Metal-Organic Frameworks-Based Hybrid Materials for Environmental Sensing and Monitoring

Metal-Organic Frameworks-Based Hybrid Materials for Environmental Sensing and Monitoring

Edited by
Ram K. Gupta, Tahir Rasheed, Tuan Anh Nguyen, Muhammad Bilal

CRC Press
Taylor & Francis Group
Boca Raton London New York

CRC Press is an imprint of the
Taylor & Francis Group, an **informa** business

First edition published 2022
by CRC Press
6000 Broken Sound Parkway NW, Suite 300, Boca Raton, FL 33487-2742

and by CRC Press
4 Park Square, Milton Park, Abingdon, Oxon, OX14 4RN

CRC Press is an imprint of Taylor & Francis Group, LLC

Library of Congress Cataloging-in-Publication Data
Names: Gupta, Ram K., editor. | Rasheed, Tahir, editor. | Nguyen, Tuan Anh
(Chemist), editor. | Bilal, Muhammad (Professor of bioengineering), editor.
Title: Metal-organic frameworks-based hybrid materials for environmental
sensing and monitoring/edited by Ram K. Gupta, Tahir Rasheed, Tuan Anh Nguyen, Muhammad Bilal.
Description: First edition. | Boca Raton : CRC Press, 2022. | Includes bibliographical references and index.
Identifiers: LCCN 2021051679 (print) | LCCN 2021051680 (ebook) |
ISBN 9781032024530 (hardback) | ISBN 9781032036007 (paperback) |
ISBN 9781003188148 (ebook)
Subjects: LCSH: Environmental monitoring. |
Metal-organic frameworks. | Electrochemical sensors.
Classification: LCC QH541.15.M64 M48 2022 (print) |
LCC QH541.15.M64 (ebook) | DDC 363.7/063–dc23/eng/20211221
LC record available at https://lccn.loc.gov/2021051679
LC ebook record available at https://lccn.loc.gov/2021051680

ISBN: 978-1-032-02453-0 (hbk)
ISBN: 978-1-032-03600-7 (pbk)
ISBN: 978-1-003-18814-8 (ebk)

DOI: 10.1201/9781003188148

Typeset in Times
by Newgen Publishing UK

Contents

Preface

With an unprecedented population boom and rapid industrial development, environmental pollution has become a severe problem for the ecosystem and public health. A large number of environmental pollutants, for example, heavy metals, organic compounds, and toxic gases are associated with potential health risks. Classical techniques for sensing and determining environmental contaminants often require complex pretreatments, expensive equipment, and longer testing times. Therefore, new and state-of-the-art sensing technologies possessing the advantages of excellent sensitivity, rapid detection, ease of use, and suitability for in situ, real-time, and continuous monitoring of environmental pollutants, are highly desirable.

Metal-organic frameworks (MOFs) are a fascinating class of crystalline porous materials consisting of metal ions (or clusters) and organic ligands linked by coordination bonds. They offer a facile means for providing structural diversity, flexibility, high porosity, large surface areas, and a wide range of pore sizes. These properties of MOFs make them superior to commercial porous materials, such as activated carbon, silica, and zeolites. Moreover, the physicochemical properties of the MOFs can be engineered during the synthesis, for example, by changing parameters such as the nature of the metals, ligands, and temperature conditions. As such, MOFs have been applied for diverse functional applications, including hydrogen storage, contaminant sorption (removal of inorganic, and organic contaminants from water during purification), separation of volatile organic compounds, sorbents for analytical purposes (for example, in-field sampling or trapping of formaldehyde in air or water and on-line solid-phase extraction of polycyclic aromatic hydrocarbons from water samples), heterogeneous catalysis, chemical sensors for environmental and biological species, proton conduction and drug delivery, for example, the release of niacin/vitamin B3.

With the progress of MOF research, many redox-active MOFs have been synthesized by the judicious selection of the electroactive metal ions and/or organic functional groups. Due to the unique properties and wide range of redox activities, MOFs can be applied in various fields like electrocatalysis, energy storage devices, and electrochemical sensors. The potential of the MOF-based composites has also been realized as ideal hosts for functional materials (like conducting nanoparticles). These composites are demonstrated to have superior electrocatalytic/electrochemical sensing properties over their pristine forms. Accordingly, various MOF composite-based platforms have been developed as efficient electrochemical sensors for environmental and biochemical targets.

The main purpose of this book is to emphasize the recent advances in MOF-based electrochemical sensors for the detection of environmental contaminants, including heavy metals, organic compounds, and toxic gases. The designed chapters are focused on the fundamental principles, mechanisms, and promising applications of MOF-based hybrid materials for electrochemical, sensing. Current limitations and challenges of MOF-based sensors and an outlook of the future direction of this emerging sensing material are also outlined.

Ram K. Gupta, Ph.D.
Associate Professor
Department of Chemistry
Kansas Polymer Research Center
Pittsburg State University
Pittsburg, Kansas, United States

About the Editors

Dr. Ram K. Gupta is an Associate Professor at Pittsburg State University. Dr. Gupta's research focuses on conducting polymers and composites, green energy production and storage using biowastes and nano materials, optoelectronics and photovoltaics devices, organic-inorganic hetero-junctions for sensors, bio-based polymers, flame-retardant polymers, bio-compatible nanofibers for tissue regeneration, scaffold and antibacterial applications, corrosion inhibiting coatings, and bio-degradable metallic implants. Dr. Gupta has published over 235 peer-reviewed articles, made over 300 national, international, and regional presentations, chaired many sessions at national/international meetings, edited many books, and written several book chapters. He has received over two and a half million dollars for research and educational activities from many funding agencies. He is serving as Editor-in-Chief, Associate Editor, and editorial board member of numerous journals.

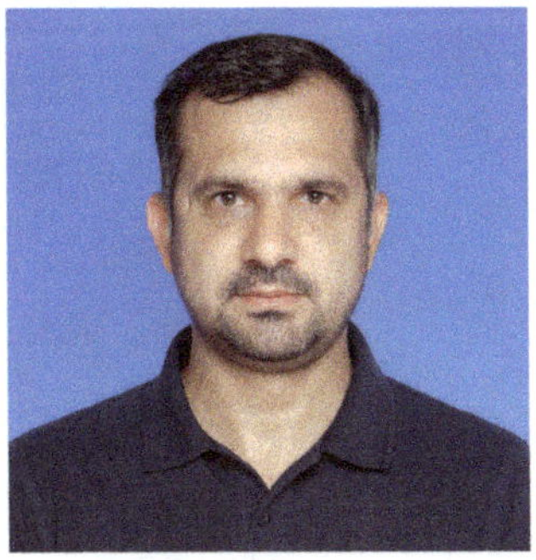

Dr. Tahir Rasheed is currently working as a res earcher at The Interdisci plinary Research Center for Advanced Materials, King Fahd University of Petroleum and Minerals, Saudi Arabia. He received his Ph.D. and degree in Polymer Chemistry and postdoctoral research from the School of Chemistry and Chemical Engineering, Shanghai Jiao Tong University, China. Dr. Rasheed has published more than 130 scientific contributions in the form of Research, Reviews, Book Chapters and Editorial type scientific articles in various areas of Science and Engineering. His research interests focus on multiple disciplines including controllable synthesis, characterization and self-assembly of polymeric materials, polymers and polymer based composites, nanomaterials and nanocomposites, hybrid nanocomposites, carbon materials, energy storage materials, with special emphasis on their potential applications in the field of sensing and biosensing, electrocatalysis, degradation and quantification of various emerging pollutants, solid polymer electrolytes, gel polymer electrolytes for flexible zinc based batteries and electrodes in energy storage devices (namely, LI-S and ZIBs, and so on).

Tuan Anh Nguyen has completed his BSc in Physics from Hanoi University in 1992, and his Ph.D. in Chemistry from Paris Diderot University (France) in 2003. He was a Visiting Scientist at Seoul National University (South Korea, 2004) and the University of Wollongong (Australia, 2005). He then worked as a Postdoctoral Research Associate and Research Scientist at Montana State University (USA), 2006-2009. In 2012, he was appointed as Head of the Microanalysis Department at the Institute for Tropical Technology (Vietnam Academy of Science and Technology). His research focuses on advanced nanomaterials and applied nanotechnology. His research activities include smart coatings, conducting polymers, corrosion and protection of metals/concrete, antibacterial materials, and smart sensors/devices. He has managed 4 Ph.D. theses as thesis director and 3 are in progress; He is Editor-In-Chief of 'Kenkyu Journal of Nanotechnology & Nanoscience' and Founding Co-Editor-In-Chief of 'Current Nanotoxicity & Prevention'. He is the author of 4 Vietnamese books and Editor of over 52 Elsevier/9 CRC Press books.

Dr. Muhammad Bilal is presently working as an Associate Professor in the School of Life Science and Food Engineering, Hua iyin Institute of Technology, Huaian, China. He accomplished his Ph.D. from Shanghai Jiao Tong University with a specialization in Bioengineering and Applied Biotechnology. He has published more than 400 scientific contributions in the form of research, reviews, book chapters and editorial type scientific articles in various areas of science and engineering. He has an H-index = 50 along with more than 9000 citations. He has edited several special issues and books and serves as a scientific reviewer in numerous peer-reviewed journals. Dr. Bilal has a collaborative network with national and international institutes/universities across the globe. His research interests include environmental biotechnology/bioengineering, nanotechnology, bio-catalysis, enzyme engineering, immobilization, chemical modifications and industrial applications of microbial enzymes, bioremediation of hazardous and emerging pollutants, liquid and solid waste management-valorization of agro-industrial wastes and biomaterials for bioenergy.

List of Contributors

Darko Kwabena Adu
Department of Pharmaceutical Chemistry
College of Health Sciences, University of KwaZulu-Natal
Westville Campus, Durban, South Africa

Nityananda Agasti
Department of Chemistry
Deen Dayal Upadhyaya College
University of Delhi, New Delhi, India

Muhammad Siddique Ahmad
Bahauddin Zakeriya University
Multan, Pakistan

Naseer Ahmad
HEJ Research Institute of Chemistry
International Center for Chemical and Biological Sciences, University of Karachi
Karachi, Pakistan

John Alake
Department of Pharmaceutical Chemistry
College of Health Sciences, University of KwaZulu-Natal
Westville Campus
Durban, South Africa

Fahad Ali
Division of Analytical Chemistry
Bahauddin Zakariya University
Multan, Pakistan

Jazib Ali
Centre for Hybrid and Organic Solar Energy
Department of Electronic Engineering, University of Rome Tor Vergata
Rome, Italy

Giaan A. Álvarez-Romero
Universidad Autónoma del Estado de Hidalgo
Área Académica de Química
Carretera Pachuca-Tulancingo, Mineral de la Reforma, Hidalgo, México

Saima Anjum
Department of Chemistry
The Govt. Sadiq College Women University
Bahawalpur, Pakistan

Dicky Annas
Department of Chemistry
Chemistry Institute for Functional Materials
Pusan National University
Busan, Republic of Korea

Karim Asadpour-Zeynali
Department of Analytical Chemistry
Faculty of Chemistry, University of Tabriz
Tabriz, Iran

Ghulam Abbas Ashraf
Department Of Physics
Zhejian Normal University
Zhejiang, China

Baoqing Bai
School of Life Science
Shanxi University
Taiyuan, China

Selva Balasubramanian
Centre for Nanotechnology & Advanced Biomaterials
SASTRA Deemed University
Thanjavur, Tamil Nadu, India
and
School of Electrical & Electronics Engineering
SASTRA Deemed University
Thanjavur, Tamil Nadu, India

Rafia Batool
Department of Biochemistry
Bahauddin Zakariya University
Multan, Pakistan

Nabakrushna Behera
School of Chemistry
Sambalpur University
Jyoti Vihar, Sambalpur, Odisha, India

Tankadhar Behera
School of Chemistry
Sambalpur University
Jyoti Vihar, Sambalpur, Odisha, India

Asnake Lealem Berhanu
Department of Chemistry
Punjabi University
Patiala, India

Prism Bhardwaj
Material Application Research Lab
Department of Nano Sciences and Materials, Central University of Jammu
Jammu, India

Shikha Bhogal
Department of Chemistry
Punjabi University
Patiala, India

John Bosco Balaguru Rayappan
Centre for Nanotechnology & Advanced Biomaterials
SASTRA Deemed University
Thanjavur, Tamil Nadu, India
and
School of Electrical & Electronics Engineering
SASTRA Deemed University
Thanjavur, Tamil Nadu, India

Ruchika Chauhan
Department of Pharmaceutical Chemistry
College of Health Sciences, University of KwaZulu-Natal
Westville Campus, Durban, South Africa

J. Ángel Cobos-Murcia
Universidad Autónoma del Estado de Hidalgo
Área Académica de Ciencias de la Tierra y Materiales
Carretera Pachuca-Tulancingo,, Mineral de la Reforma, Hidalgo, México

Raúl Colorado-Peralta
Facultad de Ciencias Químicas
Universidad Veracruzana
Orizaba, Veracruz, México

J. Antonio Cruz-Navarro
Universidad Autónoma del Estado de Hidalgo
Área Académica de Química
Carretera Pachuca-Tulancingo,
Mineral de la Reforma, Hidalgo, México
and
Facultad de Ciencias Químicas
Universidad Veracruzana
Orizaba, Veracruz, México

Gopal Das
Centre for the Environment
Indian Institute of Technology Guwahati
Assam, India
and
Department of Chemistry
Indian Institute of Technology Guwahati
Assam, India

Akash Deep
CSIR-Central Scientific Instrument Organisation
Chandigarh, India
and
Academy of Scientific and Innovative Research
Ghaziabad, India

Felipe de Souza
Kansas Polymer Research Center
Pittsburg State University
Pittsburg, Kansas, USA

Ambika Devi
Material Application Research Lab
Department of Nano Sciences and Materials, Central University of Jammu
Jammu, India

Engin Er
Department of Analytical Chemistry
Faculty of Pharmacy, Ankara University
Ankara, Turkey

Muhammad Umer Farooq
Division of Analytical Chemistry
Bahauddin Zakariya University
Multan, Pakistan
and
State Key Laboratory of Chemical Engineering
East China University of Science and Technology
130 Meilong Road, Shanghai, China

Batool Fatima
Division of Analytical Chemistry
Bahauddin Zakariya University
Multan Pakistan
and
Department of Biochemistry
Bahauddin Zakariya University
Multan, Pakistan

Arnab Ghosh
Centre for the Environment
Indian Institute of Technology Guwahati
Assam, India

Soujanya Ghosh
Department of Biotechnology
KIIT University
Bhubaneswar, India

Atal A.S. Gill
Department of Pharmaceutical Chemistry
College of Health Sciences, University of KwaZulu-Natal
Westville Campus, Durban, South Africa

Priyanshu Goel
CSIR-Central Scientific Instrument Organisation
Chandigarh, India
and
Academy of Scientific and Innovative Research
Ghaziabad, India

Caixia Guo
School of Life Science
Shanxi University
Taiyuan, China

Ram K. Gupta
Kansas Polymer Research Center
Pittsburg State University
Pittsburg, Kansas, USA
and
Department of Chemistry
Pittsburg State University
Pittsburg, Kansas, USA

Arpan Hazra
Technische Universität Dresden
Institut für Anorganische Chemie
Bergstraße 66, Dresden, Germany

Fabiola Hernández-García
Universidad Autónoma del Estado de Hidalgo
Área Académica de Ciencias de la Tierra y Materiales
Carretera Pachuca-Tulancingo, Mineral de la Reforma, Hidalgo, México

Shamim Ahmed Hira
Department of Chemistry
Chemistry Institute for Functional Materials
Pusan National University
Busan, Republic of Korea

Dilshad Hussain
HEJ Research Institute of Chemistry
ICCBS, University of Karachi
Karachi, Pakistan

M. Irfan Hussain
Institute of Powder Metallurgy and Advanced Ceramics
School of Material Science and Engineering, University of Science and Technology Beijing
Beijing, China

Blessing Wisdom Ike
Department of Pharmaceutical Chemistry
College of Health Sciences, University of KwaZulu-Natal
Westville Campus, Durban, South Africa

Sanha Jang
Department of Chemistry
Chemistry Institute for Functional Materials
Pusan National University
Busan, Republic of Korea

Supritha M Karekuladh
Department of Chemistry
Bengaluru City University, Central College Campus
Bengaluru, India

Rajshekhar Karpoormath
Department of Pharmaceutical Chemistry
College of Health Sciences, University of KwaZulu-Natal
Westville Campus, Durban, South Africa

Stefan Kaskel
Technische Universität Dresden
Institut für Anorganische Chemie
Bergstraße 66, Dresden, Germany

Harmeet Kaur
CSIR-Central Scientific Instrument Organisation
Chandigarh, India
and
Academy of Scientific and Innovative Research
Ghaziabad, India

Kuldeep Kaur
Department of Chemistry
Mata Gujri College
Fatehgarh Sahib, India

Muhammad Usman Khan
National Key Laboratory of Tunable Laser Technology
Institute of Optoelectronics
Department of Electronics Science and Technology, Harbin Institute of Technology
Harbin, China
and
Department of Physics
The University of Lahore
Sub-campus Sargodha, Pakistan

Naeem Akhtar Khan
IRCBM
COMSAT University
Islamabad, Lahore Campus, Pakistan
and
HEJ Research Institute of Chemistry
ICCBS, University of Karachi
Pakistan

Santimoy Khilari
Department of Chemistry
Guru Ghasidas Vishwavidyalaya
Bilaspur, Chhattishgarh, India

Arockia Jayalatha Kulandaisamy
School of Electrical & Electronics Engineering
SASTRA Deemed University
Thanjavur, Tamil Nadu, India

Jitendra Kumar
Nuclear Agriculture and Biotechnology Division
Bhabha Atomic Research Centre
Trombay
Mumbai, India
and
Homi Bhabha National Institute
Anushakti Nagar
Mumbai, India

Pawan Kumar
Materia
l Application Research Lab
Department of Nano Sciences and Materials, Central University of Jammu
Jammu, India

Sandeep Kumar
Department of Chemistry
Punjabi University
Patiala, India

Jianping Li
College of Chemistry and Bioengineering
Guilin University of Technology
Guilin, China
and
College of Environmental Science and Engineering
Guilin University of Technology
Guilin, China

Yang Liu
Department of Chemistry
Tsinghua University
Beijing, China

R. Mahendran
Centre of Excellence in Nonthermal Processing
Indian Institute of Food Processing Technology
Pudukkottai Road, Thanjavur, India

Saadat Majeed
Division of Analytical Chemistry
Bahauddin Zakaryia University
Multan, Pakistan
and
Institute of Chemical Sciences
Bahauddin Zakariya University
Multan, Pakistan

Sabahat Majeed
Department of Biosciences
COMSAT University
Islamabad, Pakistan

Mir Reza Majidi
Department of Analytical Chemistry
Faculty of Chemistry, University of Tabriz
Tabriz, Iran

Ashok Kumar Malik
Department of Chemistry
Punjabi University
Patiala, India

Pandurangappa Malingappa
Department of Chemistry
Bengaluru City University, Central College Campus
Bengaluru, India

Mayank
Shobhaben Pratapbhai Patel
School of Pharmacy & Technology Management
SVKM's NMIMS University
Mumbai, India

L. Humberto Mendoza-Huizar
Universidad Autónoma del Estado de Hidalgo
Área Académica de Química
Carretera Pachuca-Tulancingo, Mineral de la Reforma, Hidalgo, México

Yongjuan Meng
College of Chemistry and Bioengineering
Guilin University of Technology
Guilin, China

Archana Mishra
Nuclear Agriculture and Biotechnology Division
Bhabha Atomic Research Centre
Trombay
Mumbai, India
and
Homi Bhabha National Institute
Anushakti Nagar
Mumbai, India

Sunita Mishra
CSIR-Central Scientific Instrument Organisation
Chandigarh, India
and
Academy of Scientific and Innovative Research
Ghaziabad, India

Arjun A Mohan
School of Materials Science and Engineering
National Institute of Technology, Calicut
Calicut, Kerala, India

Sumit Mohapatra
School of Chemistry
Sambalpur University
Jyoti Vihar, Sambalpur, Odisha, India

Irshad Mohiuddin
Department of Chemistry
Punjabi University
Patiala, India

Abrar Mohyuddin
Department of Chemistry
The Emerson University
Multan, Pakistan

Ahad Mokhtarzadeh
Immunology Research Center
Tabriz University of Medical Sciences
Tabriz, Iran

Soumya Mukundan
Nuclear Agriculture and Biotechnology Division
Bhabha Atomic Research Centre
Trombay
Mumbai, India

Saravanan Nagappan
Department of Chemistry
Chemistry Institute for Functional Materials
Pusan National University
Busan, Republic of Korea

Muhammad Najam-ul-Haq
Division of Analytical Chemistry
Bahauddin Zakariya University
Multan, Pakistan
and
Institute of Chemical Sciences
Bahauddin Zakariya University
Multan, Pakistan

Sayed Tayyab Raza Naqvi
Division of Analytical Chemistry
Bahauddin Zakariya University
Multan, Pakistan

Zondi Nate
Department of Pharmaceutical Chemistry
College of Health Sciences, University of KwaZulu-Natal
Westville Campus, Durban, South Africa

Anish R Nath
School of Materials Science and Engineering
National Institute of Technology, Calicut
Calicut, Kerala, India

Noel Nesakumar
Centre for Nanotechnology & Advanced Biomaterials
SASTRA Deemed University
Thanjavur, Tamil Nadu, India
and
School of Chemical & Biotechnology
SASTRA Deemed University
Thanjavur, Tamil Nadu, India

Kang Hyun Park
Department of Chemistry
Chemistry Institute for Functional Materials
Pusan National University
Busan, Republic of Korea

Suma B Patri
Department of Chemistry
Bengaluru City University, Central College Campus
Bengaluru, India

P Abdul Rasheed
Department of Biological Sciences and Engineering
Indian Institute of Technology Palakkad
Palakkad, Kerala, India

Sufian Rasheed
HEJ Research Institute of Chemistry
International Center for Chemical and Biological Sciences, University of Karachi
Karachi, Pakistan

Ummama Saeed
Department of Biochemistry
Bahauddin Zakariya University
Multan, Pakistan

Isha Saini
Department of Applied Science
School of Engineering and Technology, Sushant University
Gurugram, Haryana, India

Verónica Salazar-Pereda
Universidad Autónoma del Estado de Hidalgo
Área Académica de Química
Carretera Pachuca-Tulancingo, Mineral de la Reforma, Hidalgo, México

Raghabendra Samantaray
Department of Biotechnology
KIIT University
Bhubaneswar, India
and
Department of Chemical Technology
KIIT University
Bhubaneswar, India

Sipun Sethi
School of Chemistry
Sambalpur University
Jyoti Vihar, Sambalpur, Odisha, India

Sharanyakanth P S
Centre of Excellence in Nonthermal Processing
Indian Institute of Food Processing Technology
Pudukkottai Road, Thanjavur, India

Amit L Sharma
CSIR-Central Scientific Instrument Organisation
Chandigarh, India
and
Academy of Scientific and Innovative Research
Ghaziabad, India

Promila Sharma
Department of Chemistry
Punjabi University
Patiala, India

Saloni Sharma
CSIR-Central Scientific Instrument Organisation
Chandigarh, India
and
Academy of Scientific and Innovative Research
Ghaziabad, India

Narinder Singh
Department of Chemistry
Indian Institute of Technology
Ropar, India

Vinamrita Singh
Department of Physics
Netaji Subhas University of Technology, East Campus
Delhi, India

Zhuo Shi
School of Life Science
Shanxi University
Taiyuan, China

Hessamaddin Sohrabi
Department of Analytical Chemistry
Faculty of Chemistry, University of Tabriz
Tabriz, Iran

Peihong Tong
College of Chemistry and Bioengineering
Guilin University of Technology
Guilin, China

Shuvendu Tripathy
Department of Physics
Vikash Group of Institutions
Bargarh Odisha, India

Hongye Wang
Department of Chemistry
Tsinghua University
Beijing, China

Xiaomin Wang
Institute of Pharmaceutical and Food Engineering
Shanxi University of Chinese Medicine
Yuci, China

Yong Wang
College of Chemistry
Nanchang University
Nanchang, China
and
Jiangxi Province, Key Laboratory of Modern Analytical Science
Nanchang University
Nanchang, China

Qin Xiao
College of Chemistry
Nanchang University
Nanchang, China

Wenyan Yan
College of Biosystems Engineering and Food Science
Key Laboratory of Agro-Products Postharvest Handling
Ministry of Agriculture
Zhejiang University
Hangzhou, China

Yukun Yang
School of Life Science
Shanxi University
Taiyuan, China

Ligang Yu
School of Life Science
Shanxi University
Taiyuan, China

Jinhua Zhang
School of Life Science
Shanxi University
Taiyuan, China

Shiyu Zhang
Department of Chemistry
Tsinghua University
Beijing, China

Juanhua Zhou
Department of Chemistry
Tsinghua University
Beijing, China

Qianfen Zhuang
College of Chemistry
Nanchang University
Nanchang, China

1 Metal-Organic Frameworks
An Introduction to Advanced Sensing Applications

Felipe de Souza[1] *and Ram K. Gupta*[1,2]
[1]Kansas Polymer Research Center, Pittsburg State University, Pittsburg, Kansas, USA
[2]Department of Chemistry, Pittsburg State University, Pittsburg, Kansas, USA

1.1 INTRODUCTION

Nano-sized materials has become a significant research area that is present in nearly all scientific fields such as chemistry, physics, materials science, engineering, medicine, and the like. Nanotechnology can be applied to electronics, drug delivery systems, biosensors, energy storage, catalysis, fuel cells, and many others. Nanomaterials can show enhanced properties in comparison to their bulky counterparts.

Nanomaterials can be classified into four categories based on their dimension: zero-dimension (0-D), one-dimension (1-D), two-dimension (2-D), and three-dimension (3-D). 0-D are materials that present all three dimensions within the nanoscale such as nanoparticles. 1-D materials have at least one dimension less than 100 nm. Applications of these include nanorods, nanowires, or nanotubes. 2-D materials have one of their dimensions out of the nanoscale. Applications of these include a single nanolayer, nanofilms, or nano-coatings. Finally, 3-D materials are usually comprised of an array of nanomaterials that are combined to have a three-dimensional structure. Application of these include nano-layered structures, nano nets which are arrays of nanotubes in the vertical plane and nanofilms in the horizontal plane, or nanosized active sites uniformly distributed, with a 3-D structure. Based on that kind of 3-D structure, was the development of metal-organic frameworks (MOFs) which are structured with a high degree of crystallinity based on coordination complexes that consist of an inorganic core, usually a transition metal, bonded with an organic ligand forming a 3D structure.

Due to the large number of materials available, there are around 2,000 types of MOFs. Such a wide variety is a consequence of several variations that can be made when one is designing their synthesis. This includes variations in starting materials, size, geometry, and functionality. The type of chemical bonding is based on the positively charged empty d orbitals in a transition metal bonding with one negatively charged organic group such as carboxylates, phenolates, phosphonates, or imidazolates for instance. Through this, robust and crystalline structures can be obtained that present a relatively high surface area that can vary from 1,000 up to 10,000 m^2/g, which is larger than other materials such as activated carbon-based compounds and zeolites. This aspect of MOFs makes them desirable for applications related to electrocatalysis, CO_2 reduction, fuel cells, and energy storage, for example. Part of the reason why MOFs can cover a wide range of applications is their tunable surface characteristics. By varying experimental conditions, their surface characteristics can be easily tuned. Through that, the pore sizes can be controlled which allows them to interact with specific molecules or biomolecules, adsorb small molecules, or even function as active sites to catalyze certain reactions. Their chemical and thermal stability can be also improved as they can be further functionalized after their synthesis. Other unique properties of MOFs can be observed when lanthanides are used, which can fluoresce when irradiated with UV light. Further, the bonding strength between the metallic and organic ligands can be redesigned to make them biodegradable. Importantly, their structure can be precisely controlled thus facilitating their interaction with guest molecules in terms of catalysis when acting as an active site or for identification of small molecules or biomolecules [1].

An example of these properties is the applications of MOFs as gas absorbents because their pore size can be controlled by means of the synthesis temperature, to selectively absorb CO_2 or CH_4 [2, 3]. Hence, MOFs can be used as sensors for many gases. Taking these properties in another direction, MOFs can also be used as effective capacitors due to their high surface area that allows an easy diffusion of ions within the frameworks, leading to high capacitance. Moreover, the frameworks can be functionalized with other elements such as N or P-based ligands to induce pseudo-capacitance, which further improves their energy storage capabilities. For this reason, the redox properties and exposed active sites allow MOFs to function as electrocatalysts for H_2O splitting to produce H_2 and O_2. MOFs have also found application in the biomedical field because, when they are on the nanoscale, they can be used as carriers to deliver drugs, contrasting agents for imaging, photo-induced therapy, or chemotherapy [4, 5].

1.2 SYNTHESIS OF MOFs

The synthesis of MOFs can be performed through several physical or chemical routes. Both routes can be subdivided into top-down or bottom-up approaches. A top-down approach is based on using a bulky starting material that is broken into

DOI: 10.1201/9781003188148-1

smaller pieces until it reaches the nanoscale. In the physical route, it can be performed through mechanical milling, for instance. The main disadvantage of this process is the wide distribution of particle sizes. Physical exfoliation is another procedure that falls within a top-down category as the bulky material can be peeled off into thinner films such as the use of adhesive tape to peel off the graphene layer from graphite. Even though this method is effective it is also tedious as it may require repetitive peelings with the tape followed by dissolution to claim the nanomaterial.

In the bottom-up approach, smaller starting materials are combined until they form a larger structure. One example of this case is the laser evaporation method in which a starting metal oxide is vaporized when exposed to a strong laser beam and condensed onto a substrate [6]. The parameters such as atmosphere composition and laser power can vary the size and property of nanoparticles. For example, magnetic nanomaterials can change their phase depending on these conditions or a larger number of defects on the surface area can be induced to enhance the catalytic activity [7, 8]. Slow solvent evaporation is a simpler precipitation method that is performed at room temperature in the presence of one or more volatile solvents to make the process cheaper against the cost of the long evaporation time [9]. Metal can also be converted from its ionic state that agglomerates into nanoparticles, which is another type of bottom-up approach. It can be performed through thermolysis of LiN_3, for example, by heating it to 400 °C which causes it to decompose, releasing N_2 gas (since its decomposition temperature is close to 370 °C). This process causes a decrease of pressure in the system which creates a driving force for the agglomeration of Li atoms that forms 5 nm nanoparticles [10].

There are several chemical routes available for the synthesis of MOFs such as co-precipitation, sol-gel, hydro or solvothermal, chemical vapor deposition, pyrolysis, microwave-assisted, microemulsion, ion exchange, intercalation, chemical exfoliation, sonochemical, mechanochemical, and electrochemical, to name a few. Co-precipitation is an easy method that can yield uniform-sized nanoparticles through the precipitation of a precursor in aqueous media. It consists of mixing two or more soluble salts and then using precipitating agents or suitable conditions to precipitate them. To avoid the formation of clusters, stirring is usually required along with a controlled temperature, as both parameters can affect the nanoparticles' sizes and shapes [6]. Slightly higher temperatures induce higher crystallinity as nanoparticles tend to have more energy to move and properly arrange. Commonly employed reducing agents are NH_3, NaOH, $NaBH_4$, $LiBH_4$, $N_2H_4 \cdot 2HCl$, $N_2H_4 \cdot H_2O$, among others which can also control the pH of the system to promote the precipitation of a specific metal. Based on that, there is a sequence of events for the synthesis of MOFs or nanomaterials. This sequence is: nucleation, growth, coarsening, agglomeration, and stabilization. Controlling the agglomeration is an important step to keep the particles within the nanoscale, otherwise undesired levels of agglomeration may occur due to the high surface energy of small nanoparticles. Hence, proper concentration of reducing or capping agents is important to control the size and shape.

A hydrothermal method is an efficient approach where the starting materials are exposed to high temperature and pressure in an autoclave to yield particles of regular size under relatively lower temperatures and a facile procedure. The variation in parameters such as solvents, pH, time, and temperature can greatly influence the MOF's properties. A solvothermal method is like a hydrothermal method except that instead of water, organic solvents such as dimethylformamide, acetonitrile, diethyl formamide, acetone, and so on, are used. It can lead to drastic changes in the morphology and in the properties of MOFs [9].

The sonochemical approach consists of ultrasonic radiation in a solution containing the starting materials that creates bubbles leading to high temperatures and pressures. When a bubble bursts, it provokes chemical excitation for the reaction to occur. Synthesis of CoS_2, CdSe, ZnSe, alloys, and oxides has taken place by this method [6]. Microwave-assisted synthesis is a convenient method as the conversion of high-frequency electromagnetic waves into heat, interacts with the solution containing the starting materials to create the driving force for the reaction. It is a method with a facile approach and a much shorter reaction time than traditional hydro or solvothermal methods. Additionally, it yields nanostructures with a narrow size distribution.

Electrochemical synthesis is based on dissolving metal ions along with organic ligands and electrolytes which are exposed to a specific voltage to carry on the reaction. This is a convenient process as it can be performed in a short time and with high reproducibility. Mechanochemical synthesis is another attractive method as chemical bonds are created with the input of mechanical force under small or no solvent content. On top of being a cheap method, it also yields different types of materials that can range from 1D, 2D, or 3D materials such as frameworks of zeolitic imidazolates [9].

1.3 CHEMISTRY AND APPLICATIONS OF MOFs

Properties such as high surface area, highly organized structure, exposed active sites, and facile functionalization are signatures of MOFs. Such properties endorse their use in several fields related to gas adsorption and their identification, sensors, batteries, supercapacitors, electrocatalysts, and biosensors/trackers. Such properties are related to the tunable chemical structure of MOFs which can be synthesized through a variety of transition metals and organic ligands. MOFs can function as adsorbents for gases which can reduce the cost since several types of gases used in the industry require highly pressurized containers and many steps for compression that make the process expensive and create the potential danger of explosions or leakage. Hence, a simpler procedure to store gases that do not require pressurized systems is desired. Also, due to the growing concern regarding large emission of greenhouse gases, it is considered important to introduce an efficient technology to address this situation. MOFs can be used as efficient nanocontainers for many gases since, by adjusting the synthesis procedure their properties can be tuned to adsorb different gases in MOFs. For example, for absorption of CO_2, MOFs structure can be changed to have polar groups such as amines or nitrogen bonded in a heterocyclic ligand. This molecular architecture, accompanied by their inherently high surface area allows MOFs to be suitable as gas adsorbents. Furukawa and his team obtained a MOF-210 synthesized from zinc (II) nitrate hexahydrate with 4,4',4"-[benzene-1,3,5-triyl-tris(ethyne-2,1-diyl)]tribenzoate (BTE) and biphenyl-4-4'-dicarboxylate (BPDC) for efficient gas adsorption and transportation [11]. The MOF-210 reached a Langmuir surface area of 10400 m^2/g and CO_2 storage of 2400 mg/g under a saturation pressure of around 50 bar. The void volume of 89% for the 210-MOF structure may be attributed to the large spacing between the Zn and the organic ligands, which yielded a large surface area for the MOF.

Methane is another important gas for energy applications. It is the main component of natural gas (95%) and has high gravimetric heat of combustion (50 MJ/Kg). Hence, synthesizing MOFs with low-cost starting materials that can adsorb high contents of CH_4 under applicable conditions, for example, around 25 °C with pressures between 35 to 850 bar are workable. For that, two aluminum-based MOFs were synthesized through the coordination reaction of aluminum nitrate with 4,4',4"-benzene-1,3,5-triyl-tribenzoic acid (H_3BTB) dissolved in N,N-dimethylformamide (DMF) at 150 °C for 4 days. The MOF's robust structure possessed empty spaces that allowed large quantities of CH_4 to diffuse in. The capture process of dangerous gases such as CO and NO is also within reach of MOFs, even though scientific work is required to achieve this in practice. The challenge of achieving appreciable CO adsorption is likely related to exposure to positively charged metal sites that can interact with the polarized CO molecule. Thus, improving these factors seems to be a suitable approach to achieve better CO adsorption. In this regard, the sorption and differentiation of gases are often challenging as these small molecules are of similar sizes and shapes to a certain degree. Yet, selectivity towards NO and O_2 has been achieved through Shimomura and his colleagues [12]. In his approach, the synthesized MOF was based on Zn centers coordinated with tetracyanoquinodimethane (TCNQ). The combination of these two components led to a dynamic framework that switched between opened and closed gates. This type of mechanism is based on a minimal concentration of gas (threshold) to start the adsorption process. This effect emerged through a concerted charge-transfer process between the host and guest molecule. It was suggested that there was a decrease in the conjugation level of TCNQ when the ligand interacted with specific gases such as O_2 or NO. Through this phenomenon, the MOF structure was selective towards these two gases. Note that the gas adsorption process takes place in MOFs due to their high surface area, which is also influenced by the ligand groups in the metal core.

The same principle of gas adsorption in MOFs can be adapted to the intercalation of ions into their highly porous structure that is composed of an organic segment that creates several empty spaces along the structure to accommodate ions. The process of ion diffusion within the structure of a material can be controlled by applying electric potential and can be used in energy storage applications such as supercapacitors. In addition, MOFs are also derived from metals such as Ni, Zn, Zr, Fe, Co which are redox-active to provide additional energy storage via pseudo-capacitance. Based on that, the structure of traditional MOFs should allow them to store charges through the diffusion of ions as well as the redox process, making them valuable materials in this area. Despite the convenience of having the potential to deliver high capacitance, MOFs can present low structural stability as well as conductibility. Hence, to address those issues, carbon-based nanomaterials such as graphene, carbon nanotubes (CNT), or others can be added because these materials possess high conductibility, flexibility, and chemical stability which can counter the drawbacks of MOFs. Thus, a composite that facilitates the kinetics of both electrons and ions can be obtained, promoting an effective method for charge delivery and storage. Xiao and his team performed liquid exfoliation of graphite (LEG) to obtain graphene nanolayers that were further functionalized with glucose derivatives through the solvothermal method [13]. This step was performed to provide chemical support for the growth of the MOF structure due to the increase of oxygen-based functions. Through that, a Ni-based MOF was grown homogeneously over the surface of functionalized graphene to create a stable 3D structure with optimized properties in the sense of electric conductibility of graphene and ionic transport of the MOF. The properties of the LEG-Ni-MOF composite surpassed those of the LEG-based materials and Ni-MOFs when used by itself. The synthetic approach to obtain the composite is shown in Figure 1.1a. During the testing of cyclic voltammetry (CV), a non-rectangular shape with peaks related to the redox process of Ni^{2+} to Ni^{3+} was observed suggesting the predominance of a pseudo-capacitance mechanism for charge storage. The composite delivered a capacitance of 913.9 F/g at 1 A/g which was about 1.5 times higher than the neat Ni-MOF. The noticeable improvement in charge storage properties was due to a few factors. Firstly, the presence of oxygen functions from the GM-LEG created a targeted site for the growth of the Ni-MOF. Because of that, a nano sandwiched structure was obtained with more exposed active sites to improve ionic adsorption and transport. Second, the LEG structure improved the electron transfer step. The combination of all these effects led to an overall improvement in the composite's capacitive properties. The CV profiles and charge-discharge curves are provided in Figure 1.1(b-c).

The presence of reactive sites and the facile adsorption of smaller molecules provide the MOFs electrocatalytic activity to act as an array of nanoreactors. This property has been widely explored for water splitting as it is considered to be a convenient and eco-friendly route to generate H_2 and O_2 which can be used as fuels. The challenge arises in using low-cost materials that can promote both a hydrogen evolution reaction (HER) and an oxygen evolution reaction (OER) with high efficiency, meaning a low overpotential process. Srinivas et al. [14] fabricated a semi-MOF-based on Ni/NiS

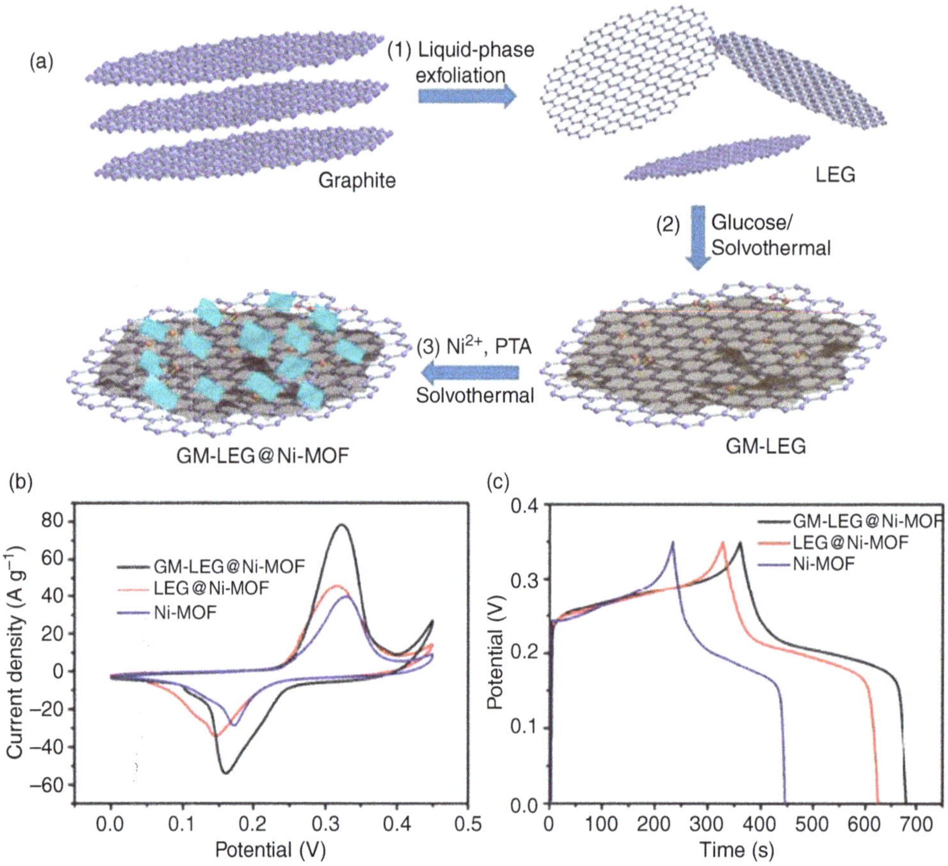

FIGURE 1.1 (**a**) Approach for the synthesis of a glucose-modified liquid exfoliated graphene adhered with Ni-MOF (GM-LEG@Ni-MOF). (**b**) CV profiles were performed at a scan rate of 10 mV/s. (**c**) Galvanostatic charge-discharge curves were performed at 1 A/g for the neat Ni-MOF, LEG@Ni-MOF, and GM-LEG@Ni-MOF. Adapted with permission from reference [13], copyright (2019), American Chemical Society.

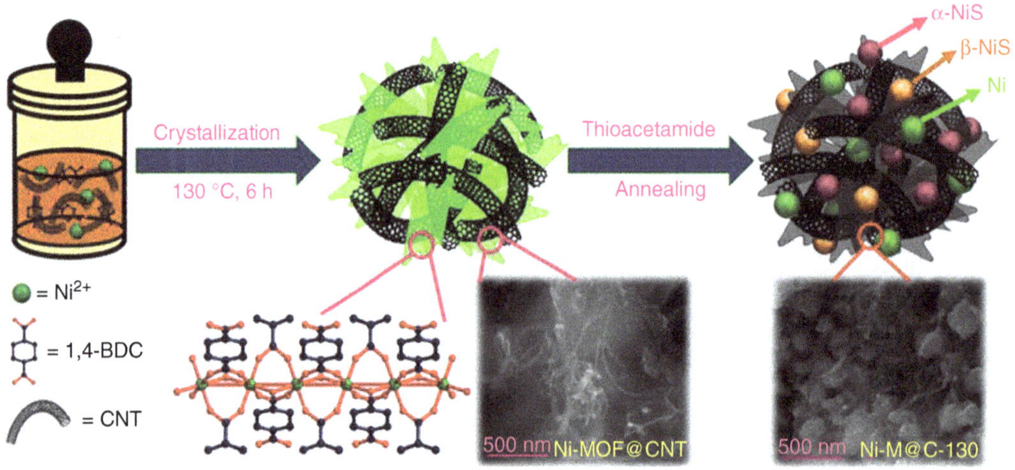

FIGURE 1.2 The fabrication process of Ni@CNT through the one-pot hydrothermal method in a mixture containing Ni^{2+} source, CNT, and 1,4-benzene dicarboxylic acid (1,4-BDC) as ligand followed by thermal treatment under the presence of a sulfur source (thioacetamide). Adapted with permission from reference [14], copyright (2021), American Chemical Society.

to promote electrocatalysis along with CNTs to improve conductivity. The composite electrocatalysts displayed low overpotential of 244 and 123 mV at η_{10}, respectively for both OER and HER. The fabrication process for the Ni-MOF@CNT is presented in Figure 1.2. This approach demonstrated the versatility for the catalytic process that MOFs can promote due to the bifunctional activity due to the presence of Ni and NiS in the same system. Yang and his team irradiated a plasma beam at low temperature and high pressure over a Co-based MOF functionalized with nitrogen-doped carbon nanotubes (NCNT) to fabricate a highly efficient electrocatalyst [15]. An increase in catalytic activity towards oxygen reduction reaction (ORR) was observed when a higher amount of pyridinic-N over pyrrolic-N was introduced into the MOF's structure. In addition, the presence of Co-N and Co-Co bonds was found to be promoting the catalysis of OER. The optimum performance was achieved when 4 min of plasma treatment was performed on MOF-NCNT. This process introduced enough defects into the structure to improve the catalysis followed by a decrease in overpotential. An overpotential of 0.35 V and satisfactory cycling stability was achieved. This improvement was observed due to the higher amount of Co-N sites along with Co-Co bonds that enabled the adsorption of O_2 that facilitated the electron transfer step and therefore the catalytic process.

MOFs are also receiving attention from the biomedical field due to their useful properties such as high porosity and large surface area that allow them to incorporate higher amounts of therapeutic agents or drugs. Additionally, the chemical versatility of MOFs can yield interesting properties, for example, there are some lanthanide-based MOFs that are fluorescent when exposed to UV light. These might work well as bio trackers. Also, their organic ligands can be functionalized with a biomolecular probe to target or identify a specific bio component allowing MOFs to function as biosensors. Due to the feasibility of these attributes, MOFs are gaining ground in the biomedical field.

1.4 MOFs FOR ENVIRONMENT SENSING AND MONITORING

The unique structure and properties of MOFs allow them to be used for many applications. In this section, applications of MOFs for environmental sensing and monitoring are provided.

1.4.1 MOFs for Toxic Chemical and Gas Sensors

Many chemicals and gases used in industrial sectors are referred to as toxic industrial chemicals (TICs) which are indispensable commodities for several chemical processes, such as the production of fertilizers, mainstream products, among many other uses. Even though these substances are widely used worldwide, their inherent toxicity is a concern for the workforce exposed to them. Hence, finding ways to not only safely handle these substances as well as recognising them as soon as there is a leak or spill, is a necessary measure. For these important tasks, several approaches have been considered to use MOFs as capture agents for TICs. One widely explored example of this is the adsorption of H_2S which has been achieved through several different types of MOF [16]. In one case Hanon and colleagues developed a MOF (MIL-47) based on V^{+4} and MIL-53 which were prepared using Al^{3+}, Cr^{3+}, Fe^{3+} as metallic centers for coordination. The transition metals were coordinated with terephthalate ligands which led to a porous structure with a diamond-like shape. The resultant MOF presented a surface area of around 1000 m^2/g which was determined by Brunauer–Emmet–Teller (BET). Another MOF developed for this group was based on Cr^{3+} metal oxide which presented an octahedral geometry with three centered Cr atoms that coordinated with O either from a trimesate named MIL-100 or terephthalate ligands named MIL-101. This MOF structure presented super-tetrahedral cavities that achieved a higher surface of 2000 m^2/g. Both MIL-100 and 101 presented the highest adsorption of H_2S (16.7 and 38.4 mmol/g, respectively) at 303.1 K under a pressure of 2 MPa. Among these MOFs, MIL-47 presented relatively lower adsorption of H_2S compared to MIL-53, 100, and 101 which was explained by the 'breathing effect'. This occurred in MIL-47 due to its highly flexible structure that causes its pores to close, likely due to a relatively stronger interaction with H_2S as the μ-OH groups can form a bridge with the metallic atoms leading to a compression of the structure. On the other hand, MIL-53, 100, and 101 presented a more rigid structure that would not deform which enabled them to display a higher uptake of the gas.

In broad terms, the adsorption mechanism of toxic gases is complex and requires the MOFs to be chemically stable to prevent a reaction with the gas and interact with it in such a way that it can perform a reversible adsorption process to allow several uses of the materials. Finally, the process should ideally be performed at room temperature or mild conditions to become applicable on a larger scale. Hence, the bond strength between metal and ligand, chemical interaction with the gas, and rigidity of MOF's structure are some of the main parameters that should be taken into consideration for the design of a MOF. MOFs can be also used for the storage of chemical warfare agents (CWAs). The chemical structure of the main CWAs is shown in Figure 1.3. Based on these issues, it deems necessary to develop ways to identify these chemicals that can pose a threat. However, it is inherently challenging as these compounds need to be detected under lower concentrations and room temperature. For that, porous carbon-based adsorbents would be employed. Yet, their chemical structure does not interact well with polar toxic chemicals which makes them inefficient in this case. Even when activated carbon is doped with ionic salts, its structure becomes less ordered which makes it unstable for detection of these polarized substances. This situation can be addressed by the use of MOFs, which possess a precise structure at the atomic level that allows them to selective adsorb a target molecule. This idea was applied by Montoro and her team who synthesized a novel Zn-based MOF with the formula Zn_4O (3,5-dimethyl-4-carboxypyrazolato) referred to as (1) and with a structure displayed in Figure 1.4 that was tested for the adsorption of different CWAs when exposed to 50% humidity conditions and which was still able to selectively promote the adsorption with CWA [19]. Two factors enabled this process. The first was the proper nanopore size that allowed it to capture the harmful chemicals. The second was the MOF's hydrophobicity as was suggested by the heat of adsorption (ΔH_{ads}) for the substance, diethyl sulfide (DES), diisopropylfluorophosphate (DIFP), and water that were -50.1, -44.8, and -17.9 kJ/mol, respectively. This effect demonstrated a higher release of energy after the

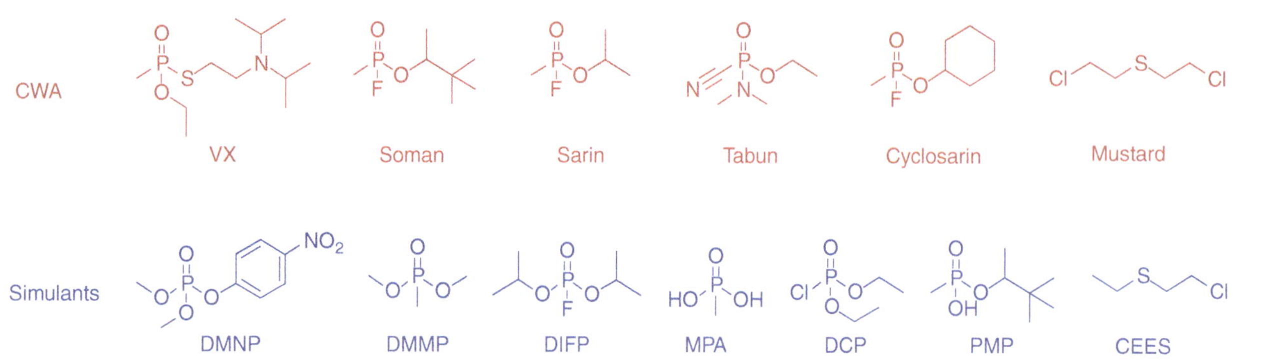

FIGURE 1.3 Chemical structure of the main CWAs and their simulant structure used for academic purposes. Adapted with permission from reference [18], copyright (2020), American Chemical Society.

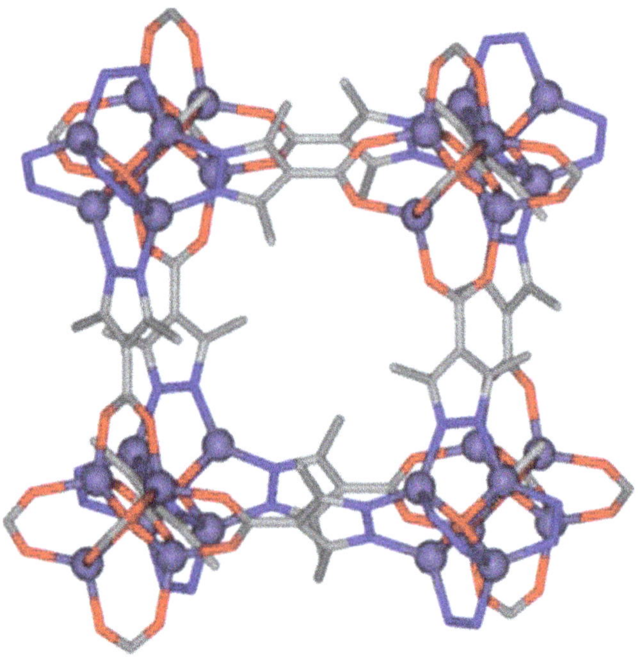

FIGURE 1.4 The predicted crystal structure for the Zn-based MOF is referred to as (1). Adapted with permission from reference [19], copyright (2011), American Chemical Society.

proper interaction of the simulated harmful compounds in comparison to water. Similarly, it also led to higher values for the partition coefficients (K_r) for the model compounds compared to water. Despite its similarity in terms of the values when compared with activated carbon or $Cu_3(btc)_2$ the Zn-based MOF presented more robusticity due to its high thermal and chemical stability arising from the bond between Zn and N and/or O. On top of that (1) can be synthesized with higher amounts through a facile method consisting of the addition of the starting materials under ethanol reflux, which is not the case for the commercialized adsorbents.

Hence, traditionally the strategies to improve the adsorption of harmful substances can be based on controlling pore size and introducing hydrophobic groups to avoid their interaction with water. Yet, to further improve the uptake of harmful substances by the MOFs their frameworks can be tuned to create open metal sites (OMS). This approach was explored by Zou et al. [20] through the synthesis of a trinuclear Zn_2Ca-based MOF along with α-PbO_2. Its structure, presented in Figure 1.5, possesses unsaturated coordinate bonds for the Zn sites which ease the MOF's interaction with their simulant CWA which was methyl phosphonic acid (MPA). The adsorption capacity for the Zn_2Ca-based MOF reached its highest at from 0.5 to 0.6 mol ratio MPA/Zn at room temperature and diluted in CH_2Cl_2/CH_3OH as shown in Figure 1.5b. Thus, the trinuclear MOF provided an adsorption density of 3.42 mmol/g of the stimulant, which is around 17 times higher than the adsorption density of goethite particles [20]. This improvement in adsorption may be attributed to the coordinately unsaturated Zn sites that allow the binding process of MPA due to its low hindrance to accommodate into the Zn exposed site. This interaction was proposed through theoretical simulation for the binding process as shown in Figures 1.5c-d.

1.4.2 MOFs for Agricultural Waste (Pesticides) Sensors

Pesticides are important chemicals used mostly to control plagues in crops to avoid the loss of crops and potential food shortage and to maintain production of vegetables at a constant pace, This is important to keep prices stable and to make sure food arrives in good quality to customers. These pesticides can cause several types of diseases in humans depending on the level of exposure. The diseases vary from different types of cancer, reproductive disorders, and many other negative health effects when these compounds find their way into the body. Hence, despite their importance in one sense, pesticides are aggressive to the body. Therefore, the identification and removal of these chemicals once they have fulfilled their function is a critical process towards providing a healthy community. MOFs can meet these requirements because of their inherent properties, for example, luminescence and their chemical tunability that allows them to be used to detect pesticides. Along those lines, a successful strategy was proposed by Seo et al. [21] to promote the removal of a pesticide named methylchlorophenoxy propionic acid (MCCP) from aqueous media. The authors synthesized a

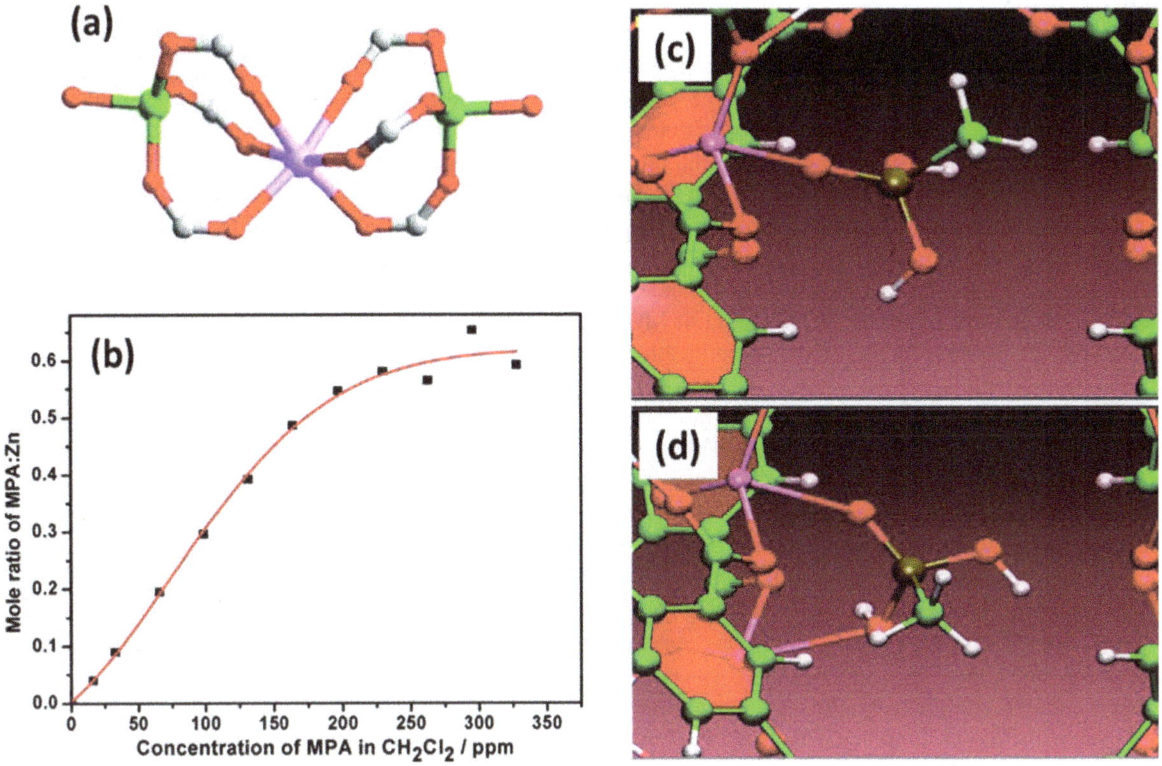

FIGURE 1.5 (**a**) Structure for the trinuclear $Zn_2Ca(CO_2)_6(H_2O)_2$ in a sole molecular building block. (**b**) Adsorption curve of methyl phosphonic acid at room temperature. Theoretical binding sites for the trinuclear MOF through (**c**) direct coordination between Zn and phosphoryl group (**d**) bridging site. Adapted with permission from reference [20], copyright (2010), American Chemical Society.

MOF named UiO-66 which was able to outperform traditional activated carbon-based materials as it presented 7.5 times higher adsorption capacity. The high adsorption capacity of the UiO-66 MOF was due to exposed aromatic rings which facilitated the π-π interaction with MCCP which also possessed an aromatic ring. Additionally, the surface area of UiO-66 was higher than that of activated carbon. These values were reflected in the maximum adsorption capacity as UiO-66 reached 370 mg/g whereas activated carbon only obtained 303 mg/g.

Another aspect that differed from the synthesized MOF was how its performance was dependent on the pH. This was because UiO-66 has an isoelectric point at a pH of 5.5. On the other hand, MCCP is neutral at pH < 3.8 and anionic above 3.8 pH. With this information, it may be noted that the optimal adsorption of MCCP over UiO-66 occurs in between a pH from 4 to 5.5 since MCCP is anionic and provides interactive forces of attraction with UiO-66 due to its positively charged surface. This behavior deviates from the properties of activated carbon materials thus showing an advantage from the use of MOFs. On top of that, UiO-66 (along with other MOFs) is reusable. Another strategy to promote the adsorption of a commonly known organophosphorus pesticide, namely, glyphosate was performed by Yang et al. [22] with their method of combining a Zr-based MOF with GO. Through that, the MOF/GO composite obtained 482.69 mg/g of adsorption capacity at pH = 4. The composite was able to adsorb glyphosate through the chemical interaction of Zr–O–P identified in the Fourier transformed infrared (FTIR) spectra at 941 cm^{-1} which appeared after MOF/GO interaction with glyphosate along with the appearance of P=O and P–O peaks at 1157 and 1075 cm^{-1}, respectively.

1.4.3 MOFs for Biosensors

The early identification of certain biomolecules that exist in the human body such as DNA fragments, RNA, antibodies (amongst others) can be the a key factor in the proper diagnosis and early treatment of disease and this can lead to a considerable increase in survival rates. Hence, developing techniques and technologies that can deliver accurate and early results is much needed. Although current approaches such as X-ray, ultrasound, positron emission tomography, computer tomography, and the like, are valuable tools for the medical field, there is still a chance of misdiagnosis and late detection of, for example, cancer which can be critical. Biosensors are a promising technology that can aid in their diagnosis. To be functional, biosensors mainly require three components: (1) recognition of bio elements such as enzymes, antibodies, or biomarkers that can respond to the targeted biomolecule, (2) emit an appreciable electrical signal that indicates the interaction between the substrate and analyte, and (3) be connected to an electronic system to gather the data. The signals such as redox processes, resistance due to electron transfer, or surface conduction are some effects that can be used in a biosensor. The fact that MOFs allow greater versatility in their synthetic approaches in terms of chemical modification, surface properties, high surface area, and the presence of active sites for immobilization of biomolecules endorses their use as biosensors. Such conditions can prompt levels of the ultralow detection limit. MOFs can identify specific species for the diagnosis of diseases on a scale of pg/mL (1×10^{-15} grams per millilitre) [23].

The mechanism for the detection of cancer, for instance, can occur based on three steps. First, the modification of MOF's surface to enhance electrochemical activity, amplification of the electrical signal, and higher bio affinity with the desired probe. Second, the MOF-based biosensors must immobilize the probe over its surface to properly interact with the bio analyte. Third, through a successful immobilization, an appreciable electrical signal from the bio analyte can be obtained. However, MOFs often lack conductibility which decreases their electrochemical activity [23]. To address that, some common strategies rely on introducing ligands to provide a stronger binding with the probe. For example, imidazole or carboxylate-based ligands can be used to functionalized MOFs. In addition, electrochemically active materials such as conducting polymers, carbon-based nanomaterials, or metal nanoparticles can be incorporated to improve their conductibility. Also, the synthesis of bimetallic MOFs is a convenient approach as the hybrid structure containing two metal centers can provide an efficient electron transfer by providing more active sites. Another way lies in synthesizing MOFs with a larger specific surface area to effectively anchor the biomolecules. Performing other processes such as annealing can also improve the porosity [24, 25]. The mechanistic process for the identification of a biomolecule can take place in four ways. First, it can occur through the specific binding of an immobilized antibody over the MOF's surface with an antigen. The second, formation of a G-quadruplet structure. These are stacked structures of guanine nitrogenated based tetrads found in DNA. Third, the capture of the electrical signal of hybridization of DNA or RNA chains. Fourth, abnormal expression of biomarkers that are found in living cancer cells. These mechanisms for the functioning of a biosensor are described in Figure 1.6.

1.4.4 MOFs for Pharmaceutical/Neurochemicals Sensors

Neurochemicals play an important role in the hormonal and chemical balance in the body which makes them responsible for certain types of behavior or disorders whenever there is a disturbance in their biochemical equilibrium. Hence, properly tracking these components is an important test for the diagnosis of specific human conditions in a similar way

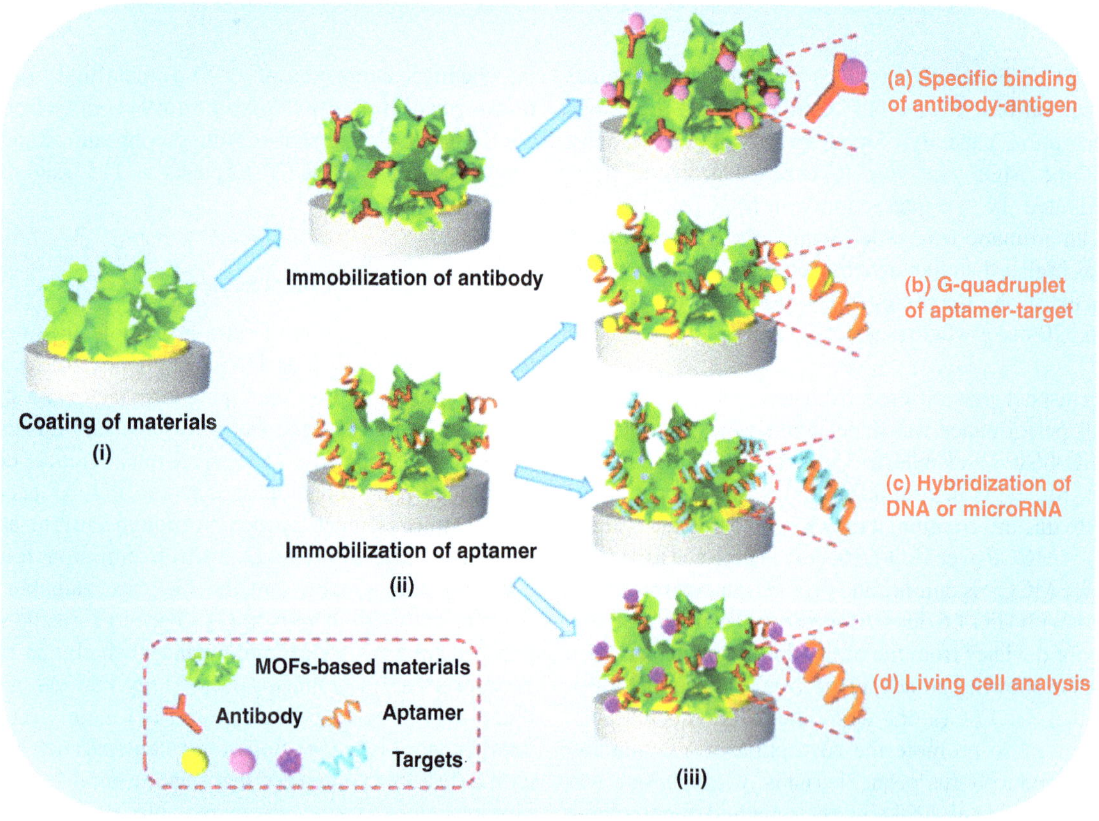

FIGURE 1.6 A general schematic for the functioning and operation of a biosensor. Adapted with permission from reference [23], copyright (2021), Elsevier.

as for biosensors. Nowadays, the main attributes that enable the applicability for neurochemical sensors are finding a straightforward synthetic procedure that does not require post-synthesis treatments, high selectivity and sensitivity towards several analytes, and simple integration of the sensor with electronic devices to make wearable or portable devices. 2D structured MOFs are convenient materials for this type of application which appears as a convenient option compared to the currently used materials such as B-doped diamonds, metallic or carbon-based nanomaterials due to their inherent properties that prompt their use in this field. The main purpose for neurochemical sensors lies in identifying important analytes such as dopamine (DA), ascorbic acid (AA), serotonin (5-HT), and uric acid (UA) which is intrinsically related to the human metabolism, physiology, mental and nutritional conditions. In the case of a chemical imbalance of DA, for instance, the person may present a clinical scenario that leans towards depression, schizophrenia, attention deficit, or hyperactivity disorder. With an imbalance in 5-HT symptoms such as problem sleeping, mood, and memory can be affected.

It is important to properly track these chemicals in the body, however, one of the recurrent obstacles is that AA and UA may cause interference in the analysis of other neurochemicals due to their chemical similarity. This issue has been addressed with the introduction of composites based on ionic liquids, polymers, carbon-based nanomaterials, and enzymes, which yielded detection limits within the range of nano to micro mols. Yet, these types of neurochemical sensors are not robust as they can deteriorate due to consecutive interaction with substances that lead to loss of the efficiency of the electrochemically active interfaces. Hence, in an approach to avoiding the need for complex post synthetical procedures, Ko and colleagues opted for the synthesis of a 2D structured MOF-based on Cu and Ni that are coordinated with 2,3,6,7,10,11-hexahydroxytriphenylene (HHTP) and 2,3,6,7,10,11-hexaiminotriphenylene (HITP) [26]. This approach proposes an alternative use of 2D MOFs since in most cases they are used as biosensors through colorimetric and luminescence analysis, drug carriers, or scaffolds instead of electroactive materials, due to their low conductivity and instability in aqueous media [27]. Yet, in this case, a high selectivity was achieved in aqueous media for DA and 5-HT showing a detection limit of 63 and 40 nm, respectively. In addition, the analysis was performed in the presence of interferents such as AA and UA, where the 2D MOF was able to differentiate as well as present considerable robustness as the electrochemical properties remained stable over 100 cycles. The layered device architecture was obtained by dispersing the MOF in water followed by drop-casting over a glassy carbon electrode (GCE). The schematics for this process are shown in Figure 1.7. This synthetical approach enabled better control of the electrocatalytic components. Also, the absence of a doping process enabled a permanent value of porosity as when a MOF is doped with an atom its surface area tends to decrease due to the obstruction of its surface area at cost to conductivity or chemical stability.

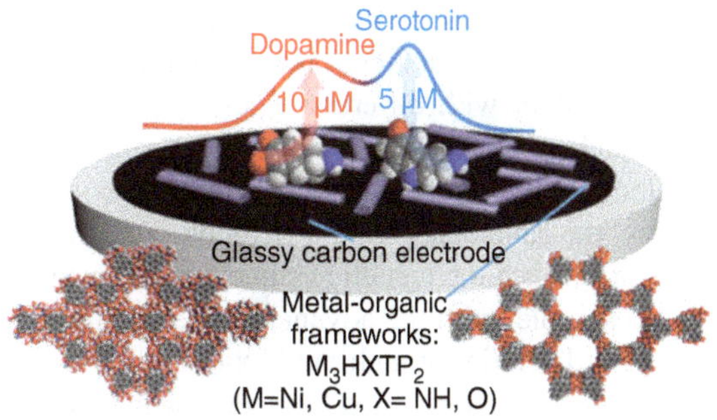

FIGURE 1.7 Schematics for a 2D layered MOF for a neurochemical sensor with high cyclic stability and sensitivity towards dopamine and serotonin. Adapted with permission from reference [26], copyright (2020), American Chemical Society.

Hence, avoiding this step favors the performance and electric signals provided by the MOF. Additionally, the coordination of metal with a conducting ligand such as hexa-substituted triphenylene improves its electrochemical performance.

1.4.5 MOFs for Viruses/Bacteria Sensors

Bacterial infection is a public health concern that takes the lives of around ten million people per year [28, 29]. The identification of pathological bacteria in the early stages is a vital procedure to avoid deaths. Currently, there are several technologies for the identification of these microbes such as polymerase chain reaction, magnetic resonance, mass spectroscopy, Raman scattering, surface-enhanced Raman scattering, electrochemical techniques, and plate culturing. However, alongside quickly identifying the bacteria, it is also necessary to eliminate them, preferably without exaggerated use of antibiotics to avoid bacterial resistance. Given that situation, MOFs can be employed for the identification as well as the elimination of bacteria. One strategy for that can be achieved through the functionalization of MOF's surface to bind specifically with a certain type of bacteria to identify it. Along with that, positively charged groups found, for example, in Ag^+, NH_4^+ have bactericidal properties. Also, boron-based compounds like boric acid are known to easily bind with hydroxyl groups found on the surface of glycolipid receptors on bacteria. With this idea, a group of researchers proposed the synthesis of a fluorescent Zr-based MOF functionalized with boric acid groups (Zr-UiO-66-B$(OH)_2$) that can bind with the glycolipids of the bacterial cell wall for biosensing [30]. Magnetic nanoparticles coupled with *E. coli* antibodies were synthesized, functioning as a capture probe that can bind selectively with the *E. coli* bacteria through antibody-antigen binding. After this process, the Zr-UiO-66-B$(OH)_2$ can latch onto the cell surface of *E. coli*. In the follow-up process, the authors added H_2O_2 to amplify the fluorescence signal intensity which is directly related to the concentration of bacteria in the system. Hence, the bacteria can be quantified. Furthermore, the Zr-UiO-66-B$(OH)_2$ can neutralize the bacteria cell wall when exposed to light illumination due to its semiconductor properties.

Another strategy that has demonstrated satisfactory performance and eco-friendliness is based on the coordination of metallic centers with different types of polysaccharides. This approach grants MOFs convenient properties such as biocompatibility, biodegradability, biosafety, and lower cost due to the abundance of the ligands. Yet, bactericidal properties arise from the metallic center such as from Ag, Zn, or Co, for example, with the advantage of avoiding the route of eliminating bacteria through antibiotics. With this insight, Qian and colleagues fabricated a biodegradable MOF with satisfactory mechanical robusticity and antibacterial properties [31]. For that, cellulose paper (CP) was grafted with carboxylic groups through a Williamson reaction. Then epichlorohydrin (ECH) was used as a crosslinker leading to a fibrous structure intercalated with hydroxyl groups that function as ligands for the metallic centers. Furthermore, the Co-based zeolitic imidazolate nanoparticles (ZIF-67) were deposited over the functionalized cellulose surface leading to the composite CP/CNF/ZIF-67. The concentration of metallic center over the cellulosic surface was directly proportional to the anti-bacterial properties. The antibacterial properties for this composite were satisfactory since low quantities of Co in the MOF (10 µg/mL) provided 80% of bacteria suppression. This effect was tested on *E. coli*. When *E. coli* was exposed to the metallic ions, it went through an ionic imbalance which further led to the decrease of its metabolism, eventually eliminating bacterial activity. Another effect was that the negatively charged polysaccharide segments can attract Ca^{2+} and Mg^{2+} from the bacteria cell wall, leading to a decrease in nutrients within the bacteria and thus eliminating it [32]. Hence, this approach shows a synergetic effect for antibacterial properties which can virtually be applied in any condition that requires a composite to present anti-bacterial properties. The facile synthetic process along with low-cost materials endorses this approach.

1.5 CONCLUSION

It is noticeable throughout this chapter that MOFs are highly versatile materials in several aspects which include synthetic approaches, a variety of starting materials, tunability of properties, and satisfactory performance in a broad range of applications. MOFs can be implemented in energy storage such as batteries and supercapacitors due to their high surface area that allows the diffusion of ions as well as rich redox properties. In addition, the inherently exposed sites of metallic centers allow MOFs catalytic activities for many applications such as water-splitting, for the production of H_2 and O_2. In another field of application, they can be used as gas adsorbents as their high porosity allows easy capture of gases and can even differentiate them. MOFs can be used to identify those gases at various concentration levels, hence working as viable sensors to detect substances that are harmful to the body. They can be used to trap and identify targeted substances such as CWA and TICs which are particularly dangerous compounds. Due to the chemical versatility of MOFs, their chemical surfaces can be functionalized to identify specific types of biomolecules that can cover the range of DNA, RNA, proteins, biomarkers for the identification of diseases to microbes such as bacteria. Through that, MOFs are indeed a valuable tool that is vital for the scientific community as a convenient and efficient material in several fields.

REFERENCES

[1]. Anand R, Borghi F, Manoli F, Manet I, Agostoni V, Reschiglian P, Gref R, Monti S (2014) Host–Guest Interactions in Fe(III)-Trimesate MOF Nanoparticles Loaded with Doxorubicin. *J Phys Chem B* 118:8532–8539

[2]. Lin Y, Kong C, Zhang Q, Chen L (2017) Metal-Organic Frameworks for Carbon Dioxide Capture and Methane Storage. *Adv Energy Mater* 7:1601296

[3]. Sumida K, Rogow DL, Mason JA, McDonald TM, Bloch ED, Herm ZR, Bae T-H, Long JR (2012) Carbon Dioxide Capture in Metal–Organic Frameworks. *Chem Rev* 112:724–781

[4]. Luo Z, Fan S, Gu C, Liu W, Chen J, Liu* BL and J (2019) Metal–Organic Framework (MOF)-based Nanomaterials for Biomedical Applications. *Curr. Med. Chem.* 26:3341–3369

[5]. Lakshmi BA, Kim S (2019) Current and emerging applications of nanostructured metal–organic frameworks in cancer-targeted theranostics. *Mater Sci Eng C 105*:110091

[6]. Kolahalam LA, Kasi Viswanath I V, Diwakar BS, Govindh B, Reddy V, Murthy YLN (2019) Review on nanomaterials: Synthesis and applications. *Mater Today Proc 18*:2182–2190

[7]. Stötzel C, Kurland H-D, Grabow J, Dutz S, Müller E, Sierka M, Müller FA (2013) Control of the Crystal Phase Composition of FexOy Nanopowders Prepared by CO2 Laser Vaporization. *Cryst Growth Des 13*:4868–4876

[8]. Kurland H-D, Grabow J, Staupendahl G, Andrä W, Dutz S, Bellemann ME (2007) Magnetic iron oxide nanopowders produced by CO_2 laser evaporation. *J Magn Magn Mater* 311:73–77

[9]. Dey C, Kundu T, Biswal BP, Mallick A, Banerjee R (2014) Crystalline metal-organic frameworks (MOFs): synthesis, structure and function. *Acta Crystallogr Sect B Struct Sci Cryst Eng Mater 70*:3–10

[10]. Fang X, Wu L (2012) Handbook of innovative nanomaterials: *From syntheses to applications*

[11]. Hiroyasu F, Nakeun K, Bok GY, Naoki A, Beom CS, Eunwoo C, Özgür YA, Q. SR, Michael O, Jaheon K, M. YO (2010) Ultrahigh Porosity in Metal-Organic Frameworks. *Science* (80-) 329:424–428

[12]. Shimomura S, Higuchi M, Matsuda R, Yoneda K, Hijikata Y, Kubota Y, Mita Y, Kim J, Takata M, Kitagawa S (2010) Selective sorption of oxygen and nitric oxide by an electron-donating flexible porous coordination polymer. *Nat Chem* 2:633–637

[13]. Xiao Y, Wei W, Zhang M, Jiao S, Shi Y, Ding S (2019) Facile Surface Properties Engineering of High-Quality Graphene: Toward Advanced Ni-MOF Heterostructures for High-Performance Supercapacitor Electrode. *ACS Appl Energy Mater 2*:2169–2177

[14]. Srinivas K, Chen Y, Wang X, Wang B, Karpuraranjith M, Wang W, Su Z, Zhang W, Yang D (2021) Constructing Ni/NiS Heteronanoparticle-Embedded Metal–Organic Framework-Derived Nanosheets for Enhanced Water-Splitting Catalysis. *ACS Sustain Chem Eng* 9:1920–1931

[15]. Yang X, Su F, Hou M, Zhang D, Dai Y, Liang F (2021) Plasma tailored reactive nitrogen species in MOF derived

carbon materials for hybrid sodium–air batteries. *Dalt Trans* 50:7041–7047

[16]. Martínez-Ahumada E, López-Olvera A, Jancik V, Sánchez-Bautista JE, González-Zamora E, Martis V, Williams DR, Ibarra IA (2020) MOF Materials for the Capture of Highly Toxic H_2S and SO_2. *Organometallics* 39:883–915

[17]. Hamon L, Serre C, Devic T, Loiseau T, Millange F, Férey G, Weireld G De (2009) Comparative Study of Hydrogen Sulfide Adsorption in the MIL-53(Al, Cr, Fe), *MIL-47(V), MIL-100(Cr), and MIL-101(Cr) Metal–Organic Frameworks at Room Temperature. J Am Chem Soc* 131:8775–8777

[18]. Islamoglu T, Chen Z, Wasson MC, Buru CT, Kirlikovali KO, Afrin U, Mian MR, Farha OK (2020) Metal–Organic Frameworks against Toxic Chemicals. *Chem Rev* 120:8130–8160

[19]. Montoro C, Linares F, Quartapelle Procopio E, Senkovska I, Kaskel S, Galli S, Masciocchi N, Barea E, Navarro JAR (2011) Capture of Nerve Agents and Mustard Gas Analogues by Hydrophobic Robust MOF-5 Type Metal–Organic Frameworks. *J Am Chem Soc* 133:11888–11891

[20]. Zou R, Zhong R, Han S, Xu H, Burrell AK, Henson N, Cape JL, Hickmott DD, Timofeeva T V, Larson TE, Zhao Y (2010) A Porous Metal–Organic Replica of α-PbO2 for Capture of Nerve Agent Surrogate. *J Am Chem Soc* 132:17996–17999

[21]. Seo YS, Khan NA, Jhung SH (2015) Adsorptive removal of methylchlorophenoxypropionic acid from water with a metal-organic framework. *Chem Eng J* 270:22–27

[22]. Yang Q, Wang J, Zhang W, Liu F, Yue X, Liu Y, Yang M, Li Z, Wang J (2017) Interface engineering of metal organic framework on graphene oxide with enhanced adsorption capacity for organophosphorus pesticide. *Chem Eng J* 313:19–26

[23]. Zhang S, Rong F, Guo C, Duan F, He L, Wang M, Zhang Z, Kang M, Du M (2021) Metal–organic frameworks (MOFs) based electrochemical biosensors for early cancer diagnosis in vitro. *Coord Chem Rev* 439:213948

[24]. Wang M, Hu M, Hu B, Guo C, Song Y, Jia Q, He L, Zhang Z, Fang S (2019) Bimetallic cerium and ferric oxides nanoparticles embedded within mesoporous carbon matrix: Electrochemical immunosensor for sensitive detection of carbohydrate antigen 19-9. *Biosens Bioelectron* 135:22–29

[25]. Guan Q, Zhou L-L, Li Y-A, Li W-Y, Wang S, Song C, Dong Y-B (2019) Nanoscale Covalent Organic Framework for Combinatorial Antitumor Photodynamic and Photothermal Therapy. *ACS Nano* 13:13304–13316

[26]. Ko M, Mendecki L, Eagleton AM, Durbin CG, Stolz RM, Meng Z, Mirica KA (2020) Employing Conductive Metal–Organic Frameworks for Voltammetric Detection of Neurochemicals. *J Am Chem Soc* 142:11717–11733

[27]. Campbell MG, Dincă M (2017) Metal–Organic Frameworks as Active Materials in Electronic Sensor Devices. *Sensors* 17

[28]. Váradi L, Luo JL, Hibbs DE, Perry JD, Anderson RJ, Orenga S, Groundwater PW (2017) Methods for the detection and identification of pathogenic bacteria: past, present, and future. *Chem Soc Rev* 46:4818–4832

[29]. Brochado AR, Telzerow A, Bobonis J, Banzhaf M, Mateus A, Selkrig J, Huth E, Bassler S, Zamarreño Beas J, Zietek M, Ng N, Foerster S, Ezraty B, Py B, Barras F, Savitski MM, Bork P, Göttig S, Typas A (2018) Species-specific activity of antibacterial drug combinations. *Nature* 559:259–263

[30]. Zuo W, Liang L, Ye F, Zhao S (2021) An integrated platform for label-free fluorescence detection and inactivation of bacteria based on boric acid functionalized Zr-MOF. *Sensors Actuators B Chem* 345:130345

[31]. Qian L, Lei D, Duan X, Zhang S, Song W, Hou C, Tang R (2018) Design and preparation of metal-organic framework papers with enhanced mechanical properties and good antibacterial capacity. *Carbohydr Polym* 192:44–51

[32]. Aguado S, Quirós J, Canivet J, Farrusseng D, Boltes K, Rosal R (2014) Antimicrobial activity of cobalt imidazolate metal–organic frameworks. *Chemosphere* 113:188–192

2 Introduction to Metal-Organic Frameworks

Arpan Hazra[1] *and Stefan Kaskel*[2]

[1]Technische Universität Dresden, Institut für Anorganische Chemie, Bergstraße 66, Dresden, Germany

[2]Technische Universität Dresden, Institut für Anorganische Chemie, Bergstraße 66, Dresden, Germany

2.1 INTRODUCTION

Until the mid-1990s, there were mainly two kinds of microporous materials, zeolites and porous carbons, which were considered as the main classical porous compounds. However, at that time, the scientists were also inspired to conceive different strategies for building new materials due to the tedious synthetic methods and lack of tunable pores in the so-called classical porous solids. The last few decades have witnessed a considerable amount of upsurge in the field of metal-organic frameworks (commonly known as MOFs) which represents an unconventional class of porous materials constructed by the infinite and periodic connections between the organic struts and metal nodes. The use of organic bridging ligands in forming porous material of interest at the moment was developed long before. The approach of the 'node and spacer' concept to generate metal-organic coordination polymers was first introduced by Robson et al. in the early 1990s [1]. They conceived the simple strategy of coordinating linear ditopic organic molecules such as 4,4′-bipyridine with the transition metal ions, giving rise to 1D chains, 2D sheets, or 3D periodic nets. In the early 2000s, only a handful of MOF examples could easily fit in one chapter of a handbook on porous solids with the discussion of relatively low porosity materials. However, currently, more than 20,000 MOFs are reported with some of the structures having 90% porosity and a high surface area extending beyond 7000 m^2/g [2; 3]. The combination of porosity, significantly high surface area, proper distribution of pore sizes, and a wide variety of metal-ligand choices have emerged with metal-organic frameworks as the most promising and reliable substitute to the zeolites and carbon-based porous materials. It is also worth mentioning that sometimes the porous properties of MOF samples are even better than zeolites and activated carbon materials. Other than their porous nature, MOFs have also potential importance in exchange[4], separation [5], heterogeneous catalysis [6], and sensing phenomena [7], and so forth.

2.2 BUILDING UNITS OF MOFS

2.2.1 Concept of Metal Nodes in the Formation of the Framework Structure

The periodic structure of the MOFs in different directions emerges by attaching the metal ions as nodes and with organic linkers as connecters. The degree of dimensionality of these crystalline materials is exclusively established by the coordination environment (coordination number) of the metal nodes which determines the ultimate topology of the system. In this regard, transition metal ions are of popular choice as they offer connections in different directions depending upon the combination of their oxidation state and coordination number. Various structural architectures of MOFs are offered by these metal centers as they can easily adopt various geometries like linear, T- or Y-shaped, tetrahedral, square-planar, square pyramidal, trigonal-bipyramidal, octahedral, trigonal-prismatic, pentagonal-bipyramidal, and their distorted forms (Figure 2.1). Having d^{10} electronic configuration, Ag(I) and Cu(I) are very flexible and thus offer a wide range of coordination spheres (linear, trigonal, tetrahedral, square-planar, square-pyramidal) depending on the synthetic conditions like solvent, temperature, pH, and so on. On the contrary, the transition metal ions with other electronic configurations possess defined coordination environments. For example, Ni(II) and Pt(II) adopt square planar coordination, Co(II) prefer octahedral geometry and Cu(II) forms Jahn–Teller distorted octahedral coordination geometry (eventually square planar or square-pyramidal). MOFs containing Zn(II) and Cd(II) as metal nodes are of paramount interest as they extend a variety of coordination numbers and geometries owing to their d^{10} electronic configuration. The coordination environment of the metal nodes also varies depending upon the reaction conditions, solvents, organic ligands, and counter anions [7]. Other than the transition metal ions, lanthanide metals are another popular choice because of the capability to adopt higher coordination numbers (varying from 7 to 10), leading towards the formation of MOFs having unusual network structures.

2.2.2 Concept of Secondary Building Unit (SBU)

Contrary to the above discussion of single-metal nodes, the framework structure constructed by the polynuclear metal-oxygen cluster, commonly termed as the secondary building unit (SBU) profiles the series of robust crystalline materials with programmable structures and properties. Multidentate organic ligands containing carboxylate moieties as binding sites have the potential to aggregate the metal ions into M–O–C

DOI: 10.1201/9781003188148-2

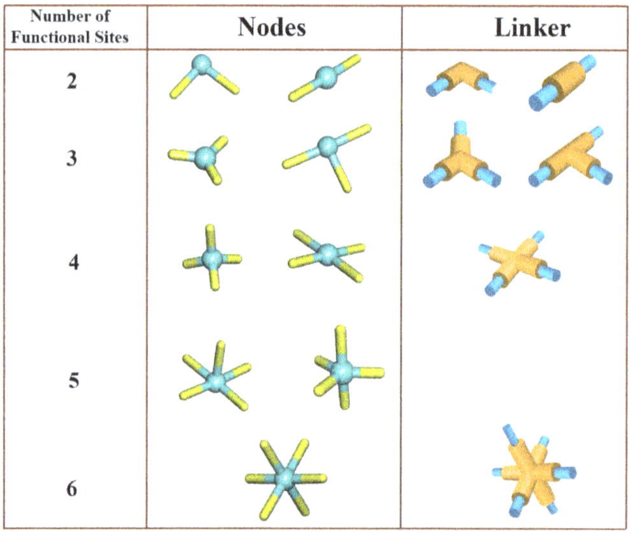

FIGURE 2.1 Schematic representation of the constituents of metal-organic frameworks (MOFs) with different coordination environments of the metal centers and different geometry of the organic linkers.

clusters or SBUs and thus allow the formation of more rigid frameworks. Compared to the MOFs having a single metal node, the position of the metal ions in the SBU are locked by the carboxylate moieties which introduces stability into the system. One can easily generate structurally and chemically stable extended networks by stitching these rigid SBUs with organic linkers of different lengths and denticity. Another advantage of having SBUs in the framework is that they offer a neutral framework precluding the presence of counter anions in the cavity which subsequently increases the porosity of the system. One of the early examples of SBU in MOF structure is found in MOF-3 [8] where a trinuclear Zn-carboxylate cluster can be considered as an octahedral SBU (Figure 2.2a) and a primitive 3D cubic net topology is generated by linking these units with BDC linkers [8]. It was found that the solvent methanol molecules are also connected to the metal centers of the trinuclear cluster and structural integrity remains intact even after removing them. The outstanding stability of MOF-5 in certain conditions is attributed to the presence of truncated tetrahedral-shaped $Zn_4O(–COO)_6$ SBU (Figure 2.2c) [9]. The BDC linkers are connected to the 6-c nodes to form the 3D structure and eliminate all strain at the center of the SBU leading to exceptional mechanical and architectural stability. A similar type of SBU is also present in MOF-177[10] where the same $Zn_4O(–COO)_6$ cluster is connected to the 1,3,5-benzene tricarboxylic acid to form a net with mixed (6,3)-coordination. Another type of $Zn_4O(–COO)_6$ SBU is also found in the CPM-7[11] MOF where three out of 6 carboxylates are chelating two metal ions and the others act as monodentate bindings (Figure 2.2d). Except for the $Zn_4O(–COO)_6$ cluster, CPM-7 also has two other types of SBU ([$Zn_3(OH)$] trimer, and Zn(II) monomer) in the same framework structure which is not discussed here.

The above examples act as the turning point in this field by demonstrating the importance of the SBU approach in synthesizing stable porous MOFs and subsequently, a numerous array of different polynuclear metal clusters have served as SBUs in many framework structures. One of them is a paddlewheel SBU, which is formed by the bridging of two metal centers (for example, Zn(II) or Cu(II)) through three or four carboxylate units (Figure 2.2b). A planar $[M_2L_2]_n$ grid is generated by the connection of a paddle wheel with a linear linker, which can be further connected to the solvent molecules to form stacked 2D porous layers [12] or can be interconnected by ditopic pillars like dabco (1,4-diazabicyclo[2.2.2]octane) to generate a 3D framework [13]. MOF structure containing paddle wheel units displays rigidity (HKUST-1) [14] as well as a remarkable degree of flexibility (DUT-8(Ni), DUT-49) [15; 16].

To introduce more stability in the structure, high valence metal ions, such as Cr(III), Fe(III), and Zr(IV) are also used as their small, hard ions, tend to form a stronger bond with the hard oxygen centers of the carboxylate moiety compared to the larger, soft ions like Zn(II) or Cd(II). This concept was employed in synthesizing MIL-101[17] where the small cluster $Cr_3(\mu_3\text{-}O)$ are connected to six carboxylates from a different BDC linker to create a highly porous structure (pore sizes range from 29 to 34 Å, and Langmuir surface area 5900 m^2g^{-1}) with $[Cr_3(\mu_3\text{-}O)(COO)_6]$ SBUs (Figure 2.2e). It was found that MIL-101 demonstrates high chemical stability in various solvents, pH, and even remains stable for several months in the open air. Extending the concept of using higher valence metal ions, Lillerud et al. reported UiO-66, the highest coordination reported for a MOF, where the SBU is a 12-connected $Zr_6O_4(OH)_4(COO)_{12}$ unit containing Zr(IV) metal ions (Figure 2.2f) [18]. Except for 12 connected SBUs, Zr also offers 8 connected (PCN-222) (Figure 2.2h) and 6 connected (PCN-224) (Figure 2.2g) nodes with Zr_6O_8 units. However, it is also important to note that it is hardly possible to predict the mechanisms of crystal formation and solution equilibria while synthesizing Zr-MOFs. It always requires tuning the reaction conditions over many experiments to obtain a good single crystalline product, instead of getting microcrystalline powders. A subtle change in reaction conditions sometimes yields MOFs with different connectivity [19; 20]. There are also examples of metal-carboxylate infinite chains which are further connected by the organic backbone to generate the 3D framework structures. One such example is MIL-53, where the axial positions of the hexa-coordinated Al(III) centers are filled with the four oxygen atoms from different carboxylates and the equatorial position with an OH group to create the $AlO_4(OH)_2$ octahedra [21]. Such $AlO_4(OH)_2$ units are interconnected to form infinite trans chains (Figure 2.2i) and the pore dimensions of the 3D architecture depends upon the nature of the inserted organic backbone. A similar kind of 1D chain like SBU is also found in MOF-74[22] where the metal ions are coordinated to the carboxylate and hydroxy groups to form helical M–O–C rods of $[M_2O_2](COO)_2]$ units (Figure 2.2j). Each octahedral metal center is connected with three carboxylates, two hydroxy groups, and the remaining coordination site is completed by a terminal DMF ligand. There are many other SBUs reported in the MOF structure

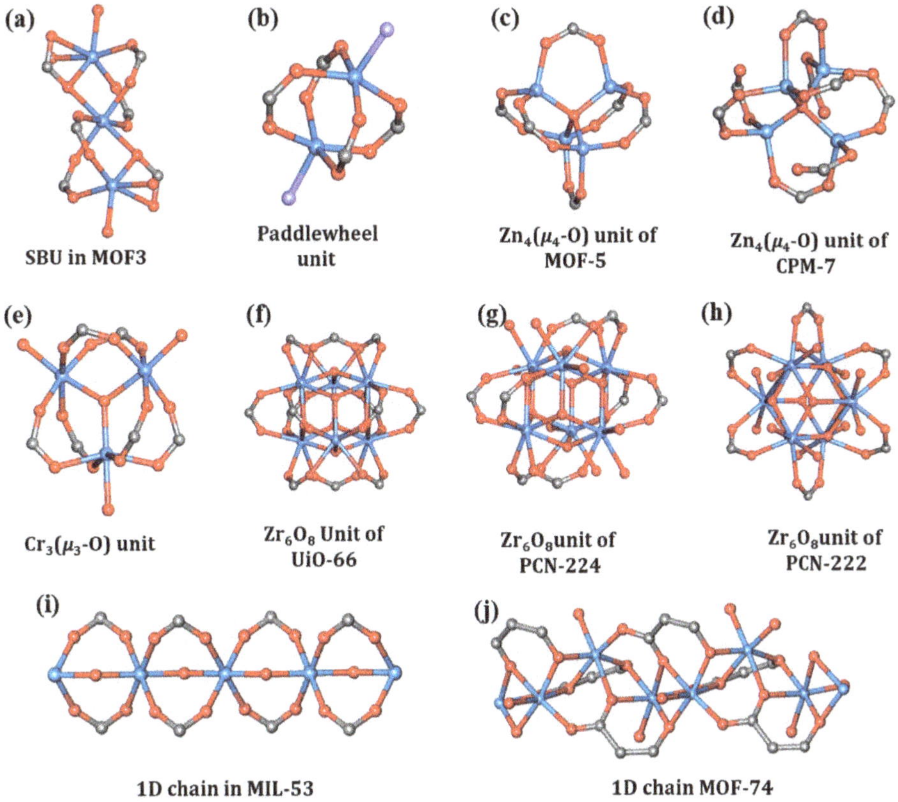

FIGURE 2.2 Some examples of popular SBUs to generate highly porous and/or flexible MOFs with points of extension ranging from 3 to 12. [Colour code: Red – Oxygen, Blue – Metal ions and Grey – Carbon]

which could not be covered in this chapter. For further reading on different kinds of SBUs, the readers are encouraged to read the review article by Yaghi et al.[23; 24]

2.2.3 Organic linkers or Spacers

The popular choice of employed organic ligands is of those with relatively high symmetry (Figure 2.3). They are synthesized from rigid unsaturated hydrocarbon fragments and make an immense contribution to the chemical and mechanical stability of the framework structure. Here it is worth mentioning, that the geometry, length, charges, and the number of binding sites control the pore size (the micro as well as mesopores) and the surface area of the overall framework material. Linkers with rigid organic backbones are typically synthesized by the coupling reactions like Suzuki coupling, Sonogashira coupling, Heck reaction, and so on. Such synthetic procedures are well established and facilitate the straightforward production of linkers of high yield and purity. There are enormous choices of the organic linkers and from which one can easily tune the porosity of MOFs from ultramicropore (<0.7 nm) to mesopore (2–50 nm).

Furthermore, the organic ligands can be classified, not only by the pillar lengths, but also according to their charge, namely (i) anionic, (ii) cationic, and (iii) neutral.[25-27] (Figure 2.4) Anionic ligands are functionalized with –COOH, –OH, or $-SO_3H$ groups whereas the neutral linkers contain primarily pyridyl functionality. Cationic ligands are very rare and hence have been used in limited cases. Depending on the pore size, the majority of MOFs are in the micropore region, although several mesoporous framework structures have also been reported which are sometimes unstable upon evaporation of the solvent molecules. There are several special procedures (solvent exchange, supercritical CO_2, and so forth.) to remove the guest molecules without destroying the mesoporous framework structures.

The organic linkers typically employed for the MOF synthesis generally possess two, three, four, six, eight, or twelve ligation points and they are generally referred to as ditopic, tritopic, tetratopic, and so on (Figure 2.5). In this regard, it is noteworthy that the denticity of a particular ligand is not the only factor to define the network topology, rather it is also important to consider the points of extension of the binding sites. For example, tetrahedral and square planar ligands generate completely different network topologies despite having the same number of ligation sites. Different shaped linkers with the combination of different SBUs render the formation of frameworks of various topologies. Changing the organic backbones and adding functionality, one can easily obtain a plethora of ligands, and the finest example is the isoreticular synthesis of MOF-5 and analogous series [28]. The geometry, functionality and even coordination sites of the linkers can be altered by the post-synthetic modification (PSM) [29] of the framework. Several PSM approaches are

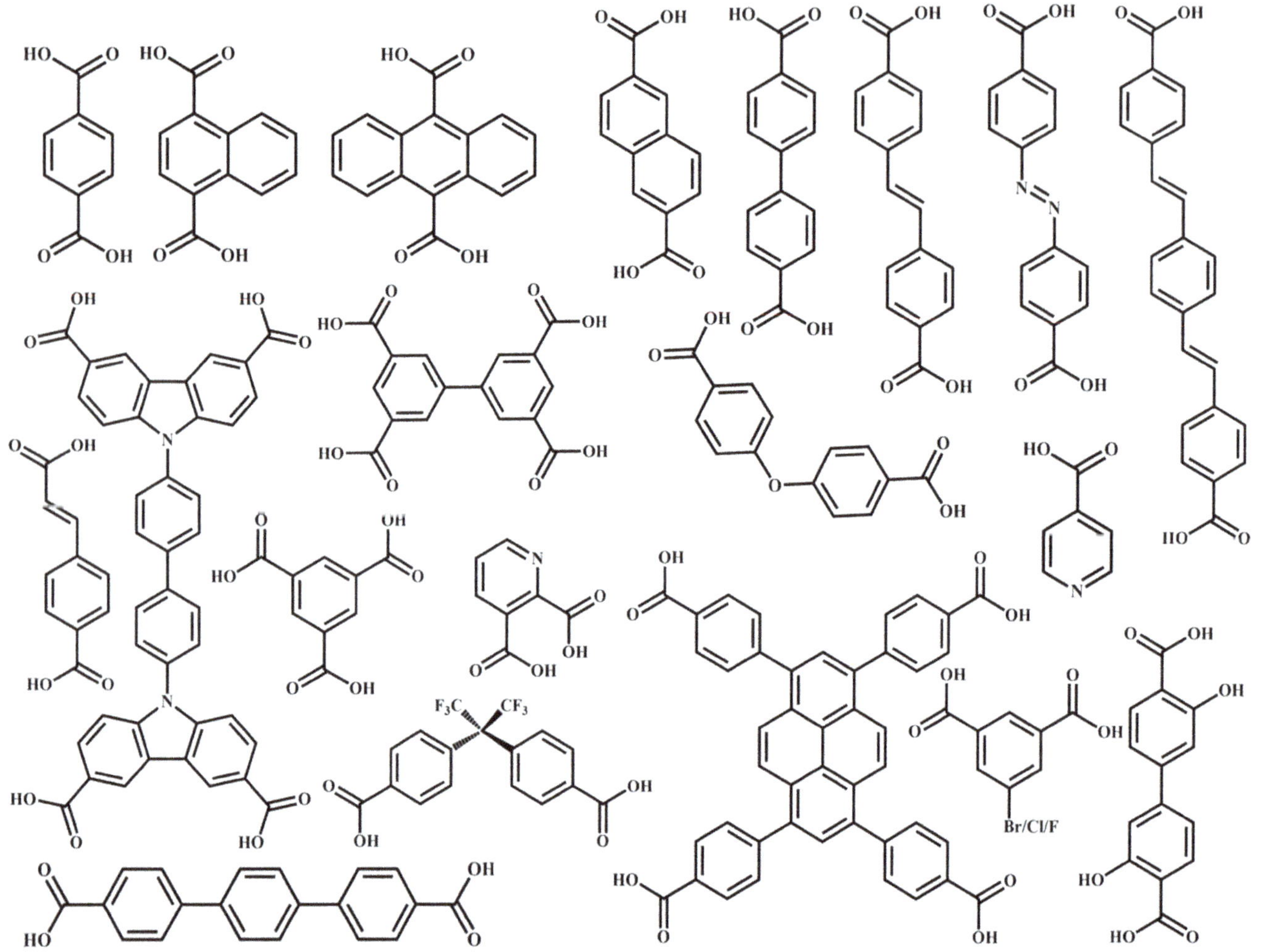

FIGURE 2.3 Commonly used anionic dicarboxylic organic ligands for the construction of porous Metal-organic frameworks.

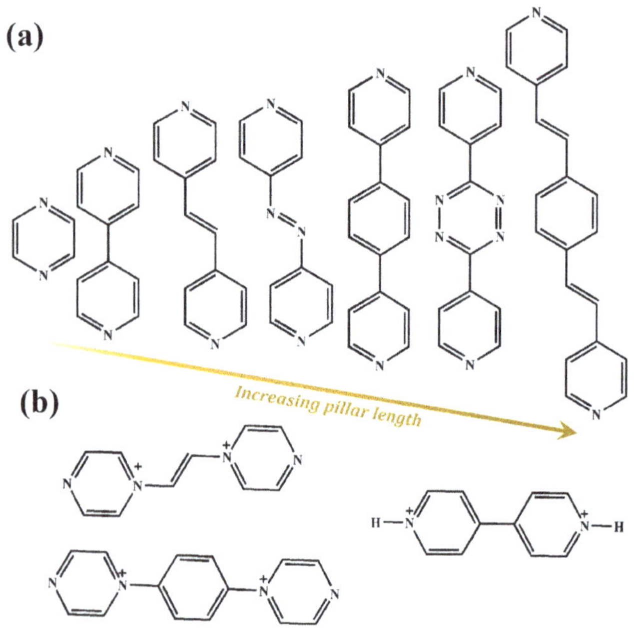

FIGURE 2.4 Various examples of (a) neutral ligands, and (b) cationic ligands. The neutral ligands are often called spacers as they are used to increase the dimensionality of the frameworks.

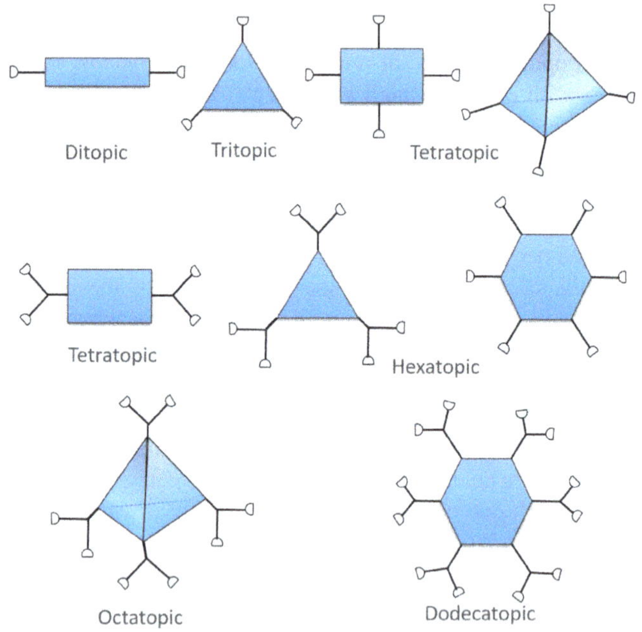

FIGURE 2.5 Figure illustrates the basic geometries of symmetrical organic linkers used in MOF synthesis ranging from 2 to 12 coordination abilities. Here the number coordination ability is described by the term topicity.

reported in the literature which describes procedures like a chemical reaction, incorporation of metal centers in the ligand backbone, photo, and heat-induced reaction, and so forth [30]. Incorporation of a chiral center in the ligand backbone including direct synthesis, post-synthesis, and chiral inductive synthesis produces homochiral porous materials which show excellent performance in the field of enantioselective adsorption, and separation [31].

2.3 FLEXIBILITY AND SOFT NATURE OF MOFs

The above discussion makes it clear that metal-organic frameworks are described as porous solids which are comprised of both a highly ordered crystalline grid and structural transformability. The porosity of MOFs is defined by Kitagawa such that the framework structure must possess at least one crystalline phase where the empty space is occupied by the guest molecules and the framework must demonstrate reproducible guest occlusion. Depending on the stability and transformability of the framework with respect to guest removal and vice versa, MOFs are categorized into three different classes namely: 1st, 2nd, and 3rd generation porous materials. The stability of the 1st generation compounds is directly related to the presence of guest molecules and exhibits irreversible structural collapse upon guest removal (Figure 2.6). These types of polymers consist of metal ions with d^{10} configuration (for example, Cu(I)) and the framework structures disintegrate with respect to guest elimination due to the weak metal-ligand bond strengths [32]. They do not show any permanent porosity. To overcome this situation, carboxylate-based aromatic building blocks have been employed in 2nd generation frameworks. This approach increases the robustness of the framework by achieving stronger metal-ligand bonds. The 2nd generation structures retain structural integrity and do not show any structural collapse or transformation upon guest removal (Figure 2.6) [33]. In contrast to the 1st generation, the 2nd generation frameworks are more stable and always depict higher porosity and specific surface area with respect to the conventional porous solids. Last but not least, the most interesting 3rd generation frameworks render flexibility and are dynamic in nature [34]. The main feature of these types of material is that they provide reversible structural transformations in response to the guest molecules and sometimes to external stimuli like light, heat, pressure, electric field, and so on. In terms of their elastic/inelastic crystalline transitions and softness in the structure, they are referred to as flexible MOFs or *f*-MOFs. One of the prototypical *f*-MOFs is MIL-53[35] which exists in an open pore phase (*op*), demonstrating a structural transformation to a contracted/close pore state (*cp*) upon the extraction of the guest molecules. The thermodynamically stable *cp* phase can reversibly expand into an open pore state (*op*) by adsorbing different guest molecules and sometimes as a function of only temperature without the interference of any guest [36]. The consecutive *op*-*cp* and vice versa transformation of *f*-MOFs with respect to the guest loading is described as 'breathing' behavior. In such a 'breathing' process in *f*-MOFs, the *op*-*cp* transformation resembles the exhalation process of human lungs. On the other hand, *cp*-*op* transformation can be compared to the inhalation process where the stabilization of the *op* phase originates from the enhanced host-guest and guest–guest interactions [37]. It has also been noted that some *f*-MOFs also undergo a metastable phase (sometimes they are referred as nanopore phase or *np* phase) during *op*-*cp* and vice versa transformation [38]. These short-lived intermediate states are very difficult to structurally characterize by employing common characteristic techniques and hence molecular-level explanations are often rare in this case. Prof. Kitagawa has also introduced the concept of the 4th generation framework. This kind of framework is the improved version of 3rd generation materials which includes (1) Hierarchy and Hybridization (double-H), (2) Anisotropy and Asymmetry (double-A), and most importantly (3) Disorder and Defect (double-D).[39] Now, if we can control these factors all together, an arena of new research fields will emerge where we will have a better fundamental understanding of *f*-MOF structures[40] to develop excellent materials for storage, catalytic and electronic applications.

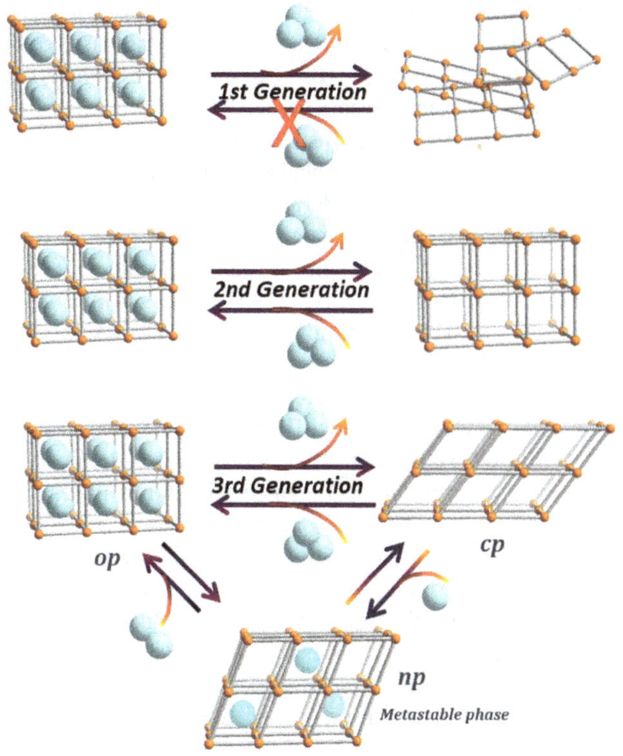

FIGURE 2.6 Schematic representation of first, second, and third-generation microporous Metal-organic frameworks.

2.4 GENERAL APPLICATION OF MOFs

The chemistry of MOF synthesis with an enormous choice of metal clusters and organic linkers has enabled the building of a library of new materials, many of which find interesting applications in the laboratory and in industry. The community is witnessing a collection of research articles being published every year on the synthesis, structures, and different applications of MOFs. However, the question that always arises is whether any breakthrough discovery has been made or the whole field is curiosity-driven in academic research, since the last critical breakthrough was accomplished decades ago. The answer is quite straightforward if we meticulously observe the continuous development accomplished during the last decade. Some of them were beyond our imagination 15 years ago.

The first thought that comes to our mind about MOFs is their porous property. Yaghi *et.al* [41] reported the highest surface area of 6240 m^2g^{-1} in MOF-210 in 2010 and before that the maximum value reported was 5200 m^2g^{-1} for UMCM2 (Figure 2.7) [42]. In the next couple of years, the surface area of MOF material was increased to 7100 m^2g^{-1}, reported by Farha et al. for MOF NUC-110 [2]. Later, Kaskel and coworkers reported a new porous MOF named DUT-60 exhibiting the current world record of 7836 m^2g^{-1} with a pore volume of 5.02 cm^3g^{-1}[3]. It is worth noticing that almost 90% of the material possesses space with a density of just 0.187 g cm^{-3}. It also extrapolates the fact that the inner surface area of one gram of such MOFs can cover a total football field whereas the traditional porous materials like zeolite and graphene typically show a surface area of 800 and 3000 m^2g^{-1}, respectively.

Realizing the availability of high empty volume, numerous attempts have been devoted to applying MOFs as storage materials for fuel gases like H_2 and hydrocarbons (for example, methane, ethane, and ethylene) at ambient conditions. The first potential application of MOFs as hydrogen storage was reported by Yaghi et al. in 2003 [43] and after that, a plethora of MOF structures have been reported with hydrogen sorption data at cryogenic temperatures. Currently, many MOFs with hydrogen adsorption at room temperature are also reported [29; 44]. It is also envisioned that the introduction of open/coordinatively unsaturated metal sites into the MOF backbone can lead towards higher host-guest interaction, resulting in improved sorption capacity at ambient conditions.

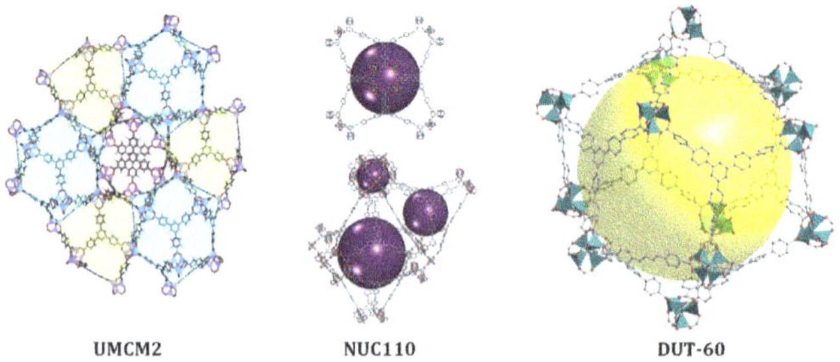

FIGURE 2.7 3D supramolecular architecture of some popular and highly porous MOFs offering more than 90% of void present in the structure. The porosity evolved from 5200 m^2g^{-1} to 7800 m^2g^{-1} in the last 10 years. DUT-60 offers the highest surface area in the MOF library till date. The figures UMCM-2 and NUC-110 are adapted with permission from [42, 2] respectively. Copyright (2009) and (2012). Copyright The Authors, some rights reserved; exclusive licensee [ACS]. Distributed under a Creative Commons Attribution License 4.0 (CC BY) https://creativecommons.org/licenses/by/4.0/

The high surface area of MOFs has been employed for the potential application of high-density methane storage via physisorption [45]. At 298 K, HKUST-1[46] shows an uptake amount of 267 mL g^{-1} at 65 bar whereas MOF-519 and 520[47] have a volumetric capacity of 279 and 231 mL g^{-1} under the same conditions. There have been intensively studied during the past decades, MOFs like Ni-MOF-74, PCN-14, DUT-8(Cu), DUT-51(Zr), and DUT-49 which displayed remarkable methane storage capacities at real applicable conditions [47-49]. Moreover, it is noteworthy that a vessel containing one of these MOFs can store at least twice the amount of methane that could be stored in an empty vessel [41] which is now commercially used by BASF in the automobile industry. The MOF has also established its potential for reversible adsorption and selective capture of carbon dioxide from the atmosphere and flue gas. Carbon dioxide having a large quadrupole moment depicts stronger interaction with the host framework compared to hydrogen and methane. The metal carboxylates for instance MOF-177[50] and MIL-101[51] are showing a huge carbon dioxide capacity having the capacity to take in CO_2 of 33.5 mmol g^{-1} (298K, 4.2 MPa) and 40 mmol g^{-1} (303 K, 5 MPa), respectively. In this context, we can presume that MOF, saturation occurs with large micropores at a comparatively much higher pressure than zeolites, which is also interesting for the recovery of CO_2 at high pressures from a gas stream. It is quite fascinating that MOF-177 and MOF-200 can store 9 and 17 times more CO_2 at 35 bar pressure compared to the empty cylinder without MOF [24]. The study of the porous nature of MOF has also been employed in the selective capture of hydrocarbons and toxic molecules since the use of traditional materials always compensates for the high energy demands. Programmable pore aperture and the advantages of reticular chemistry allow fine-tuning of pore size and functionalization of framework structure which promote MOF materials suitable for hydrocarbon separation compared to other traditional porous solids. The separation of acetylene is one of the most important aspects of the plastic industry as it decreases the catalyst activity. There are several reports where a MOF acts as a separating agent to purify ethylene from ethylene/acetylene mixtures containing 1% acetylene [52; 53]. Separations of hydrocarbon mixtures with similar kinetic diameter like propane/propylene have been successfully achieved by microporous MOFs [54].

Besides their porous nature, MOFs hold great potential for diverse applications like catalysis [55] and sensing [56], drug delivery [57; 58], and magnetism. In this chapter, we have summarized the fundamentals of metal-organic frameworks, which include different approaches to obtain programmable pore architectures for different applications. We have discussed the different building units of MOFs like metal nodes, organic struts and the importance of metal-carboxylic SBUs to establish thermal and chemical stability into the framework structures. Further, important well-studied applications of its porous nature have been pointed out in the above discussions. Subsequent chapters of this book will give the readers a precise understanding of structural design and electrochemical sensing, strategies to improve sensitivity and selectivity, and further merits of MOFs as sensors.

REFERENCES

[1]. Gable RW, Hoskins BF, and Robson R (1990) A new type of interpenetration involving enmeshed independent square grid sheets. The structure of diaquabis-(4,4′-bipyridine)zinc hexafluorosilicate. *J. Chem. Soc., Chem. Commun.* (23): 1677–1678.

[2]. Farha OK, Eryazici I, Jeong NC, Hauser BG, Wilmer CE, Sarjeant AA, Snurr RQ, Nguyen ST, Yazaydın AÖ, and Hupp JT (2012) Metal–Organic Framework Materials with Ultrahigh Surface Areas: Is the Sky the Limit? *J. Am. Chem. Soc.* 134 (36):15016–15021.

[3]. Hönicke IM, Senkovska I, Bon V, Baburin IA, Bönisch N, Raschke S, Evans JD, and Kaskel S (2018) Balancing Mechanical Stability and Ultrahigh Porosity in Crystalline Framework Materials. *Angew. Chem. Int. Ed.* 57 (42): 13780–13783.

[4]. Zhao X, Bu X, Wu T, Zheng S-T, Wang L, and Feng P (2013) Selective anion exchange with nanogated isoreticular positive metal-organic frameworks. *Nat. Commun.* 4 (1):2344.

[5]. Lin R-B, Xiang S, Zhou W, and Chen B (2020) Microporous Metal-Organic Framework Materials for Gas Separation. *Chem* 6 (2):337–363.

[6]. Bavykina A, Kolobov N, Khan IS, Bau JA, Ramirez A, and Gascon J (2020) Metal–Organic Frameworks in Heterogeneous Catalysis: Recent Progress, New Trends, and Future Perspectives. *Chem. Rev.* 120 (16):8468–8535.

[7]. Fang X, Zong B, and Mao S (2018) Metal–Organic Framework-Based Sensors for Environmental Contaminant Sensing. *Nano-Micro Letters* 10 (4):64.

[8]. Li H, Davis CE, Groy TL, Kelley DG, and Yaghi OM (1998) Coordinatively Unsaturated Metal Centers in the Extended Porous Framework of $Zn_3(BDC)_3{\cdot}6CH_3OH$ (BDC = 1,4-Benzenedicarboxylate). *J. Am. Chem. Soc.* 120 (9):2186–2187.

[9]. Yaghi OM, O'Keeffe M, Ockwig NW, Chae HK, Eddaoudi M, and Kim J (2003) Reticular synthesis and the design of new materials. *Nature* 423 (6941):705–714.

[10]. Chae HK, Siberio-Pérez DY, Kim J, Go Y, Eddaoudi M, Matzger AJ, O'Keeffe M, Yaghi OM, Materials D, and Discovery G (2004) A route to high surface area, porosity, and inclusion of large molecules in crystals. *Nature* 427 (6974):523–527.

[11]. Bu F, Lin Q, Zhai Q, Wang L, Wu T, Zheng S-T, Bu X, and Feng P (2012) Two Zeolite-Type Frameworks in One Metal–Organic Framework with $Zn_{24}@Zn_{104}$ Cube-in-Sodalite Architecture. *Angew. Chem. Int. Ed.* 51 (34):8538–8541.

[12]. Li H, Eddaoudi M, Groy TL, and Yaghi OM (1998) Establishing Microporosity in Open Metal–Organic Frameworks: Gas Sorption Isotherms for Zn(BDC) (BDC = 1,4-Benzenedicarboxylate). *J. Am. Chem. Soc.* 120 (33):8571–8572.

[13]. Dybtsev DN, Chun H, and Kim K (2004) Rigid and Flexible: A Highly Porous Metal–Organic Framework with Unusual Guest-Dependent Dynamic Behavior. *Angewandte Chemie International Edition* 43 (38):5033–5036.

[14]. Chui SS-Y, Lo SM-F, Charmant JPH, Orpen AG, and Williams ID (1999) A Chemically Functionalizable Nanoporous Material $[Cu_3(TMA)_2(H_2O)_3]_n$. *Science* 283 (5405):1148–1150.

[15]. Klein N, Herzog C, Sabo M, Senkovska I, Getzschmann J, Paasch S, Lohe MR, Brunner E, and Kaskel S (2010) Monitoring adsorption-induced switching by 129Xe NMR spectroscopy in a new metal–organic framework Ni_2(2,6-ndc)$_2$(dabco). Phys. *Chem. Chem. Phys.* 12 (37):11778–11784.

[16]. Krause S, Bon V, Senkovska I, Stoeck U, Wallacher D, Többens DM, Zander S, Pillai RS, Maurin G, Coudert F-X, and Kaskel S (2016) A pressure-amplifying framework material with negative gas adsorption transitions. *Nature* 532 (7599):348–352.

[17]. Férey G, Mellot-Draznieks C, Serre C, Millange F, Dutour J, Surblé S, and Margiolaki I (2005) A Chromium Terephthalate-Based Solid with Unusually Large Pore Volumes and Surface Area. *Science* 309 (5743):2040–2042.

[18]. Cavka JH, Jakobsen S, Olsbye U, Guillou N, Lamberti C, Bordiga S, and Lillerud KP (2008) A New Zirconium Inorganic Building Brick Forming Metal Organic Frameworks with Exceptional Stability. *J. Am. Chem. Soc.* 130 (42):13850–13851.

[19]. Bai Y, Dou Y, Xie L-H, Rutledge W, Li J-R, and Zhou H-C (2016) Zr-based metal–organic frameworks: design, synthesis, structure, and applications. *Chem. Soc. Rev.* 45 (8):2327–2367.

[20]. Bosch M, Zhang M, and Zhou H-C (2014) Increasing the Stability of Metal-Organic Frameworks. *Adv. Chem.* 2014:182327.

[21]. Loiseau T, Serre C, Huguenard C, Fink G, Taulelle F, Henry M, Bataille T, and Férey G (2004) A Rationale for the Large Breathing of the Porous Aluminum Terephthalate (MIL-53) Upon Hydration. *Chem. Eur. J.* 10 (6):1373–1382.

[22]. Rosi NL, Kim J, Eddaoudi M, Chen B, O'Keeffe M, and Yaghi OM (2005) Rod Packings and Metal–Organic Frameworks Constructed from Rod-Shaped Secondary Building Units. *J. Am. Chem. Soc.* 127 (5):1504–1518.

[23]. Tranchemontagne DJ, Mendoza-Cortés JL, O'Keeffe M, and Yaghi OM (2009) Secondary building units, nets, and bonding in the chemistry of metal–organic frameworks. *Chem. Soc. Rev.* 38 (5):1257–1283.

[24]. Furukawa H, Cordova KE, O'Keeffe M, and Yaghi OM (2013) The Chemistry and Applications of Metal-Organic Frameworks. *Science* 341 (6149):1230444.

[25]. Cheng J-W, Zheng S-T, and Yang G-Y (2007) A Series of Lanthanide–Transition Metal Frameworks Based on 1-, 2-, and 3D Metal–Organic Motifs Linked by Different 1D Copper(I) Halide Motifs. *Inorg. Chem.* 46 (24):10261–10267.

[26]. Horike S, Kishida K, Watanabe Y, Inubushi Y, Umeyama D, Sugimoto M, Fukushima T, Inukai M, and Kitagawa S (2012) Dense Coordination Network Capable of Selective CO_2 Capture from C1 and C2 Hydrocarbons. *J. Am. Chem. Soc.* 134 (24):9852–9855.

[27]. Higuchi M, Nakamura K, Horike S, Hijikata Y, Yanai N, Fukushima T, Kim J, Kato K, Takata M, Watanabe D, Oshima S, and Kitagawa S (2012) Design of Flexible Lewis Acidic Sites in Porous Coordination Polymers by using the Viologen Moiety. *Angew. Chem. Int. Ed.* 51 (33):8369–8372.

[28]. Eddaoudi M, Kim J, Rosi N, Vodak D, Wachter J, O'Keeffe M, and Yaghi OM (2002) Systematic Design of Pore Size and Functionality in Isoreticular MOFs and Their Application in Methane Storage. *Science* 295 (5554):469–472.

[29]. Kalaj M, and Cohen SM (2020) Postsynthetic Modification: An Enabling Technology for the Advancement of Metal–Organic Frameworks. *ACS Cent. Sci.* 6 (7):1046–1057.

[30]. Mandal S, Natarajan S, Mani P, and Pankajakshan A (2021) Post-Synthetic Modification of Metal–Organic Frameworks Toward Applications. *Adv. Funct. Mater*. 31 (4):2006291.

[31]. Lu Y, Zhang H, Zhu Y, Marriott PJ, and Wang H (2021) Emerging Homochiral Porous Materials for Enantiomer Separation. *Adv. Funct. Mater.* 31 (25):2101335.

[32]. Kitagawa S, and Munakata M (1993) Molecular architecture of copper (I) coordination polymers towards crystal lattice design. *Trends Inorg. Chem* 3 (5):437–462.

[33]. Horike S, Shimomura S, and Kitagawa S (2009) Soft porous crystals. *Nat. Chem*. 1 (9):695–704.

[34]. Schneemann A, Bon V, Schwedler I, Senkovska I, Kaskel S, and Fischer RA (2014) Flexible metal–organic frameworks. *Chem. Soc. Rev*. 43 (16):6062-6096.

[35]. Bourrelly S, Llewellyn PL, Serre C, Millange F, Loiseau T, and Férey G (2005) Different Adsorption Behaviors of Methane and Carbon Dioxide in the Isotypic Nanoporous Metal Terephthalates MIL-53 and MIL-47. *J. Am. Chem. Soc*. 127 (39):13519–13521.

[36]. Liu Y, Her J-H, Dailly A, Ramirez-Cuesta AJ, Neumann DA, and Brown CM (2008) Reversible Structural Transition in MIL-53 with Large Temperature Hysteresis. *J. Am. Chem. Soc*. 130 (35):11813–11818.

[37]. Boutin A, Coudert F-X, Springuel-Huet M-A, Neimark AV, Férey G, and Fuchs AH (2010) The Behavior of Flexible MIL-53(Al) upon CH_4 and CO_2 Adsorption. *J. Phys. Chem. C* 114 (50):22237–22244.

[38]. Hazra A, van Heerden DP, Sanyal S, Lama P, Esterhuysen C, and Barbour LJ (2019) CO_2-induced single-crystal to single-crystal transformations of an interpenetrated flexible MOF explained by in situ crystallographic analysis and molecular modeling. *Chem. Sci*. 10 (43):10018–10024.

[39]. Kitagawa S (2017) Future Porous Materials. *Acc. Chem. Res*. 50 (3):514–516.

[40]. Evans JD, Bon V, Senkovska I, Lee H-C, and Kaskel S (2020) *Four-dimensional metal-organic frameworks. Nat. Commun*. 11 (1):2690.

[41]. Furukawa H, Ko N, Go YB, Aratani N, Choi SB, Choi E, Yazaydin AÖ, Snurr RQ, O'Keeffe M, Kim J, and Yaghi OM (2010) Ultrahigh Porosity in Metal-Organic Frameworks. *Science* 329 (5990):424–428.

[42]. Koh K, Wong-Foy AG, and Matzger AJ (2009) A Porous Coordination Copolymer with over 5000 m^2/g BET Surface Area. *J. Am. Chem. Soc*. 131 (12):4184–4185.

[43]. Rosi NL, Eckert J, Eddaoudi M, Vodak DT, Kim J, O'Keeffe M, and Yaghi OM (2003) Hydrogen Storage in Microporous Metal-Organic Frameworks. *Science* 300 (5622):1127–1129.

[44]. García-Holley P, Schweitzer B, Islamoglu T, Liu Y, Lin L, Rodriguez S, Weston MH, Hupp JT, Gómez-Gualdrón DA, Yildirim T, and Farha OK (2018) Benchmark Study of Hydrogen Storage in Metal–Organic Frameworks under Temperature and Pressure Swing Conditions. *ACS Energy Lett*. 3 (3):748–754.

[45]. Tian T, Zeng Z, Vulpe D, Casco ME, Divitini G, Midgley PA, Silvestre-Albero J, Tan J-C, Moghadam PZ, and Fairen-Jimenez D (2018) A sol–gel monolithic metal–organic framework with enhanced methane uptake. *Nat. Mater.* 17 (2):174–179.

[46]. Peng Y, Krungleviciute V, Eryazici I, Hupp JT, Farha OK, and Yildirim T (2013) Methane Storage in Metal–Organic Frameworks: Current Records, Surprise Findings, and Challenges. *J. Am. Chem. Soc*. 135 (32):11887–11894.

[47]. Gándara F, Furukawa H, Lee S, and Yaghi OM (2014) High Methane Storage Capacity in Aluminum Metal–Organic Frameworks. *J. Am. Chem. Soc*. 136 (14):5271–5274.

[48]. Mason JA, Veenstra M, and Long JR (2014) Evaluating metal–organic frameworks for natural gas storage. *Chem. Sci*. 5 (1):32–51.

[49]. Klein N, Hoffmann HC, Cadiau A, Getzschmann J, Lohe MR, Paasch S, Heydenreich T, Adil K, Senkovska I, Brunner E, and Kaskel S (2012) Structural flexibility and intrinsic dynamics in the $M_2(2,6\text{-ndc})_2(\text{dabco})$ (M = Ni, Cu, Co, Zn) metal–organic frameworks. *J. Mater. Chem*. 22 (20):10303–10312.

[50]. Saha D, Bao Z, Jia F, and Deng S (2010) Adsorption of CO_2, CH_4, N_2O, and N_2 on MOF-5, MOF-177, and Zeolite 5A. *Environ. Sci. Technol*. 44 (5):1820–1826.

[51]. Zhang Y, Su W, Sun Y, Liu J, Liu X, and Wang X (2015) Adsorption Equilibrium of N_2, CH_4, and CO_2 on MIL-101. *J. Chem. Eng. Data* 60 (10):2951–2957.

[52]. Hu T-L, Wang H, Li B, Krishna R, Wu H, Zhou W, Zhao Y, Han Y, Wang X, Zhu W, Yao Z, Xiang S, and Chen B (2015) Microporous metal–organic framework with dual functionalities for highly efficient removal of acetylene from ethylene/acetylene mixtures. *Nat. Commun*. 6 (1):7328.

[53]. Denny MS, Moreton JC, Benz L, and Cohen SM (2016) Metal–organic frameworks for membrane-based separations. *Nat. Rev. Mater.* 1 (12):16078.

[54]. Cadiau A, Adil K, Bhatt PM, Belmabkhout Y, and Eddaoudi M (2016) A metal-organic framework–based splitter for separating propylene from propane. *Science* 353 (6295):137–140.

[55]. Ma L, Falkowski JM, Abney C, and Lin W (2010) A series of isoreticular chiral metal–organic frameworks as a tunable platform for asymmetric catalysis. *Nat. Chem*. 2 (10):838–846.

[56]. Kreno LE, Leong K, Farha OK, Allendorf M, Van Duyne RP, and Hupp JT (2012) Metal–Organic Framework Materials as Chemical Sensors. *Chem. Rev*. 112 (2):1105–1125.

[57]. Horcajada P, Gref R, Baati T, Allan PK, Maurin G, Couvreur P, Férey G, Morris RE, and Serre C (2012) Metal–Organic Frameworks in Biomedicine. *Chem. Rev.* 112 (2):1232–1268.

[58]. Thorarinsdottir AE, and Harris TD (2020) Metal–Organic Framework Magnets. *Chem. Rev*. 120 (16):8716–8789.

3 Recent Developments in MOF-Polymer Composites

Vinamrita Singh[1] *and Isha Saini*[2]
[1] Department of Physics, Netaji Subhas University of Technology, East Campus, Delhi, India
[2]Department of Applied Science, School of Engineering and Technology, Sushant University, Gurugram, Haryana, India

3.1 INTRODUCTION

With the growth of population and industrial development, the problem of environmental pollution has also grown. There are numerous types of water and air pollutants including heavy metals, organic dyes, and noxious gases, which are harmful to human health. In this context, numerous materials with extraordinary features have been developed for environmental monitoring and sensing. These include metals, metal oxides, semiconducting materials, carbon-based materials, and polymers. The past few decades have witnessed a growth in the study of various metal-organic frameworks (MOFs) as sensing materials for environmental remediation. MOFs consist of inorganic metal nodes (called secondary building units (SBUs)) and organic linkers connected to form three-dimensional porous material. The first reported MOF structure was Zn-based MOF-5 having a huge specific surface area of 2900 m^2/g and 60% porosity [1]. Afterward, MOFs based on appropriate combinations of metal ions and organic ligands were developed and they possessed improved structural, electrical, optical, and catalytic properties. MOFs display various exceptional features such as the availability of huge surface area, the ability to modify pore size, and the presence of a large number of active sites, which leads to diverse applications. The porosity, exceptional storage, and high catalytic properties of MOFs are utilized in gas separation, catalysis, energy storage, production devices, and so forth.

Despite many extraordinary properties, MOFs usually have low conductivity and lack chemical stability. To mitigate the shortcomings, MOFs have been combined with wide-ranging functional materials to obtain composite materials with enhanced properties and new functionalities. The integration of polymers into MOFs is promising for developing composite materials since the favorable properties of both materials can be combined. MOFs bring in metallic components, distinct porosity, and defined architectures, while the polymers provide easy processability, mechanical strength, and chemical stability. The MOF-polymer composites depict enhanced gas selectivity and adsorption capacity due to which these composites are increasingly being investigated for gas separation, detoxification, biomedical, and energy storage systems [2] [3].

This chapter provides an overview of the development of MOF-polymer composites. The structures and properties of the composites are discussed in detail along with the synergistic effects of the composites for environmental applications. In this chapter, we hope to address the challenges and provide a direction for further research efforts.

3.2 STRATEGIES TO SYNTHESIZE MOF-POLYMER COMPOSITES

Controlled integration of polymers and MOFs creates advanced functional composites that exhibit interesting properties in comparison to the individual components. Diverse approaches for the fabrication of MOF-polymer composites have been exploited by different research groups. An overview of the synthesis techniques is given in Figure 3.1.

3.2.1 Polymerization within MOFs

Polymerization within the confined spaces (nanochannels) of MOFs allows the control of the environment around the polymers, their molecular weight, stereoregularity, polymer sequence, and their chain arrangement. Unique host guest synergies are observed upon encapsulating functional polymers into MOFs. The first example of polymerization in the nanochannels of [M_2(1,4-benzenedicarboxylate)$_2$(triethyl enediamine)] (M = Cu^{2+} or Zn^{2+}) was performed by Uemura et al. [4] and is presented in Figure 3.2. Polymerization of the styrene (St) monomer was induced by heating the mixture with a radical initiator 2,2-azobis(isobutyronitrile) to construct the final composite of MOFs with polystyrene. XRD data of the composite confirms the retention of the pore structure of the MOFs during polymerization. For the analysis of the resulting polymer, the MOF structure was decomposed to obtain PSt. Due to the effective protection of the polymer inside the nanochannels, the propagating radical was free from side reactions and the recovered PSt showed homogenous molecular weight distribution as compared to a polymer synthesized using traditional techniques.

It is challenging to regulate the stereoregularity (tacticity) in radical polymerization since it is quite difficult to control the environment around propagating radical species. It can only be achieved by incorporating polar solvents or Lewis acids into the reaction medium which aids the stereospecific chain growth. In a recent study by Hwang et al. [5], free radical and reversible addition-fragmentation chain-transfer (RAFT)

DOI: 10.1201/9781003188148-3

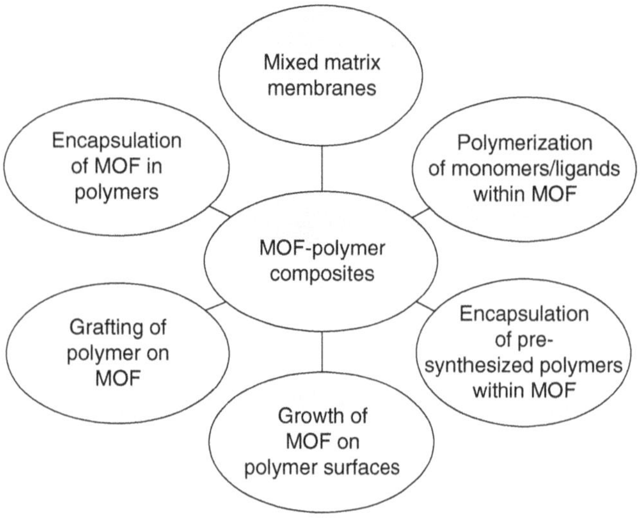

FIGURE 3.1 Overview of MOF-Polymer synthesis techniques.

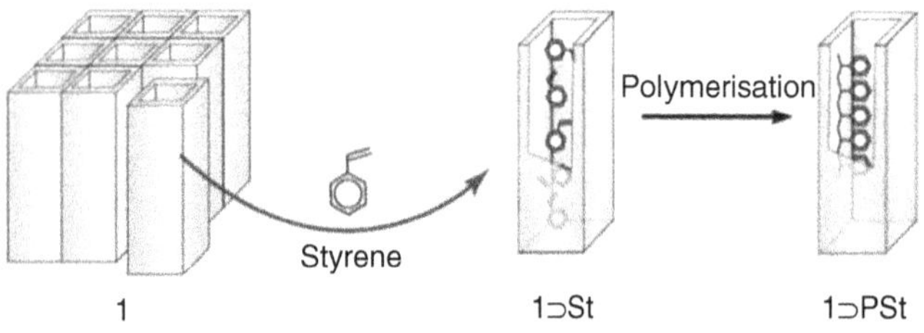

FIGURE 3.2 Schematic representation of the polymerization of St in the nanochannels of $[Zn_2(bdc)_2(ted)]_n$. Reproduced with permission from reference [4], Copyright (2005) The Royal Society of Chemistry.

polymerization of vinyl acetate (Vac), vinyl propionate (VPr), and vinyl butyrate (Vbu) in $[Zn_2$(benzene-1,4-dicarboxylic acid$)_2$triethylenediamine$]_n$ MOF was initiated. A decreasing dispersity and an increasing isotacticity were observed when the monomer size increased. No polymeric products were obtained when the bulky VPr monomer was used which could be due to the deprived monomer mobility. When the RAFT technique was used, the tacticities were the same as in free radical polymerization, however, the disparities were considerably lower. In a study by Schmidt et al. functionalized MOFs and atom-transfer radical polymerization (ATRP) techniques were adopted. The surface-initiated ARGET (activators regenerated by electron transfer) ATRP of various monomers in the nanochannels of Zn_2(benzene-1,4-dicarboxylate$)_2$-(1,4-diazabicyclo[2.2.2]octane) leads to a controlled polymer structure with constant chain length and protected end-chain functionality [6].

The control of polymerization reactions can also be achieved by changing electrochemical parameters (applied potential or applied current density). However, there are only a few reports on the electropolymerization of polymers inside MOFs. In the work by Qiu et al. [7] microporous highly conductive polyaniline polymer having a surface area of 986 m^2/g was synthesized. Polyaniline-supported HKUST-1 thin films were used as a template for electrochemical synthesis. In a similar work by Ruben et al. [8] electrochemical polymerization of monosubstituted 1-hexyne monomers has been successfully carried out in nanochannels of surface-mounted Cu(4-benzenedicarboxylate) MOF (SURMOF-2). Oxidative polymerization is yet another technique to synthesize π-conjugated highly conducting polymers such as polythiophene inside of MOF pores. This technique helps to obtain well-ordered chain alignment of unsubstituted polythiophene inside the 1D channels of [La(BTB)]n (BTB = 1,3,5-benzenetrisbenzoate) [9].

3.2.1.1 Polymerization of Ligands

The alternative approach to obtain composites by polymerization is by polymerizing the ligands (placed at specific positions) with one another to generate polymer chains within the framework or polymers may be formed from ligands by interacting with a cross-linker present inside the MOFs. Vittal et al. [10] used a photochemical [2+2] cycloaddition reaction to polymerize the conjugated diene ligand. The slip stacked assembly of C=C bonds in 1,4-bis[2-(4-pyridylethenyl] benzene (bpeb) ligands in a six-fold interpenetrated Zn(II) MOF with diamondoid topology undergoes [2+2] cycloaddition to transform into $[Zn$(poly-bppcb)(bdc)$]_n$ (poly-bppcb = 1,3-(4,4-bipyridyl)-2-phenylcyclobutane) polymer in which tetrahedral geometry is preserved around the Zn^{2+}. In another work, an infinite slip stacked assembly can also be obtained by the tetrahedral Zn(II) atoms in the CDS topology in place of diamondoid topology [11]. A syndiotactic polymer ligand was obtained in a $[Zn_2$(bpeb)(obc$)_2]_n$ (obc = 4,4'-oxybis-(benzoic acetate)) MOF with a pillared layer structure by the same research group [12].

Sada et al. [13] fabricated a Zn-based MOF by treating Aztpdc with Zn^{2+} using solvothermal method followed by the polymerization with CL2 with Cu(I) as a catalyst for click reaction to perform copper(I)-catalyzed azide-alkyne cycloaddition (CuAAC). The resulting ZnAzMOF was dissolved completely in an acidic solution to form a linear polymer. The same research group [14] described a box-like structure in which a MOF layer with reactive azide functional ligands is present over an unfunctionalized cubic MOF. CuAAC was performed on a tetra alkyne compound resulting in structures having an outer layer of crosslinked MOF and an inner MOF core. A boxlike cubic polymer structure was obtained after the removal of the MOF.

Polymer gels can also be obtained by a post-synthetic click reaction between MOF ligands and incorporated cross-linkers. Woll et al. prepared polymer gel coatings, called surface-grafted gels (SURGELs) by converting highly ordered epitaxially grown surface-mounted MOFs (SURMOFs) through crosslinking of alkyne and azido group-based linkers [15]. A state-of-the-art technique to convert a surface mounted MOF (SURMOF) into an interwoven polymer textile using a unique technique called, 'molecular weaving' is presented by Wang et al. [16]. The 2D polymer textiles are produced using liquid-phase epitaxy in which the

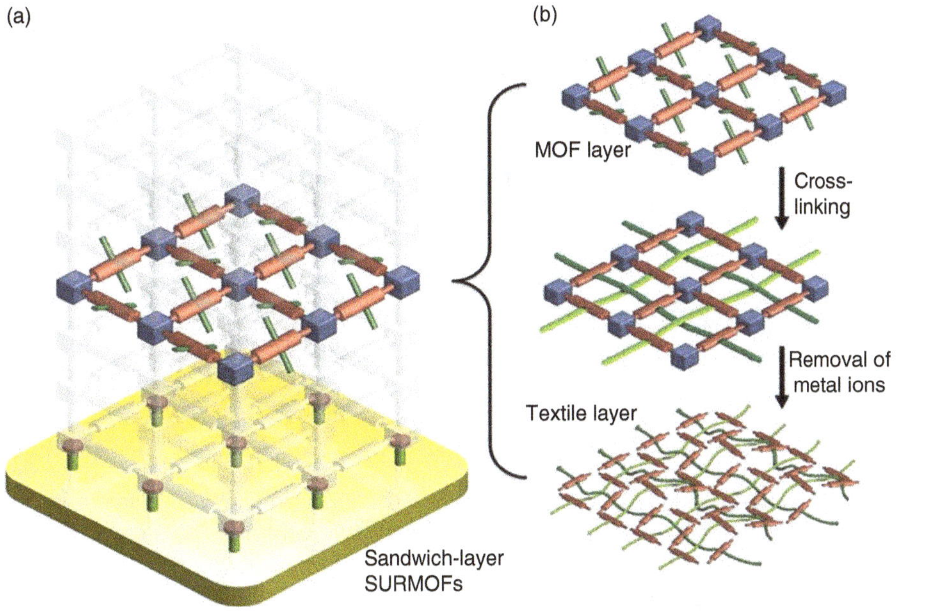

FIGURE 3.3 Schematic illustration of (a) heteroepitaxial sandwich-layer surface-mounted MOF (SURMOF) system and (b) the formation procedure of molecular weaving in the active MOF layer embedded between two sacrificial layers. Reproduced from reference [16], Copyright (2017) Nature Communications Chemistry. The article was printed under a CC-BY license.

quadritopic linkers are sandwiched between sacrificial MOF thin films to form a multi-heteroepitaxial crystalline system (Figure 3.3). The linkers were converted via Glaser-Hay coupling, which is a reaction involving the coupling of cyclic molecules with terminal alkynes to form bisalkyne molecules that subsequently yielded linear, interwoven polymer chains in a 2D array.

3.2.2 Growth of MOFs on Polymer Surfaces

There is significant progress in the fabrication of core-shell MOF-Polymer structures. Lee et al. have demonstrated a solvothermal method for the preparation of Polystyrene core@ZIF-8 shell particles. By the interaction of carboxylate groups present on the polystyrene surface and Zn^{2+} ions, the ZIF-8 MOF grows on the surface of the polystyrene spheres. A highly unique hollow ZIF-8 structure was obtained after removing polystyrene cores by etching [17]. In a similar study by Li et al. the layer-by-layer self-assembly technique was utilized to fabricate polystyrene (PS)@MIL-100(Fe) and hollow MIL-100(Fe) spheres. Sulfone groups were prepared on the surface of PS which then interacted with metal ions for the growth of a MOF. The core-shell PS@MIL-100(Fe) structures were fabricated by a stepwise growth cycle. Finally, hollow MOF spheres were obtained by removing the polystyrene core using dimethylformamide solution [18].

A bottom-up approach in which polymer ligands transform into crystalline frameworks after the incorporation of metal ions leads to the formation of fascinating MOF-polymer composites, named 'PolyMOFs'. In this field, Cohen et al. have done pioneering work. The first polyMOFs were synthesized by using a linear polymer containing 1,4-benzenedicarboxylic acid (H_2bdc) units by varying different lengths of methylene spacers. The ligands self-assemble into a MOF-5 $[Zn_4O(bdc)_3]_n$ structure by incorporation of Zn^{2+} ions. The typical MOF-5 has cubic geometry, however, with the use of polymeric ligands, spherical crystals were obtained [19]. In the following reports, Zr^{4+}-based UiO-66 framework polyMOFs with a unique morphology and hierarchical porosity were synthesized with different spacer lengths in the polymer backbone between bdc units. The polymers with precise end-groups and higher molecular weights were obtained by acyclic diene metathesis (ADMET) polymerization, but with reduced surface area [20]. To address the issue of surface area, the isoreticular polyMOFs with enhanced pore volumes and surface areas were prepared [21].

3.2.3 Encapsulation of Polymer Chains into MOF Structures

The in-situ polymerization method has been used for the fabrication of MOF-polymer composites but still the exact control of molecular weight and number of polymers pose a huge challenge. To address this limitation, the direct incorporation of pre-synthesized polymers into MOFs is an effective strategy. Pastore et al. [22] reported a highly versatile approach for MOF-polymer hybridization by post-synthetic exchange of ligands between MOF crystals and ligand functionalities on a poly(amic acid) (PAA) backbone to help integrate polymer chains into the framework architecture (Figure 3.4). The composites have been synthesized using three different MOFs namely: ZIF-8, MOF-5, and UiO-66. It was noticed that ligand exchange was not observed in UiO-66 because of the lower exchange rates of the Zr^{4+} carboxylate

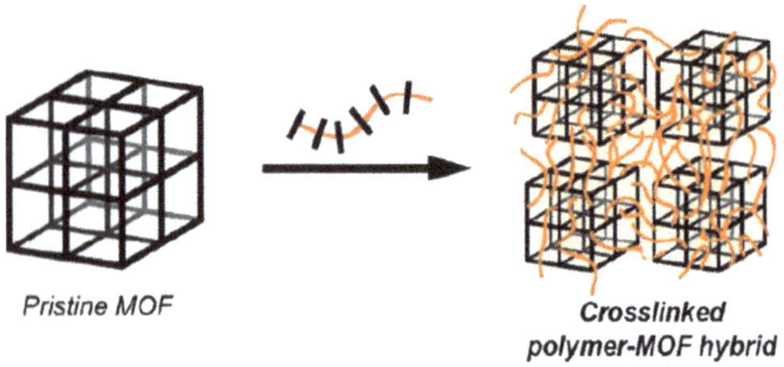

FIGURE 3.4 Schematic of direct integration through post-synthetic ligand exchange to form a cross-linked polymer-MOF network with preserved porosity. Reproduced with permission from reference [22], Copyright (2018) American Chemical Society.

bonds. Pristine MOF-5 is highly unstable in water, however, when integrated into a polymer, remarkable structural stability was observed with no degradation for at least 20 days with 99.8% retention of porosity as observed using N_2 adsorption experiments. This can be attributed to the restricted entry within the MOF pores posed by the large size of PAA chains.

Uemura et al. [23] utilized melt processing to form MOF-polymer composites. Polyethylene glycol (PEG) and poly(methylpropylsilane) have been integrated within MOFs by heating pre-synthesized polymers and MOFs beyond the polymer melting point. This technique can be used with low molecular weight polymers since high molecular weight polymers have lower diffusivity into the pores. For the incorporation of high molecular weight polymers, Matzger et al. [24] reported a solution-based approach in which poly(3-hexylthiophene) (P3HT) was introduced inside a Zn-MOF by dipping it into a polymer-dissolved chlorobenzene solution. This direct encapsulation of polymers inside the MOF gives good control over the polymer amount, however, pre-synthesized polymers can have a pore-blocking effect and it is difficult to control a homogenous distribution of the polymer. Often, polymers with highly branched structures create a restriction on diffusion inside MOF pores.

3.2.4 Polymer Grafted MOFs

In grafting strategies to synthesize MOF-polymer composites, polymers are attached to the surface of MOFs via covalent attachment. A large number of polymers were grafted onto the surface of a MOF to tackle the issues of dispersibility, hydrophilicity, protection against the environment, or stability against chemicals. There are two techniques to graft polymers onto MOFs: 'grafting-to' and 'grafting-from' strategies. The grafting-to approach involves the covalent attachment of reactive groups onto the polymers. In the first study of grafting by Sada et al.[25] an active ester end-functionalized thermoresponsive polymer poly(N-isopropylacrylamide) was grafted to the surface of UiO-66 which is functionalized using amine groups. The grafted polymer was utilized to switch from the ON state (coil state) to the OFF state (globule state) as a function of temperature, in other words, the MOF

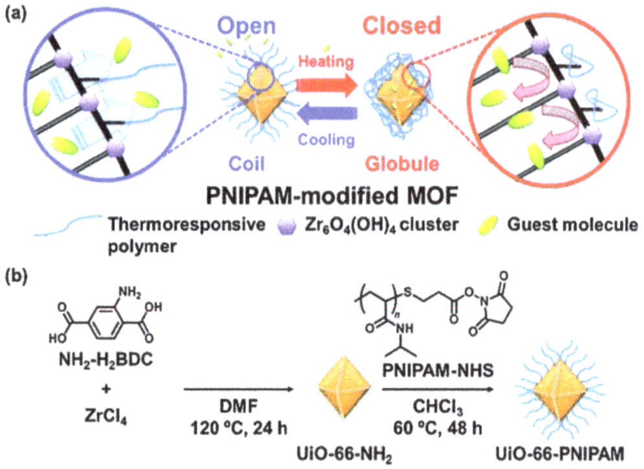

FIGURE 3.5 (a) Schematic image of controlled release using MOF tethering PNIPAM. (b) Preparation method for MOF tethering PNIPAM (UiO-66-PNIPAM). Reproduced with permission from reference [25], Copyright (2015) The Royal Society of Chemistry.

surface can be open or closed according to temperature, leading to the defined control of guest release (Figure 3.5). In another study, Wuttke et al. [26] tried to control the external surfaces of MOFs while retaining the porosity by utilizing unsaturated ligands on the MOF surfaces. A highly stable and carboxylic acid groups attached MIL-100(Fe) MOF was selectively functionalized with two different kinds of polymer, a hydrophilic polyethylene glycol (PEG) 5000 polymer, used for stability and shielding of nanoparticles and another polymer Stp10-C, which is an oligoamino amide by a reaction using carbodiimide. The resulting composites have enhanced stability in an aqueous medium.

In the grafting-from technique, the polymer is grown from initiator groups functionalized surfaces. In an approach given by Li et al. [27] for the grafting-from technique, a hybrid architecture of UiO-66-NH_2@PMAA was prepared using ATRP, click chemistry, and post-synthetic modification. The polyanionic coating aids in improving the MOFs dispersion in water. Matzger et al. [28] synthesized a layer of ATRP initiator modified IRMOF-3 on a MOF-5. A polymethylmethacrylate layer was grafted from the MOF surface to form polymershell@MOFshell@MOFcore architecture. In the resulting structure, the polymerization takes place on the outer shell only, which helps in the retention of the high surface area during polymerization.

3.2.5 Encapsulation of MOFs into Polymers

Coating MOFs with an organic polymer layer can also be done through non-covalent attachment where polymer molecules having an end group can attach to the MOF particles through electrostatic attraction or hydrogen bonding. Wu et al. explored the non-covalent attachment of MIL-101-NH_2(Fe) MOF and fluorescent dyes and studied the surface modification of the MOF with comb-shaped copolymers. The binding of dye molecules on a MIL-101-NH_2(Fe) surface is validated by the creation of surface hydrophobic assemblies [29]. The porous MOF-polymer composites can also be obtained by using high internal phase emulsions (HIPEs). Schwab et al. demonstrated for the first time the incorporation of a HKUST-1 MOF within a monolithic porous polyHIPE. A direct crystallization of HKUST-1 takes place inside the polymer, which is validated by the appearance of homogenous light blue color on the monoliths. Crystals with sizes ranging from 0.2 µm to 10 µm were obtained [30]. The simultaneous crystallization of a MOF and polymerization of monomers can also occur to obtain MOF (HKUST-1)-polyHIPE composites [31].

Fu et al. [32] fabricated poly(MAA-co-EDMA) monoliths incorporated with a UiO-66 MOF via thermally initiated polymerization. The appearance of characteristic peaks of UiO-66 in XRD and FTIR spectra confirms their presence in the monolith. An increase in BET surface area from 192.6 m^2/g for monoliths to 321.4 m^2/g for MOF-incorporated monoliths was observed and the presence of sphere-shaped MOF particles inside monoliths was confirmed using SEM micrographs. Zhang et al. [33] devised a technique to coat the MOF surface with a thin layer of polydimethysiloxane (PDMS). Different types of MOF were used for the study, namely, $[Zn(bdc)(ted)_{0.5}]\cdot 2DMF\cdot 0.2H_2O$ with pillared square-grid $Zn_2(COO)_4N_2$ clusters, MOF-5 with $Zn_4O(COO)_6$ clusters, and HKUST-1 with paddle wheel $Cu_2(COO)_4$ centers. The thermal degradation of a PDMS coating on a MOF surface leads to deposition of volatile silicone molecules, which consequently cross-link to form a hydrophobic silicone coating. All the coated MOFs display excellent retention of porosity and crystal structure, and the hydrophobic layer prevents the interaction between metal ions and water molecules, thereby increasing water stability and protecting the catalytic activity of the coated MOF.

3.2.6 Mixed Matrix Membranes

Another very advantageous synthesis method of MOF-polymer composite is in the form of membranes known as mixed matrix membranes (MMMs) which are composites of MOF embedded polymeric matrices. Generally, polymeric membranes have been used in various industrial applications such as filters, reverse osmosis (RO), water treatment, heavy metal extraction, CO_2 separation, and the like. The major requirement for a membrane is excellent permeability and good selectivity along with exceptional physical features. However, membranes are characterized by Robeson's upper bound that means highly selective membranes will have diminished permeability or vice versa. Among the various porous materials, a MOF, due to its well-defined pore architecture has been considered to be an ideal material for membrane applications. Duan et al. [34] reported a MOF-based MMM thin film of ZIF-8 and polyamide matrix for a reverse osmosis membrane. It was observed that the addition of only 0.4 wt% of ZIF-8 led to an increase of 162% in water permeance compared to a pristine polyamide membrane while maintaining 98% NaCl rejection. An example of high MOF loading in MMM was displayed by Cohen et al. [35] that uses UiO-66, UiO-66-NH_2, MIL-53, MIL-101, HKUST-1, and ZIF-8 with loading as high as 67 wt % in PVDF. Maspoch et al. presented MIL-88A/PVDF-based self-folding films which can undergo shape transformation and can be used in encapsulation, robotics, and harvesting applications [36].

Defect-free MMMs were synthesized by Tien-Binh et al. [37]. The poor polymer and filler interaction was addressed by the cross-linking of polymers of intrinsic microporosity (PIM-1) with UiO-66-NH_2. The in situ chemical reactions between 1,4-dicyanotetrafluorobenzene monomer and amine groups of UiO-66(Zr) leads to the grafting of PIM-1 on the MOF surface, which in turn leads to enhancement of adhesion and gas separation capabilities.

3.3 PROPERTIES OF MOF-POLYMER COMPOSITES

The MOF-polymer composites possess unique properties that have expanded their scope of application. The mixing of MOFs with polymers can be carried out either to enhance their stability or conductivity or to extract the desired optical and thermal properties. For example, the incorporation of PEDOT in the 1D channels of non-conductive MIL-101 (Cr) increased the conductivity to 1.1×10^{-3} S/cm, which was then used for chemiresistive sensing of NO_2 [38]. The mechanical strength of MOFs can be enhanced by adding polymers. The aspect ratio of the fillers strongly determines the strength of the synthesized composite. When a zirconium-based MOF, PCN-222, was employed as a particle filler in PMMA, the mechanical strength increased significantly with an increase in the particle aspect ratio as well as the concentration (up to 0.5 wt.%), while the chemical properties remained the same [39]. The metal-organic ligand bonds primarily determine the thermal strength of the MOFs. Upon the addition of polymers, it has been found that the thermal stability of the composite can be enhanced. In applications like fire retardants, certain MOF-polymer composites show good stability [40]. The MOF-polymer composites have also been utilized for photocatalytic degradation. Most of the MOFs like MOF-5, UiO-66(Zr), and MIL-125(Ti) absorb in the UV range, limiting their application in photocatalytic activity. Combining MOFs with suitable polymers results in lowering of the bandgap and an increase in the absorption.

3.4 APPLICATIONS OF MOF-POLYMER COMPOSITES

3.4.1 For Hydrogen Storage

Hydrogen is a clean fuel alternative for the automotive industry as it is considered environmentally benign and hence its efficient storage is crucial for promoting the hydrogen economy. It is well established that porous MOFs and polymers having high surface areas can store hydrogen via physisorption and shows faster kinetics for a reversible process. The PIM-1-MIL-101(Cr) composites show enhanced surface area, pore structure, and H_2 adsorption ability

TABLE 3.1
depicts the H_2 uptake of some of the MOF-polymer composites

MOF-Polymer Composite	Pore Area (m^2/g)	Surface Area (m^2/g)	Pore Volume (cm^3/g)	H_2 uptake (wt%)	Ref.
PIM-1/UiO-66 (Zr)/ ZTC (80 wt%)	1,668	1767	0.83	1.65	[43]
PIM-1/80 wt%UiO-66(Zr)	1014	1054	0.48	1.22	[43]
CF/PIM-1/ UiO-66(Zr)	645	768	0.43	1.05	[42]
MIL-101/PIM-1	2290	2347	1.2	1.73	[41]
MIL-101/PIM-1	-	1415	0.83	1.11	[45]

with an increase in MOF concentration without any pore blockage [41]. However, the composite has a lower thermal conductivity, which is an essential component for faster heat dissipation in H_2 storage systems. PIM-1/UiO-66(Zr) with immobilized carbon foam (CF) composite gives 1.05 wt% H_2 adsorption capacity at 77 K, but the thermal conductivity reduces from 0.2915 W/mK for UiO-66(Zr) to 0.1189 W/mK for the CF/PIM-1/UiO-66(Zr) composite [42]. To address the problem, zeolite templated carbons (ZTC), PIM-1, and the UiO-66(Zr) MOF were merged into a composite material. ZTC in the composite aids to enhance thermal conductivity and stability. The composites have shown improved hydrogen storage capabilities with a negligible loss and retention of the inherent properties of pristine ZTC [43].

Another unique approach for synthesizing the MOF-polymer composite material for H_2 storage is based on 3-D printing a composite of acrylonitrile butadiene styrene (ABS) and the MOF-5. The results of H_2 adsorption and desorption isotherms showed the hydrogen storage capacity of composite is somewhat similar to the pristine MOF-5 indicating that the performance of the MOF is not degraded in the 3-D printed matrix. Hence, this technique opens up many avenues to prepare polymer-MOF composites for storage applications using environmentally friendly synthesis techniques [44].

3.4.2 For CO_2 Storage/Adsorption

MOF-polymer-based composites have demonstrated great potential for selective CO_2 capture. Hong et al. [46] successfully fabricated beads of diamine-functionalized MOF-polymer composite for indoor CO_2 removal. Firstly, MOF powder was added to different concentrations of poly(vinylidene fluoride) (PVDF) polymer to obtain composite beads via the phase inversion process. After that, for the amine functionalization process, MOF-PVDF composite beads were dissolved in a solution containing 1-ethylpropane-1,3-diamine (epn) to obtain an epn-MOF-polymer composite. The CO_2 isotherms of the formed composite beads were obtained and the beads with 40% PVDF concentration showed an excellent CO_2 adsorption ability without structural degradation after exposure to 1000 ppm CO_2 at 60% relative humidity. Hence, a combination of MOF powders with hydrophobic polymeric binders such as PVDF proved to be a facile technique to effectively capture CO_2.

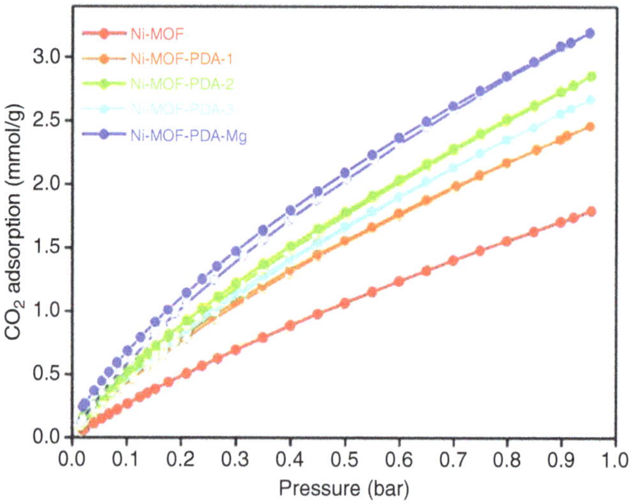

FIGURE 3.6 CO_2 adsorption isotherms of Ni-MOF (red), Ni-MOF-PDA-1 (orange), Ni-MOF-PDA-2 (green), Ni-MOF-PDA-3 (cyan), and Ni-MOF-PDA-Mg (royal blue) measured at 298 K. Reproduced with permission from reference [47] Copyright (2018) The Royal Society of Chemistry.

Ni-MOF/PDA (polydopamine) composite prepared via a double solvent method improves CO_2 adsorption. The isotherms of N_2 adsorption-desorption indicate complete retention of porosity of the composites after the addition of PDA (Figure 3.6) [47]. The results obtained from CO_2 adsorption isotherms show an increase of CO_2 uptake capacities from 1.7 mmol/g for Ni-MOF to 2.45 mmol/g for Ni-MOF-PDA-1 to 2.85 mmol/g for Ni-MOF-PDA-2. However, with further increase in PDA concentration capacity does not increase which could be either due to the pore-blocking effect of the polymer, making amine groups inaccessible or may be due to the external polymerization which limits the surface area by forming a dense layer. To further enhance the adsorption capacity of the composite Mg^{2+} was added to the Ni-MOF-PDA-2 composite. The BET surface area obtained is 1526 m^2/g for Ni-MOF-PDA-Mg, which is less than that of the original composite but the CO_2 adsorption capacity Mg^{2+} added composite is increased to 3.2 mmol/g which is approximately 12% higher than the starting composite.

Recently, many gas industries are focusing on MMMs for CO_2 gas separation. The MOFs are chemically flexible, therefore it is easier to selectively adsorb certain molecular species from a gaseous mixture, which leads to enhancement of the separation performance of the membranes. A lot of research has been concentrated on the study of the CO_2 separation of MMM-containing MOFs. For example, amine-functionalized UiO-66 MOF and PIM-1 MMMs show enhanced selectivity for CO_2 capture without affecting permeability [48].

3.4.3 Water Treatment

3.4.3.1 Photodegradation of Organic Pollutants/Dyes

MOFs have recently been used as adsorbents for the removal of hazardous water pollutants and the adsorption of dyes because of their inherent characteristics. Liu et al. were the first to research the water purification of MOF-polymer composites [49]. A 3D printed acrylonitrile butadiene styrene (ABS) template was used to coat a Cu-BTC MOF (or HKUST-1). In the study, methylene blue (MB) dye was utilized as a model compound to assess decontamination activity. The efficiency of dye removal reached 93.3% and 98.3% in only 10 min for the composites with concentrations of 10 mg/L and 5 mg/L, respectively, and it is maintained above 58% and 68% even after five cycles. This is the first report on 3D printed material for dye extraction. The main advantage of this 3D printed technique is that it does not require the restructuring of MOF powders in the form of, say, pellets.

In another report by Brahmi et al. [50], photopolymerization was used for the synthesis MOF-polymer composites, which are used as catalysts for the photodegradation of acid black pollutants from water. Cu-HKUST-1 and Cr-MIL-53 MOFs were incorporated in an acrylate monomer trimethylolpropane triacrylate (TMPTA) followed by polymerization. The structural characterization revealed successful MOF infiltration within the polymeric matrix with the preservation of pristine structures. The pollutant removal from water reaches 96% and 90%, respectively for Cr-MIL-53/polymer (2%) and HKUSTI-1/polymer (2%) composites under UV lamp irradiation for 30 min, indicating a higher performance of MOF/polymer composites than the traditional titanium dioxide-polymer composites which reduces to only 35% after 60 mins.

Dong et al. [51] used the grafting method to fabricate PDMAEMA-g-UiO-66-NH_2 (PDMAEMA = poly(2-dimethylaminoethyl methacrylate) composite via the photopolymerization of methylmethacrylate monomer. This approach is quite useful for preparing polymer brushes on the surface of MOFs under environment-friendly conditions. The prepared composite successfully removed organic dye from water.

In another study, a water-stable composite of Fe-BTC (MOF-100) (BTC:1,3,5-benzenetricaboxylate) with varying quantities (28, 38, and 42 mass %) of PDA was developed by Queen et al. [52]. The composite displays fast and selective elimination of huge amounts of heavy metals (Hg^{2+} and Pb^{2+}) from water which could be attributed to the high density of the heavy metal removing groups on the backbone of polymeric material. Despite the enhanced performance of the composites, few of the fine composite particles present in the solution lead to a loss in the active material. To address the issue, the same group used a structuring method to convert the fine powders into Fe-BTC/PDA composite beads using biodegradable poly(acrylic acid) (PAA) and sodium alginate monomers which are cross-linked using Ca^{2+} ions. It was observed that only a gram of the resulting composite beads can neutralize more than 10 L of water containing highly toxic Pb^{2+} with a concentration of 600 ppb which might be due to the selectivity of the composite for Pb^{2+} as compared to other less toxic ions (Figure 3.7) [53]. The composite also demonstrated tremendous performance in the removal of palladium ions and shows an adsorption capacity of 498 mg of Pd^{2+}.

3.4.3.2 Water Capture

Recently, a lot of interest has been generated in developing pathways for water collection and generation. MOFs have been explored as promising adsorbing materials for atmospheric water harvesting due to their exceptional properties. Not many reports are available highlighting the use of MOF-polymer composites for water harvesting. Zhao et al. reported the use of radical polymerization of N-isopropyl acrylamide (NIPAM) within Cr-MIL-101 for water capture. The hydrophilic to the hydrophobic character of the resulting composite can be controlled by temperature variation, hence, such types of composites are excellent materials for water adsorption. MOF-polymer composites with 38 wt.% of polymer capture 440 wt% of water under a relative humidity of 96% at 25°C, which is by far the best reported value for any water-capture adsorbent [54]. Similarly, the adsorption performance of Ni-MOF-PDA increases from 25.1 wt% for Ni-MOF to 44.1 wt%, for Ni-MOF-PDA (21.1%). Further increase in PDA contents leads to a decrease in adsorption, which can be due to the pore filling effect [47]. Table 3.2 presents the application of some of the MOF-polymer composites for water treatment1.

3.4.4 Sensing Applications

A few years ago, the MOF-based sensors usually focused on the changes in the luminescent properties of MOF powders and only a few reports exist on electrical detection methods. However, with the rapid progress in the field of synthesis methods for MOF-based composites, especially polymer-based composites, and enhancement in the conductivity of MOFs, development in this area has picked up a new pace. A lot of research is dedicated to the development of MOF-polymer-based sensors with high selectivity and response time. In one of the first studies by Smet et al. [60], matrimid polyimide/NH_2-MIL-53(Al) composite was fabricated for capacitive sensor application. The NH_2-MIL-53(Al) is present in the form of nanoparticles inside polymeric film, which aids in increasing its intrinsic affinity. The composite film was coated on a planar electrode and the measurements were done to note the changes in the capacitance and impedance under fixed potential. The results indicate an increase of 2.5% in the capacitance of NH_2-MIL-53(Al)/matrimid (20 wt%) when exposed to 5000 ppm of methanol. The device was further studied for selectivity for different analytes and was exposed to 20000 ppm of water, ethanol, methanol, and 2-propanol. The response time slowed down for adsorbed ethanol and 2-propanol samples, which can be due to the reduced diffusion because of large molecular sizes. The study has opened a new avenue for the usage of MOF-polymer composites in a large number of sensor applications with improved selectivity and sensitivity.

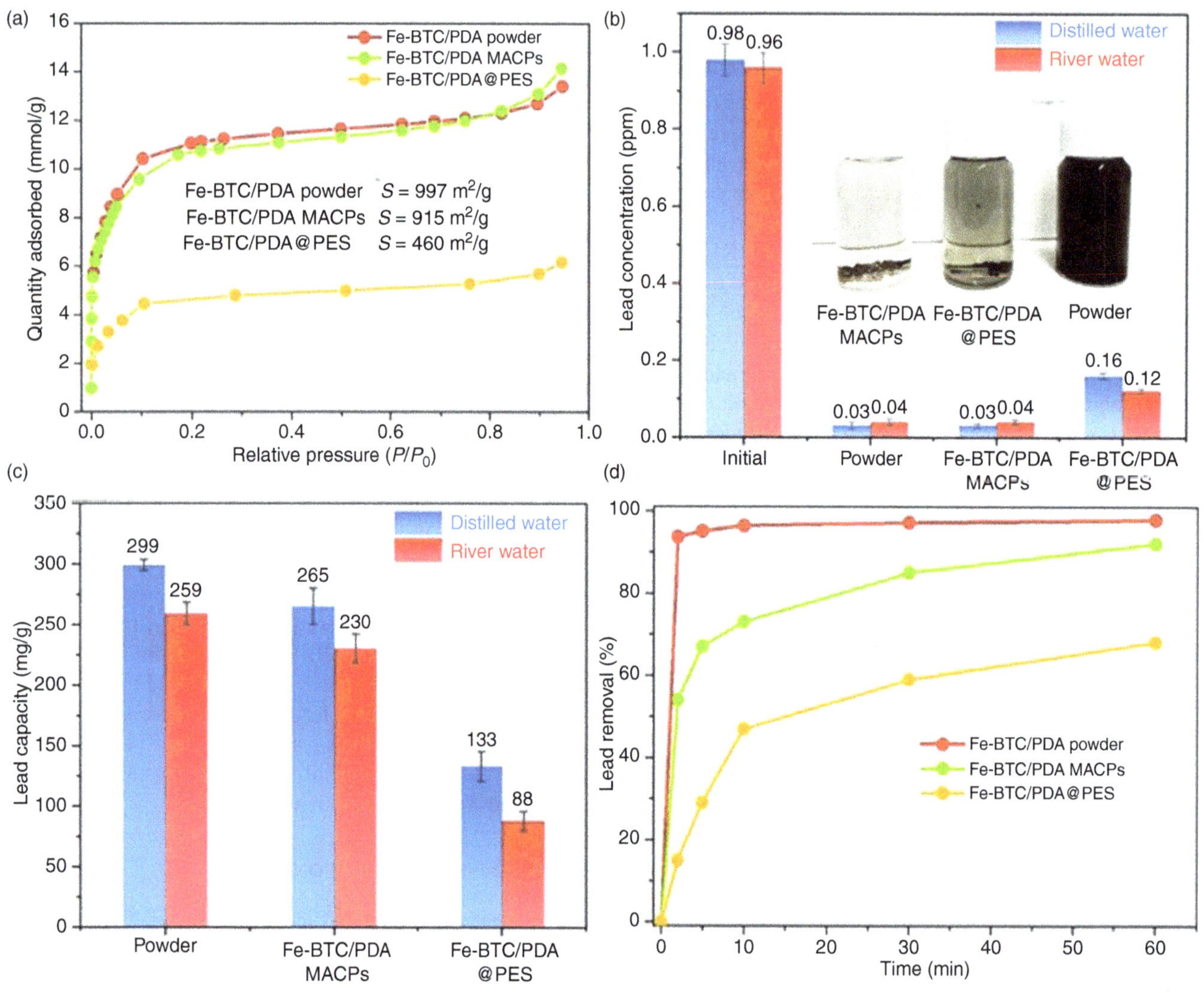

FIGURE 3.7 (a) N_2 adsorption isotherms of Fe-BTC/PDA powder, Fe-BTC/PDA beads, and the control Fe-BTC/PDA@PES beads. (b) Concentration of Pb^{2+} in water before and after treatment with three different samples. (c) Lead adsorption capacity (mg/g) at 750 ppm Pb^{2+} with a different water source. (d) Lead adsorption kinetics spiked with 1 ppm of Pb^{2+} in distilled water. Reproduced with permission from reference [53] Copyright (2020) American Chemical Society.

Furthermore, MOF-polymer composite can be used as an electrochemical sensor for the detection of heavy metal ions also. A core-shell composite of UiO-66-NH_2 MOF and polyaniline (PANI) was devised for the detection of Cd^{+2} ions. Both UiO-66-NH_2 and PANI have an abundance of amine groups that act as binding sites for Cd ions. The UiO-66-NH_2-PANI composite was coated on the electrode and electrochemical behavior was studied by cyclic voltammetry (CV) and electrochemical impedance measurements. The results showed enhanced electrochemical sensitivity. The detection limit of 0.3 μg/L and a wider linear range for Cd ions were achieved, which is by far the lowest as compared to any other electrochemical method. It was suggested that the formation of conductive pathways leads to the efficient transfer of electrons between solution and electrode [61]. In another study by Liu et al. [62], an in-situ copolymerization strategy was used to incorporate a MOF-453 having active groups, into the polymeric matrix for Fe^{+3} detection. A copolymer of polyurethane acrylate was grafted onto the surface of the MOF. Fluorescence spectroscopy was utilized to evaluate the sensing capabilities of the MOF-polymer composite for metal ion detection. In comparison with the MOF without the polymer, the luminescence intensity of the composite was notably quenched when treated with Fe^{+3} ions which can be ascribed to enhanced secondary nitramide interactions.

3.4.5 Detoxification of Chemical Warfare Agents

Chemical warfare agents (CWAs), such as blister and nerve agents are highly toxic organophosphonate compounds that can lead to death by inhalation. Despite the ban on their development, there are still recent reports of the use of CWAs on the battlefield and this poses a threat to mankind. For this reason, the reduction of CWAs is of paramount

TABLE 3.2
Application of some of the MOF-polymer composites for water treatment

MOF	Polymer	Application	Reference
MIL-53(Cr)	trimethylolpropane triacrylate	Acid black removal	[50]
HKUST-1(Cu)	trimethylolpropane triacrylate	Acid black removal	[50]
MIL-101(Cr)	N-isopropylacrylamide	Water-capture	[54]
UiO-66-NH2	poly(2-dimethylaminoethyl methacrylate)	Organic Dye removal	[51]
MIL-53(Al)	poly(vinylidene fluoride	Organic pollutants degradation	[55]
MIL-100 (Fe)	Polydopamine	Heavy metal extraction	[52]
MIL-100(Fe)	Polydopamine & Poly(acrylic acid)	Heavy metal extraction	[53]
HKUST-1	polyetherimide	Iodine Capture & recovery	[56]
HKUST-1	polyvinyl difluoride	Iodine Capture & recovery	[56]
HKUST-1	polyethersulfone	Iodine Capture & recovery	[56]
ZIF-8	Polyurethane	Iodine Capture & recovery	[57]
MOF-808	Polyacrylonitrile (PAN)	Heavy metal extraction	[58]
Fe-BTC	Poly-p-phenylenediamine	Gold ion extraction	[52]
ZIF-8/Gelatin	polyvinyl difluoride	Rhodamine B dye removal	[59]
Ni-MOF	polydopamine	Water adsorption	[47]
UiO-66-NH_2	Poly butyl methacrylate	Heavy metal extraction	[19]

importance. Mostly Zr^{4+} based MOFs have been explored to catalytically degrade CWAs. Cohen et al. [63] described the covalent attachment of UiO-66-NH_2 to a linear polyamide fiber (PA-66) using post-synthetic polymerization. The catalytic performance of the resulting PA-66-MOF composite was evaluated against the dimethyl 4-nitrophenyl phosphate (DMNP) nerve agent simulant. It was shown that the covalent attachment within the material aids more in enhanced activity against degradation of the nerve agent than that of the material in which MOF particles are physically entrapped in the polymer. In a series of studies by Parsons et al. utilization of atomic layer deposition (ALD) was made for layering a metal oxide on polymer fibers for providing a nucleation site for MOF synthesis. Al_2O_3 was deposited on polypropylene fibers for the growth of HKUST-1 and the resulting composite displays improved activity for the absorption of NH_3 gas [64]. In another work, Zhao et al. [65] used TiO_2 as the nucleation coating layer for the development of Zr-based MOFs, namely, UiO-66, UiO-66-NH_2, and UiO-67 on polyamine (PA-6) fibers. To assess the catalytic activity of MOF-PA-6 composites for CWA degradation, the hydrolysis of DMNP and nerve agent soman (also known as GD) was performed. The composite effectively degraded both DMNP and GD with half-lives as short as 7.3 min and 2.3 min, respectively. Later the same research group demonstrated the chemical association of MOF crystals (UiO-66-NH_2) on polypropylene (PP) nonwoven mats using different metals oxides (ZnO, TiO_2, Al_2O_3) for nucleation sites. The composites lead to swift catalytic degradation of the DMNP agent with composite having ZnO layer showing a half-life of less than 5 min [66].

Another strategy to obtain MOF-polymer composite for the detoxification of CWAs was developed by Farha et al. [67] by hydrogenizing the buffer using a basic polymer such as polyethyleneimine (PEI). The composite of NU-1000 (NU-Northwestern University) and PEI can hydrolyze the DMNP simulant as well as nerve agents GD and venomous agent X (VX) Half-lives of 1.8 min for DMNP, 4.8 mn for GD, and 12.4 min for VX were noted. After successful catalytic activity in the water, a portable composite of MOF catalyst, polymer buffer, and cellulose for solid-state application was fabricated. The results obtained indicated complete removal of DMNP simulant implying that the composites can be easily employed in the production of solid materials like cleaning mats, masks, and the like.

Peterson et al. [68] also reported a new strategy to obtain hierarchical composites by layering MOF-polymer films for CWA removal. The process involves sandwiching a layer of MOF between the top and bottom layers of robust polymers for the formation of a composite known as 'MOFwich' as shown in Figure 3.8. Polymers such as polystyrene-block-polyisoprene-block polystyrene (SIS) and polystyrene-block-poly(ethylene-ran-butylene)-block-polystyrene (SEBS) were used as encasing layers (having low MOF loadings) for active layer polymers such as SIS, SEBS, poly(ethylene oxide), poly(methyl methacrylate), polystyrene and PVDF (with high MOF loadings). Using this approach and varying polymer or MOF structures as well as composition or order of layers, a large number of composites for intended applications can be fabricated. Different composites demonstrate exquisite properties like providing a reactive barrier to CWAs, which makes them highly suitable for protection applications. Very recently Pellejero et al. [69] reported functionalization of a 3D printed ABS layer with a ZIF-8 MOF for the successful capture of harmful gases. The fabricated ABS/ZIF-8 composite exhibits promising adsorption characteristics for dimethyl methyl phosphonate with an efficiency of 15.5%. Hence the composites exhibit a great potential for the elimination of toxic gas.

3.4.6 Other Catalytic Applications

Recently, MOF-polymer composites are highly valued as catalysts for several reactions such as the Knoevenagel

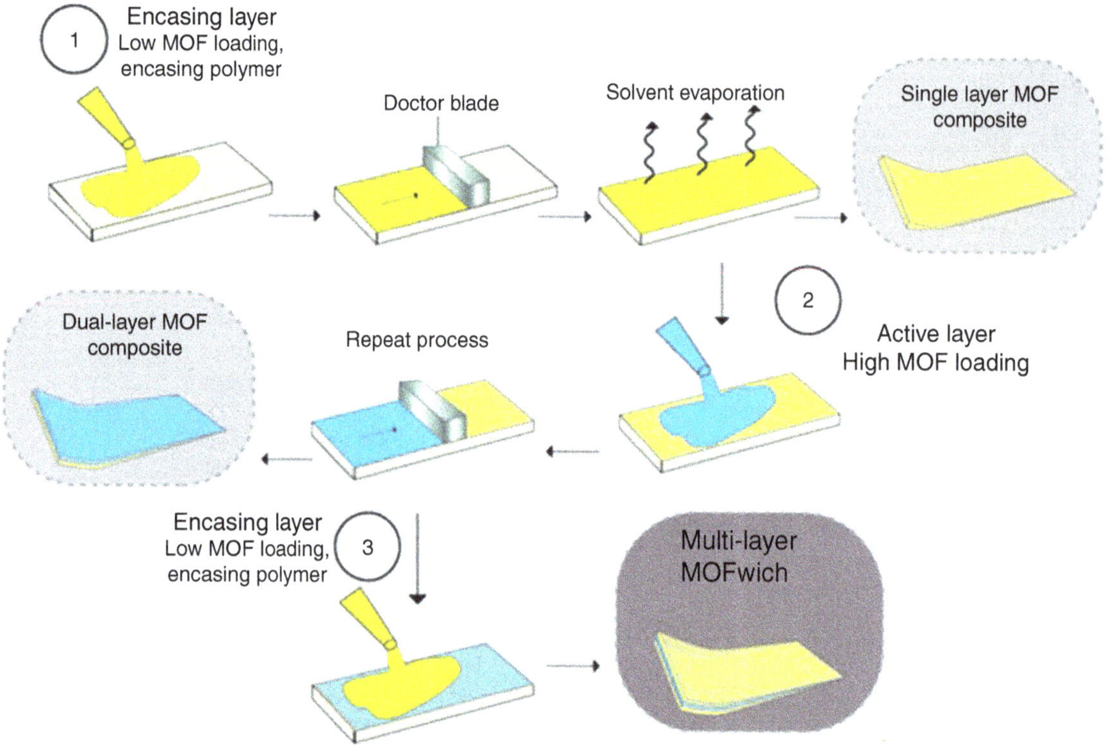

FIGURE 3.8 Strategy for fabricating the multilayer MOFwich structure. Reproduced with permission from reference [68] Copyright (2018) American Chemical Society.

condensation reaction. The Knoevenagel condensation reaction is a well-known organic reaction for the formation of carbon-carbon bonds that are essential for producing unsaturated acids, which in turn are used for fabricating a large number of products such as insecticides, pesticides, chemicals, cosmetics, and so forth. This reaction is catalyzed by organic and inorganic bases, which pose a significant threat to the environment. Hence to reduce the use of these bases, MOF-polymer composites are being fabricated and utilized. Zhao et al. [70] made use of the site isolation strategy to separate two different polymers with dissimilar properties in different porous regions in a MOF to fabricate a bifunctional catalyst. The acidic polymer, polystyrene sulfonate, and a basic polymer, polymethyl aminopyridine were formed by in-situ polymerization within the pores of Cr-MIL-101. These two types of MOF-polymer catalysts bearing isolated active sites were used to catalyze tandem reactions, namely decentralization by an acid catalyst and Knoevenagel condensation reaction of benzaldehyde glycol acetal. The simultaneous use of both acidic and basic MOF-polymer composites leads to faster catalytic conversions using mild reaction conditions. Cohen et al. [71] have fabricated layered MMM using MIL-101-NO_2 and ZIF-8 to perform a two-step catalytic reaction. At first, the top layer of MIL-101-NO_2 was used as an acid catalyst to convert a benzaldehyde dimethyl acetal into benzaldehyde and in the second reaction, a ZIF-8 MMM layer, which acts as a base catalyst, was used for the Knoevenagel condensation reaction of benzaldehyde with malononitrile. After passing through this layered MMM a conversion of 95% of the reactants to the final product was achieved.

3.5 CONCLUSION AND FUTURE OUTLOOK

MOF-polymer composites are a promising class of materials having a diverse range of applications. Some highly distinctive approaches outline the interactions and synergies between various MOFs and polymers. This research field is so vast and there are endless possibilities for their unique combination through encapsulation of polymers into the MOF, polymer grafting, polymers templating MOF, mixed matrix membranes, and so on. It has been observed that not only new MOF architectures with controlled structures have been built using polymers but also a diverse range of polymers with well-defined stereo structures can be prepared using MOFs. However, the scalability and processability of the synthesized polymers are still a challenge that needs to be addressed. There is also a need for a well-defined study to evaluate the kinetics of self-assembly of MOF structures and computational modeling of the composites. MOF-polymers have also shown excellent functionality in MMMs, yet there is a continuous need for research on hybridization and functionalization of these structures to achieve high permeability and selectivity. Many of the remarkable applications of MOF-polymer composites reported have been performed in the past few years only. Hence, significant advancements in this niche area are expected in the next few years as the resulting composites offer rich multitudes of crystal structures. There is potential to discover the new and exciting chemical and physical phenomena that hold great promise for excellent commercial applications.

REFERENCES

[1]. O. M. Y. Hailian Li, Mohamed Eddaoudi, M. O'Keeffe, "Design and synthesis of an exceptionally stable and highly porous metal-organic framework," *Nature*, vol. 402, pp. 276–279, 1999.

[2]. X. Ma, Y. Chai, P. Li, B. Wang, "Metal-Organic Framework Films and Their Potential Applications in Environmental Pollution Control," *Acc. Chem. Res.*, vol. 52, pp. 1461–1470, 2019.

[3]. A.-Z. C. Jun Zhong, Ranjith Kumar Kankala, Shi-Bin Wang, "Recent Advances in Polymeric Nanocomposites," *Eng. Polym. Chem. Complexity, Vol. I*, vol. 11, pp. 37–69, 2014.

[4]. T. Uemura, K. Kitagawa, S. Horike, T. Kawamura, S. Kitagawa, M. Mizuno, K. Endo, "Radical polymerisation of styrene in porous coordination polymers," *Chem. Commun.*, no. 48, pp. 5968–5970, 2005.

[5]. J. Hwang, H. C. Lee, M. Antonietti, B. V. K. J. Schmidt, "Free radical and RAFT polymerization of vinyl esters in metal-organic-frameworks," *Polym. Chem.*, vol. 8, no. 40, pp. 6204–6208, 2017.

[6]. S. Begum, Z. Hassan, S. Bräse, M. Tsotsalas, "Polymerization in MOF-Confined Nanospaces: Tailored Architectures, Functions, and Applications," *Langmuir*, vol. 36, no. 36, pp. 10657–10673, 2020.

[7]. C. Lu, T. Ben, S. Xu, S. Qiu, "Electrochemical synthesis of a microporous conductive polymer based on a metal-organic framework thin film," *Angew. Chemie - Int. Ed.*, vol. 53, pp. 6454–6458, 2014.

[8]. S. Klyatskaya, A.B. Kanj, C. Molina-Jirón, S. Heidrich, L. Velasco, C. Natzeck, H. Gliemann, S. Heissler, P. Weidler, W. Wenzel, C. Bufon, L. Heinke, C. Wöll, M. Ruben, "Conductive Metal-Organic Framework Thin Film Hybrids by Electropolymerization of Monosubstituted Acetylenes," *ACS Appl. Mater. Interfaces*, vol. 12, pp. 30972–30979, 2020.

[9]. M. W. A. MacLean, T. Kitao,T. Suga, M. Mizuno, S. Seki, T. Uemura, S. Kitagawa, "Unraveling Inter- and Intrachain Electronics in Polythiophene Assemblies Mediated by Coordination Nanospaces," *Angew. Chemie - Int. Ed.*, vol. 55, pp. 708–713, 2016,

[10]. I. H. Park, A. Chanthapally, Z. Zhang, S. S. Lee, M. J. Zaworotko, J. J. Vittal, "Metal-organic organopolymeric hybrid framework by reversible [2+2] cycloaddition reaction," *Angew. Chemie Int. Ed.*, vol. 53, no. 2, pp. 414–419, 2014.

[11]. I. H. Park, A. Chanthapally, H. H. Lee, H. S. Quah, S. S. Lee, J. J. Vittal, "Solid-state conversion of a MOF to a metal-organo polymeric framework (MOPF) via [2+2] cycloaddition reaction," *Chem. Commun.*, vol. 50, pp. 3665–3667, 2014.

[12]. I. H. Park, R. Medishetty, H.-H. Lee, C. E. Mulijanto, H. S. Quah, S. S. Lee, J. J. Vittal "Formation of a Syndiotactic Organic Polymer Inside a MOF by a [2+2] Photo-Polymerization Reaction," *Angew. Chemie - Int. Ed.*, vol. 54, pp. 7313–7317, 2015.

[13]. S. Anan, Y. Mochizuki, K. Kokado, K. Sada, "Step-Growth Copolymerization Between an Immobilized Monomer and a Mobile Monomer in Metal–Organic Frameworks," *Angew. Chemie - Int. Ed.*, vol. 58, pp. 8018–8023, 2019.

[14]. T. Ishiwata, A. Michibata, K. Kokado, S. Ferlay, M. W. Hosseini, K. Sada, "Box-like gel capsules from heterostructures based on a core-shell MOF as a template of crystal crosslinking," *Chem. Commun.*, vol. 54, pp. 1437–1440, 2018.

[15]. M. Tsotsalas, J. Liu, B. Tettmann, S. Grosjean, A. Shahnas, Z. Wang, C. Azucena, A. Carlos, M. Addicoat, T. Heine, J. Lahann, J. Overhage, B. Stefan, H. Gliemann, C. Wöll, "Fabrication of highly uniform gel coatings by the conversion of surface-anchored metal-organic frameworks," *J. Am. Chem. Soc.*, vol. 136, pp. 8–11, 2014.

[16]. Z. Wang, A. Błaszczyk, O. Fuhr, S. Heissler, C. Wöll, M. Mayor, "Molecular weaving via surface-templated epitaxy of crystalline coordination networks," *Nat. Commun.*, vol. 8, pp. 1–8, 2017.

[17]. H. J. Lee, W. Cho, M. Oh, "Advanced fabrication of metal–organic frameworks: Template-directed formation of polystyrene@ZIF-8 core–shell and hollow ZIF-8 microspheres," *Chem. Commun.*, vol. 48, no. 2, pp. 221–223, 2012.

[18]. A. L. Li, F. Ke, L. G. Qiu, X. Jiang, Y. M. Wang, X. Y. Tian, "Controllable synthesis of metal-organic framework hollow nanospheres by a versatile step-by-step assembly strategy," *CrystEngComm,* vol. 15, no. 18, pp. 3554–3559, 2013.

[19]. Z. Zhang, H. T. H. Nguyen, S. A. Miller, S. M. Cohen, "PolyMOFs: A Class of Interconvertible Polymer-Metal-Organic-Framework Hybrid Materials," *Angew. Chemie - Int. Ed.*, vol. 54, pp. 6152–6157, 2015.

[20]. S. Ayala, Z. Zhang, S. M. Cohen, "Hierarchical structure and porosity in UiO-66 polyMOFs," *Chem. Commun.*, vol. 53, pp. 3058–3061, 2017.

[21]. G. E. M. Schukraft, S. Ayala, B. L. Dick, S. M. Cohen, "Isoreticular expansion of polyMOFs achieves high surface area materials," *Chem. Commun.*, vol. 53, pp. 10684–10687, 2017.

[22]. V. J. Pastore, T. R. Cook, J. Rzayev, "Polymer – MOF Hybrid Composites with High Porosity and Stability through Surface-Selective Ligand Exchange," vol.30, pp.8639–8649, 2018.

[23]. T. Uemura, N. Yanai, S. Watanabe, H. Tanaka, R. Numaguchi, M. T. Miyahara, Y. Ohta, M. Nagaoka, S. Kitagawa, "Unveiling thermal transitions of polymers in subnanometre pores," *Nat. Commun.*, vol. 1, pp. 1–8, 2010.

[24]. K. Koh, A. G. Wong-Foy, A. J. Matzger, "Coordination copolymerization mediated by Zn4O(CO 2R)6 metal clusters: A balancing act between statistics and geometry," *J. Am. Chem. Soc.*, vol. 132, pp. 15005–15010, 2010.

[25]. S. Nagata, K. Kokado, K. Sada, "Metal-organic framework tethering PNIPAM for ON-OFF controlled release in solution," *Chem. Commun.*, vol. 51, pp. 8614–8617, 2015.

[26]. A. Zimpel, T. Preiß, R. Röder, H. Engelke, M. Ingrisch, M. Peller, J. O. Rädler, E. Wagner, T. Bein, U. Lächelt, S. Wuttke, "Imparting Functionality to MOF Nanoparticles by External Surface Selective Covalent Attachment of Polymers," *Chem. Mater.*, vol. 28, pp. 3318–3326, 2016.

[27]. Y. Li, J. Liu, K. Zhang, L. Lei, Z. Lei, "UiO-66-NH2@ PMAA: A Hybrid Polymer-MOFs Architecture for Pectinase Immobilization," *Ind. Eng. Chem. Res.*, vol. 57, pp. 559–567, 2018.

[28]. K. A. McDonald, J. I. Feldblyum, K. Koh, A. G. Wong-Foy, A. J. Matzger, "Polymer@MOF@MOF: 'grafting from' atom transfer radical polymerization for the synthesis of hybrid porous solids," *Chem. Commun.*, vol. 51, pp. 11994–11996, 2015.

[29]. S. Liu, L. Zhai, C. Li, Y. Li, X. Guo, Y. Zhao, C. Wu, "Exploring and exploiting dynamic noncovalent chemistry for effective

surface modification of nanoscale metal-organic frameworks," *ACS Appl. Mater. Interfaces*, vol. 6, pp. 5404–5412, 2014.

[30]. M. G. Schwab, I. Senkovska, M. Rose, M. Koch, J. Pahnke, G. Jonschker, S. Kaskel, "MOF@PolyHIPEs," *Adv. Eng. Mater.*, vol. 10, pp. 1151–1155, 2008.

[31]. C. Le Calvez, M. Zouboulaki, C. Petit, L. Peeva, N. Shirshova, "One step synthesis of MOF-polymer composites," *RSC Adv.*, vol. 6, pp. 17314–17317, 2016.

[32]. Y. Y. Fu, C. X. Yang, X. P. Yan, "Incorporation of metal-organic framework UiO-66 into porous polymer monoliths to enhance the liquid chromatographic separation of small molecules," *Chem. Commun.*, vol. 49, pp. 7162–7164, 2013.

[33]. W. Zhang, Y. Hu, J. Ge, H. Jiang, S. Yu, *"A Facile and General Coating Approach to Moisture/Water-Resistant Metal – Organic Frameworks with Intact Porosity,"* 2014.

[34]. J. Duan, Y. Pan, F. Pacheco, E. Litwiller, Z. Lai, and I. Pinnau, "High-performance polyamide thin-film-nanocomposite reverse osmosis membranes containing hydrophobic zeolitic imidazolate framework-8," *J. Memb. Sci.*, vol. 476, pp. 303–310, 2015.

[35]. M. S. Denny S. M. Cohen, "In Situ Modification of Metal-Organic Frameworks in Mixed-Matrix Membranes," *Angew. Chemie - Int. Ed.*, vol. 54, pp. 9029–9032, 2015.

[36]. J. Troyano, A. Carné-Sánchez, J. Pérez-Carvajal, L. León-Reina, I. Imaz, A. Cabeza, D. Maspoch, "A Self-Folding Polymer Film Based on Swelling Metal–Organic Frameworks," *Angew. Chemie - Int. Ed.*, vol. 57, no. 47, pp. 15420–15424, 2018.

[37]. N. Tien-Binh, D. Rodrigue, and S. Kaliaguine, "In-situ cross interface linking of PIM-1 polymer and UiO-66-NH2 for outstanding gas separation and physical aging control," *J. Memb. Sci.*, vol. 548, no. September 2017, pp. 429–438, 2018,

[38]. B. Le Ouay, M. Boudot, T. Kitao, T. Yanagida, S. Kitagawa, and T. Uemura, "Nanostructuration of PEDOT in Porous Coordination Polymers for Tunable Porosity and Conductivity," *J. Am. Chem. Soc.*, vol. 138, no. 32, pp. 10088–10091, 2016,

[39]. X. Yang, B.L. Bonnett, G. A. Spiering, H. D. Cornell, B. J. Gibbons, R. B. Moore, E. J. Foster, A. J. Morris "Understanding the Mechanical Reinforcement of Metal-Organic Framework-Polymer Composites: The Effect of Aspect Ratio," *ACS Appl. Mater. Interfaces*, Jun. 2021

[40]. J. Zhang, Z. Li, X. L. Qi, D. Y. Wang, "Recent Progress on Metal–Organic Framework and Its Derivatives as Novel Fire Retardants to Polymeric Materials," *Nano-Micro Lett.*, vol. 12, pp. 1–21, 2020.

[41]. L. Y. Molefe, N. M. Musyoka, J. Ren, H. W. Langmi, P. G. Ndungu, R. Dawson, M. Mathe, "Synthesis of porous polymer-based metal–organic frameworks monolithic hybrid composite for hydrogen storage application," *J. Mater. Sci.*, vol. 54, pp. 7078–7086, 2019.

[42]. L. Y. Molefe, N. M. Musyoka, J. Ren, H. W. Langmi, M. Mathe, and P. G. Ndungu, "Effect of Inclusion of MOF-Polymer Composite onto a Carbon Foam Material for Hydrogen Storage Application," *J. Inorg. Organomet. Polym. Mater.*, vol. 31, pp. 80–88, 2021.

[43]. L. Y. Molefe, N. M. Musyoka, J. Ren, H. W. Langmi, M. Mathe, P. G. Ndungu, "Polymer-Based Shaping Strategy for Zeolite Templated Carbons (ZTC) and Their Metal Organic Framework (MOF) Composites for Improved Hydrogen Storage Properties," *Front. Chem.*, vol. 7, pp. 1–10, 2019.

[44]. M. C. Kreider, M. Sefa, J. A. Fedchak, J. Scherschligt, M. Bible, B. Natarajan, N. N. Klimov, A. E. Miller, Z. Ahmed, M. R. Hartings, "Toward 3D printed hydrogen storage materials made with ABS-MOF composites," *Polym. Adv. Technol.*, vol. 29, pp. 867–873, 2018.

[45]. M. Tian, S. Rochat, K. Polak-Kraśna, L. T. Holyfield, A. D. Burrows, C. R. Bowen, T. J. Mays "Nanoporous polymer-based composites for enhanced hydrogen storage," *Adsorption*, vol. 25, pp. 889–901, 2019.

[46]. C. S. H. jinkyoung Park, Yun Seok Chae, DongWon Kang, Minjung Kang, Jong Hyeak Choe, Saemi Kim, Jee Yeon Kim, Yong Won Jeong, "Shaping of a metal organic framework-Polymer Composite and its CO2 Adsorption Performances from Humid Indoor Air," *Appl. Mater. interfaces*, vol. 13, pp. 25421–25427, 2021.

[47]. L. Peng, S. Yang, D. T. Sun, M. Asgari, W. L. Queen, "MOF/polymer composite synthesized using a double solvent method offers enhanced water and CO2 adsorption properties," *Chem. Commun.*, vol. 54, pp. 10602–10605, 2018.

[48]. B. Ghalei et al. "Enhanced selectivity in mixed matrix membranes for CO2 capture through efficient dispersion of amine-functionalized MOF nanoparticles," *Nat. Energy*, vol. 2, no. 7, 2017,

[49]. Z. Wang, J. Wang, M. Li, K. Sun, C. J. Liu, "Three-dimensional printed acrylonitrile butadiene styrene framework coated with Cu-BTC metal-organic frameworks for the removal of methylene blue," *Sci. Rep.*, vol. 4, pp. 4–10, 2014.

[50]. C. Brahmi, M. Benltifa, C. Vaulot, L. Michelin, F. Dumur, F. Millange, M. Frigoli, A. Airoudj, F. Morlet-Savary, L. Bousselmi, J. Lalevée, "New hybrid MOF / polymer composites for the photodegradation of organic dyes, *Eur. Polym. J.,* vol 154, pp. 110560, 2021.

[51]. L. Hou, L. Wang, N. Zhang, Z. Xie, and D. Dong, "Polymer brushes on metal-organic frameworks by UV-induced photopolymerization," *Polym. Chem.*, vol. 7, no. 37, pp. 5828–5834, 2016

[52]. D. T. Sun, L. Peng, W. S. Reeder, S. M. Moosavi, D. Tiana, K. B. David, O. Emad, W. L. Queen, "Rapid, Selective Heavy Metal Removal from Water by a Metal-Organic Framework/Polydopamine Composite," *ACS Cent. Sci.*, vol. 4, pp. 349–356, 2018.

[53]. S. Yang, L. Peng, O. A. Syzgantseva, O. Trukhina, I. Kochetygov, A. Justin, D. T. Sun, H. Abedini, M. A. Syzgantseva, E. Oveisi, G. Lu, W. L. Queen, "Preparation of Highly Porous Metal-Organic Framework Beads for Metal Extraction from Liquid Streams," *J. Am. Chem. Soc.*, vol. 142, pp. 13415–13425, 2020.

[54]. A. Karmakar, P. G. M. Mileo, I. Bok, S. Peh, J. Zhang, H. Yuan, G. Maurin, D. Zhao, s"Thermo-Responsive MOF/Polymer Composites for Temperature-Mediated Water Capture and Release," *Angew. Chemie*, vol. 132, pp. 11096–11102, 2020.

[55]. L. Hu, Y. Zhang, W. Lu, Y. Lu, H. Hu, "Easily recyclable photocatalyst Bi2WO6/MOF/PVDF composite film for efficient degradation of aqueous refractory organic pollutants under visible-light irradiation," *J. Mater. Sci.*, vol. 54, pp. 6238–6257, 2019.

[56]. B. Valizadeh, T. N. Nguyen, B. Smit, K. C. Stylianou, "Porous Metal–Organic Framework@Polymer Beads for Iodine Capture and Recovery Using a Gas-Sparged Column," *Adv. Funct. Mater.*, vol. 28, pp. 1–6, 2018.

[57]. E. M. Mahdi, A. K. Chaudhuri, J. C. Tan, "Capture and immobilisation of iodine (I2) utilising polymer-based ZIF-8 nanocomposite membranes," *Mol. Syst. Des. Eng.*, vol. 1, no. 1, pp. 122–131, 2016.

[58]. J. E. Efome, D. Rana, T. Matsuura, C. Q. Lan, "Insight Studies on Metal-Organic Framework Nanofibrous Membrane Adsorption and Activation for Heavy Metal Ions Removal from Aqueous Solution," *ACS Appl. Mater. Interfaces*, vol. 10, pp. 18619–18629, 2018.

[59]. Y. Guo, X. Wang, P. Hu, X. Peng, "ZIF-8 coated polyvinylidenefluoride (PVDF) hollow fiber for highly efficient separation of small dye molecules," *Appl. Mater. Today*, vol. 5, pp. 103–110, 2016.

[60]. S. Sachdeva, D. Soccol, D. J. Gravesteijn, F. Kapteijn, E. J. R. Sudhölter, J. Gascon, L. De Smet, "Polymer-Metal Organic Framework Composite Films as Affinity Layer for Capacitive Sensor Devices," *ACS Sensors*, vol. 1, pp. 1188–1192, 2016.

[61]. Y. Wang, L. Wang, W. Huang, T. Zhang, X. Hu, J. A. Perman, S. Ma, "A metal-organic framework and conducting polymer based electrochemical sensor for high performance cadmium ion detection," *J. Mater. Chem. A,* vol. 5, pp. 8385–8393, 2017..

[62]. C. Sen Liu, M. Chen, J. Y. Tian, L. Wang, M. Li, S. M. Fang, X. Wang, L. M. Zhou, Z. W. Wang, M. Du, "Metal–Organic Framework Supported on Processable Polymer Matrix by In Situ Copolymerization for Enhanced Iron(III) Detection," *Chem. - A Eur. J.*, vol. 23, pp. 3885–3890, 2017.

[63]. M. Kalaj, M. S. Denny, K. C. Bentz, J. M. Palomba, S. M. Cohen, "Nylon–MOF Composites through Postsynthetic Polymerization," *Angew. Chemie - Int. Ed.*, vol. 58, pp. 2336–2340, 2019.

[64]. J. Zhao, M. D. Losego, P. C. Lemaire, P. S. Williams, B. Gong, S. E. Atanasov, T. M. Blevins, C. J. Oldham, H. J. Walls, S. D. Shepherd, M. A. Browe, G. W. Peterson, G. N. Parsons, "Highly Adsorptive, MOF-Functionalized Nonwoven Fiber Mats for Hazardous Gas Capture Enabled by Atomic Layer Deposition," *Adv. Mater. Interfaces*, vol. 1, 2014.

[65]. J. Zhao, D. T. Lee, R. W. Yaga, M. G. Hall, H. F. Barton, I. R. Woodward, C. J. Oldham, H. J. Walls, G. W. Peterson, G. N. Parsons, "Ultra-Fast Degradation of Chemical Warfare Agents Using MOF–Nanofiber Kebabs," *Angew. Chemie - Int. Ed.*, vol. 55, pp. 13224–13228, 2016.

[66]. D. T. Lee, J. Zhao, G. W. Peterson, G. N. Parsons, "Catalytic 'mOF-Cloth' Formed via Directed Supramolecular Assembly of UiO-66-NH2 Crystals on Atomic Layer Deposition-Coated Textiles for Rapid Degradation of Chemical Warfare Agent Simulants," *Chem. Mater.*, vol. 29, pp. 4894–4903, 2017.

[67]. S. Y. Moon, E. Proussaloglou, G. W. Peterson, J. B. DeCoste, M. G. Hall, A. J. Howarth, J. T. Hupp, O. K. Farha, "Detoxification of Chemical Warfare Agents Using a Zr6-Based Metal–Organic Framework/Polymer Mixture," *Chem. - A Eur. J.*, vol. 22, pp. 14864–14868, 2016.

[68]. G. W. Peterson, A. X. Lu, M. G. Hall, M. A. Browe, T. Tovar, T. H. Epps, "MOFwich: Sandwiched Metal-Organic Framework-Containing Mixed Matrix Composites for Chemical Warfare Agent Removal," *ACS Appl. Mater. Interfaces*, vol. 10, pp. 6820–6824, 2018.

[69]. I. Pellejero, F. Almazán, M. Lafuente, M. A. Urbiztondo, M. Drobek, M. Bechelany, A. Julbe, L. M. Gandía, "Functionalization of 3D printed ABS filters with MOF for toxic gas removal," *J. Ind. Eng. Chem.*, vol. 89, pp. 194–203, 2020.

[70]. J. H. Zhao, Y. Yang, J. X. Che, J. Zuo, X. H. Li, Y. Z. Hu, X. W. Dong, L. Gao, X. Y. Liu, "Compartmentalization of Incompatible Polymers within Metal–Organic Frameworks towards Homogenization of Heterogeneous Hybrid Catalysts for Tandem Reactions," *Chem. - A Eur. J.*, vol. 24, pp. 9903–9909, 2018.

[71]. M. S. Denny, M. Kalaj, K. C. Bentz, S. M. Cohen, "Multicomponent metal-organic framework membranes for advanced functional composites," *Chem. Sci.*, vol. 9, pp. 8842–8849, 2018.

4 MOFs Metal Oxide-based Nanocomposites

M. Irfan Hussain[1], Ghulam Abbas Ashraf[2], Muhammad U. Farooq[3], Muhammad Siddique Ahmad[4], Jazib Ali[5] and Muhammad Usman Khan[6,7]

[1]Institute of Powder Metallurgy and Advanced Ceramics, School of Material Science and Engineering, University of Science and Technology Beijing, Beijing, China

[2]Department of Physics, Zhejian Normal University, Zhejiang, China

[3]State Key Laboratory of Chemical Engineering, East China University of Science and Technology, 130 Meilong Road, Shanghai, China

[4]Bahauddin Zakariya University, Multan, Pakistan

[5]Centre for Hybrid and Organic Solar Energy (CHOSE), Department of Electronic Engineering, University of Rome Tor Vergata, Rome, Italy

[6]National Key Laboratory of Tunable Laser Technology, Institute of Optoelectronics, Department of Electronics Science and Technology, Harbin Institute of Technology, Harbin, China

[7]Department of Physics, The University of Lahore, Sub-campus Sargodha, Pakistan

4.1 INTRODUCTION

MOFs offer great potential, owing to their exceptional porosity, specific surface area, and promise for gas storage, adsorption, elimination of toxic effluents, and numerous applications as catalysts [1]. Numerous researchers are continuously working on how to design less expensive, more stable, exceptional mechanically strong MOFs with improved performance and more functionality by altering their processing or chemistry for specific uses [2]. From a sensing perspective, their porous structure and specific surface areas are promising for providing a possible strategy for allowing a large number of ions/clusters of molecules to enter the framework, resulting in improved framework-guest interactions. MOFs often exhibit a high degree of functional and structural tunability, making them suitable for a variety of useful applications, for example, energy storage, gas filtration and separation, chemical sensing, catalysis, and drug delivery [3].

Interestingly, MOF-derived nanocomposites with controllable synthesis have boosted clean energy utilization. The emerging understanding of noble metals (Ag, Au, Pd, and Pt as well as Ag-Au alloy) and metal oxides (MnO_2, CeO_2, TiO_2, ZnO, Fe_2O_3, Cu_2O, and WO_3) has opened interesting avenues for research in photocatalytic areas, photovoltaic panels, drug delivery, surface improved, and numerous other fields [4]. One of the most significant aspects influencing the material characteristics of nanohybrids is based on results showing an excellent development of nanocomposites. By purposeful manipulation such as tailored self/external doping via metal oxides, which offer a viable solution to obtaining a controllable surface and metal chalcogenides to produce a variety of nanostructures such as ornamented metal oxide-NPs. Typically, adjustable energy band positions with noble metal/metal oxide core/shell, nanoarrays achieved, and tunable porosities are available using noble metal-metal oxide nanostructures. A moderate strength of coordination bond can be achieved through metal/metal oxide yolk/shells for various applications [5].

Herein, we focused on the surface texture's anisotropy and possible applications that can all be greatly enhanced by integrating MOFs with appropriate materials. Hence studying the domain pivotal role of metal oxide and more elaborate functionalities with hybrid nanostructures, their efficiencies were increased towards the future development of high-performance nanocomposites for better control of particle size and structure-property application [6]. The state-of-the-art progress of these materials demonstrates significantly improved catalytic sensitivity and stability for future development to mitigate against environmental effluents and to support clean renewable energy applications.

4.2 PROPERTIES OF MOF-BASED METAL OXIDE NANOCOMPOSITES

MOFs are prominent due to their structural diversity and the remarkable advances in their various functionalities through the appropriate selection of metal ions and organic ligands. Various studies available in the literature have already demonstrated many synthesis methods for tuning the chemistry, particle size, flexibility, and stability of MOF structure [2]. Moreover, these coordination polymers can be further improved by post-synthetic modification for further tuning of the properties by coating, substituting, or removal of the ligand or metal ion from the structure. The mechanical properties of MOFs can be modified by introducing malleable ligands, modulating the

DOI: 10.1201/9781003188148-4

strength of host-guest interactions, considering multi-metallic frameworks, and manipulating the size of crystals [7].

Most noteworthy in designing MOFs is how to enhance their stability for sensing and storage applications. MOF polymers have sparked interest because of their exceptional stability with adjustable energy bands, with tuned well-arranged structures and with the introduction of metal oxide and MOF derivatives. They can be widely applied in various fields such as adsorption, catalysts, and sensing. With controllable synthesis, the MOFs stabilities in their chemical, thermal, hydrothermal, and mechanical properties have shown significant development in recent years[8]. Exciting progress with tuned structure has improved the chemical stability of MOF polymers, with metal clusters having higher valence, such as Cr^{3+}, Fe^{3+}, and Zr^{4+} being used, together with soft ligands including triazolates, imidazolates, and tetrazolates [9]. Furthermore, heterocyclic molecules containing nitrogen atoms can be utilized in the preparation of MOFs together with divalent metal clusters like Zn^{2+}, or Co^{2+} to increase the stability. The thermal stability of MOFs shows degradation when metal-ligand bonds break and the organic linkers are combusted [10]. The potential can be improved by the coordination of oxy-anion ligands with high valence metal ions. Extending the hydrothermal stability of MOFs refers to the stability of materials in the existence of moisture at higher temperatures. New advancements are being unveiled for adding the hydrophobic functional groups into the MOF framework. Mechanical stability in MOF materials is directly linked to their high porosity and metal oxide allows improved sensing features [11]. Mechanical stability is crucial in MOFs that have solvent-filled pores in comparison to vacant pores. Furthermore, extended tritonic linkers relying on alkyne instead of phenylene units have been reported to improve the number of adsorption sites and large surface area of 7140 m^2/g by metal anchoring to confer large storage space [12].

4.2.1 Features of MOFs

MOFs have potential and have emerged as an extensive class, especially Zeolites, which are attractive to use in catalysts. In 1995, Yaghi was the first inventor, who gave the name to MOFs through a synthesized copper-4, 4-bipryridal complex that demonstrated prolonged metal-organic coupling [13]. For example, MOFs are members of the coordinating polymer group with organic-inorganic hybrid crystalline porous features. When compared to conventional coordination polymers, MOFs are more interesting for -dimensional crystalline (1D, 2D, and 3D) systems with porous characteristics. Driven by this trend, there has been an exponential increase in interest in the creation of MOFs in the literature.

The composition of (MOF-MOx) provides a desirable platform for idealized structures properties, in the selective determination of environmental systems including energy storage systems with good electrode materials, adsorption, photocatalytic water splitting, photocatalytic CO2 reduction, heterogeneous catalysis, chemical analysis, gas storage, and gas sensors. MOFs are coordination networks through repeating linked entities such as individual chains, loops, and single atoms linked to formed diatomic chains [14]. MOFs are well-known for geometrically imparting functionality through the use of carefully chosen linkers and metal nodes. Multifunctional MOFs incorporating different synthetic methodologies to tailor the framework's chemistry, stability, particle size, and flexibility have already been emphasized across several extensive reviews. MOFs can also be 'post-synthetically modified' to fine-tune their attributes by metal-organic complexes, organic ligands, or removing linker or node components from the framework [15]. In particular, the surface chemistry of the MOF13 can be modified by controlling crystallographic phases and crystallite size/morphology. This distinct class of porous materials' capacity to modify such features is a distinguishing strength since it allows for strategic control over host-guest chemistry in thin-film synthesis with metal exchange for energy storage applications [16].

4.2.1.1 Structure of MOFs

MOFs are constructed by organic linkers and build a network structure with a significant class of compounds comprising metal ions and clusters to reveal unique types of dimensions namely (ID, 2D, and 3D) structures [17]. The structure is best understood in terms of interactions between-metal clusters and organic linkers which generate a 1D expansion throughout the polymer within the MOF structure, and which may pave the way for smaller molecules possibly occupying the voids. In a 1D structure, the mixed inorganic-organic layers are assigned a symbol best described by the formation of different morphology: nanorods, nanotubes, and nano-shuttles, which shows the weak interaction existing between layers, based on the coordination influence of materials and their orientation [18]. Ligand modification consists of layers that are superimposed from either previous edges or staggered type layers that are stacked due to weak contact between layers. Modification of the ligands constituting the layers can govern how to channel interiors stack and function. In 2D MOF layer coordination polymers with mixed inorganic structures give high enhanced stability, which can accommodate guest species through the spaces found amongst the grids of layers and in between the layers. In the case of 3D MOFs, the structural frameworks of inorganic hybrids have high porosity and stability as a result of the spreading of coordination bonds in three dimensions and more diverse coordination geometries are achieved [19]. This functionality has been introduced with geometries (between metal ions and organic ligands) that facilitate non-covalent p-p stacking and hydrogen bonds. These interactions are responsible for converting the framework in MOFs to an infinitely dimensional network, as well as controlling their strength and direction. Moreover, the metal ion geometries have a good effect on the structures of MOF polymers.

Metal sites (also known as unsaturated metal sites or available metal sites) in MOFs have a significant impact on their adsorption capabilities. Transition metals, for example, Zn, Cu, Co, Mn, and others, are frequently used as center connectors. The metal in the MOF structure commonly acts as

Lewis' acids, allowing the coordinated organic ligands to be activated for subsequent chemical transformations [20]. Even some MOF materials have partially positive charges on metal sites with sophisticated morphologies and versatile structures that can boost adsorption capability depending on the number of oxidation states. The most commonly used transition metals produce varied geometries [21]. Each metal's most prevalent geometries are listed here,

a. The electron arrangement of copper in zero-valent metal $2d^{10}4S1$ [Ar]: the geometries of Cu-nodes are frequently distorted square-planar and octahedral. Because the Dz^2 orbital is full, but the dx^2-Y^2 orbital is only incompletely filled, Jann-Teller distortion occurs, resulting in octahedral geometries with two longer bonds and four short bonds.
b. The zero-valent metal's co-electron configuration is [Ar] $3d^74S^2$: Cobalt (II) node geometries are typically octahedral and tetrahedral.
c. The zero-valent metal's Ni-electron configuration is [Ar]$3d^84S^1$: the most common Ni configurations with coherent bond lengths are octahedral and square-planar.
d. The zero-valent metal's Mn-electron configuration is [Ar] $3d^54S^2$: Mn (II) compounds occur in a variety of coordination modes and configurations in MOFs, with octahedral being one of the most prevalent.
e. The zero-valent metal's Zn-electron configuration is [Ar] $3d^{10}4S^2$: compounds containing Zn (II) frequen tly have octahedral or tetrahedral configurations.

The organic linkers being used in coordination polymers to link the metal nodes are typically substituent organic molecules. The aromatic rings in the framework of the linkers sustain the structural stability of the dynamic and direct the topology of the framework. MOFs can have porous structures with pore volume and surface area.

4.2.1.2 MOF-based Nanocomposites

MOFs are porous materials in which organic ligands connect metal ions or groups. Their unique features, such as wide surface area, customizable pore geometry and shape, active sites, and high chemical and thermal durability, allow them to be used in a variety of photocatalytic applications, gas storage, drug delivery, absorption, and detection for diverse molecules [22]. Making composites with MOF alterations might make them the best choice for detecting various objects, including biomolecules, anions, organic compounds, cations, gas molecules, and so on.

As a result, a variety of MOF composite-based platforms have been designed to address this issue and effectively engage as electrochemical sensors for genetic and environmental targets. Noble-metal-NPs (like Ag, Pd, Au, and Pt) are expected to be used in detecting, imaging, cancer therapy, optical computing, and catalysis due to their exceptional physicochemical characteristics. However, their high surface energy causes aggregation, which limits the above-mentioned functionality [23]. As a result, researchers mix them with MOFs to circumvent this barrier, using various strategies. The resulting composites can be used as an effective sensor for recognizing diverse compounds.

4.2.2 Metal Oxide

Nanomaterials have sparked a lot of interest because of their minuscule and quantum phenomena. In comparison to their bulk counterparts, they demonstrate innovative and substantial mechanical, electrical, magnetic, and optical and magnetic properties. Metal-oxide-NPs are a type of nanomaterial that has been mass-produced on a massive scale and is gaining popularity in domains such as chemistry, physics, medicine, and biology. They have a large surface area, unusual adsorptive characteristics, surface flaws, and quick diffusivities [24].

Metal oxide monocrystals are superstructures made up of integrated metal oxide nanoparticles with configurable electrical, magnetic, and optical properties that could be beneficial in a wide range of uses from optoelectronics to catalysis. In this chapter, we report literature based on metal oxide monocrystals and evolve them into novel nanostructured materials with two metals. Many metal oxide mesocrystals (for example, ZnO, TiO_2, CuO, and Fe_2O_3) have been produced. Thus far, the most popular approach to making mesocrystal materials is to use hydrothermal/solvothermal treatments, which require the solution to be kept at a specified temperature and annealed for a certain amount of time [25]. The resultant precipitates are then gathered and contaminants rinsed away, and in some circumstances, more annealing is required. New approaches, such as microwave-assisted hydrothermal approaches, bioinspired approaches, and titration, have been recently developed to streamline and optimize product innovation [26].

4.2.2.1 Zinc Oxide

ZnO is an inorganic compound with wide bandgap promising features and is extensively employed in numerous materials owing to its remarkable properties such as good transparency, strong luminescence, and better electron mobility at ambient temperatures. Among the nanoparticles, ZnO is an n-type semiconductor with a versatile emerging application in electronic and optoelectronic devices [27]. Possessing significant exciton binding energy of 60 meV, it extensively uses in transparent electrodes in liquid crystal and a broad bandgap (E_g) of 3.2-3.4 eV. The structural optimization confirms their hexagonal wurtzite and cubic zincblende with a mix of ionic and covalent structures. Alternate structures can emerge, through 'zinc mixture' (the phrase comes when mixed with a solution like ZnS, because of cubic or hexagonal formations), including 'rock salt' (which has a cubed phase), although these are not thermodynamically stable phases under ambient circumstances [27]. A hexagonal ZnO wurtzite structure with lattice constants provides fundamental insight into crystal arrangements c= 0.52065 nm, a= 0.3296 nm and with a ratio c/a = 1.602 belonging to the P63mc space group comprised unit cell with lattice constants, more elaborate precise existence of symmetric a= 0.3296 nm and c= 0.52065 nm and a ratio c/a = 1.602. However, studies using structures depict a space group that is distinguished by the presence of two interconnected sublattices of O^{2-} and Zn^{2+} ions, that occupy a well-defined crystal structure such as a tetrahedral of four oxygen ions. As the symmetry of a Zn arrangement with oxygen is shown in the c-axis, cation, and anion polar planes are abundant in O^{2-} and Zn^{2+}, accordingly [28]. This synchronization illustration paves the way for polar symmetry as an O^{2-} and Zn^{2+} tetrahedral structure along the hexagonal axis. The symmetric structure development of 1D ZnO crystallites is accountable for this imbalance along the c-axis. The wurtzite structure with four-faced dismissals can be stabilized with durable: polar Zn ended (0001) and an equal number of Zn atom with O atom, as O terminated $(000\bar{1})$ facets, maintain a lattice symmetry property as well as non-polar $(10\bar{1}0)$ facets respectively [29]. Figure 4.1 shows a graphic illustration of the polar and non-polar facets, in both cases an atomic model of the ZnO hexagonal wurtzite structure, as well as non-polar and polar facets.

4.2.2.2 Titanium Dioxide (TiO_2)

TiO_2 is an n-type semiconductor that can be found in amorphous forms or in three crystalline forms: Rutile, tetragonal anatase, and orthorhombic brookite. The structure is Rutile, with both anatase and brookite becoming rutile if heated. The phase change from amorphous form to anatase comes with different properties lying between 300 and 500 °C, with additional brookite with an orthorhombic crystalline form and rutile tuned anisotropy occurring between 600 and 1000 °C. The corner-sharing twelve and six atoms per unit cell make up the tetragonal structures of rutile and anatase, correspondingly [32]. Notably, the Ti atom is correlated with corner shared six O atom, including both phases. Similar extend linked of O atom tend to correlate with three Ti atoms across both phases. Additionally, the deformed structure of TiO_6 octahedron come with corner shared linking, of two Ti-O atoms and tetra certain O-Ti-O bond angles shifted tilt 90° (larger distortion in anatase than rutile). Anatase and rutile crystal structures are made up of chains of TiO6 octahedra that share four sides in anatase and two in rutile. Brookite, for example, features deformed TiO6 octahedra with three edges. Furthermore, brookite has eight orthorhombic cell formula units with interatomic lengths and O-Ti-O bond angles comparable to anatase and rutile, but it has six distinct Ti-O bonds [33].

4.2.2.3 Copper Oxides

CuO is semiconducting and attracted particular attention, owing to the versatility of the monoclinic structure. It is covalent in characteristics with an average small size for synthesis with wet-chemical techniques. Cuprous oxide (Cu_2O), a reddish material with a coordinated oxygen atom, and cupric oxide (CuO) or copper (II) oxide, a black material,

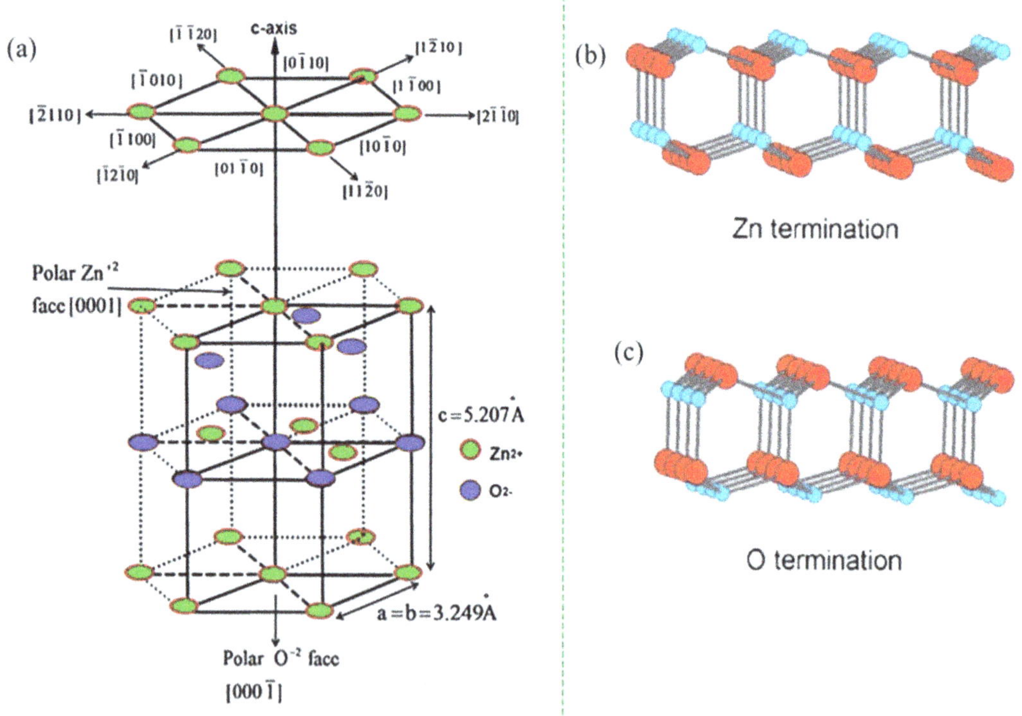

FIGURE 4.1 (a) Ionic positions of Zn^{2+} and O^{2-} in the lattice crystal comprised the ZnO hexagonal wurtzite structure from reference [30]; (b) and (c) ZnO wurtzite facets from the side, as other lattice symmetry illustrations along (0001) Zn and $(000\bar{1})$ O terminations, Adapted with permission from [31]. Copyright (2021) Royal Society of Chemistry.

are the two common copper oxides. It possesses p-type semiconductors and is a significant product of copper mining with a bulk direct E_g of 2-2.17 eV. Both cooper oxides have fine/narrow E_g of 1.2 eV (bulk) for CuO, which is extracted from solids. As the heat treatment causes negatively charged vacancies, commonly attributed to the p-type character of both oxides [34]. At normal temperature, Cu_2O has good conduction about 100 $cm^2 V^{-1}s^{-1}$, resulting in an exceptional expansion with different micrometer ranges. Cu_2O excitons have been discovered with high stability as (10 μs) after photoexcitation.

It is noteworthy that a cubic structure with the Pn3m (224) has been observed with the lattice parameter of the space group and 0.42696 nm. In addition, its unit cell is built of six atoms, with four copper atoms forming a face-centered cubic lattice and two oxygen atoms creating a BCC lattice pattern [35]. Concerning the copper sublattice, the oxygen atoms are located inside specific tetrahedral interstitial islands. These interstitial islands are the preferred nucleation with copper to tetrahedrally coordinate oxygen, whereas copper is linearly coordinated, implying that corner bond sharing occurs with two oxygen atoms. This orientation indicates a complicated monoclinic tenorite crystalline formation corresponding to the C2/c (15) space group, promoting the crystalline pattern values of 0.46833 nm, 0.34208 nm, and c=0.51294 nm [36].

4.2.3 Metal Oxide Nanocomposite

Metal oxides (MOs) have been a hot topic in this study because of their high modulus and strength at far higher temperatures than standard polymeric engineering materials, making them suitable for a wide range of applications. Individual MOs, cannot meet all of the requirements for developing new technologies and solving the world's most pressing problems. MOs, as an example, have limited applicability due to their intrinsic brittleness and low fracture toughness, besides promising biocompatibility and extensive use in catalysts. They provide a new avenue for researchers to develop new strategies for strengthening and stiffening MOs[37]. Striving to combine two or more materials results in a composite with properties that outperform the separate components. Most of the studies that exploited metal oxides, exhibit a wide range of functional features, including ferroelectricity, ferromagnetism, superconductivity, multi-ferroelectricity, and ionic conduction, have received a lot of interest in recent decades. A complicated hierarchy of contradictory interactions involving spin, orbital, charge, and lattice degrees of freedom provide enhancements to these unusual physical features [38]. Innovative characterization methods such as neutron scattering, electron microscopy, imaging, spectroscopy, and synchrotron scattering at high magnetic fields, and others have allowed us to investigate the encounters between these degrees of freedom, as well as the effects of strain, defect, and interface on features.

The form, size, dispersion, interaction, and alignment of CNTs are all crucially indispensable for the additional nucleation of nanocomposite (NC). The alignment of the carbon nanotubes, the homogeneity of the composite, nanotube aspect ratio, nanotube matrix adhesion, and the volume percentage of nanotubes can all have a significant impact on the composites' efficiency. Transition metal oxides are among the most major categories. Transition metal oxides have undergone much research in order to be widely employed in a variety of fields. Apart from their well-known usefulness, the following are some of the most important features of transition metal oxides for pioneering use in the ORR: (1) On the surfaces of transition metal oxides, numerous hydroxyl groups are introduced that were more functionalized by DNA and RNA (genetic materials) for biological catalysts; (2) Transition metal oxides feature strong interactions due to their crystalline structure, which might avoid metal particle agglomeration and keep metal particle sizes modest; (3) Transition metallic elements have much greater alkaline resistance to corrosion in the electromechanical atmosphere than noble metal and carbon-based composites and lead to the remarkable preservation of the high oxidation state. Novel improved material-based transition metal oxides have been generated in the last decade thanks to superior nanocomposite catalysts. Unraveled hierarchically porous frameworks, electrically conductive systems, and heterogeneous growth on carbonaceous substrates (micro/mesoporous carbon, graphene, and carbon nanotubes) with transition metal oxides assistance have shown significant electrocatalytic output in the ORR, particularly in comparison to unsupported counterparts [39].

4.2.4 Structure of Metal-Oxide Nanocomposite

Environmentally persistent structural and chemical transformation in metal oxide nanocomposite, is indispensable, based on quantitative and measurable composition, binary metal oxide nanocomposites' structure, and temperature, to the physicochemical activities that take place during the detection of reducing gases. The productivity and specificity of sensors based on these composites are considered [40]. The link between conductivity processes and sensor function in composites is established. It is highlighted that electron transmission across metal oxide components with different work functions, which leads to mutual charging of these components, plays a critical role [41]. In composite materials comprised of metal oxides with different electronic and chemical properties, the principles of chemical and electronic sensitization of the sensor response are reported in various literature. It is mentioned that the method by which composite materials are obtained plays an essential role. The influence of tiny clusters of one oxide on the surfaces of nanoparticles of other components generated during impregnation synthesis of composites is investigated. Systems depending on various types of metal oxides, as well as core-shell composite nanofibers, are studied. It is demonstrated that the selectivity/sensitivity of a sensor model for various chemical substances may be modified by modifying the type of the components and their relative placement in the nanofibers [42].

4.2.4.1 Core-Shell Nanostructure

Core-shell technology is leading edge and becoming a new trend in the synthesis of nanomaterials, owing to their versatile use in making designs and geometries with enhanced properties and more functional utilization in analytical science. Core-shell nanostructure is of ever-increasing importance in combining, for example, single-core/shell, different multishell properties into hybrid NPs for different functionalities in single NPs with other new and sophisticated multicores/multishells improving its interior functionalities. Wrapping metal NPs in semiconductor shells are an effective way to prevent metal NP agglomeration and deterioration caused by harsh chemicals. Adopting promising solvothermal and hydrothermal methods results in Au–TiO_2 core-shell structures [43]. This approach generally employs an autoclave process and provides a compatible strength for the desired TiO_2 and Au precursors, which show a solution of ethyl ether, glycerol, and solvothermal processes. The resulting solution was then distributed in a water system and hydrothermally processed.

The TiO_2 shell prevents AuNP agglomeration and leaching. Another exciting progress is of Au/TiO_2 core/shell NPs being utilized as star-like block copolymers and metal moieties of precursors once more. Successful results from a sandwich-like structure consisting of SiO_2 core comprised of a layer of Au nanoparticles were obtained using the sol-gel method, and a doped TiO_2 encapsulated shell provided stability. The design of a standard TiO_2–Au core-shell structure, in which Au NPs are a poor interaction between surface TiO_2 and thus unstable during heat treatment and photocatalysis makes use of sandwich architectures that provide additional stability to the Au NPs [45]. Further extended synthesis of core-shell Au@TiO_2 NPs comes with the dissolution of TiF_4, a reddish-brown blend of $HAuCl_4$ solution, sodium citrate, and ascorbic acid solution. Later on, the dispersed solution was transferred to stainless-steel autoclaves fixed with Teflon. The reaction was carried out for 48 hours at 180°C. Multi-core metal oxide/noble metal shell nanocomposites are a commonly classified different combined multi-component for metal oxide/ noble metal shell/core nanostructure that may be manufactured using a variety of templates (carbon, polystyrene, and SiO_2 spheres) and follow a similar route. PVP-functionalized carbon nanosphere templates were used to make multi-Pd core/hollow CeO_2 core/shell nanocomposites, as shown in Figure 4.2 [44]. The final calcination step in the air allows for the hydrothermal treatment of Ce^{3+}, thus different functionalities can be integrated to eliminate the carbon template and PVP stabilizer. Due to the specific design Pd/CeO_2 core/shell nanostructures are formed due to this process [46].

4.2.4.2 Yolk-Shell Nanostructure

Yolk-shell structures have remarkable features for hollow and core-shell structures such as core@ void@ shell configuration

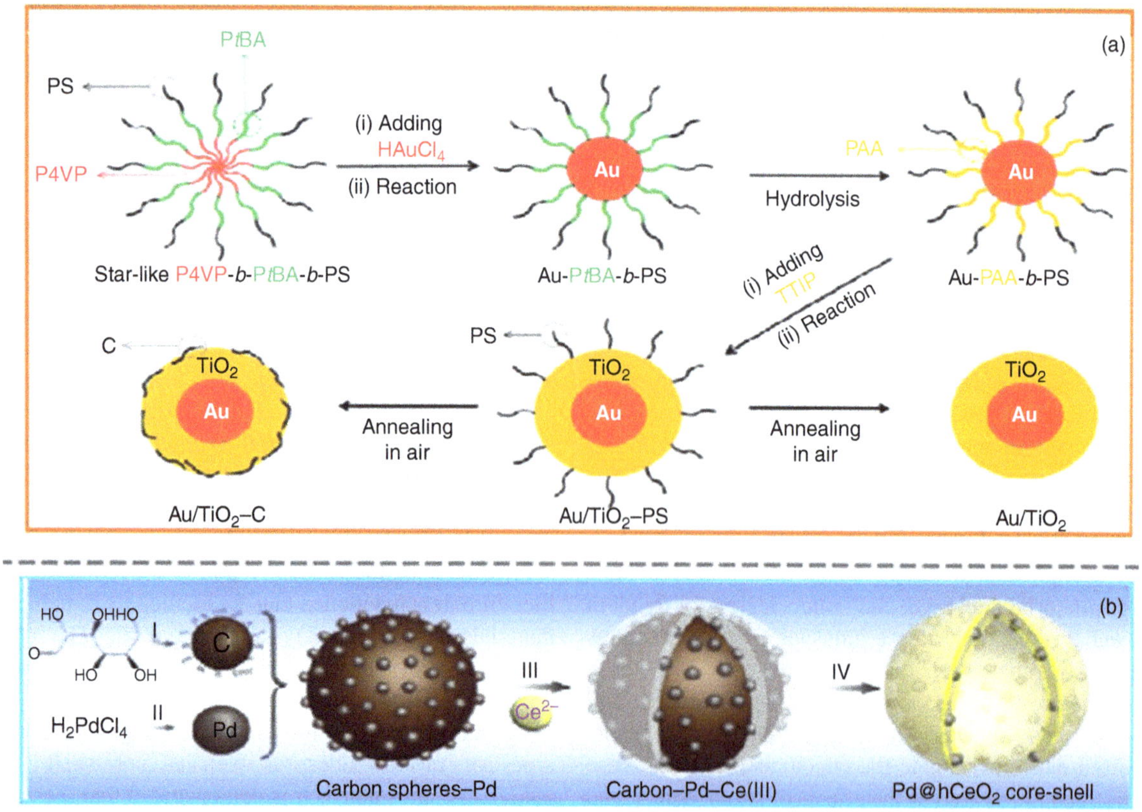

FIGURE 4.2 (a) The synthesis procedure for (PS) Au/TiO_2 core/shell NPs and (b) schematic showing the inner layer encapsulated of core/shell of Pd/CeO_2 nanocomposite. Adapted with permission from [44]. Copyright © 2015 American Chemical Society

and NM/MO yolk/shell nanostructures which makes these nanostructures ideal, due to promising shell functionally, moveable cores, and they are good candidates for various applications. Wrap metal cores within SiO_2 shells, make yolk/shell nanohybrids by precisely eliminating a piece of the noble metal core of premade core/shell NPs [47].

Because it is insulating, the SiO_2 hollow shell can only be utilized as a stabilizer. The inclusion of multifunctional MOs as shells, like Fe_3O_4, ZnO, and NiO, provides additional features and functions to the technical design. While most MOs corrode during the etching procedure, producing yolk/shell nanostructures using other MO shells is more difficult than SiO_2. As a consequence, SiO_2 is commonly employed as a sacrificial hard template in the synthesis of NM/MO yolk/shell nanostructures. Au/ZrO_2 yolk/shell catalysts with great heat resistance have also been created using SiO_2 as a heterogeneous catalyst [48]. By employing Au/SiO_2 nanospheres as expendable templates, researchers were able to effectively construct Au/SnO_2 yolk/shell nanospheres with better sensing performance [49]. A common carbon is employed as a hard template to make noble metal/TiO_2 yolk/shell nanostructures in the same way as SiO_2.

4.3 FUNDAMENTAL APPLICATION OF MOF METAL-OXIDE-BASED NANOCOMPOSITES

Due to their wide range of commercial and technical uses, MOFs, and metal oxides have been at the forefront of research over the past two decades. To date, a variety of review articles on MOFs and metal oxides have been released, emphasizing their uses and characteristics. MOFs are used in gas separation, catalysis, storage, medication delivery, the development of materials for heat pumps and absorption refrigerators, pollutant removal, and the manufacture of optical and luminous materials. MOFs are being used in device applications as adsorbent-based chemical sensors with a NO_2 gas sensor technique. MOFs surpass zeolites throughout in terms of pore specific volume and pore size, making them ideal as efficient catalysts in a variety of processes, such as those involving bulky molecules that cannot permeate zeolite pores [50].

MOFs are more adaptable than zeolites since they contain organic structures. One of the most important applications of TiO_2 and ZnO is removing pollutants from effluent. Another function for ZnO NPs is in agriculture, where they are used to treat a disease known as Grasserie in silkworms and Rice Weevil in rice [51]. ZnO has also been utilized in various consumer applications, including cosmetics and skincare products, and is noted for its antibacterial properties. Metal oxides are very efficient in oxidation products and microbial eradication. ZnO NPs have piqued the interest of researchers in biological applications, including scanning, medication delivery, gene therapy, and biosensors.

4.3.1 Sensor Application

Monitoring systems for evaluating the quality of water and air that use semiconductor metal oxides are gaining much attention because they enhance the sensitivity and selectivity of existing sensing devices while also allowing simultaneous measurements of multiple parameters with a serious response. Metal oxides offer a broad array of applications in gas sensors, with numerous studies demonstrating the advantages of incorporating these materials into such devices, as well as their overall gas detecting efficiency of size features [52]. Sensing technology is widely used in a variety of industries, as well as in residential settings. Some categories include the automobile industry, indoor environmental quality control, and greenhouse gas measurement, amongst others. The most important parameters of sensor devices are sensitivity, long-term stability, selectivity, operating temperature, energy consumption, reversibility, low humidity requirement, and production price. To be economically successful, these devices must be steady during function, with a consistent and repeatable signal over time, with the ability to accurately detect a particular gas even in mixes of multiple gases [53].

Gas sensors were divided into two groups by Liu et al. based on their sensing approaches: (a) methods based on variations in electrical characteristics, and (b) methods based on variations in other properties. The variation in electrical characteristics is used in semiconductor metal oxide gas detection. Gas sensors made of TiO_2, SnO/SnO_2, ZnO, CuO/Cu_2O, WO_3, and V_2O_5 are extensively used to detect flammable, oxidizing gases, and reduce these, with sensing relying mostly on resistance change reactions in the targeted gases. The most often utilized semiconducting metal oxides in industrial gas sensors are SnO_2 and WO_3. In particular, reducing gases including such as H_2, H_2S, CO, NH_3, SO_2, HCHO, and CH_4 cause a rise in n-type semiconductor conductivity and a reduction in p-type semiconductor conductivity, whereas oxidizing gases have the reverse effect [54].

4.3.1.1 Gas Adsorption and Storage Applications

Toxic gases are a serious global issue all around the world. The use of an adsorption method is one of the most effective techniques to remove harmful gases. Adsorptive elimination of hazardous gases has been the subject of numerous studies. MOFs have been recommended for use in gas removal due to their increased surface area and unique properties [55]. Furthermore, gas storage has seen increasing popularity year after year, and MOFs have been employed in gas storage systems because of their unique qualities, such as a large surface area and pore volume. MOFs have been widely researched for gas storage purposes, such as H_2 and CH_4, which are two examples of alternate energy supplies for future automobiles; however, their efficient use is still a hurdle for the automotive sector [56].

4.3.2 Catalyst Applications

Surface and strain-driven lattice distortion, change in electronic state density, and oxidation-induced charge redistribution are all features of nanomaterials that may increase catalytic activity. Significant advancements in synthetic chemistry have resulted in the availability of nanomaterials with shapes, well-defined sizes, structures, crystal facets, and compositions that open up many possibilities for designing catalysts with high activity, for energetically difficult reactions, excellent selectivity of valuable chemicals, and long-term reliability in harsh reaction conditions.

4.3.2.1 Localized Surface Plasmon Resonance (LSPR)

Inside NMs, LSPR is defined as the collective oscillation of free electrons caused by the incoming light. When the incident light frequency equals the natural frequency of free electrons oscillating against the restoring force of positive nuclei, the oscillation amplitude achieves a maximum [57]. The LSPR effect of NMs may easily increase the light-absorption spectrum ranges of TiO_2 and ZnO to visible and even near-infrared areas, which is a direct benefit of NM–MOs. Using Au nanorods as antennas, Liu et al. discovered broad absorption spectra of TiO_2 by adjusting the aperture of Au nanorods. The plasmon resonance absorptions of nanostructures may be controlled from 630 to 810 nm [58].

4.3.2.2 Photocatalytic CO_2 Reduction

Photocatalytic CO_2 reduction is a good option for reducing greenhouse gases created by human manufacturing processes, as well as another technique for transforming light energy into chemical energy. Photo-reduction of CO_2 requires both energy and a hydrogen supply, in contrast to photo-catalytic H_2 production, which demands solar energy and water. Due to increased competition between them in the presence of moisture, inhibiting the H_2 production process is an effective technique for improving CO_2 photo-reduction [59]. Since the discovery by Inoue's team of photoelectron catalytic CO_2 reduction over semiconductors. The photocatalytic CO_2 conversion of powders in water has been focused on ongoing research. In the photocatalytic CO_2 reduction system, charge separation and transportation are two critical components in solar energy harvesting [60]. Deposited NMs served as cocatalysts and were shown to enhance the accuracy of photocatalytic CO_2 reduction due to better charge separation, instigated CO_2 activation, and the existence of active catalytic sites for the reduction reaction, which is comparable to removal efficiency and H_2 formation. Figure 4.3 illustrates the mechanism of the reduction of systematic photocatalytic CO_2 conversion depiction with a mechanism of adsorption and interaction, and activity curve illustration of CuPt–TiO_2 for the production of CH_4 [61].

4.4 CONCLUSION AND PERSPECTIVE

In this chapter, some of the most effective ways to use MOFs in electrochemical with exceptional morphological structure and their potential applications have been discussed. MOF characteristics such as porosity, durability, and synthetic selectivity provide suitable design criteria for addressing specific constraints in electrode and electrolyte innovation. MOFs will be distinguished from porous carbons, polymers, and inorganic materials in the future by identifying, exploiting, and expanding methods to support charge storage and transit. Given their potential, there is still a lot to learn about how to use MOFs effectively in energy storage systems. Design methodologies used in polymers, ionic liquids, carbons, and solid inorganic compounds can be used to explore and create new MOF designs with enhanced storage capacity. Moreover, MOF composite materials are largely unexplored, and their exploration will almost certainly uncover novel parameter

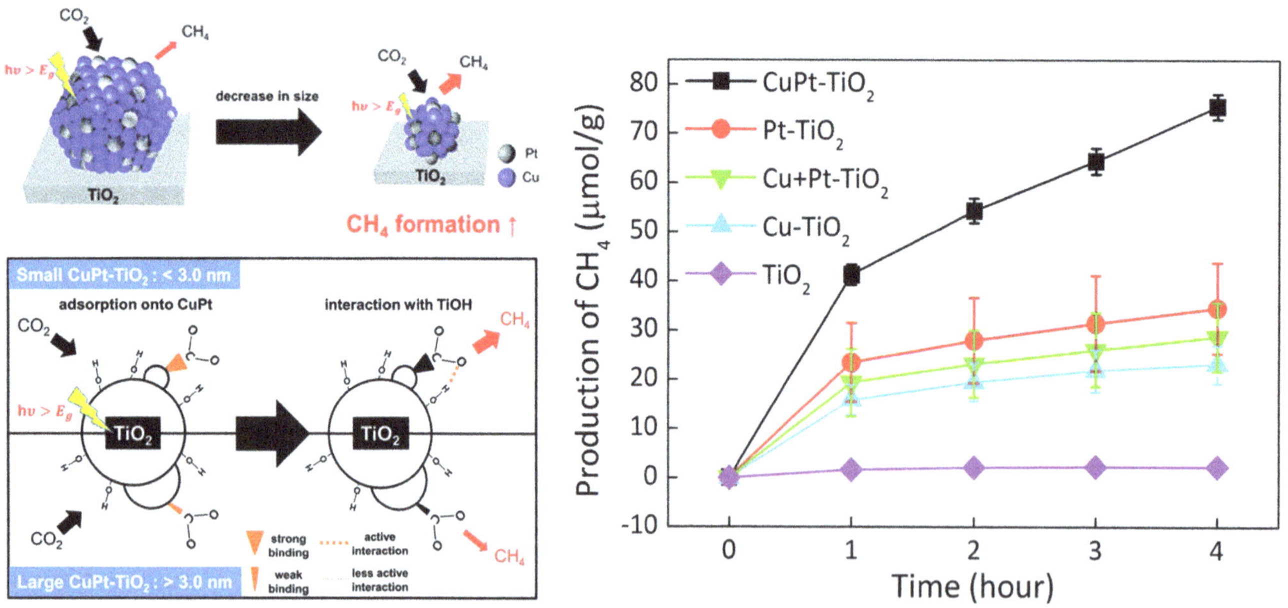

FIGURE 4.3 Systematic photocatalytic CO_2 conversion depiction with the mechanism of adsorption and interaction activity of CuPt–TiO_2. Adapted with permission from [61]; an open-access article licensed under a Creative Commons Attribution-NonCommercial 3.0 Unported Licence.

spaces at the junction of traditional soft/hard materials and MOFs. Thus, theoretical and practical knowledge gained from MOF-based devices will be critical for developing next-generation materials for emerging technologies such as modular and flexible electronics, solid-state electrolytes, and enhanced energy storage devices in mild and extreme conditions.

This chapter presents an explanation of the promise of enhanced properties of noble metal-metal oxide nanocomposites with various nanostructures such as noble metal decorated metal oxide nanoarrays, noble metal/metal oxide core/shell, and yolk/shell nanostructures, and Janus noble metal-metal oxide nanostructures in a variety of catalytic applications. Photocatalytic degradation of environmental pollutants and photocatalytic CO_2 reduction are two applications for these materials. Finally, the issues and future research directions for these materials are discussed. We anticipate that this review will be useful to both professionals and beginners interested in the design of noble metal-metal oxide hybrid nanostructures for a variety of catalytic applications.

REFERENCES

[1]. Li, H., et al. *Metal–Organic Frameworks as Metal Ion Precursors for the Synthesis of Nanocomposites for Lithium-Ion Batteries.* 2020. **132**(12): p. 4793–4799.

[2]. Song, Y., et al. *Metal/metal oxide nanostructures derived from metal–organic frameworks.* 2015. **5**(10): p. 7267–7279.

[3]. Wei, Q., et al. *Spindle-like Fe2O3/ZnFe2O4 porous nanocomposites derived from metal-organic frameworks with excellent sensing performance towards triethylamine.* 2020. **317**: p. 128205.

[4]. Singh, B.K., S. Lee, and K.J.R.M. Na, *An overview on metal-related catalysts: metal oxides, nanoporous metals and supported metal nanoparticles on metal organic frameworks and zeolites.* 2020. **39**(7): p. 751–766.

[5]. Dang, S., Q.-L. Zhu, and Q.J.N.R.M. Xu, Nanomaterials derived from metal–organic frameworks. 2017. **3**(1): p. 1–14.

[6]. Kaneti, Y.V., et al. *Nanoarchitectured design of porous materials and nanocomposites from metal-organic frameworks.* 2017. **29**(12): p. 1604898.

[7]. He, X., et al. *Rapid formation of metal–organic frameworks (MOFs) based nanocomposites in microdroplets and their applications for CO2 photoreduction.* 2017. **9**(11): p. 9688–9698.

[8]. Salunkhe, R.R., Y.V. Kaneti, and Y.J.A.n. Yamauchi, Metal–organic framework-derived nanoporous metal oxides toward supercapacitor applications: progress and prospects. 2017. **11**(6): p. 5293–5308.

[9]. Zhang, L., et al. *NH2-MIL-53 (Al) metal–organic framework as the smart platform for simultaneous high-performance detection and removal of Hg2+.* 2019. **58**(19): p. 12573–12581.

[10]. Du, X., et al. *Controlled Zn2+-triggered drug release by preferred coordination of open active sites within functionalization indium metal organic frameworks.* 2017. **9**(34): p. 28939–28948.

[11]. Nur Atiqah, S., *Design, and fabrication of solar light responsive new metal organic frameworks for photocatalysis/ Nur Atiqah Surib.* 2018, University of Malaya.

[12]. Sharmin, E. and F. Zafar, *Introductory chapter: metal organic frameworks (MOFs), in Metal-organic frameworks.* 2016, IntechOpen.

[13]. Furukawa, H., et al. *The chemistry and applications of metal-organic frameworks.* 2013. **341**(6149).

[14]. Lu, W., et al. *Tuning the structure and function of metal–organic frameworks via linker design.* 2014. **43**(16): p. 5561–5593.

[15]. Halder, A. and D.J.C. Ghoshal, *Structure and properties of dynamic metal–organic frameworks: a brief account of crystalline-to-crystalline and crystalline-to-amorphous transformations.* 2018. **20**(10): p. 1322–1345.

[16]. Baumann, A.E., et al. *Metal-organic framework functionalization and design strategies for advanced electrochemical energy storage devices.* 2019. **2** (1): p. 1–14.

[17]. Derakhshandeh, P.G., et al. *Ce (III)-Based Frameworks: From 1D Chain to 3D Porous Metal–Organic Framework.* 2019. **19**(12): p. 7096–7105.

[18]. Zhai, T., et al. *Rational hetero-interface design of Fe3N@ Ni2Co-LDHs as highly efficient electrocatalyst for oxygen evolution reaction.* 2021. **853**: p. 157353.

[19]. Chakraborty, G., et al. *Two-Dimensional Metal-Organic Framework Materials: Synthesis, Structures, Properties and Applications.* 2021. **121**(7): p. 3751–3891.

[20]. Yu, D., et al. *A combined experimental and quantum chemical study of CO2 adsorption in the metal–organic framework CPO-27 with different metals.* 2013. **4**(9): p. 3544–3556.

[21]. Gangu, K.K., et al. *A review on contemporary metal–organic framework materials.* 2016. **446**: p. 61–74.

[22]. He, L., et al. *Core-shell noble-metal@ metal-organic-framework nanoparticles with highly selective sensing property.* 2013. **125**(13): p. 3829–3833.

[23]. Guo, J., et al. *Advanced photocatalysts based on metal nanoparticle/metal-organic framework composites.* 2021. **14**(7): p. 2037–2052.

[24]. Roduner, E.J.C.S.R., *Size matters: why nanomaterials are different.* 2006. **35**(7): p. 583–592.

[25]. Pedhekar, R., et al. *Low temperature H2S gas sensor based on Fe2O3 modified ZnO-TiO2 thick film.* 2015. **3**: p. 219.

[26]. Dey, C., et al. *Crystalline metal-organic frameworks (MOFs): synthesis, structure, and function.* 2014. **70**(1): p. 3–10.

[27]. Coleman, V.A., and C. Jagadish, *Basic properties and applications of ZnO, in Zinc oxide bulk, thin films and nanostructures.* 2006, Elsevier. p. 1–20.

[28]. Wen, B., Y. Huang, and J.J.J.T.J.o.P.C.C. Boland, *Controllable growth of ZnO nanostructures by a simple solvothermal process.* 2008. **112**(1): p. 106–111.

[29]. Tong, Y., et al. *Growth of ZnO nanostructures with different morphologies by using hydrothermal technique.* 2006. **110**(41): p. 20263–20267.

[30]. Kumar, R., et al. *Zinc oxide nanostructure-based dye-sensitized solar cells.* 2017. **52**(9): p. 4743–4795.

[31]. Chang, J. and E.R.J.C. Waclawik, *Facet-controlled self-assembly of ZnO nanocrystals by non-hydrolytic aminolysis and their photodegradation activities.* 2012. **14**(11): p. 4041–4048.

[32]. Yanagisawa, K. and J.J.T.J.o.P.C.B. Ovenstone, *Crystallization of anatase from amorphous titania using the hydrothermal technique: effects of starting material and temperature.* 1999. **103**(37): p. 7781–7787.

[33]. Allen, N.S., et al. *The effect of crystalline phase (anatase, brookite and rutile) and size on the photocatalytic activity of calcined polymorphic titanium dioxide (TiO2).* 2018. **150**: p. 31–36.

[34]. Murali, D.S., et al. *Synthesis of Cu2O from CuO thin films: Optical and electrical properties.* 2015. **5**(4): p. 047143.

[35]. Meyer, B., et al. *Binary copper oxide semiconductors: From materials towards devices.* 2012. **249**(8): p. 1487–1509.
[36]. Su, D., et al. *CuO single crystal with exposed {001} facets-A highly efficient material for gas sensing and Li-ion battery applications.* 2014. **4**(1): p. 1–9.
[37]. Wu, Y., *Sophisticated construction of Au islands on Pt–Ni: an ideal trimetallic nanoframe catalyst, in Controlled Synthesis of Pt-Ni Bimetallic Catalysts and Study of Their Catalytic Properties.* 2016, Springer. p. 93–111.
[38]. Mühlberg, M.J.J.o.M.C.A., *Retraction: Recent advances of metal–metal oxide nanocomposites and their tailored nanostructures in numerous catalytic applications.* 2020. **8**(30): p. 15189–15189.
[39]. Bhasarkar, J.B., B.A. Bhanvase, and V.B. Pawade, *Future Scope of Various Nanostructured Metal Oxides for Sustainable Energy Sources, in Multifunctional Nanostructured Metal Oxides for Energy Harvesting and Storage Devices.* 2020, CRC Press. p. 309–319.
[40]. Oh, S.-H. and G.B.J.T.J.o.P.C.A. Hoflund, *Chemical state study of palladium powder and ceria-supported palladium during low-temperature CO oxidation.* 2006. **110**(24): p. 7609–7613.
[41]. Gerasimov, G., et al. *Effect of the composition and structure of metal oxide nanocomposites on the sensor process when detecting reducing gases.* 2017. **91**(9): p. 1609–1620.
[42]. Bian, Z., et al. *A nanocomposite superstructure of metal oxides with effective charge transfer interfaces.* 2014. **5**(1): p. 1–9.
[43]. Bian, Z., et al. *In situ encapsulation of Au nanoparticles in mesoporous core–shell TiO2 microspheres with enhanced activity and durability.* 2009(25): p. 3789–3791.
[44]. Zheng, D., et al. *Unconventional route to hairy plasmonic/semiconductor core/shell nanoparticles with precisely controlled dimensions and their use in solar energy conversion.* 2015. **27**(15): p. 5271–5278.
[45]. Zhang, Q., et al. *A highly active titanium dioxide based visible light photocatalyst with nonmetal doping and plasmonic metal decoration.* 2011. **50**(31): p. 7088–7092.
[46]. De Rogatis, L., et al. *Embedded phases: a way to active and stable catalysts.* 2010. **3**(1): p. 24–42.
[47]. Kamata, K., Y. Lu, and Y.J.J.o.t.A.C.S. Xia, *Synthesis, and characterization of monodispersed core– shell spherical colloids with movable cores.* 2003. **125**(9): p. 2384–2385.
[48]. Güttel, R., M. Paul, and F.J.C.c. Schüth, *Ex-post size control of high-temperature-stable yolk–shell Au,@ ZrO2 catalysts.* 2010. **46**(6): p. 895–897.
[49]. Li, X., X. Fu, and H.J.P.C.C.P. Yang, *Preparation, and photocatalytic activity of eccentric Au–titania core–shell nanoparticles by block copolymer templates.* 2011. **13**(7): p. 2809–2814.
[50]. Falcaro, P., et al. *Application of metal and metal oxide nanoparticles@ MOFs.* 2016. **307**: p. 237–254.
[51]. Kandasamy, S. and R.S.J.J.C.P.R. Prema, *Methods of synthesis of nano particles and its applications.* 2015. **7**(3): p. 278–285.
[52]. Koo, W.-T., et al. *Heterogeneous sensitization of metal–organic framework driven metal@ metal oxide complex catalysts on an oxide nanofiber scaffold toward superior gas sensors.* 2016. **138**(40): p. 13431–13437.
[53]. Kreno, L.E., et al. *Metal–organic framework materials as chemical sensors.* 2012. **112**(2): p. 1105–1125.
[54]. Nunes, D., et al. *Metal oxide nanostructures: synthesis, properties, and applications.* 2018: Elsevier.
[55]. Song, F.Z., et al. *Metal–organic framework templated porous carbon-metal oxide/reduced graphene oxide as superior support of bimetallic nanoparticles for efficient hydrogen generation from formic acid.* 2018. **8**(1): p. 1701416.
[56]. Keskin, S., D.S.J.I. Sholl, and E.C. Research, *Assessment of a metal– organic framework membrane for gas separations using atomically detailed calculations: CO2, CH4, N2, H2 mixtures in MOF-5.* 2009. **48**(2): p. 914–922.
[57]. Linic, S., P. Christopher, and D.B.J.N.m. Ingram, *Plasmonic-metal nanostructures for efficient conversion of solar to chemical energy.* 2011. **10**(12): p. 911–921.
[58]. Liu, L., S. Ouyang, and J.J.A.C. Ye, *Gold-nanorod-photosensitized titanium dioxide with wide-range visible-light harvesting based on localized surface plasmon resonance.* 2013. **125**(26): p. 6821–6825.
[59]. Low, J., J. Yu, and W.J.T.j.o.p.c.l. Ho, *Graphene-based photocatalysts for CO2 reduction to solar fuel.* 2015. **6**(21): p. 4244–4251.
[60]. Tang, L., et al. *Unique homo–heterojunction synergistic system consisting of stacked BiOCl nanoplate/Zn–Cr layered double hydroxide nanosheets promoting photocatalytic conversion of CO2 into solar fuels.* 2018. **54**(40): p. 5126–5129.
[61]. Lee, S., et al. *Low-coordinated surface atoms of CuPt alloy cocatalysts on TiO2 for enhanced photocatalytic conversion of CO2.* 2016. **8**(19): p. 10043–10048.

5 Synthesis and Applications of MOFs
Chalcogenide-based Nanocomposites

Shikha Bhogal[1], *Irshad Mohiuddin*[1], *Sandeep Kumar*[1], *Promila Sharma*[1], *Asnake Lealem Berhanu*[1], *Kuldeep Kaur*[2] *and Ashok Kumar Malik*[1]

[1]Department of Chemistry, Punjabi University, Patiala, India

[2]Department of Chemistry, Mata Gujri College, Fatehgarh Sahib, India

5.1 INTRODUCTION

Metal-organic frameworks (MOFs) are porous materials in which metal nodes are linked through organic linker molecules to form crystalline structures of varying dimensionalities [1]. As a result of the diverse structures and intriguing characteristics (like high specific surface area, tailor-made structure, and adjustable porosity) of MOFs, they have recently been applied in several applications including catalysis, gas adsorption, gas separation, drug delivery, magnetism, and sensing [2, 3]. Despite their outstanding features, they still face some challenges like poor chemical stability which hampers their use in practical applications. Much effort has been aimed at developing novel MOF structures with desired properties to widen the field of their application. The fusion of MOFs with other functional materials has been explored to improve their characteristic properties and yield novel properties [4]. The outstanding properties of many such composite materials have led to their application in a variety of fields including catalysis, sensing, and adsorption [5]. To date, MOF composites have been successfully prepared with a variety of materials including carbon nanomaterials, metal nanoparticles, metal oxides, and chalcogenides to achieve better performance in intended applications compared to individual components [6].

Recently, metal chalcogenide nanoparticles (like metal sulfides, selenides, and tellurides) have drawn significant interest by virtue of their interesting optoelectronic properties which include band gap mediated processes and localized surface plasmon resonance properties [7]. The hybridization of metal chalcogenide nanoparticles with MOF hybrids is an attractive approach to engineering the properties of both materials for desired applications. MOFs can furnish tunable pore size, large specific surface area, and multiple functional active sites to integrate with metal chalcogenides. In addition to acting as a host matrix for chalcogenides, it also prevents their aggregation to aid satisfactory dispersion. On the other hand, chalcogenides can improve the poor stability, conductivity, and electrocatalytic activity of the MOFs [8]. The improvement of the component properties and the generation of new attributes in the hybrid materials have driven researchers to synthesize MOF-chalcogenide through the immobilization of metal chalcogenides into MOFs.

MOFs have also been used as precursors/templates for the synthesis of metal chalcogenides for a few years [9]. The porous structure of MOFs and the presence of abundant metal ions make them favorable candidates for the derivation of metal chalcogenides. The MOF-derived metal chalcogenides retain the high surface area and permanent porosity of MOFs while significantly enhancing their conductivity and stability [10]. The outstanding features of derived metal chalcogenides make them promising candidates for electrochemical applications. Considering the promising aspects of metal-organic frameworks modified metal chalcogenide (MOFs@MC) nanocomposites, this chapter discusses the different methods for their synthesis along with their electrocatalytic applications. In addition to this, the electrochemical application of MOF derived metal chalcogenides has also been explored. Finally, the prospects of MOFs-chalcogenide nanocomposites for future applications have been explored.

5.2 STRUCTURE OF MOF@METAL CHALCOGENIDE COMPOSITES

Because of the different assembly methods of metal chalcogenides and MOFs, the structure of MOF@MC nanocomposites can be divided into three types: core-shell structure, sandwich structure, and within the pore structure.

5.2.1 Core-shell Structure

MOFs can be fabricated on pre-formed metal chalcogenides. This method often necessitates the use of a capping agent or surfactant to promote the heteronucleation of the MOF on the chalcogenide surface. The core-shell structure of CdTe@ZIF-8 nanocomposites was reported to exhibit oxidase activity [11]. For the synthesis of the nanocomposite, CdTe was first prepared and then stabilized by polyvinyl pyrrolidine (PVP) to form CdTe@PVP. The CdTe@PVP was further dissolved in methylimidazole and zinc-nitrate solution and stirred for 40 minutes to obtain CdTe@ZIF-8 core-shell nanocomposites.

DOI: 10.1201/9781003188148-5

5.2.2 Sandwich Structure

Metal chalcogenides may be integrated into MOF thin films to produce a sandwich-type structure. To produce the sandwich structure, a thin film of MOF is grown which is then spin-coated with chalcogenides and the cycle is repeated multiple times. A CdSe/ZIF-8 composite with sandwich structure was synthesized by Zhang et al. [12]. They obtained a thin film of ZIF-8 by immersing the glass substrate in a solution of zinc nitrate and 2-methyl-imidazole for 40 min. The glass substrate had hydroxyl groups and the PVP capped CdSe was then spin-coated onto the ZIF-8 thin film which was followed by the growth of MOF [12]. The cycle was repeated multiple times which produced the multilayer CdSe/ZIF-8 hybrid sandwich-type structure (Figure 5.1).

5.2.3 Within the Pore Structure

The immobilization of the solution or vapor phase of metal chalcogenide precursors into the MOF pores followed by the formation of metal chalcogenide nanoparticles from the precursors gives rise to the pore structure. The CdSe/MOF nanocomposite with CdSe encapsulated into the pores of MIL-101- (Al)-NH_2 was reported via a stepwise injection reaction [13]. As the source of Cd, a precursor solution of CdO and stearic acid in octadecane was injected into the MOF while a Se precursor solution containing tributylphosphine in octadecane was injected into the MOF as the source of Se. The infusion of precursors into the cavity was followed by heating, which produced the CdSe nanoparticles immobilized into the pores of the MOF (Figure 5.2).

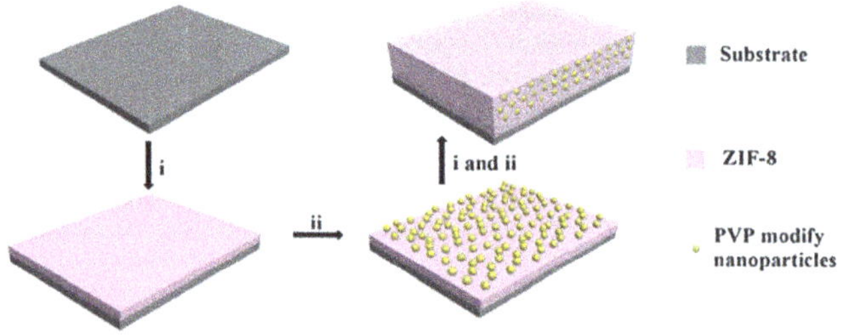

FIGURE 5.1 Fabrication procedure of CdSe/ZIF-8 hybrid thin films. Adapted with permission from [12]. Copyright (2014) RSC.

5.3 GENERAL SYNTHETIC METHODS OF MOF@METAL CHALCOGENIDE NANOCOMPOSITES

For immobilizing metal chalcogenides into MOFs, four techniques have been established: bottle-around ship, ship-in-bottle, photochemical deposition, and direct binding (Figure 5.3).

5.3.1 Bottle Around Ship Method

This method involves the assembly of MOFs around metal chalcogenides that have already been synthesized. It is also called, the template synthesis method and involves two main steps: i) Production of metal chalcogenides of homogeneous shape and size and their stabilization using capping agents or surfactants; ii) Dispersion of pre-formed metal chalcogenides in a solution containing the organic frameworks, which assemble around the chalcogenides. The metal chalcogenides are intercalated inside the MOF matrix, and their particle size is not constrained by the MOF pore size. The controlled overgrowth of MOFs on chalcogenides, on the other hand, is difficult because of the large interfacial energy barrier between the two particles [8].

Maspoch et al. (2009) encapsulated CdSe/ZnS into Zn(bix) (bix = 1,4-bis(imidazole-1-ylmethyl)benzene)) using the bottle around ship method to yield MOF@MC nanocomposite [14]. The pre-formed CdSe/ZnS nanoparticles were mixed with Zn (II) and bix linker to entrap the chalcogenide in the Zn(bix) matrix. In addition to the chalcogenide, they also encapsulated fluorescein and iron oxide nanoparticles to give multi-modal fluorescence and magnetic separation of the particles, respectively. The surface modification of chalcogenides is necessary to prevent aggregation and synthesize well-dispersed metal chalcogenides. A core-shell metal chalcogenide, CdSe/CdS/$Cd_{0.5}Zn_{0.5}S$/ZnS was synthesized using two-step ligand ((octadecylamine and 2,2,4-trimethylpentyl)-phosphinic acid)) modification [15]. These core-shell chalcogenides provided better passivation and

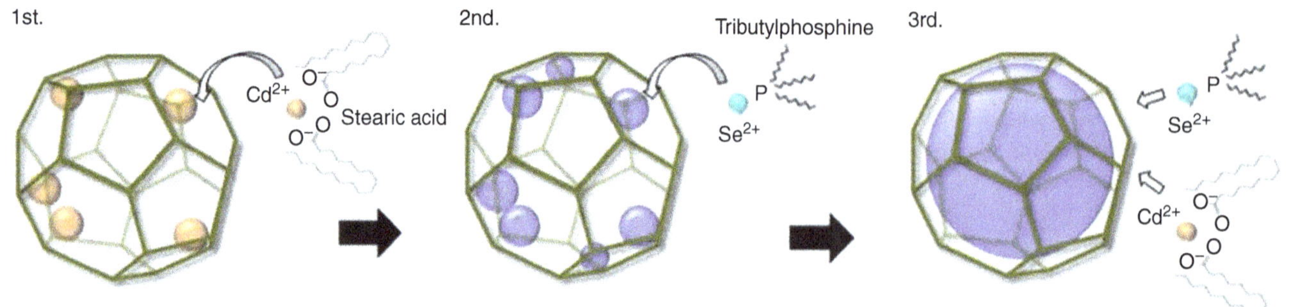

FIGURE 5.2 Schematic illustration of encapsulation of CdSe nanoparticles into the pores of MOF. Adapted with permission from [13]. Copyright (2014) RSC.

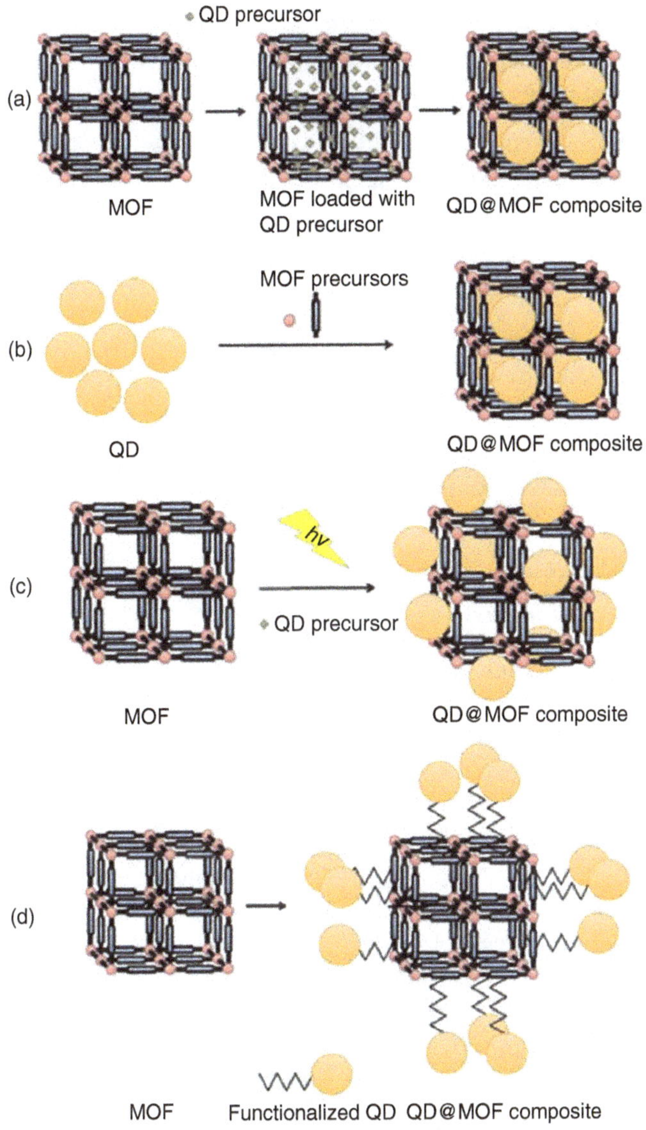

FIGURE 5.3 Different synthetic methods for MOF@ metalchalcogenide composite a) Ship in bottle method; b) Bottle around ship method; c) Photochemical deposition; d) Direct binding. Adapted with permission from [7]. Copyright (2016) Elsevier.

modification of the lattice for improved crystallization. The ligands present on the chalcogenides were then successfully exchanged by pyridine and 5-amino-1-pentanol (AP) to form metal chalcogenides that are highly dispersible. The AP-capped chalcogenides were then added into the solution containing $Zn(NO_3)_2$ and 1,4-benzenedicarboxylic acid at 100°C for 10 h to form composites. The authors also studied the solvent effect on the synthesis of composites using two different solvents which were dimethylformamide (DMF) and diethyl formamide (DEF). The particle size distribution was found to vary in the two solvents. In DEF, the distribution of chalcogenide particle size was found to be homogenous while in DMF aggregation of the particles was observed. The possible reason for the stability of AP capped chalcogenides in DEF was considered to be the presence of bulkier ethyl groups in DEF in comparison to methyl groups in DMF.

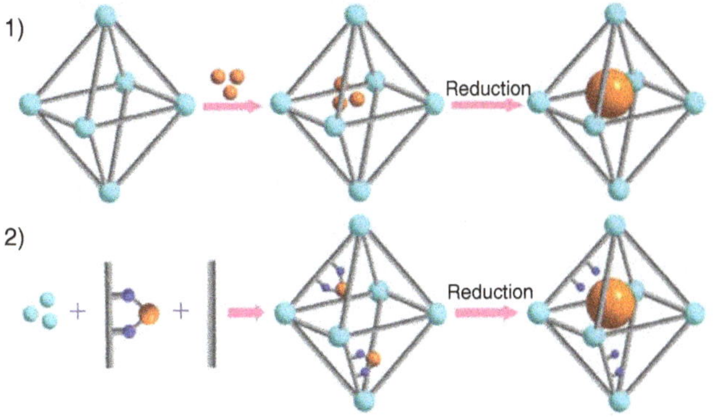

FIGURE 5.4 Post assembly incorporation (1) and pre-assembly incorporation of chalcogenides (2) into MOF. Adapted with permission from [17]. Copyright (2014) Elsevier.

5.3.2 Ship in Bottle Method

In this method, immobilization of metal chalcogenides can be done inside the MOF which must be small enough to penetrate through the pores of the MOF. The assembly of metal chalcogenides in the MOF matrix can be achieved by both pre-assembly and post-assembly incorporation [8]. In the post-assembly method, metal chalcogenide nanoparticles are embedded into an already formed MOF whereas, in the pre-assembly method, chalcogenides are introduced by being anchored to the linker before the assembly of the MOF (Figure 5.4). However, some challenges need to be resolved i) The stability of a MOF under the synthesis conditions of metal chalcogenides is an issue of concern as it can degrade; ii) Formation of chalcogenide particles larger than the pore size can cause a partial degradation of the structure; iii) The chalcogenides can get deposited on the external surface of MOF crystals; iv) The exact size, shape of the chalcogenide particles along with their location in the MOF framework are difficult to control. Wakaoka et al. reported the encapsulation of CdSe within the pores of MOF via ship in bottle method [13]. The CdSe (~2 nm) particles were encapsulated into the pores of MIL-101- (Al)-NH_2 which were functionalized by terminal primary amine groups. MIL-101(Al)-NH_2 contains two cage-type pores of diameters 2.9 and 3.4 nm. These pores of MOF were found to be ideal for encapsulation of CdSe for the following reasons: a) pore size was suitable for the encapsulation of CdSe particles; b) cage type and zero-dimensional pore configuration prevented the self-aggregation of CdSe;

c) functionalization of pores with NH_2 helped the infiltration of Cd precursors into the pores which resulted into the dense configuration of nanoparticles [16]. The MIL-101-(Al)-NH_2@CdSe was obtained by the addition of Se precursor to the reaction mixture containing MOF and Cd precursor solution (CdO and stearic acid) at 230°C.

CdS encapsulated MIL-101 was synthesized for photocatalytic hydrogen production [18]. Firstly, the CdS was embedded into MIL-101 at different concentrations (0, 5, 10, 20, and 50% by weight) to form CdS/MIL-101 composite. Afterward, Pt was photo deposited on CdS/MIL-101 composite

Pt to obtain Pt@CdS/MIL-101. The photocatalytic activity for H_2 production was performed under visible light. The hydrogen evolution activity was found to be dependent on the weight % of CdS and maximum activity was observed at a 10% (w/w) concentration of CdS. The hydrogen production activity of CdS/MIL-101 was compared with CdS/MOF-5 and CdS/MCM-41. Of all the three materials, the hydrogen production activity was found to be lowest for CdS/MCM-41 although it possesses a mesoporous structure. The study proved the role of MOF materials in achieving superior performance.

5.3.3 Photochemical Deposition

This method consists of the formation and deposition of metal chalcogenide on the surface of the metal-organic framework assisted by light. The chalcogenides are generated on the surface of the MOF from the metallic precursors through photoreduction at an appropriate redox potential. Shen et al. decorated CdS nanorods on the surface of UiO-66(NH_2) by the photo deposition technique [19]. The photo deposition was carried out by suspending UiO-66(NH_2) in ethanol containing S_8 and cadmium chloride and irradiating under visible light for selected hours (2, 4, 6, to 8 hours). The increase in irradiation time from 2 to 6 h was found to increase the amount of photo deposition and more CdS nanorods were found to be decorated on the surface of UiO-66(NH_2) surface in a relatively dispersed state at the end of 6 hours. However, after 8 hours, the CdS particles started to aggregate to a larger degree. The electrochemical measurements indicated that during the photo-deposition of CdS on UiO-66(NH_2) surface Cd^{2+} ions of the precursor were photo reduced to Cd^0 which then reacted with S_8 to yield CdS. A ternary hybrid of MOF, chalcogenide, and reduced graphene oxide (RGO), UiO-66/CdS/1%RGO in which the previously synthesized MOF was modified by CdS prepared through photo deposition method was reported by Lin et al. [20].

5.3.4 Direct Binding

This method involves the surface modification of MOFs with chalcogenides using a suitable linking group. In this method, the surface groups of the synthesized metal chalcogenides are successively exchanged with a suitable linking agent. This facilitates the direct binding of the chalcogenides with MOFs either through specific coordinate bonds or through non-specific interactions like the physisorption process. Jin et al. (2013) reported the binding of CdSe/ZnS core-shell QDs on the surface of the porphyrin-based MOFs through coordination interactions. The surface amine groups of the chalcogenide quantum dots were believed to play a role in the direct binding with MOF through amine-Zn coordination interactions. Table 5.1 shows different synthetic methods for the MOF@MC composite.

TABLE 5.1
Different synthetic methods for the MOF@MC composite

Order	MOF	Metal chalcogenide	Ref
A] Bottle around ship method			
1)	Zn(1,4-bis(imidazole-1-ylmethyl)benzene)	ZnS/CdSe	[14]
2)	Zn(BDC)	CdSe/CdS/$Cd_{0.5}Zn_{0.5}S$/ZnS	[15]
3)	MOF-5	CdSe/CdS/ZnS	[21]
4)	EuOF	CdSe	[22]
5)	ZIF-8	CdTe	[23]
B] Ship in bottle method			
6)	MIL-101-(Al)-NH_2	CdSe	[13]
7)	MIL-101	CdS	[18]
8)	ZAVCl	CdS	[24]
9)	MIL-101(Fe)	CdS	[25]
C] Photochemical deposition			
10)	UiO-66-NH_2	CdS	[19]
11)	UiO-66	CdS	[20]
D) Direct binding			
12)	PorphyrinMOF	CdSe/ZnS	[26]

5.4 ELECTROCATALYTIC APPLICATIONS OF MOF@METAL CHALCOGENIDE COMPOSITES

The consumption of energy for human activities is experiencing an exponential rise. Most of the demand is met by the use of resources like oil and coal which account for a large amount of CO_2 emission in the environment and other related issues. As a result, much effort has been expended in the pursuit of green and renewable energy sources. Electro-catalysts have attracted considerable interest as a result of their technological potential to produce clean energy. Pt catalysts are frequently used in electrochemistry, however, due to their cost; non-Pt catalysts must be developed. MOF@MC nanocomposites are now being considered as good candidates for the development of electrocatalysts for clean and efficient energy production alternatives owing to their economic and environmental feasibility.

5.4.1 Electrocatalytic Water Splitting

The heterogeneous water-splitting reaction by the MOF@MC nanocomposites is an area of recent interest. The electrochemical splitting of water is a zero-emission technology as it produces only water and hydrogen. The water-splitting reaction involves the two main half-reactions: the hydrogen evolution reaction (HER) occurring at the cathode and the oxygen evolution reaction (OER) occurring at the anode.

5.4.1.1 Hydrogen Evolution Reaction

The HER is a two-electron transfer reaction. It involves the two main steps: a) generation of H* on the surface of the electrode; b) combination of H* with a proton or with other H* to form H_2 [27]. The MOF@MC composite electrocatalyst is known to decrease the overpotential in HER. The CdS loaded ZAVCl-MOF was utilized as the photocatalyst for the H_2 production using Pt and ethanol as a co-catalyst and sacrificial reagent respectively [24]. The ethanol (sacrificial reagent) consumed photo-generated holes to prevent the photocatalyst from corrosion. The binary composite CdS@ZAVCl with size 7-9 nm showed a hydrogen efficiency of 398-418 μmol/hg under visible light. Under the irradiation of UV light, the CdS@ZAVCl showed more efficiency as a photocatalyst with H_2 production of 500-510 μmol/hg. This is because UV light contains photons of the visible and UV region. The number of photons that could be absorbed by the CdS composite is considerably higher in comparison to only CdS. The increased light absorption increases the number of photogenerated electron-hole pairs as a result of which the hydrogen production rate increases. Carbon dots (CDs) have been considered as promising alternative co-catalysts to avoid the utilization of noble metals. Meng et al. (2019) immobilized the CDs and CdS into MIL-101. The obtained composite CD/CdS@MIL-101 exhibited stronger photocatalytic activity of hydrogen evolution (14.66 μmol h^{-1}) than the CdS and CdS@MIL-101 [28].

A ternary composite, UiO-66/CdS/RGO with superior performance in catalytic hydrogen evolution in comparison to individual components has been reported [20]. The composite was found to exhibit a superior photocatalytic hydrogen evolution rate which was higher than commercial CdS (13.8 times higher), UiO-66/CdS (1.2 times higher), and P-25/CdS/RGO (2 times higher). in comparison to lone CdS and CdS/RGO composite thus proving the role of MOF in enhancing the performance. The enhanced performance of CdS in the presence of RGO and MOF was attributed to the synergistic effect of the components. The porous MOFs provided higher surface area and numerous catalytic sites which enhanced light absorption significantly. The main contribution of RGO seemed to be the acceleration of the electron emission process and inhibition of charge recombination.

5.4.1.2 Oxygen Evolution Reaction

The oxygen evolution reaction involves the evolution of O_2 from the electrochemical oxidation of water at the anode [29]. Presently, MOF@MC nanocomposites are of interest in OER reactions as they show high OER performance. A two-dimensional Ni-BDC@NiS was synthesized as a synergistic electrocatalyst which showed high activity during OER [30]. The Ni-BDC@NiS composite catalyst showed better performance with lower overpotential in OER in comparison to the NiS based catalyst. The hybridization of NiS with Ni-BDC was believed to enhance electron transport and expand the active species pool.

A Ni-BDC MOF modified Ni_3S_2/NiS hollow nanostructure was reported for OER applications [31]. The nanocomposite showed better electro-catalytic performance with low over-potential (298 mV). The electrocatalytic performance was better in comparison to the single components and the reference sample RuO_2. This was because of the high porosity and conductivity of Ni_3S_2/NiS and the coordinative unsaturation of the Ni-BDC shell. Zhang et al. reported a Ti@TiO_2@CdS composite with ZIF-67 for the oxygen evolution [32]. The Ti@TiO_2@CdS/ZIF-67 composite electrode exhibited a small Tafel slope (42 mV dec^{-1}) and low over-potential in addition to high stability in basic media.

5.4.2 Oxygen Reduction Reaction

A composite material, CuS/Cu-MOF has been explored for the electrocatalytic ORR [33]. The variation in the amount of CuS was found to affect electrical conductivity which increased with increases in the amount of CuS. The kinetic current density showed by the electrocatalyst was 6.63, 6.76, 10.1, and 11.3 mA cm^{-2} for 5.3, 8.8, and 28 wt% nano-CuS loaded composites respectively. These values were found to be higher than both Cu-BTC (3.46 mA cm^{-2}) and nano-CuS (3.36 mA cm^{-2}) materials. These results indicated the better performance activity of nano-CuS@Cu-MOF in comparison to its counterparts.

5.5 MOF DERIVED METAL CHALCOGENIDES

MOFs have recently been established to be viable templates for the synthesis of metal chalcogenides, particularly metal sulfides, and metal selenides. This section summarises recent progress in the synthesis of MOF-derived metal chalcogenides and their electrochemical applications.

5.5.1 MOF-Derived Metal Sulfides

The majority of previously reported works concentrated on the production of nanostructured noble metals and metal oxides derived from MOFs. The distinctive physical and chemical properties along with a broad scope of applications of metal sulfides has necessitated the development of metal sulfide nanostructures of definite shapes. MOFs form promising precursors/templates to develop multifunctional nanomaterials with hollow or porous structures. MOFs especially Prussian Blue Analogues (PBA) have markedly been utilized as precursors for the synthesis of porous metal sulfide nanomaterials through sulfidation [34]. Ni-Co PBA analogue precursors were used as templates to synthesize well-defined NiS nano frames with tunable pore sizes (Figure 5.5). The transformation of MOF to NiS nano frames was achieved through an anisotropic chemical etching/anion exchange method using Na_2S as a sulfur source at elevated temperatures [35]. The formation of well-defined NiS nano frames was facilitated due to differential reactivity of the edges and the plane surface of the precursor nano cubes is found to

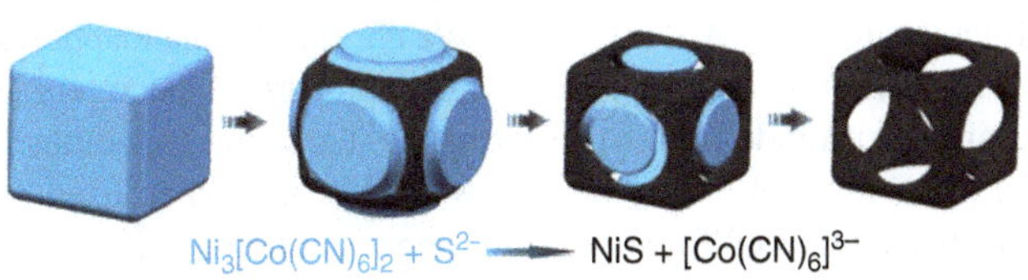

FIGURE 5.5 Schematic illustration of the formation of NiS nanoframes. Adapted with permission from [35]. Copyright (2015) Wiley Online Library.

be the key factor for the formation of NiS nano frames. The NiS nano frames were found to show high electrocatalytic activity for the hydrogen evolution reaction. The phenomenal HER performance of NiS was attributed to the open three-dimensional porous structure with a high specific surface area which provided a large contact area between electrode and electrolyte resulting in increased electrocatalytic performance.

Nitrogen-doped NiFe-sulphides/carbon composites constructed on carbon cloth (N-NiFe-S/C@CC) catalyst were synthesized by a solvothermal method followed by carbonization and sulfurization process [36]. A single MOF precursor with –surface NH_2 groups was applied as a source of both carbon and nitrogen for the synthesis of catalyst which greatly simplified the doping method. The synthesized N-NiFe-S/C@CC catalyst possessed excellent O_2 evolution with an over-potential of 232 mV at a current density of 10 mA cm^{-2} and exhibited outstanding stability in alkaline solution. The outstanding performance of the catalyst could be ascribed to a) electronic interactions between the constituent elements (Fe, Ni, and S); b) good conductivity arising due to carbon and nitrogen doping; c) stronger mechanical binding as a result of direct growth on carbon.

Cobalt sulfide (CoS) electroactive materials find applications for energy storage purposes and MOFs can be used to derive these sulfide particles. Zeolitic imidazolate framework (ZIF) was used as a template for the synthesis of amorphous CoS nanocages [37]. The CoS nanocages were synthesized by refluxing ZIF-67 and thioacetamide in ethylene glycol. The MOF ZIF-67 acted both as template and Co precursor while thioacetamide acted as a sulfur precursor. The formation of CoS nanoparticles took place by reaction of Co^{2+} ions dissolving from the surface of ZIF-67 with S2- ions from thioacetamide. As the reaction continued, a separation started appearing between the ZIF-67 and the CoS shell. Finally, the ZIF-67 cores were thoroughly consumed to form well-defined hollow truncated rhombic dodecahedral nanocages of CoS. The specific capacitance of the CoS nanocages was calculated to be 1812 F g^{-1} and 1201 F g^{-1} at a scan rate of 5 mV s^{-1} and 50 mV s^{-1}, respectively. The specific capacitance obtained from the galvanostatic discharge curves were 1475 F g^{-1} and 932 F g^{-1} at a current density of 1 A g^{-1} and 10 A g^{-1}, respectively. The results demonstrated the high performance and utility of the developed CoS in capacitive energy storage applications.

Hollow NiS_2/ZnS nanocomposites were synthesized from Ni/Zn-BDC (BDC = 1,4-benzenedicarboxylate) precursor and thioacetamide as a sulfur precursor [38]. The procedure involved the synthesis of the bimetallic MOF by the solvothermal method followed by chemical sulfidation which transformed the MOF into NiS_2/ZnS hollow nanospheres via an anion exchange reaction. The prepared chalcogenide material provided high performance (specific capacitance of 1198 F g^{-1} at a current density of 1 A g^{-1}) when used as electrode material for super capacitance applications.

Copper sulfide has been considered as an advantageous anode material for a sodium-ion battery (SIB) owing to its resource abundance, high electronic conductivity, and low cost. However, its use as anode material in SIB is hindered by its poor cycling stability and lower rate performance in comparison to some other materials. The poor electrochemical performance of the copper sulfide material in low voltage processes as a result of pulverization and volume expansion of the copper sulfide electrode is another limiting factor. The electrical conductivity of CuS can be improved by combining it with porous conductive materials like electroactive carbon materials. Liu et al. (2020) have proposed a MOF-based strategy to synthesize N-doped carbon encapsulated CuS nanoparticles (CuS@N-C) by a carbonization and sulfidation process [39]. CuS@N-C exhibits some interesting features obtained from Cu-MOFs precursors: a) N-doped sites provide more conductivity to the composite; b) the porous structure aids in alleviating large volume stress arising during long term discharge-charge processes; c) MOF derived carbon improves structural stability and electrical conductivity of the composite. All these features made the CuS@N-C excellent anode material for SIBs. The composite material exhibited a prominent rate capability (259.4 mAh g^{-1} at 5 A g^{-1}), excellent long-life stability (300.2 mAh g^{-1} at 5 A g^{-1}) over 1200 cycles.

5.5.2 MOF-derived Metal Selenides

Besides metal sulfides, MOFs have also been utilized as precursors to obtain metal selenide nanostructures. The synthesis of metal selenides generally involves pyrolysis of the MOF precursor at high temperatures and corresponding selenization in an inert atmosphere [34]. $CoSe_2$ nanoparticles affixed on the carbon fiber paper ($CoSe_2$/CF) have been developed by pyrolysis of ZIF-67 MOF which was grown in-situ followed by selenization with ZIF-67 [40]. For in-situ syntheses of MOF, the solution containing Co^{2+} ions and 2-methylimidazole (organic linker) was passed over the CF paper for several cycles which lead to the development of ZIF-67. Electrostatic interactions with the negatively charged –COOH groups on the CFs allowed Co^{2+} to be easily absorbed. Se vapors in the tube furnace then transformed the MOF nanocrystals into $CoSe_2$ nanoparticles by pyrolysis and selenization. The bifunctional catalytic activity of $CoSe_2$ particles in both the OER and HER, is capable of functioning as anode and cathode for an overall water splitting reaction. At a voltage of 1.63 V, a water electrolyzer with two CoSe/CF electrodes produced a water-splitting current of 10 mA cm^{-2}. A binary metal selenide (ZnSe/CoSe) composite encased in an N-doped carbon polyhedral connected with carbon nanotubes (ZCS@NC/CNTs) was prepared via in-situ pyrolysis and selenization at a moderate temperature [41]. The N-doped polyhedral can store electrolytes and facilitate charge transfer across the electrode/electrolyte junction. When applied as an anode material in LIBs, the composite ZCS@NC/CNTs exhibited high capacity (873 mAh g^{-1}), incredible rate ability, and high cycling steadiness. This performance could be attributed to the porous structure, stability and conductivity of the network, and shorter diffusion channels for Li^+ and electron transport. Similarly, core-shell MOF (ZIF-8@ZIF-67) derived N-doped yolk-shell carbon nanocage packed with ZnSe and $CoSe_2$ nanodots composite material was used as

TABLE 5.2
Applications of MOF-derived metal chalcogenides

Order	MOF precursor	Metal chalcogenide	Application	Ref
A] Metal sulphides				
1)	Ni-Co PBA	NiS	Hydrogen evolution	[35]
2)	Ni-Co-BTC	Ni-Co-S	Hydrogen evolution	[44]
3)	Ni-Fe MOF	$NiS-Fe_3O_4$	Oxygen evolution	[45]
4)	Ni-Fe-MOF	N-Ni/Fe-S	Oxygen evolution	[36]
5)	ZIF-9	C-CoS	Oxygen reduction	[46]
6)	ZIF-67	CoS	Supercapacitor	[37]
7)	Ni/Zn-BDC	NiS_2/ZnS	Supercapacitor	[38]
8)	Cu-MOF	N doped CuS	Sodium ion battery	[39]
9)	Co-ZIF-8	$ZnS/CoS/CoS_2$@N-dopedcarbonnanoparticles	Lithium-ion battery	[47]
B] Metal selenides				
10)	Ni-MOF	NiSe	Hydrogen evolution	[43]
11)	ZIF-67	$CoSe_2$	Hydrogen evolution and oxygen evolution	[40]
12)	ZIF-8@ZIF-67	$ZnSe/CoSe_2$	Lithium-ion battery	[42]
13)	ZIF-8/67	ZnSe/CoSe	Lithium-ion battery	[41]

anode for LIBs/SIBs[42]. The ZnSe-NC@$CoSe_2$-NC material exhibited high electrochemical performance with a reversible storage capacity of 171 mA g^{-1} after 600 cycles at a current density of 0.2 A g^{-1} in LIBs and 308.5 mA g^{-1} after 150 cycles at a current density of 0.1 A g^{-1} in SIBs.

Ni-Se nanohybrids with various shapes have received a lot of interest and have been studied as effective electrocatalysts for HER. Among all the morphologies, materials with coral morphology have gained much attention. A MOF-derived coral-like NiSe@N-doped carbon (NiSe@NC) was reported via in-situ selenation of Ni-based MOF for the HER electrocatalytic performance [43]. The NiSe@NC-600 showed outstanding electrolytic performance for HER at all pH values. The NiSe@NC-600 nanohybrid achieved low overpotentials of 123, 250, and 300 mV in 0.5 M H_2SO_4 (pH = 0), 1.0 M KOH (pH = 14), and 1.0 M phosphate buffer solution (pH = 7) solution respectively to show a current density of 10 mA cm^{-2}. Table 5.2 shows applications of MOF-derived metal chalcogenides.

5.6 CONCLUSION

MOFs have gained much attention for their wide applications in different areas. The inherent defects like poor conductivity and less thermal stability restrict their full potential. The encapsulation of metal chalcogenides with MOFs greatly enhances their stability and improves conductivity. Furthermore, the MOF@MC nanocomposites combine the advantageous features of both the components to strengthen the existing properties and generate novel attributes as a result of the synergistic effect of both the components. The MOF@MC hybrid materials exhibit remarkable features for use as electrocatalysts for energy storage applications. Despite the advantages, some challenges need to be addressed. Although many significant efforts have been made for immobilizing chalcogenide particles within the MOFs, the accessibility of MOF pores to metal chalcogenides has not been thoroughly explored. Most of the MOFs have high diffusion resistance which hinders the approach of reactants to the encapsulated active sites. This can be solved by the encapsulation of chalcogenides into the yolk-shell MOFs. A thorough understanding of the interactions between MOFs and chalcogenides is needed to explore the full potential of the MOF@MC nanocomposites. The integration of several chalcogenides and other nanomaterials inside a single framework is also predicted to open up a new avenue in the field of advanced materials. In conclusion, the development of MOF@MC composite materials has opened up a plethora of new possibilities for the design of functional materials with diverse characteristics and uses.

REFERENCES

[1]. F. Seidi, M. Jouyandeh, M. Taghizadeh, A. Taghizadeh, H. Vahabi, S. Habibzadeh, K. Formela, M.R. Saeb, Metal-organic framework (MOF)/epoxy coatings: A review, *Materials*, 13 (2020) 2881.

[2]. S. Bhogal, K. Kaur, A.K. Malik, C. Sonne, S.S. Lee, K.-H. Kim, Core-shell structured molecularly imprinted materials for sensing applications, *TrAC Trends in Analytical Chemistry*, 133 (2020) 116043.

[3]. M. Kaur, S. Kumar, S.A. Younis, M. Yusuf, J. Lee, S. Weon, K.-H. Kim, A.K. Malik, Post-Synthesis Modification of Metal-Organic Frameworks Using Schiff Base Complexes for Various Catalytic Applications, *Chemical Engineering Journal*, 423 (2021) 130230.

[4]. F.Y. Yi, R. Zhang, H. Wang, L.F. Chen, L. Han, H.L. Jiang, Q. Xu, Metal–organic frameworks and their composites: synthesis and electrochemical applications, *Small Methods*, 1 (2017) 1700187.

[5]. J. Yu, C. Mu, B. Yan, X. Qin, C. Shen, H. Xue, H. Pang, Nanoparticle/MOF composites: preparations and applications, *Materials Horizons*, 4 (2017) 557–569.

[6]. L. Chen, X. Zhang, X. Cheng, Z. Xie, Q. Kuang, L. Zheng, The function of metal–organic frameworks in the application of MOF-based composites, *Nanoscale Advances*, 2 (2020) 2628-2647.

[7]. J. Aguilera-Sigalat, D. Bradshaw, Synthesis and applications of metal-organic framework–quantum dot (QD@ MOF) composites, *Coordination Chemistry Reviews*, 307 (2016) 267–291.

[8]. L. Chen, R. Luque, Y. Li, Controllable design of tunable nanostructures inside metal–organic frameworks, *Chemical Society Reviews*, 46 (2017) 4614–4630.

[9]. Q. Wang, D. Astruc, State of the art and prospects in metal–organic framework (MOF)-based and MOF-derived nanocatalysis, *Chemical reviews,* 120 (2019) 1438–1511.

[10]. T. Ma, H. Li, J.-G. Ma, P. Cheng, Application of MOF-based materials in electrochemical sensing, *Dalton Transactions*, 49 (2020) 17121-17129.

[11]. K. Wang, N. Li, J. Zhang, Z. Zhang, F. Dang, Size-selective QD@ MOF core-shell nanocomposites for the highly sensitive monitoring of oxidase activities, *Biosensors and Bioelectronics*, 87 (2017) 339–344.

[12]. W. Zhang, G. Lu, S. Li, Y. Liu, H. Xu, C. Cui, W. Yan, Y. Yang, F. Huo, Controlled incorporation of nanoparticles in metal–organic framework hybrid thin films, *Chemical Communications*, 50 (2014) 4296-4298.

[13]. T. Wakaoka, K. Hirai, K. Murayama, Y. Takano, H. Takagi, S. Furukawa, S. Kitagawa, Confined synthesis of CdSe quantum dots in the pores of metal–organic frameworks, *Journal of Materials Chemistry C*, 2 (2014) 7173-7175.

[14]. I. Imaz, J. Hernando, D. Ruiz-Molina, D. Maspoch, Metal–organic spheres as functional systems for guest encapsulation, *Angewandte Chemie International Edition*, 48 (2009) 2325–2329.

[15]. D. Buso, J. Jasieniak, M.D. Lay, P. Schiavuta, P. Scopece, J. Laird, H. Amenitsch, A.J. Hill, P. Falcaro, Highly luminescent metal–organic frameworks through quantum dot doping, *Small,* 8 (2012) 80–88.

[16]. X. Gu, Z.-H. Lu, H.-L. Jiang, T. Akita, Q. Xu, Synergistic catalysis of metal–organic framework-immobilized Au–Pd nanoparticles in dehydrogenation of formic acid for chemical hydrogen storage, *Journal of the American Chemical Society*, 133 (2011) 11822–11825.

[17]. L. Chen, H. Chen, R. Luque, Y. Li, Metal– organic framework encapsulated Pd nanoparticles: towards advanced heterogeneous catalysts, *Chemical Science*, 5 (2014) 3708–3714.

[18]. J. He, Z. Yan, J. Wang, J. Xie, L. Jiang, Y. Shi, F. Yuan, F. Yu, Y. Sun, Significantly enhanced photocatalytic hydrogen evolution under visible light over CdS embedded on metal–organic frameworks, *Chemical communications*, 49 (2013) 6761–6763.

[19]. L. Shen, S. Liang, W. Wu, R. Liang, L. Wu, CdS-decorated UiO–66 (NH_2) nanocomposites fabricated by a facile photodeposition process: an efficient and stable visible-light-driven photocatalyst for selective oxidation of alcohols, *Journal of Materials Chemistry A*, 1 (2013) 11473–11482.

[20]. R. Lin, L. Shen, Z. Ren, W. Wu, Y. Tan, H. Fu, J. Zhang, L. Wu, Enhanced photocatalytic hydrogen production activity via dual modification of MOF and reduced graphene oxide on CdS, *Chemical Communications*, 50 (2014) 8533–8535.

[21]. P. Falcaro, A.J. Hill, K.M. Nairn, J. Jasieniak, J.I. Mardel, T.J. Bastow, S.C. Mayo, M. Gimona, D. Gomez, H.J. Whitfield, A new method to position and functionalize metal-organic framework crystals, *Nature communications,* 2 (2011) 1–8.

[22]. R. Kaur, A. Paul, A. Deep, Nanocomposite of europium organic framework and quantum dots for highly sensitive chemosensing of trinitrotoluene, *Forensic science international*, 242 (2014) 88–93.

[23]. G. Lu, S. Li, Z. Guo, O.K. Farha, B.G. Hauser, X. Qi, Y. Wang, X. Wang, S. Han, X. Liu, Imparting functionality to a metal–organic framework material by controlled nanoparticle encapsulation, *Nature chemistry*, 4 (2012) 310–316.

[24]. S. Saha, G. Das, J. Thote, R. Banerjee, Photocatalytic metal–organic framework from CdS quantum dot incubated luminescent metallohydrogel, *Journal of the American Chemical Society*, 136 (2014) 14845-14851.

[25]. F. Ke, L. Wang, J. Zhu, Facile fabrication of CdS-metal-organic framework nanocomposites with enhanced visible-light photocatalytic activity for organic transformation, *Nano Research*, 8 (2015) 1834–1846.

[26]. S. Jin, H.-J. Son, O.K. Farha, G.P. Wiederrecht, J.T. Hupp, Energy transfer from quantum dots to metal–organic frameworks for enhanced light harvesting, *Journal of the American Chemical Society*, 135 (2013) 955–958.

[27]. Y. Xue, S. Zheng, H. Xue, H. Pang, Metal–organic framework composites and their electrochemical applications, *Journal of Materials Chemistry A*, 7 (2019) 7301–7327.

[28]. X.-B. Meng, J.-L. Sheng, H.-L. Tang, X.-J. Sun, H. Dong, F.-M. Zhang, Metal-organic framework as nanoreactors to co-incorporate carbon nanodots and CdS quantum dots into the pores for improved H_2 evolution without noble-metal cocatalyst, *Applied Catalysis B: Environmental*, 244 (2019) 340–346.

[29]. T. Wu, X. Liu, Y. Liu, M. Cheng, Z. Liu, G. Zeng, B. Shao, Q. Liang, W. Zhang, Q. He, Application of QD-MOF composites for photocatalysis: Energy production and environmental remediation, *Coordination Chemistry Reviews*, 403 (2020) 213097.

[30]. P. He, Y. Xie, Y. Dou, J. Zhou, A. Zhou, X. Wei, J.-R. Li, Partial sulfurization of a 2D MOF array for highly efficient oxygen evolution reaction, *ACS applied materials & interfaces*, 11 (2019) 41595–41601.

[31]. J. Wang, H.C. Zeng, A Hybrid Electrocatalyst with a Coordinatively Unsaturated Metal–Organic Framework Shell and Hollow Ni_3S_2/NiS Core for Oxygen Evolution Reaction Applications, *ACS applied materials & interfaces*, 11 (2019) 23180–23191.

[32]. T. Zhang, J. Du, H. Zhang, C. Xu, In-situ growth of ultrathin ZIF-67 nanosheets on conductive Ti@ TiO_2/CdS substrate for high-efficient electrochemical catalysis, *Electrochimica Acta*, 219 (2016) 623–629.

[33]. K. Cho, S.H. Han, M.P. Suh, Copper–organic framework fabricated with CuS nanoparticles: Synthesis, electrical conductivity, and electrocatalytic activities for oxygen reduction reaction, *Angewandte Chemie*, 128 (2016) 15527–15531.

[34]. Y.V. Kaneti, J. Tang, R.R. Salunkhe, X. Jiang, A. Yu, K.C.W. Wu, Y. Yamauchi, Nanoarchitectured design of porous materials and nanocomposites from metal-organic frameworks, *Advanced materials,* 29 (2017) 1604898.

[35]. X.Y. Yu, L. Yu, H.B. Wu, X.W. Lou, Formation of nickel sulfide nanoframes from metal–organic frameworks with enhanced pseudocapacitive and electrocatalytic properties, *Angewandte Chemie,* 127 (2015) 5421–5425.

[36]. Y. Shen, L. Wang, P. Jiang, W.S.V. Lee, J. Xue, Metal-Organic-Framework-Derived Nitrogen-Doped Hybrid Nickel-Iron-Sulfide Architectures on Carbon Cloth as Efficient Electrocatalysts for the Oxygen Evolution Reaction, *ChemElectroChem*, 6 (2019) 2741–2747.

[37]. Z. Jiang, W. Lu, Z. Li, K.H. Ho, X. Li, X. Jiao, D. Chen, Synthesis of amorphous cobalt sulfide polyhedral nanocages for high performance supercapacitors, *Journal of Materials Chemistry A*, 2 (2014) 8603–8606.

[38]. G.-C. Li, M. Liu, M.-K. Wu, P.-F. Liu, Z. Zhou, S.-R. Zhu, R. Liu, L. Han, MOF-derived self-sacrificing route to hollow NiS_2/ZnS nanospheres for high performance supercapacitors, *RSC advances*, 6 (2016) 103517–103522.

[39]. X. Liu, X. Li, X. Lu, X. He, N. Jiang, Y. Huo, C. Xu, D. Lin, Metal-organic framework derived in-situ nitrogen-doped carbon-encapsulated CuS nanoparticles as high-rate and long-life anode for sodium ion batteries, *Journal of Alloys and Compounds*, 854 (2021) 157132.

[40]. C. Sun, Q. Dong, J. Yang, Z. Dai, J. Lin, P. Chen, W. Huang, X. Dong, Metal–organic framework derived $CoSe_2$ nanoparticles anchored on carbon fibers as bifunctional electrocatalysts for efficient overall water splitting, *Nano Research*, 9 (2016) 2234-2243.

[41]. J. Jin, Y. Zheng, L.B. Kong, N. Srikanth, Q. Yan, K. Zhou, Tuning ZnSe/CoSe in MOF-derived N-doped porous carbon/CNTs for high-performance lithium storage, *Journal of Materials Chemistry A*, 6 (2018) 15710–15717.

[42]. X. Hu, X. Liu, K. Chen, G. Wang, H. Wang, Core–shell MOF-derived N-doped yolk–shell carbon nanocages homogenously filled with ZnSe and $CoSe_2$ nanodots as excellent anode materials for lithium-and sodium-ion batteries, *Journal of Materials Chemistry A*, 7 (2019) 11016–11037.

[43]. Z. Huang, J. Liu, Z. Xiao, H. Fu, W. Fan, B. Xu, B. Dong, D. Liu, F. Dai, D. Sun, A MOF-derived coral-like NiSe@NC nanohybrid: an efficient electrocatalyst for the hydrogen evolution reaction at all pH values, *Nanoscale*, 10 (2018) 22758–22765.

[44]. C. Wu, Y. Du, Y. Fu, W. Wang, T. Zhan, Y. Liu, Y. Yang, L. Wang, MOF-derived formation of nickel cobalt sulfides with multi-shell hollow structure towards electrocatalytic hydrogen evolution reaction in alkaline media, *Composites Part B: Engineering*, 177 (2019) 107252.

[45]. K. Srinivas, Y. Chen, B. Wang, B. Yu, X. Wang, Y. Hu, Y. Lu, W. Li, W. Zhang, D. Yang, Metal–Organic Framework-Derived NiS/Fe_3O_4 Heterostructure-Decorated Carbon Nanotubes as Highly Efficient and Durable Electrocatalysts for Oxygen Evolution Reaction, *ACS Applied Materials & Interfaces*, 12 (2020) 31552-31563.

[46]. F. Bai, H. Huang, C. Hou, P. Zhang, Porous carbon-coated cobalt sulfide nanocomposites derived from metal organic frameworks (MOFs) as an advanced oxygen reduction electrocatalyst, *New Journal of Chemistry*, 40 (2016) 1679–1684.

[47]. W. Cheng, H. Di, Z. Shi, D. Zhang, Synthesis of ZnS/CoS/CoS_2@ N-doped carbon nanoparticles derived from metal-organic frameworks via spray pyrolysis as anode for lithium-ion battery, *Journal of Alloys and Compounds*, 831 (2020) 154607.

6 Merits of Selecting Metal-Organic Frameworks as Sensors

Harmeet Kaur[1,2], *Amit L Sharma*[1,2] *and Akash Deep*[1,2]
[1]CSIR-Central Scientific Instrument Organisation (CSIR-CSIO), Chandigarh, India
[2]Academy of Scientific and Innovative Research (AcSIR), Ghaziabad, India

6.1 INTRODUCTION

6.1.1 Metal-Organic Frameworks

MOFs are a class of crystalline porous compounds formed through the regular linking of inorganic building units (positively charged metal ions) with organic ligands. One-, two- or three-dimensional structures are formed through repetitive coordination bonding of metal ions and organic ligands (Figure 6.1). These materials manifest large-scale physical and chemical tunability upon astute selection of metal ions and organic ligand combinations [1].

The first synthesized MOF was a cationic MOF with nickel (Ni) as a metal ion and pyridine as an organic ligand [3]. Since the pioneering work by Yaghi and Li, numerous categories of metal ions including transition metals, alkali earth metals, p-block elements, actinides have been explored for synthesizing MOFs. The possible molecular geometries - linear, planar, pyramidal, octahedral, bipyramidal, and so on, are governed by the coordination number of the constituent metal ion. The use of diverse organic ligands enables varied sizes, geometries, and properties of synthesized MOFs. Commonly used organic ligands in MOF synthesis include carboxylates, phosphates, nitrates, and amines. The choice of organic linker for MOF synthesis is governed by the presence of Lewis basic sites and the relative angularity of these binding sites. Thereby, the crystal structure of a MOF is governed by the coordination geometries, sizes, and angularities of the ligand and metal ions. One of the structural classifications of MOFs is on the basis of secondary building units (SBUs). The underlying geometry of a MOF is identified on the basis of the SBUs attached. Commonly used SBUs in MOF synthesis is shown in Figure 6.2. Spatial arrangements along with suitable modifications in these SBUs govern the specific applicability of corresponding MOF material. Additionally, the open coordination sites in any MOF, enable opportunities of binding analytes or other metal ions which render the material suitable for application in sensing, optoelectronics, and so forth [4].

With the large variety of available metal-ligand combinations, more than 90,000 different MOFs have been synthesized and over 500,000 predicted to date [6]. Notable properties of MOFs encompass high surface area (up to 8000 m^2/g), tunable porosity (pore size ≈ 14-98 Å), low density (0.124 g/cm^3), and highly stable luminescence [7]. These remarkable properties, coupled with the attribute of design flexibility, facilitates assimilation of distinct functionalities in MOF materials through surface functionalization or through the incorporation of guest moieties, formation of MOF composites, or derivatives. The application avenues of these materials include sensing, gas storage, molecule separation, catalysis, drug delivery, optoelectronics, energy storage, and conversion, to name but a few. Some of the commonly used MOFs being used in the storage of gases like CH_4, NO, CO_2, H_2 include MILs, IRMOFs, ZIFs on account of their high porosities. HKUST-1, MOF-5, IRMOF-3 have been reported to be efficient in the separation of inorganic gases like ammonia and chlorine as well as toxic compounds including benzene, benzothiophene, and toluene. Attributes of structural flexibility, large surface area, and chemical tunability, enable the use of a large number of MOFs in sensing applications through varied sensing mechanisms including but not limited to chemical, optical, electrochemical, magnetic, mechanical, and ferroelectric sensing [8].

6.1.2 Engineering Nanomaterials for Sensing Application

A sensor is a device that detects a variable quantity and converts it into a readable signal. The measured stimuli could be physical, chemical, mechanical, optical, biological, and so forth. These devices are everywhere, for example, the thermometers and smartphones in our homes, the speedometers in our vehicles, and the thermostats in our buildings. While previously, these devices used to be bulky and expensive, the advent of micro and nanotechnologies has enabled smarter, smaller, and much more efficient sensors. Continuous advancements in sensing technology are necessary to cater for the needs of the increasing population and associated issues, for example, the increasing number of diabetes patients in the world calls for cheaper and more efficient blood glucose monitors in the near future.

Novel nanomaterials have expanded the use of sensors to unchartered fields like wearable devices (insulin, glucose detectors), implantable sensors (neuro implants, cochlear implants), and so forth. Nanomaterials find multifold applications in sensor applications as active sensing elements, transducing elements, or in the sensing circuitry in the form of nanowires. Many of these sensors offer better sensitivity, specificity, response time, and stability in comparison to the conventional sensors on account of high chemical tunability and

DOI: 10.1201/9781003188148-6

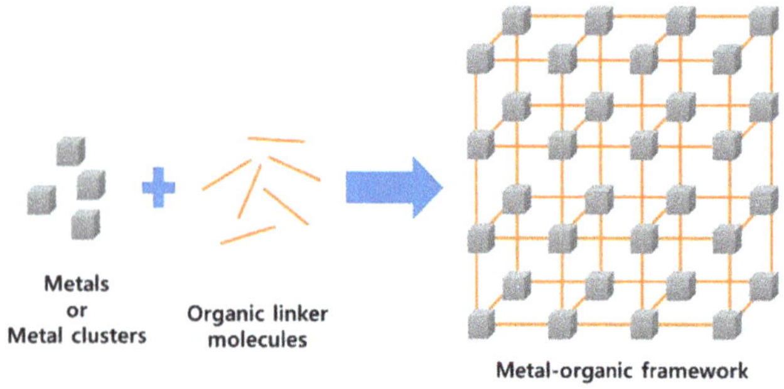

FIGURE 6.1 Schematic of MOF structure. Adapted with permission from [2]. Copyright (2020) Copyright the Authors, some rights reserved; exclusive licensee [MDPI]. Distributed under a Creative Commons Attribution License 4.0 (CC BY).

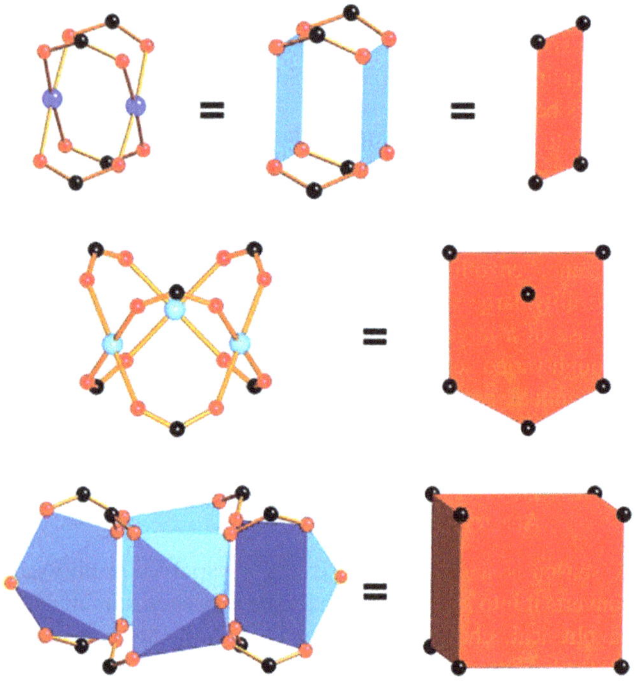

FIGURE 6.2 MOF SBUs with different geometries: (a) two planar, (b) three planar, (c) four square planar, (d) octahedron, (e) trigonal prism. Adapted with permission from [5]. Copyright (2012) ACS Publications.

surface to volume ratios enabling these to operate at the scale of natural biological and chemical processes. Numerous classes of nanomaterials including carbon and graphene have been used in various sensing contexts such as industry, environment, military, aerospace, and so on. [9]. MOFs, on account of their optical, electronic, and mechanical transduction properties have been employed in realizing high sensitivity sensors for gases, heavy metals, pesticides, explosives, aflatoxins, environmental remediation, and the like. MOF materials evince many attributes showing them to be suitable to be employed in sensors either as active sensing elements or transduction elements. The following section provides an insight into the properties of MOFs featuring their suitability for sensing applications.

6.2 ATTRIBUTES OF MOFS FOR SENSING APPLICATION

The sensing material is a key element of a sensor device. Following its interactions with the target analytes, the selectivity of the device is determined. The presence of unsaturated coordination metal sites and active functional groups on MOFs, results in specific interactions and thereby, specific identification and detection. The flexibility of MOFs enables a dynamic response to stimuli including temperature, pH, guests, pressure, and so forth. The attributes of MOFs that make them suitable for wide scale application in sensing avenues are discussed herein.

6.2.1 Synthetic Tunability

MOFs are formed through reticular coordination bonding of metal clusters and organic bridging ligands. Mostly, the metal clusters are formed in-situ while the linkers are pre-formed. The structure of the MOF is dependent upon the connectivity and geometry of the linker. The size, shape, or other properties of a MOF for an intended application can be tuned by tailoring the length, geometry, or functional groups on the ligand. The common tuning methods include solvent-assisted linker exchange, trans-metalation for MOF functionalization, and non-bridging ligand replacement. The solvent-assisted linker exchange focuses on substitution reactions for desired metal-ligand bond formations. Diverse classes of linker geometries including heterocyclic, ditopic, tritopic, hexatropic, octatopic, desymmetrized, and so on, have been explored for synthesizing MOFs and to tune their structure and functionality. Stimuli-responsive MOFs have been reported through the use of flexible linkers. Post synthetic modification enables the introduction of guest moieties with desired attributes. Desired functional groups are generally attached to the linkers upon post-modification. Recently, template-directed synthesis of MOFs has also been introduced wherein the MOFs act as hosts to template the chemical reactions between the guest species [10, 11].

6.2.2 High Surface Area

High surface area in MOFs enables adsorption and separation of a mixture of gases at ambient conditions. These display much high surface areas in comparison to activated carbon and zeolites. The highest surface area reported for a MOF structure to date is 14600 m^2/g [12]. MOFs display efficient results as sensing materials for various gases including ammonia, carbon disulfide, sulfur dioxide, and volatile organic compounds (VOCs) like ketones, aldehydes, alcohols, and so forth. [13]. The high surface area and presence of numerous active sites enable a large number of host-guest interactions and thus act as efficient sensing platforms. Various attributes of MOFs including tunable pore sizes, high porosity, and synthetic tunability, enable highly selective and sensitive MOF-based sensors. Figure 6.3 shows the timeline of evolution of MOF geometry over the years.

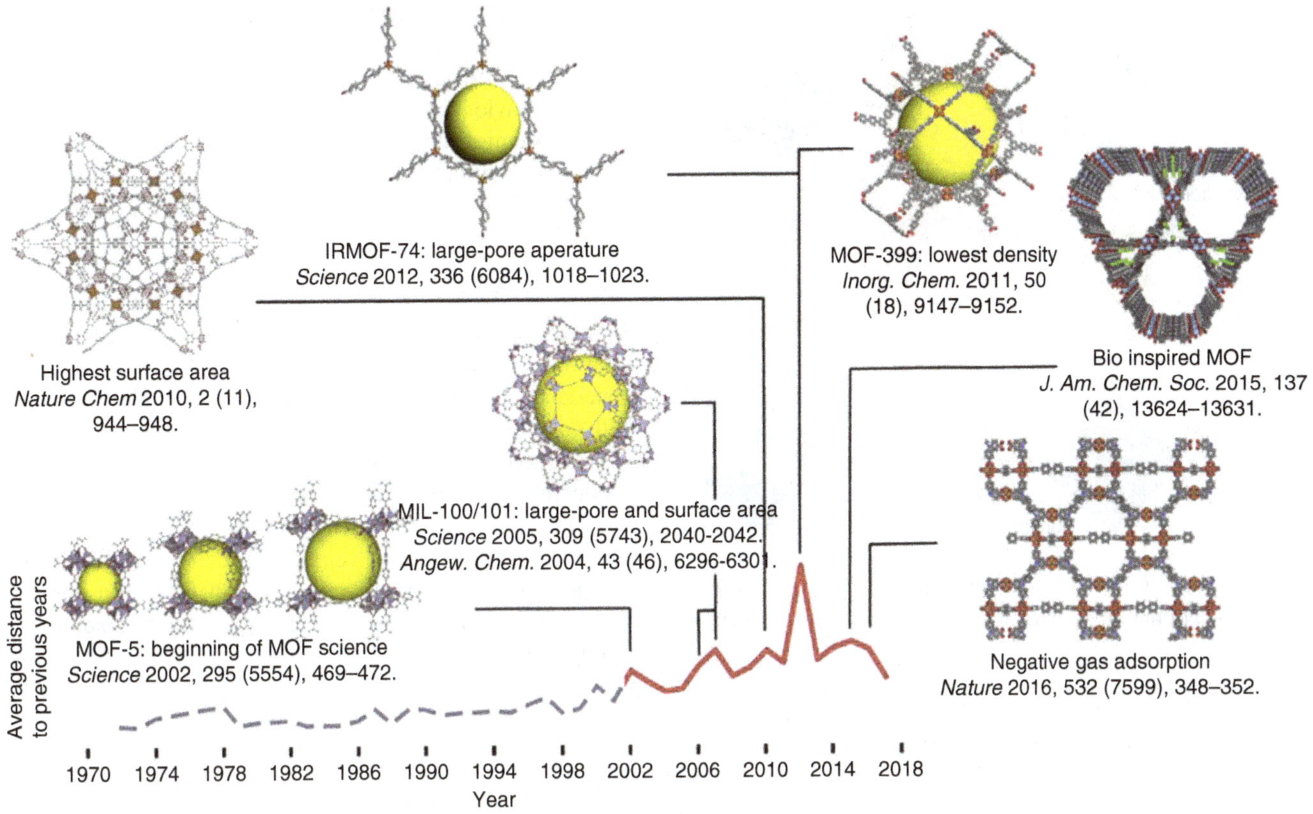

FIGURE 6.3 Evolution of MOF geometry. Adapted with permission from [6]. Copyright (2020) Copyright the Authors, some rights reserved; exclusive licensee [Springer Nature]. Distributed under a Creative Commons Attribution License 4.0 (CC BY).

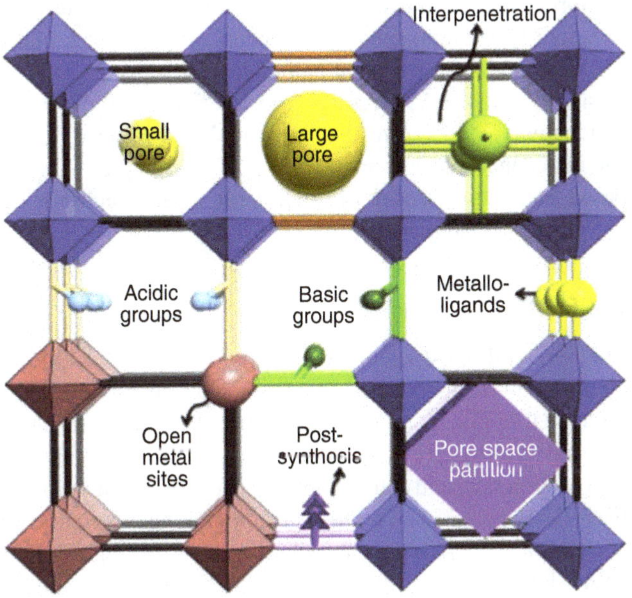

FIGURE 6.4 Various processes associated with porosity of MOFs during sensing. Adapted with permission from [17]. Copyright (2020) Elsevier.

6.2.3 Ultra-high Porosity

High porosity in MOFs enables interactions with various analytes and thus a readily measurable response to various physicochemical properties. With porosity (fraction of void volume to total volume) up to 90% of total volume and pore volume of 4.4 cm^3g^{-1}, MOFs display great sensing potential for various materials [14, 15]. A large number of research reports on MOF-based chemical sensors in literature can be attributed to the favorable attributes of these materials including: i) the presence of many active sites on account of high porosity which accelerates the host-guest interactions, ii) high levels of preconcentration of target molecules enabling high sensitivity, iii) tunability of pores enabling incorporation of functional groups thus increasing chemical selectivity and sensitivity. The host analytes adsorbed into the MOF pores could interact with the framework with any of the interactions namely hydrogen bonding, van der walls interactions, π-π stacking interactions, halogen bonding, and so on. [16]. Figure 6.4 highlights the various processes associated with the porosity of MOFs during sensing.

6.2.4 Stable and Tunable Luminescence

Luminescent MOFs are being increasingly used as sensing platforms for gases, ions, biomolecules, and so on. In comparison to inorganic material (gold nanoparticles, graphene oxide, MoS_2) based sensing platforms, MOFs offer more diverse and efficient chemical sensors on account of large-scale tunability. Moreover, on account of their intrinsic biocompatibility and biodegradability, these are also useful biosensing platforms. MOFs are multifunctional luminescent materials with either one or both of the constituents (metal ion, organic linker) augmenting the energy transfer process for

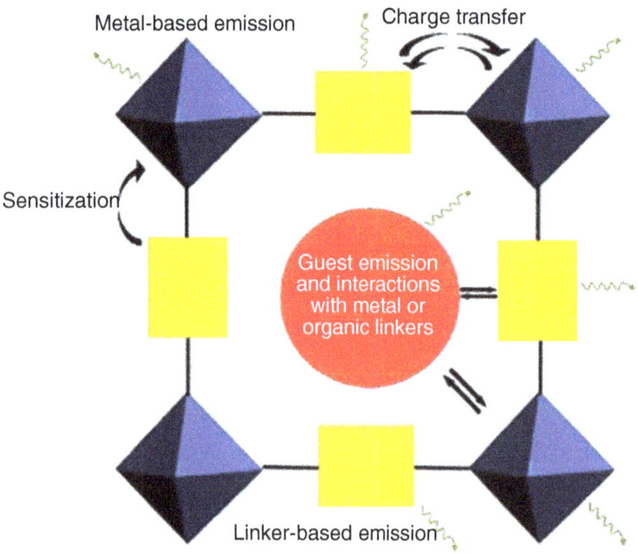

FIGURE 6.5 Luminescence phenomena in MOFs. Adapted with permission from [20]. Copyright (2009) Royal Society of Chemistry.

emission. There is the additional prospect of the introduction of luminescence through the introduction of luminescent guest molecules through doping. At present, around 1300 luminescent MOFs are known and many more are being explored on account of the interesting luminescent behavior of these materials. The high degree of porosity and structural tunability which enable the unusual luminescent behavior is not seen in the case of conventional phosphors. The ability to incorporate guest molecules within a MOF enables the realization of new luminescent MOFs with very high emission intensities at desirable emission wavelengths. Generation of luminescence in MOFs is broadly attributed to one of the following emission phenomena (Figure 6.5):

i) **Metal-based emission**: many metal ions emit luminescence at particular wavelengths, for example, lanthanide ions emit very sharp luminescence attributable to forbidden transitions (electric dipole selection rule). Additionally, the proximity of these metal ions to an organic fluorophore results in luminescence intensity enhancement on account of the antenna effect.
ii) **Linker induced emission**: some organic linkers absorb energy in UV and visible regions and subsequently emit either directly or through charge transfer to metal ions, for example, pyrene based organic linkers emit blue light
iii) **Emission due to adsorption of luminophores**: entrapment of a luminescent guest molecule inside the MOF pores enables luminescence from a non-emissive MOF as well as the tuning of emission of an already luminescent MOF through host-guest interactions, for example, guest triggered aggregation-induced emission in case of silver chalcogenolate MOFs.
iv) **Emission due to exciplex formation**: π–π interactions between adjacent conjugated linkers or between a linker and a guest molecule can produce an excited complex that typically exhibits broad luminescence [18, 19]. Also, many MOFs mimic enzyme-like behavior and catalyze particular non-emissive analytes to emissive behavior while the rest of the homologs are unaffected. This turn-on sensing mechanism forms the basis of many MOF-based biosensors.

6.2.5 Charge Transport Properties

Although MOFs are inherently insulating, several approaches have been developed to introduce conductivity in these materials. The commonly used strategies include: i) Doping with conductive guest molecules, and ii) Composite formation with conducting polymers. The doping of MOFs with conductive guest molecules enables their interactions with open metal sites in the MOFs, resulting in the creation of conductive pathways for electron transfer. Dopants employed are generally small-sized molecules that form π-bonding with linkers and result in increased electronic conductivity through the redox process. During composite formation, the combination of MOFs with conductive molecules like graphene, polyaniline, and the like, result in synergistic interactions integrating the redox properties of MOFs with the electron transfer properties of the conductive molecules. The charge transport in MOFs is governed by spatial or energetic orbital overlapping and an increase in this overlapping improves the overall charge mobility. The electroactive MOFs are increasingly being used in electrochemical sensing platforms due to the dual attributes of enhanced intrinsic selectivity and augmented electrochemical activity especially for applicability in pollutant and molecular sensing [21, 22]. Besides enhanced electrical conductivity MOFs, low conductivity MOFs have also been employed in active layers for impedance or capacitive sensing, for example, a MOF-based capacitive sensor on an interdigitated electrode for detection of hydrogen disulfide (H_2S) [23].

6.2.6 Biocompatibility

The pre-and post-synthetic functionalization of MOFs along with size modulation to nano ranges enable unlimited biological applications of these materials as these result in reduced cytotoxicity, augmented colloidal stability, efficient degradation rate, and cellular uptake. These properties provide the MOF materials with an edge in biological applications over conventional nanocarriers like silica, zeolites, or lipids. MOFs could be synthesized within permissible pharmacokinetic limits by using biocompatible components. Large porosity and surface area in MOFs ensure higher loading capacity in MOFs that are useful for the encapsulation of biological moieties for sensing or drug delivery applications. Also, labile metal-ligand bonds in MOFs ensure that not only the molecule is delivered to target tissue for sensing or drug delivery but also get promptly degraded and eliminated from the body before endogenous accumulation. Additionally, MOFs can overcome the challenges of metal nanoparticle

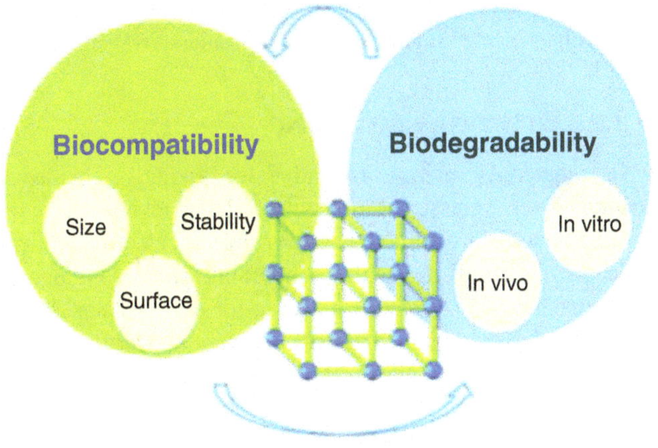

FIGURE 6.6 Biocompatibility and Biodegradability in MOFs. Adapted with permission from [26]. Copyright (2021) Royal Society of Chemistry.

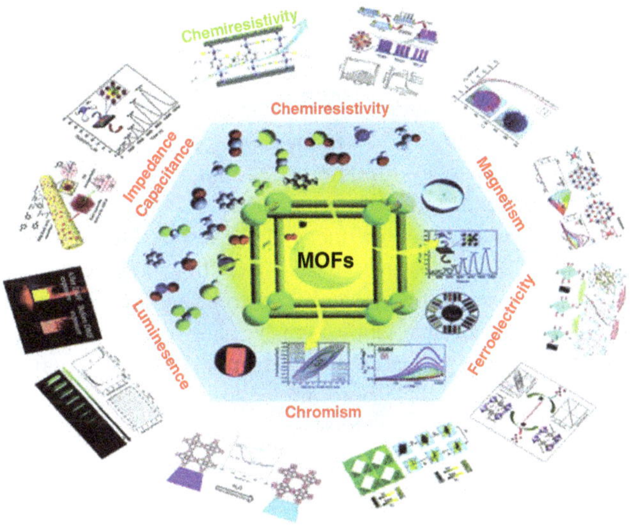

FIGURE 6.7 Sensing mechanisms in MOFs. Adapted with permission from [13]. Copyright (2020) Royal Society of Chemistry.

aggregation and bio-conjugation experienced in conventional sensing methods. The high adsorption capacity, a large surface area along with the high porosity of MOF materials protect the metal nanoparticle from aggregation and aid in preconcentrating the target at the electromagnetic field area. Thus, MOFs have been widely explored in biosensing and as bio carriers for imaging and diagnosis. Biocompatibility in MOFs is governed by the size, surface, and stability of the MOF material (Figure 6.6) [24, 25].

6.2.7 Mechanical Stability

Mechanical stability in MOF materials is very important considering that their applications involve repetitive pressure and temperature variations and capillary forces wielded by guest molecules. Mechanical stability basically measures the capability of the porous material to base its structure and porosity upon application of mechanical load. MOF-based mechanical sensors commonly employ quartz crystal microbalances as substrates upon which thin films of MOFs are grown. Such devices are employed to sense changes in the mass of the MOF thin films upon uptake or release of the analyte [27, 28]. HKUST-1 coated over a QCM substrate has been designed as a humidity sensor wherein the adsorption and desorption of water molecules in the MOF is measured in terms of change in mass [29].

6.2.8 Magnetic Functionality

The introduction of magnetic functionalities in MOFs is a relatively new area of research. Magnetism could be introduced into MOFs through appropriate choice of organic linkers, functional nodes, their bonding, or through encapsulation of magnetic guest moieties. For MOFs where the coordination framework itself is magnetic, judicious chemical design is necessary to attain the desired magnetic properties. The incorporation of functional molecules into MOF channels results in the formation of hybrid functional MOFs which could include: i) a magnetic guest in a non-magnetic framework, ii) a non-magnetic guest in a magnetic framework, or iii) a magnetic guest in a magnetic framework. Thus, intelligent selection of molecules is very important for the desired application of magnetic MOFs. A new class of stimuli-responsive materials has been designed wherein the magnetic properties are tuned according to the presence of molecular species in the MOF pores [30].

6.3 SENSING MECHANISMS IN MOF-BASED SENSORS

Sensing of an analyte involves transduction of the signal measured upon adsorption of analyte or reaction of the analyte with the MOF. MOFs, on account of their high surface areas and the presence of functional groups enable synergistic interactions with guest molecules resulting in optical, mechanical, electric, and chemical transduction signals. The commonly employed sensing mechanisms in MOFs (Figure 6.7) are discussed here.

6.3.1 Optical Sensing

On account of their unique luminescence behavior, MOFs have been extensively employed as optical sensors. The common transduction phenomenon includes quenching, luminescence enhancement, and photoinduced emission on account of guest adsorption. Host-guest interactions enable cogent electron-donor/acceptor orbitals overlap. Thus, the materials which are most efficiently sensed include electron acceptors like common explosives and electron donors like amines. In addition, the redox potential of the acceptor or donor units could also be tuned which could tune the luminescence. Besides luminescence enhancement and quenching, another sensing mechanism involves the formation of a unique emitting exciplex by adsorption of analyte within the framework,

Forster resonance energy transfer, photoinduced electron transfer, excited-state proton transfer, and intramolecular charge transfer [31].

Luminescence quenching is categorized into static and dynamic quenching. While in the case of static luminescence quenching the decay lifetime of the material does not vary with the concentration of the quencher, it is different in the presence and absence of quencher in the case of dynamic quenching. Luminescence-based quenching has been used to sense various biomolecules, nitroaromatic molecules, nitrofuran antibiotics, and so on. Forster resonance energy transfer (FRET) process involves the transfer of energy from excited molecules to acceptor molecules. Thus, the emission of donor molecules upon de-excitation is quenched. The phenomenon is ascribed to dipole interactions between the donor and acceptor molecules. FRET is governed by the quantum yield of the donor, the separation distance between acceptor and donor, the spectral overlap between the emission spectrum of donor and absorption spectrum of the acceptor, and the relative orientation of acceptor and donor dipoles. It is a readily used technique for sensing biomolecules using MOFs.

Photoinduced electron transfer (PET) involves the transfer of an electron between excited molecules. The excited state molecules act as a redox system that donates or captures electrons from other molecules. PET is governed by the distance between donor and acceptor molecules as well as the matching of their reduction and oxidation potentials. After, the transfer of an electron from donor to acceptor, the deactivation process is non-radiative. Thus, the PET receptors and excited luminophores are selected according to their redox potentials to maximize luminescence quenching. Excited-state proton transfer (ESPT) is a proton transfer process of photoacids (molecules acting as strong acids or bases in an excited state) to solvents. The emission spectrum of photoacids is dependent upon the solvent and its pH. A few examples of PET-based optical sensors include Zr-MOF-based fluorescent switches and Tb-MOF-based protamine sensors [32, 33].

Intramolecular charge transfer (ICT) is the process of tuning emission intensity as well as wavelength through the electron process. Both electron donating and electron accepting groups contribute to ICT. With the presence of both groups on a molecule, dipoles are formed. In the excited state, in the presence of different solvents and operating conditions, the dipole moment varies followed by the change in luminescence. Different environmental contaminants and solvents like toluene, chloroform, isopropanol have been sensed using ICT-based luminescence quenching [34]. Color change is one of the most basic means of sensing any phenomenon. Solvatochromism is the shift in absorption wavelength of material upon change of solvent and the shift is correlated with the polarity of the solvent. Blue shifting of absorption band has been reported in the case of Cu-MOF upon the shift of solvent from low to high polarity (water to chloroform) [35]. Apart from the solvatochromic effect, solvent-dependent color changes in MOF absorption have also been ascribed to changes in coordination environment, for example, change in absorption of Co-MOF upon absorption of different vapors due to change in geometry from octahedral to tetrahedral [35].

6.3.2 Electrochemical Sensing

MOF materials evince oxidative or radiative behavior towards diverse classes of molecules. The attached functional groups adsorb target molecules for further electrochemical processes. Varied molecules, including nitrites, glucose, hydrogen peroxide, metal ions, and so forth. have been sensed using MOFs on electrochemical sensing platforms. Direct electrochemical sensing of molecules involves limitations of poor selectivity, weak interference, and low sensitivity. MOFs on account of their high surface areas, porosity, and the presence of multiple functional sites equip a large number of electroactive sites for oxidation and reduction besides aiding in charge transfer between the electrode and molecules. The electrochemical interactions of MOFs with organic molecules include electrostatic interactions, π-π interactions, and donor-acceptor interactions [36, 37]. Figure 6.8 shows various areas of MOF-based electrochemical sensing.

Properties of large-scale tunability and high surface area in MOFs render their suitability as signal probes for electrochemical sensing. The sensing process commonly involves redox-active metals or ligands, electrochemical affinity towards redox substrates, electroactive guests loaded MOFs. One or more of these processes govern the movement of redox molecules towards the electrodes which are continually examined through an impedance spectroscopy technique. Conjugation of signal probes with nanomaterials like CNTs, graphene has shown promising results in electrochemical sensing. Unique catalytic and electronic properties of MOFs enable the enhancement of signals through the conjugation

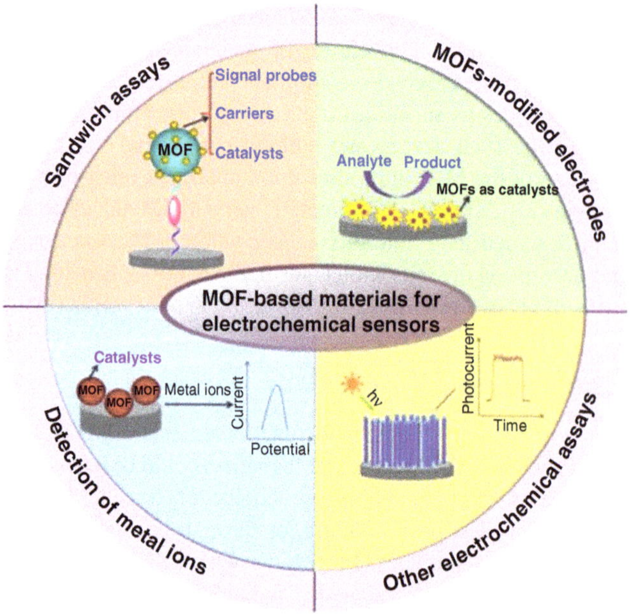

FIGURE 6.8 Electrochemical sensing using MOFs. Adapted with permission from [38]. Copyright (2018) John Wiley and Sons.

of MOFs or MOF-derived materials on signal probes. The presence of a large number of redox metal ions ensures high sensitivity in MOF-based electrochemical immunosensors. As an example, a Cu-MOF-based electrochemical immunoassay for sensing C-reactive protein has been reported with higher efficiency and smaller response time in comparison to the conventional methods [39].

Different metal ions like Cu^{2+}, Zn^{2+}, Pb^{2+} are conventionally used as electroactive indicators in sensing various molecules since these evince unique voltammetric responses to these. To enhance the electrochemical signals, these metal ions are generally doped or encapsulated into various nanoparticles via time-consuming methods. MOFs prove to be very efficient carriers for these metal ions due to their porosity and capability for the attachment of functional groups. An antibiotics detector has been designed with UiO-66-NH_2 as a carrier for Pb^{2+} and Cd^{2+} ions [40]. Functional biomolecules like aptamers, enzymes, antibodies react with specific molecules and are widely used as biosensors. Functional groups like -COOH and -NH_2 couple with biomolecules or adsorb biomolecules like single-stranded DNA through hydrogen bonding, π-π stacking, or electrostatic interactions.

6.3.3 Magnetic Sensing

Magnetic response-dependent signal, in other words, the variation in magnetic properties of MOFs in response to the analyte is proposed as a suitable alternative to optical sensing since it eliminates non-radiative quenching. Although only a few works have been reported on magnetic MOF-based sensing devices, these are still being seen as appealing candidates for future MOF-based sensing devices [41]. MOF-based magnetic sensing devices are broadly categorized into spin crossover MOFs and single-molecule magnets (Figure 6.9).

Most of these devices are based on spin-crossover MOFs which are comprised of transition metal ions with electronic configurations ranging from d^4-d^7, for example, Fe-MOF. These metal ions exist in high and low states and thus respond to external stimuli in terms of changes in these states and physical properties subsequently. The spin states also alter upon the incorporation of guest molecules indicating suitability for molecular sensing and switching. Single-molecule magnets (SMMs) comprised of magnetic metal complexes like Fe^{2+}, and Mn^{2+} exhibit utility in magnetic memories, spintronic devices, and recently, magnetic switches. Although, the SMMs are inherently highly sensitive to stimuli like light, pressure, and external guests these materials lack stability and repeatability which limits their usage in sensing applications. So, the use of these metal nodes in the MOF framework has been suggested to improve stability. Upon a change in guest molecules or formation or breakage of chemical bonds, the magnetism of these MOFs is tuned enabling their use as magnetic switches or sensors. Change in guest molecules or solvent molecules alters the distance between metal ion and ligand which could lead to changes in the coordination mode of the metal or crystal lattice subsequently varying spin states of magnetism [30]. A sensor for diaminobenzene has been developed by encapsulating carbon-coated cobalt nanoparticles in a Zn-MOF. Combining magnetic and optical detection methods, this sensor paves a way forward for the development of lab on chip microfluidics-based sensing devices [41].

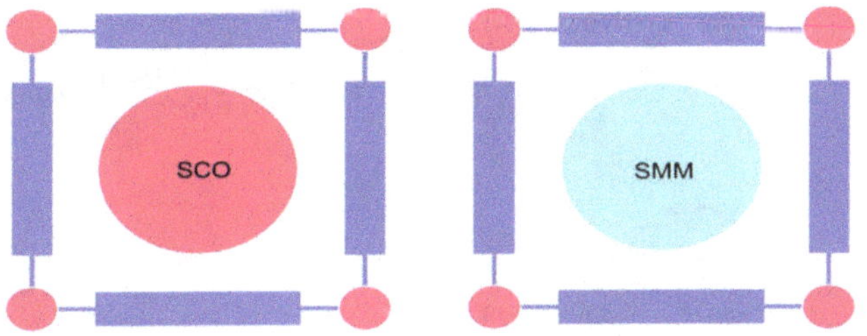

FIGURE 6.9 Schematic of magnetic functionalities in MOFs. Adapted with permission from [30]. Copyright (2018) Copyright the Authors, some rights reserved; exclusive licensee [Royal Society of Chemistry]. Distributed under a Creative Commons Attribution License 4.0 (CC BY).

6.3.4 Chemiresistive Sensing

The unique characteristics of MOFs enable their application in chemiresistive sensors which display a change in electrical resistance upon chemical reaction or adsorption of an analyte. The surface reaction or adsorption of the guest molecule results in hole or electron transfer leading to a change in resistance or conductance. Due to the large variety in chemical structures of MOFs and their derivatives, MOF-based chemiresistors employ diverse sensing mechanisms. The change in resistance is governed by the sensing material. In the case of metal nodes or organic ligands acting as absorption sites, the absorbed analytes on the MOF surface donate or accept electrons from the MOF, leading to a change in MOF resistance. The change in resistance or conductivity could also depend on the redox reactions of transition metal-based MOFs. Also, the volume change in a MOF upon gas uptake varies the number of electrons hopping between the MOFs leading to a change in resistance. The high surface area in MOFs implies higher surface interactions with analytes or a higher gas receptiveness in comparison to porous materials thereby giving a higher transduction signal and more efficient sensor response. On account of selective gas penetration and adsorption by each MOF these sensors offer very high selectivity [42].

In the case of MOF derivates or composites, the sensing characteristics are dependent on the majority phase. So, a different sensing behavior is seen in the presence of different components like metal oxides, nitrides, sulfides, or others. Metal oxide-based gas sensors involve a surface reaction between the chemisorbed oxygen and target molecules. At higher temperatures, oxygen molecules from air get absorbed on the surface of metal oxides and trap the electrons in these metal oxides, forming accumulation or depletion layers. Upon chemical reactions between the analyte molecules, the thickness of the hole accumulation layer or electron depletion layer varies, causing a change in resistance. Similarly, in the case of other chemical components, specific interactions with analytes govern the resistance change mechanism, sensitivity, selectivity, and response time of the sensor [42]. Thus, the attributes of high porosity, surface area, and electrical conductivity prove to be very beneficial for the application of MOFs as chemiresistive sensors. Selective gas adsorption of MOF membranes ensures highly selective sensors. Additionally, through intelligent control of MOF morphology, chemical structure, functionalization, and use of catalysts, highly efficient MOF-based chemiresistors could be designed.

6.3.5 Ferroelectric Sensing

Ferroelectric crystals are materials that evince spontaneous electric dipole moments due to the separation of negative and positive centers of charge even in absence of any electric field (Figure 6.10). Crystalline structure is imperative for ferroelectricity. MOFs, on account of their crystalline nature and attributes of tunability and porosity have inherent, or post introduced ferroelectric functionalities. The interactions of polar guest molecules or of the hydrogen bonds of MOFs bestow the ferroelectric properties on the MOF materials [43]. On being subjected to an external stimulus, these materials undergo single-crystal to single-crystal transformations which makes them suitable for application in ferroelectric switching.

MOF-based ferroelectric switches have been developed using a chiral 3d MOF which has externally induced guest water columns. The sensing process involves the adsorption or desorption of guest molecules followed by single-crystal to single-crystal transformation resulting in ferroelectric switching [44]. Reports on MOF-based ferroelectric sensing are limited in number to date on account of critical requirements of polar metal nodes and hydrophilic cavities. Moreover, efficient ferroelectric sensing requires large-scale single crystals, which is a limitation in MOFs to date. But the attributes of chemical tunability and porosity enables MOFs to act as host for other ferroelectric materials. Such combinations of the ferroelectricity of guests with useful properties of MOFs could lead to the realization of highly efficient ferroelectric sensors. For example, ethanol encapsulation in pores of an Mn-based MOF led to the introduction of ferroelectricity in the MOF [43]. On account of an unlimited choice of metal clusters, linkers, and guest molecules, multiple functionalities could be integrated into a single MOF, creating multiferroic MOFs. Jain and co-workers report an iron-based MOF showing both ferroelectricity and ferromagnetism [45]. With the ideas and feasibility demonstrated, new MOF-based ferroelectric compounds and sensing devices could be tailored in the near future with the convenience of a rich knowledge of MOF chemistry.

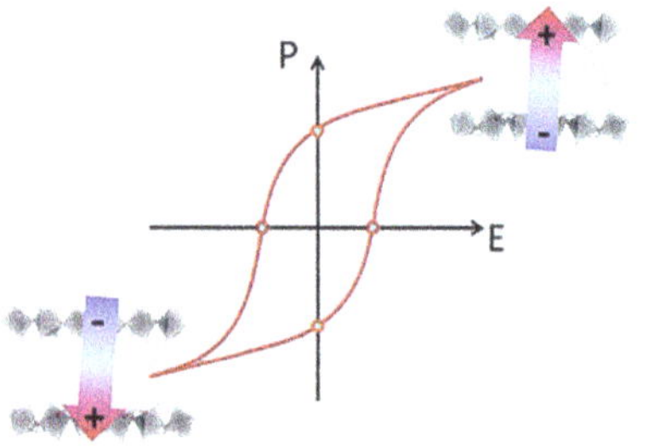

FIGURE 6.10 Ferroelectricity in MOFs. Adapted with permission from [43]. Copyright (2016) John Wiley and Sons.

6.3.6 Electronic Sensing

Most of the work on MOFs to date is focused on their chemical and physical attributes only and their application in electronic devices is a relatively unexplored avenue. But with the development of 1d, 2d MOFs and fabrication of stable thin films of these MOFs, a realization of MOF-based electronic devices has garnered attention. Since electronic sensors are subjected to harsh operating conditions like high or very low temperatures, corrosive analytes, humidity stability of the sensing material is a prerequisite. MOFs on account of their high physical and chemical stability prove to be the right choice as electronic sensing materials enabling highly stable and repeatable sensing devices.

Sensors generally face the issue of false detection due to interference issues during analyte detection. High selectivity in sensors is extremely important as well as difficult to attain. MOFs, on account of their nano-scale pore size show high selectivity since these allow diffusion of smaller-sized analytes and eliminate the rest of molecules with different sizes or shape conformations. Moreover, the presence of unsaturated coordination metal sites and functional groups on MOFs, specific interactions between MOFs and analytes take place resulting in selective sensing. Additionally, the high porosity in MOFs coupled with the fact that adsorption-desorption is a reversible process, ensures faster response time in MOF-based electronic sensors. The changes in mass and physical properties of MOFs upon analyte adsorption are converted into readable signals for quantitative detection. In electronic sensing, MOFs are employed as, i) mass-loaded layers, ii) filtration layers, iii) electronic function layers, or iv) optical sensitive layers (Figure 6.11) [46].

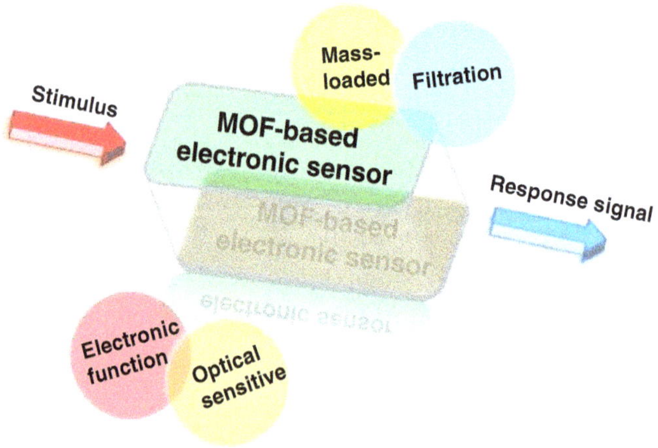

FIGURE 6.11 Various mechanisms in MOF-based electronic sensors. Adapted with permission from [46]. Copyright (2021) John Wiley and Sons.

MOFs which exhibit selective adsorption of analytes but no physical/chemical interactions with the analytes are commonly employed as mass-loaded layers in electronic sensors. The mass-loaded layers are connected to crystal microbalances, microcantilevers, or surface acoustic wave sensors. The change in mass upon adsorption of analyte, subsequently reflects in the form of a change in frequency of the quartz crystal or microbalance or other mass sensitive device which is then sent as an electric signal. Due to their selective adsorption, MOFs have been used to adsorb the target analyte and filter out the interferants or vice versa. So, these materials are employed as filtration layers in many electronic sensing devices. Metal oxidation sensors are highly selective sensors but have a limitation of degradation due to moisture exposure. The use of a layer of ZIF-8 MOF, which absorbs H_2 molecules selectively, solved the limitation of degradation of the metal oxidation sensors [47]. Inherently, MOFs display insulating behavior on account of a low degree of overlap between the d orbitals of metal ions and p orbitals of linkers. The hindered charge transport results in conductivity < 10^{-10} S cm^{-1}. Sensors with MOFs as electronic layers thus do not work on change in conductivity but rather measure the variation in inductance, capacitance, or work function. A rare-earth MOF has been employed as an electronic layer for the capacitive sensing of H_2S. The sensor displayed high sensitivity, a low limit of detection, and high stability [23].

MOFs show excellent optical characteristics in response to external stimuli like guest molecules, temperature, light, pressure, and so forth. The use of MOFs as optically sensitive layers in sensing devices is based on the electronic or optical readout of host-guest interactions. With the changes in analyte, the optical transduction platforms respond in the form of a change in intensity or wavelength of the optical signal which is further quantified to determine the quantity and quality of the analyte. The transduction and measurement units employed in MOF-based optical sensors include spectrophotometers, lasers, filters, waveguides, and so on. An optical CO_2 sensor has been designed with ZIF-8 MOF as an optical waveguide. The refractive index of ZIF-8 changes with the presence of CO_2 molecules. The ZIF-8 incorporated waveguide sensor showed high sensitivity and stability [48]. Thus, MOFs have been used in electronic sensors on account of their high selectivity, sensitivity, reproducibility, the limit of detection, and stability.

6.4 CONCLUSION AND OUTLOOK

MOFs can be rightly called one of the most propitious sensing materials of the future due to their excellent sensing characteristics on account of favorable attributes of tunability, high porosity, large surface area, stability, high luminescence. Highly selective and sensitive optical, electronic, ferroelectric, electrochemical, and magnetic sensors have been developed using MOFs, MOF composites, and MOF-derived materials. Although the field is rapidly expanding, commercialization of these sensors remains challenging. The fabrication of MOF thin films and membranes is still in the early stages. The development of robust, stable, long lifetime sensor devices is difficult with the current levels of success in the fabrication of MOF-based devices. More importantly, the high cost of bulk production of MOFs further hampers their practical usage. But with the continuous research thrust and advancements in avenues of crystal structure predictions of MOFs using computational tools as well as MOFtronics, MOF-based sensors may be available in markets soon.

REFERENCES

[1]. Y.-R. Lee, J. Kim, W.-S. Ahn, Synthesis of metal-organic frameworks: A mini review, *Korean Journal of Chemical Engineering* 30(2013) 1667–80.

[2]. D.Y. Heo, H.H. Do, S.H. Ahn, S.Y. Kim, Metal-organic framework materials for perovskite solar cells, *Polymers* 12(2020) 2061.

[3]. O. Yaghi, H. Li, Hydrothermal synthesis of a metal-organic framework containing large rectangular channels, *Journal of the American Chemical Society* 117(1995) 10401–2.

[4]. N. Stock, S. Biswas, Synthesis of metal-organic frameworks (MOFs): routes to various MOF topologies, morphologies, and composites, *Chemical Reviews* 112 (2012) 933–69.

[5]. M. O'Keeffe, O.M. Yaghi, Deconstructing the crystal structures of metal–organic frameworks and related materials into their underlying nets, *Chemical Reviews* 112 (2012) 675–702.

[6]. S.M. Moosavi, A. Nandy, K.M. Jablonka, D. Ongari, J.P. Janet, P.G. Boyd, Y. Lee, B. Smit, H. Kulik Understanding the diversity of the metal-organic framework ecosystem, *Nature Communications* 11(2020) 1–10.

[7]. W. Xia, *Fabrication of Metal–Organic Framework Derived Nanomaterials and Their Electrochemical Applications*, 2018, Springer, Singapore.

[8]. H.-C. Zhou, J.R. Long, O.M. Yaghi, Introduction to metal–organic frameworks, *Chemical Reviews* 112(2012) 673–4.

[9]. A. Tuantranont, *Nanomaterials for sensing applications: introduction and perspective, A. Tuantranont, Applications of Nanomaterials in Sensors and Diagnostics*, 2012, Springer, Berlin, Heidelberg.

[10]. C. Wang, D. Liu, W. Lin, Metal–organic frameworks as a tunable platform for designing functional molecular materials, *Journal of the American Chemical Society* 135(2013) 13222–34.

[11]. L. Feng, K.-Y. Wang, J. Powell, H.-C. Zhou, Controllable synthesis of metal-organic frameworks and their hierarchical assemblies, *Matter* 1(2019) 801–24.

[12]. O.K. Farha, I. Eryazici, N.C. Jeong, B.G. Hauser, C.E. Wilmer, A.A. Sarjeant, R.Q. Snurr, S.T. Nguyen, A.O. Yazaydin, J. Hupp, Metal–organic framework materials with ultrahigh surface areas: is the sky the limit? *Journal of the American Chemical Society* 134(2012) 15016–21.

[13]. H.-Y. Li, S.-N. Zhao, S.-Q. Zang, J. Li, Functional metal–organic frameworks as effective sensors of gases and volatile compounds, *Chemical Society Reviews* 49.17 (2020) 6364–401.

[14]. M. Alhumaimess, Metal–organic frameworks and their catalytic applications, *Journal of Saudi Chemical Society* 24(2020) 461–73.

[15]. Y.-S. Bae, D. Dubbeldam, A. Nelson, K.S. Walton, J.T. Hupp, R. Snurr, Strategies for characterization of large-pore

metal-organic frameworks by combined experimental and computational methods, *Chemistry of Materials* 21(2009) 4768–77.

[16]. T. Wittmann, C. Tschense, L. Zappe, C. Koschnick, R. Siegel, R. Stäglich, BV Lotsch, J Senker, Selective host–guest interactions in metal–organic frameworks via multiple hydrogen bond donor–acceptor recognition sites, *Journal of Materials Chemistry A* 7(2019) 10379–88.

[17]. R.-B. Lin, S. Xiang, W. Zhou, B. Chen, Microporous metal-organic framework materials for gas separation, *Chem* 6(2020) 337–63.

[18]. K. Müller-Buschbaum, F. Beuerle, C. Feldmann, MOF based luminescence tuning and chemical/physical sensing, *Microporous Mesoporous Materials* 216(2015) 171–99.

[19]. Y. Liu, X.-Y. Xie, C. Cheng, Z.-S. Shao, H.-S. Wang, Strategies to fabricate metal–organic framework (MOF)-based luminescent sensing platforms, *Journal of Materials Chemistry C* 7(2019) 10743–63.

[20]. M.D. Allendorf, C.A. Bauer, RK Bhakta, RJT Houk, Luminescent metal–organic frameworks, *Chemical Society Reviews* 38(2009) 1330–52.

[21]. B.A. Johnson, A.M. Beiler, B.D. McCarthy, S. Ott, Transport Phenomena: Challenges and Opportunities for Molecular Catalysis in Metal–Organic Frameworks, *Journal of the American Chemical Society* 142(2020) 11941–56.

[22]. C. Winkler, E. Zojer, Strategies for controlling through-space charge transport in metal-organic frameworks via structural modifications, *Nanomaterials* 10(2020) 2372.

[23]. O. Yassine, O. Shekhah, A.H. Assen, Y. Belmabkhout, K.N. Salama, M. Eddaoudi, H2S sensors: fumarate-based fcu-MOF thin film grown on a capacitive interdigitated electrode, *Angewandte Chemie* 128(2016) 16111–5.

[24]. S. Keskin, S. Kızılel, Biomedical applications of metal organic frameworks, *Industrial Engineering Chemistry Research* 50(2011) 1799–812.

[25]. R. Ricco, C. Pfeiffer, K. Sumida, C.J. Sumby, P. Falcaro, S. Furukawa, N R Champness, C Doonan, Emerging applications of metal–organic frameworks, *CrystEngComm* 18(2016) 6532–42.

[26]. N. Singh, S. Qutub, N. Khashab, Biocompatibility and biodegradability of metal organic frameworks for biomedical applications, *Journal of Materials Chemistry B* (2021).

[27]. S.M. Moosavi, P.G. Boyd, L. Sarkisov, B. Smit, Improving the mechanical stability of metal–organic frameworks using chemical caryatids, *ACS Central Science* 4(2018) 832–9.

[28]. A.J. Howarth, Y. Liu, P. Li, Z. Li, T.C. Wang, J.T. Hupp, O.K. Farha, Chemical, thermal and mechanical stabilities of metal–organic frameworks, *Nature Reviews Materials* 1(2016) 1–15.

[29]. L. Kosuru, A. Bouchaala, N. Jaber, M. Younis, Humidity detection using metal organic framework coated on QCM, *Journal of Sensors* (2016).

[30]. G.M. Espallargas, E. Coronado, Magnetic functionalities in MOFs: from the framework to the pore, *Chemical Society Reviews* 47(2018) 533–57.

[31]. Ü. Anik, S. Timur, Z. Dursun, Metal organic frameworks in electrochemical and optical sensing platforms: a review, *Microchimica Acta* 186(2019) 1–15.

[32]. B. Gui, Y. Meng, Y. Xie, J. Tian, G. Yu, W. Zeng, G. Zhang, S. Gong, C. Yang, D. Zhnag, Tuning the Photoinduced Electron Transfer in a Zr-MOF: Toward Solid-State Fluorescent Molecular Switch and Turn-On Sensor, *Advanced Materials* 30(2018) 1802329.

[33]. F. Qu, X. Li, X. Lv, J. You, W. Han, Highly selective metal–organic framework-based sensor for protamine through photoinduced electron transfer, *Journal of Materials Science* 54(2019) 3144–55.

[34]. X. Fang, B. Zong, S. Mao, Metal–organic framework-based sensors for environmental contaminant sensing, *Nano-Micro Letters* 10(2018) 1–19.

[35]. L.E. Kreno, K. Leong, O.K. Farha, M. Allendorf, R.P. Van Duyne, J. Hupp, Metal–organic framework materials as chemical sensors, *Chemical Reviews* 112(2012) 1105–25.

[36]. C.-S. Liu, J. Li, H. Pang, Metal-organic framework-based materials as an emerging platform for advanced electrochemical sensing, *Coordination Chemistry Reviews* 410(2020) 213222.

[37]. S. Kempahanumakkagari, K. Vellingiri, A. Deep, E.E. Kwon, N. Bolan, K.-H. Kim, Metal–organic framework composites as electrocatalysts for electrochemical sensing applications, *Coordination Chemistry Reviews* 357(2018) 105–29.

[38]. L. Liu, Y. Zhou, S. Liu, M. Xu, The applications of metal– organic frameworks in electrochemical sensors, *ChemElectroChem* 5(2018) 6–19.

[39]. T.-Z. Liu, R. Hu, X. Zhang, K.-L. Zhang, Y. Liu, X.-B. Zhang, R.Y. Bai, D. Li, Y.-H. Yang, Metal–organic framework nanomaterials as novel signal probes for electron transfer mediated ultrasensitive electrochemical immunoassay, *Analytical Chemistry* 88(2016) 12516–23.

[40]. X. Dou, K. Sun, H. Chen, Y. Jiang, L. Wu, J. Mei, Z. Ding, J. Xie, Nanoscale Metal-Organic Frameworks as Fluorescence Sensors for Food Safety, *Antibiotics* 10(2021) 358.

[41]. R. Ricco, L. Malfatti, M. Takahashi, A.J. Hill, P. Falcaro, Applications of magnetic metal–organic framework composites, *Journal of Materials Chemistry A* 1(2013) 13033–45.

[42]. W.-T. Koo, J.-S. Jang, I.-D. Kim, Metal-organic frameworks for chemiresistive sensors, *Chem* 5(2019) 1938–63.

[43]. K. Asadi, M.A. van der Veen, Ferroelectricity in metal-organic frameworks: characterization and mechanisms, *Eur. J. Inorg. Chem* 2016(2016) 4332–44.

[44]. Z. Sharifzadeh, K. Berijani, A. Morsali, Chiral metal–organic frameworks based on asymmetric synthetic strategies and applications, *Coordination Chemistry Reviews* 445(2021) 214083.

[45]. P. Jain, A. Stroppa, D. Nabok, A. Marino, A. Rubano, D. Paparo, M. Matsubara, H. Nakotte, M. Fiebig, S. Picozzi, Switchable electric polarization and ferroelectric domains in a metal-organic-framework, *npj Quantum Materials* 1(2016) 1–6.

[46]. L.T. Zhang, Y. Zhou, S.T. Han, *The Role of Metal–Organic Frameworks in Electronic Sensors*, Angewandte Chemie (2021).

[47]. W.-T. Koo, S. Qiao, A.F. Ogata, G. Jha, J.-S. Jang, V.T. Chen, I.-D. Kim, R. Penner, Accelerating palladium nanowire H2 sensors using engineered nanofiltration, *ACS Nano* 11(2017) 9276–85.

[48]. B. Chocarro-Ruiz, J. Pérez-Carvajal, C. Avci, O. Calvo-Lozano, M.I. Alonso, D. Maspoch, L. Lechuga, A CO2 optical sensor based on self-assembled metal–organic framework nanoparticles, *Journal of Materials Chemistry* A 6(2018) 13171–7.

7 MOFs as Sensors
Methods and Merits

Selva Balasubramanian[1,2], *Noel Nesakumar*[1,3], *Arockia Jayalatha Kulandaisamy*[2] *and John Bosco Balaguru Rayappan*[1,2]

[1]Centre for Nanotechnology and Advanced Biomaterials (CeNTAB), SASTRA Deemed University, Thanjavur, Tamil Nadu, India

[2]School of Electrical and Electronics Engineering (SEEE), School of Electrical & Electronics Engineering (SEEE), SASTRA Deemed University, Thanjavur, Tamil Nadu, India

[3]School of Chemical and Biotechnology (SCBT), SASTRA Deemed University, Thanjavur, Tamil Nadu, India

7.1 INTRODUCTION

Metal-organic frameworks (MOFs), a unique class of crystalline as well as porous materials have been the subject of intensive research in the last few decades. MOFs are defined by the construction of metal ions or clusters covalently linked by organic ligands to open frameworks that show exceptional features such as large surface area, permanent porosity, and high pore volume [1]. The inorganic metal complexes are often termed secondary building units (SBUs) and the bridging organic linkers are labeled as spacers [2]. Interestingly, their physical, chemical, and structural properties; important characteristics of MOFs can be tuned during synthesis [3]. In addition, the post-synthetic modification (PSM) approach has emerged as a versatile tool to design functionalized MOFs, where multi-functional groups can be introduced to MOFs without losing their crystallinity [4].

Tunable characteristics of MOFs have been tailored to a wide range of applications including gas storage/separation, catalysis, drug delivery, gas sensing, proton conduction, photonics, solar cells, and supercapacitors. Among these applications, MOFs have drawn more interest in the sensing of various analytes, either in the gas or in the liquid phase. In MOF-based sensing, a change in their intrinsic properties upon exposure to target analytes has been considered as a sensing signal. However, the low electrical conductivity or the nearly insulating nature of several MOFs has limited its potential to an extent, especially in electrical and electrochemical sensing platforms. Recent developments of 2D conductive MOFs have experienced unprecedented growth in the research domain of MOF-based electronic sensors [5]. Aromatic subunits in organic linkers of MOFs make most of them to readily luminesce under UV or visible excitation which makes them useful for optical sensing [6].

Intriguing framework architecture with a defined crystal structure, choice of omni-chemical functional groups, luminescencent nature, biodegradability, and non-toxic profiles are the main motives for exploiting MOFs for sensing applications especially for chemical and biosensing applications [7]. The potential of MOFs also extends to the electrochemical sensing domain, where effective mass/electron transfer and a large number of accessible active sites are taken into account to achieve highly sensitive electrochemical signals [8]. Highly porous natured MOFs have superior sorption properties, allowing interactions with other molecules [9]. Tunable pore sizes and the large surface area of MOFs support enhancement in the sensitivity and selectivity of the target sensing analytes [9]. Furthermore, permanent porosity and pore confinement provide a plethora of active sites to accelerate surface host-guest interactions [5].

Besides the synthesis of robust MOFs, various research efforts have been undertaken to fabricate MOF-based devices for a broad range of applications. Over the past decade, very many research studies have been published on MOF-based devices and this implies that the trend will grow further [10]. With this background, this chapter contextualizes the research studies by providing background information on the critical features of MOFs as active sensing elements. Furthermore, the challenges of and prospects for selecting MOFs for desirable target detection are also discussed in detail. The final section of this chapter summarises the various MOF sensors reported for various applications and discusses the implications for future research in the fabrication of MOFs for next-generation sensors.

7.2 INFLUENTIAL PARAMETERS OF MOFs FOR SENSING

The approach of MOF-based sensing has received much more interest in electronic sensors technology due to the increase in demand for efficient sensing performance parameters such as high selectivity, sensitivity, and stability in multi-various domains. Along these lines, it is important to discuss the factors that influence and predetermine those sensing performances. Hence, a detailed discussion of the factors such as secondary building units, open metal sites, porosity, which need to be considered before selecting the MOFs for the fabrication of sensors is discussed in this section.

DOI: 10.1201/9781003188148-7

7.2.1 Secondary Building Units

SBUs are molecular complexes composed of metal cluster entities, stitched together by multidentate organic ligands, where the SBUs act as rigid vertices in a framework through strong coordination bonds with organic moieties [11]. SBUs are a fundamental approach to the emergence of the reticular chemistry concept, and they are regarded as a significant milestone in the discovery of MOFs with permanent porous features [12]. Without a doubt, SBUs are the leading contenders for designing MOFs with exceptional properties. For example, the large rigid vertices coordinated by organic struts tend to form structurally stable MOFs while also contributing to their mechanical, and chemical stabilities. Moreover, it has a significant impact on the pore size and porosity of MOFs structures. Because the substitution of a group of vertices to a vertex of a framework net results in open structures with an optimal pore volume, this process is named as decoration [13].

Additionally, the decoration of such SBUs affords directionality and acts as a geometric basis for the resulting MOF structure [11]. SBU geometry is strong because of its coordination number, metal ions coordination geometry, and the characteristics of functional groups, which results in the formation of different shapes/types of cavities such as triangle (three points), square paddlewheel (four points), octahedron, and trigonal prism (six points) [1] and it's up to twelve points of extension [14] as shown in Figure 7.1. The points of extension represent the possible connections among the two metal clusters through organic linkers [11]. Also, SBUs have a more significant influence on the luminescent and electronic properties of MOFs that could fit a particular sensing application [7].

7.2.2 Open Metal Sites

Open metal sites (OMSs) or coordinatively unsaturated metal sites (CUS) are a pervasive feature of MOF compounds generated through the exclusion of solvent molecules, which are coordinated with the metal clusters [15,16]. These are carried out by thermal or chemical activation processes, [15] whereas the metal ions considered for the generation of OMSs should be mostly a part of metal SBUs [17]. The as-generated OMSs act as Lewis acid centers as well as adsorption sites [15] and have led to gas storage/separation applications of MOFs [18]. OMSs features enable reversible bonding with the target analytes, promoting high selective gas adsorption for sensing applications [16]. The presence of OMSs in MOFs leads to enhanced interactions between metal sites and guest molecules, which act as active sites for interaction with various gaseous molecules, including CO, CO_2, H_2, and NO [17]. In particular, OMSs of MOFs have a greater affinity to CO_2 *via* dipole–quadrupole interaction, which results in high selectivity. On the other hand, OMSs provide selective coordination sites for electron-rich functional groups like amines (amine grafting), which enhance selectivity for CO_2 due to their high basicity [19].

7.2.3 Porosity

Porosity is a feature that makes MOFs a promising contender for sensing applications that possess superior sorption characteristics. The appropriate selection of metal units and organic ligands paves the way to tune the pore size and spatial cavity arrangement of the framework [1]. The use of extended organic linkers for framework construction is one of the fundamental strategies for designing MOFs with ultrahigh surfaces. The large pore aperture of MOF crystals allows surface modification to be performed with more functionality without compromising their porosity [1]. However, the exceptional porosity features of MOFs are hindered due to their instability under harsh operating conditions [20] and the susceptibility of MOFs to moisture, which limits the potential

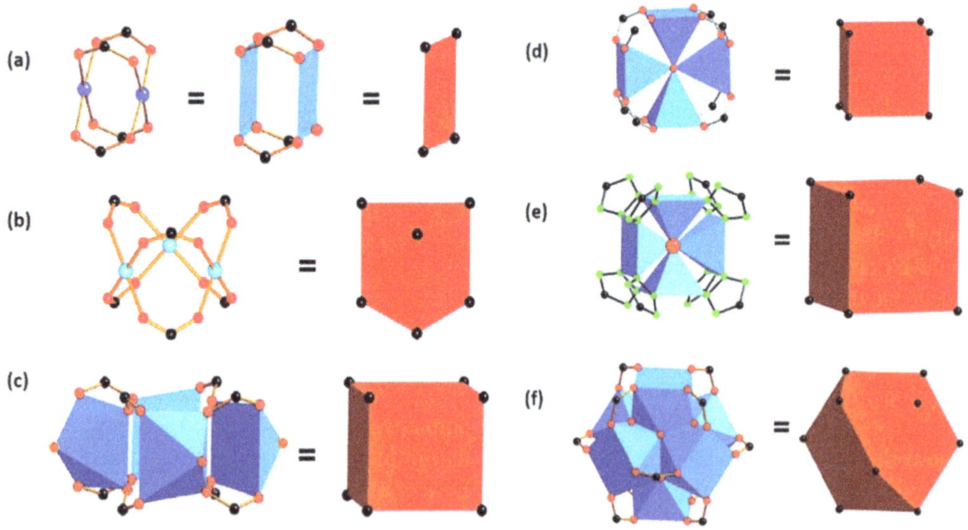

FIGURE 7.1 SBUs in MOFs with (a) two, (b) three, (c) four square planar, (d) and (e) eight, and (f) twelve points of extension. Adapted with permission from [14]. Copyright (2011) American Chemical Society.

to realize them as sensing elements in a real-time scenario. Additionally, interpenetration of the framework was found in MOFs constructed using an extended ligand approach, which has an adverse effect on stability, and also possessing an unstable nature during the removal of guest molecules [21]. Hence, a proper choice of length and flexibility of the organic linker, a topology that cannot build interpenetrating structures, and structural stability of MOFs upon guest molecules from the pores have to be considered while designing the ultrahigh porous MOFs with exceptional stability [22].

7.3 STABILITY OF MOF-BASED SENSORS

Operational stability of the sensors is an important parameter as the sensing element undergoes repetitive exposure to analytes, humidity, a harsh chemical environment for a prolonged time, and high temperature, which can result in material poisoning, structural deterioration, and sensing inhibition, and so forth. [16]. Hence, it is indispensable to evaluate and screen the chemical, thermal and mechanical stability of a sensing element. In this context, the aforementioned three categories of material capabilities, as well as their significance, are described in detail in this section.

7.3.1 Chemical Stability

MOFs are typical fragile materials that experience a lack of stability upon exposure to water molecules or air moisture, and they can degrade over a while, even at room temperature [23]. Thus, the constructed MOFs should be stable and have strong coordination bonds to preserve their structural integrity. Moreover, the chemical stability of MOFs strongly depends on the metal ion charge density, metal ion/nodes connectivity, basicity, and hydrophobicity of the exposed organic ligands, and so on [24]. The chemical and thermodynamic stability of MOFs under various operating conditions are decided by the bond strength between the metal nodes and the organic ligands [20,24]. The reinforcement of the coordination bond between the metal-ligands is an underlying concept of stable MOF construction, which represents the interaction between Lewis acid and Lewis base [23]. In addition, kinetically inert metal clusters could endow MOFs with exceptional chemical stability [23]. Thus, an amalgamation of hard Lewis acid and hard Lewis base or soft Lewis acid and soft Lewis base has to be chosen to construct stable MOFs.

From the case studies, MOFs synthesized with higher valency metal ions and carboxylate-derived ligands show a relatively stable nature in the acidic regime while lacking their stability in a basic medium. Similarly, MOFs resulting from low valency metal ions and azolate-derived ligands show admirable robustness in basic solutions, whereas they readily degrade in acids [20]. The observed difference in stability was attributed to their affinity towards protons and hydroxide ions [20]. Henceforth, the choice of MOFs for a targeted application must be considered with respect to their operating environment. To overcome these limitations, many research strategies such as post-synthetic exchange (PSE), PSM, and composite fabrication were developed to improve the stability of MOFs [24] and efforts are continuing to address the challenges.

7.3.2 Thermal Stability

The thermal stability of MOFs is an inevitable and highly desirable feature for industrial applications. Thermal stability refers to MOF compounds' irreversible ability to maintain their physical and chemical structures, even when compared to those at high temperatures [23]. In general, the phenomenon of metal node-ligand bond breakage accompanied or followed by organic spacer combustion is commonly referred to as thermal degradation of MOFs [25]. It occurs commonly due to the release or combustion of guest molecules upon heating above their decomposition temperature [23]. From the earlier case studies, it can be observed that the type of metal ions plays a vital part in defining the MOFs' thermal stability. In particular, metal ions with higher oxidation states and in their most stable states can result in MOFs with higher thermal stability. In other words, high-valence metal centers (trivalent or tetravalent metal ions) are more stable than divalent or monovalent metal ions [23]. Secondly, the SBUs with higher nuclearity possess greater thermal stability relative to the MOFs constructed from the same metal ions and similar organic ligands [23].

The thermal stability of MOFs is also determined by the nature of the organic linker. Most of the MOFs' architectures consist of oxygenated linkers because nearly all metal atoms tend to coordinate with oxygen species with different strengths [23]. However, while estimating stability, carboxylate linkers display lesser stability than phosphate and sulfate-based linkers. Moreover, the carboxylate linkers will undergo decarboxylation upon heating. In terms of MOFs' hydrothermal stability, again, the nature of the SBUs is an indispensable one. The ligand exchange constant of SBUs has an adverse effect on the stability of MOFs. For example, Cr^{3+} metal ions having a smaller ligand exchange value of 2.4×10^{-6} s^{-1} display a higher stability than Fe^{3+} of 1.6×10^{2} s^{-1} [20]. Moreover, the increase in internal hydrogen bonding in a framework can enhance hydrothermal stability [26,27]. From all of these above discussions, it may be concluded that a MOF structure having metal ions of high-valence stable oxidation states, short aromatic linker, and a strong metal-spacer interaction will result in higher thermal stability. Nevertheless, it is indispensable to consider all other parameters such as porosity and functionality with reference to the specific interest of target analyte detections.

7.3.3 Mechanical Stability

The mechanical stability of MOFs is another crucial parameter for industrial and practical applications. Most MOFs, however, exhibit less mechanical stability and undergo irreversible amorphization as a result of phase changes or partial collapse of pores within the MOF structure [20,25]. The destruction caused by the capillary force during the removal of guest molecules (in other words, anchored solvents) is also attributed to the structural collapse of the framework. Supercritical CO_2 activation is an effective method of combating structural degradation, in which the strongly associated solvent in MOFs is exchanged with low tension liquid CO_2, followed by removal under supercritical conditions. In such circumstances, the surface tension of the solvent and capillary forces is zero [23]. Another possible reason for the loss of mechanical stability is due to the availability of largely accessible pores which inherently weakens the strength of MOFs with a concomitant loss to porosity [23]. For instance, an irreversible amorphization was observed in ZIF-8 on compression beyond 0.34 GPa and loss of porosity occurs with increasing pressures up to 1.2 GPa [28]. In such cases, the use of dense structures with minimal porosity provides greater resistance against deformation and destruction under external stress [23,25]. Considering all of the described strategies, one could conclude that the negative impact on the mechanical stability of MOFs is limited, implying that their potential for applications can be expanded.

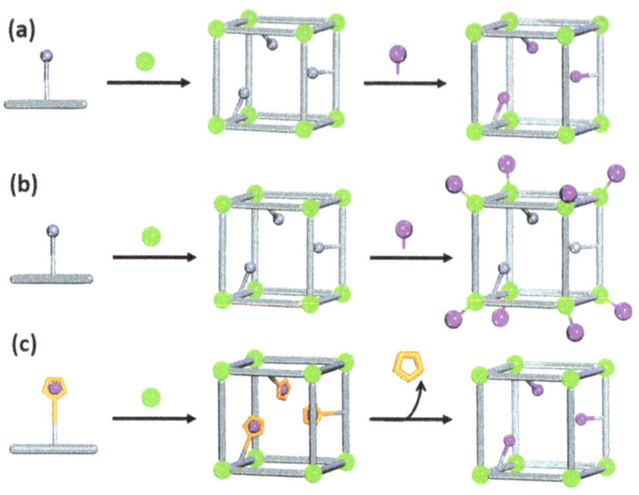

FIGURE 7.2 Schematic illustration of (a) covalent PSM, (b) dative PSM and PSD. Adapted with permission from [31]. Copyright (2011) American Chemical Society.

7.4 FUNCTIONALIZATION OF MOFs FOR ENHANCED SENSING

The tuning of chemical functionalities in MOFs can be done by functionalization through the introduction of various functional organic groups, generating active sites like Lewis acid/bases [29] for a target-specific application. The terminology 'Functionalization' for MOFs is quite different from others and has been classified according to the timeframe in which it will be functionalized before or after synthesis. Pre-synthetic functionalization involves introducing previously functionalized organic ligands during the initial stage of the synthesis reaction, whereas the subjected functional group is devoted to the structural component of the framework [30]. In post-synthesis functionalization, the surface of MOFs is functionalized with a non-structural component through covalent, dative, and post-synthetic deprotection strategies [31] (Figure 7.2). To be precise, covalent means the attachment of an organic moiety, most probably a phenyl or amine group, to the existing organic ligands of the constructed MOFs. Secondly, the dative represents the functional group's attachment to the metal node *via* a dative bond and this is only possible if the unsaturated metal sites are available within the framework. Finally, the post-synthetic deprotection (PSD) reactions constitute an exclusion or modification of the existing functional group in the MOF's structure to create active sites for catalysis or separation applications [30]. However, these approaches necessitate careful consideration of the reaction conditions and comprehensive characterization studies on the modified MOFs to ensure their physiochemical properties and resultant structure without any degradation.

7.4.1 Defect Engineering

Despite the advantage of the MOF functionalization process, the addition of other metal nodes to a parental framework would result in phase separation due to impurity creation and an option of multifunctional linker synthesis is quite complex [32]. To overcome this, one straightforward approach to controlling the physicochemical characteristics of MOFs is to create defects in their structure. Defects are the missing or dislocated atoms or ions in a periodic arrangement of the crystalline framework, and defect engineering is the controlled introduction of such defect sites [33]. To date, two synthetic routes have been reported for the preparation of defective MOFs: (i) de novo synthesis, and (ii) post-synthetic treatment. Among these, the modulation approach that comes under the de novo synthetic route is the most common method to create defects, in which considerable amounts of acids typically monocarboxylic acids, were added, along with organic linkers during the MOF synthesis [33]. The added monocarboxylic acids slow down the MOF crystallization process [33] and compete with linker molecules to coordinate with metal nodes in the synthesis environment [32] and, in turn, defect formation (that is, the missing-cluster defects and missing-linker defects). As a result, electron localization, chemical bond cleavage/formation, and lattice distortion occur in a framework and enhance its catalytic performance [32].

Another critical strategy is post-synthetic treatment, which involves the heterogeneous treatment of the native MOFs through modified acids or linkers. From this perspective, the research study by Wang et al. [32] demonstrated the fine-tuning of active sites in the UiO-66 (UiO- University of Oslo) MOF for cyclohexanone conversion. The addition of trifluoroacetic acid (TFA) during the synthesis of two batches of MOFs resulted in the existence of both the structural defects (missing-cluster and missing-linker) and caused an efficient conversion profile of 62-72% towards cyclohexanone. However, to distinguish the dominance of missing-cluster or missing-linker defects on the conversion profile, PSE was adopted, whereas the synthesized MOFs were immersed in benzene dicarboxylic acid (BDC) linker solution. During immersion, unsaturated metal nodes, called OMSs, were re-coordinated with BDC linkers and resulted in lower catalytic activity with a conversion profile of 22-18%. From this, it was found that the loss of open metal sites due to the missing-linker defect reduces catalytic performance, and thus fine-tuning the missing-linker defect rather than the missing cluster defect will increase catalytic activity. In summary, the engineering of defects accompanied by acid modulation has led to fine-tuning the formation of effective OMSs and thus converting the MOFs for a wide range of catalytic applications. However, there is still a long way to go in the area of defective MOFs to improve their heterogeneous catalytic performance by fine-tuning the type and number of defects and also helping us to know their advanced fundamental knowledge.

7.5 APPLICATIONS OF MOFs IN SENSING

In general, the working principle of sensors is based on signal transduction mechanisms that arise from the changes in the electrical, optical, or mechanical properties of the sensing element in its interaction with the analytes [34]. Based on the working principle, various classes of MOFs have been developed to be used in a wide range of sensing domains, including chemiresistive, colorimetric, luminescent, ferroelectric, and magnetic sensors [5]. Reversible adsorption, large surface area, diverse coordination geometry, tunable chemical functionalities, and porosity make them unmatched sensing elements in sensor platforms to detect heavy metal ions, toxic anions, volatile organic compounds (VOCs), CWAs, and organic pollutants. In this context, the progress of MOF sensors based on their sensing principles and the recent emerging findings on their development have been discussed in this section.

7.5.1 Chemiresistive Sensing

Chemiresistive sensing in MOF-based sensors involves the sorption process between gas molecules and functional materials and the resulting change in the surface resistance or conductance of the sensing element. The uniqueness of MOFs lies in the fact that they possess different sensing mechanisms in their chemiresistive behavior: i) redox reaction between the active-metal sites/functional organic groups and target gas molecules results in changes in conductivity, and ii) structural changes upon adsorption of guest molecules changes their conductivity [5,35]. The sensitivity of MOF-based sensing depends on the binding affinity of the analyte to the active sites of MOF, thus representing stronger binding that leads to lower detection limits (LOD). Another inevitable parameter in sensing is selectivity, which is used to distinguish between specific analytes. The selectivity of MOFs can be achieved either by fine-tuning the pore aperture sizes or *via* specific chemical interactions of the analytes with the MOF internal surface: i) hydrogen bonding, ii) π-π interactions, and iii) Mulliken-type electron donor-acceptor interactions [6].

To date, several studies have been explored based on the chemiresistive sensing behavior of MOFs towards volatile organic compounds (VOCs) and toxic industrial gas detection. At first, Chen et al. [36] demonstrated the chemiresistive sensing of Co-based ZIF-67 MOF to detect formaldehyde with a linear response in the range of 5-50 ppm at an elevated temperature 150 °C and it was attributed to the large surface of 1832.2 $m^2\,g^{-1}$. Then, Mello et al. [37] investigated the sensing characteristics of UiO-66-NH_2 MOF for the detection of acidic gases like SO_2 in the argon atmosphere. The conductivity of UiO-66-NH_2 MOF increased upon interaction with all the above-mentioned gases and showed a higher response of 21.6 ± 2.7%, to 10 ppm of SO_2 at 150°C. The reason for a higher response was ascribed to the fact that the high acidity of SO_2 aids a charge transfer complex with amine functionalities more than the other two acidic gases. Nevertheless, these research works have shown appreciable sensing features, slow response-recovery time, elevated operating temperature, low electrical conductivity and the requirement for enhanced sensitivity hinders its potential in the chemiresistive sensing platform. As a result, numerous research efforts have been undertaken to develop highly stable and conductive MOFs for chemiresistive sensors.

With this background, Campbell et al. [38] first reported that the 2D conductive MOF $Cu_3(HITP)_2$ (HITP = 2,3,6,7,10,11-hexaiminotriphenylene), exhibited a bulk conductivity 0.2 S/cm and detected 0.5 to 10 ppm of ammonia at room temperature. On the other hand, Yao et al. [39] have first reported the controllable growth of $Cu_3(HHTP)_2$ nanofilms in various thicknesses using the spray layer-by-layer (LbL) liquid phase epitaxial method. The fabricated $Cu_3(HHTP)_2$ nanofilm of 20 nm thickness displays higher sensitivity to ammonia among other interference gases and has shown an average resistance change of 129% to 100 ppm with a LOD of 0.5 ppm. Also, the fabricated nanofilms possess an excellent reproducibility of response (88.4%) with the long-term stability of 3 months.

Smith et al. [40] demonstrated the direct self-assembly of $Cu_3(HHTP)_2$ and $Ni_3(HITP)_2$ MOFs on a shrinkable polymeric film made of polystyrene is shown in Figure 7.3a. At first, the graphite interdigitated electrode (IDE) was made on the polymeric film and the MOFs were deposited over the polymeric film through the solution-phase synthesis approach. The fabricated sensors demonstrated an appreciable response to the 80 ppm of tested gas molecules, namely NH_3, NO, and H_2S, with the exemption of $Ni_3(HITP)_2$ MOF sensor which has not shown any noticeable response to NH_3. In a subsequent study, the same approach was applied to cotton fabrics, in which $Ni_3(HITP)_2$ and $Ni_3(HTTP)_2$ MOFs modified textile chemiresistive sensor arrays were developed (see Figure 7.3b) and tested against NO and H_2S [41]. In sensing performances, $Ni_3(HITP)_2$ MOF showed a greater response ($\Delta G/G_0$) of 81% ±6% to 80 ppm of NO, whereas $Ni_3(HTTP)_2$ possessed a higher response of 97% ± 2% towards H_2S of 80 ppm. Overall, the development of these conductive MOFs sensors on flexible media suggests the evolution of next-generation MOF-based flexible electronic sensors.

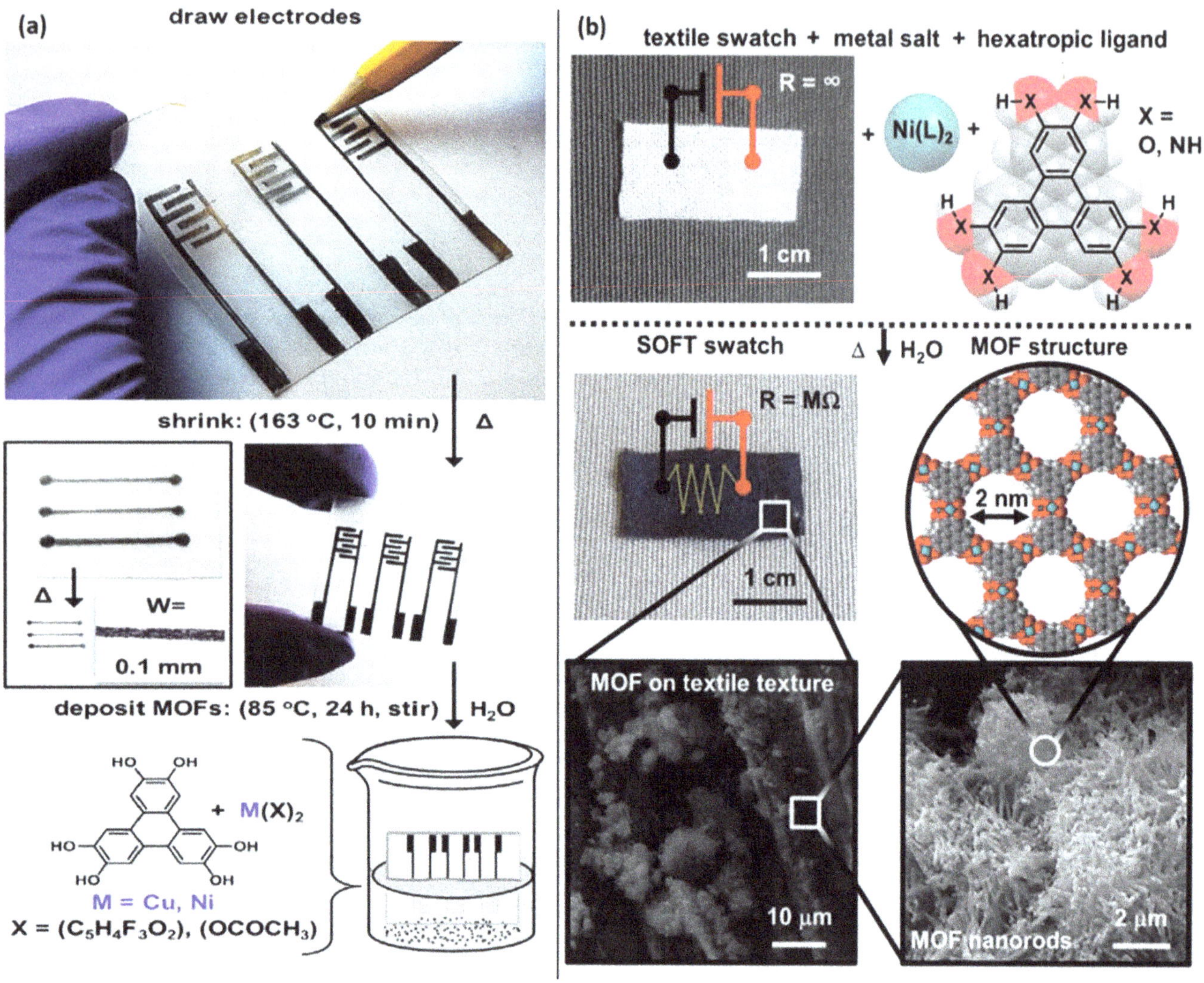

FIGURE 7.3 Illustration of chemiresistive MOF sensors on flexible substrates, (a) polymer sheet, and (b) cotton fabrics. Adapted with permission from [40,41]. Copyright (2016, 2017) American Chemical Society.

7.5.2 Luminescence Sensing

Luminescence in MOFs can originate from the building components: i) ligand-based luminescence (π-conjugation), ii) metal-centered luminescence, iii) charge transfer luminescence, and iv) guest-host framework interactions [5,42]. In detail, the extended π-conjugation system features the spin-ALLOWED radiative transition (fluorescence) and the spin-forbidden radiative transition (phosphorescence). The phenomenon of effective quenching by d^{10} transition-metal ions with unpaired electrons [43] and lanthanide metal ions with exceptional 4f-4f transitions and coordination with high absorption struts (antenna molecules) are solely attributed to metal-centered emission. The charge-transfer process is referred to as charge transfer luminescence, and it is made up of '*metal to ligand charge transfer (MLCT), the ligand to metal charge transfer (LMCT), metal to metal charge transfer (MMCT), and ligand to ligand charge transfer (LLCT)*' [5]. It is worth noting that none of these interactions are mutually exclusive; the emission pathway of one interaction can competitively coexist with that of another. For luminescent-based MOF (LMOF) sensors, changes in their spectroscopic features can be denoted as sensing signals, and they predominantly occur *via* two effects: quenching and guest interactions. Moreover, the intensity of both these effects was strongly influenced by the nature of the host-guest interactions [6].

The inherent porosity in LMOFs facilitates a natural habitat for analyte species and increases the probabilities of guest-host interactions through abundant interactions such as π-π stacking interactions, van der Waals interaction, and hydrogen bonding [5]. Moreover, systematic control of size and chemical environment in pores (polarizability, polarity, hydrophobicity, and acidity) allows selective capture of targeted analytes. In addition, the choice of functional groups with Lewis-acidic/-basic sites and the availability of OMSs further enhances the selective detection in sensing applications [42]. For example, an LMOF $[Eu(pdc)_{1.5}(dmf)]\cdot(DMF)_{0.5}\text{-}(H_2O)_{0.5}$ with Lewis basic sites has shown preferential binding with various metal ions [44]. On the other hand, LMOFs based on Zn and Mg compounds have selective sensing of NH_3 (at 100°C) due to the presence of OMSs [45]. Notably, $Mg(H_2DHBDC)$ has

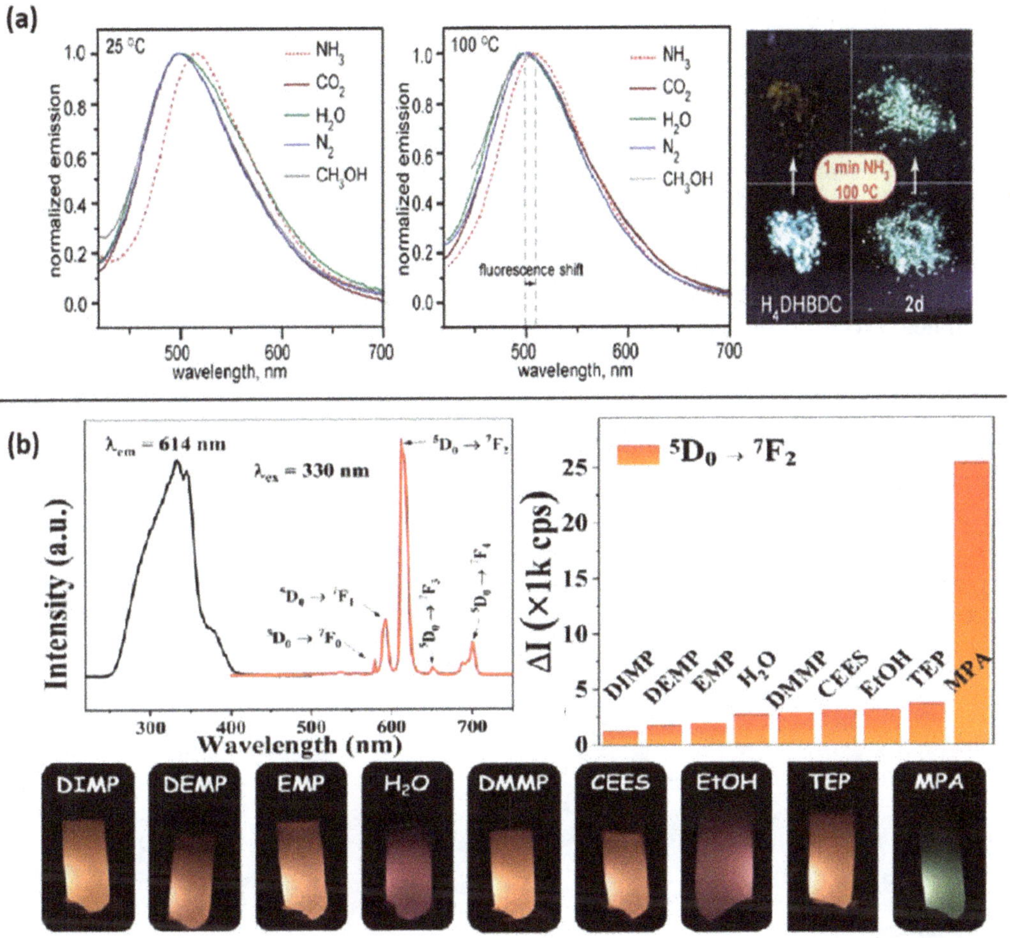

FIGURE 7.4 Emission spectra and photographs of LMOFs, (a) Mg (H_2DHBDC) and (b) Eu modified UiO-67 (Hf). Adapted with permission from [45,46] Copyright (2013, 2018) American Chemical Society.

shown a reversible sensing response to NH_3 after evacuation without any additional heating, which suggests the design of reversible NH_3 sensors (see Figure 4a). The diverse structures and unique characteristics of LMOFs have made them promising contenders for sensing VOCs, explosives, solvents, and biomolecules.

For the detection of CWAs, Lian et al. [46] demonstrated the luminescent sensing performance of Europium (Eu) modified UiO-67 (Hf)-bpydc MOF in which selective reactivity against methanephosphonic acid (MPA) was observed and displayed a high sensitivity with a LOD of 0.4 ppm. From the emission spectra, it was observed that the emission of MOF-MPA dispersed samples had dropped quickly, whereas the rest of the MOF-CWAs dispersed samples did not show any significant changes in fluorescence intensity as shown in Figure 7.4b. The shift in emission spectra was attributed to the charge transfer between the luminescence centers of MOF and MPA molecules, generating a weak fluorescent emission. In total, luminescent-based sensing displays reliable performance against the sensing of a wide range of target molecules. However, further research studies are required to know the interaction of MOFs with analytes because, in some cases, the signal from LMOFs gets lost in response to the sorption of analyte molecules [6]. Hence, exploration of the signal-transduction mechanism is recommended for the optimization of the sensing signatures of MOF-based luminescent sensors.

7.5.3 Electrochemical Sensing

Electrochemical sensing measures the change in output electrical signals caused by the reduction-oxidation reactions of analytes [34]. Although the reactions occur only at the electrode surface, an approach to developing an active electrode for electrochemical sensing is to immobilize the electrocatalytic materials on the electrode surface, resulting in the development of highly active electrochemical sensors [47]. The highly porous and large surface area features of MOFs inherently allow the guest molecules to diffuse into the framework, which intrinsically affords sensitivity for electrochemical detection [48]. In addition, fine-tuning the size and shape of pores for the analyte molecules helps achieve specific target interactions. Also, the incorporation/modification of chemical functionalities in the framework will also offer specific chemical interactions, namely π-π interactions, hydrogen bonding, and van der Waals interaction, between the MOFs and analyte molecules [48]. In this interest, the surface modification of working electrodes with active MOFs towards the electrochemical sensing of various analytes and its performance efficiency are briefly discussed in this sub-section.

Guo et al. [49] demonstrated an effective electrochemical sensor for detecting Pb^{2+} ions, in which the surface of the glassy carbon (GC) electrode was modified with an as-prepared NH_2-MIL-53 (Cr) MOF, MIL stands for Materials of Institute Lavoisier. In this study, the MOF modified sensor demonstrated excellent sensitivity to lead (Pb^{2+}) ions, with a linear increase in oxidation peak current as the concentration of Pb^{2+} ions increased between 0.4 and 80 μM, and a LOD of 30.5 nM as shown in Figure 7.5a. In addition, the developed sensor displayed selective detection of Pb^{2+} ions, in the presence of other ions such as Zn^{2+}, Hg^{2+},Cu^{2+}, and Cd^{2+} ions, to the same reaction medium (see Figure 7.5b). In terms of organic compound sensing, Dong et al. [50] fabricated a $Cu_3(BTC)_2$ modified carbon paste electrode (CPE) for the sensitive detection of 2,4-dichlorophenol (2,4-DCP). The modification of $Cu_3(BTC)_2$ to CPE enhanced the total adsorption capacity of 2,4-DCP as a result of an increase in effective surface area and led to a linear increase in oxidation peak current for the concentration in the range of 0.04-1.0 μM and exhibited a LOD of 9 nM (see Figure 7.5c and d).

Thus, the catalytic efficiency of MOFs has demonstrated that they are a potential candidate for electrochemical sensing applications. Furthermore, the rational design of porosity, tunable chemical functionalities, and specific target recognitions via multifarious interactions have met the requirements of the electrochemical sensing platform. Nevertheless, there are certain challenges such as low-conductivity, sluggish charge transport, and the chemical instability of most MOFs in the aqueous medium that have hindered their practical use for electroanalytical purposes. In this scenario, it is anticipated that the emergence of research strategies in the future will address these challenges and lead to extensive development in MOF-based electrochemical sensing applications.

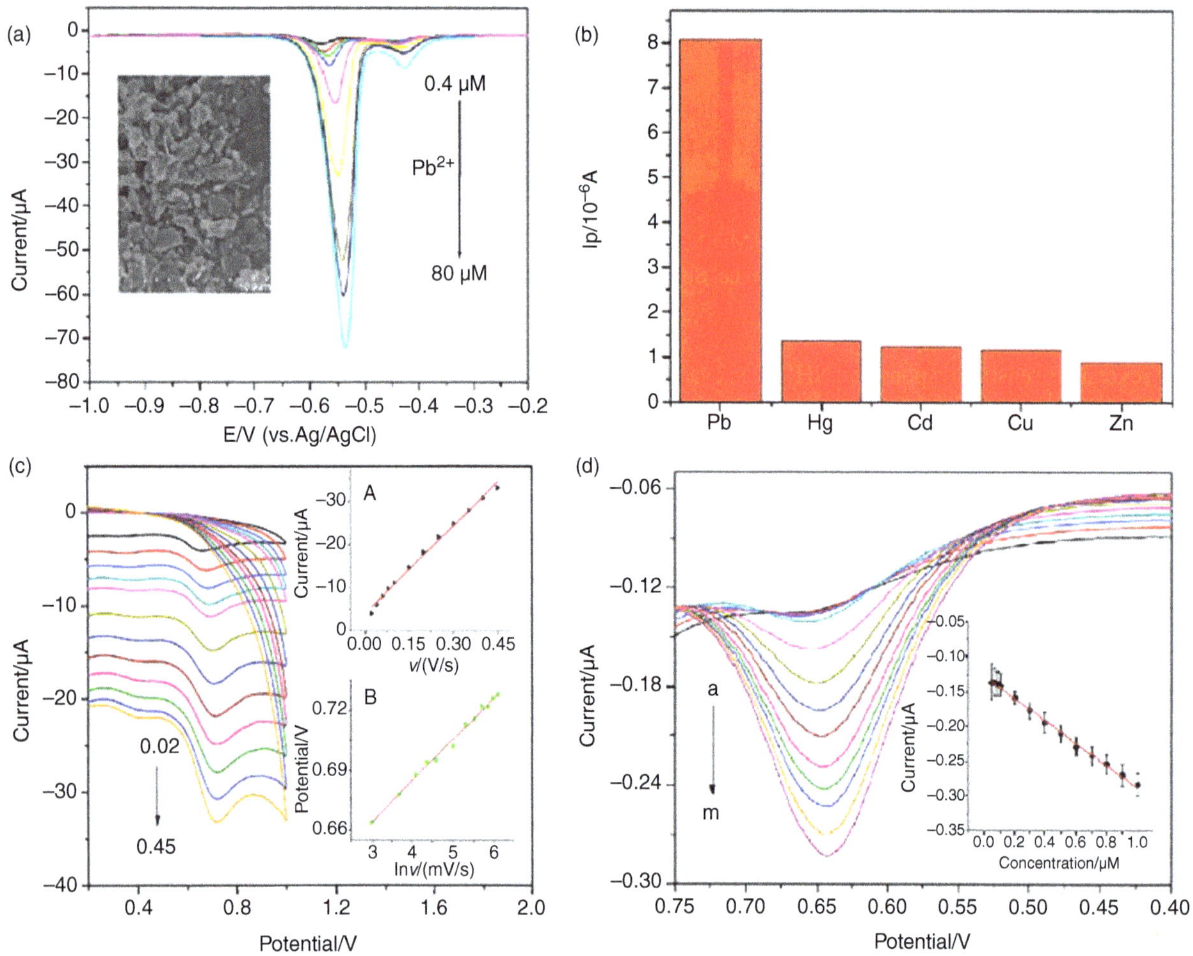

FIGURE 7.5 Electrochemical performance of MOFs, (a) Linear response and (b) changes in peak current of NH_2-MIL-53 (Cr) to various metal ions, (c) Cyclic voltammograms, and (d) DPV curves of $Cu_3(BTC)_2$-CPE to 2,4-DCP. Adapted with permission from [49,50] Copyright (2016, 2015) Elsevier.

7.6 CONCLUSION AND FUTURE PERSPECTIVES

With an interest in the development of MOF sensors for real-time applications, significant efforts have been made to move from controlled to well-organized synthesis MOFs for the development of electronic sensors. The topics discussed in this chapter have been selected to illustrate the reliable parameters that influence both the stability and the sensing performance of MOFs. Also, the performance of MOF sensors for multiple analytes such as VOCs, CWAs, metal ions, and organic pollutants has been narrated based on their sensing mechanism. From existing reports, it is known that properties of MOFs, namely porosity, surface area, SBU geometry, functional groups, and defect sites, have their own merits in enhancing the performance of the resultant sensing devices. It is indeed important to mention the synthesis and design strategies of MOFs such as the modulation approach, functionalization, and PSM because of their greater dominance in the construction of stable MOFs to tailor the chemical functionalities for the realization of MOF-based sensors.

Overall, MOFs' rational design and multifunctionality have enabled them to achieve superior sensing performance in a wide range of analyte detection. Irrespective of working principles, MOF-based sensors have shown promising results in the quantification of analyte species in trace amounts. However, many improvements to existing technologies need to be addressed to enhance the sensitivity and selective detection of target species. Moreover, a well-patterned fabrication approach is required to facilitate the effective and uniform distribution of MOFs on the supporting substrate. Hence, a synergism between MOF synthesis with functionality for the detection of specific targets and feasible fabrication technologies are highly desirable to broaden their applicability in a wide range of sensing applications. Meanwhile, certain reports on the development of MOF sensors on flexible substrates like polymer membranes and cotton fabrics have fuelled the growth of these sensors for next-generation electronic devices. This approach has also discovered the utility of MOFs in wearable and flexible technologies, where the surface modification of garments with sensing capability can respond to an environmental stimulus. With the aid of advancement in developing robust MOFs and excellent fabrication technologies, one could expect the advent of wearable electronic textile (e-textile) sensors in the future.

ACKNOWLEDGMENTS

The authors sincerely thank & acknowledge the Department of Science and Technology, New Delhi for the funding support DST/INT/JSPS/P-313/2020 and SR/FST/ET-I/2018/221 (C). One of the authors, Selva Balasubramanian thanks the Council of Scientific and Industrial Research, New Delhi, for the Senior Research Fellowship (09/1095(0058)/2020 EMR-I). We also wish to acknowledge SASTRA Deemed University, Thanjavur for extending the infrastructure support to carry out this work.

REFERENCES

[1]. E. Sharmin, F. Zafar, Introductory Chapter: Metal Organic Frameworks (MOFs), F.Z. and E. Sharmin (Ed.), *Met. Org. Fram.*, 2016. IntechOpen, Rijeka.

[2]. K. Kim, M. Banerjee, M. Yoon, S. Das, Chiral Metal-Organic Porous Materials: Synthetic Strategies and Applications in Chiral Separation and Catalysis, M. Schröder (Ed.), *Funct. Met. Fram. Gas Storage, Sep. Catal.*, 2010, Springer, Berlin Heidelberg.

[3]. S. Keskin, Molecular Simulations for Adsorption-Based CO_2 Separation Using Metal Organic Frameworks, S.E. Z. Fahmina (Ed.), *Met. Fram.*, 2016, IntechOpen, Rijeka.

[4]. A.D. Burrows, L.K. Cadman, W.J. Gee, H. Amer Hamzah, J. V Knichal, S. Rochat, Tuning the Properties of Metal–Organic Frameworks by Post-synthetic Modification, S.N. Hermenegildo García (Ed.), *Met. Fram. Appl. Sep. Catal.*, 2018, Wiley, Weinheim.

[5]. H.-Y. Li, S.-N. Zhao, S.-Q. Zang, J. Li, Functional metal–organic frameworks as effective sensors of gases and volatile compounds, *Chem. Soc. Rev.*, 2020, 49, 6364–6401.

[6]. L.E. Kreno, K. Leong, O.K. Farha, M. Allendorf, R.P. Van Duyne, J.T. Hupp, Metal–Organic Framework Materials as Chemical Sensors, *Chem. Rev.*, 2012, 112, 1105–1125.

[7]. P. Kumar, A. Deep, K.-H. Kim, Metal organic frameworks for sensing applications, *TrAC Trends Anal. Chem.*, 2015, 73, 39–53.

[8]. C.-S. Liu, J. Li, H. Pang, Metal-organic framework-based materials as an emerging platform for advanced electrochemical sensing, *Coord. Chem. Rev.*, 2020, 410, 213222.

[9]. Y. Li, A.-S. Xiao, B. Zou, H.-X. Zhang, K.-L. Yan, Y. Lin, Advances of metal–organic frameworks for gas sensing, *Polyhedron,* 2018, 154, 83–97.

[10]. R. Ricco, M.J. Styles, P. Falcaro, MOF-based devices for detection and removal of environmental pollutants, Ghosh (Ed.), *Met. Fram. Environ. Appl.*, 2019, Elsevier, Cambridge.

[11]. A. Schoedel, O.M. Yaghi, Reticular Chemistry of Metal–Organic Frameworks Composed of Copper and Zinc Metal Oxide Secondary Building Units as Nodes, S. Kaskel (Ed.), *Chem. Met. Fram. Synth. Charact. Appl.*, 2016, Wiley, Weinheim.

[12]. M.J. Kalmutzki, N. Hanikel, O.M. Yaghi, Secondary building units as the turning point in the development of the reticular chemistry of MOFs., *Sci. Adv.*, 2018, 4, eaat9180.

[13]. M. Eddaoudi, D.B. Moler, H. Li, B. Chen, T.M. Reineke, M. O'Keeffe, O.M. Yaghi, Modular Chemistry: Secondary Building Units as a Basis for the Design of Highly Porous and Robust Metal–Organic Carboxylate Frameworks, *Acc. Chem. Res.*, 2001, 34, 319–330.

[14]. M. O'Keeffe, O.M. Yaghi, Deconstructing the Crystal Structures of Metal–Organic Frameworks and Related Materials into Their Underlying Nets, *Chem. Rev.*, 2012, 112, 675–702.

[15]. J.N. Hall, P. Bollini, Quantification of Open-Metal Sites in Metal–Organic Frameworks Using Irreversible Water Adsorption, *Langmuir,* 2020, 36, 1345–1356.

[16]. A. Chidambaram, K.C. Stylianou, Electronic metal–organic framework sensors, *Inorg. Chem. Front.*, 2018, 5, 979–998.

[17]. Ü. Kökçam-Demir, A. Goldman, L. Esrafili, M. Gharib, A. Morsali, O. Weingart, C. Janiak, Coordinatively unsaturated metal sites (open metal sites) in metal–organic frameworks:

design and applications, *Chem. Soc. Rev.*, 2020, 49, 2751–2798.
[18]. L.-C. Lin, J. Kim, X. Kong, E. Scott, T.M. McDonald, J.R. Long, J.A. Reimer, B. Smit, Understanding CO_2 Dynamics in Metal–Organic Frameworks with Open Metal Sites, *Angew. Chemie Int. Ed.,* 2013, 52, 4410–4413.
[19]. M. Mohamedali, H. Ibrahim, A. Henni, Application of Metal–Organic Frameworks (MOFs) for CO_2 Separation, S.N. Hermenegildo García (Ed.), *Met. Fram. Appl. Sep. Catal.*, 2018, Wiley, Weinheim.
[20]. S. Yuan, L. Feng, K. Wang, P. Jiandong, M. Bosch, C. Lollar, Y. Sun, J. Qin, X. Yang, P. Zhang, Q. Wang, L. Zou, Y. Zhang, L. Zhang, Y. Fang, J. Li, H.-C. Zhou, Stable Metal-Organic Frameworks: Design, Synthesis, and Applications, *Adv. Mater.,* 2018, 30 , 1704303.
[21]. Y.C. and S. Ma, Mesoporous Metal-Organic Frameworks, C.M.L. Leonard R. MacGillivray (Ed.), *Met. Framew. Mater.*, 2014, John Wiley & Sons, Ltd, Chichester.
[22]. H. Furukawa, X. Sun, Extended Linkers for Ultrahigh Surface Area Metal–Organic Frameworks, S. Kaskel (Ed.), *Chem. Met. Fram. Synth. Charact. Appl.*, 2016, Wiley, Weinheim.
[23]. G. Mouchaham, S. Wang, C. Serre, The Stability of Metal–Organic Frameworks, in: S.N. Hermenegildo García (Ed.), *Met. Fram. Appl. Sep. Catal.*, 2018, Wiley, Weinheim.
[24]. M. Ding, X. Cai, H.-L. Jiang, Improving MOF stability: approaches and applications, *Chem. Sci.,* 2019, 10, 10209–10230.
[25]. A.J. Howarth, Y. Liu, P. Li, Z. Li, T.C. Wang, J.T. Hupp, O.K. Farha, Chemical, thermal and mechanical stabilities of metal–organic frameworks, *Nat. Rev. Mater.*, 2016, 1, 15018.
[26]. W. Li, A. Thirumurugan, P.T. Barton, Z. Lin, S. Henke, H.H.-M. Yeung, M.T. Wharmby, E.G. Bithell, C.J. Howard, A.K. Cheetham, Mechanical Tunability via Hydrogen Bonding in Metal–Organic Frameworks with the Perovskite Architecture, *J. Am. Chem. Soc.,* 2014, 136, 7801–7804.
[27]. N. ul Qadir, S.A.M. Said, H.M. Bahaidarah, Structural stability of metal organic frameworks in aqueous media – Controlling factors and methods to improve hydrostability and hydrothermal cyclic stability, *Microporous Mesoporous Mater.*, 2015, 201, 61–90.
[28]. K.W. Chapman, G.J. Halder, P.J. Chupas, Pressure-Induced Amorphization and Porosity Modification in a Metal–Organic Framework, *J. Am. Chem. Soc.*, 2009, 131, 17546–17547.
[29]. J. Li, T. Musho, J. Bright, N. Wu, Functionalization of a Metal-Organic Framework Semiconductor for Tuned Band Structure and Catalytic Activity, *J. Electrochem. Soc.*, 2018, 166, H3029–H3034.
[30]. C.F. Cogswell, Z. Xie, S. Choi, Tuning of Metal–Organic Frameworks by Pre- and Post-synthetic Functionalization for Catalysis and Separations, S.N. Hermenegildo García (Ed.), *Met. Fram. Appl. Sep. Catal.*, 2018, Wiley, Weinheim.
[31]. S.M. Cohen, Postsynthetic Methods for the Functionalization of Metal–Organic Frameworks, *Chem. Rev.*, 2012, 112, 970–1000.
[32]. J. Wang, L. Liu, C. Chen, X. Dong, Q. Wang, L. Alfilfil, M.R. AlAlouni, K. Yao, J. Huang, D. Zhang, Y. Han, Engineering effective structural defects of metal–organic frameworks to enhance their catalytic performances, *J. Mater. Chem. A.*, 2020, 8, 4464–4472.
[33]. S. Dissegna, K. Epp, W.R. Heinz, G. Kieslich, R.A. Fischer, Defective Metal-Organic Frameworks, *Adv. Mater.,* 2018, 30, 1704501.
[34]. X. Fang, B. Zong, S. Mao, Metal–Organic Framework-Based Sensors for Environmental Contaminant Sensing, *Nano-Micro Lett.,* 2018, 10 .
[35]. W.-T. Koo, J.-S. Jang, I.-D. Kim, Metal-Organic Frameworks for Chemiresistive Sensors, *Chem.*, 2019, 5, 1938–1963.
[36]. E.-X. Chen, H. Yang, J. Zhang, Zeolitic Imidazolate Framework as Formaldehyde Gas Sensor, *Inorg. Chem.*, 2014, 53, 5411–5413.
[37]. M.E. DMello, N.G. Sundaram, A. Singh, A.K. Singh, S.B. Kalidindi, An amine functionalized zirconium metal–organic framework as an effective chemiresistive sensor for acidic gases, *Chem. Commun.*, 2019, 55, 349–352.
[38]. M.G. Campbell, D. Sheberla, S.F. Liu, T.M. Swager, M. Dincă, Cu_3(hexaiminotriphenylene)$_2$: an electrically conductive 2D metal-organic framework for chemiresistive sensing., *Angew. Chem. Int. Ed. Engl.,* 2015, 54 , 4349–4352.
[39]. M.-S. Yao, X.-J. Lv, Z.-H. Fu, W.-H. Li, W.-H. Deng, G.-D. Wu, G. Xu, Layer-by-Layer Assembled Conductive Metal–Organic Framework Nanofilms for Room-Temperature Chemiresistive Sensing, Angew. *Chemie Int. Ed.*, 2017, 56, 16510–16514.
[40]. M.K. Smith, K.E. Jensen, P.A. Pivak, K.A. Mirica, Direct Self-Assembly of Conductive Nanorods of Metal–Organic Frameworks into Chemiresistive Devices on Shrinkable Polymer Films, *Chem. Mater.,* 2016, 28, 5264–5268.
[41]. M.K. Smith, K.A. Mirica, Self-Organized Frameworks on Textiles (SOFT): Conductive Fabrics for Simultaneous Sensing, Capture, and Filtration of Gases, *J. Am. Chem. Soc.*, 2017, 139, 16759–16767.
[42]. Z. Hu, B.J. Deibert, J. Li, Luminescent metal–organic frameworks for chemical sensing and explosive detection, *Chem. Soc. Rev.,* 2014, 43, 5815–5840.
[43]. M.D. Allendorf, C.A. Bauer, R.K. Bhakta, R.J.T. Houk, Luminescent metal–organic frameworks, *Chem. Soc. Rev.*, 2009, 38, 1330–1352.
[44]. B. Chen, L. Wang, Y. Xiao, F.R. Fronczek, M. Xue, Y. Cui, G. Qian, A luminescent metal-organic framework with Lewis basic pyridyl sites for the sensing of metal ions., *Angew. Chem. Int. Ed. Engl.,* 2009, 48, 500–503.
[45]. N.B. Shustova, A.F. Cozzolino, S. Reineke, M. Baldo, M. Dincă, Selective Turn-On Ammonia Sensing Enabled by High-Temperature Fluorescence in Metal–Organic Frameworks with Open Metal Sites, *J. Am. Chem. Soc.*, 2013, 135, 13326–13329.
[46]. X. Lian, B. Yan, Trace Detection of Organophosphorus Chemical Warfare Agents in Wastewater and Plants by Luminescent UIO-67(Hf) and Evaluating the Bioaccumulation of Organophosphorus Chemical Warfare Agents, *ACS Appl. Mater. Interfaces.,* 2018, 10, 14869–14876.
[47]. C.-H. Chuang, C.-W. Kung, Metal–Organic Frameworks toward Electrochemical Sensors: Challenges and Opportunities, *Electroanalysis.,* 2020, 32, 1885–1895.
[48]. L. Liu, Y. Zhou, S. Liu, M. Xu, The Applications of Metal–Organic Frameworks in Electrochemical Sensors, *ChemElectroChem.,* 2018, 5, 6–19.
[49]. H. Guo, D. Wang, J. Chen, W. Weng, M. Huang, Z. Zheng, Simple fabrication of flake-like NH2-MIL-53(Cr) and its application as an electrochemical sensor for the detection of Pb2+, *Chem. Eng. J.*, 2016, 289, 479–485.
[50]. S. Dong, G. Suo, N. Li, Z. Chen, L. Peng, Y. Fu, Q. Yang, T. Huang, A simple strategy to fabricate highly sensitive 2,4-dichlorophenol electrochemical sensor based on metal organic framework Cu3(BTC)2, *Sensors Actuators B Chem.,* 2016, 222, 972–979.

8 Strategies to Improve Sensitivity and Selectivity of MOF-based Sensors

Arnab Ghosh[1] *and Gopal Das*[1,2]
[1]Centre for the Environment, Indian Institute of Technology Guwahati, Assam, India
[2]Department of Chemistry, Indian Institute of Technology Guwahati, Assam, India

8.1 INTRODUCTION

Over the past few years, metal-organic frameworks (MOFs) have been reported widely by researchers in the domain of electrochemistry. Compared to most conventional porous materials, MOFs usually possess a much higher surface area ranging between 1000-10000 m^2 g^{-1}. MOF-based materials have been explored in sensing analytes for diverse fields such as biomedicine, environmental, and in the food field in recent years [1,2]. MOF derivatives such as metals, metal oxides/ hydroxides, metal sulfides, metal phosphides, carbons, or their composites have been employed as promising electrochemical sensors in the liquid and gas phases. MOFs proved to be promising electrochemical sensors as [3,4] (a) MOFs possess coordinatively unsaturated metal sites, well-arranged periodic structures, and eccentric structural advantages such as high surface area, tunable pore size (via alteration of the length of the organic ligand), and facile functionalization. These unique features bestow MOFs with superior catalytic capacity, establishing an effective coating material for electrocatalytic electrodes used in sensing platforms. (b) the large surface area and high porosity of MOFs favors the high-efficiency mass transfer of the analytes, which effectively amplifies the signal response and improves the detection sensitivity. (c) MOF-based support matrices possess channels and cavities of specific sizes and shapes, which shows good selectivity towards target analytes through size exclusion effects. Based on the above features, MOFs have been explored as electrocatalysts in sensing various small molecules like H_2O_2, glucose, dopamine, heavy metals, and so on [5-7]. The incorporation of MOFs with functional groups confers them with additional new properties and multi-functionality, therefore resulting in the development of new electrochemical sensors with enhanced performances. For example, the incorporation of MOFs with biomolecules such as enzymes, antibodies, and aptamers extends the present applications of the pristine MOFs into the biosensing platforms by exploring the prepared MOF-bio composites in the detection of large molecules, proteins, biomarkers, DNA, and the like [8-11]. Furthermore, the introduction of carbon nanomaterials to the pristine MOFs enhances their conductivity and dispersibility, hence boosting the electroactivity of the sensors [12-14]. In another approach, under controlled chemical/thermal treatment, pristine MOFs and MOF composites can be further converted to metallic and/or carbonaceous nanomaterials and their respective composites [15]. Such MOF-derived materials feature the advantages of MOFs, namely, large surface area and high porosity and the merits of metal/ carbons such as good conductivity and plenty of active metal sites, making them promising electroactive materials for electrochemical sensing applications.

8.2 DESIGN CONSIDERATIONS FOR MOF-BASED ADVANCED ELECTROCHEMICAL SENSING APPLICATIONS

Most MOFs possess large surface areas, porous structures, easily tunable surface properties, molecular size-selective ability, and the like [16]. However, the exploration of MOFs for electrochemical sensing applications requires specific properties such as water stability, electrical conductivity, electro-chemical activity, and biocompatibility [1,13]. Therefore, it is of significant importance to characterize and discuss the properties mentioned above of MOF-based materials before their application in electrochemical sensing platforms. The fabrication of reliable electrochemical sensors requires attention to specific parameters such as sensitivity/specificity, selectivity, accuracy, stability/ robustness, reproducibility, and so forth [14]. MOFs play a vital role in this context because of their various structures and compositions that directly influence the above-discussed parameters (Figure 8.1).

Some important considerations to improve the sensitivity of MOF-based electrochemical sensors are uniformly distributed and easily accessible active sites and efficient mass/electron transfer [13]. Sensors with desired sensitivities could be effectively obtained through controllable synthesis of nano-MOFs, meso-MOFs, 2D MOFs, and smart MOF composites of conductive functional materials, such as carbon nanomaterials and metal nanoparticles [5,17-19]. A stable and reproducible electrochemical sensor could be constructed with uniformly immobilized electroactive materials such as 2D MOF thin films or 3D self-supported MOF film with uniform and controlled film thickness to prevent any depilation from the electrodes [20,21]. To

DOI: 10.1201/9781003188148-8

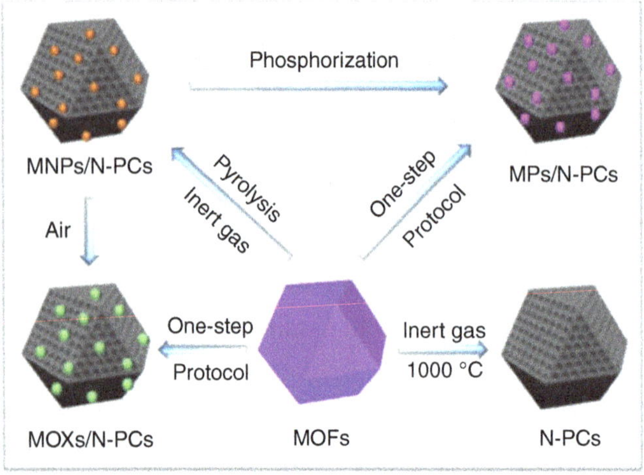

FIGURE 8.1 MOF-derived materials for electrochemical sensing applications. Reproduced with permission from [4]. Copyright 2020, Elsevier.

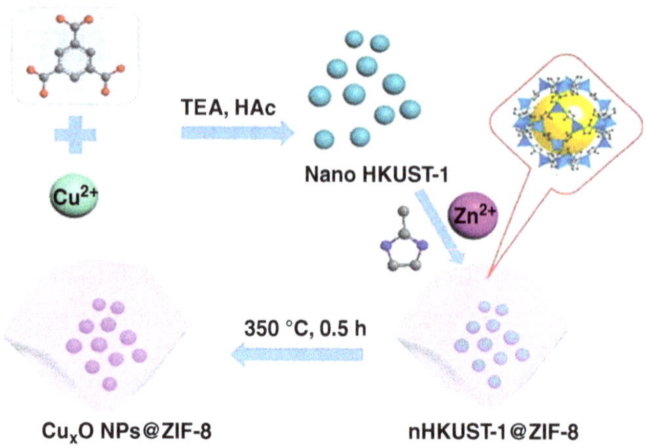

FIGURE 8.3 Pictorial representation of the preparation of Cu_xO NPs@ZIF-8. Reproduced with permission from [22]. Copyright 2020, American Chemical Society.

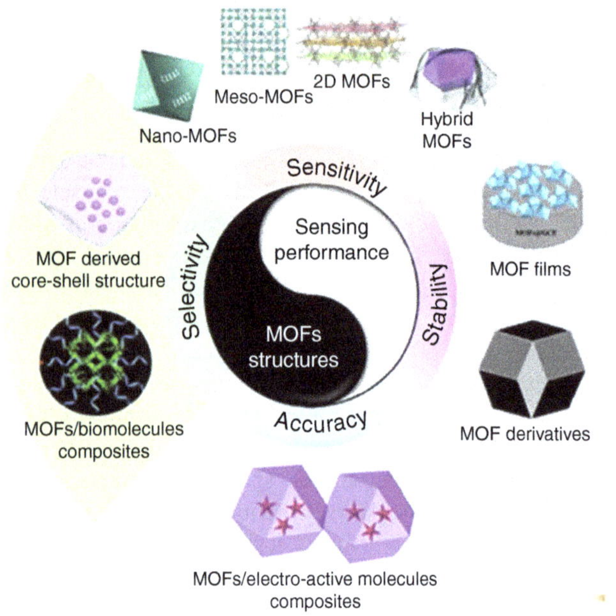

FIGURE 8.2 Structure performance relationship of MOF-based electrochemical sensors. Reproduced with permission from [4]. Copyright 2020, Elsevier.

improve the selectivity of the sensor, it is essential to develop MOF-based materials with core-shell structures, or defect structures, or MOFs functionalized with specific recognition elements like chiral cavities, imprinted polymer molecules, or biomolecules, for example, enzymes, aptamers, or antibodies with tailor-made molecular recognition functions [10,22]. The reliability and accuracy of the ratiometric electrochemical sensors can be obtained by introducing redox-active small organic molecules, which provide a reference signal [23,24]. The design and selection of MOF-based electrochemical sensors should be based on comprehensive consideration of properties such as sensitivity, selectivity, stability/robustness, accuracy, reliability, reproducibility, and so on [4]. The above parameters interact with one another and sometimes result in MOFs with contradictory properties. For instance, the hierarchical porous structures of mesoporous MOFs improve the sensitivity of the sensors by facilitating mass/electron transfer [25], whereas the drawback of such an electrochemical sensor is the poor molecular sieving effect. Therefore, the rational design of the MOFs should be based on their major application by keeping a balance of the multiple requirements' optimal sensing performance (Figure 8.2).

8.3 STRATEGY TO IMPROVE SELECTIVITY

Despite the significant development of MOF materials, selectivity in sensing most analytes remains a considerable challenge. In this context, to improve selectivity, the natural merit of MOFs such as the molecular sieving effect could be integrated with innovative design techniques such as defect design and specific recognition elements (molecular imprinting polymers and biomolecules) [1,10,22].

8.3.1 Core-Shell Structures

The specific pore sizes of the MOFs impart intrinsic molecular size-selective abilities, due to which MOFs possess excellent potential in the selective absorption of the analytes. Additionally, the size exclusion effects play a significant role in the detection of the specific analytes [26]. Core-shell heterostructures of the MOF-based materials have been reported to demonstrate good selectivity in analyte detection. For instance, Zeng et al. [22] fabricated a core-shell heterostructure of Cu_xO NPs@ZIF-8 via the thermal decomposition of $Cu_3(BTC)_2$@ZIF-8 composites taking into account the different thermal stability of the two MOFs (Figure 8.3). As a result, the uniformly dispersed small Cu_xO

NPs cores derived from $Cu_3(BTC)_2$ were found inside the ZIF-8 shell. Such a core-shell structure allows the passage of small H_2O_2 molecules while hindering the movement of the larger molecules. The fabricated sensor displayed excellent electrochemical sensing of the H_2O_2 molecules with high selectivity in the presence of the other interferents like dopamine, uric acid, ascorbic acid, amino acid, and so forth. In contrast, the conventional Cu_xO NPs devoid of the protecting MOF structures failed to display anti-interference activity.

8.3.2 Defect Design

It has been reported that defect structures such as plenty of oxygen vacancies played an essential role in enhancing the catalytic performance of the nanomaterials while effectively reducing the oxidation potential of the analytes [27]. The electrochemical sensors working at low negative potentials display the advantage of good anti-interference activity and high specificity in analyte detection. Wu et al. demonstrated the use of oxygen vacancies in the design of highly selective electrochemical sensors. CeO_{2-x}/C nanorods with plenty of oxygen vacancies were obtained using Ce-MOF as a template and applied for indirect uric acid determination at very low working potentials (-0.4 V vs. SCE) [28]. Physical conditions such as high temperature and low oxygen partial pressures supported the formation of oxygen vacancies in CeO_{2-x}/C nanorods, resulting in abundant Ce^{3+} active sites for high-efficiency electrocatalysis. In contrast to the conventional CeO_2 nanoparticles obtained via combustion in the atmosphere, CeO_{2-x}/C nanorod-based uric acid sensor demonstrated a much stronger CV response and a superior anti-interference activity towards glucose and ascorbic acid.

8.3.3 Recognition Element Incorporation

Shape and size alone are not always sufficient to construct a sensor with high specificity. Mostly, the incorporation of specific recognition elements into the MOFs proves to be a helpful strategy. For instance, in MOF-based bio composites, MOFs are integrated with functional biomolecules like enzymes, antibodies, and nucleic acids that impart high specificity and selectivity to detect the corresponding targets. In MOF-bio composite-based sensing platforms, a majority of MOF-bio composites are obtained via surface adsorption and covalent bonding. On this basis, Qiu et al. divided the strategies for bio composite formation into four types: surface adsorption, covalent attachment, pore infiltration, and in-situ encapsulation [1]. Strong binding occurs in the covalent attachment-based method among the interaction forces, while weak interactions like van der Waals forces, hydrogen bonding, and π-π stacking form the other three strategies.

8.3.3.1 Surface Adsorption

Surface adsorption is the oldest and most straightforward method of biomolecule immobilization. It is a physical method that utilizes weak binding forces such as hydrogen bonds, van der Waals forces, ionic interactions, and/or hydrophobic interactions to anchor the biomolecules on the MOF surface [29,30]. Generally, such forces are weak in comparison to a covalent bond. For example, a hydrogen bond has a low energy content of about 20 kJ mol^{-1} compared to a covalent bond with an energy content of 200-500 kJ mol^{-1}. In a general process, the pre-synthesized MOFs are mixed via simple stirring (for minutes to hours) with the biomolecules under mild reaction conditions to produce the MOF bio composites [31]. It is a simple technique devoid of any synthetic condition like temperature/solvent and only requires the MOF as a supporting matrix to immobilize the biomolecules. In surface adsorption, a high number of MOFs can be explored as supporting matrices to immobilize the biorecognition elements [1].

8.3.3.2 Covalent Attachment

Covalent attachment is a chemical method of biomolecule immobilization through a covalent bond. Although covalent bonding is a tedious process, it can produce immobilized biomolecules strongly bound to MOFs' surfaces. The strong affinity of the covalent linkages restrains the immobilized biomolecules leaching from the MOFs [8,32]. Typically, covalent bonding requires a linking or bridging molecule which anchors onto a MOF's surface and must possess a different reactive group capable of reacting with the biomolecules. An example of such a linking molecule is the crosslinking polymer, in other words, glutaraldehyde, consisting of two aldehyde groups, one at either end of its $(CH_2)_3$ moiety. At neutral pH, the amino groups on the biomolecules can react with the aldehyde groups, where one end of the glutaraldehyde may be attached to the MOF, and the other to the biomolecule [31]. Further, some MOF composites like noble metal MOF composites can be employed as a support to immobilize the biomolecules. In such a case, the covalent attachment is achieved by the formation of chemical bonds between the sulfhydryl groups on the biomolecules and the noble metal nanomaterials [33].

8.3.3.3 Pore Infiltration

In pore infiltration strategy, the mesoporous MOFs act as a support matrix where the biomolecules are adsorbed into the cavity of the MOFs instead of on the MOF surface [34]. This technique of biomolecule immobilization offers several advantages [35,36]. Firstly, the large pore size and mesoporous (2-50 nm) cavities of the MOF demonstrate the high loading capacity of the biomolecules. Secondly, the MOF acts as a protective layer around the biomolecule, thereby preventing its leaching and preserving activity.

8.3.3.4 In-Situ Encapsulation

The in-situ encapsulation technique introduces the bio-recognition element during the MOF synthesis process for the in-situ generation of the MOF bio composite [37]. In the in-situ encapsulation strategy, the MOFs are synthesized under mild conditions, for example, a ZIF series of MOFs, where the activity of the biomolecules is preserved [38,39]. The MOF protective shell allows the biomolecules to retain their activity under harsh reaction conditions. Additionally, during in-situ encapsulation, biomolecules larger than the pore size of the MOF can be encapsulated, thereby preventing further leaching of the biomolecules from the MOFs. Based on the use of the co-precipitating agents, the in-situ encapsulation strategy is divided into co-precipitation and biomimetic mineralization [8]. In the co-precipitation method, the biomolecules coated with co-precipitating agents like polyvinyl pyrrolidone (PVP) in-situ encapsulated into MOFs, co-precipitating agents protect and help in dispersing the biomolecules in the solution. Contrastingly, in the biomimetic mineralization method, the biomolecules are directly mixed with the MOF precursors, devoid of any co-precipitating agents to obtain biomolecule@ MOF bio composites.

8.4 STRATEGY TO IMPROVE SENSITIVITY

The sensitivity of an electrochemical sensor mainly depends on the electroactive area of the electrode and the mass transfer efficiency of the analytes [13]. It is widely known that particle reduction generates nanomaterials with high surface areas, and therefore the electroactive surface area is also increased [40]. In addition, MOFs featuring hierarchical pores and unique structures like 2D MOF nanosheets and hybrid structures enhance the accessibility of the electrode area, thereby promoting the easy diffusion of the analytes, thus enabling the electron/ mass transfer process [41,42]. Therefore, improving the sensitivity of the development of the sensor of nano-scale MOFs with unique and hierarchical porous structures can be considered an effective approach (Figure 8.4).

8.4.1 Nano-MOFs

MOF particles in the nano-size (5-500 nm) region, called nano-MOFs have displayed impressive success towards enhancing structural/functional features compared to conventional MOFs with sizes in micrometers. Nano-MOFs possess a higher number of exposed active sites, which offers better mass diffusion of the analytes to and from the active sites and therefore improves the electrocatalytic activity and detection sensitivity. Wang et al. have demonstrated that by reducing the MOF size from 1.5 μm to 300 nm the electrocatalytic performance of the NH_2-Ni MOF increased by 2.7-fold [33]. Nano-scale MOFs such as Ni-MOFs and Cu-MOFs have been reported to be excellent electrocatalysts for analyte detection [43-46]. The nano-MOF-based electrochemical sensors display higher sensitivity and ultra-low detection limits (nM or pM concentrations), in comparison to their bulk analogs.

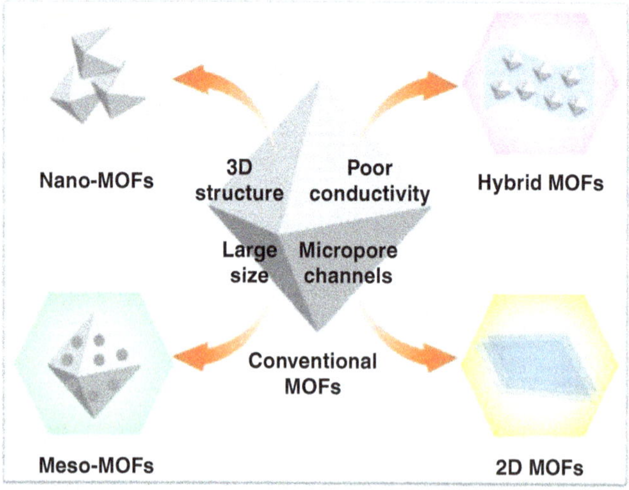

FIGURE 8.4 Strategy to improve the sensitivity of MOF-based electrochemical sensing materials. Reproduced with permission from [4]. Copyright 2020, Elsevier.

8.4.2 Hierarchical Porous MOFs

Hierarchical porous MOFs (HP MOFs) combine the merits of both mesoporous/macroporous materials and microporous MOFs, thus improving the sensitivity of the electrochemical sensor by facilitating rapid mass diffusion/electron transfer [47,48]. However, the facile fabrication of the HP MOFs remains a significant challenge. Some traditional approaches to obtain HP MOFs (featuring intrinsic mesopores and/or mesoporous channels) are ligand exchange, ligand extension, and ligand mixing. However, complicated synthesis of the extended ligand could be a significant drawback. In contrast, a more facile strategy could be constructing HP MOFs through crystal defect formation or assembly [49]. Jiang et al. reported a controllable synthesis of HP MOFs with tunable porous features via a modulator-induced defect design approach [50]. However, such a strategy resulted in the collapse of some MOF structures during the removal of guest molecules. Further, Liu et al. simplified the synthetic approach and developed a template-free self-assembly simple strategy to obtain extrinsic HP Cu-BTC MOFs [51]. The prepared mesoporous MOF exhibited tunable pore size from 26 to 72 nm due to varying degrees of packing of Cu-BTC nanoparticles due to a simple change of synthesis temperature.

8.4.3 2D MOFs

Compared to conventional 3D bulk layered MOFs, 2D MOFs, particularly 2D MOF nanosheets, possess large surface area, easy diffusion, and more available active sites for the analytes, therefore improving the response sensitivity of the sensors [52]. Different methods such as the top-down approach via ultrasonication/liquid exfoliation, the surfactant-assisted bottom-up synthetic method, and the layer-by-layer growth method have been investigated to fabricate 2D MOF nanosheets [53]. Nevertheless, the fabrication of ultra-thin MOF nanosheets with dimension thickness sub-10 nm remains a significant challenge. Li et al. reported a facile solvothermal method for the synthesis of Ni-MOF nanosheets with a uniform thickness of ~8 nm. They employed it for the effective detection of ascorbic acid (AA). Further, Zang et al. developed a surfactant-assisted synthesis method for obtaining high yields of ultrathin bimetallic M-TCPP (Fe)

(M = Co, Cu, or Zn) 2D MOF nanosheets (~5 nm thickness) [53,54]. When suspended in ethanol, a typical Tyndall effect was observed, which suggests the colloidal structure of the prepared MOF nanosheets. The as-synthesized MOF nanosheets demonstrated fast and better sensing responses towards H_2O_2 compared to natural heme proteins. They could be explored in real-time monitoring of the trace H_2O_2 released by the live cells [5].

8.4.4 Hybrid MOFs

As discussed above, tuning of the topological properties improves the electrochemical sensing ability of the pristine MOF-based sensors. However, these pristine MOF-based electrochemical sensors still suffer from some drawbacks such as poor conductivity, low electrocatalytic abilities, and high aggregation. An effective strategy to overcome these limitations is fabricating MOF-based hybrid structures, namely, carbon-based hybrid MOFs with superior conductivity, or noble metal nanoparticle-based hybrid MOFs, to derive a synergistic catalytic performance [55]. MOF-based hybrid nanostructures offer a short route between each component, promote electron transfer, and deter the aggregation of the individual components, thereby providing more available active sites. For instance, Zheng et al. constructed ultrathin 2D MOF M-TCPP (M= Cu, Co, and Ni) nanofilms of 1-3 nm thickness and corresponding 2D MOF nanosheets with 6-10 nm thickness [17]. Further, these nanofilms and nanosheets were incorporated with CNTs and GO to prepare MOF-based hybrid nanostructures. The hybrid M-TCPP nanofilm/CNT demonstrated a significant response towards reducing H_2O_2 and exhibited an ultralow detection limit up to nM concentrations, in comparison to as-synthesized M-TCPP nanofilm.

8.5 CONCLUSION AND PERSPECTIVES

Overall, this chapter summarizes the different strategies to improve the selectivity and sensitivity of the MOF-based electrochemical sensing materials. It has been demonstrated that the development of nano-MOFs with unique structures (like 2D MOF nanosheets) and hierarchical porous features (like mesoporous MOFs) effectively enhances the electroactive surface area and promotes electron/mass transfer, thus improving the sensitivity of detection. Additionally, fabrication of MOFs with conductive functional material like carbon nanomaterial/metal nanoparticles facilitates the uniform distribution and exposure of the active sites, which effectively improves the sensor's sensitivity. Moreover, the incorporation of biomolecules like enzymes, antibodies, and aptamers with specific molecular recognition functions endows the MOF-based sensors with high selectivity. Therefore, to design a MOF-based electrochemical sensor with high sensitivity and selectivity, the above-discussed aspects must be considered comprehensively. Despite significant efforts and progress, the practical application of MOF-based electrochemical sensors is still at a nascent stage. In this exciting field of research, there is still plenty of room for innovation. Soon, with the joint efforts of researchers from multi-disciplinary fields, one could expect many significant breakthroughs.

REFERENCES

[1]. Q. Qiu, H. Chen, Y. Wang and Y. Ying, Recent advances in the rational synthesis and sensing applications of metal-organic framework biocomposites, *Coord. Chem. Rev.,* 2019, **387**, 60–78.

[2]. H. Wang, W. P. Lustig and J. Li, Sensing and capture of toxic and hazardous gases and vapors by metal-organic frameworks, *Chem. Soc. Rev.,* 2018, **47**, 4729–4756.

[3]. Y. Xu, Q. Li, H. Xue and H. Pang, Metal-organic frameworks for direct electrochemical applications, *Coord. Chem. Rev.,* 2018, **376**, 292–318.

[4]. C. Sen Liu, J. Li and H. Pang, Metal-organic framework-based materials as an emerging platform for advanced electrochemical sensing, *Coord. Chem. Rev.,* 2020, **410** , 213222.

[5]. Y. Wang, M. Zhao, J. Ping, B. Chen, X. Cao, Y. Huang, C. Tan, Q. Ma, S. Wu, Y. Yu, Q. Lu, J. Chen, W. Zhao, Y. Ying and H. Zhang, Bioinspired Design of Ultrathin 2D Bimetallic Metal-Organic-Framework Nanosheets Used as Biomimetic Enzymes, *Adv. Mater.,* 2016, **28**, 4149–4155.

[6]. Z. Wang, M. Gui, M. Asif, Y. Yu, S. Dong, H. Wang, W. Wang, F. Wang, F. Xiao and H. Liu, A facile modular approach to the 2D oriented assembly MOF electrode for non-enzymatic sweat biosensors, *Nanoscale,* 2018, **10**, 6629–6638.

[7]. W. Ling, G. Liew, Y. Li, Y. Hao, H. Pan, H. Wang, B. Ning, H. Xu and X. Huang, Materials and Techniques for Implantable Nutrient Sensing Using Flexible Sensors Integrated with Metal–Organic Frameworks, *Adv. Mater.,* 2018, **30**, 1–9.

[8]. C. Doonan, R. Riccò, K. Liang, D. Bradshaw, and P. Falcaro, Metal-Organic Frameworks at the Biointerface: Synthetic Strategies and Applications, *Acc. Chem. Res.,* 2017, **50**, 1423–1432.

[9]. H. An, M. Li, J. Gao, Z. Zhang, S. Ma and Y. Chen, Incorporation of biomolecules in Metal-Organic Frameworks for advanced applications, *Coord. Chem. Rev.,* 2019, **384**, 90–106.

[10]. S. Kempahanumakkagari, V. Kumar, P. Samaddar, P. Kumar, T. Ramakrishnappa and K. H. Kim, Biomolecule-embedded metal-organic frameworks as an innovative sensing platform, *Biotechnol. Adv.,* 2018, **36**, 467–481.

[11]. K. Lu, T. Aung, N. Guo, R. Weichselbaum and W. Lin, Nanoscale Metal-Organic Frameworks for Therapeutic, Imaging, and Sensing Applications, *Adv. Mater.,* 2018, **30**, 1–20.

[12]. S. K. Bhardwaj, N. Bhardwaj, R. Kaur, J. Mehta, A. L. Sharma, K. H. Kim and A. Deep, An overview of different strategies to introduce conductivity in metal-organic frameworks and miscellaneous applications thereof, *J. Mater. Chem. A,* 2018, **6**, 14992–15009.

[13]. S. Kempahanumakkagari, K. Vellingiri, A. Deep, E. E. Kwon, N. Bolan and K. H. Kim, Metal–organic framework composites as electrocatalysts for electrochemical sensing applications, *Coord. Chem. Rev.,* 2018, **357**, 105–129.

[14]. Y. Xue, S. Zheng, H. Xue and H. Pang, Metal-organic framework composites and their electrochemical applications, *J. Mater. Chem. A,* 2019, **7**, 7301–7327.

[15]. K. Y. Zou and Z. X. Li, Controllable Syntheses of MOF-Derived Materials, *Chem. - A Eur. J.,* 2018, **24**, 6506–6518.

[16]. A. Kirchon, L. Feng, H. F. Drake, E. A. Joseph, and H. C. Zhou, From fundamentals to applications: a toolbox for robust and multifunctional MOF materials, *Chem. Soc. Rev.,* 2018, **47**, 8611–8638.
[17]. W. Bai, S. Li, J. Ma, W. Cao and J. Zheng, Ultrathin 2D metal-organic framework (nanosheets and nanofilms)-based: X D-2D hybrid nanostructures as biomimetic enzymes and supercapacitors, *J. Mater. Chem. A,* 2019, **7**, 9086–9098.
[18]. Y. Liu, Y. Zhang, J. Chen and H. Pang, Copper metal-organic framework nanocrystal for plane effect nonenzymatic electro-catalytic activity of glucose, *Nanoscale,* 2014, **6**, 10989–10994.
[19]. L. Hua, Z. Hui, Y. Sun, X. Zhao, H. Xu, Y. Gong, R. Chen, C. Yu, J. Zhou, G. Sun and W. Huang, Oxygen vacancy enriched hollow cobaltosic oxide frames with ultrathin walls for efficient energy storage and biosensing, *Nanoscale,* 2018, **10**, 21006–21012.
[20]. Y. Zhou, C. Li, Y. Hao, B. Ye and M. Xu, Oriented growth of cross-linked metal-organic framework film on graphene surface for non-enzymatic electrochemical sensor of hydrogen peroxide in disinfectant, *Talanta,* 2018, **188**, 282–287.
[21]. C. H. Su, C. W. Kung, T. H. Chang, H. C. Lu, K. C. Ho and Y. C. Liao, Inkjet-printed porphyrinic metal–organic framework thin films for electrocatalysis, *J. Mater. Chem. A,* 2016, **4**, 11094–11102.
[22]. J. Yang, H. Ye, F. Zhao and B. Zeng, A Novel CuxO Nanoparticles@ZIF-8 Composite Derived from Core-Shell Metal-Organic Frameworks for Highly Selective Electrochemical Sensing of Hydrogen Peroxide, *ACS Appl. Mater. Interfaces,* 2016, **8**, 20407–20414.
[23]. S. Li, Y. Duan, S. Lei, J. Qiao, G. Li and B. Ye, A new electrochemical sensing strategy for echinacoside based on an original nanocomposite, *Sensors Actuators, B Chem.*, 2018, **274**, 218–227.
[24]. J. Zhang, D. Wang and Y. Li, Ratiometric Electrochemical Sensors Associated with Self-Cleaning Electrodes for Simultaneous Detection of Adrenaline, Serotonin, and Tryptophan, *ACS Appl. Mater. Interfaces*, 2019, **11**, 13557–13563.
[25]. Z. Wang, T. Liu, Y. Yu, M. Asif, N. Xu, F. Xiao and H. Liu, Coffee Ring–Inspired Approach toward Oriented Self-Assembly of Biomimetic Murray MOFs as Sweat Biosensor, *Small,* 2018, **14**, 1–11.
[26]. Z. Zhao, J. DIng, R. Zhu and H. Pang, The synthesis and electrochemical applications of core-shell MOFs and their derivatives, *J. Mater. Chem. A,* 2019, **7**, 15519–15540.
[27]. J. Ren, M. Ledwaba, N. M. Musyoka, H. W. Langmi, M. Mathe, S. Liao and W. Pang, Structural defects in metal–organic frameworks (MOFs): Formation, detection and control towards practices of interests, *Coord. Chem. Rev.,* 2017, **349** , 169–197.
[28]. B. Peng, J. Cui, Y. Wang, J. Liu, H. Zheng, L. Jin, X. Zhang, Y. Zhang and Y. Wu, CeO_2-: X/C/rGO nanocomposites derived from Ce-MOF and graphene oxide as a robust platform for highly sensitive uric acid detection, *Nanoscale,* 2018, **10**, 1939–1945.
[29]. H. S. Wang, Metal–organic frameworks for biosensing and bioimaging applications, *Coord. Chem. Rev.,* 2017, **349** , 139–155.
[30]. E. Gkaniatsou, C. Sicard, R. Ricoux, J. P. Mahy, N. Steunou and C. Serre, Metal-organic frameworks: A novel host platform for enzymatic catalysis and detection, *Mater. Horizons,* 2017, **4**, 55–63.
[31]. P. K. Robinson, Enzymes: principles and biotechnological applications, *Essays Biochem.*, 2015, **59**, 1–41.
[32]. J. Cui, S. Ren, B. Sun and S. Jia, Optimization protocols and improved strategies for metal-organic frameworks for immobilizing enzymes: Current development and future challenges, *Coord. Chem. Rev.,* 2018, **370**, 22–41.
[33]. Z. Wang, P. Dong, Z. Sun, C. Sun, H. Bu, J. Han, S. Chen and G. Xie, NH2-Ni-MOF electrocatalysts with tunable size/morphology for ultrasensitive C-reactive protein detection via an aptamer binding induced DNA walker-antibody sandwich assay, *J. Mater. Chem. B,* 2018, **6**, 2426–2431.
[34]. D. Feng, T. F. Liu, J. Su, M. Bosch, Z. Wei, W. Wan, D. Yuan, Y. P. Chen, X. Wang, K. Wang, X. Lian, Z. Y. Gu, J. Park, X. Zou and H. C. Zhou, Stable metal-organic frameworks containing single-molecule traps for enzyme encapsulation, *Nat. Commun.,* 2015, **6**, 1–8.
[35]. Y. Chen, V. Lykourinou, C. Vetromile, T. Hoang, L. J. Ming, R. W. Larsen, and S. Ma, How can proteins enter the interior of a MOF? investigation of cytochrome c translocation into a MOF consisting of mesoporous cages with microporous windows, *J. Am. Chem. Soc.,* 2012, **134**, 13188–13191.
[36]. V. Lykourinou, Y. Chen, X. Wang, L. Meng, T. Hoang, L. Ming, R. L. Musselman and S. Ma, Immobilization of MP-11 into a Mesoporous Metal–Organic Framework, MP-11@mesoMOF: A New Platform for Enzymatic Catalysis, *J. Am. Chem. Soc.,* 2011, **133**, 10382–10385.
[37]. Y. Cui, B. Li, H. He, W. Zhou, B. Chen and G. Qian, Metal-Organic Frameworks as Platforms for Functional Materials, *Acc. Chem. Res.,* 2016, **49**, 483–493.
[38]. J. Mehta, N. Bhardwaj, S. K. Bhardwaj, K. H. Kim and A. Deep, Recent advances in enzyme immobilization techniques: Metal-organic frameworks as novel substrates, *Coord. Chem. Rev.,* 2016, **322**, 30–40.
[39]. K. Liang, R. Ricco, C. M. Doherty, M. J. Styles, S. Bell, N. Kirby, S. Mudie, D. Haylock, A. J. Hill, C. J. Doonan and P. Falcaro, Biomimetic mineralization of metal-organic frameworks as protective coatings for biomacromolecules, *Nat. Commun.,* 2015, **6**, 4–11.
[40]. M. B. Majewski, H. Noh, T. Islamoglu and O. K. Farha, NanoMOFs: Little crystallites for substantial applications, *J. Mater. Chem. A,* 2018, **6**, 7338–7350.
[41]. J. Duan, Y. Li, Y. Pan, N. Behera and W. Jin, Metal-organic framework nanosheets: An emerging family of multifunctional 2D materials, *Coord. Chem. Rev.,* 2019, **395**, 25–45.
[42]. D. Liu, D. Zou, H. Zhu and J. Zhang, Mesoporous Metal–Organic Frameworks: Synthetic Strategies and Emerging Applications, *Small,* 2018, **14**, 1–40.
[43]. X. Q. Wu, J. G. Ma, H. Li, D. M. Chen, W. Gu, G. M. Yang and P. Cheng, Metal-organic framework biosensor with high stability and selectivity in a bio-mimic environment, *Chem. Commun.,* 2015, **51**, 9161–9164.
[44]. X. Gao, S. DelaCruz, C. Zhu, S. Cheng, D. Gardner, Y. Xie, C. Carraro and R. Maboudian, Surface functionalization of carbon cloth with cobalt-porphyrin-based metal organic framework for enhanced electrochemical sensing, *Carbon N. Y.,* 2019, **148**, 64–71.
[45]. N. S. Lopa, M. M. Rahman, F. Ahmed, T. Ryu, J. Lei, I. Choi, D. H. Kim, Y. H. Lee and W. Kim, A chemically and electrochemically stable, redox-active, and highly sensitive metal

azolate framework for non-enzymatic electrochemical detection of glucose, *J. Electroanal. Chem.,* 2019, **840**, 263–271.

[46]. P. Arul and S. A. John, Size controlled synthesis of Ni-MOF using polyvinylpyrrolidone: New electrode material for the trace level determination of nitrobenzene, *J. Electroanal. Chem.,* 2018, **829**, 168–176.

[47]. X. Chen and Q. Zhang, Recent advances in mesoporous metal-organic frameworks, *Particuology,* 2019, **45**, 20–34.

[48]. H. V. Doan, H. Amer Hamzah, P. Karikkethu Prabhakaran, C. Petrillo and V. P. Ting, *Hierarchical Metal–Organic Frameworks with Macroporosity: Synthesis, Achievements, and Challenges*, Springer Singapore, 2019, vol. 11.

[49]. H. Huang, J. R. Li, K. Wang, T. Han, M. Tong, L. Li, Y. Xie, Q. Yang, D. Liu and C. Zhong, An in-situ self-assembly template strategy for the preparation of hierarchical-pore metal-organic frameworks, *Nat. Commun.,* 2015, **6**, 1–8.

[50]. G. Cai and H. L. Jiang, A Modulator-Induced Defect-Formation Strategy to Hierarchically Porous Metal–Organic Frameworks with High Stability, *Angew. Chemie - Int. Ed.,* 2017, **56**, 563–567.

[51]. Y. Cao, Y. Ma, T. Wang, X. Wang, Q. Huo and Y. Liu, Facile Fabricating Hierarchically Porous Metal-Organic Frameworks via a Template-Free Strategy, *Cryst. Growth Des.,* 2016, **16**, 504–510.

[52]. M. Zhao, Y. Huang, Y. Peng, Z. Huang, Q. Ma and H. Zhang, Two-dimensional metal-organic framework nanosheets: Synthesis and applications, *Chem. Soc. Rev.,* 2018, **47**, 6267–6295.

[53]. S. Zhao, Y. Wang, J. Dong, C. T. He, H. Yin, P. An, K. Zhao, X. Zhang, C. Gao, L. Zhang, J. Lv, J. Wang, J. Zhang, A. M. Khattak, N. A. Khan, Z. Wei, J. Zhang, S. Liu, H. Zhao and Z. Tang, Ultrathin metal-organic framework nanosheets for electrocatalytic oxygen evolution, *Nat. Energy,* 2016, **1**, 1–10.

[54]. M. Zhao, Y. Wang, Q. Ma, Y. Huang, X. Zhang, J. Ping, Z. Zhang, Q. Lu, Y. Yu, H. Xu, Y. Zhao and H. Zhang, Ultrathin 2D Metal-Organic Framework Nanosheets, *Adv. Mater.,* 2015, **27**, 7372–7378.

[55]. M. Muschi and C. Serre, Progress and challenges of graphene oxide/metal-organic composites, *Coord. Chem. Rev.*, 2019, **387**, 262–272.

9 MOF Composites as Catalysts for Electrochemical Sensors

J. Antonio Cruz-Navarro[1,3], *L. Humberto Mendoza-Huizar*[1], *Verónica Salazar-Pereda*[1], *J. Ángel Cobos-Murcia*[2], *Fabiola Hernández-García*[2], *Raúl Colorado-Peralta*[3] *and Giaan A. Álvarez-Romero*[1]

[1]Universidad Autónoma del Estado de Hidalgo. Área Académica de Química. Carretera Pachuca-Tulancingo, Mineral de la Reforma. Hidalgo. México

[2]Universidad Autónoma del Estado de Hidalgo. Área Académica de Ciencias de la Tierra y Materiales. Carretera Pachuca-Tulancingo, Mineral de la Reforma. Hidalgo. México

[3]Facultad de Ciencias Químicas, Universidad Veracruzana, C.P. 94340 Orizaba, Veracruz, México

9.1 INTRODUCTION

Metal-organic frameworks (MOFs) are metal coordinated polymers obtained through the coordination of metal cluster nodes with organic ligands to generate unlimited 3D networks with multiple coordination topologies and environments [1]. In various reports, MOFs have been described as having a tetrahedral or octahedral coordination environment and interesting crystal shapes [2]. In recent years, MOFs have drawn attention due to the limitless applications of their porous structure. Specific molecules can be selectively adsorbed onto the pores of the MOF, without modifying the structure [3]. The application of MOFs and their derivatives in electroanalytical chemistry is a new and growing topic. Most of the work reported on MOFs in this field is devoted to energy applications [4]. However, at present, more and more attention is been directed towards evaluating the electrocatalytic performance of the transition metal MOF and their composites towards specific molecules in order to create smart electrochemical sensors.

Nowadays, the use of electrochemical sensors is becoming the main way to quantify important analytes (such as bio-molecules, drugs, nitroaromatic compounds, heavy metals, and the like). These sensors are modified electrodes that can be integrated into potentiostats or portable devices to detect analytes in real-time by using electroanalytical methods such as voltammetry or chronoamperometry. Electrochemical sensors have wide versatility, repeatability, extraordinary detection limits, and high sensitivity. The basis of electrochemical sensors is to use electroactive materials to modify the electrode surface and directly participate in the selective electro-oxidation or electro-reduction of molecules in the presence of specific electrolytes. Various types of materials (nanomaterials, porphyrin complexes, organic polymers, and the like) are being evaluated as electrode modifiers for the selective redox electrocatalysis of diverse molecules. In this sense, metal-organic frameworks have proven to be excellent candidates because they provide an extraordinary inner surface area (> 6000 m^2/g) [3], chemical resistance, electroactive behavior, selective absorption of molecules, and they improve the detection limit, sensitivity, and stability of electrochemical sensors [5]. Despite having interesting properties, their low overlap between electronic states and frontier orbits in MOFs induces low electron conductivity [6,7]. As a result, the use of pure MOFs for electrochemical applications is hardly reported [8-10]. To solve these conductivity problems, pristine MOFs are combined with nanomaterials or carbon nanomaterials to obtain highly conductive composites that can be used as active materials in electrochemical sensors, for energy production, or in storage devices [11-12]. The electrical analysis applications of MOFs are new and rapidly expanding. Therefore, in this chapter, we have summarized the use of MOF composites as catalysts in electrochemical sensing applications, hoping that the information found here will inspire further studies in this interesting field. The first part of this section is a general overview of MOF composites, their application in the fabrication of modified electrodes, and their evaluation as electroanalytical sensors. The second part concerns the main role of MOFs as electrocatalysts and their proposed redox mechanisms. Finally, in the third part, we converse about the examples of MOF composites used in electrochemical sensors for sensing biomolecules and water pollutants.

9.2 MOF COMPOSITES FOR ELECTROANALYTICAL APPLICATIONS

MOF composites are solid products composed of two or more components, which can maintain their performance and contribute to the entire system [13]. Generally, for electroanalysis applications, composite materials are manufactured by combining innovative materials with high conductivity and redox properties. These materials can increase the analytical signal, reduce the work potential, and induce a selective catalyst for the analyte, even in the presence

DOI: 10.1201/9781003188148-9

of other similar molecules. By recognizing the great potential that MOFs can offer in electroanalytical platforms, researchers began to synthesize diverse hybrid MOF composites based on carbon nanomaterials such as CNTs, fullerenes, graphite, graphene, graphene oxide, or reduced graphene oxide [14], amongst others. These smart composites combine the highly porous properties of MOFs with the high-electron conductivity and high-density active sites of the other materials which provide the bulk electrical conductivity inside the modified electrode [11] and which moderate the deficiency of each component in the composite, resulting in synergistic effects and attractive functions [15].

Other types of nanomaterials such as noble metal nanoparticles and transition metal-oxide nanoparticles are also used in the production of redox-active composites based on MOFs. In these NPs/MOF composites, the NPs are anchored inside the porous frameworks to provide selectivity and high sensitivity to the sensors. In addition, the main function of the nanoparticles in the MOF composite is to mediate the oxidation or reduction reactions towards specific analytes whereas the MOFs work only as support matrices.

9.3 ELABORATION OF ELECTROCHEMICAL SENSORS BASED ON MOF COMPOSITES: AN OVERVIEW

As mentioned before, electrochemical sensors are electrodes containing MOF composites or other materials that mediate the redox catalysis of diverse molecules in the modified electrode's surface. The challenge that researchers must face during the elaboration of electrochemical sensors based on MOF composites is how to load the MOF material on an accurate electrode. In literature, most of the reported MOF-based electrochemical sensors involve the use of glassy carbon electrodes (GCEs) due to their low cost, high conductivity and wide electrochemical window. The most common GCE modification process with MOF composites is carried out by the drop-coating process. In this process, the composite is dispersed in a solvent and the suspension obtained is cast on the polished surface of the electrode to create a layer that acts as a catalyst in electrochemical measurements. During the modification process, it is important to control the surface coverage concentration (Γ) of the MOF composite. Low values of Γ may cause high resistance in the electrochemical system related to the scarce conductivity of MOF materials and deficiencies in the electrochemical response.

Another kind of electrode used to immobilize MOF-based composites is carbon paste electrodes (CPEs). These electrodes are extensively employed in electrochemical sensors and are directly fabricated by using a wet paste fabricated from graphite powder, a binder agent, and a pristine MOF or a MOF composite [10,16]. The components are blended in a specific ratio to produce the paste which is packed into a plastic tube to create the electrode. Unlike GCE, the active surface of CPEs can be renewed several times, allowing the presence of fresh layers of MOF-composites. The same carbon paste electrode can be used in diverse experiments, therefore it is not necessary to prepare and modify new electrodes for each measurement. To evaluate the capabilities of the modified electrodes as electrochemical sensors, researchers evaluate the electrodes containing the pristine MOFs and the composites by cyclic voltammetry. The evaluation is carried out in the presence or absence of specific analytes at certain concentrations considering the effect of the pH and the effect of the proportion of MOF or non-MOF materials in the electrode. The evaluation aims to identify the best parameters to increase the current intensity of the peaks related to the electrocatalytic process of the analyte. Then, the effect of the analyte concentration is tested using the best features previously optimized. As a result, diverse analytical parameters such as the limit of detection (LOD), sensitivity, and linear range are improved during the design of electrochemical sensors based on MOF composites.

9.4 REDOX REACTION INDUCED BY METAL NODES IN MOF COMPOSITES: THE KEY ROLE OF THE METAL CENTER

Composites of MOFs, based on Zn(II) Fe(II), Co(II), Ni(II) and Cu(II) exhibit electrocatalytic activity in the presence of basic electrolytes (usually NaOH or KOH 0.1 M), where their metallic centers play a significant role in mediating redox reactions of organic molecules. In this respect, obtaining highly reactive oxyhydroxide species (MOOH, M = Fe(III), Co(III), Ni(III), or Cu(III)) from the metallic clusters in the MOF catalyze the analyte's oxidation or reduction at the electrode's surface in alkaline media, which helps to enhance current densities. The electro-oxidation process with MOF materials can be explained in terms of the Fleishman mechanism based on an M(III) species [16,17], which is represented in the following equations:

$$\text{M(II)-MOF} + 2\text{OH}^- \rightarrow \text{M(II)-MOF-(OH)}_2 \quad (1)$$

$$\text{M(II)-MOF-(OH)}_2 + \text{OH}^- \leftrightarrows \text{M(III)-MOF-OOH} + \text{H}_2\text{O} + e^- \quad (2)$$

$$\text{M(III)-MOF-OOH} + \text{analyte} \rightarrow \text{M(II)-MOF} + \text{P}_{\text{ox}} + e^- \quad (3)$$

In Eq. (1), Cu-BTC is coordinated with hydroxyl ions in the metallic center of the cluster node in the MOF, then, in Eq. (2), the MOF interacts with another hydroxyl ion to produce the high valence M(III) species. Then, in Eq. (3), the analyte is oxidized via the M(III) species to produce the hypothetical oxidation products (P_{ox}), and Cu(II)-BTC is newly regenerated. This process is easily observed during the electrochemical characterization of MOFs by using cyclic voltammetry, where the presence of anodic and cathodic peaks represent the redox pair [M(II)/ M(III)] expected for the MOF. Fleischmann's mechanism was initially proposed to explain the methanol electro-oxidation process with electrodes containing nickel or copper oxide. However, it has been extensively revised to explain the short-chain alcohols' electro-oxidation with coordination complexes based on Co(II), Ni(II) or Cu(II). In addition, in the last two years, Fleischmann's mechanism has been revised to explain the electro-oxidation process of glucose on non-enzymatic electrochemical sensors based on the MOF's composites. Other oxidation processes for other kinds of molecules catalyzed by MOFs have a similar explanation and will be discussed elsewhere in this chapter.

The changes in the MOF coordination environment during the electro-oxidation of analytes are still unclear [10, 16]. All reported information about MOFs as electrocatalysts in sensors only describes their electrochemical characterization and their application. However, there are no existing evaluations of changes of MOF structure during the electrocatalysis process. DFT calculations for [$(HOC_2H_4)_2$dtoaCu] in the presence of ethanol have demonstrated changes in the coordination environment due to strong interaction between the metallic center and the hydroxyl group, which suggests that the electro-oxidation reaction is carried out by the loss of an H from the hydroxyl group and is mediated by the metallic center [18]. In electro-reduction processes, the MOF's redox pair acts inversely. In this respect, the redox pair obtained could be M(II)/M(I). Most of the papers reporting the electro-reduction with MOFs are focused on the catalyst of hydrogen peroxide. In the following section, we will discuss the mechanism related to this process.

9.5 ELECTROCHEMICAL SENSORS BASED ON PRISTINE MOFs AND MOF COMPOSITES

At present, there is an increase in the use of electrochemical sensors based on MOFs for the electrochemical quantification of several kinds of analytes such as biomolecules, heavy metals and water pollutants. The results obtained with these electrodes are exceptional, presenting low detection limits and high sensitivities. New proposals of electrochemical sensors based on MOFs are monthly reported in important journals, and these proposals can be explored in the design of wearable or mobile analyzers. Nowadays, the electrochemical sensors based on MOFs are mainly focused on biomedical and environmental applications. In the next section, we discuss the examples of MOF-based electrochemical sensors and their application in these fields.

9.5.1 Biomedical Applications: Electrochemical Sensing of Molecules with Biological Importance

Research on electrochemical sensors for the rapid and exact quantification of biomolecules is growing continuously, and all projects are focused on their application in clinical chemistry and medical sciences. In this respect, most innovative designs involve the use of nanomaterials or inorganic materials in conjunction with enzymes to obtain smart electrochemical sensors that display low detection limits and high sensitivities. Some examples of MOF materials have been usefully used to immobilize enzymes in electrochemical sensors and increase their selectivity. Despite the excellent results obtained with enzyme-based electrodes, there is an enormous interest in the applicability of MOF composites and MOF derivatives with redox-active behavior for the design of non-enzymatic electrochemical sensors with enhanced analytical parameters. Several reports indicate that MOF materials have electrocatalytic activity towards glucose, hydrogen peroxide, uric acid, and ascorbic acid.

9.5.1.1 Detection of Glucose

The constant increase of diabetes mellitus cases in the world has led researchers to design more reliable and accurate sensors for the exact quantification of glucose using small devices. There are reported diverse proposals for smart and wearable devices based on enzymes for detecting glucose in biological fluids [19, 20]. Nonetheless, the implementation of enzymes on the electrode surface, their cost and their instability during electrochemical measurements make them a non-viable option in electrocatalysis. On the other hand, metallic oxides have been used widely to create non-enzymatic glucose sensors. Despite their efficiency for the determination of glucose in neutral media, some of these kinds of electrodes cannot be used for long-term analysis due to their instability [21]. In this respect, metal-organic frameworks based on Co(II), Ni(II) and Cu(II) represent novel and cheap materials for designing non-enzymatic electrodes due to their redox-active behavior in alkaline media. Nonetheless, despite the intensive research for new sensing platforms based on MOFs and increasing their performance, the poor conductance of MOFs limits their application on non-enzymatic sensors. Consequently, diverse types of nanomaterials are used in combination with MOFs to create smart composites with high redox activity towards glucose. For example, MOFs can be used as solid support to immobilize nanoparticles and produce a high conductive composite. In this respect, Yang [22] reported the use of AgNPs@Co-MOF composite that was loaded in a GCE to create a non-enzymatic electrochemical sensor that displayed high selectivity to glucose even in the presence of strong interferences such as ascorbic acid, urea, and sugar molecules. It was evident that the presence of AgNPs in framework increased the current intensity of the MOF. In the absence of glucose, the Co(III)/Co(II) redox appeared at +0.24 and -0.55 V correspondingly and increased in the presence of glucose, indicating moderate electrochemical catalytic activity toward glucose oxidation. The high analytical performance of the electrode showed an LOD and a sensitivity of 1.32 μM and 0.135 μA cm^{-1}. Carbon nanotubes have also been used in the design of MOF composites for non-enzymatic sensors. CNTs have several features such as good conductivity and high surface area that make them attractive for MOF composites. Diverse authors [23-25] report the use of CNTs@MOF composites in the design of electrodes for amperometric detection of glucose. The sensors have displayed detection limits in the micromolar order (from 0.4 to 4.6 μM) and excellent selectivity and sensitivity. In comparison with the results obtained with pristine MOFs, the current intensity of MOF composites is enhanced as a result of the presence of carbon nanotubes.

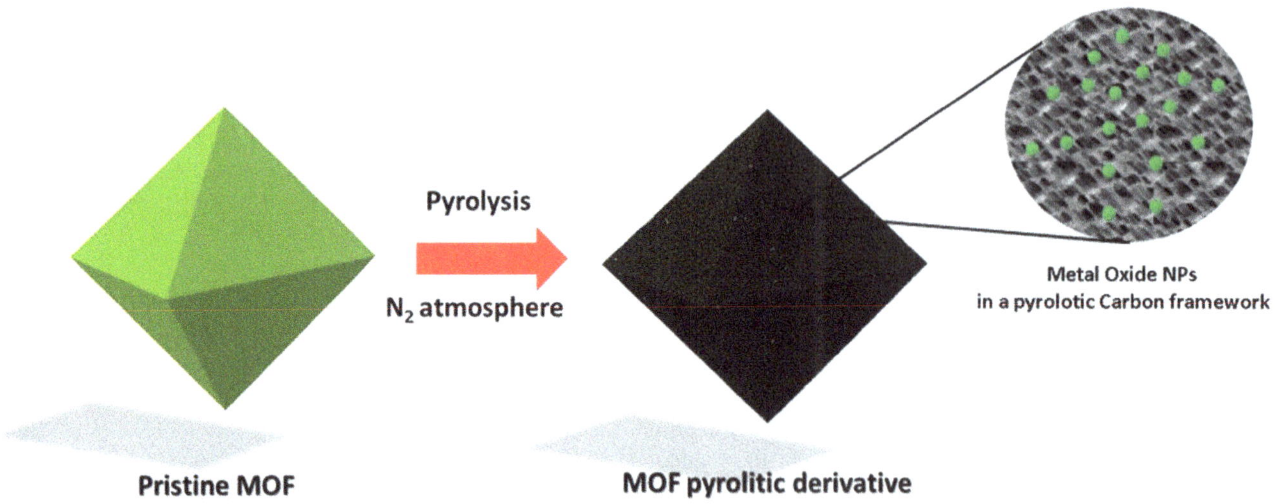

FIGURE 9.1 Schematic exemplification of MOF pyrolysis to obtain redox-active pyrolytic frameworks.

At present, there is a tendency to use MOFs as innovative sacrifice platforms to obtain redox-active nanoparticles of mono or bimetallic oxides embedded in pyrolytic carbon frameworks (Figure 9.1). Several authors have reported the implementation of these composites on non-enzymatic electrochemical sensors for glucose [26-31]. These composites improve mass and electron transfer processes during glucose oxidation due to their elevated surface area. For these purposes, the pyrolytic carbon framework must keep the original MOF structure [32]. The obtained pyrolitic composites are loaded directly on the surface of GCEs and act as an efficient redox catalyst for glucose electro-oxidation, increasing the anodic peak current intensity in comparison to electrodes based only on MOF materials. For example, Archana [29] and coworkers reported the use of a CuNi-BTC MOF as a sacrifice material to obtain CuO/NiO spherical nanoparticles embedded in hierarchical carbon. The composite loaded in a sellotape film displayed an enhanced current intensity (0.6 mA) in the presence of 5.0 mM glucose in comparison with the electrode modified only with the bimetallic MOF that displayed a low current intensity (0.25 mA). The value of the sensitivity is 586.7 µA and the LOD value is 37 nM. These results are superior compared to those presented only with MOF composites.

As represented in equations 4 and 5, and Figure 9.2, the glucose oxidation mechanism on the surface of MOF non-enzymatic sensors is based on the formation of metal oxyhydroxides (MOOH) that catalyze the reaction:

$$\text{M(II)-MOF-(OH)}_2 + \text{OH}^- \leftrightharpoons \text{M(III)-MOF-OOH} + \text{H}_2\text{O} + \text{e}^- \quad (4)$$

$$\text{M(III)-MOF-OOH} + \text{glucose} \rightarrow \text{M(II)-BTC} + \text{glucoronic acid} \quad (5)$$

In equation 4, the M(II) inside the MOF is oxidized in alkaline to produce highly reactive M(III) species that generate the presence of an anodic peak current in voltammograms; then,

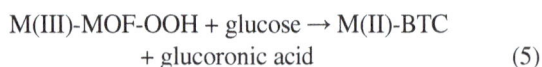

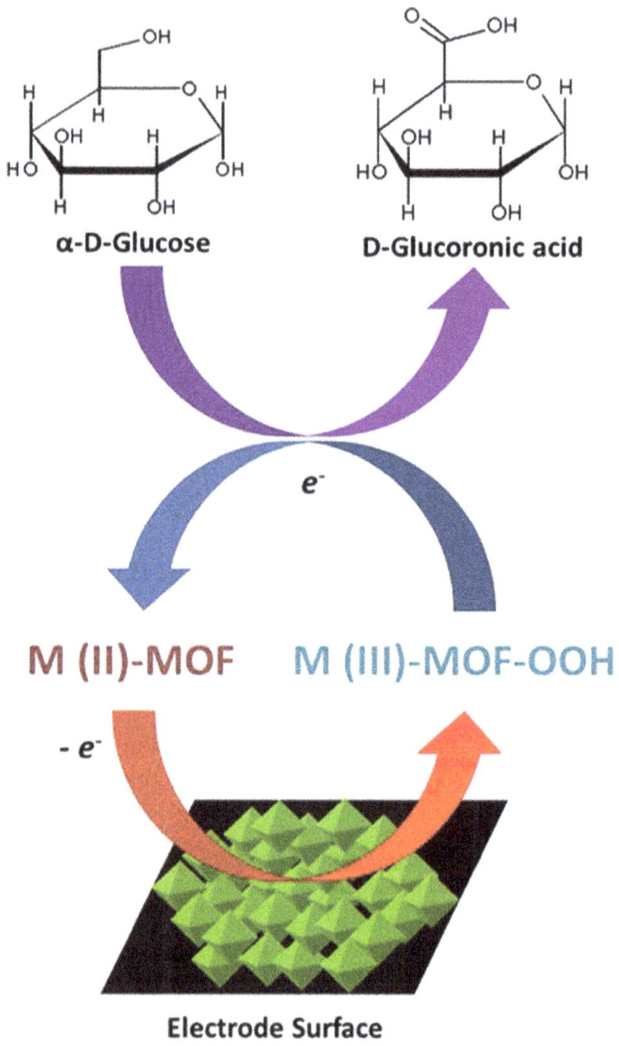

FIGURE 9.2 Graphic representation of the oxidation mechanism of glucose in the presence of high valence M(III)-MOF species, where M represents Co, Ni or Cu ions.

in equation 5, the M(III) species oxidize glucose ($C_6H_{12}O_6$) to gluconolactone ($C_6H_{12}O_7$), this process increases the anodic peak current intensity and after that, M(II) is regenerated again.

9.5.1.2 Detection of Hydrogen Peroxide

Hydrogen peroxide (H_2O_2) is a small molecule that plays a significant role in the defense of the organism. However, the presence of high amounts in the body induces damage to cell proliferation or Alzheimer's disease. H_2O_2 is also a by-product obtained as a result of the oxidation of glucose by oxidase enzymes and is employed on indirect quantification of glucose by titrimetry or fluorometry. Recently, there is an increasing interest in the design of hydrogen peroxide sensors for the indirect electrochemical detection and quantification of glucose. Nevertheless, this application requires the use of enzyme-based electrochemical sensors, which are expensive, unstable and difficult to calibrate. Nowadays, there is a growing interest in the design of non-enzymatic sensors for the detection of H_2O_2 in alkaline media, and materials based on transition metals are playing an important role as active materials in enzyme-free electrodes. MOF materials containing transition metals such as Fe(III), Co(II), Ni(II), Cu(II), allow the selective electrocatalysis of hydrogen peroxide in alkaline and neutral media. These MOFs are used to create conductive composites with metallic gold or silver NPs [33-35] where the NPs mediate the electrocatalysis of hydrogen peroxide. In the electrode surface, hydrogen peroxide is decomposed into H_2O and O_2 in the presence of the metallic nanoparticles, and after that, an electro-reduction of O_2 yields H_2O as the main product. In the composite, the main function of the MOFs is capturing H_2O_2 in their structure. Other composites used on non-enzymatic peroxide sensors are elaborated with carbon-based materials like CNTs [36] or reduced graphene oxide [37-38] to enhance selectivity and cathodic current intensity. Additionally, pristine MOFs have been successfully applied in GC or carbon paste electrodes for detecting hydrogen peroxide.

In the literature about MOF-based electrochemical sensors for H_2O_2 detection, the electrocatalytic processes for hydrogen peroxide are based on its reduction. In this case, the formation of a redox pair (M(III)/M(II) or M(II)/M(I)) inside the MOF mediates the electro-reduction reaction (Figure 9.3). In the mechanism, cathodic sweep reduces M(III) or M(II) to M(II) or M(I), then H_2O_2 reacts with the metallic reduced species to produce H_2O in the cathodic potential. Finally, the reduced metallic species are oxidized via H_2O_2 to their initial state. By cyclic voltammetry, this process is usually perceived as two peaks (anodic and cathodic), the intensity of which tends to be modified in the presence of diverse peroxide concentrations. For example, Yang and coworkers [37] reported the use of Fe(III)-MOF as a catalyst in a hydrogen peroxide sensor. According to the authors, the electro-reduction process is carried out as follow:

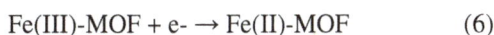

$$\text{Fe(III)-MOF} + e^- \rightarrow \text{Fe(II)-MOF} \quad (6)$$

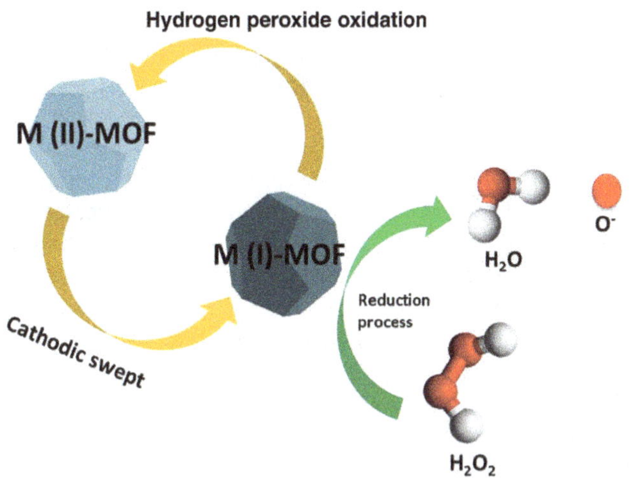

FIGURE 9.3 Graphical representation of the hydrogen peroxide reduction via M(II)-MOF

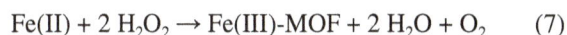

$$\text{Fe(II)} + 2\,H_2O_2 \rightarrow \text{Fe(III)-MOF} + 2\,H_2O + O_2 \quad (7)$$

In equation (6), the Fe(III)-MOF is electrically reduced to Fe(II)-MOF during the cathodic sweep; then, in equation (7), the Fe(II)-MOF reduces the H_2O_2 and returns to its original oxidation state Fe(III). Other examples of these mechanisms are reported by Naseri [39] and Meing [40]. In both works, Cu(II)-MOFs are used as active materials on non-enzymatic amperometric sensors for hydrogen peroxide detection. The redox couple Cu(II)/Cu(I) obtained induce the reduction of H_2O_2 as follow:

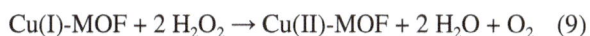

$$\text{Cu(II)-MOF} + e^- \rightarrow \text{Cu(I)-MOF} \quad (8)$$

$$\text{Cu(I)-MOF} + 2\,H_2O_2 \rightarrow \text{Cu(II)-MOF} + 2\,H_2O + O_2 \quad (9)$$

In this case, in equation (8), Cu(II) is reduced to Cu(I) at the cathodic sweep and then in equation (9), H_2O_2 oxidizes Cu(I) to produce Cu(II) again whereas H_2O_2 is reduced to water and oxygen.

9.5.1.3 Detection of other Biomolecules

Most of the electrochemical sensors based on MOFs are focused on glucose or hydrogen peroxide sensing applications. However, due to the redox-active properties of MOFs, there are several reports about MOF-based electrochemicals for quantifying important biomolecules like amino acids, cysteine, dopamine, steroids, among others. An interesting example is an electrochemical sensor based on the composite Pt NPs/MIL-101[41] for the simultaneous quantification of dopamine, uric acid and xanthine by differential pulse voltammetry. In the sensor, the MIL-101 MOF acts as a host for the metallic nanoparticles and helps in the recognition of the molecules, whereas the nanoparticles increase conductivity and mediate the oxidation of the three analytes. Another example is the modification of a NACP film electrode with the hybrid composite Ni-BDC/AuNPs/

MWCNTs [42]. The modified electrode was employed in the chronoamperometric determination of dopamine from living C6 cells, presenting a linear range of 50 nM to 15μM, and high sensitivity, with a value of 1250 mA M^{-1} cm^{-2}. The authors did not report the mechanism; nonetheless, it is probably that the Ni-BDC MOF mediated the electro-oxidation of dopamine whereas the AuNPs and the carbon nanotubes increased the conductivity. These results are superior compared with those results obtained with a sensor based on a pristine bimetallic Fe_2NiMIL-88B that showed a sensitivity of 124.7 μA mM^{-1} cm^{-2}. For detecting estradiol, Duan and coworkers reported the use of the composite MIL-53/CNTs in combination with a polymer to create a highly sensitive electrode towards estradiol oxidation at 0.2 V presenting a low detection limit of 6.19 $\times 10^{-9}$ μM. A similar combination of molecular imprinted polymer and a MOF-5/CNTs composite was used in an electrochemical sensor for detecting L and D-cysteine enantiomers, with a detection limit of 6 $\times 10^{-9}$ μM. In both cases, the oxidation mechanism was not reported.

9.5.2 Environmental Applications: Electrochemical Sensing of Molecules with Biological Importance

Rapid analysis of the presence of organic compounds and heavy metals in water using inexpensive techniques is an important topic for environmental sciences. Several compounds such as heavy metal salts, pharmaceuticals or phenolic compounds are commonly found in waste water and even in tap water. These pollutants are mainly introduced in drain systems as a result of industrial activity in alimentary, chemical or pharmaceutical industries. Their presence in water effluents represents a threat to ecosystems and a serious environmental problem due to their elevated toxicity and accumulation in living beings. A typical analysis for pollutant determination in water requires costly instrumental methods such as atomic absorption spectroscopy, chromatography or mass spectrometry. As a result, there is an increase in the development and design of electrochemical sensors focused on the quantification of heavy metals and specific organic compounds. Diverse materials such as molecular impressed polymers and nanocomposites are used as active compounds in these electrochemical sensors, and at present, materials based on transition metals are being explored.

9.5.2.1 Detection of Heavy Metals

Recently, MOF-composites are being incorporated in electrodes for the selective or simultaneous detection of As(III) [43], Cd(II) [44], Cu(II) [45], and Pb(II) [46,47]. Most works in the literature reported the use of MOF composites based on CNTs and graphene derivatives incorporated on carbon paste electrodes. The exclusive selection of carbon paste electrodes is due to their absorptive properties that facilitate to pre-concentrate heavy metal ions on the electrode surface. As a result, diverse electrochemical techniques such as DPV, chronoamperometry and stripping voltammetry can be easily used for efficient quantification of the pre-concentrated heavy metals. An interesting example of the application of a MOF composite for heavy metal detection is reported by Sangeetha and coworkers[44]. In their work, a composite of MOF-177 and polytetrafluoroethylene was incorporated in a carbon paste electrode for the simultaneous quantification of Cd(II) and Pb(II). Chronoamperometric evaluation of the composite in a pH 7 phosphate buffer and the presence of both heavy metal ions in solution indicated an enhancement of the anodic current intensity when Cd(II) or Pb(II) were present in the electrolyte. The linear range obtained was 10-120 μM whereas the LOD was 0,004 and 0,03 μM for Cd(II) and Pb(II) respectively [44].

Another interesting example is the elaboration of a NH_2-$Cu_3(BTC)_2$ MOF and its incorporation on a GCE for detecting Pb(II). The NH_2-$Cu_3(BTC)_2$/GCE showed a strong interaction between the amino groups in the MOF and the Pb(II) ions, which allow the pre-concentration the metal ions. The determination of Pb(II) was performed by anodic stripping voltammetry at 0.6 V in pH 4.5, presenting a low detection limit of 5 nM. Similar results and behavior were obtained by Wang and coworkers [45]. In their work, they also reported the use of an amino-MOF (NH_2-UiO-66) functionalized with ferrocene carboxylic acid and reduced graphene oxide in a GCE. The modified electrode was used for the detection of Cd(II), Pb(II), and Cu(II) by anodic stripping voltammetry. The MOF showed the same interaction of amine groups as an LOD of 8.5, 0.6, and 0.8 nM for Cd(II), Pb(II), and Cu(II) respectively. In all works related to the determination of heavy metals by using MOF-based electrodes, authors did not report the electro-oxidation mechanism; nonetheless, a redox pair in MOFs may mediate the electrocatalysis. Despite the advantages that MOFs could offer in electroanalysis, the determination of heavy metals by using MOF-composite-based electrodes is a new trend, and at present, few reports can be found in the literature.

9.5.2.2 Detection of Aromatic Compounds

Electrochemical sensors based on metal-coordinated materials for the direct quantification of phenolic compounds in aqueous media are extensively reported in the literature [48]. MOF-composite-based electrodes have taken an interesting role in the recent advance of sensors applied to environmental sciences and specifically, in the determination of organic pollutants in water. The absorptive properties of MOFs and their composites are essential for the selective or simultaneous oxidation or reduction of these compounds. Diverse MOF composites have been employed on sensors for detecting hydroquinone and catechol, which are compounds widely used in the elaboration of dyes, cosmetics and pharmaceuticals. The determination of these compounds is difficult due to their similarity and MOF-based electrochemical sensors have presented excellent results in the simultaneous determination of both aromatic compounds. For example, Zhan and coworkers [49] reported the synthesis of a pyrolytic framework composite derived from ZIF-67 and polyacrylonitrile. The resulting composite (C-ZIF-67/PAN-800) was loaded in GCE and evaluated by DPV in the presence of catechol and hydroquinone. The electrode showed a linear range of 1-120 μM and 1-200 μM for hydroquinone and catechol respectively, and the detection limit of 1.0 μM for both compounds. More efficient results were presented by Yang and coworkers [50]. In their work, a composite elaborated with Ni-BTC and MoS_2 sheets was used to produce MoS_2/$Ni(OH)_2$ that was loaded in a GCE. The modified electrode showed a strong activity toward both molecules, presenting linear ranges of 1-500 μM and 2-400 μM, and detection limits of 0.43 and 0.48 μM for hydroquinone and catechol respectively.

Overall, MOF composites show promising results with detection limits in the order of micro and nanomoles, thus allowing the detection of the presence of organic compounds in small concentrations. The constant improvement of this class of electrochemical sensors is creating a new pathway for the electroanalytical chemistry in environmental sciences.

9.6 CONCLUSIONS

The recent incorporation of MOFs in the electrochemical field is new, attractive, and promising. As demonstrated before, the presence of active metal centers in their framework is the key for the electrocatalytic mechanism in the presence of alkaline or acidic media, which has benefited the development of new electrochemical sensors with the capability of the selective electrocatalyst toward diverse molecules with biological or environmental interest. The use of pristine MOFs in electrochemical application is scarce; the main reason is the poor conductivity of these materials. Therefore, current research on the electrochemical application of MOFs is focused on the design of hybrid composites elaborated with carbon nanomaterials or metal NPs to obtain highly conductive and reactive composites that could be applied to electrochemical sensors or energy storage devices. Other composite materials elaborated from MOFs result from direct pyrolysis to obtain carbonaceous frameworks embedded with metallic nanoparticles or metallic oxides derived from the metal clusters. These materials are often found on glucose or hydrogen peroxide electrochemical sensors; nonetheless, researchers are exploring their application for sensing other kinds of molecules. The efficiency of MOF-based composites has been demonstrated in their incorporation on the non-enzymatic electrochemical sensor and other kinds of sensor that display limits in detection of nano and micromolar order.

REFERENCES

[1]. Zhou, H.-C.; Long, J. R.; Yaghi, O. M. Introduction to Metal–Organic Frameworks. *Chem. Rev.* **2012**, *112* (2), 673–674.

[2]. Kalmutzki, M. J.; Hanikel, N.; Yaghi, O. M. Secondary Building Units as the Turning Point in the Development of the Reticular Chemistry of MOFs. *Sci. Adv.* **2018**, *4* (10), 9180.

[3]. Yaghi, O. M.; Kalmutzki, M. J.; Diercks, C. S. Building units of MOFs. In *Introduction to Reticular Chemistry*; John Wiley & Sons, Ltd: Chichester, UK, **2019**, pp 57–81.

[4]. Wang, H.; Zhu, Q.-L.; Zou, R.; Xu, Q. Metal-Organic Frameworks for Energy Applications. *Chem* **2017**, *2* (1), 52–80.

[5]. Liu, L.; Zhou, Y.; Liu, S.; Xu, M. The Applications of Metal–Organic Frameworks in Electrochemical Sensors. *ChemElectroChem* **2018**, *5* (1), 6–19.

[6]. Morozan, A.; Jaouen, F. Metal Organic Frameworks for Electrochemical Applications. *Energy Environ. Sci.* **2012**, *5*, 9269–9290.

[7]. Jaouen, F.; Morozan, A. Metal-Organic Frameworks: Electrochemical Properties. In *Encyclopedia of Inorganic and Bioinorganic Chemistry*; John Wiley & Sons, Ltd: Chichester, UK, **2014**; pp 1–24.

[8]. Zhao, Y.; Song, Z.; Li, X.; Sun, Q.; Cheng, N.; Lawes, S.; Sun, X. Metal Organic Frameworks for Energy Storage and Conversion. *Energy Storage Mater.* **2016**, *2*, 35–62.

[9]. Liu, W.; Yin, X.-B. Metal–Organic Frameworks for Electrochemical Applications. *TrAC Trends Anal. Chem.* **2016**, *75*, 86–96.

[10]. Cruz-Navarro, J. A.; Hernandez-Garcia, F.; Alvarez Romero, G. A. Novel Applications of Metal-Organic Frameworks (MOFs) as Redox-Active Materials for Elaboration of Carbon-Based Electrodes with Electroanalytical Uses. *Coord. Chem. Rev.* **2020**, *412*, 213263.

[11]. Chuang, C. H.; Kung, C. W. Metal–Organic Frameworks toward Electrochemical Sensors: Challenges and Opportunities. *Electroanalysis.* **2020**, *32* (9), 1885–1895.

[12]. Liu, C. Sen; Li, J.; Pang, H. Metal-Organic Framework-Based Materials as an Emerging Platform for Advanced Electrochemical Sensing. *Coord. Chem. Rev.* **2020**. *410*, 213222

[13]. Li, S.; Huo, F. Metal-Organic Framework Composites: From Fundamentals to Applications. *Nanoscale.* **2015**, pp 7482–7501.

[14]. Kempahanumakkagari, S.; Vellingiri, K.; Deep, A.; Kwon, E. E.; Bolan, N.; Kim, K. H. Metal–Organic Framework Composites as Electrocatalysts for Electrochemical Sensing Applications. *Coord. Chem. Rev.* **2018**, *357* , 105–129.

[15]. Yi, F.-Y.; Zhang, R.; Wang, H.; Chen, L.-F.; Han, L.; Jiang, H.-L.; Xu, Q. Metal-Organic Frameworks and Their Composites: Synthesis and Electrochemical Applications. *Small Methods* **2017**, *1* (11), 1700187.

[16]. Cruz-Navarro, J. A.; Mendoza-Huizar, L. H.; Salazar-Pereda, V.; Cobos-Murcia, J. Á.; Colorado-Peralta, R.; Álvarez-Romero, G. A. Progress in the Use of Electrodes Modified with Coordination Compounds for Methanol Electro-Oxidation. *Inorg. Chim. Acta* **2021**, *520*, 120293.

[17]. Fleischmann, M.; Korinek, K.; Pletcher, D. The Kinetics and Mechanism of the Oxidation of Amines and Alcohols at Oxide-Covered Nickel, Silver, Copper, and Cobalt Electrodes. *J. Chem. Soc. Perkin Trans. 2* **1972**, No. 10, 1396.

[18]. Yang, L.; Kinoshita, S.; Yamada, T.; Kanda, S.; Kitagawa, H., Tokunaga, M., Ishimoto, T.; Ogura, T.; Nagumo, R.; Miyamoto, A.; Koyama, M. A Metal-Organic Framework as an Electrocatalyst for Ethanol Oxidation. *Angew. Chemie Int. Ed.* **2010**, *49* (31), 5348–5351.

[19]. Luo, X.; Shi, W.; Liu, Y.; Sha, P.; Chu, Y.; Cui, Y. A Smart Tongue Depressor-Based Biosensor for Glucose. *Sensors* **2019**, *19* (18), 3864.

[20]. Vargas, E.; Teymourian, H.; Tehrani, F.; Eksin, E.; Sánchez-Tirado, E.; Warren, P.; Erdem, A.; Dassau, E.; Wang, J. Enzymatic/Immunoassay Dual-Biomarker Sensing Chip: Towards Decentralized Insulin/Glucose Detection. *Angew. Chemie Int. Ed.* **2019**, *58* (19), 6376–6379.

[21]. Arul, P.; Abraham J., S. Electrodeposition of CuO from Cu-MOF on Glassy Carbon Electrode: A Non-Enzymatic Sensor for Glucose. *J. Electroanal. Chem.* **2017**, *799*, 61–69.

[22]. Liu, Y.; Shi, W.-J.; Lu, Y.-K.; Liu, G.; Hou, L.; Wang, Y.-Y. Nonenzymatic Glucose Sensing and Magnetic Property Based on the Composite Formed by Encapsulating Ag Nanoparticles in Cluster-Based Co-MOF. *Inorg. Chem.* **2019**, *58* (24), 16743–16751.

[23]. Zhang, X.; Xu, Y.; Ye, B. An Efficient Electrochemical Glucose Sensor Based on Porous Nickel-Based Metal Organic Framework/Carbon Nanotubes Composite (Ni-MOF/CNTs). *J. Alloys Compd.* **2018**, *767*, 651–656.

[24]. Wang, F.; Chen, X.; Chen, L.; Yang, J.; Wang, Q. High-Performance Non-Enzymatic Glucose Sensor by Hierarchical Flower-like Nickel(II)-Based MOF/Carbon Nanotubes Composite. *Mater. Sci. Eng. C* **2019**, *96*, 41–50.

[25]. Wu, L.; Lu, Z.; Ye, J. Enzyme-Free Glucose Sensor Based on Layer-by-Layer Electrodeposition of Multilayer Films of Multi-Walled Carbon Nanotubes and Cu-Based Metal Framework Modified Glassy Carbon Electrode. *Biosens. Bioelectron.* **2019**, *135*, 45–49.

[26]. Arul, P.; Abraham John, S. Electrodeposition of CuO from Cu-MOF on Glassy Carbon Electrode: A Non-Enzymatic Sensor for Glucose. *J. Electroanal. Chem.* **2017**, *799*, 61–69.

[27]. Zhang, L.; Ye, C.; Li, X.; Ding, Y.; Liang, H.; Zhao, G.; Wang, Y. A CuNi/C Nanosheet Array Based on a Metal–Organic Framework Derivate as a Supersensitive Non-Enzymatic Glucose Sensor. *Nano-Micro Lett.* **2018**, *10* (28), 1–10.

[28]. Chen, J.; Xu, Q.; Shu, Y.; Hu, X. Synthesis of a Novel Au Nanoparticles Decorated Ni-MOF/Ni/NiO Nanocomposite and Electrocatalytic Performance for the Detection of Glucose in Human Serum. *Talanta* **2018**, *184*, 136–142.

[29]. Archana, V.; Xia, Y.; Fang, R.; Gnana kumar, G. Hierarchical CuO/NiO-Carbon Nanocomposite Derived from Metal Organic Framework on Cello Tape for the Flexible and High Performance Nonenzymatic Electrochemical Glucose Sensors. *ACS Sustain. Chem. Eng.* **2019**, *7* (7), 6707–6719.

[30]. Feng, Y.; Xiang, D.; Qiu, Y.; Li, L.; Li, Y.; Wu, K.; Zhu, L. MOF-Derived Spinel $NiCo_2O_4$ Hollow Nanocages for the Construction of Non-enzymatic Electrochemical Glucose Sensor. *Electroanalysis* **2020**, *32* (3), 571–580.

[31]. Ma, X.; Tang, K.L.; Yang, M.; Shi, W.; Zhao, W. Metal–Organic Framework-Derived Yolk–Shell Hollow Ni/NiO@C Microspheres for Bifunctional Non-Enzymatic Glucose and Hydrogen Peroxide Biosensors. *J. Mater. Sci.* **2021**, *56*, 442–456.

[32]. Yang, L.; Zeng, X.; Wang, W.; Cao, D. Recent Progress in MOF-Derived, Heteroatom-Doped Porous Carbons as Highly Efficient Electrocatalysts for Oxygen Reduction Reaction in Fuel Cells. *Adv. Funct. Mater.* **2018**, *28* (7), 1704537.

[33]. Dang, W.; Sun, Y.; Jiao, H.; Xu, L.; Lin, M. AuNPs-NH2/Cu-MOF Modified Glassy Carbon Electrode as Enzyme-Free Electrochemical Sensor Detecting H_2O_2. *J. Electroanal. Chem.* **2020**, *856*, 113592.

[34]. Zhang, T.; Xing, Y.; Song, Y.; Gu, Y.; Yan, X.; Lu, N.; Liu, H.; Xu, Z.; Xu, H.; Zhang, Z.; Yang, M. AuPt/MOF-Graphene: A Synergistic Catalyst with Surprisingly High Peroxidase-like Activity and Its Application for H_2O_2 Detection. *Anal. Chem.* **2019**, *91* (16), 10589–10595.

[35]. Arul, P.; John, S. A. Silver Nanoparticles Built-in Zinc Metal Organic Framework Modified Electrode for the Selective Non-Enzymatic Determination of H_2O_2. *Electrochim. Acta* **2017**, *235*, 680–689.

[36]. Rani, S.; Sharma, B.; Malhotra, R.; Kumar, S.; Varma, R. S.; Dilbaghi, N. Sn-MOF@CNT Nanocomposite: An Efficient Electrochemical Sensor for Detection of Hydrogen Peroxide. *Environ. Res.* **2020**, *191*, 110005.

[37]. Yang, S. Facile Synthesis of Fe-MOF/RGO Nanocomposite as an Efficient Electrocatalyst for Nonenzymatic H2O2 Sensing. *Int. J. Electrochem. Sci.* **2019**, *14* , 7703–7716.

[38]. Li, C.; Wu, R.; Zou, J.; Zhang, T.; Zhang, S.; Zhang, Z.; Hu, X.; Yan, Y.; Ling, X. MNPs@anionic MOFs/ERGO with the Size Selectivity for the Electrochemical Determination of H_2O_2 Released from Living Cells. *Biosens. Bioelectron.* **2018**, *116* , 81–88.

[39]. Naseri, M.; Fotouhi, L.; Ehsani, A. Nanostructured Metal Organic Framework Modified Glassy Carbon Electrode as a Highly Efficient Non-Enzymatic Amperometric Sensor for Electrochemical Detection of H_2O_2. *J. Electrochem. Sci. Technol.* **2018**, *9* (1), 28–36.

[40]. Meng, W.; Xu, S.; Dai, L.; Li, Y.; Zhu, J.; Wang, L. An Enhanced Sensitivity towards H_2O_2 Reduction Based on a Novel Cu Metal–Organic Framework and Acetylene Black Modified Electrode. *Electrochim. Acta* **2017**, *230*, 324–332.

[41]. Zhang, L.; Li, S.; Xin, J.; Ma, H.; Pang, H.; Tan, L.; Wang, X. A Non-Enzymatic Voltammetric Xanthine Sensor Based on the Use of Platinum Nanoparticles Loaded with a Metal-Organic Framework of Type MIL-101 (Cr). Application to Simultaneous Detection of Dopamine, Uric Acid, Xanthine and Hypoxanthine. *Chinese Chem. Lett.* **2018**, *2*, 4–13.

[42]. Shu, Y.; Lu, Q.; Yuan, F.; Tao, Q.; Jin, D.; Yao, H.; Xu, Q.; Hu, X. Stretchable Electrochemical Biosensing Platform Based on Ni-Mof Composite/Au Nanoparticle-Coated Carbon Nanotubes for Real-Time Monitoring of Dopamine Released from Living Cells. *ACS Appl. Mater. Interfaces* **2020**, *12* (44), 49480–49488.

[43]. Baghayeri, M.; Ghanei-Motlagh, M.; Tayebee, R.; Fayazi, M.; Narenji, F. Application of Graphene/Zinc-Based Metal-Organic Framework Nanocomposite for Electrochemical Sensing of As(III) in Water Resources. *Anal. Chim. Acta* **2020**, *1099*, 60–67.

[44]. Sangeetha, S.; Krishnamurthy, G. Fabrication of MOF-177 for Electrochemical Detection of Toxic Pb(II) and Cd(II) Ions. *Bull. Mater. Sci.* **2020**, *43* (1), 29.

[45]. Wang, X.; Qi, Y.; Shen, Y.; Yuan, Y.; Zhang, L.; Zhang, C.; Sun, Y. A Ratiometric Electrochemical Sensor for Simultaneous Detection of Multiple Heavy Metal Ions Based on Ferrocene-Functionalized Metal-Organic Framework. *Sensors Actuators B Chem.* **2020**, *310*, 127756.

[46]. Wang, Y.; Wu, Y.; Xie, J.; Hu, X. Metal-Organic Framework Modified Carbon Paste Electrode for Lead Sensor. *Sensors Actuators, B Chem.* **2013**, *177* , 1161–1166.

[47]. Wang, Y.; Ge, H.; Wu, Y.; Ye, G.; Chen, H.; Hu, X. Construction of an Electrochemical Sensor Based on Amino-Functionalized Metal-Organic Frameworks for Differential Pulse Anodic Stripping Voltammetric Determination of Lead. *Talanta* **2014**, *129*, 100–105. I

[48]. Cruz-Navarro, J. A.; Hernández-García, F.; Mendoza-Huizar, L. H.; Salazar-Pereda, V.; Cobos-Murcia, J. Á.;

Colorado-Peralta, R.; Álvarez-Romero, G. A. Recent Advances in the Use of Transition-Metal Porphyrin and Phthalocyanine Complexes as Electro-Catalyst Materials on Modified Electrodes for Electroanalytical Sensing Applications. *Solids* **2021**, *2* (2), 212–231.

[49]. Zhang, M.; Li, M.; Wu, W.; Chen, J.; Ma, X.; Zhang, Z.; Xiang, S. MOF/PAN Nanofiber-Derived N-Doped Porous Carbon Materials with Excellent Electrochemical Activity for the Simultaneous Determination of Catechol and Hydroquinone. *New J. Chem.* **2019**, *43* (9), 3913–3920.

[50]. Yang, W.; Guo, H.; Fan, T.; Zhao, X.; Zhang, L.; Guan, Q.; Wu, N.; Cao, Y.; Yang, W. MoS_2/Ni$(OH)_2$ Composites Derived from in Situ Grown Ni-MOF Coating MoS_2 as Electrode Materials for Supercapacitor and Electrochemical Sensor. *Colloids Surfaces A Physicochem. Eng. Asp.* **2021**, *615*, 126178.

10 Recent Advancement and Challenges in MOF-based Electrochemical Sensors

Sharanyakanth P S[1] *and R. Mahendran*[1]
[1]Centre of Excellence in Nonthermal Processing, Indian Institute of Food Processing Technology (IIFPT), Pudukkottai Road, Thanjavur, India

10.1 INTRODUCTION

The swift growth of science and the world economy has also carried various problems related to air and water pollution [1]. In addition, air quality is a major concern confronted globally [2]. With trading globalization, developing countries are at higher risk of gas hazards [3], followed by developed countries who require safety standards to face these environmental issues [4]. According to recent reports, hazardous gases such as acetone, toluene, formaldehyde, and others carry malignant diseases. Due to the constant volatilization of these gases, it is impossible to evade contact with them [5]. Additionally, patients who are weak or ill are often made worse by humidity (water vapor) in the air [6]. These unseen dangers have a negative impact and this ensures significant global attention. Effective action must be taken to ensure environmental safety. This situation gives rise to the need for the development of smart sensors that are capable of sensing hazardous gases/humidity, and other components for environmental sensing and monitoring.

Common environmental sensors include electrochemical sensors, mass-sensitive sensors, fluorescent sensors, quartz crystal microbalance sensors, and so forth [7, 8]. The mass sensing materials coated on the sensors adsorb the target compounds such as metal oxides, carbon materials, organic polymers, and the like. However, due to the lack of porous structure in these materials, the adsorption site is unable to absorb a large number of target components. To improve the adsorption rate and sensitivity of the sensors, porous materials such as metal-organic frameworks (MOFs) in hybrid forms are used [9]. MOFs are crystalline materials self-assembled from the organization of polydentate ligands to metal clusters [10]. The properties and structure of MOFs can be distinctively altered by customizing ligands and metal clusters in addition to their facile synthesis methods. However, certain MOFs are hindered by chemical and physical limitations despite these advantageous properties, leading to reduced performance. For instance, the crystalline structure of certain MOFs is degraded at higher temperatures or ambient moisture conditions [11, 12]. In addition to this, MOFs are limited by expensive linkers and the high cost of a metal node, making mass production currently impractical. Moreover, some MOFs' functions are constrained by physical limitations in conductivity, steric effects, or porosity, amongst others [13]. Hybridizing current MOFs with external components gives rise to novel materials with features similar to both native MOFs and material supplemented MOFs. These hybrid MOF materials classically outperform their native materials and have exhibited promising results in the field of sustainable energy [14], food packaging [15], catalysis [16], proton conductivity [17], detoxification [18], drug delivery [19], gas storage and separation [20], sensing and lighting [21], amongst other fields.

The application of the MOFs as smart sensors is wide enough due to its interesting framework architecture properties such as ultrahigh porosity, stable framework, pore-volume, selectivity, internal surface area, thermal stability, exponentially high specific area, non-toxic nature, chemical stability, tunable porosity, chemical functionality, and luminescent nature [22]. In particular, various excellent properties (biodegradability, the accessibility of surplus functional groups, luminescence, and non-toxicity) are an important driving force for exploring MOFs in biosensing fields [23]. Therefore, this chapter is organized to highlight the challenges faced by MOFs to be effective sensors for environmental sensing and monitoring, along with future perspectives.

10.2 STRUCTURAL CHALLENGES AND FUTURE PERSPECTIVE

MOFs are self-assembled with organic linkers and metal nodes. The organic linkers serve as bridging molecules, and the metal nodes are used as connecting pairs in the formation of the framework [24]. Therefore, the ligands (organic linkers) and metal ions used to synthesize MOFs are proposed as the 'primary building blocks.' The choice of metal nodes during the synthesis of MOFs is wide open and includes alkali metal ions, rare-earth metals, transition metal ions (Fe^{3+}, Cu^{2+}, Zn^{2+}, and Cr^{3+}), and alkali-earth metal ions [25]. In addition, other variables such as coordinate numbers, coordinate geometries, and oxidation states should also be considered during the synthesis process. One of the significant components to

DOI: 10.1201/9781003188148-10

consider in the formation of MOFs is the coordination geometry system (in other words: square-planner, cubic, T- or Y- shaped, tetrahedral, octahedral, or paramidal). The selection of organic linkers depends on binding sites' relative angularity and Lewis number [25]. Organic ligands are linear spacers and they form divergent sites with a metal ion to procedure a crystal geometry [26, 27]. Apparently, the spacer linearity is broken upon the construction of the spacer-node-spacer unit. Hence, ligand angularity, size, metal clusters, and coordination geometry are necessary components to define the crystal structure of MOFs.

While classifying the MOF structures, the concept of SBU (secondary building units) was advised as a shaping concept [28]. An SBU is necessary for the identification and design, and describes the underlying geometry of the MOFs. These features help in enabling the sensing nature (namely, the electronic and luminescent properties) of a MOF to be custom-made for a specific application through SBU modification and its absolute spatial arrangements. In addition, open coordination sites present in a MOF can bind analytes and metal ions reversibly with a strong absorbing antenna molecule for an efficient energy transfer. This process can produce a sensor transduction mechanism. Finally, MOFs' chemical structural flexibility can dramatically change in the local coordination environment, which helps luminescent properties directly fit into the sensing application [29]. An outline of the literature available [30, 31] indicates that the synthesis of MOFs is primarily accomplished by non-aqueous techniques such as microwave, mechanochemical, electrochemical, hydro/solvothermal, high-throughout, layer by layer, and ultrasonic synthesis [13, 22]. The bottom-up approach is the most common strategy for the synthesis of MOFs in which organic molecules and metal ions are combined directly. As MOF synthesis methods are well established in the laboratory, the following parameters need to be considered: (1) organic linkers type, (2) metal type, (3) solvent medium, (4) synthesis condition, and (5) the synthesis method to active high yield by preventing large quantities of impurities.

10.3 CHALLENGES IN BASIC PROPERTIES AND MECHANISM OF MOFs FOR SENSING

The MOFs have emerged as a promising tool for environmental sensing and monitoring applications through sensing various gases, explosives, small molecules, and solvents that are not safe [31, 32]. For instance, Lei et al. [33] have explored the viability of LMOFs (luminescent MOFs) for several sensing applications. However, the sensitivity of hybrid MOF-based detection mainly depends on the method of sensing used for signal transduction. Notably, lower detection limits are realized by facilitating analytes' inclusion in the sensing material. In addition, the collective effects of the analyte transport dynamics within the MOF and higher loading capacity due to higher volume to surface ratio can enhance MOF sensitivity.

Though the selective process of luminescent MOF-based sensing has been proposed, the theory is not yet sufficiently developed to fully report all of the required properties. Kreno et al. [34] have reviewed MOF progress in various sensing applications and suggest that MOFs are normally coupled with various signal transduction techniques such as interferometry, micro-electrochemical systems, microcantilever-based methods, surface acoustic waves, and quartz crystal microbalances. Based on the MOFs, combining the above-mentioned diverse signal transduction methods holds great promise for miniaturized sensor development. In addition, MOF-based thin-film methods show promise in developing future generation chemical sensors [34]. Still, during the technology up-scale, the bulk molecules of MOFs are assumed to face certain problems. For example, bulky MOFs' poor solubility is an obstacle during the processing of related devices, and to overcome this hurdle, nano-scale MOFs have been proposed as a pragmatic alternative that can offer a processability solution, possible functionality, and stable luminescence [35].

LMOFs' popularity is increasing because of a few principal factors such as lower detection limits up to unimolecular level, production of signals visible to the naked eye, and the direct application of powdered material without film fabrication (or any other treatment). In addition, the structural framework of the LMOF (in other words, porous crystalline) offers sorptive properties with a molecular specificity at a higher degree. However, some limitations such as many quenching pathways, the possible rate of quenching, counterproductive porosity, and medium stability have been identified in LMOFs [36]. As the modification in the MOF properties is identified, changes in significant physicochemical properties such as absorption, mass, electrical condition, and optical properties are reflected in MOFs. Hence, these properties can be used during signal trapping generated at any given sensing event. Among these properties, the optical mechanism has been used most intensively for analyte sensing. The luminescent properties of a few developed MOFs are shown to generate a useful signal to sense small molecules, explosives, amines, gas, and other materials [37, 38]. Besides, an LMOF has been developed for biosensing applications [39]. The capability for sensing is a comparatively new research venture, nonetheless, the information available points to the emergence of MOFs to lead next-generation sensors.

10.4 REQUIREMENTS FOR AN IDEAL SMART SENSOR

The meticulous configuration of a sensor for a specific application may be based on the requirements and nature of the focus. Yet, it is beneficial to set the properties that one wants in an ideal sensor for biochemical and chemical species. By knowing the individual subsystem requirement, the ideal sensor system requirements can be analyzed in which sensing material, sample introduction, transduction principle, and implementation and data analysis are major requirements of individual subsystems. Potyrailo, R.A et al. [40] report that the overall system requirement and its performance are affected significantly by every individual subsystem's design.

For instance, with an exact type of sensing material, sample introduction, data analysis, and transduction principle chosen for sensor signal generation, a desired dynamic range for the sensor can be gained.

Practically, the potentials of an ideal sensor are often evaluated differently following its application. Factors such as adequate long-term stability, resolution, and high reliability are at the top of the priority list for industrial sensor operators, whereas technology maturity and size are at the bottom of the list [41]. First responders face difficulty in ergonomic design and a low false-positive rate. On the other hand, the cost of disposable sensors is focused on by medical users. Definite requirements for in-vivo medical sensors include minute size and blood compatibility [40]. Resistance to drift-free performance, gamma radiation during sterilization, and cost are the critical requirements for sensors in disposable bioprocess components. Moreover, continuous monitoring importance also differs for each application [40].

10.5 CHALLENGES AND PERSPECTIVE OF ELECTROCHEMICAL SENSORS FOR ORGANIC MOLECULES

Even though hybrid MOF sensors have exhibited exceptional potential for detecting ions and organic molecules along with radiation detection, there is much need for improvement. In most of the literature reports, luminescent MOFs are fabricated using traditional methods. Nevertheless, noteworthy progress has been made with two particular advances. Firstly, the work of supplementary signal transduction approaches has given access to non-luminescent MOFs to avoid no observable sensing signal display. Various methods mentioned here, including colloidal crystals, interferometry, microcantilever devices, localized surface plasmon resonance (LSPR), surface acoustic wave (SAW) devices, and quartz crystal microbalance (QCM), can be adapted to any MOF material, reducing concerns about the structure and properties of specific MOFs' materials. In addition, micro-electro-mechanical system (MEMS) devices specifically lend themselves to miniature portable sensor development.

Secondly, growth in MOF thin-film techniques has been vital in aiding the fabrication of sensor devices. When the only solvothermal synthesis of MOF produced free-standing crystals, the availability of signal transduction was limited to a few approaches, and as we adopt new methods for producing these structures on the substrates, we can access MOF structures of different scale lengths and create a better interface between the support surface and the MOFs. Despite the growing list of hybrid MOF materials, the aim of highly selective recognition is still unrealized for many analytes. It is improbable that a great level of specificity would be achieved only by size or shape selectivity. In a few cases, design tricks can be used to fit sophisticated identified elements into MOFs. For instance, reports on chiral frameworks recognizing only a single enantiomer in a combination and a few instances of preferential (albeit) enantiomer sorption could be imagined [42]. MOFs with exposed metal centres could bind certain molecules of gas while the rest bind to atmospheric interferents [43].

Although exceptions doubtless will be created, routinely discovering or designing MOF materials that react very selectively to a particular analyte will likely prove impossible or difficult for the foreseeable future. To overcome this, the construction of an array-based device comprising various distinct materials (in MOFs) would be a superficially attractive substitute that has been widely explored with chemosensory polymers [44]. In principle, every analyte interacts with the series in a different way to produce a unique fingerprint. In reality, the effectiveness of this approach is limited since the 'unique fingerprint' varies with analyte concentration. Additionally, the nose or array addresses only identification problems and not signal problems caused due to interferants. It is essential to recognize that vapor or gas sorption by a conventional polymer is a solubility issue and, in many cases, analyte shape and size are not important. In fact, strategically positioning numerous analyte identification functionalities within the uniform and well-defined voids is impossible with conventional polymers.

Computational ranking and MPF screening may be a better approach, where the catalog of materials can be created by data mining or by combining nodes and linkers in silico to form both hypothetical and existing MOFs. Although screening of electrostatically and structurally simple molecules like methane can readily be accomplished [45], a higher degree of motion is possessed by large molecules or molecules exhibiting added complex electrostatics and these require highly complex simulations that inevitably will be slower. Finally, another useful concept for refining selective detection in hybrid MOF smart sensors is to couple analytical techniques providing molecular fingerprints or vibrational spectroscopy with MOFs. While infrared spectroscopy (Raman to a lesser extent) has been used in MOF characterization, the characterization/observation of guest molecules present inside MOFs via these methods has been underutilized [46, 47]. Analytes are detectable without sorbent assistance by vibrational spectroscopy, however, the limit of detection is improved by MOFs via molecular-specific preconcentration. Specifically, if preconcentration arrives proximal to (within a few nanometers) a material such as nanostructured silver or gold then it can support surface-enhanced Raman scattering (SERS).

10.6 CHALLENGES OF MOF-BASED VOLATILE ORGANIC COMPOUND AND GAS SENSORS

MOF-based volatile organic compound (VOC) and gas sensors with different chemical or physical responses are comparatively independent. However, MOF-based multifunctional materials which contain two or more physical (or chemical) properties are coupled or exist together, such as porosity, chirality, conductivity, luminescence, ferroelectricity, magnetism, and so forth, would enable the development of new sensor types. These targeted materials are produced by integrating functional ligands, metal nodes, and guest molecules (with different properties). Until now, few MOFs that are multifunctional with different properties such as luminescence-ferroelectricity, luminescence- electricity, and luminescence-magnetism, have been reported [48]. But, in most of these MOFs, different properties are not correlated; they just exist together. Therefore, further studies must design MOF-based smart switches and sensors that couple more than two properties and evaluate their application as barcode sensors and multi anti-counterfeiting uses.

Even after achieving excellent sensing performance for some MOF-based VOC and gas sensors, and although the field is expanding rapidly, many challenges must be overcome to meet commercial application standards. Firstly, the major challenge is to manufacture sensing materials that are MOF-based into workable devices. Thin films of MOF-based and MOF membrane development have come near to practical application, however, this field is still in its budding stages. Secondly, MOF stability in a different environment (temperature, pressure, pH condition, humidity, and so on) is a key issue to be considered each time for real-time applications. Continuous efforts are essential for designing strong MOF sensors that are capable of withstanding high performance during multiple cycles. Thirdly, although constant improvement is seen in MOF sensing performance under laboratory conditions, there is a need to make them appropriate for real-world commercial applications. Lastly, widespread practical application is sustainably hampered by the high cost of MOFs due to large-scale production. These challenges are addressed only by great and collective efforts, but we accept that MOFs have true potential applications as VOC and gas sensors.

10.7 CHALLENGES OF ELECTROCHEMICAL SENSORS AS SMART SENSORS

MOFs have been emerging as capable materials for electrochemical sensing in their original form and as templates or robust matrices to design hybrid MOFs (MOF-based material) for electrochemical sensing that act as smart sensors. Recent studies on MOF-based material designed for electrochemical sensing shows that three types of material are required for effective sensing: MOF-based composites, MOF-derived functional materials, and pristine MOFs [49, 50]. Among these materials, enhancement in electrochemical sensing performance and electron conductivity is focused on by MOF-based composites. Additionally, the suitability of MOF-based composites for sensing is due to the following reasons: (1) the strong framework/skeletons of MOFs offer a perfect platform to integrate with functionally active species, (2) the tunable porosity together with the pore number and pore size is an advantage for the pre-enrichment and immersion of analytes, thus improving the selectivity and sensitivity,(3) synergetic effects cause an augmentation in electrochemical sensing when functional species and MOFs interact with each other.

Yet, unsolved questions still exist around the use of MOF-based materials for electrochemical sensing. For example, the application of MOFs as electrochemical sensors is hindered due to their poor conductivity. Additionally, inadequate stability has also reduced their performance. Lastly, the electrochemical sensing mechanism is obscure because of fewer efforts under in-situ characterization and theoretical calculation. Hence, a few suggestions and strategies can be implemented in future research to overcome these issues. Firstly, 2D MOFs have excellent conductivity so can be considered to be suitable candidates for the matrix. Secondly, some suitable post-synthesis approaches can be taken to increase the MOFs stability or the redox activities of MOF-based materials can be enhanced by introducing redox ligands. Finally, the electrochemical test can include some in-situ characterization in its procedure to evaluate the interaction between host and guest and further theoretical calculations can be used to explore the complete discrimination mechanism. Thus, with suitable design and MOF-based material integration, future research can generate even more groundbreaking results and create a path for MOF-based materials to be broadly applied in a pragmatic application as smart sensors for environmental sensing and monitoring.

10.8 CONCLUSION

Based on this comprehensive study, MOFs have emerged as effective tools for various sensing applications, especially for aromatic compounds, vapors, small molecules, explosives, and gases. The primary characteristic of such detection is MOFs' label-free nature (biomolecules such as aptamers, and enzymes are not labeled) with greater sensitivity. Therefore, MOFs with luminescence-based sensing molecules are an appropriate solution. In contrast, conducting MOFs with electrochemical sensors is relatively new and offers a good opportunity in a highly fascinating research area. The properties of hybrid MOFs that are established to work effectively as smart sensors for environmental sensing and monitoring can be summarized as porosity, stable framework, pore-volume, high surface area, chemical and thermal stability, the viability of building a crystal structure, luminescence, chemical functionality, and toxic nature. As such, several traditional practices that have been used for various products like quantum dots, graphenes, organic dyes, and carbon nanotubes can be replaced by a potent tool, MOF. Because of the constant research efforts engaged in optimizing novel synthesized material, hybrid MOF-based smart sensors for environmental sensing and monitoring are ready for commercialization. Shortly, MOFs development will open a new track for success in various industrial applications.

REFERENCES

[1]. Bolisetty., Sreenath., Peydayesh, M., Mezzenga, R. (2019) Sustainable technologies for water purification from heavy metals: review and analysis. *Chem Soc Rev* 48.2:463–487.

[2]. Heft-Neal, S., Burney, J., Bendavid, E., Burke, M. (2018) Robust relationship between air quality and infant mortality in Africa. *Nature* 559:254–258.

[3]. Liu, J., Mooney, H., Hull, V., Davis, S. J., Gaskell, J., Hertel, T., J. Lubchenco, K. C., & Seto, P. Gleick, C. K. (2015) Systems integration for global sustainability. *Science* (80) 347:1258832.

[4]. Wang, H., Zhang, Y., Zhao, H., Lu, X., Zhang, Y., Zhu, W., & McElroy, M. B. (2017) Trade-driven relocation of air pollution and health impacts in China. *Nat Commun* 8:1–7.

[5]. Han, D., Zhao, M. (2020) Facile and simple synthesis of novel iron oxide foam and used as acetone gas sensor with sub-ppm level. *J Alloys Compd* 815:152406.

[6]. Pang, Y., Jian, J., Tu, T., Yang, Z., Ling, J., Li, Y., & Ren, T. L. (2018) Wearable humidity sensor based on porous graphene network for respiration monitoring. *Biosens Bioelectron* 116:123–129.

[7]. Brown, M. D., Schoenfisch, M. H. (2019) Electrochemical nitric oxide sensors: principles of design and characterization. *Chem Rev* 119:11551–11575.

[8]. Schroeder, V., Swager, T. M. (2018) Translating catalysis to chemiresistive sensing. *J Am Chem Soc* 140:10721–10725.

[9]. Wang, L. (2020) Metal-organic frameworks for QCM-based gas sensors: a review. *Sensors Actuators A Phys* 111984.

[10]. Moghadam, P. Z., Li, A., Wiggin, S. B., Tao, A., Maloney, A. G., Wood, P. A., & Fairen-Jimenez, D. (2017) Development of a Cambridge Structural Database subset: a collection of metal–organic frameworks for past, present, and future. *Chem Mater* 29:2618–2625.

[11]. Nguyen, J. G., Cohen, S. M. (2010) Moisture-resistant and superhydrophobic metal– organic frameworks obtained via postsynthetic modification. *J Am Chem Soc* 132:4560–4561.

[12]. Yang, S. J., Park, C. R. (2012) Preparation of highly moisture-resistant black-colored metal organic frameworks. *Adv Mater* 24:4010–4013.

[13]. Stock, N., Biswas, S. (2012) Synthesis of metal-organic frameworks (MOFs): Routes to various MOF topologies, morphologies, and composites. *Chem Rev* 112:933–969.

[14]. Cao, X., Tan, C., Sindoro, M., Zhang, H. (2017) Hybrid micro-/nano-structures derived from metal–organic frameworks: preparation and applications in energy storage and conversion. *Chem Soc Rev* 46:2660–2677.

[15]. Sharanyakanth, P. S., Mahendran, R. (2020) Synthesis of metal-organic frameworks (MOFs) and its application in food packaging: A critical review. *Trends Food Sci Technol* 104:102–116.

[16]. Aijaz, A., Xu, Q. (2014) Catalysis with metal nanoparticles immobilized within the pores of metal–organic frameworks. *J Phys Chem Lett* 5:1400–1411.

[17]. Li, X. M., Dong, L. Z., Li, S. L., Xu, G., Liu, J., Zhang, F. M., & Lan, Y. Q. (2017) Synergistic conductivity effect in a proton sources-coupled metal–organic framework. *ACS Energy Lett* 2:2313–2318.

[18]. Liu, Y., Howarth, A. J., Hupp, J. T., Farha, O. K. (2015) Selective photooxidation of a mustard-gas simulant catalyzed by a porphyrinic metal–organic framework. *Angew Chemie* 127:9129–9133.

[19]. Rojas, S., Carmona, F. J., Maldonado, C. R., Horcajada, P., Hidalgo, T., Serre, C., & Barea, E. (2016) Nanoscaled zinc pyrazolate metal–organic frameworks as drug-delivery systems. *Inorg Chem* 55:2650–2663.

[20]. Kumar, K. V., Preuss, K., Titirici, M. M., Rodríguez-Reinoso, F. (2017) Nanoporous materials for the onboard storage of natural gas. *Chem Rev* 117 :1796–1825.

[21]. Buru, C. T., Li, P., Mehdi, B. L., Dohnalkova, A., Platero-Prats, A. E., Browning, N. D., & Farha, O. K. (2017) Adsorption of a catalytically accessible polyoxometalate in a mesoporous channel-type metal–organic framework. *Chem Mater* 29:5174–5181.

[22]. Furukawa, H., Cordova, K. E., O'Keeffe, M., Yaghi, O. M. (2013) The chemistry and applications of metal-organic frameworks. *Science* (80) 341:6149.

[23]. Wriedt, M. (2012) Systematic investigations on magneto-structural correlations of copper (II) coordination polymers based on organic ligands with mixed carboxylic and nitrogen-based moieties. *Dalt Trans* 41:4207–4216.

[24]. Burnett, B.J., Barron, P. M., Choe, W. (2012) Recent advances in porphyrinic metal–organic frameworks: materials design, synthetic strategies, and emerging applications. *CrystEngComm* 14:3839–3846.

[25]. Kitagawa, S., Kitaura, R., Noro, S. I. (2004) Functional porous coordination polymers. *Angew Chemie Int Ed* 43: 2334–2375.

[26]. Guo, X., Zhu, G., Sun, F., Li, Z., Zhao, X., Li, X., & Qiu, S. (2006) Synthesis, structure, and luminescent properties of microporous lanthanide metal–organic frameworks with inorganic rod-shaped building units. *Inorg Chem* 45:2581–2587.

[27]. Gustafsson, M., Bartoszewicz, A., Martín-Matute B, Sun, J., Grins, J., Zhao, T., & Zou, X. (2010) A family of highly stable lanthanide metal–organic frameworks: structural evolution and catalytic activity. *Chem Mater* 22:3316–3322.

[28]. Eddaoudi, M., Moler, D.B., Li, H., Chen, B., Reineke, T. M., O'keeffe, M., & Yaghi, O. M. (2001) Modular chemistry: secondary building units as a basis for the design of highly porous and robust metal–organic carboxylate frameworks. *Acc Chem Res* 34:319–330.

[29]. Tranchemontagne, D. J., Mendoza-Cortés, J.L., O'Keeffe, M., Yaghi, O.M. (2009) Secondary building units, nets and bonding in the chemistry of metal–organic frameworks. *Chem Soc Rev* 38:1257–1283.

[30]. Stock, N., Biswas, S. (2012) Synthesis of metal-organic frameworks (MOFs): routes to various MOF topologies, morphologies, and composites. *Chem Rev* 112:933–969.

[31]. Jin, L.N., Liu, Q., Sun, W.Y. (2014) An introduction to synthesis and application of nanoscale metal–carboxylate coordination polymers. *CrystEngComm* 16:3816–3828.

[32]. Hu, Z., Deibert, B.J., Li, J. (2014) Luminescent metal–organic frameworks for chemical sensing and explosive detection. 43(16). *Chem Soc Rev* 43:5815–5840.

[33]. Lei, J., Qian, R., Ling, P, L... Cui, and H. Ju. (2014) Design and sensing applications of metal–organic framework composites. *TrAC Trends Anal Chem* 58:71–78.

[34]. Kreno, L.E., Leong, K., Farha, O. K., Allendorf, M., Van Duyne, R. P., & Hupp, J. T. (2012) Metal–organic framework materials as chemical sensors. *Chem Rev* 112:1105–1125.

[35]. Falcaro, P., Ricco, R., Doherty, C. M., Liang, K., Hill, A. J., & Styles, M. J. (2014) MOF positioning technology and device fabrication. *Chem Soc Rev* 43:5513–5560.

[36]. Chang, Z., Yang, D. H., Xu, J., Hu, T. L., & Bu, X. H. (2015) Flexible metal-organic frameworks: Recent advances and potential applications. *Adv Mater* 27:5432–5441.

[37]. Wei, Z., Gu, Z. Y., Arvapally, R. K., Chen, Y. P., McDougald Jr, R. N., Ivy, J. F., & Zhou, H. C. (2014) Rigidifying

fluorescent linkers by metal–organic framework formation for fluorescence blue shift and quantum yield enhancement. *J Am Chem Soc* 136:8269–8276.

[38]. Hu, Z., Huang, G., Lustig, W. P., Wang, F., Wang, H., Teat, S. J., & Li, J. (2015) Achieving exceptionally high luminescence quantum efficiency by immobilizing an AIE molecular chromophore into a metal–organic framework. *Chem Commun* 51:3045–3048.

[39]. Jung, S., Kim, Y., Kim, S. J., Kwon, T. H., Huh, S., & Park, S. (2011) Bio-functionalization of metal–organic frameworks by covalent protein conjugation. *Chem Commun* 47:2904–2906.

[40]. Potyrailo, R.A., Mirsky, V.M. (2008) Combinatorial and high-throughput development of sensing materials: the first 10 years. *Chem Rev* 108:770–813.

[41]. Taylor, R.F., Schultz, J.S. (1996) Handbook of chemical and biological sensors, *CRC Press*.17 (4):319–319.

[42]. Vaidhyanathan, R., Bradshaw, D., Rebilly, J. N., Barrio, J. P., Gould, J. A., Berry, N. G., & Rosseinsky, M. J. (2006) A family of nanoporous materials based on an amino acid backbone. *Angew Chemie Int Ed* 45:6495–6499.

[43]. Murray, L. J., Dinca, M., Yano, J., Chavan, S., Bordiga, S., Brown, C. M., & Long, J. R. (2010) Highly selective and reversible O_2 binding in Cr_3(1, 3, 5-benzenetricarboxylate) 2. *J Am Chem Soc* 132:7856–7857.

[44]. Rakow, N.A., Suslick, K.S. (2000) A colorimetric sensor array for odour visualization. *Nature* 406:710–713.

[45]. Wilmer, C. E., Leaf, M., Lee, C. Y., Farha, O. K., Hauser, B. G., Hupp, J. T., & Snurr, R. Q. (2012) Large-scale screening of hypothetical metal–organic frameworks. *Nat Chem* 4:83.

[46]. FitzGerald, S. A., Allen, K., Landerman, P., Hopkins, J., Matters, J., Myers, R., & Rowsell, J. L. (2008) Quantum dynamics of adsorbed H_2 in the microporous framework MOF-5 analyzed using diffuse reflectance infrared spectroscopy. *Phys Rev B* 77:224301.

[47]. Vitillo, J. G., Regli, L., Chavan, S., Ricchiardi, G., Spoto, G., Dietzel, P. D., & Zecchina, A. (2008) Role of exposed metal sites in hydrogen storage in MOFs. *J Am Chem Soc* 130:8386–8396.

[48]. Li, H. Y., Zhao, S. N., Zang, S. Q., Li, J. (2020) Functional metal–organic frameworks as effective sensors of gases and volatile compounds. *Chem Soc Rev* 49:6364–6401.

[49]. Ma, T., Li, H., Ma, J. G, Cheng, P. (2020) Application of MOF-based materials in electrochemical sensing. *Dalt Trans* 49:17121–17129.

[50]. Qiu, Q., Chen, H., Wang, Y., Ying, Y. (2019) Recent advances in the rational synthesis and sensing applications of metal-organic framework biocomposites. *Coord Chem Rev* 387:60–78.

11 MOF-based Electrochemical Sensors for Toxic Anions

Naseer Ahmad[1], Sufian Rasheed[1], Batool Fatima[2], Saadat Majeed[3], Abrar Mohyuddin[4], Muhammad Najam-ul-Haq[3] and Dilshad Hussain[1]

[1]HEJ Research Institute of Chemistry, International Center for Chemical and Biological Sciences, University of Karachi, Karachi, Pakistan

[2]Department of Biochemistry, Bahauddin Zakariya University, Multan, Pakistan

[3]Institute of Chemical Sciences, Bahauddin Zakariya University, Multan, Pakistan

[4]Department of Chemistry, The Emerson University, Multan, Pakistan

11.1 INTRODUCTION

Water pollution is now a global issue across the world. Death and disease rates world-wise increased due to the consumption of polluted water, and almost 14000 people die every day because of polluted water consumption. Developing and developed countries both come under the influence of polluted water. Different factors are involved in the contamination of water such as precipitation, soil type, climate, geology, vegetation, groundwater, flow conditions, and human activities. But industry, agriculture, and municipal sector are the main contributors over other factors [1]. Rapid industrialization in the last few years has become a major source of heavy metal ions, small toxic inorganic anions (SO_4^-, NO^-, Ac^-, NO_2^-, IO_3^-, CO_3^-, ClO_4^-, I^-, PO_4^-, $H_2PO_4^-$, Br^-, Cl^-, SCN^-, or $Cr_2O_7^-$), and organic molecules (pharmaceuticals, endocrine disruptors, personal care products, detergents, organic dyes, pesticides, and common industrial organic wastes) finding their ways into the water. Different industries such as textile, paper, leather, food, makeup products, and medical dispose of their waste material into the water without some of the necessary treatment which causes water pollution. Contaminants pollute the groundwater and under-groundwater and it becomes unhealthy for drinking and domestic use. These substances pollute the water and this ultimately has diverse effects on plants, humans, animals, and aquatic life [2]. In agricultural production, various types of chemicals (fertilizer, pesticides, and the like.) are used to meet the need of a huge population. Unpolluted water is essential for life on earth. Growing demand for well-timed and precise environmental pollution checking and control involves new sensing techniques with exceptional performance, in other words, extraordinary selectivity, high sensitivity, and good reliability. Currently used sensing procedures, such as GC-MS, LC-MS, HPLC, Fluorimetry, enzymatic biosensor, and colorimetry depend on chromatography. Some of the problems associated with these procedures are complex sample preparation, cost, time, expensive equipment, and toxic chemicals. The development of a precise, easy-to-use, timesaving, rapid detection, simple, unbroken environmental pollutants measurement, and low-cost method for the measurement of various nature of analytes, therefore, is immediately required [3].

Sensor devices mainly comprise of a sensing unit which interacts with the analyte of interest and a transducer unit which ultimately transforms the chemical or physical change into a signal from electrical/optical. The working mechanism of the transducer and sensing unit sensors are classified into numerous types. In a sensor, vital sensing features comprise of high sensitivity, good selectivity, low response time, reusability, durable stability, and cost. To develop a sensor for monitoring environmental pollutants, several types of materials are designed and employed, such as metal oxide nanoparticles, carbon-based nanomaterials, quantum dots, polymers, porous nanomaterials, metal nanoparticles, and semiconducting nanomaterials. The enhancement of new sensing materials with admirable properties significantly supports the need for sensor research and applications [4]. In the past decade, 2D materials attained the special attention of scientists due to their excellent properties. Metal-organic frameworks are researched more relative to other materials due to their chemical and functional designability [5].

MOF synthesis and applications in different fields started in the early 1990s, by Yaghi, Moore, Robson, and Zaworotko. Metal-organic frameworks (MOFs) are types of crystal-like polymers which are made up of metal ions and organic ligands by coordination chemistry. MOFs are crystalline where a cluster of ions or metal cations known as (nodes) are attached with organic 'strut' or 'linker' ions or molecules (carboxylates, phosphonates, and N-donor groups). These materials contain a large degree of tuneability and structural diversity with good physical and chemical properties. A large number of metal cations, organic linkers, photosynthetic modification, and structural motifs make it possible to design MOF materials which different combinations (more than 20,000 MOF nanomaterials reported in the literature) [6]. MOF materials are matched with zeolites because of their high degree of crystallinity, high surface area, and wide-ranging porosity. Due to these similarities, MOFs, and zeolite materials are often employed in similar applications such as separation, catalysis, and gas storage [7]. MOF materials

DOI: 10.1201/9781003188148-11

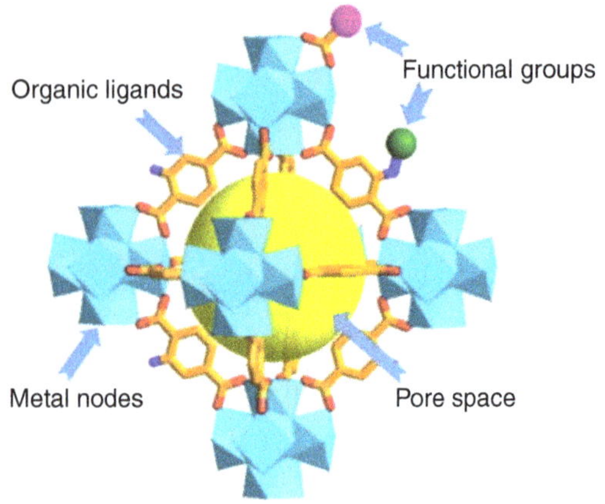

FIGURE 11.1 Schematic representation of MOF structure. Adapted with permission from [8], Copyright (2019) Elsevier.

have high chemical tailorability because of several functional groups in their framework. These properties make them more valuable materials for the bulk of applications in comparison to other porous materials such as zeolites, mesoporous silica, and active carbon. Until now, the different composite structures like MOF/biomolecules, MOF/metal/MOx NPs, MOF/polyoxometalates (POMs), MOF/polymers, MOF/enzymes, MOF/quantum dots, and MOF/carbon nanotubes have been manufactured and utilized for various applications such as magnetic application, proton conductivity, electrical conductivity, photocatalysis, gas adsorption, heterogeneous catalysis, separation, and medicinal drug delivery. A general structure of MOF materials is given below in Figure 11.1 [8]. Numerous other researchers have been developing the uses of MOFs in chemical and electrochemical sensing because of their strange porosity, exceptional physical and chemical stability, and capacity to integrate vastly conjugated aromatic ligands in the structure of the framework [9].

In this chapter, we explain MOF nanomaterials, nanostructured MOFs, MOF composites, and enzyme mimicking MOFs for the electrochemical sensing of toxic anions. As an electrochemical sensor, MOF-based materials show good advantages because of active metal sites. Composites of MOF materials with other materials also enhance the electrochemical sensing capability. Different natures of negative anions such as sulfates, chlorides, arsenates, fluorides, phosphates, cyanides, carbonates and bicarbonates become parts of water from numerous sources and pollute the water. These anions have significant effects on every type of life. MOF-based electrochemical sensors are used to detect these toxic inorganic anions electrochemically. The toxicity of these inorganic ions is discussed in detail and lastly, MOF-based electrochemical sensors are prescribed to monitor their concentration levels in water to make it safer and sustainable. Different pure MOF-based, and composite-based sensors are prescribed.

11.2 TOXICITY OF ANIONS

Inorganic anions are another group of toxic pollutants in water in addition to metal ions and organic molecules. These anions are also carcinogenic and toxic. Numerous technologies have been developed to nullify their presence in water to prevent their diverse effects on life [10]. Some major toxic inorganic anions, their source of production, and their health effects are discussed below.

11.2.1 Arsenates

Arsenic is mainly present in two forms: arsenite As(III) (H_3AsO_3), or arsenate As(V) (H_2AsO^{4-} and $HAsO_4^{2-}$). Arsenic becomes part of the environment through human activities and natural activities such as abiotic and biologically catalyzed weathering reactions. Because of these reactions, the arsenic level increases in the water. These two forms of arsenic are inorganic and considered more toxic than organic forms. According to the world health organization (WHO) arsenic levels, if increased above 10 μg L^{-1} stimulate toxicity and cause serious health issues: cancer in the kidney and bladder, cancer of the skin and lungs. The toxicity of arsenic in humans mainly relies on its valence state. For example, the arsenic (III) form is more toxic than arsenic (V). The arsenic (III) form has the capability to be retained for longer in the human body. To prevent its toxicity, different technologies are designed to eliminate it from water and to monitor its concentration level in water. Different instruments are used, including electrochemical sensors which are easy to use, reliable, selective, and more sensitive [11].

11.2.2 Fluorides

The mixing of fluorides in groundwater is a basis of severe human health issues. Fluorides are present in water with iron, beryllium, and aluminum as fluoride ions. It is due to the elimination of mineral deposits and industrial waste which contains fluoride into bodies of water. Fluorine-based inorganic compounds are employed in industry for aluminum fabrication, in glass fiber, and steel utilized as a flux, sodium hexafluorosilicate, fluorosilicic acid, and sodium fluoride NaF is applied in water fluoridation treatment. Fluorides in the human body within range are also essential for the usual mineralization of skeletons and the formation of the dental coating. A fluoride amount in drinking water is beneficial at levels up to 0.7 mg/L but is toxic above 1.5 mg/L, according to the WHO. If fluoride is present in excess in the body it may cause dental and skeletal fluorosis. Fluorosis is widely spread across 20 developed and developing countries. Some other health effects which may result from Fluorides are arthritis, brain damage, cancer, Alzheimer's syndrome, infertility, osteoporosis, brittle bones, and thyroid disorder. A large fluoride concentration affects some other biological molecules such as carbohydrates, lipids, vitamins, proteins, and the mineral metabolism. The most common methods to eliminate fluoride from polluted water are bone char, clay, dolomite, precipitation with lime, ion exchange, reverse osmosis, nanofiltration, and electrodialysis. Monitoring their concentrations to prevent toxic effects, MOF-based electrochemical sensors are utilized [12].

11.2.3 Chlorides

Chloride-based ions are present in natural water. Its amount in water is variable depending on river flow and local area. Naturally, its presence in water is very low but currently, human activities increase its quantity in water. These human activities are industrial discharge, coal mining, washing out of rocks, and the use of different mixtures to decrease the level of ice on roads. Due to these activities, the concentration of chloride ions has increased from the limit defined by WHO and these ions become a source of different problems associated with the environment and life on earth. Their high concentration proves to be toxic and causes severe issues: kidney failure, loss of body fluids from severe vomiting, diarrhea, diabetic coma, high fever sweating, and high levels of blood sodium chloride [13].

11.2.4 Sulfates

Sulfate anions are present in water due to some natural resources and play a crucial role in biochemical cycles. But recent industrialization and urbanization cause an increase in sulfur content in water. The increase in sulfate concentration in water is not only an issue for human and ecological life but it also affects the carbonate weathering, global carbon cycle evolution, and erosion processes. If humans ingest these anions in too large an amount they might suffer from severe health issues such as diarrhea, gastrointestinal disorders, dehydration, and so on. Within the water environment, sulfate ions are converted into toxic compounds which results in the loss of necessary metal ions in aquatic plants. It also disturbs ecohydrological functions. Human activities that are sources of sulfates are industrial wastewater, infiltration, fertilizers, synthetic detergents, and mining drainage. So, to prevent its toxic effects appropriate steps should be taken to protect the environment [14].

11.2.5 Phosphates and nitrates

Phosphates and nitrates are essential for the normal growth of plants. Too much human activity increases the level of these ions which has significant effects on the ecosystem and human health. Phosphate and nitrate sources in various types of fertilizers are used in the agricultural field to enhance crop production. Unfortunately, unused amounts of these ions run off and contaminate the water. Their high concentration in water boosts algae growth. Excess growth of algae dramatically affects aquatic life. Light penetration is decreased, the bacteria life cycle is disturbed, and the amount of dissolved oxygen, essential for aquatic life, is changed. This results in negative consequences. Aquatic life is more sensitive towards the concentration of these anions. To prevent their hazardous effects on human life and aquatic life it is essential to take steps for their ultimate removal from water and to develop the types of electrochemical sensors to monitor these ion levels continuously [15].

11.2.6 Cyanide

On the other hand, the cyanide ion (CN^-) is fatal and has been associated with many human deaths. If its concentration is higher than the normal value defined by WHO, it causes numerous health effects. Early indications include headache, fast heart rate, dizziness, shortness of breath, and vomiting [16]. Cyanide anions cause toxicity by inhibiting the growth of cells, which results in cell death and these ions have a high affinity toward iron which is present in many enzymes. By the combination of these ions with iron present in various enzymes, the result is hypoxia. So, to prevent its toxicity it should be made clear that the water released by industry and other sources is free of cyanide [17].

11.2.7 Carbonates and Bicarbonates

Carbonate and bicarbonate ions are involved in many physical, environmental, and industrial processes. These ions are important because these ions control the blood pH value. By maintaining the pH of blood these ions prevent acidosis. On the other hand, CO_2 levels in oceans are also maintained by these ions, which also prevent acidification. These ions have a crucial role in biomineralization. If we consider this on an industrial scale these ions have lots of applications such as toothpaste abrasive, food, energy storage materials, food additives, and in the engineering of materials such as glass and cosmetic items. These ions should be present in water and in a manageable amount. In excess, these ions prove to be toxic and affect nature in many ways. High concentration levels of these ions may cause metal-induced toxicity in humans [18].

11.3 METAL-ORGANIC FRAMEWORK-BASED SENSING DEVICES

In recent years, research has grown considerably in the synthesis and utilization of MOFs. MOFs can also be termed as porous coordination polymers which are long crystalline structures having coupling units, namely, metal clusters or centers coordinated by organic ligands as a result of their electron-donating ability. Due to distinct pore size and higher volume of pores, MOFs have very high surface areas which makes them very significant materials in various research areas such as drug delivery, gas storage, separation and removal of pollutants, and sensing. Moreover, the microporous structures of MOFs can be adjusted from angstroms to nanometers, depending on the nature of the organic linkers, Further, endless combinations can be produced through a wide number of metal ions and organic linkers. Thus, by varying the organic linkers and coordination nodes, their structural design can be tailored to specific target molecules. Notably, diverse framework functionality can be synthesized from metal centers (for example, catalysis, electricity, magnetism) or organic ligands (for example, chirality, fluorescence, and luminescence) or a mixture of both, beyond the available porosity (Figure 11.2) [19].

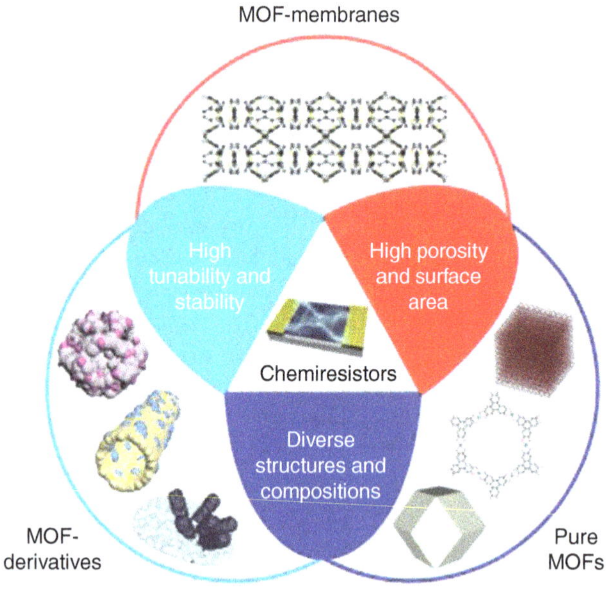

FIGURE 11.2 Characteristics of MOF-based sensing devices. Adapted with permission from [19], Copyright (2019) Elsevier.

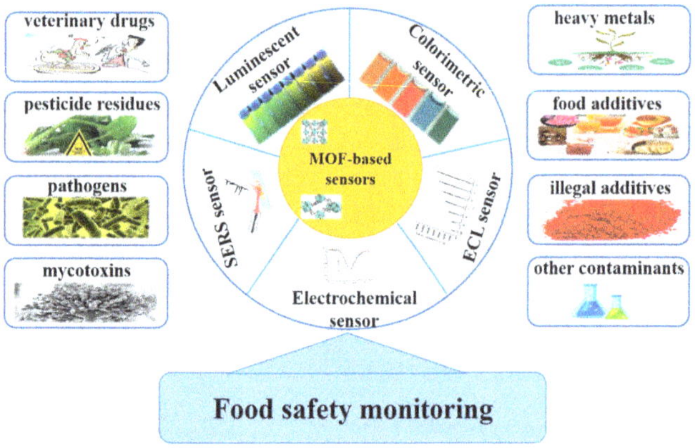

FIGURE 11.3 Different types of MOF-based sensing devices. Adapted with permission from [20], Copyright (2021) Elsevier.

11.4 CLASSIFICATION OF METAL-ORGANIC FRAMEWORK (MOF)-BASED SENSING DEVICES

We found three key concepts that can be utilized to measure or detect the presence of particular analytes, leading to three categorizations of MOF-based sensing devices. In the first category of MOF-based sensors, fluctuations in optical characteristics refer to spectroscopic measurements of aspects such as luminescence, refractive index, phosphorescence, and fluorescence. In the second category, the primary techniques used to detect mechanical stress fluctuations are surface acoustic wave (SAW) and quartz crystal microbalance (QCM) technologies. Finally, electrochemical-based devices are primarily concerned with changes in conducting properties, such as impedance or amperometric fluctuations. MOF-based sensors are shown in Figure 11.3 [20].

11.4.1 MOF-BASED OPTICAL SENSORS

The tunable properties and variable porous structures of the metal-organic frameworks make them choice candidates for use in optical sensing devices. The flexible nature of MOF structures in particular, allows them to be quickly modified resulting in more efficient results for these sensing devices. Optical sensing devices comprised of MOFs are rapid responsive, cost-effective, simple, and portable which can be utilized in a wide number of areas especially for the detection of heavy metals and toxic oxoanions in an aqueous environment [21]. In contrast to other spectroscopic approaches, the uses of optical sensing techniques are straightforward and the need for specialized apparatus is eliminated because the results may be recognized with the 'naked eye.' The phenomenon of optical sensing is the induced color change of the analytes where the sensing responses in terms of selectivity, sensitivity, and rapid response-time are generated by the interactions of an immobilized indicator chromogen 'molecular probe' with analyte 'cations'. The capacity to integrate chromophore probes from bulk to nanoscale materials as sensing receptors has recently attracted attention in the creation of flexible chemosensors for the identification of various species such as neutral and charged organic/inorganic anions and metal cations [22].

Sahat A. et al. fabricated a Zr-based MOF (UiO-66) sensor with well-defined particle morphology, size, and microspore geometry, for the optical sensing and elimination of ultra-traces of certain hazardous metal ions, namely, Bi(III), Zn(II), Pb(II), Hg(II) and Cd(II) [23]. The UiO-66 sensor that was designed was used to detect certain ultra-trace hazardous metal ions visible to the naked eye. The novel sensor had a high sensitivity as well as selectivity for a wide variety of detectable metal-metal ions up to 10-10 mol dm^3 in solution, as well as a fast analyte uptake response, namely, within seconds. The designed optical sensor was found to be stable, inexpensive, and simple to fabricate, and beneficial for the fast detection as well as for the effective removal of ultra-traces of hazardous metal analytes from an aqueous environment. Further classification of optical MOF-based sensors is discussed below.

11.4.1.1 MOF-based Luminescence Sensors

A large amount of research work is centered on the luminescent properties of MOFs as sensing materials. The phenomenon of luminescence is also known as the generation of light by sources/methods except heating and usually, it occurs when excited singlet state electrons return to the ground state through photon emissions. The luminescence phenomenon of MOFs occurss in two ways. Either it is quenched or suppressed as the analyte is absorbed. This phenomenon is also termed as, the turn off mechanism or it can be enhanced upon analyte absorption, which is also known as, the turn on mechanism [24]. The enhancement in luminescence, quenching/ suppression, or shifting of the emission wavelength can be used to determine the quantitative and qualitative study of the analytes. MOF-based sensing materials are potential multifunctional luminescent sensors since they can emit a luminescence signal from both organic as well as inorganic components. Other dimensional luminous capabilities may be added to MOFs through metal-ligand electron transfer luminescence. Moreover, certain guest molecules included in MOFs can generate or cause additional luminescence properties [25].

A considerable amount of research work has been focused on MOFs as luminescence sensing materials in literature. Luminescence sensor MOFs are widely used for the detection of inorganic hazardous anions such as phosphate ions (PO_4^{3-}), ClO^-, CN^-, and F^-. Qian et al. designed Tb (III)-based metal-organic framework, TbNTA1 (NTA = nitrilotriacetate) having an asymmetric cavity (C3v). The designed MOF-based quenching photoluminescent sensor was found to be highly selective in the detection of trace amounts of PO_4^{3-} anions in water. Other inorganic anions present in the water did not affect the luminescent properties of TbNTA1 and only PO_4^{3-} incorporation in TbNTA1 led to significant quenching luminescence impact [26]. This study was the first to describe the use of MOF sensors for selective detection of PO_4^{3-} anions in water. Subsequently, many other MOF-based luminescent sensors were designed with different combinations of metal clusters and organic linkers for the detection of inorganic anions. In addition to the application of luminescent MOF (LMOF) materials towards the sensing of inorganic anions they are widely used in the detection of toxic heavy metal ions, organic molecules, and bioorganic molecules and they have demonstrated excellent results. The reason for such excellent multifunctional sensing properties in an LMOF is that its sensing capabilities could readily mediate and be achieved through the rational selection of chromophoric organic units and metal atoms. In addition to the introduction of robust aromatic groups or conjugated electron systems, several uncoordinated functional sites like O, S, and N atoms can also be introduced to modify the intensities and luminescence emission region and the interaction of the LMOF for specific analyte detection could be improved. Li et al. designed cage like SBUs for two cluster-based microporous MOFs that included coordinated sites of unsaturated metal(II) [and Open Metal(II)]. Both materials showed extremely selective CO_2 absorption under ambient circumstances over CH_4 and N_2. More significantly, MOF-1 offers an excellent luminescent sensor, based on substantial luminescence improvement or quenching for a selective Ba^{2+} and Cu^{2+} ions test [27]. Recently, a luminescent europium organic framework was designed by Liu et al. for selective detection of UO_2^{2+} and Cu^{2+} ions. The LOD toward Cu^{2+} was found to be 17.2 μg/L where the maximum tolerable concentration in drinking water is 2 mg/L (as defined by EPA). As for UO_2^{2+} ions, the LOD is found to be 309.2 μg/L, which could be utilized for the detection of uranium in polluted areas. Additionally, the prepared LMOF was found to be highly selective detection of monovalent, divalent as well trivalent states of these cations via a turn-off mechanism [28].

11.4.1.2 MOF-based Fluorescent Sensors

MOF-based fluorescent sensing materials have received a lot of interest due to their adaptability and sensing diversification. In the recent 10 years, fluorescent MOF sensors have gained increasing attention in monitoring with different features (for example, easy operation, high sensitivity, and quick response time). Most significantly, the non-fluorescent material may be identified by combining a second or third chemical that has indirect fluorescent characteristics [29], considerably expanding the variety of analytes. However, when measuring fluorescence intensity, the influences of concentration, surroundings, and excitation light intensity lead to low detection accuracy for monochromatic fluorescence MOF sensors. To address this shortcoming, another fluorescence signal is used to build a ratiometric fluorescence MOF (RFMOFs) sensor. Because the emission intensities at two of the wavelengths are independent of the abovementioned interfering circumstances, RF sensors may overcome the limitations of single fluorescence sensing by dual-emission self-calibration and accomplish accurate detection. Zhang [30], for example, described an RFMOFs sensor used to detect H_2S. The ratio signal aids in the exclusion of interference, resulting in a reduced limit of detection 5.45 μM that is comparable to certain other previously fluorescence MOF sensors for H_2S detection. Consequently, MOF-based sensors have higher sensitivity and selectivity than fluorescent sensors without the amplification effect of MOFs, and dual-emission sensors generate more accurate results than MOF-based single signals. As the intensity of the dual emission may be utilized as an internal standard and exhibits self-reference qualities, interference from surrounding factors can be successfully avoided. RFMOFs sensing has gained rapid advancement in recent years due to the benefits listed above [31].

11.4.2 MOF-based Electrochemical Sensors

An electrochemical sensor is a device that is used to transform the electrochemical data into the appropriate analytical signal. The basic working principle of the electrochemical sensor includes the series of redox reactions of specific analytes in the electrochemical system. Usually, the electrochemical measurement is carried out by using a three electrode system in which one is the working electrode, one is the counter, and one is the reference electrode. The number of analytes participating in the reaction may be calculated from the measurement of the electric potential, current, or any other electrical signals. Electrochemical sensing techniques are particularly suitable in comparison to other existing traditional approaches because of their simple and low-cost analytical systems. Moreover, electrochemical sensors are typically capable of detecting a wide range of substances, including neutral, organic, inorganic, and ionic molecules, as well as multiple analytes at the same time [32]. Amperometric techniques are widely utilized for environmental sensing applications (inorganic anions, volatile organic compounds, toxic ions, and gases).

The application of MOF-based electrochemical sensors to detect various environmental contaminants have grown in popularity over the last several decades. The stable MOFs with excellent electrical conductivity have been successfully developed and synthesized with the rapid progress of synthesis techniques. However, most of the MOFs still exhibit weak electrical conductivity and somewhat low stability in an aqueous environment due to the fact that the coordination bonds have a reversible nature. Furthermore, MOFs typically have a size in micrometers that leads to a low adhesive affinity between the electrode surface and MOF. These shortcomings of MOFs limit their use in electrochemical sensing platforms [33]. The key to fabricating effective electrochemical signals is to synthesize MOFs with strong redox activity and electrical conductivity while retaining their distinctive pore structure which can be achieved by designing MOFs with a combination of other highly electric conductive materials as discussed above. MOF-based electrochemical sensors have been extensively applied for the detection of numerous environmental analytes such as inorganic anions, ions, organic molecules, volatile organic compounds, gases, and biomolecules [34].

11.5 SENSING OF VARIOUS TOXIC ANIONS BY MOF-BASED ELECTROCHEMICAL SENSORS

Anions such as sulfates, nitrite, chlorides, arsenates, fluorides, phosphates, cyanides, carbonates, bicarbonates, and bromates have a health hazard effect on living beings when they are present beyond their specified limits in aqueous media as discussed earlier. The application of MOF-based sensors in an aqueous environment is limited due to the poor conductivity and instability of MOFs in aqueous solutions so very little research work has been carried out for the sensing of the above-mentioned ions in the water phase utilizing MOF-based electrochemical sensing devices. In contrast, other MOF-based techniques especially luminescence and fluorescence sensing are mostly utilized for this purpose due to excellent results with high selectivity and sensitivity and less use of auxiliary devices. However, few research groups worked to overcome the drawbacks associated with MOF-based electrochemical sensing materials, in other words, by combing MOFs with highly conductive materials like macroporous carbon, graphene, and metallic nanoparticles to fabricate novel MOF composites with enhanced electrochemical properties.

The chloride ion (Cl^-) is a common anion found in waste water from landfills, the leather industry and animal farms. Excessive concentrations of Cl ions invariably result in water contamination, plant growth inhibition, salinization of soil, and pipeline corrosion. Therefore, a rapid and effective technique for their detection in an aqueous medium is essential and for which MOF-based electrochemical sensors have been utilized. Qianyun et al. fabricated an electrochemical sensor for the detection of Cl^- ions by MOF composite with silver nanoparticles [35]. They impregnate Ag^+ ions into the pores of ZIF-8 and after reduction treatment with $NaBH_4$ the Ag NPs/ZIF-8 composite was coated on a glassy carbon electrode. Furthermore, the sensor's analytical performance was good, including selectivity, stability, and repeatability, which is important for the use of MOFs in electrochemical sensing. Recently, Wenbiao et al. used a strategy to detect as well as remove chloride ions from aqueous solvent by preparing conductive chlorine ion-imprinted polymer threaded in metal-organic frameworks [36]. In-situ polymerization in UiO-66 cages of polypyrrole (PPy) was accomplished for electrochemically selective separation of the chloride ions (Cl^-) from the waste water using a conductive ionic imprinted polymer film (Cl-IIP@UiO-66) electrode. The prepared MOF-based electrochemical electrode was highly specific because it was designed to have pore sizes according to the size of Cl^- ions (6.64 Å). Other ions such as F^-, Br^-, SO_4^{2-}, PO_4^{3-} do not affect the sensitivity and selectivity of the prepared sensor due to the different sizes of these ions to the size of the pore of prepared MOF. This kind of material can be designed for the detection and selective separation of other toxic anions in water.

Ashraf M. M et al. [37] designed an electrochemical voltammetric-based sensor for cyanide inorganic anions sensing. To selectively detect the cyanide ions in water, they synthesized a modified carbon paste electrode. A copper metal-based metal-organic framework was utilized for the modification of the carbon paste electrode. The electrochemical performance of the newly synthesized electrode was examined by using various voltammetry techniques such as differential pulse voltammetry, cyclic voltammetry, and impedance-based spectroscopy. All of these electrochemical readings were taken at pH=7 by using 0.25M KCl. The modified carbon paste electrode with a MOF gives good selectivity toward cyanide toxic ions under the influence of different interferences and is used successfully to determine the cyanide concentrations in various types of water samples. The synthesis of electrodes and sensing analysis of cyanides ions are shown below in Figure 11.3. In optimized conditions, the anodic peak current (Ipa) declines linearly in the range of 1.87–25 μM and with a limit of detection (LOD) of 0.60 μM (at S/N = 3). This electrode exhibited better results toward some real water samples but the high concentration of sulfide, EDTA, and thiocyanate interfere with the performance of the electrode (Figure 11.4).

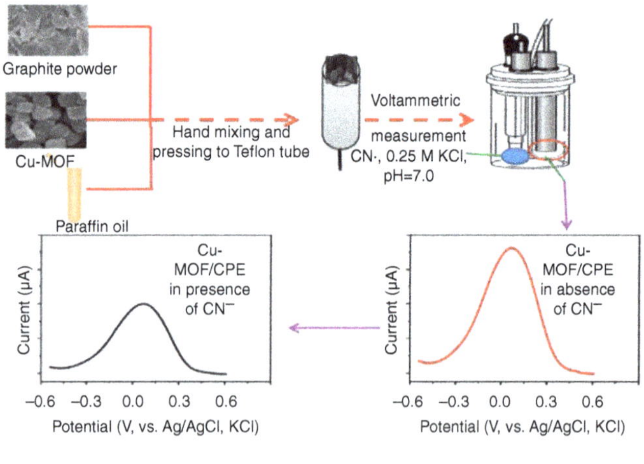

FIGURE 11.4 MOF-based electrochemical sensor for the detection of cyanide ions. Adapted with permission from [37], Copyrights (2018) Elsevier.

Nitrite has been used extensively in the food industry as a preservative but the excess nitrite intake is, nevertheless, damaging to human health. In addition, it is widely recognized that nitrite is commonly associated with urinary tract infections upon its appearance in human urine. Porphyrin MOFs offer a large electro-catalytic site for the oxidation reaction nitrite and can further enhance nitrite sensitivity and detection limits. A recent research group has succeeded in synthesizing and dispersing nanocrystals for water-stable zirconium-based porphyrinic MOF (MOF-525). Thus, the MOF-525 suspensions can be a suitable alternative for the production of MOF thin electrochemically adjustable films for nitrite sensing using the injection jet printing technique [38]. Chung-Wei et al. fabricated thin coatings on glass substrates via a solvothermal method on a zirconium-based porphyrin metal-organic frame (MOF–525) [39]. The resulting film MOF-525 is electrochemically turntabled in an aqueous solution and demonstrates electrocatalytic activity to oxidize nitrite. A cyclic voltammetry technique is used for the mechanism for electro-catalytic nitrite oxidation in the MOF-525 thin layer. Amperometry is used to study the redox mechanism of the thin film MOF-525 of the KCl aqueous solution. The amperometric nitrate sensor is the thin film MOF-525. The linear range of 20–800 μM, the sensitivity of 95 μA/mM-cm^{2,} and detection limit 2,1 μM were found for the synthesized MOF-525.

Because of the health and environmental consequences, the detection of bromate, a category B-2 carcinogen designated by the International Agency for Research on Cancer (IARC), is of significant importance. Among many analytical techniques for bromate, identification by electrochemical approaches, due to their eco-friendly methods, quick response times, frequently effective cost, and viability for developing portable sensors, are regarded as attractive. In this context, significant studies have been investigated to develop direct electrochemistry-based materials that might be used in the electrochemical detection of bromate. Erbin S et al. [40] developed a simple electrochemical plating technique (PTA) to synthesize the electroactive metal-organic framework film (NENU-3) on a copper electrode in an acid electrolyte. The prepared film electrodes had high electrocatalytic activity for bromate reduction and could be utilized successfully for amperometric bromate detection. By chronoamperometry, the suggested sensor has a large linear range (0.05–72.74 mM) and a low LOD (12 μM) under optimal conditions. Furthermore, the films have good electrochemical stability and anti-interference performance in the bromate detection phase. The electrochemical plating process described has been proved to be a reliable and efficient method for fabricating MOF films on conductive substrates for bromate detection.

11.6 CONCLUSION

In this chapter, we discussed the various causes of environmental pollution especially due to inorganic anions. We have identified the sources of these hazardous anions and also discussed their toxicity. MOF-based sensors are promising materials and demonstrated outstanding performance for both the ultra-trace level detection and separation of these anions. As metal-organic frameworks have distinct pore size, pore volumes as well as high surface area, they have multi-functional sensing capability. MOF-based sensors are classified mainly into three classes, namely, optical, sonochemical, and electrochemical sensors. Optical MOF sensors work as a function of changes in optical properties, sonochemical MOF sensors deal with mass change while electrochemical MOF sensors work as a function of change in conductive properties. Environmental pollutants analytes are being detected by different types of MOF-based sensors. Gas sensing is mostly performed by sonochemical MOF sensors especially QCM and SAW. The detection of heavy metal ions and inorganic anions is mostly achieved with optical sensors like luminescence and fluorescence MOF sensors. Electrochemical MOF sensors are mainly utilized for trace level detection of biomarkers such as glucose, dopamine, H_2O_2, and the like. However, the research area of application MOF sensors in the electrochemical detection of ions in aqueous media by electrochemical MOF sensors is still in its infancy because of the instability and poor conductivity of MOFs in the aqueous phase. It is, for this reason, the sensing of inorganic toxic anions is mainly performed with luminescence and fluorescence-based sensors while these drawbacks can be overcome by making the incorporation of MOFs by guest materials having high conducting properties such as activated carbon, graphene, and metallic nanoparticles. MOF-based electrochemical sensors require a lot of future work to improve the sensing performance of analytes in the aqueous environment.

REFERENCES

[1]. Chaudhry, F. N., & Malik, M. F. (2017). Factors affecting water pollution: a review. *J Ecosyst Ecography*, *7*(225), 1–3.

[2]. Gao, L., Jiao, C., Chai, H., Ren, Y., Zhang, G., Yu, H., & Tang, L. (2020). A highly sensitive multifunctional Eu-MOF sensor with pentacarboxylate for fluorescence detecting acetone, Cu^{2+}, and Cr_2O7^{2-}, and electrochemical detection of TNP. *Journal of Solid-State Chemistry, 284*, 121199.

[3]. Amini, A., Kazemi, S., & Safarifard, V. (2020). Metal-organic framework-based nanocomposites for sensing applications–A review. *Polyhedron, 177*, 114260.

[4]. Fang, X., Zong, B., & Mao, S. (2018). Metal-organic framework-based sensors for environmental contaminant sensing. *Nano-micro letters*, *10*(4), 1–19.

[5]. Qiu, Q., Chen, H., You, Z., Feng, Y., Wang, X., Wang, Y., & Ying, Y. (2020) Shear Exfoliated Metal-Organic Framework Nanosheet-Enabled Flexible Sensor for Real-Time Monitoring of Superoxide Anion. *ACS applied materials & interfaces*, *12*(5), 5429–5436

[6]. Tranchemontagne, D. J.; Mendoza-Cortes, J. L.; O'Keeffe, M.; Yaghi, O. M. *Chemical Society reviews* 2009, 38, 1257

[7]. Kreno, L. E., Leong, K., Farha, O. K., Allendorf, M., Van Duyne, R. P., & Hupp, J. T. (2012). Metal-organic framework materials as chemical sensors. *Chemical Reviews*, *112*(2), 1105–1125.

[8]. Jiao, L., Seow, J. Y. R., Skinner, W. S., Wang, Z. U., & Jiang, H. L. (2019). Metal–organic frameworks: Structures and functional applications. *Materials Today, 27*, 43–68.

[9]. Sk, M., Banesh, S., Trivedi, V., & Biswas, S. (2018). Selective and sensitive sensing of hydrogen peroxide by a boronic acid functionalized metal-organic framework and its application in live-cell imaging. *Inorganic chemistry*, *57*(23), 14574–14581.

[10]. Deng, S. Q., Mo, X. J., Zheng, S. R., Jin, X., Gao, Y., Cai, S. L., & Zhang, W. G. (2019). Hydrolytically stable nanotubular cationic metal–organic framework for rapid and efficient removal of toxic oxo-anions and dyes from water. *Inorganic chemistry*, *58*(4), 2899–2909.

[11]. Malakar, A., Singh, R., Westrop, J., Weber, K. A., Elofson, C. N., Kumar, M., & Snow, D. D. (2021). Occurrence of arsenite in surface and groundwater associated with a perennial stream located in Western Nebraska, USA. *Journal of Hazardous Materials*, *416*, 126170.

[12]. John, Y., David, V. E., & Mmereki, D. (2018). A comparative study on removal of hazardous anions from water by adsorption: a review. *International Journal of Chemical Engineering, 2018*, 3975948.

[13]. Vorobiev, E. V., Usova, E. V., & Orkhova, Y. V. (2021, April). Analysis of Sources of Anthropogenic Pollution of the Transboundary River, the Serevsky Donets, Based on the Dynamics of the Anion Composition (Nitrites, Nitrates, Sulfates, Chlorides, Phosphates) in 2007-2016. In *IOP Conference Series: Earth and Environmental Science* (Vol. 720, No. 1, p. 012059). IOP Publishing.

[14]. Wang, H., & Zhang, Q. (2019). Research advances in identifying sulfate contamination sources of water environment by using stable isotopes. *International journal of environmental research and public health*, *16*(11), 1914.

[15]. Ribet, S. M., Shindel, B., Dos Reis, R., Nandwana, V., & Dravid, V. P. (2021). Phosphate Elimination and Recovery Lightweight (PEARL) membrane: A sustainable environmental remediation approach. *Proceedings of the National Academy of Sciences*, *118*(23).

[16]. Silpcharu, K., Soonthonhut, S., Sukwattanasinitt, M., & Rashatasakhon, P. (2021). Fluorescent Sensor for Copper (II) and Cyanide Ions via the Complexation–Decomplexation Mechanism with Di (bissulfonamido) spirobifluorene. *ACS Omega*. *6*, 25, 16696–16703.

[17]. Mehta, R., Paul, K., & Luxami, V. (2021). 1-Oxo-1H-phenalene-2, 3-carbonitride Based Sensor for Selective Detection of Cyanide ions in Industrial Waste. *Journal of Molecular Structure, 1234*, 130077.

[18]. Verma, S., Ravichandiran, V., & Ranjan, N. (2021). Selective, pH-sensitive, 'turn on' fluorescence sensing of carbonate ions by a benzimidazole. *Spectrochimica Acta Part A: Molecular and Biomolecular Spectroscopy*, *255*, 119624.

[19]. Koo, W. T., Jang, J. S., & Kim, I. D. (2019). Metal-organic frameworks for chemiresistive sensors. *Chemistry*, *5*(8), 1938–1963.

[20]. Cheng, W., Tang, X., Zhang, Y., Wu, D., & Yang, W. (2021). Applications of metal-organic framework (MOF)-based sensors for food safety: Enhancing mechanisms and recent advances. *Trends in Food Science & Technology, 112*, 268–282.

[21]. Gao, Y., Qi, Y., Zhao, K., Wen, Q., Shen, J., Qiu, L., & Mou, W. (2018). An optical sensing platform for the dual channel detection of picric acid: the combination of rhodamine and metal-organic frameworks. *Sensors and Actuators B: Chemical, 257*, 553–560.

[22]. Chandra Rao, P., & Mandal, S. (2018). Europium-based metal–organic framework as a dual luminescence sensor for the selective detection of the phosphate anion and Fe^{3+} ion in aqueous media. *Inorganic chemistry*, *57*(19), 11855–11858.

[23]. Shahat, A., Hassan, H. M., & Azzazy, H. M. (2013). Optical metal-organic framework sensor for selective discrimination of some toxic metal ions in water. *Analytica chimica acta, 793*, 90–98.

[24]. Zhang, M., Feng, G., Song, Z., Zhou, Y. P., Chao, H. Y., Yuan, D., & Zhao, D. (2014). Two-dimensional metal–organic framework with wide channels and responsive turn-on fluorescence for the chemical sensing of volatile organic compounds. *Journal of the American Chemical Society*, *136*(20), 7241–7244.

[25]. Fang, X., Zong, B., & Mao, S. (2018). Metal–organic framework-based sensors for environmental contaminant sensing. *Nano-micro letters*, *10*(4), 1–19.

[26]. Xu, H., Xiao, Y., Rao, X., Dou, Z., Li, W., Cui, Y., & Qian, G. (2011). A metal–organic framework for selectively sensing of PO43– anion in aqueous solution. *Journal of Alloys and Compounds*, *509*(5), 2552–2554.

[27]. Liu, C., & Yan, B. (2016). A novel photofunctional hybrid material of pyrene functionalized metal-organic framework with conformation change for fluorescence sensing of Cu^{2+}. *Sensors and Actuators B: Chemical, 235*, 541–546.

[28]. Liu, W., Wang, Y., Song, L., Silver, M. A., Xie, J., Zhang, L., & Wang, S. (2019). Efficient and selective sensing of Cu^{2+} and UO_2^{2+} by a europium metal-organic framework. *Talanta, 196*, 515–522.

[29]. Bigdeli, A., Ghasemi, F., Abbasi-Moayed, S., Shahrajabian, M., Fahimi-Kashani, N., Jafarinejad, S., ... & Hormozi-Nezhad, M. R. (2019). Ratiometric fluorescent nanoprobes for visual detection: Design principles and recent advances-A review. *Analytica chimica acta, 1079*, 30–58.

[30]. Zhang, X., Hu, Q., Xia, T., Zhang, J., Yang, Y., Cui, Y., & Qian, G. (2016). Turn-on and ratiometric luminescent sensing of hydrogen sulfide based on metal–organic frameworks. *ACS applied materials & interfaces*, *8*(47), 32259–32265.

[31]. Chen, L., Liu, D., Peng, J., Du, Q., & He, H. (2020). Ratiometric fluorescence sensing of metal-organic frameworks: Tactics and perspectives. *Coordination Chemistry Reviews, 404*, 213113.

[32]. Amini, A., Kazemi, S., & Safarifard, V. (2020). Metal-organic framework-based nanocomposites for sensing applications–A review. *Polyhedron, 177*, 114260.

[33]. Lu, G., Farha, O. K., Kreno, L. E., Schoenecker, P. M., Walton, K. S., Van Duyne, R. P., & Hupp, J. T. (2011). Fabrication of Metal-Organic Framework-Containing Silica-Colloidal Crystals for Vapor Sensing. *Advanced Materials*, *23*(38), 4449–4452.

[34]. Kim, K. J., Lu, P., Culp, J. T., & Ohodnicki, P. R. (2018). Metal–organic framework thin film coated optical fiber sensors: a novel waveguide-based chemical sensing platform. *ACS sensors*, *3*(2), 386–394.

[35]. Bin, Q., Wang, M., & Wang, L. (2020). Ag nanoparticles decorated into metal-organic framework (Ag NPs/ZIF-8) for electrochemical sensing of chloride ion. *Nanotechnology*, *31*(12), 125601.

[36]. Ma, W., Du, X., Liu, M., Gao, F., Ma, X., Li, Y., & Hao, X. (2021). A conductive chlorine ion-imprinted polymer threaded in metal-organic frameworks for electrochemically selective separation of chloride ions. *Chemical Engineering Journal, 412*, 128576.

[37]. Mahmoud, A. M., Mahnashi, M. H., & El-Wekil, M. M. (2021). Indirect differential pulse voltammetric analysis of cyanide at porous copper-based metal organic framework modified carbon paste electrode: Application to different water samples. *Talanta, 221*, 121562.

[38]. Chang, T. H., Kung, C. W., Chen, H. W., Huang, T. Y., Kao, S. Y., Lu, H. C., & Ho, K. C. (2015). Planar heterojunction perovskite solar cells incorporating metal–organic framework nanocrystals. *Advanced Materials*, *27*(44), 7229–7235.

[39]. Kung, C. W., Chang, T. H., Chou, L. Y., Hupp, J. T., Farha, O. K., & Ho, K. C. (2015). Porphyrin-based metal–organic framework thin films for electrochemical nitrite detection. *Electrochemistry Communications, 58*, 51–56.

[40]. Shi, E., Zou, X., Liu, J., Lin, H., Zhang, F., Shi, S., & Qu, F. (2016). Electrochemical fabrication of copper-containing metal–organic framework films as amperometric detectors for bromate determination. *Dalton Transactions*, *45*(18), 7728–7736.

12 MOF-based Electrochemical Sensors for Alkali Metal Cations

Narinder Singh[1] *and Mayank*[2]
[1]Department of Chemistry, Indian Institute of Technology, Ropar, India
[2]Shobhaben Pratapbhai Patel – School of Pharmacy & Technology Management, SVKM's NMIMS University, Vile Parle, Mumbai, India

12.1 INTRODUCTION

chemo sensors for the quantitative and qualitative determination of the analyte of interest seem vital in the modern era of advanced science and technology [1]. The analyte of interest may include a large library of components, including anions, cations, biomolecules, and markers for several types of diseases. In short, everything that is available in our local environment may behave as an analyte of interest and developing sensors for these analytes currently seems relevant. For this, chemo sensors for alkali metal ions seem particularly important because of the critical implications of these analytes in our day-to-day lives [2]. The main examples of alkali metal ions include Li^+, Na^+, K^+, Rb^+, Cs^+ and all these ions have a direct impact on our daily lives [3]. The use of lithium salt for the treatment of bipolar disorder is an age-old practice [4]. However, maintaining optimum blood concentration is crucial, as uncontrolled high Li^+ may harm nephron and neuron-related function. Therefore, quantitative Li^+ detection is vital in this case [5]. Similarly, Na^+ and K^+ are essential elements in biological systems, and deregulation of their concentration is responsible for hypertension, cardiovascular disease, and related health conditions [6-7]. The detection of Cs^+ is still considered extremely important because it is the most common contaminant in the medical and industrial sectors. Moreover, it is also associated with nuclear waste and toxic environmental pollutants [8]. Considering all of these factors, the development of sensors for the detection of alkali metal ions is currently considered to be critical [9].

Recently, analytical techniques have emerged as a highly dynamic scientific area for research and we have witnessed many significant advances. Conventional methods for metal ion detection include inductively coupled plasma mass spectrometry, atomic absorption spectrometry, atomic emission spectroscopy, and many more. However, increased understanding of various electrochemical techniques has now facilitated the development of electrochemical sensors for several forms of metal ion entities [10]. The electrochemical sensing process generally consists of three essential components, namely, a receptor that binds to the analyte, the analyte itself, and a transducer that can convert an analyte-receptor binding reaction into measurable signaling [11]. The electrode is the essential part of electrochemical sensing. However, the majority of conventional electrode materials reveal several limitations, namely, low surface area, the limited scope of surface modification, and lack of structural design. All these limitations are significant, and we have thus explored many other advanced materials with desirable physicochemical properties and improved electrochemical performances [10]. In these cases, nanotechnology has been significant and has provided us with many novel materials for selective, optimal sensitive, and accurate electrochemical sensor development. Nanoparticles are particles with very high surface areas, making them highly sensitive to the external environment under varied analyte concentrations [12].

Because of very high surface area, uniform porosity, and tunable nanoscale properties, MOFs have created plenty of opportunities for ideal sensor development [13]. These MOFs are the coordination polymers fabricated by the combination of organic ligands and metal ions [14]. In this case, multiple varieties of the organic ligand-metal ion coordination modes are possible, providing us with the option of an infinite number of possible MOF configurations (Figure 12.1). Therefore, the various porosity features possible can be designed and controlled, and by doing so, the selective entrapment of a particular type of analyte is possible. The pores present within MOFs is exceptional, as these pores can trap several forms of metal ions. Thus, MOFs can be considered ideal candidates for our sensor development [15]. Moreover, using synthetic skills, we can control the size and shape of the pores of MOFs [16] and by doing so, we can produce MOFs that will be suitable for capturing particular types of metal ions present in diverse environments. Eventually, upon metal-MOF binding, modulation in the electronic properties of the MOF is expected, and electrochemical, and other available analytical methodologies can quickly record these signatures. Therefore, the scientific community is continuously working and has successfully developed sensors for metal ion detection. Here, we discuss the role of MOFs in alkali metal detection.

12.2 THE CATION SENSING MECHANISM OF MOFs

The precisely controlled porosity within the MOF is expected to capture a particular kind of analyte selectively, say metal cations or other analyte entities. Upon the capturing of the

DOI: 10.1201/9781003188148-12

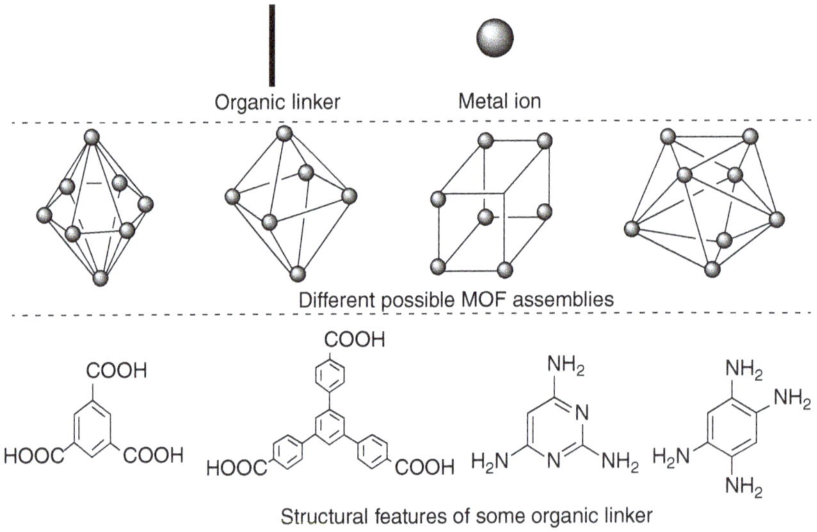

FIGURE 12.1 The general structure feature of MOF using various organic linkers and different metal ions.

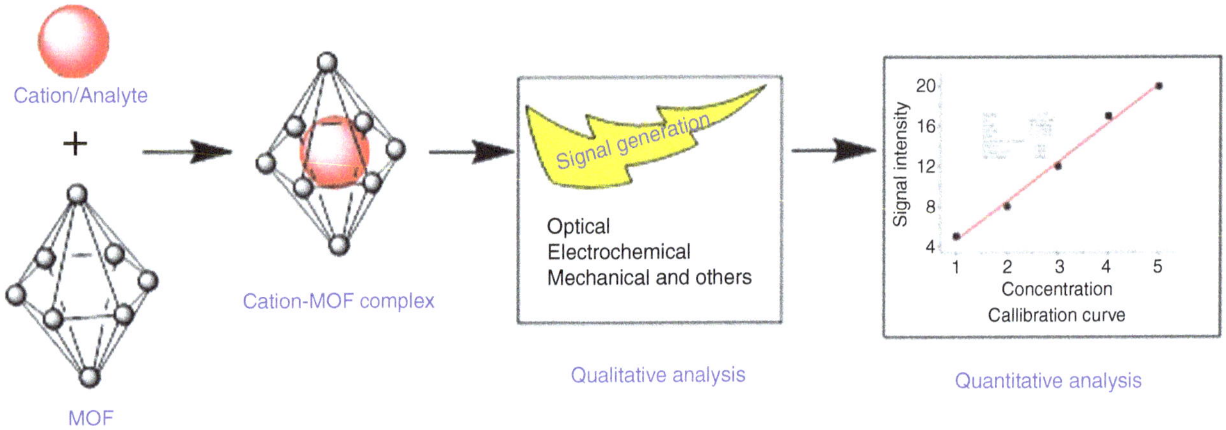

FIGURE 12.2 Diagrammatic representation of the general sensing mechanism of the MOF for cations and other analytes.

pollutant by the MOF, it is further expected that parameters of the MOF such as photoluminescence, UV-Vis absorption, cyclic voltammetry, and related electrochemical behaviors may show significant modulation. Therefore, using these signals, we can easily detect various pollutants within the environment. Qualitative and quantitative analysis for key compounds is also possible. Figure 12.2 represents the general sensing mechanism. The porous core of the MOF represents the accommodation site for the analyte entities. The analytes represent the cations, anions, biomolecules, or other entities and are termed guest molecules. Being selective for a particular analyte only that we can design, the porous core of the MOF can only accommodate a specific kind of analyte, say a specific cation in our case. Once the guest-MOF complex is formed, specific signals such as a change in optical, electrochemical, mechanical, and other similar properties are anticipated, producing scope for the qualitative analysis for one particular cation. By measuring and processing, the intensity of the signal, a calibration curve is possible, and quantitative analysis is possible. Therefore, selective capturing of only a desired analyte and ignoring others in the vicinity is crucial here. It can be achieved by MOF design, namely, by controlling the size and shape of its linked porosities. Hence, while discussing the MOF for alkali metal cation sensing, alkali metal selective MOFs are also emphasized in this chapter.

12.3 MOF-BASED ALKALI CATIONS CAPTURING: ELECTROCHEMICAL SENSING POTENTIAL

12.3.1 MOFs in Cesium Capturing

From the above discussion, we can conclude that the selective capturing of analytes is vital for sensor development. Therefore, the electrochemical sensing potential of MOFs for alkali metal cation detection can be realized by demonstrating the selective capturing of these elements by

MOF-based materials. In due course, the representative examples that are present in the literature strongly advocate possibilities for the use of MOFs in alkali metal detection, particularly by electrochemical sensing mechanisms. In a study by Thallapally et al. the successful development of MOFs with Cs^+ capturing potential has been demonstrated successfully [17]. They have designed a MOF with cation exchange potential. The smartly controlled cation exchange properties for the MOF are responsible for the selective capturing of Cs^+ along with Sr^{2+} under highly competitive environmental conditions [17]. The MOF, namely MIL-101-SO_3H, contains the SO_3H group. They hypothesized that the exchange of – SO_3H protons and Cs^+ and/or Sr^{2+} under aqueous conditions is responsible for selective capturing [17]. The structural feature of MIL-101-SO_3H includes MTN type zeolite architecture, and the – SO_3H group was found evenly distributed throughout the MOF. The MIL-101-SO_3H was also stable at varied pH ranges and capable of capturing Cs^+ under diverse pH conditions.

In a study by H. Faghihian et al. a magnetic MOF was found efficient in capturing Cs^+ under aqueous conditions [18]. The research group prepared a magnetic MOF by the magnetization and functionalization of the HKUST-1 MOF. The HKUST-1 is a well-known MOF, also known as $[Cu_3(BTC)_2]$, which contains Cu^+ as a metal entity. The magnetization and functionalization process of HKUST-1 was executed by soaking it with potassium nickel hexacyanoferrate (KNiFC) [18]. The resulting functionalized MOF was found to be very efficient in capturing Cs^+ ions under aqueous conditions. The potassium nickel hexacyanoferrate is well known for its Cs^+ adsorbing potential, and its hybridization with MOFs can be thus considered an essential strategy in MOF-based Cs^+ sensor development [18].

A study by Q. J. Pan et al. has also developed a porous anionic uranyl-organic framework (UOF) [19]. For this development, they have used tetra topic carboxylate ligand 1,3,6,8-tetrakis(p-benzoic acid)pyrene (H_4TBAPy) and D_{3h}-symmetrical triangular $[UO_2(COO)_3]^-$. By doing so, they have successfully developed two interesting compounds with very high Cs^+ adsorbing properties under aqueous conditions. The proposed binding mode is represented in Figure 12.3 below [19].

Similarly, Shuao Wang and his research group have also developed 3D uranyl organic framework material with Cs^+ capturing capabilities. The structural feature of the produced material is unique, has a very high surface area, and completely exchangeable $[(CH_3)_2NH_2]^+$ cations. Therefore, this material was found to be very efficient in capturing Cs^+ from the aqueous environment [20].

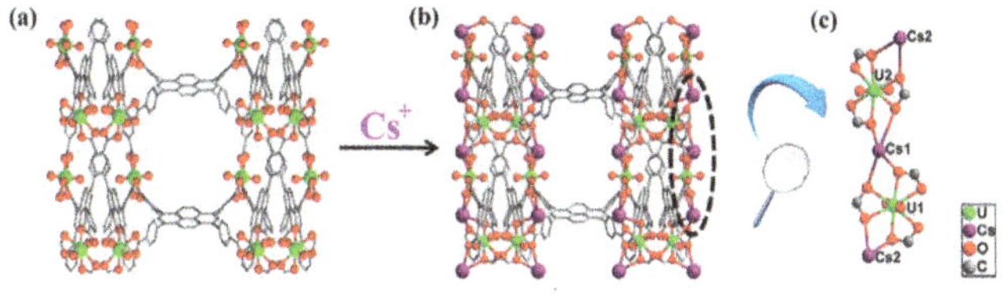

FIGURE 12.3 Pictorial representation of UOF (a) and its Cs^+ capturing mechanism of the same. [Reprinted with permission from Polyhedron. 2019, 57, 8, 4419-4426, Copyright 2019 Elsevier].

12.3.2 MOFs in Rubidium Capturing

In a research work by Wei Dai et al. HS-Fe_3O_4@MIL-53 (Al) has been developed and found to be very efficient in capturing rubidium ions [21]. This particular MOF contains sulfhydryl groups and magnetic Fe_3O_4 doped in it, and because of that, it was found efficient in capturing rubidium ions. The Rb^+ capturing mechanism of this particular MOF includes an acidic proton exchange process between the sulfhydryl group and the Rb^+ ion. It was further observed that the selectivity of Rb^+ was only moderately affected by K^+, Na^+, and Cs^+ ions [21]. It has been further observed that the adsorption kinetic was also fast, and maximum adsorption was attained in 30 min. The fast adsorption kinetic is significant, as it will allow us to develop sensors for the rapid detection of the Rb^+ cations [21].

In yet another report by Wei Dai et al. a rubidium capturing MOF has been developed by facial polydimethylsiloxane (PDMS) coating treatment on the bimetallic HKUST-1 surface. By doing so, a water-resistant MOF, namely PDMS@HKUST-1, has been designed with admirable properties [22]. The PDMS@HKUST-1, on the one hand, exhibits good wettability, and on the other hand, it also shows excellent Rb^+ adsorption capabilities. The observed adsorption process is further described by using the Freundlich adsorption model and the pseudo-second-order equation [22]. To explain the Rb^+ capturing mechanism, two mechanisms are considered. The first one is based on the sieve-threshold limiting effect (STLE). According to this, adsorbing entities only of a specific size are sieved by the passage of complementary shapes on the adsorbent material. Finally, the adsorbate is captured firmly inside the sorbent MOF material. In this, the path size of the HKUST-1 is sufficiently larger so that it can capture Rb^+ of 0.148 nm diameter. The second mechanism assumed here includes an electrostatic attraction force between MOF and the Rb^+ ion [22].

Similarly, XiaoyingYan et al. have also developed phosphomolybdic acid promoted HKUST-1 (PMA@HKUST-1) with selective Rb^+ capturing abilities [23]. Various characterization techniques such as XRD, SEM, gravimetric analysis,

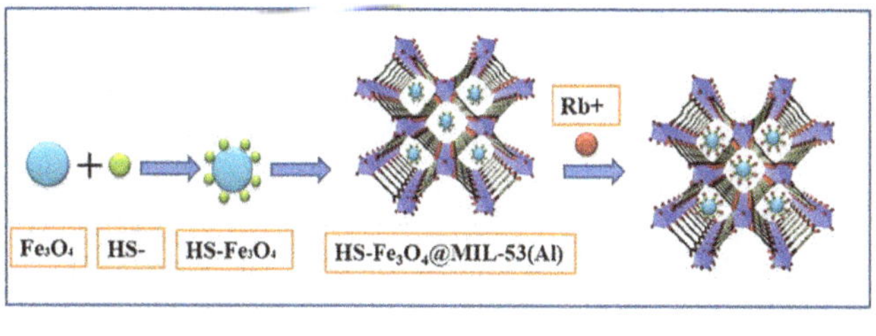

FIGURE 12.4 Pictorial representation of HS-Fe_3O_4@MIL-53(Al) and its Rb^+ capturing mechanism. Reprinted with permission from Inorg. Chem. 2018, 166, 109-114, Copyright 2018 American Chemical Society.

IR, and N_2 adsorption isotherms have revealed that the immobilization of PMA was present over HKUST-1. The final material, namely, PMA@HKUST-1, was found suitable for capturing Rb^+ ions with solid chemical affinity. The proposed mechanism by which the MOF captures Rb^+ includes Lewis acid-base interaction in between the adsorbant and the Rb^+ molecule [23]. Moreover, in this case, the pseudo-second-order model describes the adsorption kinetics, and the Freundlich model expresses the adsorption isotherm. The influence of other ions such as K^+, Na^+, and Cs^+ was found moderate, and thus selectivity towards Rb^+ was also ensured in the study [23].

Similarly, yet another report by Na Ma et al. also suggested the use of MOFs in Rb+ ion capturing in the aqueous medium [24]. In this case, the research group has developed phenol@MIL-101(Cr) by functionalization of Mil-101(Cr) with different amounts of phenol. Based on the results obtained from XRD, thermogravimetry, SEM, FTIR, and N_2 adsorption isotherm, immobilization of phenol over MIL-101(Cr) was concluded in the study [24]. The resulting MOF, namely, phenol@MIL-101(Cr), when tested for Rb^+ capturing efficiency, revealed a high Rb^+ charging tendency. The experiment was conducted at (25 °C) and under aqueous conditions. The Rb^+ capturing mechanism includes a proton exchange between the acidic proton of the phenol group and the Rb^+ ion. Moreover, phenol@MIL-101(Cr), once saturated with Rb^+ can be regenerated easily by washing with ammonium nitrate solution. By doing so, the regaining of more than 90% of the total Rb^+ capturing capacity is possible [24].

Therefore, a significant number of examples of MOFs with Rb^+ capturing properties are available in the literature. Hence, these Rb^+ capturing MOFs are ideal candidates for the development of electrochemical sensors to detect Rb^+ ions. Further, as in the majority of cases, acidic proton exchange is the primary mechanism, and that can alter the conductivity strength of the local environment. Thus, the design and fabrication of electrochemical sensors is entirely possible. Meanwhile, the defined Rb^+ capturing role of the aforementioned MOF designs in the aqueous condition remains an attractive feature.

12.3.3 MOFs in Potassium Capturing

In a study by Prof. Yu Tang et al. an interesting MOF has been fabricated, showing K^+ capturing abilities [25]. The fabrication process of the material is quite interesting as it includes Fe_3O_4 as a magnetic core on which mesoporous silica ($mSiO_2$) produces a shell-like arrangement. Subsequently, a three-dimension ligand, namely, 1,1,1-tris[[(2′-benzylaminoformyl)phenoxy]methyl] ethane (L) was allowed to adsorb it to produce a final arrangement, namely $Fe_3O4@mSiO_2@L$. The $Fe_3O_4@mSiO_2@L$ so produced was found to be very efficient in separating K^+ ions in the presence of Ca^{2+}, Mg^{2+}, Na^+, ions. In this, the mechanism of K^+ capturing includes nanoparticle-assisted MOF crystallization and the formation of $Fe_3O_4@mSiO_2@MOF$ as a hybrid material. Therefore, it is a MOF-based unique strategy for capturing K^+ under aqueous conditions [25].

12.3.4 MOFs in Lithium Capturing

In a research work by Sourav Saha et al, the development of the lithium binding MOF has been described. The structural feature of the current MOF includes 2D sheets like neutral Cu(I)-sulfonate and π-acidic naphthalenediimide (NDI) ligand entities. The current arrangement was found to be binding the guest lithium-ion because of the availability of carbonyl and uncoordinated sulfonate oxygen atoms within its structure. The recent example revealed that the rationally incorporated functional groups on the surface of the MOF and surface functionalization are critical in developing MOFs with particular metal binding affinities. Figure 12.5 represents the binding mechanism of Li^+ as a guest molecule against the developed MOF material [26].

Yet another report by Z. Liu et al. also represented the development of the MOF, namely, Zn(HBTC)(bpbenz)·H_2O (Zn-MOF-COOH) with free carboxylic acid moieties. In this, uncoordinated -COOH groups that are present can be used for Li^+ ion exchange. The Li^+ ion-exchange mechanism of the current MOF is as represented in Figure 12.6 [27].

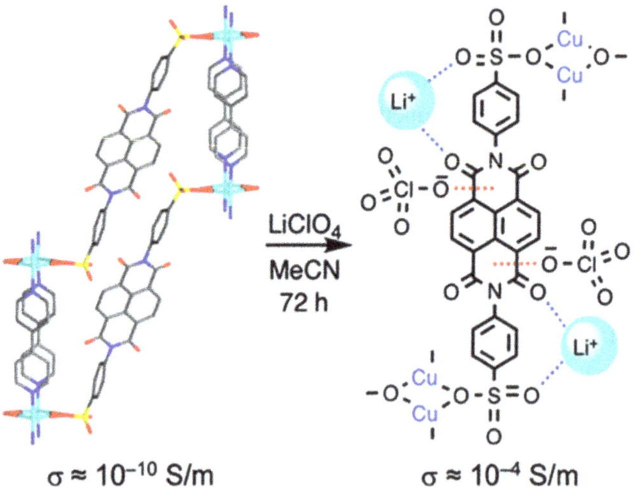

FIGURE 12.5 Pictorial representation of $Cu_2(BPY)_2(NDIDS)$ MOF and its Li^+ binding mechanism. Reprinted with permission from Inorg. Chem. 2018, 7, 5, 4619–4624, Copyright 2018 American Chemical Society.

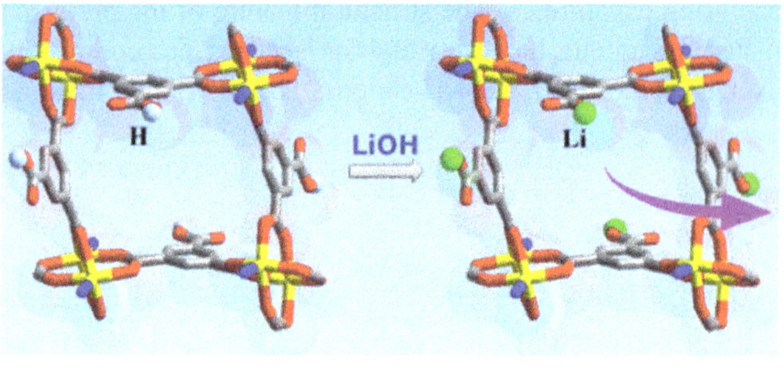

FIGURE 12.6 Pictorial representation of Zn(HBTC)(bpbenz)·H_2O (Zn-MOF-COOH) MOF and its Li^+ binding mechanism. Reprinted with permission from Langmuir. 2018, 2021, 37, 13, 3922–3928, Copyright 2021 American Chemical Society.

Based on the above discussion, we can conclude that the scientific community continuously develops MOFs with particular metal ion capturing capabilities. Examples of the alkali metal ion-specific MOFs are also available in the literature. Therefore, the development of electrochemical sensors for the detection of the alkali metal ion sensors by using a MOF-based approach is entirely possible. As per the above description, the alkali metal ion capturing mechanism of MOFs generally includes ion exchange, electrostatic interaction with a free functional group, and shape-selective entrapment of cation entities. Moreover, all these examples were found to be working well in aqueous conditions. If we capture a particular ion from the local environment, it can also affect the electron conductivity and related parameters detected by cyclic voltammetry and other associated techniques. It can therefore provide us with a way to develop sensors for alkali metal ions and associated entities.

Meanwhile, not all MOFs are perfectly selective for a particular kind of alkali metal ion. We can realize the situation from the research work by V. P. Fedin et al. where the research group has developed a MOF, namely $(H_3O)_2[Zn_4(ur)(Hfdc)_2(fdc)_4]$ [28], which can capture multiple alkali metal ions. Therefore, detection and quantification of a specific alkali cation looks to be challenging in this case. To overcome such a scenario, it is always ideal for designing and developing MOFs to capture or interact with alkali metal ions in a selective manner. Otherwise, we can also develop a method that can allow us to identify and process signals that originate from a particular analyte amongst all the available signals that originate from all the captured cations. MOFs and their application in electrochemical alkali metal sensing is not sufficiently explored, however, reports related to MOFs and their alkali metal ion capturing efficiency are sufficiently available in the literature. Meanwhile, the important report for the alkali metal capturing MOFs is summarized in Table 12.1 below.

12.4 MOF DESIGNING STRATEGY

For designing MOFs for electrochemical sensor development, two different approaches, namely, a structural approach and a functional approach, are possible. The primary purpose of the structural approach is to control the structure, topologies, and porosities of the MOF so that it should allow entrapment of only specific sized guest entities [29]. It can be further categorized into symmetry-guided designing and designing cavities of customized sizes and shapes. The fluorite topology is a particular topology that contains tetrahedral nodes and cubic nodes linked together by only one type of linkage. Therefore, if we want to develop this kind of topology within our MOF design, we can select two appropriate components, one that would represent a tetrahedral node and the second that would represent a cubic node. Therefore, by choosing the components appropriately, the size, shape, and topology of the MOF-associated pores can be controlled [29].

The second structural approach includes the design of the cavities of customized sizes and shapes. If a MOF contains cavities similar to the guest molecules, it may or may not entrap the guest molecule. However, a cavity with a similar shape and size and complimentary polarity will selectively attract and trap the guest molecule. Therefore, in this approach, we will try to select the components in such a way that the final MOF will not only provide perfectly shaped and sized cavities but also provide suitable charge distribution [29]. Therefore, while designing a MOF for alkali metal cations, the initial

TABLE 12.1
Summarization of reported MOFs, reported to show binding against alkali metal cations

Sr no	MOF name	Binding Mechanism	Binding Metal
1	MIL-101-SO_3H	Exchange of $-SO_3H$ proton and Cs^+ and/or Sr^{2+}	Cs^+
2	Magnetized and functionalized HKUST-1	Potassium nickel hexacyanoferrate functionalization, which is well known for Cs^+ adsorption	Cs^+
3	Uranyl-organic framework development, by tetratopic carboxylate ligand 1,3,6,8-tetrakis(p-benzoic acid) pyrene (H_4TBAPy) and D_{3h}-symmetrical triangular $[UO_2(COO)_3]^-$.	Adsorption	Cs^+
4	3D uranyl organic framework Based MOF	Completely exchangeable $[(CH_3)_2NH_2]^+$ cations.	Cs^+
5	HS-Fe3O_4@MIL-53(Al)	Acidic proton exchange process between the sulfhydryl group and the Rb^+ ion	Rb^+
6	PDMS@HKUST-1	Size-specific sieving, electrostatic interaction	Rb^+
7	Phenol@MIL-101(Cr),	Proton exchange between the acidic proton of the phenol group and the Rb^+ ion	Rb^+
8	Fe_3O_4@mSiO_2@L	Fe_3O_4@mSiO_2@MOFformation	K^+
9	$Cu_2(BPY)_2$(NDIDS)	Binding of Li^+ by carbonyl and sulfonate oxygen atoms	Li^+
10	Zn(HBTC)(bpbenz)·H_2O (Zn-MOF-COOH)	Li^+-H^+ exchange	Li^+

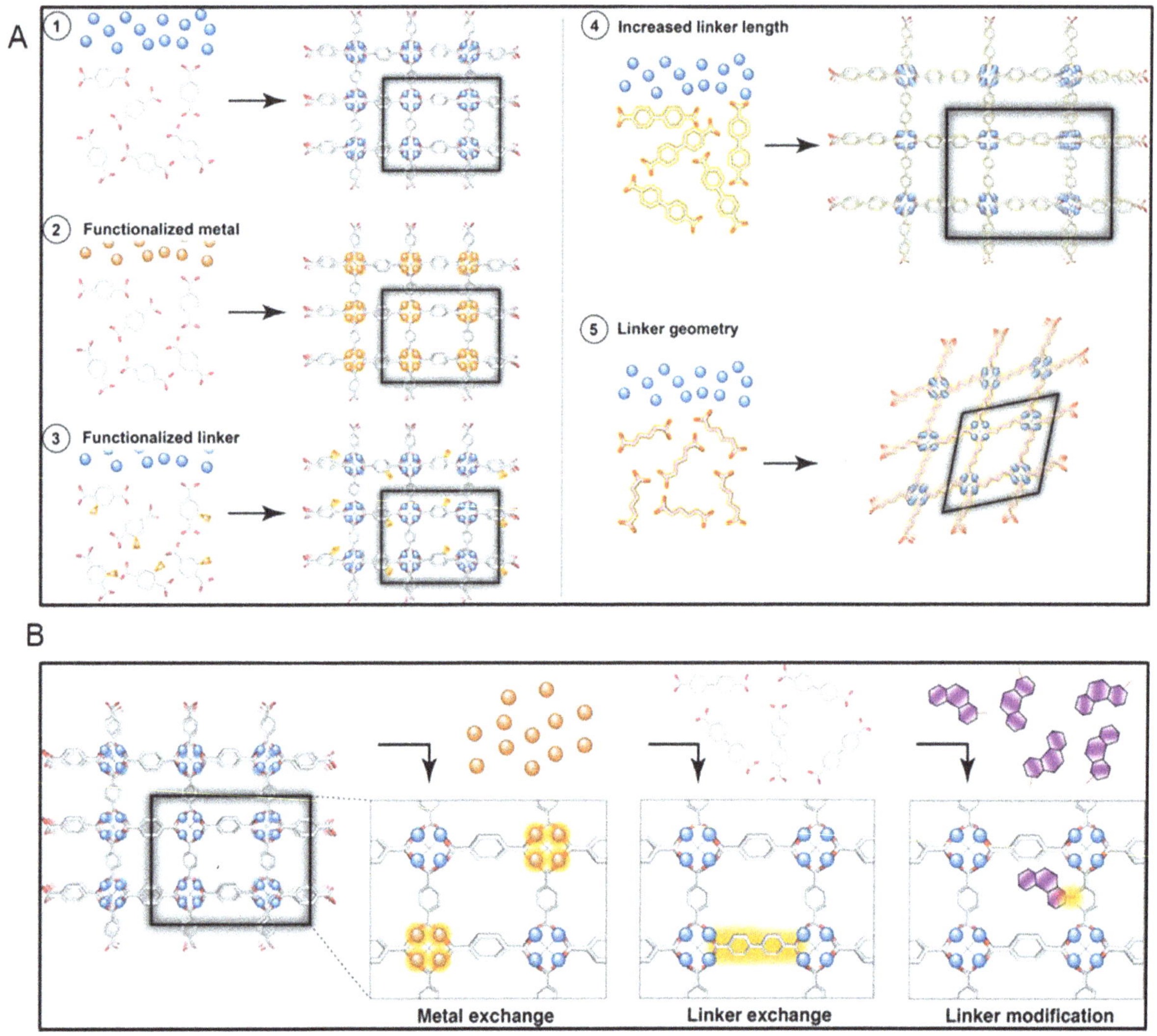

FIGURE 12.7 Pictorial representation showing the pre-synthetic (**A**) and post-synthetic approach (**B**) in MOF design. This approach is ideal for getting the desired functional groups within the final MOF structure. Reprinted with permission from Advanced Functional Materials 2020, 30, 1909062, Copyright 2020 The Authors. Published by WILEY-VCH Verlag GmbH & Co. KGaA, Weinheim.

selection of organic and metal components is a critical aspect in suitable MOF design.

The second important approach for MOF design is the functional approach, which is also categorized into two subgroups. The first subgroup includes pre synthetic approach, where the building blocks are initially functionalized with the desired functional group before its incorporation into the MOF. Therefore, we can expect selected functional groups in the MOFs prepared using pre-functionalized building block moieties [30]. The second subgroup includes post-synthetic modifications, where initially simpler building blocks are joined together to produce a MOF. Subsequently, the structural features of the MOF are modified by metal exchange, linker exchange, and linker modification-based strategies [31]. Therefore, this approach is ideal for obtaining appropriate functional groups within the final MOF structure.

As discussed above, most of the alkali metal cation capturing MOFs work via a proton exchange mechanism. Therefore, by selecting and incorporating appropriate functional groups into the MOF, either by pre- or post-synthetic modification, we can create some exchangeable protons within its cavities for capturing alkali metal ions. Therefore, considering the size and shape of the guest alkali metal cation, the cavity size of the MOF can be controlled by a structural approach, and required functionalities can be incorporated there via a functional system. In this way, designing an ideal MOF candidate for electrochemical alkali metal sensing is possible.

12.5 RECENT PROGRESS AND FUTURE CHALLENGES

The applications of MOFs in the development of sensors and biosensors is increasing day by day. Because of the pioneering work of Lan et al. we have now realized the importance and sensing applications of MOFs for various analytes, including alkali metal ions [31]. In the case of sensors, sensitivity is a critical factor, and the high binding affinity of analyte against sensing material and high loading capacity of the sensing material indicates the MOF in our case is very important for optimizing the sensitivity of the sensor probe. Because of their porosity, MOFs are associated with a very high adsorbing surface area, and thus, loading capacity is always high. If we design a MOF to provide a very high affinity towards an analyte, we can develop a sufficiently sensitive MOF-based electrochemical sensor for alkali metal cations and other analyte entities. Therefore, by using MOFs, highly sensitive sensor development is quite possible. However, providing MOFs with very high-affinity properties towards a particular alkali metal cation is currently a major challenge.

In due course, designing and developing MOFs that are perfectly selective for a particular metal ion is considered essential if we want to impart selectivity in sensor design. For this, an understanding of MOF designing strategies is necessary. It will guide us in controlling the size, shape, and functional groups of cavities to make the MOFs selective for particular alkali metal cations. Therefore, sensor-sensitivity is related to binding affinity, and sensor-selectivity is also related to MOF-cation selective binding properties.

In literature, examples of MOFs are available, showing some selectivity in binding against particular alkali metal cations. The selective binding mechanism is also documented, but most of these MOFs are not explored for their alkali metal sensing capabilities. Based on these results, in the future, it is quite possible to develop selective and sensitive MOF-based electrochemical sensors for alkali metal detection. To do so, already documented alkali metal capturing MOFs and associated mechanisms can be explored, and by rational MOF design, the selectivity and binding affinity of MOFs for a specific alkali cation can be fine-tuned. In the case of MOF-based electrochemical alkali metal sensing, no significant work is currently published. However, we cannot ignore the sensing potential of MOFs, and we are expecting a significant number of MOF-based alkali metals shortly.

12.6 CONCLUSION

Alkali metal cations are a vital component of the biological system. These are involved in maintaining cardiovascular, neurological, and enzymatical functions in the case of the living organism. Considering these facts, quantitative and qualitative determination of these cations seems very important these days. The scientific community throughout the world is working extensively to develop alkali metal sensors, and many success stories are available in the literature. Meanwhile, electrochemical sensors have several advantages over others, and the development of such sensors for alkali metal cations is also possible. The MOF is one such starting material that can be rationally designed to develop electrochemical sensors for alkali metal cations.

As per the current scenario, MOFs with selective alkali metal cation capturing potential are reported in the literature, but these are not adequately explored for their sensing potential. These MOFs, either as they are or by slight modification, can provide us with ideal electrochemical sensors for alkali metal cations. Meanwhile, several knowledge-based MOF design strategies are also available, and therefore, controlling the size, shape, morphology, and free functional groups within the cavities of the MOF is entirely possible. Consequently, we can design and develop MOFs for electrochemical sensing applications. Therefore, although not much is reported for MOFs in the electrochemical sensing of alkali metal cations, they have significant potential for such sensor development.

REFERENCES

[1]. Nandre, J.; Patil, S.; Patil, V.; Yu, F.; Chen, L.; Sahoo, S.; Prior, T.; Redshaw, C.; Mahulikar, P.; Patil, U., A novel fluorescent 'turn-on' chemosensor for nanomolar detection of Fe (III) from aqueous solution and its application in living cells imaging. *Biosensors and Bioelectronics* **2014,** *61*, 612–617.

[2]. Xia, W.-S.; Schmehl, R. H.; Li, C.-J.; Mague, J. T.; Luo, C.-P.; Guldi, D. M., Chemosensors for lead (II) and alkali metal ions based on self-assembling fluorescence enhancement (SAFE). *The Journal of Physical Chemistry B* **2002,** *106* (4), 833–843.

[3]. Perkins, J.; Gadd, G. M., Interactions of Cs+ and other monovalent cations (Li+, Na+, K+, Rb+, NH4+) with K+-dependent pyruvate kinase and malate dehydrogenase from the yeasts Rhodotorula rubra and Saccharomyces cerevisiae. *Mycological Research* **1996,** *100* (4), 449–454.

[4]. Mitchell, P. B.; Hadzi-Pavlovic, D., Lithium treatment for bipolar disorder. *Bulletin of the World Health Organization* **2000,** *78*, 515–517.

[5]. Kamenica, M.; Kothur, R. R.; Willows, A.; Patel, B. A.; Cragg, P. J., Lithium-ion sensors. *Sensors* **2017**, *17* (10), 2430.

[6]. Gao, G.; Cao, Y.; Liu, W.; Li, D.; Zhou, W.; Liu, J., Fluorescent sensors for sodium ions. *Analytical Methods* **2017**, *9* (38), 5570–5579.

[7]. Huang, C.-C.; Chang, H.-T., Aptamer-based fluorescence sensor for rapid detection of potassium ions in urine. *Chemical communications* **2008,** (12), 1461–1463.

[8]. Chen, C.; Wang, J., Removal of Pb^{2+}, Ag^{+}, Cs^{+} and Sr^{2+} from aqueous solution by brewery's waste biomass. *Journal of hazardous materials* **2008,** *151* (1), 65–70.

[9]. Özcan, E.; Çoşut, B., Fluorescent Sensing of Cesium Ions by an Amide-Linked BODIPY Dye: Synthesis and Photophysical Properties. *ChemistrySelect* **2018**, *3* (27), 7940–7944.

[10]. Wang, Y.; Wang, L.; Huang, W.; Zhang, T.; Hu, X.; Perman, J. A.; Ma, S., A metal–organic framework and conducting polymer based electrochemical sensor for high performance cadmium ion detection. *Journal of Materials Chemistry A* **2017,** *5* (18), 8385–8393.

[11]. Harper, A.; Anderson, M. R., Electrochemical glucose sensors—developments using electrostatic assembly and carbon nanotubes for biosensor construction. *Sensors* **2010,** *10* (9), 8248–8274.

[12]. Grieshaber, D.; MacKenzie, R.; Vörös, J.; Reimhult, E., Electrochemical biosensors-sensor principles, and architectures. *Sensors* **2008,** *8* (3), 1400–1458.

[13]. Cheng, W.; Tang, X.; Zhang, Y.; Wu, D.; Yang, W., Applications of metal-organic framework (MOF)-based sensors for food safety: Enhancing mechanisms and recent advances. *Trends in Food Science & Technology* **2021,** *112*, 268–282.

[14]. Moghadam, P. Z.; Rogge, S. M.; Li, A.; Chow, C.-M.; Wieme, J.; Moharrami, N.; Aragones-Anglada, M.; Conduit, G.; Gomez-Gualdron, D. A.; Van Speybroeck, V., Structure-mechanical

stability relations of metal-organic frameworks via machine learning. *Matter* **2019,** *1* (1), 219–234.

[15]. Ma, X.; Chai, Y.; Li, P.; Wang, B., Metal–organic framework films and their potential applications in environmental pollution control. *Accounts of chemical research* **2019,** *52* (5), 1461–1470.

[16]. Mian, M. R.; Redfern, L. R.; Pratik, S. M.; Ray, D.; Liu, J.; Idrees, K. B.; Islamoglu, T.; Gagliardi, L.; Farha, O. K., Precise Control of Cu Nanoparticle Size and Catalytic Activity through Pore Templating in Zr Metal–Organic Frameworks. *Chemistry of Materials* **2020**, *32* (7), 3078–3086.

[17]. Aguila, B.; Banerjee, D.; Nie, Z.; Shin, Y.; Ma, S.; Thallapally, P. K., Selective removal of cesium and strontium using porous frameworks from high level nuclear waste. *Chemical Communications* **2016**, *52* (35), 5940–5942.

[18]. Naeimi, S.; Faghihian, H., Performance of novel adsorbent prepared by magnetic metal-organic framework (MOF) modified by potassium nickel hexacyanoferrate for removal of Cs^+ from aqueous solution. *Separation and Purification Technology* **2017**, *175*, 255–265.

[19]. Ai, J.; Chen, F.-Y.; Gao, C.-Y.; Tian, H.-R.; Pan, Q.-J.; Sun, Z.-M., Porous Anionic Uranyl–Organic Networks for Highly Efficient Cs^+ Adsorption and Investigation of the Mechanism. *Inorganic Chemistry* **2018,** *57* (8), 4419–4426.

[20]. Wang, Y.; Liu, Z.; Li, Y.; Bai, Z.; Liu, W.; Wang, Y.; Xu, X.; Xiao, C.; Sheng, D.; Diwu, J.; Su, J.; Chai, Z.; Albrecht-Schmitt, T. E.; Wang, S., Umbellate Distortions of the Uranyl Coordination Environment Result in a Stable and Porous Polycatenated Framework That Can Effectively Remove Cesium from Aqueous Solutions. *Journal of the American Chemical Society* **2015,** *137* (19), 6144–6147.

[21]. Tian, N.; Dai, Y.; Liu, Q.; Dai, W., Highly efficient capture of rubidium ion by a novel HS-Fe_3O_4@MIL-53(Al) composite material. *Polyhedron* **2019,** *166*, 109–114.

[22]. Tian, N.; Gao, Y.; Wu, J.; Luo, S.; Dai, W., Water-resistant HKUST-1 functionalized with polydimethylsiloxane for efficient rubidium ion capture. *New Journal of Chemistry* **2019,** *43* (39), 15539–15547.

[23]. Dai, W.; Fang, Y.; Yu, L.; Zhao, G.; Yan, X., Rubidium ion capture with composite adsorbent PMA@HKUST-1. *Journal of the Taiwan Institute of Chemical Engineers* **2018,** *84*, 222–228.

[24]. Fang, Y.; Zhao, G.; Dai, W.; Ma, L.; Ma, N., Enhanced adsorption of rubidium ion by a phenol@MIL-101(Cr) composite material. *Microporous and Mesoporous Materials* **2017,** *251*, 51–57.

[25]. Wu, W.; Kirillov, A. M.; Yan, X.; Zhou, P.; Liu, W.; Tang, Y., Enhanced Separation of Potassium Ions by Spontaneous K^+-Induced Self-Assembly of a Novel Metal–Organic Framework and Excess Specific Cation–π Interactions. *Angewandte Chemie International Edition* **2014,** *53* (40), 10649–10653.

[26]. Panda, D. K.; Maity, K.; Palukoshka, A.; Ibrahim, F.; Saha, S., Li^+ Ion-Conducting Sulfonate-Based Neutral Metal–Organic Framework. *ACS Sustainable Chemistry & Engineering* **2019,** *7* (5), 4619–4624.

[27]. Tian, L.; Xu, X.; Liu, M.; Liu, Z.; Liu, Z., Significantly Enhancing the Lithium Ionic Conductivity of Metal–Organic Frameworks via a Postsynthetic Modification Strategy. *Langmuir* **2021,** *37* (13), 3922–3928.

[28]. Sapchenko, S. A.; Demakov, P. A.; Samsonenko, D. G.; Dybtsev, D. N.; Schröder, M.; Fedin, V. P., A Cryptand Metal–Organic Framework as a Platform for the Selective Uptake and Detection of Group I Metal Cations. *Chemistry – A European Journal* **2017**, *23* (10), 2286–2289.

[29]. Zhang, M.; Bosch, M.; Gentle III, T.; Zhou, H.-C., Rational design of metal–organic frameworks with anticipated porosities and functionalities. *CrystEngComm* **2014,** *16* (20), 4069–4083.

[30]. Ploetz, E.; Engelke, H.; Lächelt, U.; Wuttke, S., The Chemistry of Reticular Framework Nanoparticles: MOF, ZIF, and COF Materials. *Advanced Functional Materials* **2020,** *30* (41), 1909062.

[31]. Lei, J.; Qian, R.; Ling, P.; Cui, L.; Ju, H., Design, and sensing applications of metal–organic framework composites. *TrAC Trends in Analytical Chemistry* **2014,** *58*, 71–78.

13 MOF-based Electrochemical Sensors for Nitrogen Oxide/Carbon Dioxide

Raghabendra Samantaray[1,2], *Soujanya Ghosh*[1] *and Nityananda Agasti*[3]
[1]Department of Biotechnology, KIIT University, Bhubaneswar, India
[2]Department of Chemical Technology, KIIT University, Bhubaneswar, India
[3]Department of Chemistry, Deen Dayal Upadhyaya College, University of Delhi, New Delhi, India

13.1 INTRODUCTION

Increased emissions of various pollutants like carbon dioxide (CO_2), nitrogen oxides (NO_x), and matter particles that adversely impact air quality are considered and recognized as significant environmental and health risks. Nitric oxide (NO), nitrogen dioxide (NO_2) (generally referred to as NO_x), and nitrous oxide (N_2O) are acknowledged as harmful substances to have short and long-term effects on living systems. As a glaring example, nitrogen dioxide (NO_2) is a ubiquitous pollutant generated during combustion reactions [1]. Accordingly, the concentration level of NO_2 that is immediately dangerous to life is 20 ppm. This NO_2 has toxic effects on the respiratory systems of living beings and can result in a reduced life span of living organisms having prolonged exposure to the said pollutant [2]. In comparison, exposure above a level of 150 ppm for 30 mins to an hour results in fatal pulmonary edema or asphyxia and can result in rapid death. Therefore, it is urgent to detect and monitor these pollutants. Due to recent progress in material science, metal-organic frameworks (hereafter referred to as MOFs) have provided an avenue for sensing various gaseous pollutants and reactants, especially the CO_2 and NO_x class of contaminants.

Metal-organic frameworks (MOFs) are hybrid porous materials comprised of metal ions or clusters connected by organic linker molecules to form crystalline coordination polymer networks. MOFs are well known for novel properties like porosity (enormous pore volumes up to 7000 m^3/g) [3], ultralow densities [3], high surface area [4], and high chemical versatility [4]. These properties make them well-matched for sensing NO_x and CO_2 molecules. In addition, they are recognized for relatively easy chemical synthesis, they are inexpensive, and environmental-friendly as most of them are non-toxic [4]. MOFs are smart porous materials and are considered superior to zeolites in certain aspects [5]. The unprecedented porosity, structural and functional diversity make them accessible to multiple uses. It also acts as a stimuli-response material which is a requisite character for sensor materials. MOFs are composed of two major components: (a) metal nodes, a metal ion or cluster of metal ions, and (b) linkers: organic molecules. Aromatic, non-aromatic, aliphatic, and coordinating carboxylate, imidazolate, triazolates, tetrazolate, pyridine-based linkers, amino acids, and also their derivatives [6].

Since MOFs possess attractive structural features and flexibility, they have been successfully utilized for various applications in catalysis, gas separation, gas storage, drug delivery, and sensing. These applications of MOFs can be ascribed to: (a) Structural diversity arising from the synthetic routes in various MOFs through multiple possible combinations of metal centers and organic linkers leading essentially to an infinite number of variations with predictable structures; (b) the ability to tune the framework architectures and properties thereof via ligands and metal ions providing a significant advantage over other porous material such as zeolites; (c) easy synthesis and possible scaling up by applying various strategies like room-temperature synthesis, electrochemical synthesis, and the like, are possible; (d) high porosity, tunable pore sizes, easy incorporation of functional groups in MOFs. The dimensions and topology of channels can be tuned through organic synthesis by modifying the molecular structure of the organic ligand that bridges the metal ions; (e) the surface properties of channels can be altered by appending different organic substituents onto the organic ligands without distorting the original topology; and (f) the prospect of predictable alteration of organic units to provide tailored materials for applications [12,13].

For the aforementioned reasons, MOFs have overtaken other conventional materials for applications as sensor material. Here, in this chapter, we present a systematic overview of MOFs as sensor materials for sensing pollutants like CO_2 and NO_x.

13.2 MOFs FOR SENSING APPLICATIONS

The structural flexibility and tunability of MOFs make them one of the attractive materials for sensor applications. As hosts, the ability of MOFs to attract guest molecules to their pores (host-guest interaction), is an important parameter in making MOFs suitable as sensors. In comparison to other porous materials such as zeolites, MOFs are considered superior because of their structural variability, which arises from their various constituent metal centers and organic linkers. These properties are necessary for the application of MOFs as sensors for gaseous molecules like carbon dioxide (CO_2) and nitrogen oxide (NO_x) gases [4]. Selective and sensitive detection of these vapor phase molecules is needed

DOI: 10.1201/9781003188148-13

for applications in the field of industrial processes, medical diagnoses, and environmental monitoring and management [5, 8, 13]. Most of the current commercial sensors are based on organic polymers or inorganic semiconductor films. These materials detect the analytes through their changes in electrical and photophysical behavior [11]. The strength of the output signals depends on the nature and concentration of analytes.

The important parameters for suitability as a sensor material are sensitivity, selectivity, output signals, and material stability [11]. MOFs with their highly porous characters possess all these properties and can sense vapor phase molecules like CO_2 and NO_x with significant efficacy. MOFs attract gas molecules to their pores where the analytes are deposited. The detection of these vapor phase molecules can be achieved via host-guest interactions and subsequent signal transduction. The sensitivity of detection depends on the strength of binding of these gas molecules with the MOFs, the hosts. The stronger the binding (interaction), the lesser will be the signal transduction [9]. Thus, the selection of MOFs for sensing these gas molecules is a key step in the overall process. While selecting MOFs, the MOFs aperture is given utmost importance as the gas molecules are going to get adsorbed to the internal pores of MOFs, and the gas molecules should be smaller than the MOFs pore apertures [14].

13.2.1 MOFs as Electrochemical Sensors

Typically, an electrochemical sensor interacts with heat, light, sound, pressure, or movement of particles, and transmits the signal in the form of an electrical impulse [10, 14]. The electrical interactions and impulses arising from a change in the structure of the sensor material is an intrinsic property [15]. The electrochemical sensor materials possess the property of being able to sense the situational changes that occur around them and readily act on those. In an electrochemical sensor, an electrode is normally used as a transducer. The signals from the electrode solely depend on the kind of interactions with the analyte species. In usual practice, the electrochemical responses are derived from the electric potential (volt), resistance (Ohm), current (Ampere), and the capacitance (Farad) that arise as a result of chemical reactions [7]. Based on the type of interactions between the electrode and the analyte, the electrochemical methods can be classified into two types: interfacial method and non-interfacial method [15]. In an interfacial method, the electrodes directly interact with the analytes at their surfaces and produce measurable electrical signals. This can be achieved either in a static or in a dynamic mode. In the static method, electron flow is from the electrode to the analyte or from the analyte to the electrode, thus keeping the current in the system zero, whereas the dynamic method is based on the redox reactions where a significant amount of current flows from the electrode to analytes, often leading to a charge transfer reaction. In a non-interfacial method, the electrical resistance of the solution between two electrodes is measured via conductometric techniques where the direct involvement of electrodes is minimal. This technique is based on applying an alternating current and measuring the resistance as a function of the cell constant of the cell. The cell constant usually depends on the volume of the solution, surface area of the cell, and the spacing between the electrodes.

MOFs, hybrid structures of inorganic metallic nodes and organic linkers, are an ideal choice for electrochemical sensor materials in comparison to other fundamental metal oxides or organic polymeric sensors [12]. Recently, conducting, and semiconducting MOFs heve been reported to be perfect replacements for conventional electrochemical sensor materials for sensing vapor phase molecules [28]. MOFs have an enormous interconnected porous area that can be utilized as active sites and can connect to analytes. MOFs are structurally flexible and this property allows them to accommodate various chemical functionalities within their porous structure that can be used for electrochemical purposes. Thus, MOFs can be considered as superior materials for electrochemical applications in comparison to other traditional porous materials. A major concern in the application of MOFs as electrochemical materials is their stability in water as the electrochemical sensing techniques require an aqueous environment. Though most of the MOFs are relatively unstable in water systems, robust MOFs such as UiO-66 can be utilized for this purpose. Recently, there is a surge in publications on stable MOFs as electrochemical sensors [8].

13.3 SENSING NO_X MOLECULES

NO and NO_2, generally referred to as NO_x beyond the permissible limit, are harmful to living systems. Vehicular exhaust is a common source of these gases. Although nitric oxide (NO) is a toxic gas it also acts as an important messenger in biological systems such as blood pressure regulation, neural communication, and nervous system processes [17-18]. NO with low concentration can cause Alzheimer's and Parkinson's diseases, and at high concentration causes cancer [19, 20]. Therefore, detecting/monitoring NO concentration is crucial.

Though nitrite ions (NO_2^-) are found in water, food, and physiological systems [21], but, beyond a certain limit (20 ppm) [22], they pose detrimental effects on human health [23]. Also, nitrite ions react with dietary components to become transformed into secondary amines and amides, generating carcinogenic substances like N-nitrosoamines [24]. Irreversible oxidation of hemoglobin to methemoglobin can be caused due. to nitrites [25,26]. In Europe, the European Commission has introduced a series of strict regulations to detect and control such pollutants to meet the standard of higher air quality. These standards include the Euro V, which was established in 2009, and following that, the even more limiting Euro VI in 2014 [27-29] In the USA, 2008 was the year where U.S. Environmental Protection Agency established very stringent emission controls for nitric oxide, nitrogen dioxide, and nitrous oxide for every vehicle present in the nation [14,28]. Nitric oxide (NO) and nitrogen dioxide (NO_2) have been linked with increasing greenhouse gas emissions and reduction of life span due to inculcation of pulmonary distress at a young age. Children are quite a vulnerable group to these pollutants, especially nitrogen dioxide, attributed to pediatric asthma [28]. It is the reason for four million new pediatric asthma cases every year and hence needs to be brought under control quickly. Therefore, efficient and economical detection of the NO_x class of pollutants with high sensitivity is critical for many practical occasions.

Several methods of instrumental analysis such as gas chromatography, high-performance liquid chromatography, and nuclear magnetic resonance are already in place to detect the NO_x class of pollutants. However, multiple issues are plaguing these techniques, ranging from complex sample preparations to pre-treatment analysis. These techniques have limited portability, low accuracy and are expensive [14, 15, 29]. Anything that can reduce the complexity of samples is always helpful when dealing with pollutants. Hence, NO_x sensors that can address the challenges of such chromatographic methods while meeting the standards of portability, economy, and user-friendliness become an essential subject of research and invention.

MOFs can be used to overcome the limitations of conventional sensing technologies. The changes in properties of MOFs, either optical, electronic, or magnetic, on the inclusion of chemicals, have already been exploited to detect low concentrations of analytes and have shown remarkable effectiveness in advanced sensors [4]. By changing and tuning their composition, MOFs can be optimized for sensing different analytes. The ease of synthesis of MOFs is quite favorable compared to the complex, high-precision use of nanotechnological products and fabrics. By altering the composition of the MOF's metal nodes and organic linkers, the MOF can achieve selective chemical adsorption and sensing properties [7].

13.3.1 Usage of MOFs for Electrochemical Sensing of the NO_x Class of Compounds

Although there are various conventional analytical techniques like titrimetry and spectrophotometry, available for the detection of nitrite ions, it demands more efficient techniques with improved sensitivity. Electrochemical sensing is one such potential analytical technique with high sensitivity, selectivity, simple to operate, and fast and clean. Normally, an electrochemical sensor converts the electrochemical process into an analytical signal. The analytical signals can be monitored through techniques like; conductometry, potentiometry, amperometry, impedimetric, and voltammetry [32]. Electrochemical sensors for NO_x are mainly voltammetric and amperometric sensors. In electrochemical methods, the materials used for the modification of electrode surfaces play a vital role in the sensitivity of the technique. Metal-organic frameworks (MOFs) due to their high surface area and ultrahigh porosity (up to 90% free volume) [32] is a potential candidate for the modification of the electrode surface to design electrochemical sensors for detection of nitrite ions. The general principle behind designing an electrochemical sensor based on MOFs is: (a) to prepare a working electrode having a coating of MOFs on its surface; and (b) to measure the current vs potential through cyclic voltammetry in the presence of nitrite. Therefore, coating the MOFs on the surface of the working electrode is fundamental in designing an electrochemical sensor. Further, for designing an electrochemical sensor, the material has to be electroactive. In many cases, MOFs are found not to be electroactive, and such MOFs have to be modified by electroactive components to form a resultant composite or hybrid material, which can act as an electrochemical sensor. For this purpose, MOFs can be modified with electrically conductive materials like graphene, metal nanoparticles, alloy nanoparticles, and proteins. MOFs having high surface area, act as platforms for the immobilization of electrically conductive materials. Additionally, conductive materials like metal/alloy nanoparticles can be incorporated inside the porous structure of MOFs. Consequently, the MOF-based hybrid material offers a synergistic effect, which improves its electrochemical performance. Although electrochemical sensing of NO_x by MOF-based materials, is still in its nascent stage, this section presents most of the electrochemical sensors based on MOFs reported to date.

In 2015, Ho et al. demonstrated the solvothermal growth (growth in a solvent with a temperature above its boiling point) of MOF thin films composed of a Zr-MOF constructed from an active redox porphyrinic linker, MOF-525. They utilized the obtained MOF-525 thin films for electrochemical nitrite detection in neutral KCl aqueous solutions. A significant electrocatalytic activity towards nitrite was observed, attributed to the catalytically active porphyrinic units that were spatially dispersed in the porous structure of MOF 525. The MOF-525 thin film was also reported to maintain its crystal structure after electrochemical operations in the neutral aqueous electrolyte [31, 32].

In recent work, graphene quantum dots (GQD) were installed in the entire pore structure of a mesoporous Zr-MOF, PCN-222, to enable and view the long-range donor-acceptor charge transfer. The obtained Zr-MOF adsorbed with GQD shows an electrical conductivity that is two orders of magnitude higher than that of PCN-222 and can be utilized in electrochemical nitrite sensors in aqueous NaCl electrolyte. The obtained Zr-MOF adsorbed with GQD found the resulting sensing performances to outperform those of the individual GQD, individual PCN-222, and the physical mixture of GQD and PCN-222 [32].

Multiple examples of composites composed of stable MOFs and conductive materials have been designed for electroanalytical purposes. For instance, as a follow-up study, nanocomposites consisting of the same Zr-MOF, MOF-525, and graphene nanoribbons were prepared for electroanalysis. A significantly improved electrocatalytic activity towards nitrite was observed with the nanocomposites [7].

An electrochemical sensor device can be prepared by depositing MOFs on the conducting glass substrate. A thin film of MOF coated on a conducting glass substrate was used as an amperometric nitrite sensor with good sensitivity [33]. The MOF was prepared from hexa-zirconium nodes and free-base meso-tetra(4-carboxyphenyl) porphine (H_4TCPP) linkers. The thin film of the MOF was coated on the working electrode to study electrocatalytic oxidation of nitrite by cyclic voltammetry. The redox mechanism was investigated by amperometry.

Due to their porous structure, MOFs can be suitable hosts to impregnate guest molecules to form a resultant material with modified properties. MOFs, on modification by electrically conductive materials, enhance their electrochemical performance. Modification of MOFs by incorporating metal nanoparticles into the internal surface of MOFs has been an effective strategy to improve their catalytic properties. Metal nanoparticles can be grown inside the cavities of MOFs. The MOFs protect metal nanoparticles from aggregation and thus form stable composite materials, MNPs@MOF. These composites can exhibit modified electrochemical properties. Gold nanoparticles (Au NPs) are incorporated into zinc-based MOFs to produce Au-MOF. The Au-MOF has been used for the electrochemical sensing of nitrite through electrocatalytic oxidation by cyclic voltammetry [33]. Oxidation of nitrite has been electro catalyzed using a working electrode coated with Au-MOF. For surface modification of the electrode, Au-MOF solution was coated on a glassy carbon (GC) electrode to produce GC/Au-MOF. The GC/Au-MOF was used as a working electrode. In the presence of nitrite, oxidation current was significantly increased with GC/Au-MOF in comparison to GC and GC-MOF electrodes, thus suggesting the sensitivity GC/Au-MOF to nitrite. Another MOF, based on copper has been modified with Au NPs for electrochemical sensing of nitrite in water. Cu-based MOF (Cu-MOF) was drop coated on a glassy carbon electrode to produce Cu-MOF/GCE. Further, Au NPs were electrodeposited on Cu-MOF/GCE by a potentiostatic method to fabricate Cu-MOF/Au/GCE. The amperometric technique was used to study the electrocatalytic oxidation of nitrite at Cu-MOF/Au/GCE. In the presence of nitrite there was an increase in oxidation current with low anodic peak potential ($E_{pa} = 0.79V$) demonstrating the catalytic performance of Cu-MOF/Au/GCE. This can be attributed to good catalytic activity and the high electrical conductivity of Au NPs. MOFs provide a greater surface area for efficient adsorption of nitrite. Also, MOFs facilitate the formation of stable Au NPs by preventing their agglomeration. Comparison of the electrocatalytic performances of GCE, Cu-MOF/GCE, and Cu-MOF/Au/GCE reveals the role of MOF and Au NPs for the electrochemical sensing of nitrite [34]. Au NPs can be incorporated into the Cu-based MOF (Cu-MOF) to form a hybrid material, Au/Cu-MOF for the electrochemical oxidation of nitrite [35]. Au/Cu-MOF modified carbon paste electrode (CPE) was used as a working electrode, Au/Cu-MOF/CPE to study the electrochemical oxidation of nitrite by cyclic voltammetry. In the voltammogram for oxidation of nitrite with the Au/Cu-MOF/CPE electrode, the oxidation current was twice that of Cu-MOF/CPE. This suggests the role of Au NPs in accelerating the electron transfer in the Cu-MOF during electrochemical oxidation of nitrite. Amperometric measurement suggests the detection of nitrite in the range of 50.0 nM – 717.2 μM with a detection limit of 30.0 nM.

Along with metal nanoparticles, carbon nanomaterials can be integrated with MOFs for improved electrochemical performance. Copper-based MOFs modified with Au NPs and electrochemically reduced graphene oxide (ERGO) has been used to develop an electrochemical sensor for nitrite in water, milk, food, and vegetables. The suspension of MOF (Cu-TDPAT) was mixed with a suspension of graphene oxide to obtain Cu-TDPAT/ GO composites. A Cu-TDPAT/GO suspension was dropped onto a glassy carbon electrode (GCE) to form Cu-TDPAT/GO/GCE, which was further subjected to electrochemical reduction to form Cu-TDPAT/ERGO/GCE. The Cu-TDPAT/ERGO/GCE was dipped into $HAuCl_4$ solution followed by electrochemical reduction to obtain an Au/Cu-TDPAT/ERGO/GCE electrode. The electrochemical oxidation of nitrite was studied by cyclic voltammetry and electrochemical impedance spectroscopy (EIS) using Au/Cu-TDPAT/ERGO/GCE as a working electrode [51]. In the presence of nitrite, there was a high current density at oxidation peak potential 0.77V, which is much lower than bare GCE (0.903V). This suggests the electrocatalytic oxidation of nitrite by Au/Cu-TDPAT/ERGO/GCE. Comparison of catalytic performance of GCE, Cu-TDPAT/ERGO/GCE, and Au/Cu-TDPAT/ERGO/GCE, shows the synergistic effect of Au NPs, ERGO, and Cu-TDPAT for accelerated electrocatalytic oxidation of nitrite. For nitrite oxidation, the modified electrode exhibits high electrocatalytic performance. Au/Cu-TDPAT/ERGO/GCE was used for the detection of nitrite in a broad linear range (0.001–1000 $\mu mol \cdot L^{-1}$) and with a detection limit of 0.006 $\mu mol \cdot L^{-1}$. Another Cu-based MOF modified with reduced graphene oxide (Cu-MOF/rGO) has been used for the electrochemical detection of nitrite [37]. Although Cu-MOF has a large surface area and pore volume, its poor conductivity poses limitations for its electrochemical applications. Therefore, the introduction of rGO into Cu-MOF increases the conductivity and enhances the electrochemical applications of the hybrid material. The synergistic effect of Cu-MOF and rGO nanosheets enables the hybrid material to be an effective electrochemical sensor for nitrite. Sensing of nitrite can be shown, where the hybrid material oxidizes nitrite into nitrate. A suspension of Cu-MOF/rGO in ethanol was dropped onto the surface of a glassy carbon electrode (GCE) to obtain the modified electrode, Cu-MOF/rGO/GCE. The Cu-MOF/rGO/GCE electrode was used as a working electrode for electrocatalytic oxidation of nitrite by cyclic voltammetry. In the presence of nitrite, an oxidation peak was observed which was not observed in the absence of nitrite. The oxidation peak corresponds to the oxidation of nitrite into nitrate with the subsequent generation of an electron. In the presence of nitrite, a high current response with an oxidation peak at 0.76V was observed. An increase in the concentration of nitrite gives linear current increases revealing the potential of the hybrid material Cu-MOF/rGO/GCE as a nitrite sensor. Differential pulse voltammetry (DPV) shows a reduction in oxidation potential in the presence of nitrite, confirming the electrocatalytic performance of the hybrid material. Cu-MOF/rGO/GCE was found to detect nitrite in the range 3–40 000 μM with a detection limit 33 nM by amperometry.

The electrochemical oxidation of nitrite can be shown as:

$$2NO_2^- \rightleftharpoons 2NO_2 + 2e-$$

$$2NO_2 + H_2O \rightarrow NO_3^- + NO_2^- + 2H^+$$

$$NO_2^- + H_2O \rightarrow NO_3^- + 2H^+ + 2e-$$

Pd NPs incorporated in an amino-functionalized Cr-based MOF [(NH_2-MIL 101(Cr)] has been used as an electrochemical sensor for nitrite in pickles and sausage [38]. A dispersion of Pd/NH_2-MIL 101(Cr) was dropped onto the surface of a screen-printed carbon electrode (SPCE). The modified electrode Pd/NH_2-MIL 101(Cr)/SPCE was used for the determination of nitrite through electrocatalytic oxidation of nitrite by cyclic voltammetry, linear sweep voltammetry, and square wave voltammetry. In the presence of nitrite, cyclic voltammograms with unmodified SPCE and Pd/NH_2-MIL 101(Cr)/SPCE, show peaks of oxidation current 15 nA at 0.983V and 120.74 nA at 0.904V respectively. No such peak was observed in the absence of nitrite, suggesting that the peak corresponds to the oxidation of nitrite. Further, the increase in oxidation current and decrease in overpotential is due to the fast transfer of electrons and it reveals the electrocatalytic performance of the modified electrode Pd/NH_2-MIL 101(Cr)/SPCE. This can be attributed to the conductive Pd nanoparticles incorporated onto MOFs of high surface area and porosity. The oxidation current increases in the range of nitrate concentration 5 to 150 nanomolar (nM) with a detection limit of 1.3 nM.

Like metal, alloy nanoparticles can be incorporated inside MOFs to obtain hybrid material with modified electrochemical activity. Amino functionalized Zr-based MOF, UiO-66-NH_2 decorated with Au and Pd nanoparticles, produce a composite material having an application for electrochemical sensing of nitrite in sausage and pickle samples [39]. The MOF acts as a supporting platform and protecting agent for AuPd nanoparticles. The electrochemical oxidation of nitrite has been studied by cyclic voltammetry and chronoamperometry. A homogenous suspension of AuPd/UiO-66-NH_2 was dropped onto the surface of a glassy carbon electrode (GCE) to obtain AuPd/UiO-66-NH_2/GCE. For the electrochemical study, AuPd/UiO-66-NH_2/GCE was used as a working electrode. With AuPd/UiO-66-NH_2/GCE as working electrode, upon addition of nitrite, an anodic peak with high current density corresponding to the oxidation of nitrite, was observed in the voltammograms. To ascertain the electrocatalytic effect of AuPd alloy nanoparticles inside the MOFs, voltammograms were recorded with UiO-66-NH_2/ GCE, Au/UiO-66-NH_2/GCE, and Pd/UiO-66-NH_2/GCE, as working electrodes. The oxidation current density was the least with UiO-66-NH_2/GCE. The oxidation current density for Au/UiO-66-NH_2/GCE and Pd/UiO-66-NH_2/GCE, was higher than that of UiO-66-NH_2/ GCE, revealing the catalytic effect of Au and Pd nanoparticles. But the oxidation current was highest for the alloy nanoparticles, namely, AuPd/UiO-66-NH_2/GCE. This confirms the better catalytic performance of AuPd/UiO-66-NH_2/GCE composite. This can be ascribed to the synergistic effect of Au, Pd, and the MOF. To study the detection sensitivity of AuPd/UiO-66-NH_2/GCE, chronoamperometric measurement was carried out. The result shows the detection of nitrite in a concentration range 0.05 – 15666 μM with a detection limit 0.01 μM.

Electron transfer across protein is a key factor in many biological processes. Therefore, adsorption of proteins on electrodes can be an effective route to designing electrochemical biosensors. Porphyrin, having a conjugated π- system, is suitable for electron transfer and thus finds application in designing electrochemical biosensors. Thus porphyrin-based MOFs can be potential materials for the preparation of working electrodes to study the electrochemical sensing of NO and NO_2. A zirconium-based nano-mateallоporphyrinic metal-organic framework (N_{por}MOF(Fe)) has been used as an electrochemical sensor for nitric oxide (NO) [40]. The MOFs, (N_{por}MOF(Fe)) were prepared from, Zr and metalloporphyrin (TCPPFe). An aqueous solution of (N_{por}MOF(Fe)) was coated on a glassy carbon electrode (GCE). (N_{por}MOF(Fe))/GCE was used as a working electrode to test the electroreduction of NO by cyclic voltammetry. $NaNO_2$ was used as a precursor in an acidic medium to produce NO.

$$3\ HONO^+ \rightarrow H^+ + 2\ NO + NO_3^- + H_2O$$

Upon addition of $NaNO_2$, a cathodic peak appeared at −0.55V corresponding to the generation of NO and its catalytic reduction by N_{por}MOF(Fe). NO could be reduced to N_2O or NH_2OH. On increasing the concentration of $NaNO_2$, the current density at −0.55V increases, which indicates the catalytic reduction of NO by N_{por}MOF(Fe). The current density increases linearly with $NaNO_2$ in the concentration range 5 to 200 μM with a detection limit 1.3 μM.

Chen et al. have reported impregnation of graphene quantum dots (GQD) inside zirconium-based MOFs with pophyrinic linkers (PCN-222), to produce a hybrid material for the electrochemical sensing of nitrite in an aqueous solution [40, 41]. The incorporation of GQD in Zr-MOF offers donor-acceptor charge transport between electron-donating GQD and electron-accepting porphyrin linkers, resulting in an increase in electrical conductivity by a magnitude of 2 orders compared to that of pristine MOF and thus making the material suitable for electrochemical sensing applications. To prepare a working electrode with the hybrid material, GQD-PCN-222 dispersed in acetone was drop coated on fluorine-doped tin oxide (FTO) conducting glass substrate. The GQD-PCN-222 coated FTO, GQD-PCN-222 was used as a working electrode to study the electrochemical sensing of nitrite by cyclic voltammetry, amperometry, and electrochemical impedance spectroscopy. In the presence of nitrite, the voltammogram shows an increase in current with a peak corresponding to the oxidation of nitrite. This confirms the electrocatalytic activity of GQD-PCN-222. There was a linear increase in the current signal with an increase in the concentration of nitrite. Amperometric measurement suggests the detection of nitrite in a range of concentrations 40-1800 μM and a limit of detection of 6.4 μM. Another Zr-based porphyrinic MOF (MOF-525) has been used for the electrocatalytic oxidation of nitrite [40]. MOF-525 was dispersed in dimethylformamide to form ink for inkjet printing. A MicroFab JetLab4 system (Micro Fab Technologies Inc.) was used for inkjet printing of MOF-525 thin film on indium tin oxide (ITO) coated conducting glass substrate. The thin film-coated glass substrate was used as the working electrode for electrocatalytic oxidation of nitrite. In the presence of nitrite, the voltammogram shows an anodic peak at a potential of 0.9V, corresponding to the oxidation of nitrite. The current density increases by increasing the nitrite concentration. This confirms the electrocatalytic activity of MOF thin film. Amperometric measurement suggests the determination of nitrite concentration in the range from 10 to 800 μM and a detection limit of 0.72 μM. This study demonstrates the use of inkjet-printed MOF thin films for the fabrication of nitrite sensors.

Just as in the case of graphene quantum dots, graphene nanoribbons can be incorporated inside Zr-based MOF to modify its electrochemical property. Graphene nanoribbons (GNR) incorporated in a porphyrinic Zr based MOF (MOF-525) to form a nanocomposite MOF-525/GNR acts as an electrochemical sensor for nitrite [56]. MOF-525/GNR dispersed in dimethylformamide was dropped onto the surface of indium tin oxide (ITO) conducting glass substrate, which was used as a working electrode to study the electrocatalytic oxidation of nitrite by cyclic voltammetry and amperometry. In the presence of nitrite with a MOF-525/GNR/ITO working electrode, the voltammogram shows an anodic peak with high current density, corresponding to the oxidation of nitrite. However, there was no defined anodic peak obtained with MOF-525 thin film coated on ITO. This confirms the role of GNR in the electrocatalytic performance of the composite MOF-525/GNR. Although the catalytic current density was observed with GNR thin film coated on ITO, it was much smaller than that of the composite MOF-525/GNR. This further confirms the synergistic effect of MOF-525 and GNR for the electrocatalytic performance of the composite. The high surface area of porphyrinic MOF-525 and electrical conductivity of GNRs, collectively make the composite a better electrocatalyst. The interconnected GNRs between MOF crystals act as conductive bridges that facilitate charge transfer. Amperometric measurement suggests the detection of nitrite in a concentration range from100 to 2500 μM with a detection limit of 0.75 μM.

Ag NPs incorporated in a zinc-based MOF (Zinc thiosalicylate, Zn-TSA) to form a MOF composite Ag@ Zn-TSA was modified with myoglobin and ionic liquid: 1-ethyl-3-methylimidazolium tetrafluoroborate ($[BMIm]^+$ $[BF_4]^-$) to form an electrochemical biosensor for nitrite [57]. Ag@Zn-TSA solution was cast onto the surface of a carbon paste electrode (CPE) to form Ag@Zn-TSA-CPE. Further, solutions of, myoglobin and $[BMIm]^+[BF_4]^-$ were used to form a Mb/Ag@Zn-TSA/IL-CPE electrode. The large surface area, the porosity of Zn-TSA, and the available carboxyl and thiol functional groups on TSA, facilitates the immobilization of myoglobin on the MOF surface. The synergistic effect of silver and the delocalized electrons due to the conjugate structure of TSA contributes to excellent conductivity. The thin film of ionic liquid wraps the myoglobin on the surface of the MOF. The resultant effect enables the modified electrode Mb/ Ag@Zn-TSA/IL-CPE as an effective electrochemical sensor for nitrite. Electrochemical oxidation of nitrite was studied by cyclic voltammetry, electron impedance spectroscopy, and amperometry. In the presence of nitrite, oxidation peak current was increased to 0.8V suggesting the electrocatalytic effect. The detection limit for nitrite was found to be 0.5 μM in the range of 1.3 μM - 1660 μM.

13.3.2 Hypothetical Sensors that are Under Review for Application in NO_x Sensors

Another class of MOFs, called electroluminescent MOFs, have been investigated in the development of electroluminescent sensors. These cells are the simplest type of electroluminescent devices, consisting of a thin layer of conjugated polymer placed between two electrodes [42]. Electroluminescent devices can emit light by applying an electrical current or a strong electric field as electrical energy is converted into visible radiation. It is known that electroluminescent devices can be fabricated by using either organic or inorganic electroluminescent materials. MOFs are an organic-inorganic hybrid system that could form attractive electroluminescent materials with the overall advantages of both [42].

Such MOFs are promising for application in electroluminescent devices, but this has not been demonstrated in practical applications. All it would take are a few successful examples that could encourage researchers to make more conscious efforts to investigate electroluminescent MOFs for NO_x sensing applications. For instance, a new 3D microporous MOF has been constructed using a highly conjugated anthracene-based ligand, which provides high electrical conductivity and electroluminescence properties. As another example, the entrapment of perylene dianhydride molecules into MOF-5 (Zn_4O clusters with terephthalate ligand) pores with solid interaction between host and guest has been shown to yield a remarkable electroluminescent response [43, 44, 45]. Typically, deep blue electro phosphorescent behavior is observed in one of the original MOFs, whose composition was determined to be $[Cu_2(C_{39}H_{32}P_2)_2]n$ (Cu-P_6) [45]. Based on these examples, it does seem like electroluminescent sensors can be used to detect NO, NO_2, and N_2O based pollutants [46, 47].

13.4 SENSING CO_2

Excessive CO_2 emissions from the burning of fossil fuels are substantially harmful and it requires a significant effort to address this challenge. The increased CO_2 level is a primary environmental concern, and therefore it requires careful sensing and monitoring. Detecting CO_2 is a challenging task because of the highly oxidized and inert structure of CO_2. Additionally, CO_2 is always present in a mixture of gases such as O_2, NO_x, H_2 and other chemically active molecules in the atmosphere. All of these environmental gas molecules have similar resonating frequencies, and hence, detection of CO_2 by spectroscopic methods is challenging [48]. The commercially available sensors, namely nondispersive infrared (NDIR) and chemical CO_2 sensors, have disadvantages: (a) NDIR sensors are selective and have a fast response time, but their size and power consumption are detrimental to the environment; (b) chemical CO_2 sensors have poor selectivity and short and long-term sensor drift, leading to inaccurate measurements over time [50, 51]. These also express heavy power consumption and high-temperature requirements [50]. These disadvantages pave the way for new CO_2 sensors.

Polymeric porous materials are commonly used for sensing CO_2 because of their high selectivity and sensitivity. Usually, CO_2 commercial sensor materials are amine-based polymeric materials. These materials interact with CO_2 molecules reversibly to form ionic compounds such as carbamates and cations of ammonia [49]. Carbamates are unstable compounds, and upon heating, disintegrate into CO_2 molecules and free up the amine groups for reuse as sensor active sites. There are multiple advantages of amine-based sensors as they are selective, sensitive to analytes, and are used in small quantities. Other types of materials that are used for sensors are metal oxides. Metal oxide sensors work on the principle of redox reactions between the analytes and metal oxides. The redox reactions alter the conductivity at the metal oxide surface which helps to measure the quantity of analytes electronically. Other potential sensor materials are porous zeolites, carbon nanotubes, and MOFs.

13.4.1 Sensing of CO_2 by MOFs

MOFs are considered superior among the sensor materials for their exceptional structural flexibility and tunability [49, 50]. The high surface area to volume ratio is advantageous for sensing applications as it increases the chance of interaction between the analyte and the sensing material, leading to higher sensitivity. In addition, less sensing material is used to adsorb gases than the low surface area materials, thereby allowing for the miniaturization of the sensing layer [50, 52].

MOFs are studied frequently for their potential in CO_2 capture and storage, and various MOF structures have been designed to achieve high selectivity towards CO_2. Van Duyne et al. fabricated a localized surface plasmon resonance (LSPR) sensor with HKUST-1 and demonstrated the detection of CO_2 levels down to 10 % at ambient conditions [50]. Wang et al. designed a near-IR optical fiber coated with HKUST-1 that showed a limit of detection of 20 ppm for CO_2 [50]. Chocarro-Ruiz et al. reported an optical CO_2 sensor by integrating a thin transparent film of ZIF-8 onto a bimodal optical waveguide [50].

MOFs are utilized as sensor materials through their structure-function cooperative interactions with the analytes. One of the most common techniques employed for sensing CO_2 is electro-chemical impedance spectroscopy (EIS) which evaluates the proton conductivity around the MOFs. The ionic conductivity of the MOFs is altered when CO_2 binds or adsorbs onto metals or clusters of MOFs, and the change in conductivity is measured electrically to detect CO_2 and also to measure its quantity [49]. The host-guest interaction of MOFs and CO_2 are very responsive to external stimuli such as pressure, temperature, and concentration of gases, and these behaviors give a fast response, better specificity, and greater sensitivity. The detailed analysis of the co-operative phenomenon helps us to understand the sensing ability in MOFs.

Structure-function relations are more complex for the MOFs. The reason lies in the specific arrangement of the framework components, creating large pores, that can accommodate guest molecules with possible cooperative interactions [50]. The presence of foreign molecules in the pores, CO_2, or similar air pollutant NO_x, tunes the structure via co-operative behaviors, and thus the function. Major structural factors in MOFs that affect CO_2 adsorption and sensing are: (a) the symmetry and size of pores within the framework; (b) metal centers in MOFs; and (c) type and chemical nature of linkers in the frameworks. MOFs with structural flexibilities, UiO 66 (Zr) and MOF-74, are of special interest for adsorption and sensing studies. UiO 66 (Zr) compounds possess variable pores within the framework. These are large octahedral pores, which are connected through smaller windows. MOF-74, with unsaturated metal centers, has drawn much attention for its enhanced CO_2 sensing capability and has been under investigation [50]. Also, materials containing amine groups have been fascinating for sensing applications [51].

13.4.2 Electrochemical Sensing of CO_2 by MOFs

Recently, the electrochemical sensing of CO_2 by MOFs has attracted much attention. In an electrochemical sensor, the physicochemical interaction between MOFs and CO_2 is translated into electrical capacitance (electric charge storage at a certain potential), electrical resistance, and electric potential [52]. In a typical capacitive sensor, the dielectric constants are measured as a function of time. Upon adsorption of CO_2 on MOFs, the dielectric constants of MOFs are altered and the change in the dielectric constant of MOFs is measured as a function of time to estimate the quantity of CO_2 adsorbed on MOFs surfaces. As a capacitive sensor, Mg-based MOF-74 has shown excellent CO_2 sensing capability at a low limit of detection (LOD) of 200 ppm [50]. The amine-modified MOF-74 showed even better LOD values owing to the amine functionalization in MOF-74 which enhances the basicity of open metal sites in the MOFs [54].

Lately, electrical resistance-based MOF sensors or chemiresistive sensors have also been studied. These sensors usually detect the change in the resistance of the MOF materials upon adsorption of CO_2. The low conductivity of MOFs has been a major challenge in chemiresistive sensors. The linkers in MOFs, which are usually organic molecules, are nonconductors of electricity. To get better conductivity, efforts have been made to incorporate electrically conducting linkers such as 7,7,8,8- tetracyanoquinondimethane (TCNQ) and tetrathiafulvalene (TTF) into the MOF-74 structures so that they can be used as chemiresistive sensors [53]. Another chemiresistive CO_2 sensor MOF is Cu_3 (hexaamino benzene) which showed excellent sensing ability at a LOD value of 400 ppm [50]. To overcome the low conductivity in MOFs, MOF composites are considered for sensor applications. Kaskel et al. synthesized composite carbon nanotubes (CNT) with AL-MIL-53 MOFs, which allowed the electrons to flow through the composite materials. Upon adsorption of CO_2, the conductivity of the composite is altered and the change in conductivity is measured to detect the level of CO_2 [54].

The inclusion of electrically conducting linkers into MOFs has set the scene for MOFs as electrochemical sensors for vapor phase analytes. In this regard, impedance spectroscopic techniques have also been utilized for sensing analytes. The electrical impedance of MOF sensors is evaluated around the applied frequencies of electrical current upon adsorption of CO_2. MOFs with hydroxyl functionalities are known to have proton-conducting capability through the release of protons to their nanopores [51]. This behavior of MOFs is exploited to realize MOFs as CO_2 sensors. MOFs, upon adsorption of CO_2, can alter the proton conductivity via the reaction of CO_2 with protons of the MOFs. In this regard, an Rb-based MOF was found to be suitable for CO_2 sensing [52]. Another MOF, CDMOF-2 was reported to have excellent CO_2 sensing capability without compromising sensory activities even after multiple cycles of usage [50-54]. Zn-MOF-74 and Nd MO-MOF showed a very selective CO_2 adsorption ability with high proton conductivity properties [50, 51]. The fabrication of sensor composites was achieved by combining an appropriate amount of MOF powders to polyvinylpyrrolidone (PVP) [54].

MOFs were also utilized as Kelvin probes for sensing CO_2. In a typical Kelvin probe setup, the electrical signal variations, known as work function, are evaluated as a function of time. The electrical response from the sensor materials gets altered upon reaction with vapor phase analytes. The work function of a sensor is calculated by the energy requisite for the process of electron liberation from the surface of a material. MOF-74 with various metal centers as Mg, Co, Zn, Ni has been utilized as a CO_2 sensor with the help of a Kelvin probe [50]. A CO_2 molecule adsorbs onto the MOFs and alters the electronic properties of the materials, and the change is transduced to detect the CO_2 level. Post-synthetically modified, ethylene diamine functionalized MOF-74 revealed enhanced sensing behavior in comparison to the parent MOFs [50]. These studies reveal the potential of MOFs as CO_2 sensor materials.

13.5 CONCLUSION

In summary, MOFs have many advantages over other nanomaterials as crystalline molecular materials, including chemical stability, adjustable/tunable structure and property, and significant porosity. Extreme progress has been achieved in designing novel and excellent MOF-based electrochemical sensors and biosensors in the past decade. However, with the achievements of nanoscience, MOF-based electrochemical sensors show up-and-coming applications in detecting CO_2 and NO_x groups of gases and pollutants, and electrochemical-related research will experience profitable growth in the next few years. The application of MOF-based sensing devices has been explored on a minuscule scale. Sensors for NO, NO_2, N_2O and similar gases are dominated by optical, electrochemical, and electroluminescent feature-based MOFs. Both conventional methods and novel techniques (for example, interferometry, colloidal crystals, and microcantilever-based techniques) based on MOFs are being studied thoroughly to develop more efficient detection techniques for CO_2 and NO_x gases as a whole. Although some of reports are available on electrochemical sensing of NO_2 in water, food, or vegetable samples by using MOFs, adsorption of NO_2 gas in MOFs still needs further investigation.

All current technologies for CO_2 sensing have their own set of advantages and limitations that make them relevant to specific sensing applications. The costs, power consumption, and portability of such technologies play a major role in their applications. For instance, devices based on optical detection have high sensitivity. However, they get hindered by water interference. The polymer-based chemical sensors are produced quite cheap but have a shorter operating life, and their high operating temperatures limit conductometric metal-oxide sensors despite enabling the fabrication of portable instruments.

MOFs tend to be promising sensing layers for selective and sensitive CO_2 detection due to their wide range of operating temperatures (relevant for practical applications), high surface area, and tunability. Quite significantly, the possibility of tailoring the properties of MOFs enables synthetic chemists to gain control over the MOF– CO_2 interactions by fine-tuning the pore sizes and active sites of these materials through an exhaustive selection of organic, inorganic parts and advanced MOF functionalization. It is strongly believed that the potential of MOFs for chemical sensing applications justifies the research efforts to solve the hurdles mentioned above (and others like toxicity, low electrical conductivity, poor mechanical stability, and the like.) or to optimize the existing prototypes. Thus, we advocate further studies on these materials in relation not only to CO_2 and NO_X sensing but also concerning other relevant chemical sensing applications (such as H_2S, NH_3, N_2O volatile organic compounds, and explosive sensing), as any insight would be beneficial for the development of MOF-based technologies.

REFERENCES

[1]. Gielen, D.; Boshell, F.; Saygin, D.; Bazilian, M. D.; Wagner, N.; Gorini, R., The Role of Renewable Energy in the Global Energy Transformation., *Energy Strateg. Rev.*, **2019**, 24, 38–50.

[2]. Moutinho, V.; Madaleno, M.; Macedo, P., The Effect of Urban Air Pollutants in Germany: Eco-Efficiency Analysis through Fractional Regression Models Applied after DEA and SFA Efficiency Predictions., *Sustain. Cities Soc.*, **2020**, 59,102204.

[3]. Li, Z. X.; Yang, B. L.; Zou, K. Y.; Kong, L.; Yue, M. L.; Duan, H. H., Novel Porous Carbon Nanosheet Derived from a 2D Cu-MOF: Ultrahigh Porosity and Excellent Performances in the Supercapacitor Cell., *Carbon N. Y.*, **2019**, 144, 540–548.

[4]. Asghar, A.; Iqbal, N.; Noor, T., Ultrasonication Treatment Enhances MOF Surface Area and Gas Uptake Capacity., *Polyhedron,* **2020**, 181, 114463.

[5]. Sneddon, G.; Greenaway, A.; Yiu, H. H. P., The Potential Applications of Nanoporous Materials for the Adsorption, Separation, and Catalytic Conversion of Carbon Dioxide., *Adv. Energy Mater.,* **2014**, 4 (10), 1301873.

[6]. Chen, L.; Zhang, X.; Cheng, X.; Xie, Z.; Kuang, Q.; Zheng, L., The Function of Metal-Organic Frameworks in the Application of MOF-Based Composites., *Nanoscale Adv.,* **2020**, 2 (7), 2628–2647.

[7]. Kumar, P.; Kim, K. H.; Rarotra, S.; Ge, L.; Lisak, G., The Advanced Sensing Systems for NO_x Based on Metal-Organic Frameworks: Applications and Future Opportunities., *Trends Anal. Chem.,* **2020**, 122 (2), 115730.

[8]. Tang, J.; Yang, M.; Yang, M.; Wang, J.; Dong, W.; Wang, G., Heterogeneous Fe-MIL-101 Catalysts for Efficient One-Pot Four-Component Coupling Synthesis of Highly Substituted Pyrroles., *New J. Chem.,* **2015**, 39 (6), 4919–4923.

[9]. Hu, J.; Liu, Y.; Liu, J.; Gu, C., Chelation of Transition Metals into MOFs as a Promising Method for Enhancing CO_2 Capture: A Computational Study., *AIChE J.,* **2020**, 66 (2), 1–9.

[10]. Zheng, S.; Li, X.; Yan, B.; Hu, Q.; Xu, Y.; Xiao, X.; Xue, H.; Pang, H., Transition-Metal (Fe, Co, Ni) Based Metal-Organic Frameworks for Electrochemical Energy Storage., *Adv. Energy Mater.,* **2017**, 7 (18), 1–27.

[11]. Namsheer; Rout, C. S., Conducting Polymers: A Comprehensive Review on Recent Advances in Synthesis, Properties and Applications., *RSC Adv.,* **2021**, *11* (10), 5659–5697.

[12]. Sakamaki, Y.; Tsuji, M.; Heidrick, Z.; Watson, O.; Durchman, J.; Salmon, C.; Burgin, S. R.; Hassan Beyzavi, M., Preparation and Applications of Metal-Organic Frameworks (MOFs): A Laboratory Activity and Demonstration for High School and/or Undergraduate Students., *J. Chem. Educ.,* **2020**, 97 (4), 1109–1116.

[13]. Wang, H.; Lustig, W. P.; Li, J., Sensing and Capture of Toxic and Hazardous Gases and Vapors by Metal-Organic Frameworks., *Chem. Soc. Rev.,* **2018**, 47 (13), 4729–4756.

[14]. Panda, S. K.; Mishra, S.; Singh, A. K., Recent Progress in the Development of MOF-Based Optical Sensors for Fe_3., *Dalton Trans.,* **2021**, 50 (21), 7139–7155.

[15]. Zhou, T.; Cao, S.; Zhang, R.; Tu, J.; Fei, T.; Zhang, T., Effect of Cation Substitution on the Gas-Sensing Performances of Ternary Spinel MCo_2O_4 (M = Mn, Ni, and Zn) Multishelled Hollow Twin Spheres., *ACS Appl. Mater. Interfaces,* **2019**, 11 (31), 28023–28032.

[16]. Gonçalves, J. M.; Martins, P. R.; Rocha, D. P.; Matias, T. A.; Julião, M. S. S.; Munoz, R. A. A.; Angnes, L., Recent Trends and Perspectives in Electrochemical Sensors Based on MOF-Derived Materials., *J. Mater. Chem. C Mater. Opt. Electron. Devices,* **2021**, 9 (28), 8718–8745.

[17]. Hunter, R. A.; Privett, B. J.; Henley, W. H.; Breed, E. R.; Liang, Z.; Mittal, R.; Yoseph, B. P.; McDunn, J. E.; Burd, E. M.; Coopersmith, C. M.; Ramsey, J. M.; Schoenfisch, M. H., Microfluidic Amperometric Sensor for Analysis of Nitric Oxide in Whole Blood., *Anal. Chem.,* **2013**, 85 (12), 6066–6072.

[18]. Hunter, R. A.; Storm, W. L.; Coneski, P. N.; Schoenfisch, M. H., Inaccuracies of Nitric Oxide Measurement Methods in Biological Media., *Anal. Chem.,* **2013**, 85 (3), 1957–1963.

[19]. Jo, A.; Do, H.; Jhon, G.-J.; Suh, M.; Lee, Y., Electrochemical Nanosensor for Real-Time Direct Imaging of Nitric Oxide in Living Brain., *Anal. Chem.,* **2011**, 83 (21), 8314–8319.

[20]. Hajisafari, M.; Nasirizadeh, N., An Electrochemical Nanosensor for Simultaneous Determination of Hydroxylamine and Nitrite Using Oxadiazole Self-Assembled on Silver Nanoparticle-Modified Glassy Carbon Electrode., *Ionics,* **2017**, 23 (6), 1541–1551.

[21]. Hord, N. G.; Tang, Y.; Bryan, N. S., Food Sources of Nitrates and Nitrites: The Physiologic Context for Potential Health Benefits., *Am. J. Clin. Nutr.,* **2009**, 90 (1), 1–10.

[22]. Morcos, E.; Wiklund, N. P., Nitrite and Nitrate Measurement in Human Urine by Capillary Electrophoresis., *Electrophoresis,* **2001**, 22 (13), 2763–2768.

[23]. Zhou, Y.; Ma, M.; He, H.; Cai, Z.; Gao, N.; He, C.; Chang, G.; Wang, X.; He, Y., Highly Sensitive Nitrite Sensor Based on AuNPs/RGO Nanocomposites Modified Graphene Electrochemical Transistors., *Biosens. Bioelectron.,* **2019**, 146, 111751.

[24]. Yue, R.; Lu, Q.; Zhou, Y., A Novel Nitrite Biosensor Based on Single-Layer Graphene Nanoplatelet-Protein Composite Film., *Biosens. Bioelectron.,* **2011**, 26 (11), 4436–4441.

[25]. Labib, M.; Sargent, E. H.; Kelley, S. O., Electrochemical Methods for the Analysis of Clinically Relevant Biomolecules., *Chem. Rev.,* **2016**, 116 (16), 9001–9090.

[26]. Cheng, J.; Zhu, K.; Lu, H.; Yue, H.; Liu, C.; Liang, B.; Tang, S., Quantitative Relationship between CO_2 Absorption Capacity and Amine Water System: DFT, Statistical, and Experimental Study., *Ind. Eng. Chem. Res.,* **2019**, 58 (31), 13848–13857.

[27]. Achakulwisut, P.; Brauer, M.; Hystad, P.; Anenberg, S. C., Global, National, and Urban Burdens of Paediatric Asthma Incidence Attributable to Ambient NO_2 Pollution: Estimates from Global Datasets., *Lancet Planet. Heal.,* **2019**, 3 (4), e166–e178.

[28]. Anenberg, S. C.; Henze, D. K.; Tinney, V.; Kinney, P. L.; Raich, W.; Fann, N.; Malley, C. S.; Roman, H.; Lamsal, L.; Duncan, B.; Martin, R. V.; van Donkelaar, A.; Brauer, M.; Doherty, R.; Jonson, J. E.; Davila, Y.; Sudo, K.; Kuylenstierna, J. C. I., Estimates of the Global Burden of Ambient PM2:5, Ozone, and NO_2 on Asthma Incidence and Emergency Room Visits., *Environ. Health Perspect.,* **2018**, 126 (10), 1–14.

[29]. Zhou, X.; Lee, S.; Xu, Z.; Yoon, J., Recent Progress on the Development of Chemosensors for Gases., *Chem. Rev.,* **2015**, 115 (15), 7944–8000.

[30]. Tranchemontagne, D. J.; Hunt, J. R.; Yaghi, O. M., Room Temperature Synthesis of Metal-Organic Frameworks: MOF-5, MOF-74, MOF-177, MOF-199, and IRMOF-0. *Tetrahedron,* **2008**, 64 (36), 8553–8557.

[31]. Leng, K.; Sun, Y.; Li, X.; Sun, S.; Xu, W., Rapid Synthesis of Metal-Organic Frameworks MIL-101(Cr) Without the Addition of Solvent and Hydrofluoric Acid., *Cryst. Growth Des.,* **2016**, 16 (3), 1168–1171.

[32]. Zhou, H.-C.; Long, J. R.; Yaghi, O. M., Introduction to Metal-Organic Frameworks., *Chem. Rev.,***2012**, 112 (2), 673–674.

[33]. Chen, H.; Yang, T.; Liu, F.; Li, W., Electrodeposition of Gold Nanoparticles on Cu-Based Metal-Organic Framework for the Electrochemical Detection of Nitrite., *Sens. Actuators B Chem.,***2019**, 286, 401–407.

[34]. Yuan, B.; Zhang, J.; Zhang, R.; Shi, H.; Wang, N.; Li, J.; Ma, F.; Zhang, D., Cu-Based Metal–Organic Framework as a Novel Sensing Platform for the Enhanced Electro-Oxidation of Nitrite., *Sens. Actuators B Chem.,***2016**, 222, 632–637.

[35]. He, B.; Yan, D., Au/ERGO Nanoparticles Supported on Cu-Based Metal-Organic Framework as a Novel Sensor for Sensitive Determination of Nitrite., *Food Control,***2019**, 103, 70–77.

[36]. Saraf, M.; Rajak, R.; Mobin, S. M., A Fascinating Multitasking Cu-MOF/RGO Hybrid for High Performance Supercapacitors and Highly Sensitive and Selective Electrochemical Nitrite Sensors., *J. Mater. Chem. A Mater. Energy Sustain.*,**2016**, 4 (42), 16432–16445.

[37]. Ezhil Vilian, A. T.; Dinesh, B.; Muruganantham, R.; Choe, S. R.; Kang, S.-M.; Huh, Y. S.; Han, Y.-K., A Screen-Printed Carbon Electrode Modified with an Amino-Functionalized Metal Organic Framework of Type MIL-101(Cr) and with Palladium Nanoparticles for Voltammetric Sensing of Nitrite., *Mikrochim. Acta*,**2017**, 184 (12), 4793–4801.

[38]. Yang, J.; Yang, L.; Ye, H.; Zhao, F.; Zeng, B., Highly Dispersed AuPd Alloy Nanoparticles Immobilized on UiO-66-NH_2 Metal-Organic Framework for the Detection of Nitrite., *Electrochim. Acta*,**2016**, 219 (219), 647–654.

[39]. Ling, P.-H.; Zang, X.-N.; Qian, C.-H.; Gao, F., A Metal-Organic Framework with Multienzyme Activity as a Biosensing Platform for Real-Time Electrochemical Detection of Nitric Oxide and Hydrogen Peroxide., *Analyst*,**2021**, 146 (8), 2609–2616.

[40]. Kung, C.-W.; Li, Y.-S.; Lee, M.-H.; Wang, S.-Y.; Chiang, W.-H.; Ho, K.-C., In Situ Growth of Porphyrinic Metal–Organic Framework Nanocrystals on Graphene Nanoribbons for the Electrocatalytic Oxidation of Nitrite., *J. Mater. Chem. A Mater. Energy Sustain.*,**2016**, 4 (27), 10673–10682.

[41]. Dong, S.; Zhang, D.; Suo, G.; Wei, W.; Huang, T., Exploiting Multi-Function Metal-Organic Framework Nanocomposite Ag@Zn-TSA as Highly Efficient Immobilization Matrixes for Sensitive Electrochemical Biosensing., *Anal. Chim. Acta*,**2016**, 934, 203–211.

[42]. En-On, J.; Tuantranont, A.; Kerdcharoen, T.; Wongchoosuk, C., Flexible Alternating Current Electroluminescent Ammonia Gas Sensor., *RSC Adv.*, **2017**, 7 (27), 16885–16889.

[43]. Sun, X.; Wang, Y.; Lei, Y., Fluorescence Based Explosive Detection: From Mechanisms to Sensory Materials., *Chem. Soc. Rev.*, **2015**, 44 (22), 8019–8061.

[44]. Diamantis, S. A.; Margariti, A.; Pournara, A. D.; Papaefstathiou, G. S.; Manos, M. J.; Lazarides, T., Luminescent Metal-Organic Frameworks as Chemical Sensors: Common Pitfalls and Proposed Best Practices., *Inorg. Chem. Front.*, **2018**, 5 (7), 1493–1511.

[45]. Klinowski, J.; Almeida Paz, F. A.; Silva, P.; Rocha, J., Microwave-Assisted Synthesis of Metal-Organic Frameworks., *Dalt. Trans.*, **2011**, 40 (2), 321–330.

[46]. Ye, B.; Gheorghe, A.; Van Hal, R.; Zevenbergen, M.; Tanase, S., CO_2sensing under Ambient Conditions Using Metal-Organic Frameworks., *Mol. Syst. Des. Eng.*, **2020**, 5 (6), 1071–1076.

[47]. Shiozawa, H.; Bláha, M.; Valeš, V.; K Bastl, Z.; Kalbáč, M., Host - Guest Interactions in Metal-Organic Frameworks Doped with Acceptor Molecules as Revealed by Resonance Raman Spectroscopy., *J. Phys. Chem. C*, **2020**, 124 (44), 24245–24250.

[48]. Werner, R. A.; Rothe, M.; Brand, W. A., Extraction of CO_2 from Air Samples for Isotopic Analysis and Limits to Ultra High Precision Delta18O Determination in CO_2 Gas., *Rapid Commun. Mass Spectrom.*,**2001**, 15 (22), 2152–2167.

[49]. Forse, A. C.; Milner, P. J., New Chemistry for Enhanced Carbon Capture: Beyond Ammonium Carbamates., *Chem. Sci.*,**2020**, 12 (2), 508–516.

[50]. Gheorghe, A.; Lugier, O.; Ye, B.; Tanase, S., Metal–Organic Framework Based Systems for CO_2 Sensing., *J. Mater. Chem. C Mater. Opt. Electron. Devices*, **2021**.

[51]. Li, J.-R.; Kuppler, R. J.; Zhou, H.-C., Selective Gas Adsorption and Separation in Metal-Organic Frameworks., *Chem. Soc. Rev.*,**2009**, 38 (5), 1477–1504.

[52]. Yeon, J. S.; Lee, W. R.; Kim, N. W.; Jo, H.; Lee, H.; Song, J. H.; Lim, K. S.; Kang, D. W.; Seo, J. G.; Moon, D.; Wiers, B.; Hong, C. S., Homodiamine-Functionalized Metal–Organic Frameworks with a MOF-74-Type Extended Structure for Superior Selectivity of CO_2 over N_2., *J. Mater. Chem. A Mater. Energy Sustain.*,**2015**, 3 (37), 19177–19185.

[53]. Strauss, I.; Mundstock, A.; Treger, M.; Lange, K.; Hwang, S.; Chmelik, C.; Rusch, P.; Bigall, N. C.; Pichler, T.; Shiozawa, H.; Caro, J., Metal-Organic Framework Co-MOF-74-Based Host-Guest Composites for Resistive Gas Sensing., *ACS Appl. Mater. Interfaces*,**2019**, 11 (15), 14175–14181.

[54]. Stassen, I.; Dou, J.-H.; Hendon, C.; Dincă, M., Chemiresistive Sensing of Ambient CO_2 by an Autogenously Hydrated Cu_3(Hexaiminobenzene)$_2$ Framework., *ACS Cent. Sci.*,**2019**, 5 (8), 1425–1431.

14 MOF-based Electrochemical Sensors for Ammonia

Priyanshu Goel[1,2], Saloni Sharma[1,2], Akash Deep[1,2] and Sunita Mishra[1,2]
[1]CSIR-Central Scientific Instrument Organisation (CSIR-CSIO), Chandigarh, India
[2]Academy of Scientific and Innovative Research (AcSIR), Ghaziabad, India

14.1 INTRODUCTION

Our environment embraces various gases/vapors (ammonia, methane, carbon monoxide, carbon dioxide, and so on.) that directly affect the surrounding flora and fauna [1]. The production of these gases depends on volcanic eruptions, industrialization, fossil fuel combustion, and other metabolic processes [2, 3]. Sensing these gases is important in varied applications such as food safety, pharmacology, health, environmental monitoring, and so on [2]. Ammonia is one of such toxic gases which is a threat to human life. It is a colorless, pungent, and corrosive gas with high toxicity. Being such a harmful gas, it is still used in diversified fields like fertilizers, plastics manufacturing, refrigeration, dyes, pharmaceuticals, chemical industries, and so forth [4]. In our day-to-day lives, the consumption of ammonia can be easily seen in products like dairy, ice cream, soft drink manufacturing, petrochemical facilities, fruit and vegetable juice [2, 4]. Besides food industries, ammonia is well known in the automobile sector especially in diesel-based engines where it is utilized as a catalyst in the reduction of harmful NOx gases [4]. Increasing globalization and industrialization have doubled the emission of ammonia in the last ten years in our ecosystem. Direct exposure to ammonia even at a low concentration of 50–100 ppm, will cause respiratory problems which may further lead to fatal ailments [5, 6]. Ammonia forms aerosols, namely, ammonium nitrate and ammonium sulfate when gets in contact with the acid present in the air [7]. These aerosols create smog in the environment hence contributing to global warming. The National Institute of Occupational Safety and Health (NIOSH) in the United States, suggests that a safe limit for NH_3 exposure is 25 ppm for 10 h per day at the workplace [8]. This increasing threat of ammonia gas disrupts the balance in our ecosystem, promotes climate change and causes many health issues [7, 8]. Moreover, ammonia is considered one of the critical biomarkers for diagnoses of human health. Monitoring the amount of ammonia in the breath will help in analyzing many lung or renal-related diseases [5].

Thus, sensing, and accurate detection of ammonia concentration led to a growing interest in the field of gas sensors. Extensive work in measuring and monitoring the concentration of ammonia even at ultra-low concentration levels has been going on in countless fields. Most studies focus on improving the performance parameters including the sensitivity, selectivity, stability, reversibility, operating temperature, low price, and portability of a sensor. For enhancing the NH_3 gas-sensing performance, various detection techniques have been employed in literature from which the optical, solid-state, and electrochemical methods are the three most researched and studied techniques. Commercially available ammonia gas sensors are mostly based on the solid-state method and use the metal-oxide and conducting polymers as a sensing material in their fabrication [5, 9, 10]. The simplicity of design, low cost, flexible synthesis routes of metal oxide are of interest for use in the gas sensing field. Transition metal oxides like TiO_2, SnO_2, WO_3, and MoO_3 are some of the highly exploited materials for NH_3 detection [10, 11]. Metal oxide gas sensors exhibit high sensitivity, but their input power and temperature requirements are very high. Besides this, the conducting polymers, namely, polypyrrole and polyaniline follow the redox mechanism between a target gas molecule and the conducting polymer for gas detection. In this mechanism, when the polymer comes closer to the ammonia gas, the overall conductivity of the sensing system decreases, thus showing high sensitivity and selectivity towards the target molecule. However, such devices may have some issues such as swelling, sluggish reaction kinetics, instability, and irreversibility when continuously exposed to NH_3 [9, 12].

Moreover, studies in enhancing the sensor functioning of conducting polymer-based gas sensors employ the technique of modifying their structures by using nanostructured dopants. The modified structures improve the overall physical and electrical properties thus increasing the charge transfer kinetics of the redox reaction between the analyte gas and the sensing material. Also, the high surface-to-volume ratio in them improved the interactions thus resulting in high selectivity sensitivity, and quick response time. Carbon-based nanostructures like carbon nanotubes, exfoliated graphene sheets, ribbon, and thin films are examples of carbon-based nanomaterials that have been extensively studied and which show promising NH_3 sensing when used as a dopant. Another ammonia sensing mechanism that has attracted enormous attention for the past few years is optical-based detection. This technique is very much used in commercial sensors due to its low cost, high durability, wide operation range in rigorous environmental conditions and possesses high sensitivity and selectivity with fast response time.

In addition to sensing mechanisms, aspects like low energy consumption, portability, and cost-effective fabrication with high sensitivity and selectivity are some of the attractive features of

DOI: 10.1201/9781003188148-14

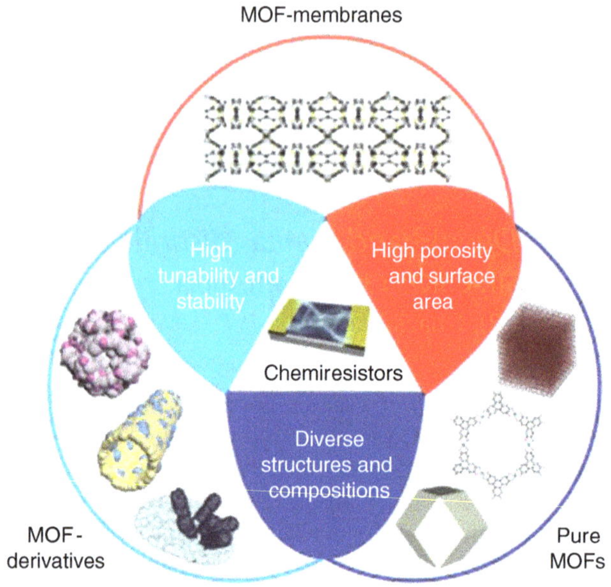

FIGURE 14.1 Illustration of various properties of MOFs. Reused from reference [13]. Copyright (2019) Elsevier

the electrochemical sensor. The electrochemical system mainly consists of three-electrodes: working, counter, and reference electrodes. The mechanism is based on the redox activity of analytes involved in the reaction. In the presence of the gas molecule the measured current or potential changes from the previous one thus showing a fast and real-time response. In the case of electrochemical sensing of NH_3, the most commonly used approach is the potentiometric method. For instance, the difference in the potential between the working electrode and the reference electrode is measured. First, the NH_3 diffuses in the system where it reacts with the electrolyte. Then, NH_3 is oxidized at the working electrode thus forming the nitrogen, hydrogen ion, and electron. The main limitation with this type of sensor is associated with the choice of the electrode and electrolyte material used (Figure 14.1)[13].

For the past two decades, much research has taken place by the scientific community which has led to the fabrication of the new materials that paved the way to designing the sensing probe from which metal-organic frameworks (MOF) have been developed. As the name suggests, MOFs are the compounds formed by the combination of metal-ligand and an organic linker, connected in such a fashion as to form a porous mesh or a framework-like structure [14, 15]. The motivation behind using MOFs in various applications lies in their excellent physical, chemical, and thermal stability with a high degree of structural tailorability [14, 16]. Additionally, MOFs have unique features such as a large specific surface area, structural flexibility, tunable pore size, high crystallinity, high catalytic activity, and easy synthesis routes that make them one of the strongest candidate materials for use in future devices [16–19]. These properties depend upon the arrangement of metal clusters and organic linkers and thus by altering these atomic arrangements MOFs can exhibit an impressive array of electrical, optical, and electrochemical properties. The first MOF was developed by Yaghi et al in 1999 and since that over 20,000 MOFs have been developed and explored in different applications like drug delivery, gas storage, separation, catalysis, and so on, by researchers all over the planet [20–22]. Due to their very high porosity, and the diversity of both the metallic centers and organic ligands, they are a highly exploited gas storage, separation, and purification application [21, 23].

Transition metal-based MOFs show high catalytic activity, thus showing high redox activity in the presence of the gas analyte molecules. Zhang et al. employed a hydrothermal approach to synthesize flaky hexagon-shaped Co-MOF cobalt(II/III) metal–organic framework (FJU-56) with aqueous Co^{2+} and tetrazolate ligands [tris-(4-tetrazolyl-phenyl)amine, H_3L] [24]. The resulting Co-MOF shows high purity, homogeneity, and superior stability as an NH_3 sensor. The as-fabricated sensor exhibits great properties for NH_3 sensing with high selectivity, a LOD of 1.38 ppm, and outstanding reversibility with a good visible color response from red to brown, when subjected to NH_3 gas [24]. Recently, our group developed a novel low-cost inkjet printing method for patterning the thin films of transition metals and Terbium (Tb) MOFs on different substrates [25]. Inkjet printing of Cr-MIL-101, Mn-BDC, Fe-MIL-101, Co-MOF-71, and Ni-BDC MOFs conveniently facilitate the homogeneous growth of MOF thin film patterns onto different substrates like paper and flexible polyethylene terephthalate (PET). The printed thin films possess intact structural and functional characteristics that remain stable for months. Further Mn- and Tb- based thin films were tested as colorimetric/fluorescent sensing strips for the detection of NH_3 vapor over a concentration range of 5–80 ppm. The results showed that Mn-BDC thin films changed to brown color from colorless and the green luminescence of the Tb-BTC was quenched when subjected to NH_3 gas [25]. Yuan et al. synthesized a Cu-MOF with micropore size through a one-pot solvothermal process [26]. The as-prepared Cu-MOF-modified carbon paste electrode exhibited outstanding redox performance towards H_2O_2 and glucose in alkaline environments. The excellent electrochemical performance suggested that the Cu-MOF was a promising candidate for electrochemical sensing [26].

Furthermore, many studies reported the MOF-based composites and derived materials showing enhanced adsorptive performance towards the removal of many gases, namely, NH_3, H_2S, NO_2, or H_2 [27–29]. Assen et al. introduced a MOF-based thin film NH_3 sensing layer comprising of a rare-earth MOF (RE-fcu-MOF) deposited on an interdigitated electrode (IDE) (Figure 14.2a) [30]. The fabricated RE-fcu-MOF sensor exhibited excellent detection for NH_3 at 1 ppm concentration, which extended to a detection limit to 100 ppb. Figure 14.2b shows the sensing results in the presence of humidity and CO_2 at room temperature. This NDC-Y-fcu-MOF-based sensor was highly stable and sensitive towards NH_3, in comparison to other reported MOFs. Also, the sensor showed high detection and selectivity towards NH_3 in comparison with CH_4, NO_2, H_2, and C_7H_8 [30]. In this chapter, the latest information and discussion are provided on the state-of-the-art sensing

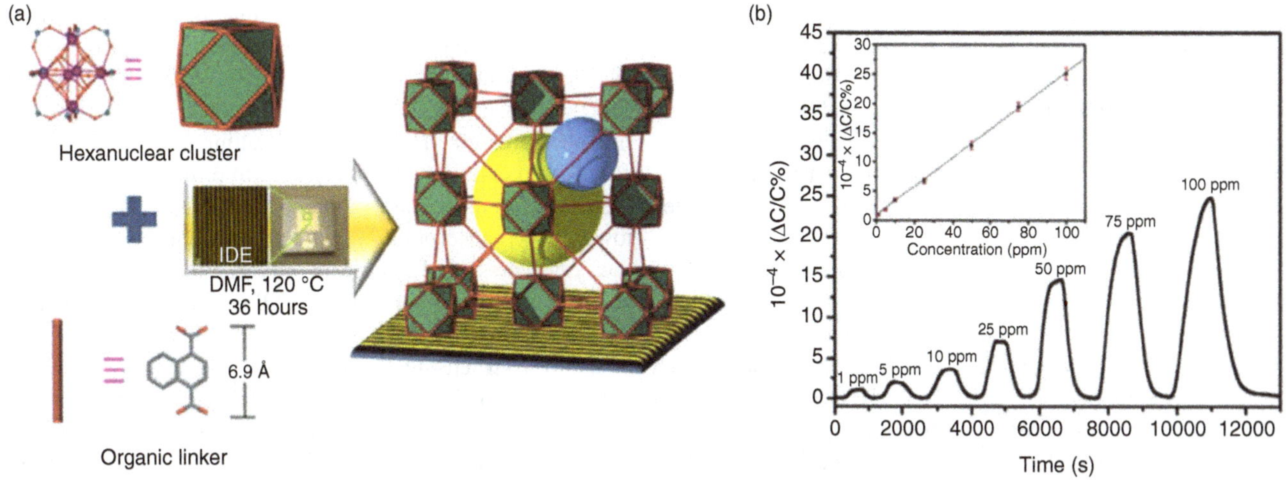

FIGURE 14.2 (a) Schematic illustration of the fabricate sensor using NDC-Y- fcu-MOF thin film on IDE substrate. (b) 1 to 100 ppm concentration detection of NH_3 Insets: plot showing the linear response of the senor with the ammonia concentration. Reused from reference [30]. Copyright (2017) American Chemical Society.

strategies with MOFs and the effects of their properties on the electrochemical performance of ammonia sensors. Furthermore, the effect of doping with different nanomaterials in the MOFs' porous structures and the use of their derivative carbon and metal oxides for sensing applications are also discussed in brief.

14.2 PRISTINE MOFS AS AN ELECTROCHEMICAL SENSORS FOR AMMONIA

In recent years, MOFs have shown great potential in the field of electrochemistry because of their controllable morphology, large pore volume, good absorbability, and high specific surface area. All such great properties feature MOFs as promising candidates in numerous electrochemical applications such as molecular sensing, gas sensing, catalysis, drug delivery, gas absorption, supercapacitors, and batteries. The metal ion or the organic ligand enhance the redox and catalytically active sites with capacity to be used as electrochemical sensors. These interactive adsorption sites, promote selective interactions between the analyte molecule and MOFs, thus enhancing the overall sensing performance. The latest progress of MOFs has made outstanding achievements in the detection of the electrochemical sensing of NH_3, H_2S, NO_2, H_2, H_2O_2, glucose, heavy metal ions, and so on [23, 27].

Nowadays many studies focus on the exploration of conductive MOFs to harness their use in many electrochemical applications. One of the widely used MOFs in electrochemical sensing is the zeolitic imidazole framework (ZIF). It is a subclass of MOFs that are primarily made with a combination of zinc or cobalt metal ions, cross-linked with imidazole linkers [5, 6]. Owing to their property of large chemical and thermal stability with huge surface area and accessible pores, they are highly explored in gas-sensing applications. Reddy et al. prepared three different Zn metal ion-based MOFs with methylimidazole (ZIF-8), nicotinic acid (Zn(NA)), and isonicotinic acid (Zn(INA)) linkers respectively as gas (ammonia, formaldehyde, ethanol, acetone, and xylene) sensing platforms [31]. The morphological and structural characteristics of the synthesized sample reveal that the MOFs have homogeneity, high porosity with high surface areas, 525 m^2g^{-1} (ZIF-8), 160.6 m^2g^{-1} (Zn(NA), and 103.6 m^2g^{-1} (Zn(INA) with average pore diameters of 1.8691 nm, 7.851 nm, and 4.5415 nm, respectively. Moreover, the fabricated sensors exhibited a stable response for NH_3 gas sensing with a LOD of 10 ppm at room temperature [31]. On the other hand, Li et al. selected two 3D Co-based MOFs, {[Co_3(mClPhIDC)2(H_2O)$_6$]·2H_2O}n and {[Co_3(p-ClPhHIDC)3(H_2O)$_3$]·6H_2O}n, to investigate their proton conductivity in the NH_3 and water vapors [32]. Their results suggested that the proton conductivity inside the MOF matrix is directly proportional to the concentration of NH_3 vapors [32]. A similar group [Ba(o-CbPhH$_2$IDC)(H_2O)$_4$]n was further synthesized in 2018 for the NH_3 gas sensor [33]. The fabrication of the device worked on the principle of reversible impedance sensing in the NH_3 gas. It contained a Ba-MOF sandwich in between the two platinum electrodes. This sensor exhibited excellent performance with high response, high selectivity, and a low detection limit of 1 ppm at 30 °C temperature and 98% of RH value. At a constant RH value when the NH_3 vapor was passed through the testing chamber, the overall sensor impedance value started decreasing and reached a constant value over time. Then, when this gas was blown out from the chamber, the impedance started to fall within 10 min and reverted to its initial value. Furthermore, this experiment was conducted 10 times and the sensors showed high reversibility and reproducibility. This work opens up wonderful opportunities for electrochemical NH_3 sensing under high RH conditions [33]. Currently, the progressive study on proton-conductive MOFs using imidazole linkers for the synthesis of MOFs and using them for the detection of NH_3 has made great achievements. Recently, Li et al.

prepared {Na[Cd(MIDC)]} using a solvothermal reaction with $Cd(NO_3)_2$ metal-ligand and a 2-methyl-1H-imidazole-4,5-dicarboxylic acid (H_3MIDC) organic linker [34]. The synthesized 3D MOF {Na[Cd(MIDC)]}n had a well-ordered one-dimensional pathway with uncoordinated carboxylate sites that provided pathways for the proton transfer which finally resulted in a change in impedance and helped in the recognition of NH_3 vapors [34].

Among the handful of electrically conductive MOFs, 2D MOFs have shown the highest conductivity values, which resulted in their in-plane charge delocalization and π- π conjugation between the sheets [35, 36]. Dinca et al. recently reported the 2D $Ni_3(HITP)_2$ (HITP=2,3,6,7,10,11-hexaiminotriphenylene) MOF, with a microporous structure and high conductivity [37]. In this work, the Ni site in the $Ni_3(HITP)_2$ MOF was substituted with a Cu ion, and the obtained material resulted in high electrical conductivity (Figure 14.3a). Further, the developed sensor highlighted a fast response towards NH_3 which is attributed to the rational tuning of metal ions in the MOF matrix. This work postulated that systematic variation in the metal center will result in a change of the overall structure of the MOF which may further lead to diverse properties and functionality. As shown in Figure 14.3b, the $Cu_3(HITP)_2$ (HITP = 2,3,6,7,10,11-hexaiminotriphenylene) displayed a high conductivity of 0.2 Scm^{-1} and sensed NH_3 within a concentration range from 0.5 to 10 ppm [37]. Meanwhile, Xu et al. reported a spray coating method for layer-by-layer assembly of electronically conductive MOFs [38]. A solution-based technique comprised of the spray-coating of metal salts and organic linkers was used to produce crystalline $Cu_3(HHTP)_2$ nanofilms (HHTP= 2,3,6,7,10,11-hexahydroxytriphenylene). These nanofilms had controllable fabrication with a thickness increase of ~2 nm per cycle. By controlling the number of spray-coating cycles, the thickness and homogeneity of $Cu_3(HHTP)_2$ thin film was easily modulated. The 2D conductive MOFs thin film showed high NH_3 sensing with a limit of detection as low as 0.5 ppm and high stability for up to 3 months. It is due to the change in electrical conductivity of MOFs that inclined it towards NH_3 adsorption [38]. These researchers, over the last few years presented 2D electrically conductive MOFs (EC-MOFs) and proton-conducting MOFs as excellent materials for electronic sensors and switches [35, 39].

Apart from sensing, the removal of NH_3 by using MOFs as adsorbents is gaining importance. NH_3 emissions constitute alkaline pollutants that react with other acidic contaminants to form particulate matter (PM 2.5) which has a hazardous impact on the environment. Li et. al, utilized series of M(BDC) (M = Cu, Zn, Cd) MOFs for NH_3 adsorption [40]. Their work suggested that 3 Dimension M(BDC) would restructure them into one-dimensional $M(BDC)(NH_3)_2$ after the adsorption of NH_3. High, repeatable adsorption capacities of 17.2, 14.1, and 7.4 mmol/g for Cu, Zn, Cd - M(BDC) were achieved [40]. Another excellent work about the absorption of NH_3 was carried out by Farha et al. in which they reported a coating of a Zn-Azolate MOF (MFU-4), on textile fiber for capturing NH_3 gas (Figure 14.4)[41]. The MFU-4 presented a large uptake capacity of 17.7 mmol/g for NH_3 at 1 bar pressure with a low limit of detection of 10.8 mmol/g at 0.05 bar pressure. The moderate synthesis conditions helped to immobilize MFU-4 on cotton textile fiber which paved the way to its use in making personal protective equipment (PPE) kits [41].

Along with the porous structure, different functionalized groups present or created by appropriate chemical or thermal

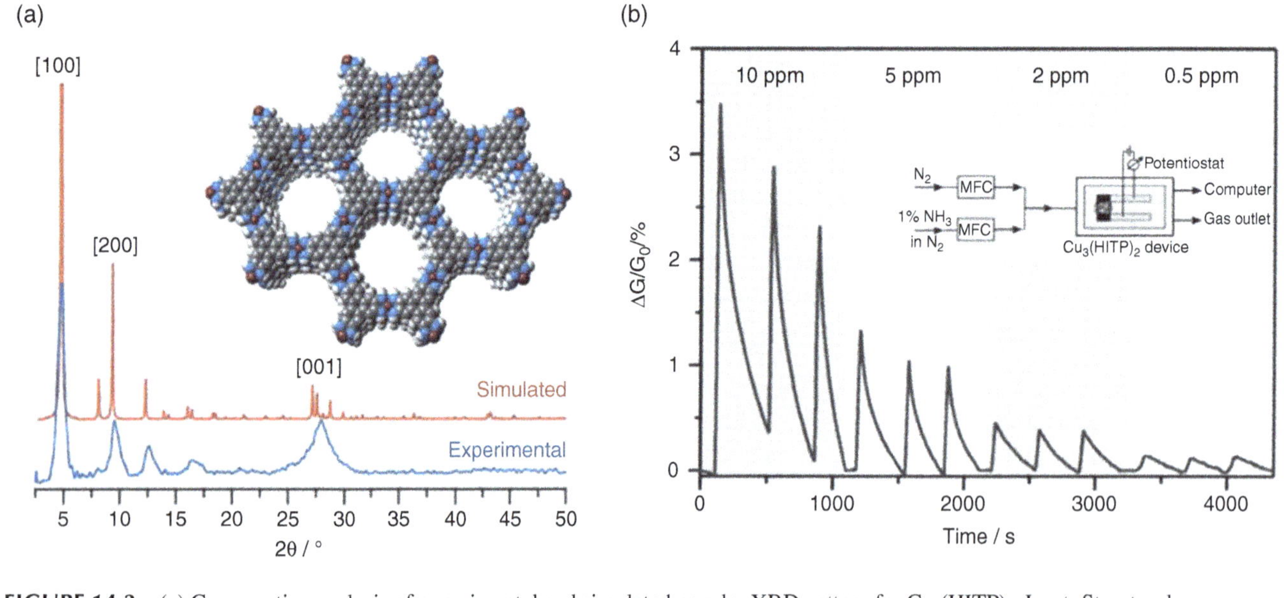

FIGURE 14.3 (a) Comparative analysis of experimental and simulated powder XRD pattern for $Cu_3(HITP)_2$. Inset: Structural arrangement of $Cu_3(HITP)_2$ with atoms having (carbon, hydrogen, nitrogen, copper) (b) $Cu_3(HITP)_2$ sensor response plot with different ammonia concentration (0.5, 2, 5, and 10 ppm). Inset: schematic depiction of the sensing module. Reproduced from reference [37]. Copyright (2015) WILEY.

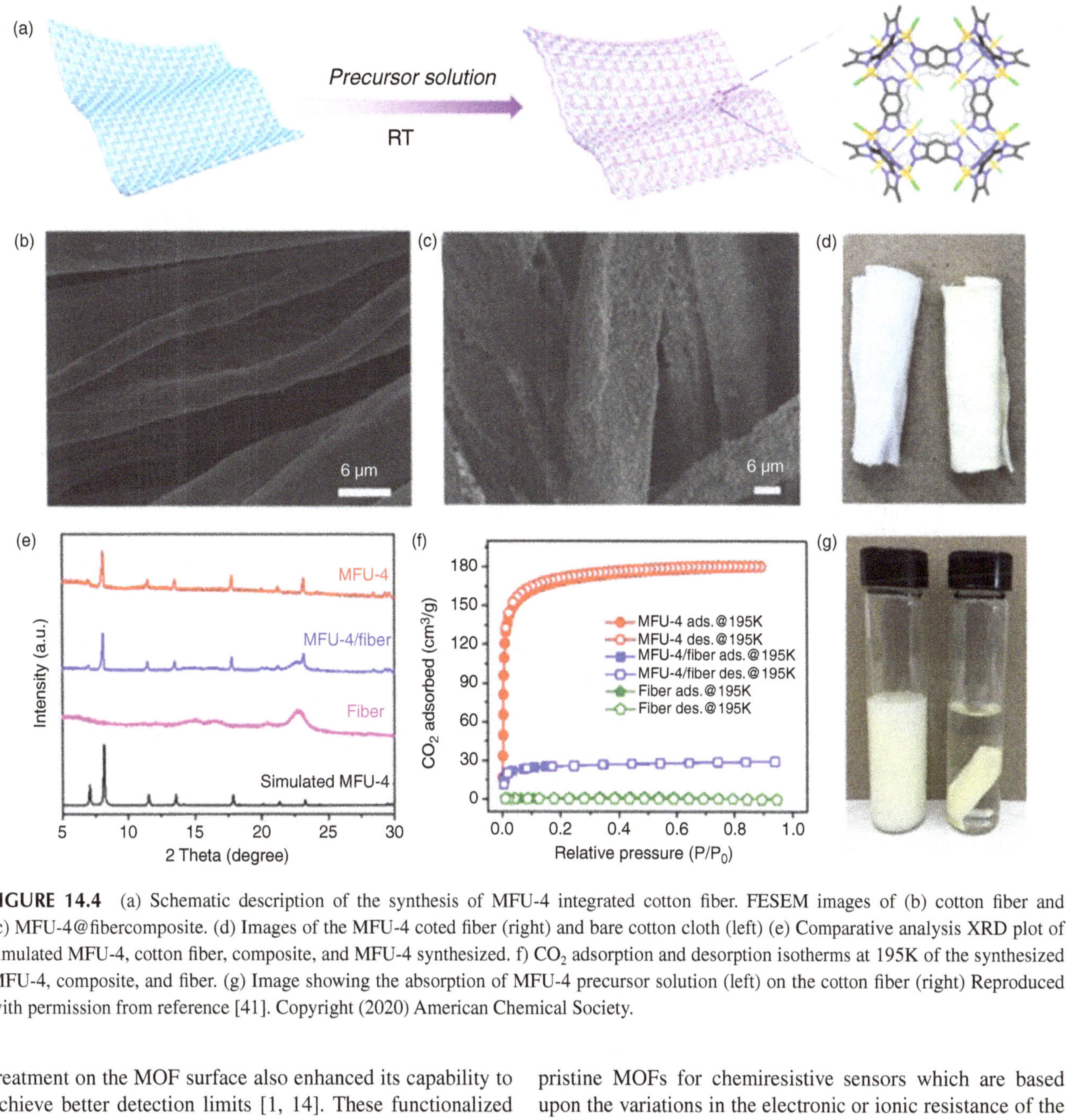

FIGURE 14.4 (a) Schematic description of the synthesis of MFU-4 integrated cotton fiber. FESEM images of (b) cotton fiber and (c) MFU-4@fibercomposite. (d) Images of the MFU-4 coted fiber (right) and bare cotton cloth (left) (e) Comparative analysis XRD plot of simulated MFU-4, cotton fiber, composite, and MFU-4 synthesized. f) CO_2 adsorption and desorption isotherms at 195K of the synthesized MFU-4, composite, and fiber. (g) Image showing the absorption of MFU-4 precursor solution (left) on the cotton fiber (right) Reproduced with permission from reference [41]. Copyright (2020) American Chemical Society.

treatment on the MOF surface also enhanced its capability to achieve better detection limits [1, 14]. These functionalized groups interact with different types of gas analyte molecules and deliver high sensitivity and selectivity to the sensor [1]. Another common way of enhancing the capability of MOFs is the incorporation of other molecules into the porous structure. These MOF composites lead to the formation of many new properties with enhanced physical adsorption forces and improved selectivity with specific interactions between the material and the gas molecules [2, 21].

14.3 MOF DERIVATIVES AND COMPOSITES AS AMMONIA SENSORS

Pure MOFs are used in various applications such as magnetic sensors, ferroelectric sensors, colorimetric sensors as well as luminescent sensors. However, there are challenges in using pristine MOFs for chemiresistive sensors which are based upon the variations in the electronic or ionic resistance of the sensing layer on interaction with the target gas [21, 23]. In the case of electrochemical sensing, various electrochemical parameters such as resistance, current, potential, impedance, and capacitance can be changed through the interaction of MOFs with target gases and VOC analytes. It boosts their potential as electrical sensing and switching materials. At ambient atmospheres, most MOFs have low electrical conductivity due to the presence of hard metal ions in MOFs and the poor orbital conjugation that hinders the flow of electrons [1, 14, 16]. These factors restrict the utilization of the porosity and ultra-large surface area of MOFs and decrease the chemical stability that results in low electrical signal transfer for chemiresistive detection of gas molecules like ammonia (NH_3) [30]. Therefore, to report this intrinsic issue in pure MOFs, researchers started focusing on the other

MOF-based derivative like porous carbon composites and metal oxides for application in electrochemical sensors. These compounds show excellent electrical sensing performance with fast redox activity. For instance, Wang et al. fabricated a highly-sensitivity NH_3 sensor based on $CoSe_2$@NC/MWCNTs nanocomposite [42]. The composite comprised of nitrogen-doped amorphous carbon-coated cobalt diselenide and multi-walled carbon nanotubes which was derived by pyrolysis of a cobalt-based MOF. For NH_3 sensing, firstly an epoxy substrate was used to deposit $CoSe_2$@NC/MWCNTs film on the interdigitated electrode. Then the microstructure was exposed to different concentrations of NH_3 gas. The characteristics of the fabricated NH_3 sensor were then studied at room temperature with a relative humidity of 43%. As shown in Figure 14.5, the $CoSe_2$@NC/MWCNTs sensor showed a response of up to 93.37% towards 10 ppm NH_3 concentration with a detection limit of 0.1 ppm. This remarkable performance of the NH_3 sensor was primarily due to the synergistic effects of Co-MOF derived $CoSe_2$@NC/MWCNTs composite and the effect of p-n heterojunction present in it. In addition, the effect of humidity and other toxic gases on sensor performance was also explored (Figure 14.5a). The results shown in Figure 14.5d revealed that with an increase in humidity of 11%, 43%, 67%, and 85% (RH), the response of the sensor deteriorated due to the presence of the OH group of water molecules that might have occupied adsorption sites under high RH conditions and hindered the reaction of NH_3 [42].

Due to the high energy bandgap, most of the MOFs behave similarly to insulators which limit their applications in many electrochemical systems. Hence, evolution in MOF-based electrochemical sensors needs some conduction pat channels in the material which can enhance the overall performance of the electrochemical system [43, 44]. Keeping this in mind, scholars have integrated MOFs with conducting polymers, carbon materials, and other metal nanoparticles. These integrated composites of MOFs retained excellent conductivity and a large surface-to-volume ratio. Recently Mirica et al. have reported the fabrication of pure semiconductive MOF (M_3HHTP_2) analogs with different metal centers (M = Fe, Co, Ni, or Cu) [20]. The synthesized material contained conductive additives of M_3HHTP_2/graphite blends in 9:1 by mass. The MOF/graphite blend revealed a significant increase in the bulk conductivity of 3.8×10^{-2} to 9.8×10^{-1} Scm^{-1} in comparison to pure MOFs (2.7×10^{-6} to 2.0×10^{-2} Scm^{-1}). The NH_3 sensing performance of M_3HHTP_2/graphite for varying concentrations (1200, 800, 80, 40, 20, 10, 5 ppm) showed a linear increase in response to NH_3 through 80 ppm exposure. The sensor exhibited a LOD of 19 ppm with a saturation limit above 80 ppm [20].

The graphene and reduced graphene oxide (rGO) showed promising performance with MOFs. Their composite possessed an outstanding conductivity and high surface area. In addition, defects over the surface and a large number of oxygen functional groups on rGO can provide active sites for the adsorption of target analytes. Recently Wang et al. [45]

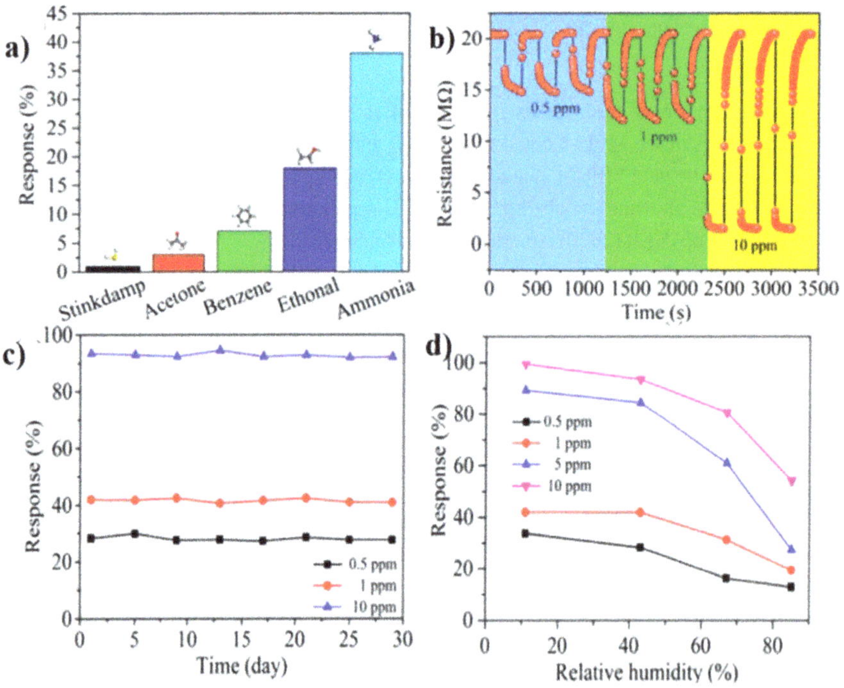

FIGURE 14.5 (a) Plot showing the selectivity of the$CoSe_2$@NC/MWCNTs-S2composite based sensor to various gases of 1 ppm concentration at RT and 43% RH.(b)Graph representing the repeatability and (c) long-term stability test conducted on 0.5, 1 and 10 ppm concentration of NH3 at RT and 43% RH. (d) RH response on$CoSe_2$@NC/MWCNTs-S2based NH_3 sensor Reproduced with permission from reference [42]. Copyright (2020) Elsevier.

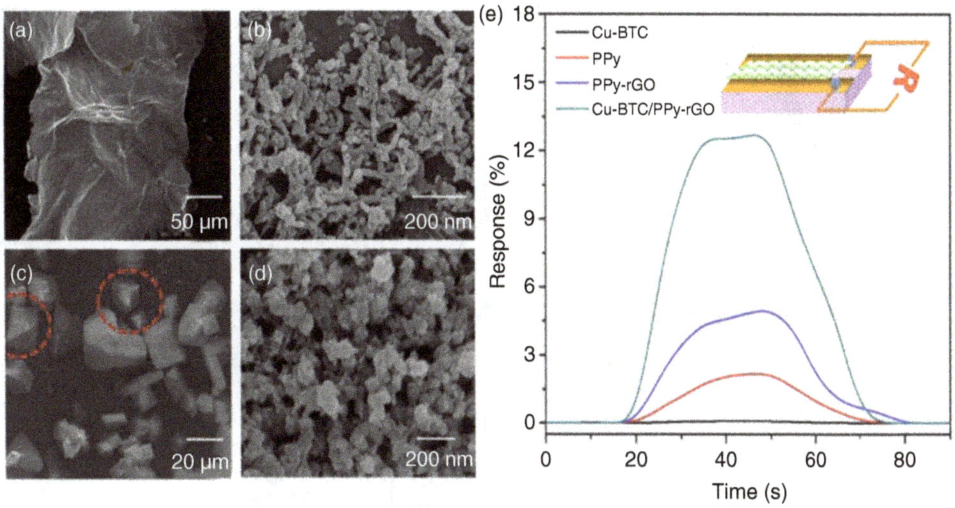

FIGURE 14.6 (a) SEM images of (a) rGO sheets, (b) PPy coated rGO, (c) cubical crystal of Cu–BTC, and (d) Cu–BTC decorated by PPy–rGO composite of (e) Electrochemical response of the prepared Cu-BTC/PPy–rGO sensor to 50ppm NH_3 at room temperature Inset: schematic diagram of the Cu-BTC/PPy–rGO based sensing module. Reproduced from reference [47]. Copyright (2018) Elsevier.

reported a NH_3 sensor fabricated using MOFs derived zinc oxide (ZnO) and reduced graphene oxide (rGO). The nanocomposite showed high linearity with excellent stability and selectivity over a sensing range of 0.5–30 ppm concentration of NH_3 gas at room temperature. Also, the sensor exhibited a quick response of 50 s with a fast recovery time of 25 s in the presence of a 30 ppm concentration of NH_3 [45]. Similar work is also reported by Sharma et. al, for the sensing of NH_3 via a zeolite imidazole framework (ZIF-67) and rGO composite [6]. This synthesized composite had a high surface area of 1080 m^2g^{-1} with excellent conductivity. The fabricated sensor offered a stable response with a LOD of 74 ppb and provide high tolerance against humidity. For a 50 ppm concentration of NH_3, the sensor displayed the average response and recovery time to be 46.5 ± 2.12 and 66.5 ± 2.12 s, respectively [6].

Within this class of MOFs, Cu-BTC (copper- benzene-1, 3, 5-tricarboxylate) is a well-studied structure for sensing purposes. This structure is also known as HKUST-1 among researchers and Basolite C300 commercially. The Cu-BTC cages contain unsaturated polar copper sites which preferentially coordinate with polarizable analyte molecules such as NO and CO, CO2, water, or alkenes. Although Cu-BTC structures have higher pore volumes and surface areas than many other MOFs, their lack of conductivity still creates a big problem for their use in electrochemical sensing. Deep et al. prepared a three-phase Cu-MOF composite consisting of SiO2, single-layer graphene, and polyaniline molecules [46]. The study demonstrated the Cu-MOF and the graphene composite were bridged by an aniline component in developing a highly efficient sensory material that has the superior conduction properties of graphene and a highly porous nature of MOF with a high surface area of 756 m2g-1. The developed sensor showed a high linear detection range of NH_3 over 1–100 ppm with a low limit of detection 0.6 ppm [46]. Another significant work highlighting this encapsulation of a conductive guest molecule in Cu-MOF for NH_3 sensing was done by Y. Yin et al. They reported the synthesis of Cu-BTC (benzene-1,3,5-tricarboxylate) on a nanocomposite of polypyrrole-nanofiber-coated reduced graphene oxide (PPy–rGO) for the detection of NH_3 [47]. SEM characterization in Figure 14.6 demonstrates PPy–rGO nanocomposites which were decorated with nanoparticles of Cu–BTC $[Cu_3(BTC)_2(H2O)_3$. A simple one-step hydrothermal process, followed by an in-situ chemical polymerization was used in the synthesis of a new MOF@G composite for sensing NH_3 gas. As shown in Figure 14.6, the synergistic effect between Cu–BTC and PPy–rGO, the MOF@G nanocomposite sensor showed a fast electrochemical response to 50 ppm of NH_3 with rapid recovery rates, high sensitivity, good reproducibility, and long-term stability [47].

Ammonia is an environmental toxin that is especially problematic for all organisms. It accumulates easily in air and aquatic systems through industrial discharge and also with animals' excretions. Moreover, people with poor nutrition have elevated levels of NH_3 in their breath as the liver or kidneys of the patients are unable to metabolize NH_3 in the body. If monitored properly, it can act as a biomarker for many lethal diseases. Recently, Banga et al. have developed a ZIF-8 based electronic nose (ZENose) [5]. As shown in Figure 14.7a, the novel electrochemical nose system comprised of a ferrocene entrapped zeolitic imidazole framework (Fc@ZIF-8) which was used to detect NH_3 levels in the breath. This breathomics NH_3 sensor showed high sensitivity, specificity, and a fast response time of 60s with a LOD of 400 ppb in real-time monitoring (Figure 14.7b, c). This work is a perfect example of the synergistic effect of MOF porosity and its guest host chemistry that leads to significant improvements in the detection and sensing level of many gases [5]. MOFs and the compounds based on them demonstrate an efficient

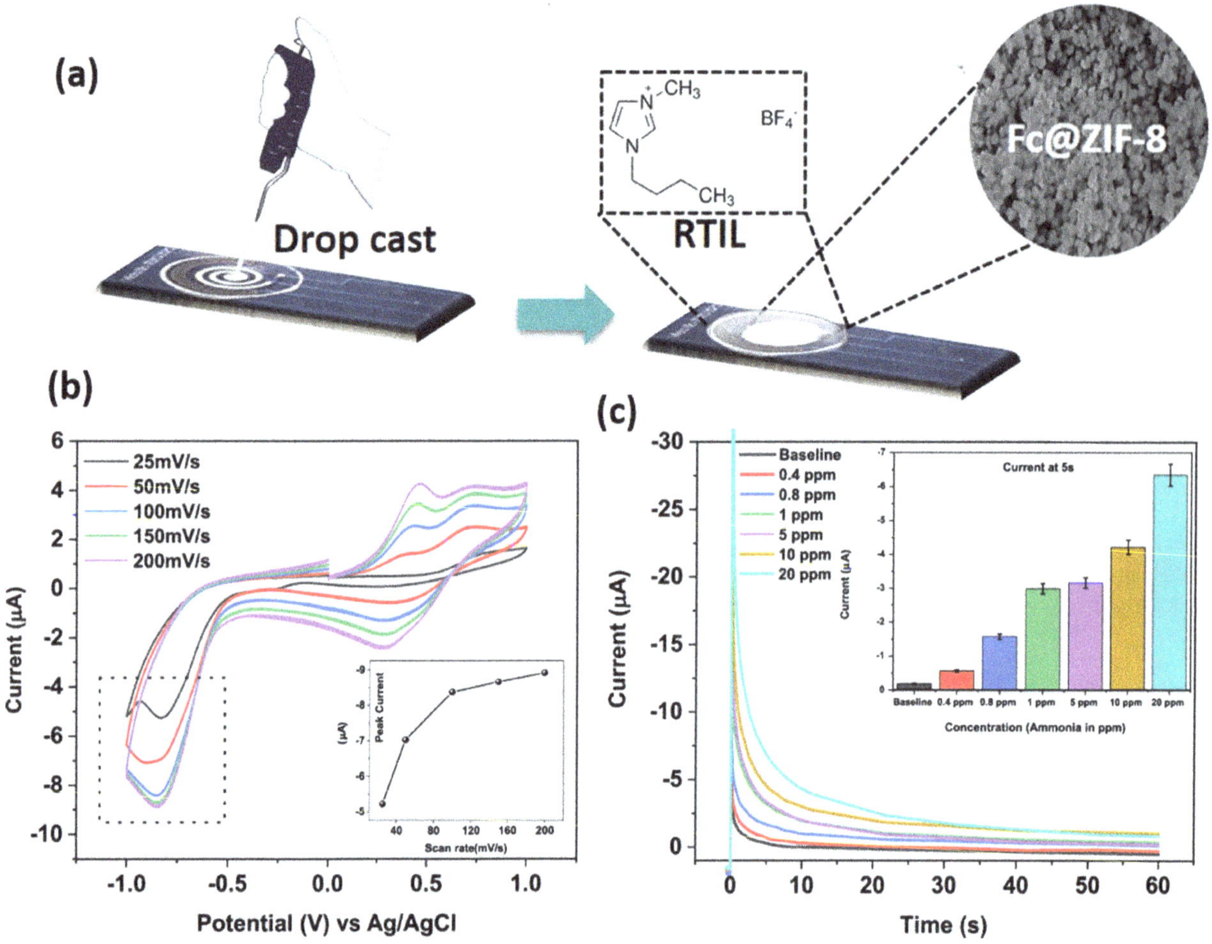

FIGURE 14.7 (a) Representation of experimental setup of Fc@ZIF based sensor (b) CV curve on a potential window of −1 to +1 V at 25 to 250 mV/s scan rate showing the NH_3 sensing at 400 ppb concentration. Inset: Current vs scan rate plot depicting the linearity in the sensing performance. c) Chronoamperometry performance ofFc@ZIF-8-modifiedsensing electrode at −0.8 V potential and for 60 s period for the sensing analyte. Inset: Peak current response for 400 ppb to 20 ppm of ammonia gas concentration. Reused with permission from reference [5]. Copyright (2021) American Chemical Society.

material for an electrochemical sensor platform that can be used for sensing NH_3 with high sensitivity and specificity. With the appropriate design and integration of MOF-based materials, innovative research in the future can be applied to the real-world applications of electrochemical sensing such as clinical diagnosis and waste treatments, health monitoring, and so on.

14.4 CONCLUSION

Due to their unique material properties, MOFs are envisaged with great potential in electrochemical gas storage, separation, and sensing material. Large surface area and tunable porosity, high catalytic activity, reversible adsorption, and tunable chemical functionalization create a wonderful opportunity for its utilization in the field of electrochemical detection and sensing. However, the usage of pristine MOFs towards electrochemical sensing always faces many challenges which can be further improved by using MOF-derived materials and MOF-based composites with other materials. For instance, transition metal-based MOFs show great potential as gas sensing materials in electrochemical ammonia sensing. The introduction of the guest NH_3 gas molecules in the host MOF matrix induces resistivity, thus decreasing the conductivity of the overall system. Further, the combination with other conductive materials, with carbon nanotubes, and graphene enhance the MOF with better stability and increased electroconductivity, thus improving the overall charge kinetics of the electrochemical system. The above-discussed host-guest chemistry directly influences the performance of the sensor. Also, the functional groups present on the MOFs, surfaces are used as recognition agents for the target analyte gas and enhance the sensing selectivity of a sensor. The research in this particular domain is still in its early stages and only a few researchers have successfully demonstrated the functionalization of MOFs in improving the performance parameters of electrochemical NH_3 sensors. On account of their considerable porous structures, very large surface areas, and thermal stability, the application of MOFs or MOF-based materials as NH_3 sensors can provide multiple benefits. These advantageous factors may include the higher loading of NH_3 gas molecules, enhanced separation of the stored gas, control on the charge redox kinetics of the electrochemical reaction and enhancing the performance parameters of the sensor.

Although MOF-based material has been very much explored in NH_3 sensing there are always some gaps in terms of selectivity, sensitivity, and durability which have to be filled by further exploration. Various research groups have worked and are still working to get a better understanding of the MOF as an NH_3 sensing platform. Combing new metal ligands with different organic linkers harnesses new properties such as plasmonic, electrical, and thermochromic properties in the MOFs. These upgraded versions of MOFs provide more opportunities for of sensor design, integration, and performance. Moreover, the working condition of the NH_3 sensor should also be optimized before designing a real-time sensor. For future research, it is stressed that MOFs and MOF-based materials possess very attractive features that should ensure that they are a promising option for highly efficient electrochemical ammonia sensors.

REFERENCES

[1]. E. Comini, Metal oxide nano-crystals for gas sensing, Anal. *Chim. Acta*. 568 (2006) 28–40.

[2]. N. Yamazoe, N. Miura, Environmental gas sensing, *Sensors Actuators B Chem*. 20 (1994) 95–102.

[3]. C. Barbu, S. Moise, I.V. Matei, Evolution of industrialisation and pollution in Craiova. the current state of the management of sustainable development, *Artic. J. Environ. Prot. Ecol.* 16 (2015) 470–478.

[4]. D. Kwak, Y. Lei, R. Maric, Ammonia gas sensors: A comprehensive review, *Talanta*. 204 (2019) 713–730.

[5]. I. Banga, A. Paul, S. Muthukumar, S. Prasad, ZENose (ZIF-Based Electrochemical Nose) Platform for Noninvasive Ammonia Detection, *ACS Appl. Mater. Interfaces*. 13 (2021) 16155–16165.

[6]. N. Garg, M. Kumar, N. Kumari, A. Deep, A.L. Sharma, Chemoresistive Room-Temperature Sensing of Ammonia Using Zeolite Imidazole Framework and Reduced Graphene Oxide (ZIF-67/rGO) Composite, *ACS Omega*. 5 (2020) 27492–27501.

[7]. J.B. Nowak, J.A. Neuman, R. Bahreini, C.A. Brock, A.M. Middlebrook, A.G. Wollny, J.S. Holloway, J. Peischl, T.B. Ryerson, F.C. Fehsenfeld, Airborne observations of ammonia and ammonium nitrate formation over Houston, *Texas, J. Geophys. Res. Atmos*. 115 (2010) 22304.

[8]. K. Clark, R. E. Crossgrove, *Acute Exposure Guideline Levels for Selected Airborne Chemicals: Volume 3*, (2003), The National Academies Press, Washington D.C.

[9]. B. Timmer, W. Olthuis, A. Van Den Berg, Ammonia sensors and their applications - A review, *Sensors Actuators*. 107 (2005) 666.

[10]. C.S. Rout, M. Hegde, A. Govindaraj, C.N.R. Rao, Ammonia sensors based on metal oxide nanostructures, *Nanotechnology*. 18 (2007) 205504.

[11]. C.S. Rout, K. Ganesh, A. Govindaraj, C.N.R. Rao, Sensors for the nitrogen oxides, NO_2, NO and N_2O, based on In_2O_3 and WO_3 nanowires, *Appl. Phys*. 85 (2006) 241.

[12]. A. Hasani, H.S. Dehsari, M.A. Lafmejani, A. Salehi, F.A. Taromi, K. Asadi, S.Y. Kim, Ammonia-Sensing Using a Composite of Graphene Oxide and Conducting Polymer, *Phys. Status Solidi – Rapid Res. Lett*. 12 (2018) 1800037.

[13]. W.-T. Koo, J.-S. Jang, I.-D. Kim, Metal-Organic Frameworks for Chemiresistive Sensors, *Chem*. 5 (2019) 1938–1963.

[14]. J.L.C. Rowsell, O.M. Yaghi, Metal-organic frameworks: A new class of porous materials, *Microporous Mesoporous Mater*. 73 (2004) 3–14.

[15]. S.K. Bhardwaj, G.C. Mohanta, A.L. Sharma, K.H. Kim, A. Deep, A three-phase copper MOF-graphene-polyaniline composite for effective sensing of ammonia, Anal.

[16]. O. Shekhah, J. Liu, R. A. Fischer, Ch. Wöll, MOF thin films: existing and future applications, *Chem. Soc. Rev.* 40 (2011) 1081–1106.

[17]. M.E. Davis, Ordered porous materials for emerging applications, *Nature*. 417 (2002) 813–821.

[18]. V. Stavila, A.A. Talin, M.D. Allendorf, MOF-based electronic and opto-electronic devices, *Chem. Soc. Rev.* 43 (2014) 5994–6010.

[19]. X. Zhu, H. Zheng, X. Wei, Z. Lin, L. Guo, B. Qiu, G. Chen, Metal-organic framework (MOF): A novel sensing platform for biomolecules, *Chem. Commun*. 49 (2013) 1276–1278.

[20]. M. Ko, A. Aykanat, M.K. Smith, K.A. Mirica, Drawing Sensors with Ball-Milled Blends of Metal-Organic Frameworks and Graphite, *Sensors* 17 (2017) 2192.

[21]. L.J. Small, M.E. Schindelholz, T.M. Nenoff, Hold on Tight: MOF-Based Irreversible Gas Sensors, *Ind. Eng. Chem. Res.* 60 (2021). 7998–8006

[22]. H. Li, M. Eddaoudi, M. O'Keeffe, O.M. Yaghi, Design and synthesis of an exceptionally stable and highly porous metal-organic framework, *Nat. 1999 4026759*. 402 (1999) 276–279.

[23]. W.-T. Koo, J.-S. Jang, I.-D. Kim, Metal-Organic Frameworks for Chemiresistive Sensors, *Chem*. 5 (2019) 1938–1963.

[24]. J. Zhang, J. Ouyang, Y. Ye, Z. Li, Q. Lin, T. Chen, Z. Zhang, S. Xiang, Mixed-Valence Cobalt(II/III) Metal–Organic Framework for Ammonia Sensing with Naked-Eye Color Switching, *ACS Appl. Mater. Interfaces*. 10 (2018) 27465–27471.

[25]. P. Goel, S. Singh, H. Kaur, S. Mishra, A. Deep, Low-cost inkjet printing of metal–organic frameworks patterns on different substrates and their applications in ammonia sensing, *Sensors Actuators B Chem*. 329 (2021) 129157.

[26]. L. Yang, C. Xu, W. Ye, W. Liu, An electrochemical sensor for H_2O_2 based on a new Co-metal-organic framework modified electrode, *Sensors Actuators B Chem*. 215 (2015) 489–496.

[27]. K. Tan, S. Zuluaga, Q. Gong, Y. Gao, N. Nijem, J. Li, T. Thonhauser, Y.J. Chabal, Competitive Coadsorption of CO_2 with H_2O, NH_3, SO_2, NO, NO_2, N_2, O_2, and CH_4 in M-MOF-74 (M = Mg, Co, Ni): The Role of Hydrogen Bonding, *Chem. Mater.* 27 (2015) 2203–2217.

[28]. G.W. Peterson, J.J. Mahle, J.B. DeCoste, W.O. Gordon, J.A. Rossin, Extraordinary NO2 Removal by the Metal–Organic Framework UiO 66-NH_2, Angew. *Chemie*. 128 (2016) 6343–6346.

[29]. G. Liu, A. Cadiau, Y. Liu, K. Adil, V. Chernikova, I.-D. Carja, Y. Belmabkhout, M. Karunakaran, O. Shekhah, C. Zhang, A.K. Itta, S. Yi, M. Eddaoudi, W.J. Koros, Enabling Fluorinated MOF-Based Membranes for Simultaneous Removal of H_2S and CO_2 from Natural Gas, *Angew. Chemie Int. Ed.* 57 (2018) 14811–14816.

[30]. A.H. Assen, O. Yassine, O. Shekhah, M. Eddaoudi, K.N. Salama, MOFs for the Sensitive Detection of Ammonia: Deployment of fcu-MOF Thin Films as Effective Chemical Capacitive Sensors, *ACS Sensors*. 2 (2017) 1294–1301.

[31]. A.J. Mohan Reddy, N.K. Katari, P. Nagaraju, S.B. Manabolu Surya, ZIF-8, Zn(NA) and Zn(INA) MOFs as chemical

selective sensors of ammonia, formaldehyde and ethanol gases, *Mater. Chem. Phys.* 241 (2020) 122357.

[32]. X. Liang, B. Li, M. Wang, J. Wang, R. Liu, G. Li, Effective Approach to Promoting the Proton Conductivity of Metal–Organic Frameworks by Exposure to Aqua–Ammonia Vapor, *ACS Appl. Mater. Interfaces.* 9 (2017) 25082–25086.

[33]. K. Guo, L. Zhao, S. Yu, W. Zhou, Z. Li, G. Li, A Water-Stable Proton-Conductive Barium(II)-Organic Framework for Ammonia Sensing at High Humidity, *Inorg. Chem.* 57 (2018) 52.

[34]. R. Liu, Y. Liu, S. Yu, C. Yang, Z. Li, and G. Li, A Highly Proton-Conductive 3D Ionic Cadmium-Organic Framework for Ammonia and Amines Impedance Sensing, *ACS Appl. Mater. Interfaces.* 11 (2019) 1713–1722.

[35]. J. Liu, X. Song, T. Zhang, S. Liu, H. Wen, L. Chen, 2D Conductive Metal–Organic Frameworks: An Emerging Platform for Electrochemical Energy Storage, *Angew. Chemie Int. Ed.* 60 (2021) 5612–5624.

[36]. X. Song, J. Liu, T. Zhang, L. Chen, 2D conductive metal-organic frameworks for electronics and spintronics, *Sci. China Chem. 2020 6310.* 63 (2020) 1391–1401.

[37]. M.G. Campbell, D. Sheberla, S.F. Liu, T.M. Swager, M. Dincă, Cu_3(hexaiminotriphenylene)$_2$: An Electrically Conductive 2D Metal–Organic Framework for Chemiresistive Sensing, *Angew. Chemie Int. Ed.* 54 (2015) 4349–4352 .

[38]. M.-S. Yao, X.-J. Lv, Z.-H. Fu, W.-H. Li, W.-H. Deng, G.-D. Wu, G. Xu, Layer-by-Layer Assembled Conductive Metal–Organic Framework Nanofilms for Room-Temperature Chemiresistive Sensing, *Angew. Chemie Int. Ed.* 56 (2017) 16510–16514.

[39]. M.G. Campbell, S.F. Liu, T.M. Swager, M. Dincă, Chemiresistive Sensor Arrays from Conductive 2D Metal–Organic Frameworks, *J. Am. Chem. Soc.* 137 (2015) 13780–13783.

[40]. Y. Chen, Y. Du, P. Liu, J. Yang, L. Li, and J. Li, Removal of Ammonia Emissions via Reversible Structural Transformation in M(BDC) (M = Cu, Zn, Cd) Metal-Organic Frameworks., *Environ. Sci. Technol.* 54 (2020) 3636–3642.

[41]. R. Cao, Z. Chen, Y. Chen, K.B. Idrees, S.L. Hanna, X. Wang, T.A. Goetjen, Q. Sun, T. Islamoglu, O.K. Farha, Benign integration of a Zn-Azolate metal-organic framework onto textile fiber for ammonia capture, *ACS Appl. Mater. Interfaces.* 12 (2020) 47747–47753.

[42]. Q. Mi, D. Zhang, X. Zhang, D. Wang, Highly sensitive ammonia gas sensor based on metal-organic frameworks-derived $CoSe_2$@nitrogen-doped amorphous carbon decorated with multi-walled carbon nanotubes, *J. Alloys Compd.* 860 (2021) 158252.

[43]. L. Liu, Y. Zhou, S. Liu, M. Xu, The Applications of Metal–Organic Frameworks in Electrochemical Sensors, *ChemElectroChem.* 5 (2018) 6–19.

[44]. Teng Ma, Huibo Li, Jian-Gong Ma, Peng Cheng, Application of MOF-based materials in electrochemical sensing, *Dalt. Trans.* 49 (2020) 17121–17129.

[45]. D. Wang, M. Chi, D. Zhang, D. Wu, Ammonia sensing properties of metal–organic frameworks-derived zinc oxide/reduced graphene oxide nanocomposite, *J. Mater. Sci. Mater. Electron.* 31 (2020) 4463–4472.

[46]. S.K. Bhardwaj, G.C. Mohanta, A.L. Sharma, K.H. Kim, A. Deep, A three-phase copper MOF-graphene-polyaniline composite for effective sensing of ammonia, *Anal. Chim. Acta.* 1043 (2018) 89–97.

[47]. Y. Yin, H. Zhang, P. Huang, C. Xiang, Y. Zou, F. Xu, L. Sun, Inducement of nanoscale Cu–BTC on nanocomposite of PPy–rGO and its performance in ammonia sensing, *Mater. Res. Bull.* 99 (2018) 152–160.

15 MOF-based Electrochemical Sensors for Hydrogen Peroxide

Saadat Majeed[1], *Muhammad Umer Farooq*[1], *Naeem Akhtar Khan*[2], *Batool Fatima*[3], *Sabahat Majeed*[4], *Saima Anjum*[5], *Muhammad Najam-ul-Haq*[1], *Fahad Ali*[1], *Sayed Tayyab Raza Naqvi*[1]

[1]Division of Analytical Chemistry, Bahauddin Zakariya University, Multan, Pakistan

[2]IRCBM, COMSAT University, Islamabad, Lahore Campus, Pakistan

[3]Department of Biochemistry, Bahauddin Zakariya University, Multan, Pakistan

[4]Department of Biosciences, COMSAT University, Islamabad, Pakistan

[5]Department of Chemistry, The Govt. Sadiq College Women University, Bahawalpur, Pakistan

15.1 INTRODUCTION

Hydrogen peroxide is a strong oxidizing agent, colorless, and weakly acidic liquid that is commercially available in different range of concentrations. It is used for various processes such as oxidizing agents and preparation of peroxygen compounds, textiles, and in the bleaching industry. It occurs in water as well as the in atmosphere that is produced by a photochemical process and the photolysis of O_3 and aldehydes, respectively. It is also present in living bodies such as blue-green algae and the bombardier beetle. It is found in humans with a concentration ranging from 300–1000 mg/m^3.

Hydrogen peroxide, one of the reactive oxygen species, is a major cause of oxidative stress that can lead to various diseases from mild to acute. These life-threatening diseases include Parkinson's disease, cancer, diabetes, myocardial infarction, atherosclerosis, chronic fatigue syndrome, and renal failure. The concentration of H_2O_2 in urine is the biomarker of oxidative stress and diagnosis of these diseases. It's very toxic in its higher concentration range in animals, plants, and bacterial cells [1]. However, its toxicity is observed to be limited at or below about 20–50 µM. In most physiological reactions, H_2O_2 is the by-product catalyzed by enzymes such as glucose oxidase, oxalate oxide, lactate oxidase, and cholesterol oxidase. Oxidizing and reducing properties of H_2O_2 enable it to be used in industrial, pharmaceutical, and environmental applications. It also acts as a signaling molecule in various biological processes. Therefore, an accurate and sensitive detection method for H_2O_2 is needed. Spectrophotometry, titrimetric, chemiluminescence, chromatography, colorimetry, fluorescence, and electrochemistry have been developed for the qualitative and quantitative determination of H_2O_2. Among these methods, the electrochemical technique has received more attention due to its ease of operation, more selectivity, high sensitivity, stability, and miniaturization. Slow kinetics, interfering agents, and overpotential can downgrade the oxidation or reduction of hydrogen peroxide [2].

The electrochemical enzymatic and non-enzymatic approach is a promising way for in vivo and in vitro detection of H_2O_2 due to its higher sensitivity, selectivity, ease of operation, fast response, low cost, and accuracy. The enzymatic sensor applications have been limited due to some disadvantages such as high cost, poor stability, complicated immobilization procedure, dependence on pH and temperature. While the development of non-enzymatic sensors for H_2O_2 has been promoted due to fast response, more stability, and wider applications [3]. Recently, various materials have been extensively used for electrode modification to make sensors for H_2O_2, such as graphene oxide, reduced graphene oxide, carbon nanotubes, nanoparticles, polymers, metal, and metal oxides. These materials can solve issues generated by enzymatic sensors as these provide a prompt electrocatalytic response, have low cost, low background noise, inertness, and a wide potential window. Metal-organic frameworks (MOFs) are a porous and network polymeric structure of organic linkers and inorganic metal nodes. MOFs emerged as an extensive crystalline class of compounds due to their structural and functional tunability, ultra-porosity, and enormous surface area. A large variety of organic and inorganic components made them interesting for application in gas storage, electro-catalysis, membranes, thin films, drug delivery, sensing, and biomedical imaging [4]. Synthesis of MOFs with different geometries is not difficult and only needs inexpensive inorganic salts and organic linkers. Different methodologies have been reported for the synthesis of MOFs such as solvothermal, sonochemical, chemical, mechanical, electrochemical, microwave-assisted, and hydrothermal methods. These methods provide high yield and good quality crystals. After the synthesis of MOFs, trapped ligands or solvents that are added during the synthesis of MOFs to affect the surface area are removed by drying or vacuum drying [5].

DOI: 10.1201/9781003188148-15

MOFs have unique properties such as chemical stability, tunable chemical functionality, good mechanical strength, and good electrical response. These materials have been investigated preferably because of their large surface area and higher porosity. This porosity allows the trapping of the guest molecules without structural modification of the MOF. The addition of metals in MOFs provides luminescent properties for biomedical applications. The applications of pure MOFs onto electrode surfaces are limited due to their poor electrical conductivity. Therefore, to resolve this poor conductivity problem and enhance the electron transfer, a new strategy is adopted which is to create a composite of MOFs with higher conductive materials such as carbon-based materials, metallic nanoparticles, polymers, and metals or metals oxides [6]. Carbon-based materials such as graphite, carbon nanotubes, and graphene are the mostly used materials due to their high mechanical strength, good electrical and chemical properties, thermal stability, low toxicity, and low cost [7].

A large variety of composites based on MOFs haves been synthesized and used for the detection of hydrogen peroxide. These composites include MOF-nanoparticles, MOF-polymers, MOF-carbon-based material, MOF-metal or metal oxide, and MOF-enzymes or biomolecules. Here, we will discuss the electrochemical sensing of H_2O_2 based on MOFs, and MOF composites. MOF composites include MOF-nanoparticles (gold, silver, iron, and platinum), carbon-based (graphene oxide, reduced graphene oxide, and carbon nanotubes), polymer based, metal-based (chromium, copper, nickel, cobalt, cerium, stannic), and biomolecules such as cytochrome C. Currently, the use of carbon-based modified electrodes is growing constantly due to their easy construction, modification, and wide range of analyte detection. Generally, carbon allotropes such as graphite and multi-walled carbon nanotubes are employed for electrode modification. The aromatic ring of six-membered and sp^2 bonding allows it to demonstrate good conductivity. Polymers provide a soft and large network with MOFs. Sensitivity and selectivity are two main factors in electrochemistry and MOFs must provide these two features. So, the ligands with conjugate systems provide these two features. Similarly, electroactive metal species can provide good selectivity for hydrogen peroxide [8].

15.2 ELECTROCHEMICAL SENSORS FOR H_2O_2 DETECTION BASED ON MOFs

A rapid and fast electrochemical sensor for the detection of hydrogen peroxide is necessary for food analysis and the pharmaceutical applications. An electrochemical sensor converts the chemical change into an electrical signal. This signal is proportional to the concentration of the analyte of interest. Transducers are classified as potentiometric, amperometric, and conductometric. In an amperometric sensor, the current resulting from oxidation and/or reduction of the analyte is measured. A constant potential is applied and a change in current depending on the concentration of the electroactive species is measured. Hydrogen peroxide is detected in two ways: enzyme-based and non-enzyme-based.

15.2.1 Enzymatic H_2O_2 Sensing

These biosensors use heme protein or horse reddish peroxidase. These are very popular due to their high selectivity, ease of operation, and more sensitivity. Redox groups present in enzymes, change their oxidation states during the reaction. These are classified into two categories, one is mediator-free biosensors in which H_2O_2 is catalyzed by natural mediators (O_2), and the other is mediated biosensors in which electron transfer mediators are used to catalyze the H_2O_2. Mediated biosensors have low detection limits from micro to nanomolar. Hexacyanoferrates, ferrocene derivatives, and phenazine methosulfate are examples of electron transfer mediators. Enzyme-based sensors work only in optimized conditions such as a narrow range of pH and temperature, showing low LOD, and high selectivity [9]. In the first instance, a hemin-based MOF was prepared, in which chitosan and reduced graphene oxide were used due to their biocompatibility with H_2O_2 and to increase the electrical conductivity. The composite was characterized with characterization tools. The sensor showed that the size of Cu-hemin MOFs was reduced whilst it maintained its crystalline structure in its composite which enhanced its electrical properties. The prepared sensor showed an LOD of 0.019 μM with a wide linear range from 0.065 to 410μM for H_2O_2 [10]. Another hemin-based sensor for H_2O_2 was synthesized in which iron was used as the metal instead of copper. This sensor had unique bioactivity just like peroxidase (Figure 15.1). In this scheme, a chitosan-rGO was used as a supporting material for synthesized Fe-hemin, and an amplified electrochemical response was obtained. The linear range was observed to be in two segments, one from 1 to 61 μM and the other from 61 to 1311 μM. The low LOD value was 0.57 μM. This was also applied for real-time monitoring of H_2O_2 in the physiological processes of living cells [11].

Usually, MOFs exhibit less enzymatic and apparent affinity for analytes in comparison to enzymes. However small cavities of MOFs provide more stability for enzymes. In this protocol, cytochrome C was immobilized onto the zeolitic imidazolate framework-8 of the MOF that was synthesized with mesoporous and microporous channels. Immobilized Cyt C modified onto the surface of a screen-printed electrode showed more affinity for substrate in comparison to native Cyt C. Here Michael's constant was reduced by up to about 50%, with increased enzymatic activity of about 128% and 1.4 times increase in sensitivity for the electrochemical quantification of H_2O_2. The modified sensor was also implemented for the residual detection of H_2O_2 in food samples such as beer and milk [12].

15.2.2 Non-enzymatic H_2O_2 Sensing

Electrochemical methods that are carried out for the detection of H_2O_2 without enzyme have their advantages such as ease of operation, no need for immobilization of the enzyme, working ability in a broad range of pH and temperature, more stability, low cost, and free of oxygen limitation. These sensors overcome the drawbacks of enzymatic sensors. Chemically modified

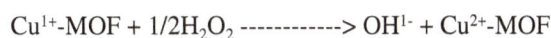

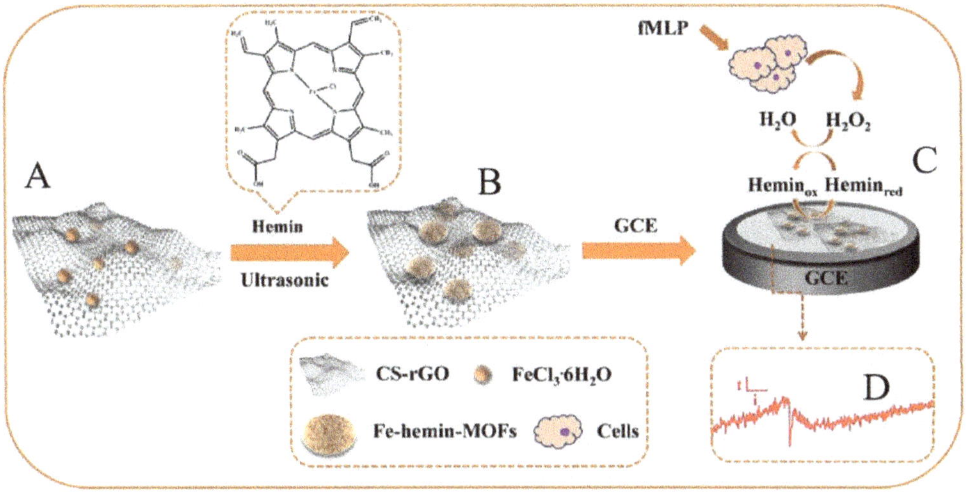

FIGURE 15.1 Synthetic route of Fe-hemin-MOFs/CS-rGO for detection of H_2O_2. Adapted with permission from [11]. Copyright (2020) Elsevier.

electrodes have low overpotential for electron transfer and less surface fouling. Various metals and metal oxides in combination with MOFs have been used for the investigation of H_2O_2, such as, copper, nickel, cobalt, cerium, chromium, and stannic. Similarly, metal nanoparticles/MOFs have been used for the quantification of H_2O_2 that include gold nanoparticles, silver nanoparticles, iron nanoparticles, and platinum nanoparticles. Some of the bimetallic MOFs have been employed for the detection of H_2O_2 [13]. In the electrochemical procedure, a redox couple initiates the reduction of H_2O_2 and converts it into water. For example, GO-HKUST-1@GCE for the sensing of H_2O_2 at neutral pH. Here H_2O_2 is converted into H_2O with Cu°/Cu^{1+} according to the following expression

$$Cu^{1+}\text{-HKUST}^{-1} + e^{-} \text{--------> } Cu^{0}\text{-HKUST}^{-1}$$

$$Cu^{0}\text{-HKUST}^{-1} + 1/2H_2O_2 \text{---------> } OH^{1-} + Cu^{1+}\text{-HKUST}^{-1}$$

In this reaction, Cu^{2+} is reduced into Cu^{1+} and Cu^0 respectively, resulting in the two alternate cathodic peaks. H_2O_2 produces hydroxyl ions while generating Cu^{1+} from Cu^0 as shown below.

$$Cu^{2+}\text{-MOF} + e^{-} \text{-----------> } Cu^{1+}\text{-MOF}$$

$$Cu^{1+}\text{-MOF} + 1/2H_2O_2 \text{------------> } OH^{1-} + Cu^{2+}\text{-MOF}$$

15.2.2.1 Carbon Hybrid Nanomaterials for Non-enzymatic H_2O_2 Sensing

Carbon nanomaterials emerged as novel materials due to their inherent properties, such as structural diversity, different allotropic forms such as graphene oxide, carbon nanotubes, fullerenes, carbon dots, and diamonds. The most used materials, graphene, and carbon nanotubes, have good optical, electronic, magnetic, chemical, thermal, and mechanical properties along with good biocompatibility for biomolecules and cells. These also possess a large surface area that is helpful for higher sensitivity to analytes. These are more stable than other nanomaterials in terms of factors such as pH, stability, ionic strength, and temperature. In this section, we discuss the use of these materials for the electrochemical sensing of H_2O_2. But CNTs are superior to graphene in terms of electron transfer, purity, easy functionalization, and cost [14]. The construction and modification of carbon-based electrodes is easy. Generally, glassy carbon and carbon paste electrodes are used for the detection of H_2O_2. The MOF composites are applied to the surfaces of these carbon-based electrodes. The modified glassy carbon electrodes are prepared by deposition of the polished surface of the electrode by MOF material. For the fabrication of carbon paste electrodes, MOF material is mixed with graphite powder in the presence of an organic compound as a linker [15].

15.2.2.1.1 Graphene Oxide-based MOF Sensors

Graphene oxide is an excellent conductive material with its flower-like hierarchal shape with high surface area, large pore size, and fast mass transfer rate. A graphene-based MOF composite was prepared by using GO and HKUST-1 through the solvothermal method. The main components in this composite were copper nitrate, GO, and benzene-1,3,5-tricarboxylic acid. The presence of GO and its morphology indicates that it can act as an effective structure-converting agent. It converts the HKUST-1 structure from octahedral to the hierarchical flower shape. Initially, GO was cut into small pieces and then converted into its reduced form to increase the conductivity, surface area, and redox activity through the solvothermal process. This composite exhibited a low LOD of 0.49 μM, with a wide linear range from 1.0 μM to 5.6 mM and a fast response time (<4s). An electrochemical assay shows a fast response with excellent selectivity for the detection of H_2O_2 [16]. In another report, carboxylated graphene oxide sheets were modified onto the electrode surface and further copper was grown onto these carboxylic functionalities to increase the density and surface area in the form of a large network. The thin film of graphene composite was used to identify the H_2O_2 that shows a LOD of 6.7×10^{-8} $molL^{-1}$ with more sensitivity having a linear range from 2.00×10^{-7} to 1.85×10^{-4} $molL^{-1}$. The increased sensitivity was due to the porous structure with increased conductivity and fast electron transfer reaction. Proper pore size and shape of modified material was the reason to attain good selectivity. This proposed strategy was also implemented with disinfectant samples for the detection of H_2O_2 [17].

15.2.2.1.2 Reduced Graphene Oxide-based MOF Sensors

Reduced graphene oxide modified composite was synthesized by the cation exchange method and reduced by an electrochemical process. This composite (MNPs@Y-1, 4-NDC-MOF/ERGO, M = Ag, Cu) contains two components, one is AgNPs@Y-1, 4-NDC-MOF/ERGO and the other is CuNPs@Y-1, 4-NDC-MOF/ERGO. Both of which exhibit good conductivity. In this process, metal nanoparticles were encapsulated into the anionic MOFs. The characterization techniques show that both are equal in size but have different constants for electron transfer rates. The first component shows a detection range from 4 to 11000 μM with a LOD of 0.18 μM. The composite shows good anti-interference performance and is the best material for the quantification of H_2O_2 in living cells. This composite shows good selectivity due to an optimized size of Y-1, 4-NDC-MOF, and a fast electron rate of rGO. This material can be used to monitor the activities of oxidase enzymes while monitoring the production of H_2O_2 and can be applied in the sensing, biosensing, biomedicine, and catalysis fields [18]. Surface modification can enhance the electrocatalytic property of composite material. This study shows the surface modification of the rht-type metal-organic framework in the form of nanosize. Further copper nanoparticles were implemented in the pores of this network structure. The as-prepared material shows good sensitivity and selectivity for H_2O_2. This composite shows good electrical conductivity due to rGO in comparison to Cu-TDPAT. This sensor shows a lower LOD value (0.17 μM) with a wide linear range (4–12000 μM) that is better in comparison to the noble metal or horseradish biosensors. The sensor also shows good stability in serum as well as in urine samples. This modified electrode can be applied for the construction of biosensors, sensors, and biofuel cells/bioelectronics devices [10].

15.2.2.1.3 Carbon Nanotube-based MOF Sensors

Although MOFs possess unique features, these cannot be used directly for the application of the sensing of H_2O_2 due to poor electrical conductivity and electrocatalytic ability. Therefore, to enhance the electrical conductivity of the MOF, carbon nanotubes are combined with MOFs. A carbon nanotube-based nickel-MOF composite was prepared by the solvothermal method. Here in this strategy, a Ni-MOF was anchored onto the network structure of carbon nanotubes. The MOFs nanoparticles of size 23 nm were homogeneously dispersed onto CNTs that show a wide linear range from 0.01 to 51.6 $mmolL^{-1}$. This study shows the low LOD of 2.1 $\mu molL^{-1}$ with a fast response time (2.5 s) is useful for the biosensing of H_2O_2. Its conductivity and chemical stability were higher in comparison to pure MOFs (Figure 15.2) [19].

In another study, multi-walled carbon nanotubes were combined with MOFs to prepare a nanocomposite that was characterized by different characterization techniques. First of all, a copper-based MOF was prepared and then immobilized onto multi-walled carbon nanotubes. This composite was deposited onto the electrode surface which shows redox behavior in phosphate buffer solution. This modified biosensor was applied for the detection of H_2O_2 and showed good stability and repeatability in water samples. The low LOD was 0.46 $\mu molL^{-1}$ with a wide linear range from 3–70 $\mu molL^{-1}$ to 70–30000 $\mu molL^{-1}$ [20]. In this strategy, carbon nanotubes@ stannic-based metal-organic framework and stannic-based

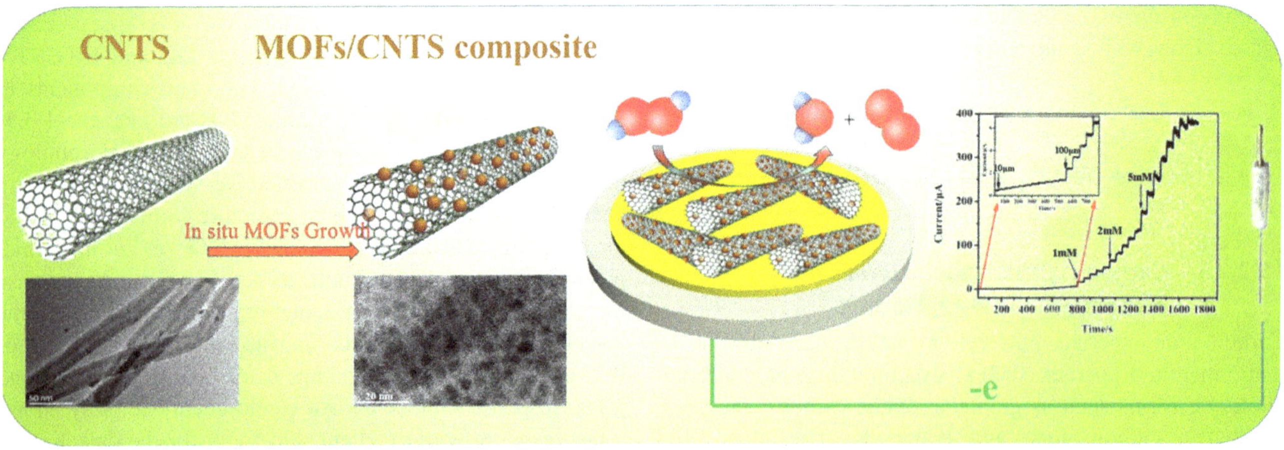

FIGURE 15.2 Schematic representation of Ni(II)-MOFs/CNTs and its electrochemical response for H_2O_2. Adapted with permission from [19]. Copyright (2016) Elsevier.

MOF composite were prepared by sonochemical, and solvothermal methods. The electroactive composites were synthesized to increase the surface area and porosity for the electrochemical response to H_2O_2. Carbon nanotubes maintain their crystalline structure upon the inclusion of the Sn-MOF. The morphology and structural composition of the MOF and its composite were characterized by characterization tools. The revealed lower LOD and linear range for this composite were ~ 4.7×10^{-3} μM and from 0.2 μM to 2.5 mM for H_2O_2 sensing [21].

15.2.2.2 Nanoparticles-based MOF Sensors for the Detection of H_2O_2

Noble metal nanoparticles have received more attention in nano photonics, sensing, solar cells, and nanomedicine as these possess large surface area, low toxicity, biocompatibility with biomolecules, stability, and varied functionality. Gold nanoparticles are non-toxic and biocompatible catalysts. These are used in biomedical fields as they offer a suitable environment for biomolecules. These are good sensing materials because of their exceptional optical as well as electric properties and ease of functionalization. Similarly, silver nanoparticles are good sensors as these have good electrical conductivity, electrocatalytic activity, and large surface area [22]. Palladium nanoparticles are important industrial catalysts. These are used in dehydrogenation, hydrogenation, hydrogen storage, fuel cell reactions, and reduction of pollutants in automobiles at low temperatures. They can work in alkaline and neutral media [23]. Platinum nanoparticles also possess a large surface area and good catalytic activity for H_2O_2. All these nanoparticles have been used for the electrochemical detection of H_2O_2 through cyclic voltammetry and amperometry.

15.2.2.2.1 Gold nanoparticle-MOF Sensors

Here, a AuNPs-NH_2/Cu-MOF composite was synthesized by using copper-based MOFs and amino-functionalized gold nanoparticles. Cu-MOFs were prepared in ionothermal conditions. This composite was deposited onto the glassy carbon electrode to quantify the H_2O_2 in human cervical cancer cells by adding a stimulant (ascorbic acid). This study shows that this sensor is highly selective and sensitive with LOD as low as 1.2 μM, with a wide linear concentration range of 5-850 μM. Sensitivity observed was 1.71 μA/cm^2 [24].

15.2.2.2.2 Silver Nanoparticle-MOF (Ag NPs-MOF) Sensors

In this study, a silver nanoparticle-based MOF was used for selective and sensitive detection of H_2O_2. First of all, a zinc MOF was prepared by the hydrothermal method and then silver nanoparticles were incorporated into it. The prepared material, AgNPs-Zn-MOF was characterized by different techniques. These results show the spherical structure of this material having a size of about 50 nm. The modified electrode was investigated by cyclic voltammetry and impedance spectroscopy after the drop coating of material (AgNPs-Zn-MOF) onto the glassy carbon electrode. Silver shows both oxidation and reduction peaks in cyclic voltammogram at + 0.20 and -0.50 V, respectively. The only response with AgNPs-Zn-MOF coated material was observed at -0.67 V. A linear response was observed upon the increase in the concentration of H_2O_2 from 1 to 5 μM. This sensor demonstrated a LOD

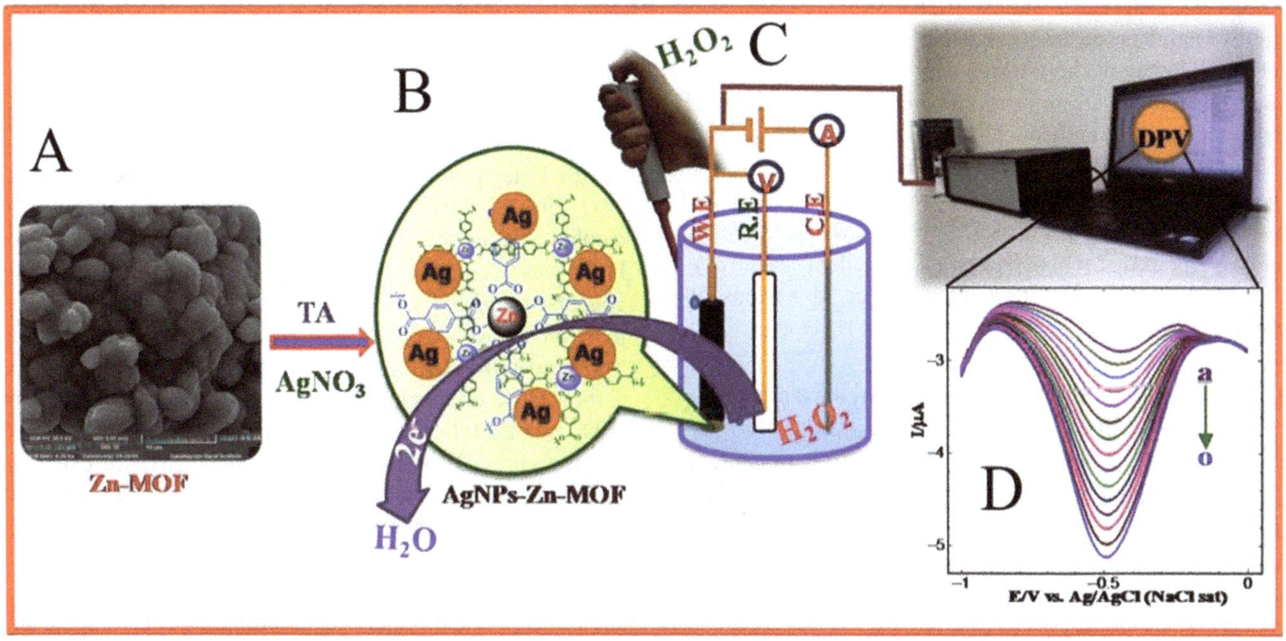

FIGURE 15.3 A schematic representation of Ag-Zn-MOF and its electrochemical response for H_2O_2. Adapted with permission from [25]. Copyright (2017) Elsevier.

value of 67 nM and was also applied for detection of H_2O_2 in serum, milk, blood, and urine [25].

15.2.2.2.3 Platinum Nanoparticle-MOF-based Sensors

Platinum nanoparticles (Pt-NPs) in combination with UiO-66 have been used for sensing H_2O_2. In this study material was prepared in the form of a core-shell. Platinum nanoparticles were synthesized separately. Then these were encapsulated in UiO-66, which acts as a host matrix having a diameter of 6 Å. The morphology of different samples was checked by electron microscopy. An increase in the concentration of Pt NPs was observed in the composite as the initially added concentration was increased. This composite shows a good crystalline structure of P tNPs in well-dispersed form. Here, sample 2 shows good electrocatalytic activity for H_2O_2 and excellent anti-interference activity for interference species. The LOD observed was very low with a linear concentration range from 5 μM to 14.77 mM. This sensor was more stable and reproducible [26].

15.2.2.2.4 Fe_2O_3 Nanoparticle-MOF Sensors

Nanoparticles can enhance sensing performance by loading enzymes through fast electron transfer. The nanocomposite of Fe_2O_3@C was synthesized by the annealing of Fe-1,3,5-benzenentricarboxylate and characterized by scanning electron microscopy, X-ray diffraction, and energy dispersive spectroscopy. Carbon paste electrodes based on Fe_2O_3@C were synthesized and immobilized by acetylcholinesterase for the investigation of H_2O_2. Electrochemical studies exhibit a LOD value as low as 1.7×10^{-7} M. This sensor can be used for biomolecules other than H_2O_2 [27].

15.2.2.3 Polymer-based MOF Sensors

MOFs unique properties enable their use as artificial enzymes for non-enzymatic applications for the detection of H_2O_2. Here, a MOF-based on nano-metalloporphyrins was used to enhance the electrochemical signal of H_2O_2. This material [NporMOF(Fe)] was coated onto the electrode surface thus increasing the surface area and providing more active sites for our analyte of interest. This material shows good electrical activity for reduction of H_2O_2 and excellent selectivity as well as sensitivity for it in the presence of various other analytes [28].

15.2.2.4 Noble Metals and their Alloy-based MOF Sensors

Noble metal composites have their use in different fields such as nano photonics, sensing, solar cells, and nanomedicine due to their low toxicity, chemical stability, high surface area, and other functionalities. Some of the noble metals and metal oxide composites have been synthesized and successfully applied for the detection of H_2O_2 [29].

15.2.2.4.1 Chromium-MOF Sensors

Here, a stable and reproducible electrochemical sensor for the non-enzymatic detection of H_2O_2 was synthesized that can work even in basic conditions, namely, NaOH. A chromium-based MOF (Cr-MOF) containing dicarboxylate groups was deposited onto the surface of the glassy carbon electrode. The MOF composite was prepared by the solvothermal method with the assistance of microwaves. After treating the Cr-MOF, no changes in crystallinity, and morphology were observed. The sensor was so stable that there was a negligible resistance in charge transfer after redox cycling. It showed good electrochemical activity in 0.1M NaOH solution. The sensitivity observed for H_2O_2 with this sensor was 11.9 μA mM^{-1} with a wide linear range (25 to 500 μM). The LOD for this modified sensor was 3.52 μM which can be used for point of care detection of H_2O_2 [30].

15.2.2.4.2 Copper-MOF Sensors

A copper-based three dimensional composite of MOF (Cu(adp)(BIB)(H2O)]n (BIB = 1,4-bis-imidazole benzene; H2adp = adipic acid) was prepared by the hydrothermal method for the detection of H_2O_2. This structure contains two-dimensional layers that are extended into channels of three-dimensional supramolecular architecture revealed by single-crystal X-ray analysis. Hydrogen bonds were observed between two-dimensional layers and the supramolecular architecture. The modified electrode exhibits a pair of redox peaks in 0.2M NaOH. A linear range from 0.1 μM to 2.75 μM along with an LOD of 0.068 μM shows a high electrocatalytic activity for H_2O_2 [31]. The strategy of increasing the conductive sites of MOFs is performed for the improvement of the conductivity of the material that enhances the electrochemical response of the analyte. A novel pillar, layered in three dimensions along with secondary building units as a paddle wheel from a copper MOF was synthesized by the hydrothermal method and a modified glassy carbon electrode showed a good electrochemical response for H_2O_2. The addition of acetylene black enhances the response to H_2O_2. A wide linear range shows high sensitivity having a LOD (0.014 mM) with Cu-MOF/AB-2%/ GCE modified electrode. It was a highly selective, faster response time (4 s) and a stable sensor for H_2O_2, and no interference was observed in the presence of alcohol, glycerin, lactose, and glucose [32]. The study of H_2O_2 was performed by a group of researchers on porous copper-based MOFs possessing large pores. The composite material shows more stability even in a basic solution. The prepared MOF was deposited onto the surface of the carbon paste electrode and the electrochemical behavior of H_2O_2 was examined. This modified electrode showed good redox signals. This non-enzymatic sensor shows excellent performance. The linear range for H_2O_2 was 0.001-0.9 mM and LOD, 1 μM [33]. Electrochemical detection of H_2O_2 was also performed by a copper-based MOF containing a layered topology having two dimensions [34, 35]. The MOF composite was synthesized by coordinating two components: cupric ions, and carboxylic groups. The layers in its structure were connected by hydrogen bonds. The MOF had good biocompatibility with cells that can be used for in vivo detection of H_2O_2.

15.2.2.4.3 Nickle-MOF Sensors

AP-Ni-MOF is a nickel-based metal-organic framework biosensor that was used for the detection of H_2O_2. In this MOF adipic acid and piperazine were used as a linker and ligand, respectively. Its morphology was confirmed by different characterization techniques such as scanning electron microscopy, X-ray diffraction, infrared spectroscopy, TGA, and so forth. There was a good electrochemical response from the composite shown by cyclic voltammetry and chronoamperometry due to its greater conductivity, 1.28×10^{-3} S/cm. The linear range of H_2O_2 and LOD were from 0.004 mM to 60mM and 0.0009 mM, respectively. Interference species change the signal of reduction peaks to less than 5%. The sensor was more stable and reproducible. The modified electrode was also used for the investigation of H_2O_2 in a real sample of lens clear solution [36].

15.2.2.4.4 Cobalt-MOF Sensors

A cobalt-based biosensor for H_2O_2 was synthesized by the hydrothermal method. This sensor (Co(pbda)(4,4-bpy)·$2H_2O$] nCo-MOF) (bpy = 4,4- bipyridine; H2pbda = 3-(pyridine-3-yloxy)benzene-1,2-dicarboxylic acid) contains electrochemical active centers for H_2O_2. It is a three-dimensional supramolecular architecture and net of unimodal-4, having a GIS topology revealed by single-crystal X-ray analysis. It shows a pair of reduction peaks in cyclic voltammetry in a basic solution of NaOH. The electrode was modified with this material and the electrochemical response was analyzed. The amperometric results demonstrated that it is a highly electrochemical active material with a linear range from 5 M to 9.0 mM, sensitivity (83.10 A mM^{-1} cm^{-2}), and LOD of 3.76 M for H_2O_2. It acts as a peroxidase and catalyzes H_2O_2 to produce hydroxyl radicals [37]. A uniform micro rod of cobalt-based metal-organic framework (Co-MOF) was prepared by a group of researchers for electrochemical quantification of H_2O_2. The composite was prepared by using 2, 5-dihydroxy terephthalic acid as an electroactive ligand. The micro rods exhibit good redox behavior in a buffer solution of acetate at pH 4. The LOD and linear range of this sensor were 0.5 μM and from 1 to 823 μM. This non-enzymatic sensor was more stable and reproducible [38].

15.2.2.4.5 Tellurium-MOF Sensors

A two-step hydrothermal strategy was applied for the synthesis of a tellurium doped zinc imidazole metal-organic framework for the sensing of H_2O_2. After characterization of the material through scanning electron microscopy, thermogravimetric analysis, X-ray diffraction, and so forth, it was also characterized with cyclic voltammetry by using a three-electrode system. Results revealed by amperometry, cyclic voltammetry, and impedance show its LOD to be as low as 60 μM with linearity up to 0.98855. The sensor shows a 101% recovery of H_2O_2 from the human serum sample, with 1000 cycle stability with no change in electrochemical response. The modified glassy carbon surface was porous and rough. An interface was developed between the MOF composite and H_2O_2 [39].

15.2.2.4.6 Bimetallic-MOF Sensors

Biomedical and environmental applications require a stable, sensitive, and selective sensor. Such a type of sensor was fabricated (for the sensing of H_2O_2) that was bimetallic and porous. The reported homeotypic bimetallic crystals (Ce1-xTbxOy) were prepared by the direct pyrolysis of CexTb1-x-MOFs precursors. The modified glassy carbon electrode with these crystals was used for the analysis of H_2O_2 in a basic medium. The sensor shows its low LOD, sensitivity, stability, 7.7 μm, 12.99 μA·mM^{-1}·cm^{-2}, and 1.6831 μA·mM^{-1} respectively with a broad linear range from 0.1 μm to 4.2 mM and fast response of less than 10s (Figure 15.4) [40].

Synthesis of a two-dimensional MOF is a facile and cost-effective process for the efficient electrochemical detection of H_2O_2. A short time is required for the fabrication of 2D-MOF through a nucleation and crystal growth method at low-temperature requirements. Here, cobalt, nickel, and NiCo-based MOFs were synthesized for the detection of H_2O_2. The oxidation potential of these non-enzymatic sensors was checked in a basic solution of KOH. The low LOD was 0.69 μM and the linear range was wide from 5 μM to 832.5 μM with fast response in KOH solution. This bimetallic MOF shows a good response in comparison to carbon-based nanomaterials. It is a highly stable, selective, and sensitive sensor [41].

The combination of 2D-MOF and Persian blue analogs resolve the problem of the poor conductivity and slow transfer of electrons in 2D-MOF for the detection of H_2O_2. The PBA/ZIF nanocomposite was successfully prepared onto the nickel foam. 2D-leaf-like zeolitic imidazolate frameworks were converted into PBAs by an in-situ method. The different MOF

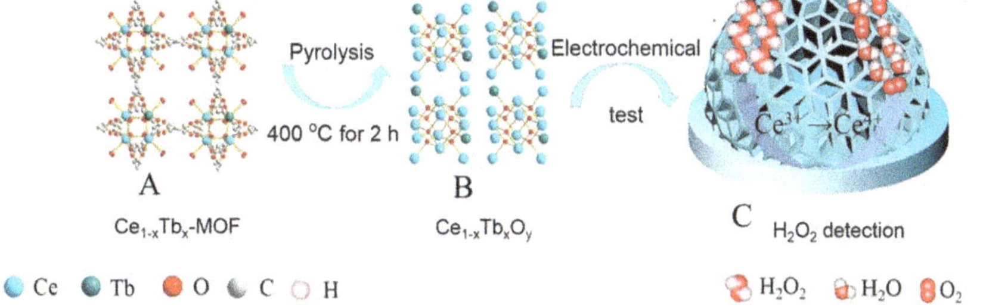

FIGURE 15.4 A stepwise synthetic route of $Ce_{0.9}Tb_{0.1}O_y$ from $Ce_{1-x}Tb_x$-MOF. Adapted with permission from [40]. Copyright (2019) Elsevier.

composites, such as FeFe-PBA/Fe-ZIF/NF, CoCo-PBA/Co-ZIF/NF, Fe/CoCo-PBA/Co-ZIF/NF, and CoFe-PBA/Co-ZIF/NF were prepared to determine the changes in conductivity and electron transfer. These composites were used for multiple functions of which H_2O_2 detection was one of the applications. The sensors showed a sensitive, selective, and stable response. Among the synthesized MOFs, CoFe-PBA/Co-ZIF/NF shows the best electrochemical response for the detection of H_2O_2. The low LOD and sensitivity were 1.08 nM, and 196 μA mM^{-1} cm^{-2} with a wide linear range from 0.2 mM to 6.0 mM [42].

15.3 COMPARISON BETWEEN DIFFERENT SENSORS, BASED ON ELECTROCHEMICAL ACTIVITY

It is understood that electrochemical sensors for the detection of H_2O_2 have got appeal due to their unique properties such as high stability, more selectivity, ease of operation, low cost, and fast response. H_2O_2 detection has been carried out via enzymatic as well as non-enzymatic methods (Table 15.1). However, carbon-based materials such as carbon nanotubes are well-preferred due to their greater sensitivity, conductivity, and active sites for the attachment of H_2O_2 in comparison to other MOF-based materials such as nanoparticles, polymers, and metals. Various MOF-based materials have been used for the detection of H_2O_2 such as enzyme-based, and non-enzyme-based. Here, we have discussed the quantification of H_2O_2 by these materials as summarized in Table 15.1. The whole study indicates that bimetallic MOFs are the most sensitive materials for the detection of H_2O_2, showing its lowest LOD (1.08 nM). The bimetallic MOF works as the most advanced sensing material for the analysis of H_2O_2. Carbon-based MOF materials are conductive and smaller in size. These are the next choice for H_2O_2 sensing after the bimetallic MOF. These have shown a LOD of 4.7 nM. If we compare the carbon-based MOFs, it can be observed in the table that carbon-based MOFs are better in comparison to MOFs, based on graphene and reduced graphene oxide that have shown a lower limit of detection 0.067 μM and 0.17 mM, respectively. Nanoparticle-based electrochemical sensing is better than MOFs, based on metal/metals, and polymers but less sensitive than carbon-based and bimetallic MOFs. These have shown the lowest LOD of 67 nM. Among the nanoparticles, the silver nanoparticles are the best in combination with a MOF. Platinum nanoparticles are the least sensitive.

Iron and gold nanoparticles are in between silver and platinum in terms of sensitivity. Polymers are porous and have large surface areas containing materials that have been studied for the detection of H_2O_2. Metal-based MOF composite has the lowest limit of detection in comparison to bimetallic MOFs. The study shows that bimetallic sensors are the best MOF materials for sensing H_2O_2. In the case of metal-based MOFs,

TABLE 15.1
Analytical Performances of MOF-based Electrochemical Sensors

Sr. No.	Type of Material	LOD	Sensitivity	Linear Range	Response Time	Ref.
1.	Cu-hemin MOFs/CS-rGO	0.019 μM		0.065–410 μM		[9]
2.	Fe-hemin- CS-rGO	0.57 μM		1 to 61 μM, 61–1311 μM		[11]
3.	Cytc@mesoZIF-8		3.84 mA·M^{-1} cm^{-2}	90-3600 μM		[12]
4.	GO-HKUST-1	0.49 μM		1.0 μM to 5.6 mM	<4s	[16]
5.	GO-Cu-MOF	0.067 μM		2.00 × 10^{-7} to 1.85 × 10^{-4} molL^{-1}		[17]
6.	MNPs@Y-1, 4-NDC-MOF/ ERGO, M = Ag, Cu	0.18 μM		4 to 11000 μM		[18]
7.	Cu-TDPAT–n-ERGO	0.17μM		4–12000 μM		[10]
8.	CNT-Ni-MOF	2.1 μM		0.01 to 51.6 mmolL^{-1}	2.5s	[19]
9.	MWCNT-Cu-MOF			3–70 μmolL^{-1} to 70–30000 μmolL^{-1}		[20]
10.	AuNPs-NH_2/Cu-MOF	1.2 μM	1.71 μA/cm^2	5 to 850 μM		[24]
11.	AgNPs-Zn-MOF	67 nM		1 to 5 μM		[25]
12.	PtNPs@UiO-66-2/GCE	3.06 μM	75.33 μA.mM^{-1}	5 μM to 14.77 mM		[26]
13.	MIL-53-CrIII-MOF	3.52 μM	11.9 μA.mM^{-1}	25 to 500 μM		[30]
14.	Cu-MOF	0.068 μM		0.1 μM to 2.75 μM		[31]
15.	Cu-MOF/AB-2%/GCE	0.014 mM		0.05 to 3 mM	4 s	[32]
16.	Cu-MOF	1 μM.		0.001 to 0.9 mM		[33]
17.	AP-Ni-MOF	0.9 μM		0.004 mM to 60 mM		[36]
18.	Co(pbda)(4,4-bpy)·$2H_2O$] nCo-MOF	3.76 μM	83.10 A.mM^{-1} cm^2	5 μM to 9.0 mM		[37]
19.	Co-MOF	0.5 μM		1 to 823 μM		[38]
20.	NiCo-MOF	0.69 μM		5 to 832.5 μM		[41]
21.	CoFe-PBA/Co-ZIF/NF	1.08 nM	196 μA.mM^{-1} cm^{-2}	0.2 to 6.0 mM		[42]
22.	Sn-MOF-CNT	4.7 nM		0.2 μM to 2.5 mM		[21]

the least electrochemically active material is a tellurium-based composite. The respective order of sensitivity among metallic and bimetallic MOFs composite is below:

CoFe-MOF > Co-MOF > Cu-MOF > Ni-MOF
> Sn-MOF > Te-MOF

15.4 CONCLUSION AND OUTLOOK

MOFs are the most advanced materials for the sensing of H_2O_2 due to the variety of functionalities and their composites that include, non-enzymatic sensors (carbon-based, metal-based, nanoparticle-based, polymer-based) and enzyme-based sensors. These have more potential for the quantification of H_2O_2 that causes oxidative stress, potentially leading to Parkinson's disease, diabetes, cardiovascular and myocardial issues, cancer, and chronic diseases. The unique features of MOFs enable their use for electrochemical sensors and biosensors. The electrochemical activity of a variety of MOF-based materials for the detection of H_2O_2 has been summarized in this chapter. Enzyme-based and non-enzyme-based sensors have their advantages and drawbacks. But the use of bimetallic MOF sensors has resolved many challenges faced by the enzyme-based sensors. Further improvement in the material should enable us to control many diseases caused by H_2O_2 through diagnosis and prognosis of the concentration of H_2O_2.

REFERENCES

[1]. K. Dhara, and D. R. Mahapatra, "Recent advances in electrochemical nonenzymatic hydrogen peroxide sensors based on nanomaterials: a review," *Journal of Materials Science,* vol. 54, pp. 12319–12357, 2019.

[2]. L. Wang, and E. Wang, "A novel hydrogen peroxide sensor based on horseradish peroxidase immobilized on colloidal Au modified ITO electrode," *Journal of Electrochemistry Communications,* vol. 6, pp. 225–229, 2004.

[3]. D. Janasek, W. Vastarella, U. Spohn *et al.*, "Ruthenium/rhodium modified gold electrodes for the amperometric detection of hydrogen peroxide at low potentials," *Journal of Analytical and Bioanalytical Chemistry,* vol. 374, pp. 1267–1273, 2002.

[4]. H. Zhou, J. R. Long, and O. M. Yaghi, "Introduction to metal–organic frameworks," *Journal of Chemical reviews,* vol. 112, pp. 673–674, 2012.

[5]. A. J. Howarth, A. W. Peters, N. A. Vermeulen *et al.*, "Best practices for the synthesis, activation, and characterization of metal–organic frameworks," *Journal of Chemistry of Materials,* vol. 29, pp. 26–39, 2017.

[6]. Y. Xu, Q. Li, H. Xue *et al.*, "Metal-organic frameworks for direct electrochemical applications," *Journal of Coordination Chemistry Reviews,* vol. 376, pp. 292–318, 2018.

[7]. X. W. Liu, T. J. Sun, J. L. Hu *et al.*, "Composites of metal–organic frameworks and carbon-based materials: preparations, functionalities and applications," *Journal of Materials Chemistry A,* vol. 4, pp. 3584–3616, 2016.

[8]. F. Wang, X. Chen, L. Chen *et al.*, "High-performance non-enzymatic glucose sensor by hierarchical flower-like nickel (II)-based MOF/carbon nanotubes composite," *Journal of Materials Science Engineering: C,* vol. 96, pp. 41–50, 2019.

[9]. A. Lindgren, T. Ruzgas, L. Gorton *et al.*, "Biosensors based on novel peroxidases with improved properties in direct and mediated electron transfer," *Journal of Biosensors and Bioelectronics,* vol. 15, pp. 491–497, 2000.

[10]. C. Li, T. Zhang, J. Zhao *et al.*, "Boosted sensor performance by surface modification of bifunctional rht-type metal–organic framework with nanosized electrochemically reduced graphene oxide," *Journal of applied materials and interfaces,* vol. 9, pp. 2984–2994, 2017.

[11]. P. Zhao, S. Chen, J. Zhou *et al.*, "A novel Fe-hemin-metal organic frameworks supported on chitosan-reduced graphene oxide for real-time monitoring of H_2O_2 released from living cells," *Journal of Analytica Chimica Acta,* vol. 1128, pp. 90–98, 2020.

[12]. C. Zhang, X. Wang, M. Hou *et al.*, "Immobilization on metal–organic framework engenders high sensitivity for enzymatic electrochemical detection," *Journal of applied materials and interfaces,* vol. 9, pp. 13831–13836, 2017.

[13]. Z. Li, T. Zhao, X. Zhan *et al.*, "High capacity three-dimensional ordered macroporous $CoFe_2O_4$ as anode material for lithium ion batteries," *Journal of Electrochimica Acta,* vol. 55, pp. 4594–4598, 2010.

[14]. J. Wang, M. Musameh, and Y. Lin, "Solubilization of carbon nanotubes by Nafion toward the preparation of amperometric biosensors," *Journal of the American Chemical Society,* vol. 125, pp. 2408–2409, 2003.

[15]. E. Sardini, M. Serpelloni, and S. Tonello, "Printed electrochemical biosensors: Opportunities and metrological challenges," *Journal of Biosensors,* vol. 10, pp. 166, 2020.

[16]. Q. Wang, Y. Yang, F. Gao *et al.*, "Graphene oxide directed one-step synthesis of flowerlike graphene@ HKUST-1 for enzyme-free detection of hydrogen peroxide in biological samples," *Journal of applied materials and interfaces,* vol. 8, pp. 32477–32487, 2016.

[17]. Y. Zhou, C. Li, Y. Hao *et al.*, "Oriented growth of cross-linked metal-organic framework film on graphene surface for non-enzymatic electrochemical sensor of hydrogen peroxide in disinfectant," *Journal of Talanta,* vol. 188, pp. 282–287, 2018.

[18]. C. Li, R. Wu, J. Zou *et al.*, "MNPs@ anionic MOFs/ERGO with the size selectivity for the electrochemical determination of H_2O_2 released from living cells," *Journal of Biosensors and Bioelectronics,* vol. 116, pp. 81–88, 2018.

[19]. M. Q. Wang, Y. Zhang, S. J. Bao *et al.*, "Ni (II)-based metal-organic framework anchored on carbon nanotubes for highly sensitive non-enzymatic hydrogen peroxide sensing," *Journal of Electrochimica Acta,* vol. 190, pp. 365–370, 2016.

[20]. E. Zhou, Y. Zhang, Y. Li *et al.*, "Cu (II)-Based MOF Immobilized on Multiwalled Carbon Nanotubes: Synthesis and Application for Nonenzymatic Detection of Hydrogen Peroxide with High Sensitivity," *Journal of Electroanalysis,* vol. 26, pp. 2526–2533, 2014.

[21]. S. Rani, B. Sharma, R. Malhotra *et al.*, "Sn-MOF@ CNT nanocomposite: An efficient electrochemical sensor for detection of hydrogen peroxide," *Journal of Environmental Research,* vol. 191, pp. 110005, 2020.

[22]. C. Welch, C. Banks, A. Simm *et al.*, "Silver nanoparticle assemblies supported on glassy-carbon electrodes for the electro-analytical detection of hydrogen peroxide," *Journal of Analytical and bioanalytical chemistry,* vol. 382, pp. 12–21, 2005.

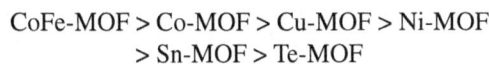

[23]. M. Pournaghi-Azar, F. Ahour, and F. Pournaghi Azar, "Simple and rapid amperometric monitoring of hydrogen peroxide in salivary samples of dentistry patients exploiting its electro-reduction on the modified/palladized aluminum electrode as an improved electrocatalyst," *Journal of Sensors and Actuators B: Chemical,* vol. 145, pp. 334–339, 2010.

[24]. W. Dang, Y. Sun, H. Jiao *et al.*, "AuNPs-NH_2/Cu-MOF modified glassy carbon electrode as enzyme-free electrochemical sensor detecting H_2O_2," *Journal of Electroanalytical Chemistry,* vol. 856, pp. 113592, 2020.

[25]. P. Arul, John, and S. Abr, "Silver nanoparticles built-in zinc metal organic framework modified electrode for the selective non-enzymatic determination of H_2O_2," *Journal of Electrochimica Acta,* vol. 35, pp. 680–689, 2017.

[26]. Z. Xu, L. Yang, and C. Xu, "Pt@ UiO-66 heterostructures for highly selective detection of hydrogen peroxide with an extended linear range," *Journal of Analytical chemistry,* vol. 87, pp. 3438–3444, 2015.

[27]. W. Wei, S. Dong, G. Huang *et al.*, "MOF-derived Fe_2O_3 nanoparticle embedded in porous carbon as electrode materials for two enzyme-based biosensors," *Journal of Sensors and Actuators B: Chemical.,* vol. 260, pp. 189–197, 2018.

[28]. P. Ling, X. Zang, C. Qian *et al.*, "A metal–organic framework with multienzyme activity as a biosensing platform for real-time electrochemical detection of nitric oxide and hydrogen peroxide," *Journal of Analyst,* vol. 146, pp. 2609–2616, 2021.

[29]. S. Chen, R. Yuan, Y. Chai *et al.*, "Electrochemical sensing of hydrogen peroxide using metal nanoparticles: a review," *Journal of Microchimica Acta,* vol. 180, pp. 15–32, 2013.

[30]. N. S. Lopa, M. M. Rahman, F. Ahmed *et al.*, "A base-stable metal-organic framework for sensitive and non-enzymatic electrochemical detection of hydrogen peroxide," *Journal of Electrochimica Acta,* vol. 274, pp. 49–56, 2018.

[31]. C. Zhang, M. Wang, L. Liu *et al.*, "Electrochemical investigation of a new Cu-MOF and its electrocatalytic activity towards H_2O_2 oxidation in alkaline solution," *Journal of Electrochemistry communications,* vol. 33, pp. 131–134, 2013.

[32]. W. Meng, S. Xu, L. Dai *et al.*, "An enhanced sensitivity towards H_2O_2 reduction based on a novel Cu metal–organic framework and acetylene black modified electrode," *Journal of Electrochimica Acta,* vol. 230, pp. 324–332, 2017.

[33]. D. Zhang, J. Zhang, R. Zhang *et al.*, "3D porous metal-organic framework as an efficient electrocatalyst for nonenzymatic sensing application," *Journal of Talanta,* vol. 144, pp. 1176–1181, 2015.

[34]. W. Ling, Y. Hao, H. Wang *et al.*, "A novel Cu-metal-organic framework with two-dimensional layered topology for electrochemical detection using flexible sensors," *Journal of Nanotechnology,* vol. 30, pp. 424002, 2019.

[35]. Y. Zhang, X. Bo, C. Luhana *et al.*, "Facile synthesis of a Cu-based MOF confined in macroporous carbon hybrid material with enhanced electrocatalytic ability," *Journal of Chemical Communications,* vol. 49, pp. 6885–6887, 2013.

[36]. B. Sherino, S. Mohamad, S. N. A. Halim *et al.*, "Electrochemical detection of hydrogen peroxide on a new microporous Ni–metal organic framework material-carbon paste electrode," *Journal of Sensors and Actuators B: Chemical,* vol. 254, pp. 1148–1156, 2018.

[37]. L. Yang, C. Xu, W. Ye *et al.*, "An electrochemical sensor for H_2O_2 based on a new Co-metal-organic framework modified electrode," *Journal of Sensors and Actuators B: Chemical,* vol. 215, pp. 489–496, 2015.

[38]. D. Zhang, J. Zhang, H. Shi *et al.*, "Redox-active microsized metal-organic framework for efficient nonenzymatic H_2O_2 sensing," *Journal of Sensors and Actuators B: Chemical,* vol. 221, pp. 224–229, 2015.

[39]. B. Fatima, D. Hussain, A. Saeed *et al.*, "Tellurium doped zinc imidazole framework (Te@ ZIF-8) for quantitative determination of hydrogen peroxide from serum of pancreatic cancer patients," *Journal of Scientific reports,* vol. 10, pp. 1–9, 2020.

[40]. L. Zhao, K. Sun, N. Youliwasi *et al.*, "Highly sensitive H_2O_2 sensor based on porous bimetallic oxide Ce1− xTbxOy derived from homeotypic Ln-MOFs," *Journal of Applied Surface Science,* vol. 470, pp. 91–98, 2019.

[41]. B. Liu, X. Wang, Y. Zhai *et al.*, "Facile preparation of well conductive 2D MOF for nonenzymatic detection of hydrogen peroxide: relationship between electrocatalysis and metal center," *Journal of Electroanalytical Chemistry,* vol. 858, pp. 113804, 2020.

[42]. C. Chen, D. Xiong, M. Gu *et al.*, "MOF-derived bimetallic CoFe-PBA composites as highly selective and sensitive electrochemical sensors for hydrogen peroxide and nonenzymatic glucose in human serum," *Journal of Applied Materials.,* vol. 12, pp. 35365–35374, 2020.

16 MOF-based Capacitive and Resistive Sensors for Hydrogen Sulfide

P. Abdul Rasheed[1], *Anish R. Nath*[2] *and Arjun A. Mohan*[1]

[1]Department of Biological Sciences and Engineering, Indian Institute of Technology Palakkad, Palakkad, Kerala, India

[2]School of Materials Science and Engineering, National Institute of Technology Calicut, Calicut, Kerala, India

16.1 INTRODUCTION

Hydrogen sulfide (H_2S) is a highly toxic, flammable gas which has a characteristic rotten egg smell. The odor detection threshold of H_2S ranges from 0.2 to 2.0 g/μgm³. H_2S is a very potent gas and can cause immediate paralysis of parts of the nervous system if inhaled in even small quantities (100-150 ppm). Very small quantities of H_2S (10 ppm) can also cause tiredness and headaches. Inhalation of this gas has also been associated with many fatal diseases like pneumonia, bronchitis, and pulmonary edema [1]. In terms of effects on the environment, H_2S easily dissipates into the air and forms sulfur dioxide and sulfuric acid which leads to acid rain. It is estimated that H_2S remains in the air for about 18 hours after dissipation which also makes it a serious pollutant.

Even though the effects of H_2S gas are highly dangerous to the body and the environment, it is an industrially relevant gas where it is used for the production of a large number of molecules such as methanethiol, ethanethiol, and thioglycolic acid. It is also used for the quantification of heavy metals and also as a precursor for the synthesis of metal sulfides. From a physiological perspective, H_2S acts as the endogenous gas transmitter in various conditions such as Down's syndrome, Alzheimer's disease, ischemia, asthma, and halitosis. H_2S is produced from cysteine by certain body enzymes where it acts as a relaxant of smooth muscles, as a vasodilator, and is also an important part of the memory forming process. It also forms important constituents of myocardial infarction and Alzheimer's disease cures. Agriculturally, it is an important gas where it contributes to the increase in storage time for post harvest fruits and vegetables [2].

It can be easily seen that the amount of H_2S is very limited within the body and any further increase in its concentration leads only to detrimental effects. This is especially true for workers on petroleum rigs where H_2S gas is released along with oil and natural gas. Since exposure to this gas is very harmful, it is highly important to develop technologies that can detect this gas before any human intervention can be made in a site where its presence is suspected. Also, some industries like sugar, paper, and pulp manufacturing plants release H_2S gas as effluent into the water. It can only be detected if H_2S is released into the atmosphere when the water temperature rises, thus inducing the characteristic rotten egg smell.

Metal-organic frameworks (MOFs) are modular structures whose structure and porosity can be controlled by controlling the metal nodes and ligands that form the structure. The modularity of the MOF is attributed to the extent of modification possible. In some cases, the MOF is modified post synthetically, and further, certain reports have outlined improvements in the electrochemical characteristics of MOFs by using redox-active linkers [3]. In addition to this, the porosity exhibited by this class of materials makes it highly possible for large surface areas to be exposed for reaction, specifically sensing applications. Owing to a plethora of desirable properties of MOFs, this material is highly desirable for the sensing of gases like H_2S (Figure 16.1). The sensing of H_2S gas is a very relevant topic owing to its large-scale applications, and also its harmful effects in large concentrations. The following sections will discuss in detail the various sensing methodologies based on MOFs adopted for the detection of H_2S.

16.2 SENSING TECHNIQUES FOR H_2S DETECTION

The sensing of H_2S has been carried out using a large variety of materials like ceramics [4], polymers [5], metal oxides [6], and precious metals [7] by different techniques such as optical, electrochemical, and mass-based (quartz crystal microbalance). Even though a wide variety of materials have been applied for the sensing of H_2S, the use of MOFs as sensor elements is highly desirable owing to the modularity that is possible. However, MOFs have been applied to the sensing of H_2S using only limited techniques, namely, luminescence, cataluminescence, capacitance, and resistance-based sensing. The scope of this chapter is primarily the application of MOFs for capacitance and resistance-based sensors. However, in order to gain a comprehensive understanding of the former techniques, an overview of the use of MOFs for other sensing techniques will also be discussed.

DOI: 10.1201/9781003188148-16

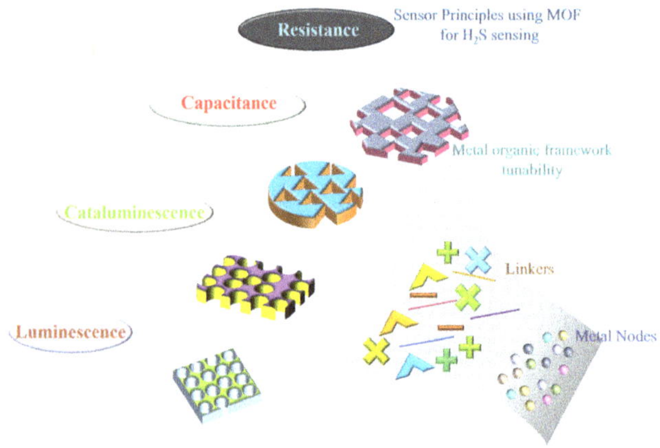

FIGURE 16.1 Schematic representation of the modularity of MOFs and their varying sensing methods for H_2S sensors.

In luminescence sensing of H_2S by MOF-based sensors, the sensing methodology is based either on an increase in luminescence intensity (Turn 'ON' sensor) or on quenching it (Turn 'OFF' sensor) based on the addition of the analyte [8]. The advantage of using MOFs for luminescence sensing is their modular nature which allows fine-tuning of the luminescence property. In addition to this, the tunable pore sizes guarantee good sensitivity and molecular diffusion, thus modulating the response time. The molecular sieving effect of MOFs can be used to reduce interference from competing molecules. MOFs are highly stable in the face of chemical attacks and also there is a huge scope for post-synthetic modification which improves the functionality of this material.

Most of the MOF-based luminescence sensors have employed the Turn 'ON' sensing mechanism for the detection of H_2S gas to avoid false positives and also for improved signal-to-noise ratio as detection is done against a dark background. The luminescence of MOFs that have been reported recently can be classified into many broad mechanisms such as MOFs containing azide/nitro groups whose fluorescence turns on when H_2S reduces these groups [9, 10], MOFs that employ the double nucleophilic character of H_2S to turn 'ON' the luminescence [11, 12], MOFs that contain a chelating unit that forms complexes with Cu^{2+} or Fe^{3+} wherein the addition of H_2S can release these ions thereby leading to luminescence enhancement [13, 14], ratiometric sensing strategy [15, 16], destruction of the MOF leading to releasing ligand or dye and the emergence of luminescence by dye or ligand [17, 18], and cataluminescence [19].

Having gone through some of the prominent sensing methodologies in this area, it can be seen that optical methods using onsite hand-held devices have not been reported. This step is highly crucial for the commercialization of the H_2S sensor. Also, most of these reports deal with turn 'ON' sense which is done in the dark atmosphere. This could lead to fluctuations even in the presence of a small amount of light. Also, the time for stabilization of the fluorescence needs to be taken into account which adds to the list of improvements to be made. However, in the case of capacitive and resistive sensors which are non-optical methods no interference in the presence of light is observed. In addition, they can be built easily using cost-effective components employing a variety of on-site reading devices which makes them more favorable for commercialization. The succeeding sections will deal with mechanisms for resistive and capacitive sensing of H_2S and a brief discussion of sensors using these mechanisms.

16.3 H_2S SENSING MECHANISM IN MOF-BASED CAPACITIVE AND RESISTIVE SENSORS

The sensing mechanism for H_2S used in MOF-based capacitive and resistive sensors depends on the adsorption of H_2S on the MOF. The most important factor affecting the adsorption performance of MOFs is adsorbate-adsorbent interaction and it is determined by the metal sites and structures of the MOFs [20]. In some of the MOFs such as MIL-47(V) and MIL-53(Cr), the hydrogen bond is the main interaction between H_2S molecules and MOFs [21]. The influence of the pore size in H_2S adsorption in the MOFs was investigated by a molecular simulation study and it was observed that the adsorbent capacity increases with increasing pore size of MOFs up to a particular temperature [22]. It was also found that extending linkers in MOFs leads to a reduction in the interaction between H_2S and the MOFs. H_2S molecules are preferentially adsorbed in the tetrahedral cages at low pressure while it is pushed to the octahedral cages at increased pressure. Another study found that the H_2S adsorption capacity of MOF-199 decreases with increasing temperature and increases with increasing pressure [23]. The main factor affecting the adsorption performance of MOF-199 is that the framework contains binding sites of copper dimers at low pressures, while the free volume of MOF-199 contributes to the adsorption capacity as well at high pressures.

Recently, different modification methods have been introduced to enhance the adsorption performance of MOFs, and amine modification is one important method among them. To modify the MOFs with an amino group, one method is to introduce amino groups into ligands to obtain the amino-tagged MOF structures and another method is directly grafting the amine reagents on the metal centers of MOFs [24, 25]. Zhang et al. used an impregnation method to modify the MOF-199 with amine groups and evaluated its H_2S adsorption performance [24]. They used triethanolamine (TEA) as the modifying agent and found that the H_2S adsorption capacity of the TEA/MOF-199 was enhanced by 65%. In addition, the density functional theory calculations have shown that the binding energies of H_2S adsorption on the TEA/MOF-199 were larger than that on bare MOF-199. This higher binding energy of TEA/MOF-199 is considered as the reason behind the improved adsorption and the enhanced binding energy of H_2S on the TEA/MOF-199 was attributed to the hydroxyl group in TEA. In another work, Joshi et al. evaluated the H_2S adsorption efficacy of parent and amine-functionalized analogs of selected MOFs such as UiO-66(Zr), MIL-125(Ti), and MIL-101(Cr) [25]. They found that linker-based amine functionalization in smaller pore MOFs significantly enhanced H_2S breakthrough time in comparison to their respective parent materials which led to enhanced H_2S adsorption.

It was reported that copper-based MOFs provide high sulfur adsorption capacity owing to the stronger interaction between the Cu^{2+} ions and the S atom [26]. Peluso et al. evaluated the H_2S adsorption on a laboratory-synthesized polymeric chromium terephthalate (MIL-101(Cr)) MOF [27]. The results showed that MIL-101 (Cr) has a high H_2S adsorption capacity at near-ambient temperature and low heat release during adsorption. Petit et al. used composites of a copper-based MOF and graphite oxide (GO) for evaluating the H_2S adsorption at ambient conditions [28]. An enhancement in H_2S adsorption was found in composite materials compared to its counterparts and it is attributed to the physical adsorption of water and H_2S in the pore space in the interface between the MOF units and the graphene layers. In addition, H_2S molecules bind to the copper centers of the MOF which leads to the formation of copper sulfide. In another work, Daraee et al. evaluated the H_2S adsorption performance of a novel MOF-TiO_2 nanocomposite (TiO_2/UiO-66) and found that the metal active sites of titanium enhance the adsorption of nano-composite in comparison to UiO-66 MOF [29].

16.4 MOF-BASED CAPACITIVE SENSORS FOR H_2S

The working principle of gas sensors is based on the variation in electrical parameters by the diffusion or chemical reaction of various gases and the sensing layer. Different transduction mechanisms commonly used for the detection of target gaseous analytes are optical, thermal, magnetic, surface acoustic waveguide (SAW), resistance, resonance, and capacitive mechanisms [30, 31]. Among these, the interdigitated electrode (IDE) capacitive transducer method is considered as the most suitable technique owing to its features such as miniature size, low-cost, simple structure, compatibility with standard complementary metal-oxide-semiconductor technology, ability to operate at room temperature, and enabling miniaturization coupled with high sensing performance features [32]. In addition, the sensor performance depends mainly on the coated or deposited sensing layer. The research on MOF-based capacitive sensing of H_2S is very limited and there is a lot of scope in this domain.

There are very few reports on IDE capacitive sensors for H_2S detection based on MOF structures. Surya et al. introduced the integration of UiO-66 (Zr) MOF and Ag_2O nanoparticles (NPs) to form Ag_2O@UiO-66 composite for the first time and deposited it on Si/SiO_2 substrate to make an IDE sensor platform for H_2S gas sensing [30]. The Ag_2O NPs were anchored on MOFs with different modified functional groups (N_3, NO_2). Based on this, three different combinations of MOF have been made, namely UiO-66(Zr) BDC, UiO-66(Zr) BDC-NO_2, and UiO-66(Zr) BDC-N_3 where BDC is 1,4-benzenedicarboxylic acid. Then the synthesized Ag_2O-UiO-66 composite materials were deposited on top of a Si/SiO_2 substrate for the fabrication of the entire IDE sensor platform and their H_2S sensing properties were investigated. From the experimental results, it was found that the UiO-66(Zr) BDC-NO_2 loaded with Ag_2O showed the highest sensitivity towards H_2S. From the capacitive response of the Ag_2O@UiO-66 (Zr)-NO_2 based sensor, it was found that the sensor can be used to detect the H_2S gas ranging from 1 ppm to 100 ppm with a limit of detection of 1 ppm. In addition, the device exhibited a near-linear response for H_2S sensing and low response time.

In another work, Yassine et al. used rare-earth (RE)-based MOFs with fcu topology (RE-fcu-MOF) for the fabrication of an IDE sensor for the detection of H_2S at room temperature [32]. This RE-fcu-MOF is isostructural MOF to the zirconium-based UiO-66 MOF [33]. They constructed RE-fcu-MOFs by the molecular building block (MBB) approach, in which the unique RE hexanuclear cluster is assembled with a series of both hetero- and homofunctional linkers and the assembly is assisted by the modulator/structural directing agent 2-fluorobenzoic acid (2-FBA). Considering the exceptional thermal and chemical stability of these RE-fcu-MOFs [34] and their high adsorption selectivity towards H_2S, they explored a new field of application of sensing of toxic gases such as H_2S. They fabricated the sensing layer of thin-film isoreticular RE fcu-MOF on capacitive IDEs. The IDE sensors were fabricated using cleanroom technologies and it was functionalized with an OH-terminated self-assembled monolayer of 11-mercaptoundecanol. They have used an in-situ crystallization method under the predetermined reaction conditions for the construction of the RE-fcu-MOF and successfully grown a homogenous fumarate-based fcu-MOF (fum-fcu-MOF) thin film in the presence of fumaric acid. They also investigated the sensing properties of fum-fcu-MOF thin film on capacitive IDEs for different types of gases/vapors. They found that the developed sensor was highly stable and it showed outstanding detection capability for H_2S at concentrations down to 100 ppb, with the limit of detection at around 5 ppb. In addition, the fum-fcu-MOF sensor showed promising selectivity towards H_2S detection in the presence of CH_4, NO_2, H_2, and C_7H_8 [32].

16.5 MOF-BASED RESISTIVE SENSORS FOR H_2S

Resistive sensors rely highly on surface reactions that demand high surface area, stability, and porosity. Apart from the surface characteristics, the sensor should possess high selectivity and sensitivity. The sensing mechanism in chemiresistive sensors is based on the surface-based reactions resulting in adsorption/desorption of gas molecules accompanied by the transfer of electrons/holes which causes variations in the resistance of the sensing materials [35–37]. The sensor displays a fall in resistance when exposed to H_2S, related to the oxidation and release of electrons to the surface by the ionized oxygen anions present on the electrode surface. The sensing can be further enhanced through the incorporation of functional groups and metal nodes. Response (S) of the resistive sensor is defined by the ratio of sensor resistance in the air (R_a) to resistance in the presence of H_2S gas (R_g). Generally, the recovery/response time (t_{resp}/t_{reco}) is demarcated as the time required by the sensors to express at least the initial 90 % of resistance upon the exposure to analyte gas or air.

Semiconducting metal oxides (SMOs) were widely investigated for the detection of target gases [38, 39]. However, the requirements for elevated temperatures result in poor selectivity and limited stability. MOFs often act as templates resulting in enhanced surface area (up to ~8,000 m^2/g) and diverse morphologies enabling them to be better contenders for chemiresistive sensor materials. MOFs with high electrical conductivity can transfer electrical signals effectively, resulting in better selectivity and sensitivity [35, 37]. By integrating MOFs with SMOs, high-performance H_2S sensors can be fabricated. By using this strategy, X. Wu et al. have fabricated a sensor using a MOF-SMO hybrid material of zeolitic imidazolate framework-8 (ZIF-8) particle-loaded ZnO nanorod for H_2S sensing [40]. The fabricated ZIF-8/ZnO nanorod (ZZO) sensor provides a rough surface established by etching of the exposed surface by H_2O and 2-methylimidazolate, resulting in enhanced surface area and more active sites. Figure 16.2(a) shows the schematic illustration of the H_2S sensor based on ZIF-8/ZnO hybrids that operate at ambient temperatures. This ZZO sensor had an optimal operating temperature of 125 °C, however, it gave a response of 52.1 % to 10 ppm H_2S at room temperature. The ZZO sensor had a detection limit of 50 ppb at 25 °C and the sensing behavior was slightly affected by humidity.

In another work, core-shell flower-like WO_3@ZIF-71 was synthesized through hydrothermal synthesis and used for sensing H_2S. The sensor exhibited a response of 19.12 to 20 ppm H_2S at 250 °C with a detection limit of 0.697 ppm [41]. Similarly, ZIF-67 MOF-derived $Co_3O_4/NiCo_2O_4$ nanocages were prepared through a stepwise template-assisted method by J. Tan et al. The sensor was fabricated on an alumina substrate through screen printing followed by aging at 400 °C. The sensor showcased a response of 57 down to 50 ppm of H_2S gas in an optimal temperature of 250 °C [42].

Different bimetallic conductive MOFs have been used as active material in the chemiresistive sensing of H_2S. In this aspect, Z. Meng et al. used nickel-based bimetallic conductive MOFs as materials for the chemiresistive sensing of H_2S gases with a limit of detection of 20-30 ppb [44]. In another work, M.K. Smith et al. formulated a sensor using template-assisted growth of Cu/Ni MOFs employing graphene as nucleation sites [45]. This chemiresistive sensor device was capable of detecting and distinguishing different gases such as NH_3, NO, and H_2S at ppm concentrations. The same group has fabricated a wearable device by incorporating two-dimensional conductive MOFs onto textile fabrics as shown in Figure 16.2(b) [43]. Organic ligands such as 3,6,7,10,11-hexahydroxytriphenylene (HHTP) or 2,3,6,7,10,11-hexaaminotriphe-nylene (HATP) with metallic nickel nodes were coated into fabrics for sensor applications. The pictorial representation of the self-organized frameworks on textiles (SOFT) devices on fabric is demonstrated in Figure 16.2(b) with the nanorod morphology of the MOF-coated fabrics. The sensor device exhibited a theoretical limit of detection of

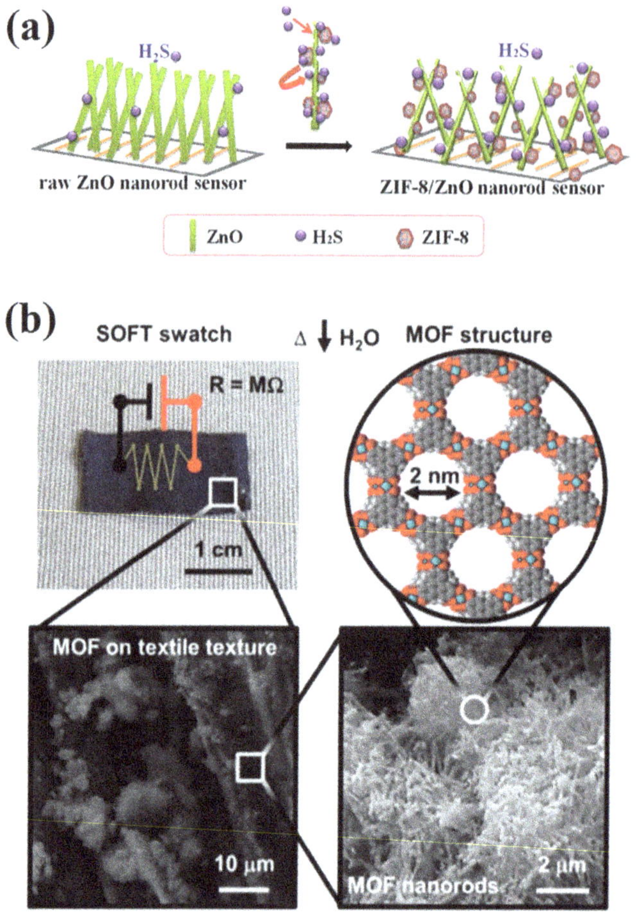

FIGURE 16.2 (a). Schematic illustrating the sensor fabrication of ZIF-8/ZnO nanorod from raw ZnO nanorods. Reprinted with permission from Ref [40] Copyright (2019) Elsevier. (b) Pictorial representation of the solvothermally fabricated self-organized frameworks on textiles (SOFT) devices on fabric swatch along with electron microscopic images showing the morphology and structure of the coated MOF. Reprinted with permission from Ref [43] Copyright (2017) Elsevier.

0.23 ppm for H_2S and showed a response increase of 65 % to 10 ppm H_2S gas.

A novel nanomaterial flexible membrane of MOF–polymer mixed-matrix has been introduced for the detection of H_2S gas at room temperature. A. Ali et al. fabricated a flexible membrane-like sensor based on a MOF-chitosan mixed-matrix for room temperature detection of H_2S [46]. A membrane was assembled over a copper foil, and the electrical measurement circuit, as well as sensor setup inside the test chamber, are shown in Figure 16.3(a). Here, the test chamber is sealed to maintain a non-humid atmosphere for performing the testing of the sensor. The sensor exhibited a sensitivity of 1 ppm at room temperature with a response time of under 8 sec. It is established that human health has a direct correlation to the quality of the air we breathe out. The exhaled air possibly contains numerous biomarkers which can indicate our health conditions even if they are present in minute traces. MOF-based sensors have been used for detecting trace levels

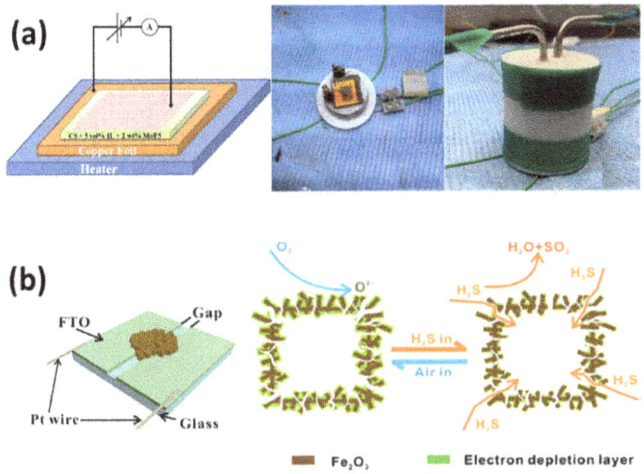

FIGURE 16.3 (a) Schematic diagram illustrating the sensor based on MOF-chitosan mixed-matrix. The electrical measurement circuit, sensor setup inside the test chamber; and picture of the sensitivity test chamber are given. Reprinted with permission from Ref [46] Copyright (2021) ACS, (b) Schematic representation illustrating the sensor fabricated on FTO along with the sensing mechanism. Reprinted with permission from Ref [36] Copyright (2017) American Chemical Society.

of gases from exhaled air. In this aspect, Yang et al. have fabricated a sensor to detect trace H_2S biomarkers in exhaled breath through a MOF-based three-dimensional inverse opal structure (3DIOS) [47]. Core-shell structures of ZnO/Au MOF were prepared and coated onto a IOS template which was further heat-treated to fabricate WO_3/ZnO@Au. The sensor was operated at an optimal working temperature of 170 °C with a response of 175 for a H_2S gas of 10 ppm. The sensor had a detection limit of 50 ppb having a response of 8.4 exhibiting the high sensitivity and limited response towards other gases like ammonia, acetone, toluene, formaldehyde, and ethanol proving its high selectivity. The sensor exhibited stability of 93% after 30 days of operation under optimal working conditions. The synergic effects of ZnO, WO_3 and Au have facilitated the enhanced activity of the H_2S sensor. ZnO is well distributed on the surface of WO_3 creating a large number of oxygen vacancies that facilitate the modification in bandgaps and leading to enhanced absorption/desorption during sensing. The core-shell structure has controlled the agglomeration of Au that acts as a catalyst for oxygen molecular dissociation.

MOF-derived multi-dimensional nanomaterials have been used to fabricate H_2S sensors with promising sensing performance. Zhou et al. synthesized 3DIOS based on Pt/ZnO using a simple self-assembly template method from ZIF-8-encapsulated Pt NPs. The polymethylmethacrylate (PMMA) spheres are impregnated with synthesized Pt@ZIF-8 and zinc precursor followed by annealing to achieve Pt/ZnO [48]. At the optimal operating temperature of 320 °C, the sensor exhibited a response of 11.2 to 1 ppm of H_2S gas with response and recovery times of 8.7 sec and 19.4 sec. In another work, MOF-derived hierarchical hollow Co_3O_4@NiO microcubes were prepared by a hydrothermal method by X. Wang et al. [49]. They have formulated sensors based on heterostructure and through the incorporation of noble metal catalysts, leading to the reduction in the activation energies for chemical reactions. Prussian blue (PB) analogs were used as templates for the synthesis of Pt functionalized cobalt/nickel oxide microcubes. Precipitation followed by a hydrothermal synthesis approach was carried out to synthesize Co/Ni oxide-based microcubes which was further impregnated with Pt to achieve Co_3O_4@NiO@Pt microcubes (CNOP). The CNOP sensor displayed a response of 250 at a working temperature of 200 °C to 100 ppm H_2S. The sensor showcased a response and recovery time of 213 s and 135 s respectively. The heterostructure without Pt loading showcased a response of 9.9 for the above conditions.

Wang et al. used Zn/Cu-BTC MOF (BTC is benzoic acid radical [50]) as a sacrificial template to obtain the ZnO/CuO composite and this ZnO/CuO composite (ZCO) was used as a sensing material for the H_2S sensor [51]. The ZCO sensor operates at an optimal working temperature of 40 °C giving a response of 393.35 to 10 ppm H_2S. At the tested conditions, the sensor showed a response time of 173 sec, however, it requires a higher temperature of 300 °C to recover to the initial condition for sensing. The limit of detection of the sensor was capped at 300 ppb and stability of over 95% for 42 days at optimal conditions. The sensing mechanism depends on factors that include the influence of Zn in delivering a surface roughness to the composite material, leading to enhancement in oxygen dissociations and the formation of the p-n junction between the n-type ZnO and p-type CuO. This leads to the Zn-O-Cu bond, resulting in stronger interactions. Similarly, a MOF-derived bamboo-like CuO/In_2O_3 heterostructure has been used for the fabrication of high-performance H_2S gas sensors with low operating temperatures [52]. In this report, a CuO/In_2O_3 heterostructure was synthesized through a two-step synthesis technique involving cation impregnation followed by calcination. The CuO/In_2O_3 sensor operates at an optimal working temperature of 70 °C giving a response of 229.3 to 5 ppm of H_2S gas.

Iron oxide (Fe_2O_3) is widely used in H_2S sensor due to its low cost and abundance in the earth's crust. The widespread applicability of these sensors is limited by the long recovery time and the unpromising detection limits along with temperature prerequisites for reaction initiations. An ideal sensor should possess high porosity and surface area along with superior thermal and chemical stabilities. K. Tian et al. fabricated a H_2S based gas sensor on an FTO glass substrate using Fe_2O_3 nano boxes as shown in Figure 16.3(b). A three-step process involving the synthesis of PB-based MOFs, followed by an ion-exchange process leading to the formation of hollow $Fe(OH)_3$ nano boxes. Hollow $Fe(OH)_3$ nano boxes were used as the template which upon calcination leads to the formation of Fe_2O_3 nano boxes (FONB). The FONBs achieved a specific surface area (SSA) of 113 m^2/g with a wider pore distribution in micro, meso, and macro ranges which promoted gas permeation and diffusion. The XPS

TABLE 16.1
A comparison of MOF-based resistive sensors for H_2S detection

Material	Operating Temperature	Response	Response/ Recovery times	Detection Limit	Ref
Co-PBA MOF-derived Co_3O_4 @ NiO microcubes	200 °C	250 @ 100 ppm	213 sec/135 sec	20 ppm	[49]
ZIF-67 MOF-derived $Co_3O_4/NiCo_2O_4$	250 °C	57 @ 50 ppm	153 sec/40 sec	10 ppm	[42]
MOF drivenγ-Fe_2O_3/rGO	Room temperature	520.73 @ 97 ppm		3 ppm	[53]
MOFmodifiedWO_3 Au@ZnO functionalized	375 °C	97 @ 10 ppm	480 sec	50 ppb	[47]
MOF-chitosan	Room temperature	96-98 % @ 50 ppm	8 sec/30 sec	1 ppm	[46]
MOF derived Fe_2O_3	200 °C	1.23 @ 0.25 ppm	~ 75 sec/~132 sec (5 ppm)	250 ppb	[36]
Bimetallic MOFs	Room temperature	98 % @ 80 ppm	-	19 ppb	[44]
Self-Organized Frameworks on Textiles (SOFT)	Room temperature	65 % @ 10 ppm	-	0.23 ppm	[43]
MOF derived ZnO/CuO composite	40 °C	393.35 @ 10 ppm	173 sec	300 ppb	[51]
MOF derived CuO/In_2O_3	70 °C	229.3 @ 5 ppm	10 sec	200 ppb	[52]
ZIF-8/ZnO	125 °C	52.1 % @ 10 ppm (25 °C)	420 sec/642 sec	50 ppb	[40]
MOF derived Pt/ZnO	320 °C	11.2 @ 1 ppm	8.7 sec/19.4 sec	25 ppb	[48]
WO_3@ZIF-71	250 °C	19.12 @ 20 ppm	118 sec/431 sec	0.697 ppm	[41]

confirmed that the material was Fe_2O_3 by the spectral peaks at 710.9 eV and 724.8 eV stating the characteristic of Fe^{3+} and peaks at 529.8 eV and 531.2 eV corresponding to O 1s [36]. Here, they tested its applicability at a temperature below 250 °C. The sensors were subjected to varying temperatures of 250 °C, 200 °C, and 50 °C, and the response-recovery curves were analyzed at 5 ppm of H_2S gas, where the sensors exhibited good repeatability. The sensor mechanism is illustrated in Figure 16.3(b) which is showing the surface reactions of the Fe_2O_3 towards the exposed air and H_2S gas. The mechanism of the sensor lies with the chemisorbed oxygen species on the Fe_2O_3 surface on exposure to air. These anionic species in reaction with the introduced H_2S gas decrease the sensor resistance through the release of electrons thereby leading to its detection. The sensor showed a promising response value at all exposed temperatures of 250 °C, 200 °C, and 50 °C. The sensor responded to 250 ppb showing a response of 1.23 which confirms the good performance at lower detection levels. The sensor exhibited a good selectivity with regards to other gases like NH_3, CO, and NO.

In another work, Zhang et al. have synthesized γ-Fe_2O_3/rGO (FoG) using MIL-88 as a sacrificial agent [53]. A hydrothermal approach was followed to synthesize MIL-88/rGO composite which, upon calcination resulted in FoG. γ-Fe_2O_3 exhibited an octahedral morphology which is well distributed over the graphene facilitating active sites for gas sensing. The XPS analysis displayed a shift in the binding energy of FoG to a higher magnitude in comparison to γ-Fe_2O_3 establishing the strong electron transfer interaction within the composite. The sensor was operated at room temperature and the concentration of H_2S was varied from 97 ppm to 2.91 ppm resulting in a response variation from 520.73 to 1.47. The FoG sensor exhibited good selectivity in comparison to other interfering gases such as NH_3, $CHCl_3$, NO, SO_2 and HCHO. The decrease in the charge transfer resistance (R_{ct}) is evident from the EIS measurements which shows the enhancement in the conductivity. In addition, the Mott-Schottky (MS) measurement revealed that the increase in rGO concentration influences increasing the conductivity, leading to the easy capture and conduction of electrons from the conduction bands. A comparison of various MOF-based resistive sensors for H_2S detection is given in Table 16.1.

16.6 CONCLUSION AND FUTURE PERSPECTIVES

The research and development on gas sensors, in particular, H_2S sensors are growing due to the need for constant vigilance in the industrial and healthcare sectors. Non-invasive detection of H_2S by semiconducting metal oxide gas sensors is often set back by the slow dynamic response and high operating temperature. The revolutionary benchmark enhancements brought by MOFs have significantly changed the outlook of electrochemical sensors. MOFs have played a vivid role through their unique properties of enhanced surface area, morphologies, and multi-responsive property. The research on MOF-based capacitive sensing of H_2S is limited and based on interdigitated electrode capacitive transducers. A promising detection limit of 5 ppb was achieved in the MOF-based capacitive H_2S sensor. However, there is a lot of scope for enhancing the sensitivity, minimizing the complexity of the sensor, practical applicability, and so forth.

Chemiresistive sensors possess an inherent advantage over other sensors due to their ease of fabrication, low power requirements, and low-temperature applications. This makes these devices represent an exciting opportunity for advancements in terms of sensor applications. A limit of detection as small as 9 ppb was achieved for chemiresistive sensors that are operational at room temperatures.

Also, there is a need for innovation to move in the direction of fabricating functional devices which can be used for the onsite detection of H_2S. Another direction that needs to be explored is the wearable sector where H_2S sensors can be used for diagnostic purposes. The use of chemiresistive and capacitive sensors employing MOFs will be highly versatile for all of the above applications. In conclusion, although this area has seen articles in the materials innovation perspective, more studies involving the use of MOFs will bring about progress in the technological paradigms too.

ACKNOWLEDGMENTS

PAR acknowledges the financial support from Ramalingaswami Fellowship, Department of Biotechnology, Government of India.

REFERENCES

[1]. Martínez-Ahumada, E.; López-Olvera, A.; Jancik, V.; Sánchez-Bautista, J.E.; González-Zamora, E.; Martis, V.; Williams, D.R.; Ibarra, I.A. MOF Materials for the Capture of Highly Toxic H_2S and SO_2. *Organometallics* **2020**, *39*, 883–915,

[2]. Carter, J.M.; Qian, Y.; Foster, J.C.; Matson, J.B. Peptide-based hydrogen sulphide-releasing gels. *Chem. Commun.* **2015**, *51*, 13131–13134.

[3]. Baumann, A.E.; Burns, D.A.; Liu, B.; Thoi, V.S. Metal-organic framework functionalization and design strategies for advanced electrochemical energy storage devices. *Commun. Chem.* **2019**, *2*, 86.

[4]. Liu, Y.; Liu, F.; Bai, J.; Liu, T.; Yu, Z.; Dai, M.; Zhou, L.; Wang, H.; Zhang, Y.; Suo, H.; et al. Direct growth of NiO films on Al_2O_3 ceramics by electrochemical deposition and its excellent H_2S sensing properties. *Sensors Actuators B Chem.* **2019**, *296*, 126619.

[5]. Duc, C.; Boukhenane, M.-L.; Wojkiewicz, J.-L.; Redon, N. Hydrogen Sulfide Detection by Sensors Based on Conductive Polymers: A Review. *Front. Mater.* **2020**, *7* , 215.

[6]. Wang, H.; Luo, Y.; Liu, B.; Gao, L.; Duan, G. CuO nanoparticle loaded ZnO hierarchical heterostructure to boost H_2S sensing with fast recovery. *Sensors Actuators B Chem.* **2021**, *338*, 129806.

[7]. Chen, J.; Guo, L.; Chen, L.; Qiu, B.; Hong, G.; Lin, Z. Sensing of Hydrogen Sulfide Gas in the Raman-Silent Region Based on Gold Nano-Bipyramids (Au NBPs) Encapsulated by Zeolitic Imidazolate Framework-8. *ACS Sensors* **2020**, *5*, 3964–3970.

[8]. Gogoi, C.; Kumar, A.; SK, M.; Biswas, S. Specific fluorescence sensing of hydrogen sulphide by an azide functionalized Zr(IV) MOF with DUT-52 topology. *Microporous Mesoporous Mater.* **2021**, *311*, 110725.

[9]. Dalapati, R.; Balaji, S.N.; Trivedi, V.; Khamari, L.; Biswas, S. A dinitro-functionalized Zr(IV)-based metal-organic framework as colorimetric and fluorogenic probe for highly selective detection of hydrogen sulphide. *Sensors Actuators B Chem.* **2017**, *245*, 1039–1049.

[10]. Nandi, S.; Banesh, S.; Trivedi, V.; Biswas, S. A dinitro-functionalized metal–organic framework featuring visual and fluorogenic sensing of H_2S in living cells, human blood plasma and environmental samples. *Analyst* **2018**, *143*, 1482–1491.

[11]. Nandi, S.; Reinsch, H.; Biswas, S. An acetoxy functionalized Al(iii) based metal–organic framework showing selective "turn on" detection of perborate in environmental samples. *Dalt. Trans.* **2020**, *49*, 17612–17620.

[12]. Guo, L.; Wang, M.; Cao, D. A Novel Zr-MOF as Fluorescence Turn-On Probe for Real-Time Detecting H_2S Gas and Fingerprint Identification. *Small* **2018**, *14*, 1703822.

[13]. Ma, Y.; Su, H.; Kuang, X.; Li, X.; Zhang, T.; Tang, B. Heterogeneous Nano Metal–Organic Framework Fluorescence Probe for Highly Selective and Sensitive Detection of Hydrogen Sulfide in Living Cells. *Anal. Chem.* **2014**, *86*, 11459–11463.

[14]. Zhu, Z.; Natarajan, V.; Wang, W.-N. The role of Fe^{3+} ions in fluorescence detection of H_2S by a bimetallic metal-organic framework. *J. Solid State Chem.* **2020**, *288*, 121434.

[15]. Kaneti, Y.V.; Zhang, J.; He, Y.-B.; Wang, Z.; Tanaka, S.; Hossain, M.S.A.; Pan, Z.-Z.; Xiang, B.; Yang, Q.-H.; Yamauchi, Y. Fabrication of an MOF-derived heteroatom-doped Co/CoO/carbon hybrid with superior sodium storage performance for sodium-ion batteries. *J. Mater. Chem. A* **2017**, *5*, 15356–15366.

[16]. Zhao, X.; Zhang, L.; Bai, J.; Wu, P.; Li, Y.; Liang, L.; Xie, L.; Wang, J. A copper-based metal-organic framework for ratiometric detection of hydrogen sulfide with high sensitivity and fast response. *Spectrochim. Acta Part A Mol. Biomol. Spectrosc.* **2020**, *243* , 118794.

[17]. Li, H.; Liang, J.; Li, H.; Zheng, X.; Tao, Y.; Huang, Z.-H.; Yang, Q.-H. Activated carbon fibers with manganese dioxide coating for flexible fiber supercapacitors with high capacitive performance. *J. Energy Chem.* **2019**, *31*, 95–100.

[18]. Ling, P.; Qian, C.; Yu, J.; Gao, F. Metal–organic framework nanosheets with flower-like structure as probes for H_2S detection and in-situ singlet-oxygen production. *Chem. Commun.* **2019**, *55*, 6385–6388.

[19]. Dong, X.; Su, Y.; Lu, T.; Zhang, L.; Wu, L.; Lv, Y. MOFs-derived dodecahedra porous Co_3O_4: An efficient cataluminescence sensing material for H_2S. *Sensors Actuators B Chem.* **2018**, *258*, 349–357.

[20]. Li, Y.-X.; Jiang, W.-J.; Tan, P.; Liu, X.-Q.; Zhang, D.-Y.; Sun, L.-B. What Matters to the Adsorptive Desulfurization Performance of Metal - Organic Frameworks? *J. Phys. Chem. C* **2015**, *119*, 21969–21977.

[21]. Hamon, L.; Leclerc, H.; Ghoufi, A.; Oliviero, L.; Travert, A.; Lavalley, J.-C.; Devic, T.; Serre, C.; Férey, G.; De Weireld, G.; et al. Molecular Insight into the Adsorption of H_2S in the Flexible MIL-53(Cr) and Rigid MIL-47(V) MOFs: Infrared Spectroscopy Combined to Molecular Simulations. *J. Phys. Chem. C* **2011**, *115*, 2047–2056.

[22]. Al-Jadir, T.M.; Siperstein, F.R. The influence of the pore size in Metal–Organic Frameworks in adsorption and separation of hydrogen sulphide: A molecular simulation study. *Microporous Mesoporous Mater.* **2018**, *271*, 160–168.

[23]. Zhang, H.-Y.; Zhang, Z.-R.; Yang, C.; Ling, L.-X.; Wang, B.-J.; Fan, H.-L. A Computational Study of the Adsorptive Removal of H2S by MOF-199. J. *Inorg. Organomet. Polym. Mater.* **2018**, *28*, 694–701.

[24]. Zhang, H.-Y.; Yang, C.; Geng, Q.; Fan, H.-L.; Wang, B.-J.; Wu, M.-M.; Tian, Z. Adsorption of hydrogen sulfide by amine-functionalized metal organic framework (MOF-199): An experimental and simulation study. *Appl. Surf. Sci.* **2019**, *497*, 143815.

[25]. Joshi, J.N.; Zhu, G.; Lee, J.J.; Carter, E.A.; Jones, C.W.; Lively, R.P.; Walton, K.S. Probing Metal–Organic Framework Design for Adsorptive Natural Gas Purification. *Langmuir* **2018**, *34*, 8443–8450.

[26]. Blanco-Brieva, G.; Campos-Martin, J.M.; Al-Zahrani, S.M.; Fierro, J.L.G. Effectiveness of metal–organic frameworks for removal of refractory organo-sulfur compound present in liquid fuels. *Fuel* **2011**, *90*, 190–197.

[27]. Peluso, A.; Gargiulo, N.; Aprea, P.; Pepe, F.; Caputo, D. Modeling Hydrogen Sulfide Adsorption on Chromium-Based MIL-101 Metal Organic Framework. *Sci. Adv. Mater.* **2014**, *6*, 164–170.

[28]. Petit, C.; Mendoza, B.; Bandosz, T.J. Hydrogen Sulfide Adsorption on MOFs and MOF/Graphite Oxide Composites. *ChemPhysChem* **2010**, *11*, 3678–3684.

[29]. Daraee, M.; Saeedirad, R.; Rashidi, A. Adsorption of hydrogen sulfide over a novel metal organic framework –metal oxide nanocomposite: TOUO-x (TiO_2/UiO-66). *J. Solid State Chem.* **2019**, *278* , 120866.

[30]. Surya, S.G.; Bhanoth, S.; Majhi, S.M.; More, Y.D.; Teja, V.M.; Chappanda, K.N. A silver nanoparticle-anchored UiO-66(Zr) metal–organic framework (MOF)-based capacitive H_2S gas sensor. *CrystEngComm* **2019**, *21*, 7303–7312.

[31]. Sapsanis, C.; Omran, H.; Chernikova, V.; Shekhah, O.; Belmabkhout, Y.; Buttner, U.; Eddaoudi, M.; Salama, K. Insights on Capacitive Interdigitated Electrodes Coated with MOF Thin Films: Humidity and VOCs Sensing as a Case Study. *Sensors* **2015**, *15*, 18153–18166.

[32]. Yassine, O.; Shekhah, O.; Assen, A.H.; Belmabkhout, Y.; Salama, K.N.; Eddaoudi, M. H_2S Sensors: Fumarate-Based fcu-MOF Thin Film Grown on a Capacitive Interdigitated Electrode. *Angew. Chemie Int. Ed.* **2016**, *55*, 15879–15883.

[33]. Xue, D.-X.; Belmabkhout, Y.; Shekhah, O.; Jiang, H.; Adil, K.; Cairns, A.J.; Eddaoudi, M. Tunable Rare Earth fcu -MOF Platform: Access to Adsorption Kinetics Driven Gas/Vapor Separations via Pore Size Contraction. *J. Am. Chem. Soc.* **2015**, *137*, 5034–5040.

[34]. Assen, A.H.; Belmabkhout, Y.; Adil, K.; Bhatt, P.M.; Xue, D.-X.; Jiang, H.; Eddaoudi, M. Ultra-Tuning of the Rare-Earth fcu-MOF Aperture Size for Selective Molecular Exclusion of Branched Paraffins. *Angew. Chemie Int. Ed.* **2015**, *54*, 14353–14358.

[35]. Koo, W.-T.; Jang, J.-S.; Kim, I.-D. Metal-Organic Frameworks for Chemiresistive Sensors. *Chem* **2019**, *5*, 1938–1963.

[36]. Tian, K.; Wang, X.-X.; Yu, Z.-Y.; Li, H.-Y.; Guo, X. Hierarchical and Hollow Fe_2O_3 Nanoboxes Derived from Metal–Organic Frameworks with Excellent Sensitivity to H_2S. *ACS Appl. Mater. Interfaces* **2017**, *9*, 29669–29676.

[37]. Li, H.-Y.; Zhao, S.-N.; Zang, S.-Q.; Li, J. Functional metal–organic frameworks as effective sensors of gases and volatile compounds. *Chem. Soc. Rev.* **2020**, *49*, 6364–6401.

[38]. Wetchakun, K.; Samerjai, T.; Tamaekong, N.; Liewhiran, C.; Siriwong, C.; Kruefu, V.; Wisitsoraat, A.; Tuantranont, A.; Phanichphant, S. Semiconducting metal oxides as sensors for environmentally hazardous gases. *Sensors Actuators B Chem.* **2011**, *160*, 580–591.

[39]. Nasiri, N.; Clarke, C. Nanostructured Chemiresistive Gas Sensors for Medical Applications. *Sensors* **2019**, *19*, 462.

[40]. Wu, X.; Xiong, S.; Gong, Y.; Gong, Y.; Wu, W.; Mao, Z.; Liu, Q.; Hu, S.; Long, X. MOF-SMO hybrids as a H_2S sensor with superior sensitivity and selectivity. *Sensors Actuators B Chem.* **2019**, *292*, 32–39.

[41]. Zhou, Y.; Zhou, T.; Zhang, Y.; Tang, L.; Guo, Q.; Wang, M.; Xie, C.; Zeng, D. Synthesis of core-shell flower-like WO_3@ZIF-71 with enhanced response and selectivity to H_2S gas. *Solid State Ionics* **2020**, *350*, 115278.

[42]. Tan, J.; Hussain, S.; Ge, C.; Wang, M.; Shah, S.; Liu, G.; Qiao, G. ZIF-67 MOF-derived unique double-shelled Co_3O_4/$NiCo_2O_4$ nanocages for superior Gas-sensing performances. *Sensors Actuators, B Chem.* **2020**, *303*, 127251.

[43]. Smith, M.K.; Mirica, K.A. Self-Organized Frameworks on Textiles (SOFT): Conductive Fabrics for Simultaneous Sensing, Capture, and Filtration of Gases. *J. Am. Chem. Soc.* **2017**, *139*, 16759–16767.

[44]. Meng, Z.; Aykanat, A.; Mirica, K.A. Welding Metallophthalocyanines into Bimetallic Molecular Meshes for Ultrasensitive, Low-Power Chemiresistive Detection of Gases. *J. Am. Chem. Soc.* **2019**, *141*, 2046–2053.

[45]. Smith, M.K.; Jensen, K.E.; Pivak, P.A.; Mirica, K.A. Direct Self-Assembly of Conductive Nanorods of Metal–Organic Frameworks into Chemiresistive Devices on Shrinkable Polymer Films. *Chem. Mater.* **2016**, *28*, 5264–5268.

[46]. Ali, A.; Alzamly, A.; Greish, Y.E.; Bakiro, M.; Nguyen, H.L.; Mahmoud, S.T. A Highly Sensitive and Flexible Metal–Organic Framework Polymer-Based H_2S Gas Sensor. *ACS Omega* **2021**, *6*, 17690–17697.

[47]. Yang, S.; Sun, J.; Xu, L.; Zhou, Q.; Chen, X.; Zhu, S.; Dong, B.; Lu, G.; Song, H. Au@ZnO functionalized three–dimensional macroporous WO_3: A application of selective H_2S gas sensor for exhaled breath biomarker detection. *Sensors Actuators B Chem.* **2020**, *324*, 128725.

[48]. Zhou, X.; Lin, X.; Yang, S.; Zhu, S.; Chen, X.; Dong, B.; Bai, X.; Wen, X.; Geyu, L.; Song, H. Highly dispersed Metal–Organic-Framework-Derived Pt nanoparticles on three-dimensional macroporous ZnO for trace-level H_2S sensing. *Sensors Actuators B Chem.* **2020**, *309*, 127802.

[49]. Wang, X.; Lu, J.; Han, W.; Yang, J.; Jiang, B.; Sun, Y.; Zhang, H.; Lu, G. Co-PBA MOF-derived hierarchical hollow Co_3O_4@NiO microcubes functionalized with Pt for superior H_2S sensing. *Sensors Actuators B Chem.* **2021**, *342*, 130028.

[50]. Dai, W.; Tian, N.; Liu, C.; Yu, L.; Liu, Q.; Ma, N.; Zhao, Y. (Zn, Ni, Cu)-BTC Functionalized with Phosphotungstic Acid for Adsorptive Desulfurization in the Presence of Benzene and Ketone. *Energy & Fuels* **2017**, *31*, 13502–13508.

[51]. Wang, X.; Li, S.; Xie, L.; Li, X.; Lin, D.; Zhu, Z. Low-temperature and highly sensitive H_2S gas sensor based on ZnO/CuO composite derived from bimetal metal-organic frameworks. *Ceram. Int.* **2020**, *46*, 15858–15866.

[52]. Li, S.; Xie, L.; He, M.; Hu, X.; Luo, G.; Chen, C.; Zhu, Z. Metal-Organic frameworks-derived bamboo-like CuO/In_2O_3 Heterostructure for high-performance H_2S gas sensor with Low operating temperature. *Sensors Actuators B Chem.* **2020**, *310*, 127828.

[53]. Zhang, C.; Zhang, S.; Yang, Y.; Yu, H.; Dong, X. Highly sensitive H_2S sensors based on metal-organic framework driven γ-Fe_2O_3 on reduced graphene oxide composites at room temperature. *Sensors Actuators, B Chem.* **2020**, *325*, 128804.

17 MOF-based Sensors for Detecting Hydrogen Sulfide

Saravanan Nagappan[1], Shamim Ahmed Hira[1], Sanha Jang[1], Dicky Annas[1] and Kang Hyun Park[1]

[1]Department of Chemistry, Chemistry Institute for Functional Materials, Pusan National University, Busan, Republic of Korea

17.1 INTRODUCTION

Metal-organic frameworks (MOFs), also known as porous coordination polymers, have been a subject of intensive research since the discovery of the first MOF structure. To date, over 90,000 different types of MOFs have been successfully synthesized, and over 500,000 are forecast [1]. MOFs can be easily synthesized using metal nodes and organic linkers to develop highly uniform micro/mesoporous crystalline solids [1]. Judiciously selected metal nodes and organic linkers have been used for tuning the porosity and functionality of MOFs for multiple applications [2]. MOFs with higher porosity and larger pore sizes have attracted considerable attention in several applications such as chemical sensing, catalysis, biomedical applications, gas storage, energy, environment-related applications, and separation, owing to the easier transport of gases and molecules through the pore channels [3]. MOFs having larger pore sizes can be easily synthesized using larger organic linkers [4]. In general, MOFs with high specific surface areas (1000 $m^2 g^{-1}$ to 10000 $m^2 g^{-1}$) can be easily synthesized using appropriate organic linkers in conjunction with suitable metal sources to form highly porous network structures with tunable chemical compositions and pore sizes [5]. For example, Wang et al. used 1,3,5-tris(4-carboxyphenyl)benzene as an organic linker to develop MOF-177 which has a high specific surface area of 5000 $m^2 g^{-1}$ and a pore diameter of 10.5 Å.[6]

Recently, MOFs have found considerable utility in the field of catalysis due to the high demand for the synthesis of various organic compounds [7]. The unique inorganic-organic hybrid structures of MOFs with easily tunable properties based on various requirements make them highly valuable to various fields of research [8]. MOFs are employed predominantly in electrochemical energy conversion and storage applications because of their enhanced electrical conductivity based on the metal source as well as the presence of heteroatoms and the large numbers of active sites in their molecular structures [2]. Recently, porous carbon materials derived from MOFs have attracted significant interest in several applications because of their highly porous network structures with higher specific surface areas, enhanced stability, improved electrical conductivity, and thermal stability [2].

Electrochemical sensors fabricated using various organic-inorganic hybrid materials have found considerable utility in the detection and sensing of various biomolecules (dopamine, glucose, uric acid, and ascorbic acid), gases (carbon dioxide, hydrogen, nitrogen, and various toxic gases), environmental pollutants (nitrophenols, nitrobenzenes, pesticides, and hydrazines), and heavy metal ions (As^{3+}, Cd^{2+}, Hg^{2+}, and Pb^{2+}) [9]. Typically, electrochemical sensors are used to detect biomolecules or gases at the parts per million (ppm) level [10]. A three-electrode set-up including working, reference, and counter electrodes separated by a thin hydrophilic membrane are used for analyte detection via amperometric or potentiometric methods based on the reduction-oxidation (redox) principle [10]. Amperometric methods are used for gas sensing, whereas potentiometric methods are used to detect solid electrolytes. Abdalla et al. briefly reviewed the design and use of various three-dimensional (3D) printed electrochemical sensors for heavy metal detection and environmental monitoring applications [11]. MOFs are widely used in electrochemical sensors because of their excellent electrical properties and sensitivity to various organic compounds, gases, and heavy metals [12]. As chemical sensors, MOFs have unique advantages compared to other nanomaterials since they can be easily functionalized, and their active metal centers and ligands are flexible to tuning, which can impart the desired properties [13].

Hydrogen sulfide (H_2S) is a highly flammable and toxic gas that smells of rotten eggs and causes severe health effects. Exposure to H_2S gas at concentrations exceeding 500 ppm can cause unconsciousness or even death [14]. The poisonous effect of H_2S gas is primarily due to the inhibition of cytochrome c oxidase, which hinders oxidative phosphorylation. Moreover, the toxicity of H_2S is comparable to that of carbon monoxide (CO) and nitric oxide (NO); it causes similar adverse effects such as tumorigenesis, rapid apnea, loss of neuroprotection, high blood pressure, loss of consciousness, inflammation, and decreased fertility [14]. The severity depends on the inhalation concentration of the gas. H_2S is generally produced during petroleum and natural gas refining and drilling, food processing, paper pulping, water

DOI: 10.1201/9781003188148-17

treatment, and coal gasification. It is also emitted from coke ovens, sewage sludge, and landfills [15]. H_2S gas is released along with other volatile sulfur compounds by *Saccharomyces cerevisiae* strains [16]. Various electrochemical sensors have been developed using carbon and graphene derivatives, metal-metal oxides, inorganic and organic materials, combinations of organic-inorganic hybrids, and dichalcogenides to detect H_2S gas [17]. Gas chromatography and spectrophotometry are typically used for detecting H_2S in air. However, these methods are very expensive [18]. Inexpensive semiconductor-based and electrochemical gas sensors have been widely applied to detect H_2S gas in air at low concentrations. In addition, an inexpensive wearable/portable gas sensor was used to monitor H_2S gas [18]. Some H_2S detection methods are very expensive owing to the need for complex instruments and procedures and highly skilled personnel. Moreover, these methods can damage tissues and are thus incompatible with biological systems thus limiting their use in H_2S sensing applications [19].

17.2 DETERMINATION OF H_2S GAS USING MOF-BASED NANOMATERIALS

Cataluminescent and luminescent techniques have been employed primarily for the detection of H_2S in gas or liquid phases using MOF-based nanomaterials [19]. Vikrant et al. briefly reviewed the important aspects of MOF-based nanomaterials and their advantages in the detection or sensing of H_2S in gases or liquids using various analytical techniques [19]. Fang et al. also reviewed MOF-based luminescent and electrochemical sensors and other types of sensors for the detection of various environmental contaminants [20]. In particular, luminescent sensors are used to determine inorganic anions, organic compounds, and heavy metal ions, whereas electrochemical sensors are used for sensing ions and organic compounds. Other sensors such as calorimetric sensors are used to detect water contaminants [20]. Luminescent sensors comprised of luminescent MOF-based nanomaterials are widely used as chemical sensors in aqueous phases. The luminescence properties of the MOF nanomaterials are determined based on the return of electrons from the excited singlet states to the ground states. Furthermore, electrochemical sensors are used to determine the redox behavior of analytes using three-electrode systems. The amount of analyte is calculated from the electric potential or electric signal and current [21]. The gas detection response ($S = R_a/R_g$) is determined from the sensor resistance of H_2S in a gas (R_g) and air (R_a).

17.3 MOF NANOMATERIALS BASED ON METALS AND METAL OXIDES FOR ELECTROCHEMICAL H_2S SENSORS

The use of various metal precursors and organic linkers for the synthesis of MOFs can have a significant impact on the structural properties of MOF-based nanomaterials, which enables the use of these materials for the detection or sensing of toxic H_2S gas and other organic compounds [22]. MOFs exhibit improved chemical and thermal stability, porosity, and functionalities compared to other materials. Metal oxides such as Fe_2O_3, CuO, ZnO, WO_3, MoO_3, SnO_2, and $BaTiO_3$, have been widely applied for H_2S gas sensing with ppm and sub-ppm detection limits [23]. We briefly discuss the various metal sources used to develop MOFs for H_2S sensors.

17.3.1 Fe AND Fe_2O_3-BASED MOFs AS H_2S SENSORS

Iron-based MOFs are promising candidates for H_2S gas sensing owing to their abundant availability, low cost, and ease of synthesis and structure-tuning. The combination of iron with the MOF structure facilitates improved gas permeation, detection, and sensing activity even at lower concentrations. Amino-functionalized iron (III)-based MILL-88, such as Fe(III)-MIL-88-NH_2, which was synthesized via a solvothermal method according to the literature reported procedure, exhibited excellent detection and sensing activity for the H_2S gas [13]. Various factors are responsible for improving the sensing behavior of the synthesized material, such as an enhanced rate of reaction between Fe(III) and H_2S in comparison to those of the reactions between the organic ligands and functional groups in the MOFs with $H_2S.Fe_2S_3$, which is readily produced by the reaction of Fe(III) and H_2S, subsequently transforming to FeS and S. FeS easily produces Fe(III) when exposed to air. Under these conditions, MOFs play a significant role in gas storage. The sizes of MOFs decrease with an increase in the specific surface area [13]. The Fe(III)-MIL-88-NH_2 based MOF sensor exhibits fluorescence upon exposure to H_2S, thus demonstrating the outstanding sensitivity of the synthesized material.

Fe_2O_3 delivers higher gas-sensing performance with the technical advantages of non-toxicity, low cost, and easy availability and synthesis, leading to the wide applications of Fe_2O_3 based nanomaterials [23]. However, metal oxides have drawbacks, such as the necessity of higher heating temperatures due to higher power consumption, insufficient detection limits, and poor selectivity. The H_2S sensing behavior of Fe_2O_3 was improved by doping with other metal oxides and noble metals. Tian et al. discussed the synthesis of a hierarchical and hollow Fe_2O_3-based MOF with a high surface area (113 m^2g^{-1}), which exhibited high selectivity, fast response, and recovery, and proved capable of detecting H_2S gas at 200 °C under the ppb sensing limit [23]. The Fe_2O_3-based MOF, which was synthesized from derivatives of Prussian blue (PB)-type MOFs via chemical etching and high-temperature calcination at 400 °C for 1 h, exhibited excellent porosity and hollow and hierarchical surface morphology [23]. The outstanding H_2S gas-sensing behavior of this MOF was due to the formation of negatively charged oxygen species (O_2^-, O^-, O^{2-}) on the surface of Fe_2O_3, which has a high resistance when exposed to air. However, Fe_2O_3 becomes conductive upon exposure to H_2S gas [23]. The porous structure facilitates the faster transport of gases through the pore channels, leading to faster sensing of H_2S, even at low concentrations (ppm or ppb). In addition, these surface structures facilitate easier switching between conductive and non-conductive behavior and faster response based on exposure to H_2S gas or air.

Zhang et al. developed a γ-Fe_2O_3/rGO gas sensor for the detection of H_2S at RT using an electrochemical method [24]. In general, γ-Fe_2O_3 gas sensors exhibit poor selectivity and require high temperatures, which are major drawbacks of these materials. These were resolved by the introduction of rGO to γ-Fe_2O_3 because rGO exhibits exceptional conductivity and has numerous active sites. The sensor developed using γ-Fe_2O_3/rGO displayed excellent H_2S gas-sensing behavior compared to the pristine γ-Fe_2O_3 sensor. The gas detection mechanism of γ-Fe_2O_3 and rGO depends on the bulk resistance and is surface-controlled [24]. The resistance of the γ-Fe_2O_3/rGO sensor immediately changes when H_2S gas is passed over it, and the resistance increases upon exposure to air, which clearly indicates the superior sensitivity of the sensor toward H_2S gas. The faster decrease in the resistance of the Fe_2O_3/rGO sensor is due to the reaction between H_2S and chemisorbed oxygen in the presence of the absorbed electrons, which facilitates improved conductivity [24]. The γ-Fe_2O_3/rGO sensor has various detection limits in ppm based on the concentration of H_2S gas (Figure 17.1a) [24]. Notably, the detection performance was slightly reduced at lower concentrations compared to higher concentrations. The sensing performance of γ-Fe_2O_3/rGO was evaluated for various gases, such as formaldehyde (HCHO), nitric oxide (NO), sulfur dioxide (SO_2), ammonia (NH_3), and chloroform ($CHCl_3$), in addition to H_2S, and the results obtained proved the higher selectivity and sensing activity toward H_2S in comparison to the other gases (Figure 17.1b) [24].

The γ-Fe_2O_3/rGO-20 composite sensor exhibited a stable sensing response toward H_2S gas at 97 ppm relative to the response time (Figure 17.1c). Furthermore, on examining the stability of the γ-Fe_2O_3/rGO-20 composite sensor at RT under various relative humidities, a dependence of the sensing performance on the humidity was revealed. The sensing performance declined with an increase in humidity owing to the adsorption of water on the surface of rGO (Figure 17.1d) [24]. Excellent H_2S gas sensitivity and selectivity were observed for the Fe_2O_3 nanomaterial synthesized from PB [25]. PB was used to modify the Fe_2O_3 nanomaterial by calcination, and the obtained nanomaterial displayed excellent sensitivity and selectivity toward H_2S compared to methane, carbon monoxide, and hydrogen gases [25]. The Fe_2O_3 nanomaterials

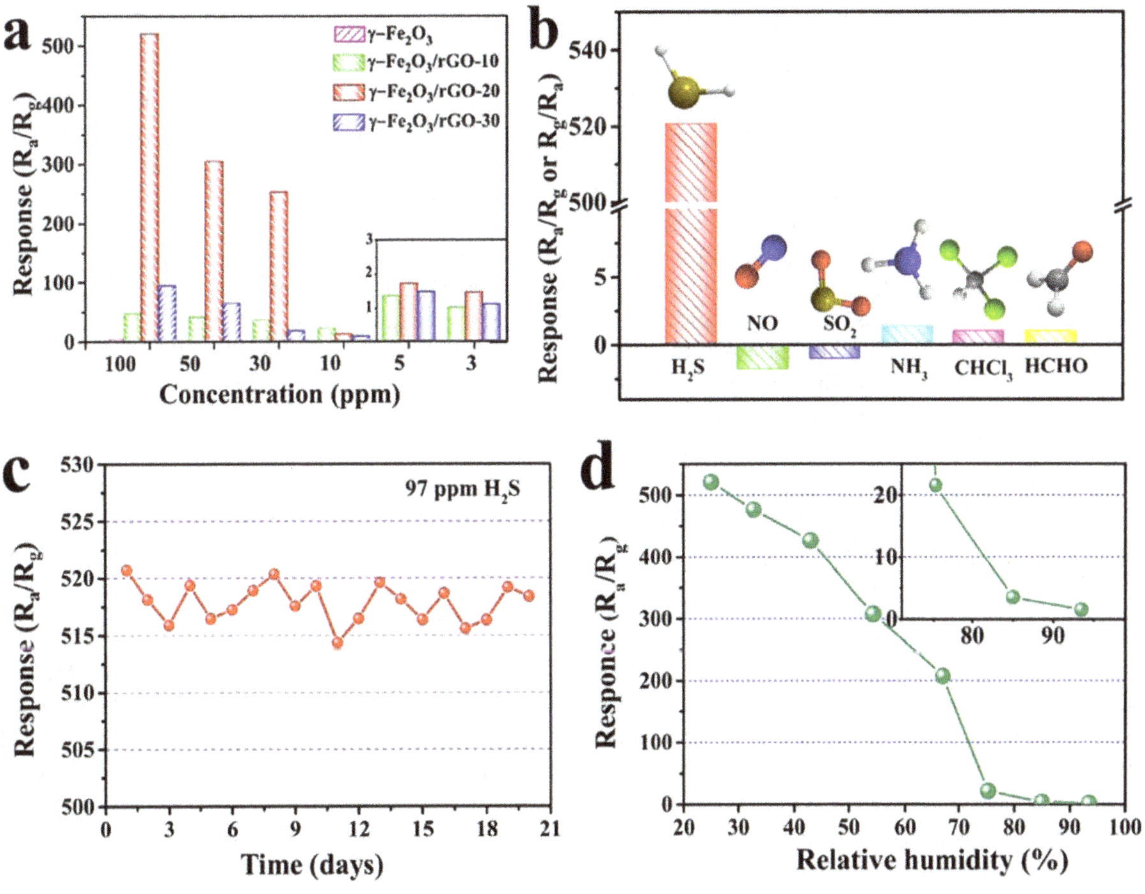

FIGURE 17.1 (a) Relationship between the concentrations of H_2S and gas response, (b) sensing responses of the γ-Fe_2O_3/rGO-20 composite sensor to 50 ppm HCHO and 97 ppm of various gases, including H_2S, NO, SO_2, NH_3, $CHCl_3$, (c) corresponding response graph (broken line) of the gas sensors based on the γ-Fe_2O_3/rGO-20 composites for H_2S (97 ppm) operated at RT in air, and (d) sensing performance of γ-Fe_2O_3/rGO-20 composites at different humidities. Reproduced with permission [24], Copyright 2020, Elsevier.

synthesized by this method existed in different morphologies, such as nanospheres, nano cubes, and nanoflowers. The nanosphere-based Fe_2O_3 nanomaterial delivered excellent H_2S sensing ability with good reproducibility in comparison to the nanotubes and nanoflowers [25]. Zhan et al. synthesized size-controlled porous Fe_2O_3 from MOFs and studied their catalytic behavior with respect to the selective oxidation of H_2S via the adsorption of H_2S on the surface of Fe_2O_3 [26]. Further reaction occurred with Fe^{3+} to produce Fe^{2+} and S. In contrast, Fe^{2+} reverted to Fe^{3+} by the diffusion of oxygen to the pore channels and was adsorbed on the catalytic surface. Here, the porous structure plays a dominant role in enhancing the catalytic activity.

17.3.2 Cu-based MOF for H_2S Sensor

Cu is one of the superior conducting materials used in various electrical, environmental, and biological applications owing to its relatively low price and easy availability. Recently, Cu-based MOFs have been widely used as biosensors for the electrochemical determination of glucose [27]. Zhao et al. synthesized a Cu-based MOF by mixing 5-hydroxyisophthalic acid (HIA) with copper nitrate trihydrate and bipyridine in an ethanol-water mixed medium [28]. The synthesized Cu-MOF exhibited luminescent detection of H_2S in a highly selective manner with a detection limit of 0.21 μM within 30 s [28]. The luminescent sensing mechanism of the synthesized Cu-MOF-based nanomaterial against different concentrations of H_2S was identified, based on the changes in the color of the sample from blue to light blue to pale green to sepia to pitchy [28]. In contrast, the fumarate-modified Cu-MOF fabricated using an interdigitated electrode (IDE) sensor via a soft lithographic technique led to improved H_2S sensing in comparison with the sensor made by the colorimetric method [29]. This IDE-based sensor has advantages such as the relatively low cost of the sensor and ease of miniaturization with excellent surface functionality. The gas-sensing behavior of the fumarate-modified Cu-MOF sensor was studied using different gases such as toluene, H_2S, hydrogen, methane, and nitric oxide in a fully automated gas-sensing instrument. The fumarate-modified Cu-MOF sensor exhibited excellent H_2S sensing ability in the range of 1 to 100 ppm as well as 100 to 1000 ppb with a higher selectivity toward H_2S in comparison with other toxic gases [29]. Furthermore, the fabricated electrode showed a detection limit of 5.4 ppb which was calculated by the root-mean-square division method. These results demonstrate the outstanding H_2S sensing behavior of the fabricated electrode compared with that of the unmodified Cu-MOF.

Ma et al. fabricated a fluorescent sensor using a heterogeneous Cu(II)-metalated non-interpenetrated 3D porous MOF, which showed superior sensitivity and selectivity toward H_2S in living cells [30]. The Cu-MOF was supported by the coordination of the reactive metal centers in the framework, which was responsible for the faster response and selectivity of the MOF toward H_2S in living cells. The fabricated fluorescent sensor follows an on/off switching mechanism controlled by its exposure to H_2S as well as air. Kakaei et al. developed a natural clay-modified Cu-MOF using sepiolite (Sep) as natural clay and terephthalate (BDC) as a ligand to generate a highly porous structure where Cu is coordinated with BDC [31]. The synthesized Sep/Cu-BDC (25 wt.%) MOF exhibits excellent H_2S adsorption with a maximum adsorption capacity of 55.13 ± 0.70 mg g^{-1} and a maximum adsorption time of 46.32 ± 1.53 min. The synthesized Sep/Cu-BDC (25 wt.%) MOF has a teal color, which becomes black upon H_2S adsorption [31]. This color change proves the superior H_2S adsorption capacity of the MOF. Moreover, the adsorption of H_2S in the Sep/Cu-BDC (25 wt.%) MOF indicated the formation of a stronger Cu-S bond due to the reaction between H_2S and CuO in the MOF structure [31].

Li et al. fabricated a bamboo-like mesoporous CuO/In_2O_3 heterostructure derived from a MOF for H_2S sensing by synthesizing indium-based microrods using indium nitrate hydrate and 1,4-benzene dicarboxylic acid (H_2BDC) in dimethyl formamide (DMF) at 140 °C for 10 min [32]. The microrods were decorated using copper (II) acetate monohydrate by the impregnation technique and calcined at 450 °C for 1 h to obtain a porous CuO/In_2O_3 heterostructure (Figure 17.2a) [32]. The p-type CuO semiconductor leads to the formation of a stable heterojunction with n-type In_2O_3. This p-n type CuO/In_2O_3 heterostructure manifested better interactions with H_2S and led to excellent sensitivity and selectivity toward toxic H_2S in comparison to other gases. The H_2S sensing mechanism of the CuO/In_2O_3 heterostructure was studied, based on its active surface area and resistance (Figure 17.2b) [32]. The calcined CuO/In_2O_3 heterostructure has a high surface area, which can accumulate a larger amount of H_2S throughout the pore channels. The CuO/In_2O_3 heterostructure has high resistance (21.2 MΩ) due to aerial exposure compared to In_2O_3 (4.3 MΩ). The resistance of the fabricated device was reduced drastically by exposure to H_2S at 70 °C owing to the formation of a CuS layer on the surface of the CuO/In_2O_3 heterostructure, which facilitated the selective detection of H_2S over other toxic gases. The absorption of electrons from the conduction band followed by their reversion was observed upon exposure to air and H_2S in parity with other metal oxides [32].

17.3.3 Zn-based MOFs for a H_2S Sensor

Zn-based MOFs are a well-known type of MOF used in various applications owing to their unique zeolitic imidazolate framework (ZIF-8) structure and ease of synthesis [15]. ZIF-8 based MOFs are widely synthesized by mixing zinc nitrate hexahydrate ($Zn(NO_3)_3.6H_2O$) or zinc acetate ($Zn(OAC)_2$), and 2-methylimidazole (Hmim) in water or alcohol(s) at RT [33]. Gao et al. discussed the important role of various metal ion-loaded nanoclusters (NCs) in MOFs and studied effective approaches for the determination of various toxic gases using such MOFs [34]. The AuNCs@ZIF-8 MOF delivered a highly sensitive and selective response to H_2S

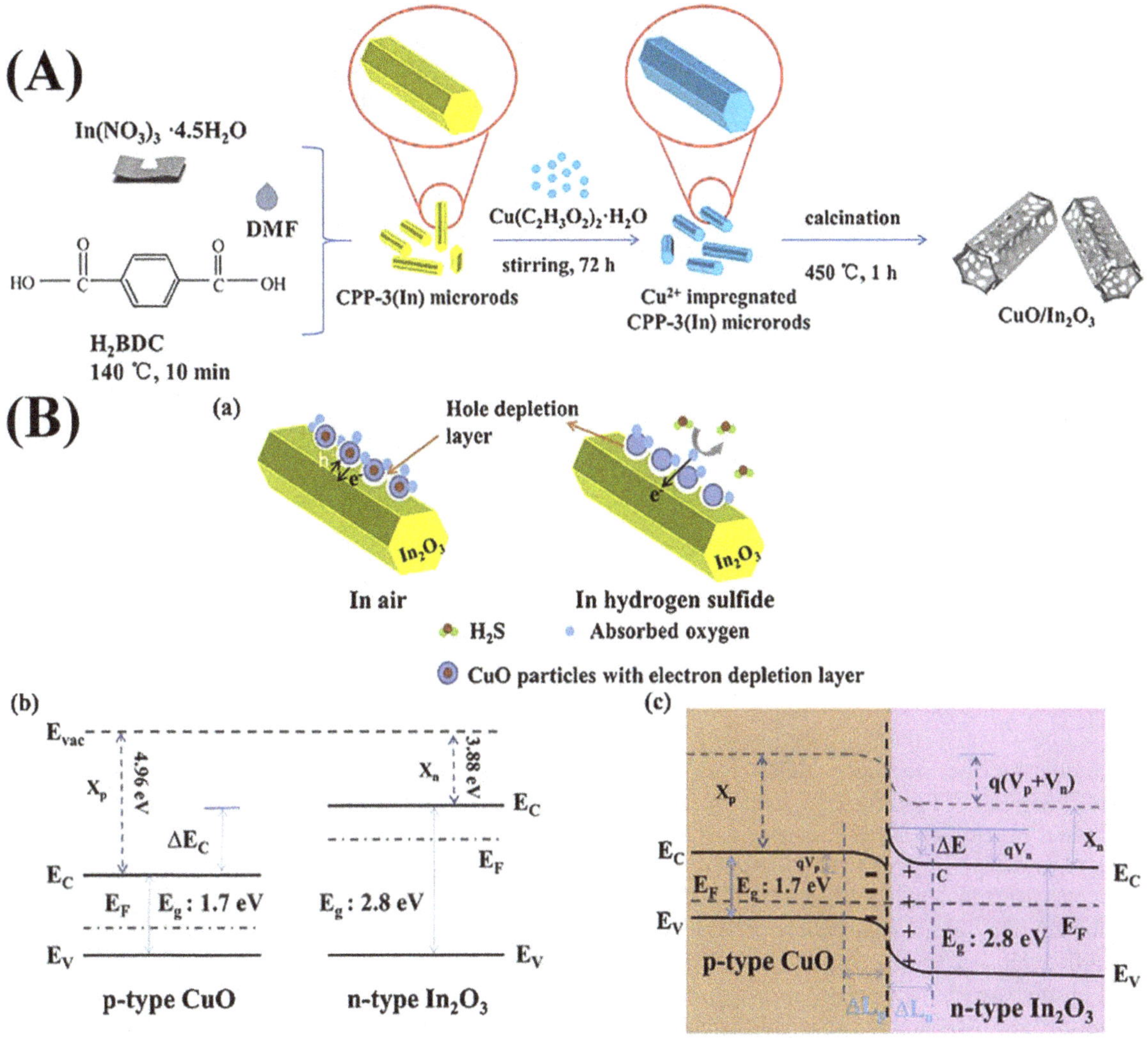

FIGURE 17.2 (A) Schematic illustration of the formation of CuO/In_2O_3. (B) Schematic of carrier transportation and gas-sensing mechanism for (a) CuO/In_2O_3 heterostructure, (b) Energy band diagram of individual CuO and In_2O_3 before contact, (c) Energy level diagram of CuO/In_2O_3 heterostructure. Adapted with permission [32], Copyright 2020, Elsevier.

over various other toxic gases (Figure 17.3) [34]. The AuNCs were well dispersed and uniformly decorated on ZIF-8 and delivered long-lived luminescence properties, high quantum yields, and selective fluorescence responses toward H_2S. Hu et al. synthesized a fluorinated $ZnFe^{III}$ hollow MOF using iron and zinc precursors with a fluorinated H_2BDC ligand. The synthesized MOF was used to fabricate a ^{19}F NMR probe for the selective and sensitive detection of H_2S as well as for use as an on/off sensor [35].

A 3D inverse opal (3DIO) ZnO sensor prepared from the MOF derivative was embedded with Pt nanoparticles [36]. The prepared sensor had an excellent H_2S response with a low detection limit of 25 ppb and exhibited highly stable and superior selectivity toward H_2S. The 3DIO Pt/ZnO sensor has a response value of 11.2 at 1 ppm which is 5.7 times higher than the sensor response in the absence of Pt [36]. MOF-SMO hybrids prepared by introducing SMOs such as ZnO to ZIF-8 also have excellent sensitivity and selectivity toward H_2S (10 ppm) at RT [37]. Moreover, the response is 15 times higher than that of the pristine ZnO nanorod-based sensor. In addition, the ZIF-8/ZnO nanorod sensor can detect H_2S at a concentration as low as 50 ppb.

17.3.4 Zr-Based MOF for a H_2S Sensor

UiO-66 is a zirconium (Zr)-based MOF that is widely used in various applications because of its highly porous structure, stability under acidic conditions as well as thermal and mechanical stress, and ease of synthesis on a large scale [38]. Li et al. designed a Zr-based fluorescent MOF sensor, UiO-66-CH=CH_2 using zirconium chloride ($ZrCl_2$) and 2-vinylterephthalic acid by a simple solvothermal approach for the sensitive and selective determination of H_2S [39]. The UiO-66-CH=CH_2 fluorescent probe exhibited a fast response to H_2S (10 s) with a detection limit of 6.46 μM with high sensitivity and selectivity. The cleavage of the vinyl and conjugated organic

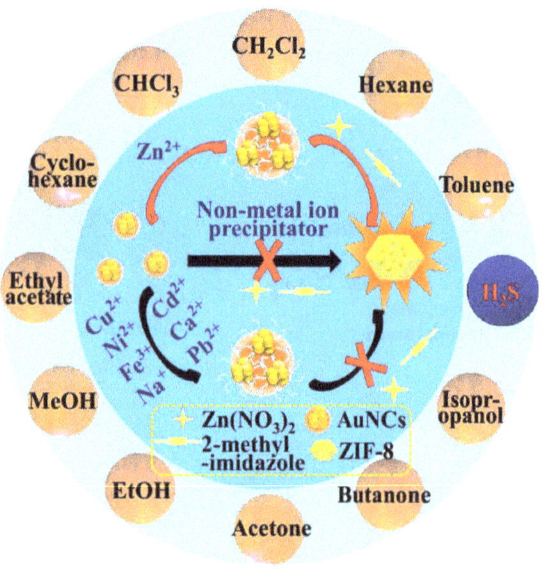

FIGURE 17.3 Illustration for fabrication of AuNCs@ZIF-8 and selective fluorescence sensing of H_2S. Reproduced with permission [34], Copyright 2018, American Chdemical Society.

groups in UiO-66-CH=CH_2 through reaction with H_2S was responsible for the fluorescence response. Guo et al. prepared a Zr-MOF, Zr(TBAPy)$_5$(TCPP) using $ZrCl_4$ as the metal source and 1,3,6,8-tetra (4-carboxylphenyl) pyrene (TBAPy) and tetrakis(4-carboxyphenyl)porphyrin (TCPP) as double linkers [40]. The synthesized material was used as a fluorescence turn-on probe for the selective detection of H_2S and was used to identify unique chemical fingerprints. Furthermore, Nagarkar et al. prepared a fluorescence turn-on probe using a nitro functionalized MOF for selective H_2S detection [41]. Exposure of the nitro functionalized MOF to H_2S gas at pH 7.4 transforms the nitro functional group to an amino group and a fluorescence turn-on response is observed. The probe displayed the lowest H_2S detection limit (188 μm) under physiological conditions. Similarly, the nitro functionalized UiO-66 MOF decorated with Ag_2O demonstrated excellent H_2S gas-sensing properties [42]. In another study, Nagarkar et al. developed an azide-functionalized MOF that exhibited outstanding sensitivity and selectivity toward H_2S along with a fast response through the interaction of the azide moiety with H_2S [43]. The maleic anhydride (MA)-functionalized MOF exhibited dual-sensing behavior toward H_2S and Cu^{2+} with detection limits of 3.3 nM and 2.6 nM for H_2S and Cu^{2+}, respectively [44].

17.3.5 Al-based MOF for a H_2S Sensor

An Al-based MOF, Christian-Albrechts-University (CAU)-10, also known as (CAU)-10-V-H, was first synthesized using a vinyl functionalized isophthalic acid (H_2IPA-V) and 1,3-benzenedicarboxylic acid (H_2IPA) in a 1:1 molar ratio [45]. The CAU-10-V-H MOF exhibited excellent H_2S sensing due to the reaction of the vinyl functional group in the MOF with H_2S. The Al-MOF also facilitated the detection of lead (Pb^{2+}) in aqueous solutions. Das et al. synthesized an azide-functionalized MOF, Al-MIL-53-N_3, and demonstrated its superior sensitivity to Fe(III) as well as H_2S [46]. Zhang et al. synthesized a highly flexible, free-standing, and stable Al-MOF using a polymer mixed-matrix membrane (MMM) with poly(vinylidene fluoride) (PVDF) [47]. The Al-MOF-MMM sensor exhibited outstanding H_2S sensitivity even at a low concentration of 92.31×10^{-9} M which was considerably superior to the H_2S sensitivity of the sensor prepared in powder form. Nitro functionalized Al-MOFs have an excellent capacity for the selective determination of H_2S, similar to nitro functionalized Zr-MOFs. The sensing response is driven by the transformation of the nitro functional group to an amino group upon reaction with H_2S [48].

17.3.6 Other MOFs as H_2S Sensors

Similar to the aforementioned Fe, Cu, Zn, Zr, and Al-based MOFs, other metal-based MOFs have also been synthesized using various approaches and used for the selective and sensitive detection of H_2S. Double-shelled Co_3O_4/$NiCo_2O_4$ MOF nanocages were prepared from ZIF-67 derivatives and their H_2S gas-sensing performance was evaluated using a four-channel material gas sensitivity system (SD101) [49]. The prepared nanocages delivered excellent H_2S sensing at 250 °C and a concentration of 100 ppm owing to their high specific surface areas. At sensing temperatures below 200 °C aerial oxygen is chemisorbed, forming a layer on the sensor surface. In contrast, increasing the sensing temperature above 200 °C activates the adsorbed O^{2-} and tunes the electrical conductivity of the Co_3O_4/$NiCo_2O_4$ MOF sensor, which further delivers selective H_2S sensing [49]. A Europium (Eu)-1,1,1-trifluoro 2,4-pentanedione (TFA)-functionalized MOF and indium (In)-MOF were also used for the selective and sensitive detection of H_2S [50]. The important results of MOF-based sensors for the selective detection of H_2S are shown in Table 17.1.

17.4 CONCLUSIONS

MOFs synthesized using various metal centers and ligands delivered excellent sensitivity and selectivity toward the detection of H_2S. Numerous studies have focused on the synthesis of various types of MOFs because of the diverse advantages of these materials, such as their highly porous and well-controlled architectures, high surface areas, metal centers, and active sites, which facilitate the selective detection of H_2S and various other toxic gases. Several types of gas-sensing methods, such as fluorescence, cataluminescence, optical, electrochemical, absorption, and surface acoustic wave techniques have been applied for H_2S gas-sensing. Recently, an electrochemical method has been adopted for sensing H_2S. In addition, a limited number of studies have been reported on the sensing of H_2S using electrochemical methods. The preparation of MOF-based nanomaterials has been widely focused on for the development of low-temperature H_2S sensors. Therefore, more studies are expected in the future

TABLE 17.1
The MOF-based Sensor for the Selective Detection of H_2S

Material	H_2S (ppm)	T (°C)	Response	t_{resp}/t_{reco} (S)	Ref.
Fe_2O_3 nanoboxes	5	50	2.58	806/1100	[23]
	5	200	2.74	75/132	[23]
	5	250	5.36	31/187	[23]
	0.25	200	1.23	145/134	[23]
Porous α-Fe_2O_3 nanoparticles	0.05	RT	1.08	~250/200	[51]
Porous α-Fe_2O_3 nanospheres	1	350	3.4	-	[52]
α-Fe_2O_3 nanoparticles	0.05	300	1.25	-	[53]
α-Fe_2O_3 microellipsoids	0.5	350	1.45	80/7	[54]
CuO microflowers	1	RT	1.4	-	[55]
3%Cu@ZnO	6.2	RT	4.72	-	[56]
ZIF-8/ZnO	10	RT	-	-	[37]
Ag@ UiO-66(Zr)-NO_2	1	Rt	200	-	[42]
Chitosan/glycerol/ WO_3	15	40	13.6	-	[57]

that will lead to the improved sensing of toxic H_2S and other gases using electrochemical and various other methods.

ACKNOWLEDGMENTS

This research was supported by the Basic Science Research Program through a National Research Foundation of Korea (NRF) grant funded by the Korean Government (MSIP) (NRF-2020R1I1A3067208).

REFERENCES

[1]. Moosavi, S. M.; Nandy, A.; Jablonka, K. M.; Ongari, D.; Janet, J. P.; Boyd, P. G.; Lee, Y.; Smit, B.; Kulik, H. J. Understanding the Diversity of the Metal-Organic Framework Ecosystem. *Nat. Commun.*, **2020**, *11* (1), 1–10.

[2]. Yang, W.; Li, X.; Li, Y.; Zhu, R.; Pang, H. Applications of Metal–Organic-Framework-Derived Carbon Materials. *Adv. Mater.*, **2019**, *31* (6), 1804740.

[3]. Li, H.; Eddaoudi, M.; O'Keeffe, M.; Yaghi, O. M. Design and Synthesis of an Exceptionally Stable and Highly Porous Metal-Organic Framework. *Nature*, **1999**, *402* (6759), 276–279.

[4]. Maya, F.; Palomino Cabello, C.; Frizzarin, R. M.; Estela, J. M.; Turnes Palomino, G.; Cerdà, V. Magnetic Solid-Phase Extraction Using Metal-Organic Frameworks (MOFs) and Their Derived Carbons. *Trends Anal. Chem.*, **2017**, *90*, 142–152.

[5]. Furukawa, H.; Cordova, K. E.; O'Keeffe, M.; Yaghi, O. M. The Chemistry and Applications of Metal-Organic Frameworks. *Science*, **2013**, *341*, 6149.

[6]. Wang, G.-H.; Lei, Y.-Q.; Song, H.-C. Evaluation of Fe_3O_4@ SiO_2–MOF-177 as an Advantageous Adsorbent for Magnetic Solid-Phase Extraction of Phenols in Environmental Water Samples. *Anal. Methods*, **2014**, *6* (19), 7842–7847.

[7]. Dhakshinamoorthy, A.; Garcia, H. Catalysis by Metal Nanoparticles Embedded on Metal–Organic Frameworks. *Chem. Soc. Rev.*, **2012**, *41* (15), 5262–5284.

[8]. Li, B.; Wen, H. M.; Cui, Y.; Zhou, W.; Qian, G.; Chen, B. Emerging Multifunctional Metal–Organic Framework Materials. *Adv. Mater.*, **2016**, *28* (40), 8819–8860.

[9]. Ramachandran, R.; Chen, T.-W.; Chen, S.-M.; Baskar, T.; Kannan, R.; Elumalai, P.; Raja, P.; Jeyapragasam, T.; Dinakaran, K.; Kumar, G. P. G. A Review of the Advanced Developments of Electrochemical Sensors for the Detection of Toxic and Bioactive Molecules. *Inorg. Chem. Front.*, **2019**, *6* (12), 3418–3439.

[10]. Hanafi, R.; Mayasari, R. D.; Masmui; Agustanhakri; Raharjo, J.; Nuryadi, R. Electrochemical Sensor for Environmental Monitoring System: A Review. *AIP Conf. Proc.*, **2019**, *2169* (1), 30011.

[11]. Abdalla, A.; Patel, B. A. 3D Printed Electrochemical Sensors. *Annu. Rev. Anal. Chem.*, **2021**, *14* (1), 19–20.

[12]. Olorunyomi, J. F.; Geh, S. T.; Caruso, R. A.; Doherty, C. M. Metal–Organic Frameworks for Chemical Sensing Devices. *Mater. Horizons*, **2021**, *8* (9), 2387–2419.

[13]. Cao, Y. Y.; Guo, X. F.; Wang, H. Highly Sensitive Luminescence Metal-Organic Framework Sensor for Hydrogen Sulfide in Aqueous Solution: A Trial of Novel Turn-on Mechanism. *Sensors Actuators B Chem.*, **2017**, *243*, 8–13.

[14]. Cao, X.; Wu, Z.; Xiong, S.; Cao, L.; Sethi, G.; Bian, J. song. The Role of Hydrogen Sulfide in Cyclic Nucleotide Signaling. *Biochem. Pharmacol.*, **2018**, *149*, 20–28.

[15]. Wan, X.; Wu, L.; Zhang, L.; Song, H.; Lv, Y. Novel Metal-Organic Frameworks-Based Hydrogen Sulfide Cataluminescence Sensors. *Sensors Actuators, B Chem.*, **2015**, *220*, 614–621.

[16]. Wang, C.; Liu, M.; Li, Y.; Zhang, Y.; Yao, M.; Qin, Y.; Liu, Y. Hydrogen Sulfide Synthesis in Native Saccharomyces Cerevisiae Strains during Alcoholic Fermentations. *Food Microbiol.*, **2018**, *70*, 206–213.

[17]. Kroll, A. V.; Smorchkov, V.; Nazarenko, A. Y. Electrochemical Sensors for Hydrogen and Hydrogen Sulfide Determination. *Sensors Actuators B Chem.*, **1994**, *21* (2), 97–100.

[18]. Llobet, E.; Brunet, J.; Pauly, A.; Ndiaye, A.; Varenne, C. Nanomaterials for the Selective Detection of Hydrogen Sulfide in Air. *Sensors*, **2017**, *17* (2), 391.

[19]. Vikrant, K.; Kumar, V.; Ok, Y. S.; Kim, K. H.; Deep, A. Metal-Organic Framework (MOF)-Based Advanced Sensing Platforms for the Detection of Hydrogen Sulfide. *Trends Anal. Chem.*, **2018**, *105*, 263–281.

[20]. Fang, X.; Zong, B.; Mao, S. Metal–Organic Framework-Based Sensors for Environmental Contaminant Sensing. *Nano-Micro Lett.*, **2018**, *10* (4), 64.

[21]. Chen, X.; Wang, Y.; Zhang, Y.; Chen, Z.; Liu, Y.; Li, Z.; Li, J. Sensitive Electrochemical Aptamer Biosensor for Dynamic Cell Surface N-Glycan Evaluation Featuring Multivalent Recognition and Signal Amplification on a Dendrimer–Graphene Electrode Interface. *Anal. Chem.*, **2014**, *86* (9), 4278–4286.

[22]. Yin, Y.; Zhang, N.; Han, J.; Liu, C.; Adimi, S.; Wen, S.; Li, X.; Ruan, S. Metal-Organic Framework Derived Core-Shell $PrFeO_3$-Functionalized α-Fe_2O_3 Nano-Octahedrons as High

Performance Ethyl Acetate Sensors. *Sensors Actuators B Chem.*, **2019**, *297*, 126738.

[23]. Tian, K.; Wang, X.-X.; Yu, Z.-Y.; Li, H.-Y.; Guo, X. Hierarchical and Hollow Fe_2O_3 Nanoboxes Derived from Metal–Organic Frameworks with Excellent Sensitivity to H_2S. *ACS Appl. Mater. Interfaces*, **2017**, *9* (35), 29669–29676.

[24]. Zhang, C.; Zhang, S.; Yang, Y.; Yu, H.; Dong, X. Highly Sensitive H_2S Sensors Based on Metal-Organic Framework Driven γ-Fe_2O_3 on Reduced Graphene Oxide Composites at Room Temperature. *Sensors Actuators B Chem.*, **2020**, *325*, 128804.

[25]. Guo, Y.; Tian, X.; Wang, X.; Sun, J. Fe_2O_3 Nanomaterials Derived from Prussian Blue with Excellent H_2S Sensing Properties. *Sensors Actuators B Chem.*, **2019**, *293*, 136–143.

[26]. Zhan, Y.; Shen, L.; Xu, C.; Zhao, W.; Cao, Y.; Jiang, L. MOF-Derived Porous Fe_2O_3 with Controllable Shapes and Improved Catalytic Activities in H_2S Selective Oxidation. *CrystEngComm*, **2018**, *20* (25), 3449–3454.

[27]. He, J.; Yang, H.; Zhang, Y.; Yu, J.; Miao, L.; Song, Y.; Wang, L. Smart Nanocomposites of Cu-Hemin Metal-Organic Frameworks for Electrochemical Glucose Biosensing. *Sci. Rep.*, **2016**, *6* (1), 36637.

[28]. Zhao, X.; Zhang, L.; Bai, J.; Wu, P.; Li, Y.; Liang, L.; Xie, L.; Wang, J. A Copper-Based Metal-Organic Framework for Ratiometric Detection of Hydrogen Sulfide with High Sensitivity and Fast Response. *Spectrochim. Acta - Part A Mol. Biomol. Spectrosc.*, **2020**, *243*, 118794.

[29]. Yassine, O.; Shekhah, O.; Assen, A. H.; Belmabkhout, Y.; Salama, K. N.; Eddaoudi, M. H_2S Sensors: Fumarate-Based Fcu-MOF Thin Film Grown on a Capacitive Interdigitated Electrode. *Angew. Chemie Int. Ed.*, **2016**, *55* (51), 15879–15883.

[30]. Ma, Y.; Su, H.; Kuang, X.; Li, X.; Zhang, T.; Tang, B. Heterogeneous Nano Metal-Organic Framework Fluorescence Probe for Highly Selective and Sensitive Detection of Hydrogen Sulfide in Living Cells. *Anal. Chem.*, **2014**, *86* (22), 11459–11463.

[31]. Kakaei, H.; Beygzadeh, M.; Golbabaei, F.; Ganjali, M. R.; Jahangiri, M.; Shahtaheri, S. J. Preparation of a Sepiolite/Cu-BDC Nanocomposite and Its Application as an Adsorbent in Respirator Cartridges for H_2S Removal. *New J. Chem.*, **2019**, *43* (29), 11575–11584.

[32]. Li, S.; Xie, L.; He, M.; Hu, X.; Luo, G.; Chen, C.; Zhu, Z. Metal-Organic Frameworks-Derived Bamboo-like CuO/In_2O_3 Heterostructure for High-Performance H_2S Gas Sensor with Low Operating Temperature. *Sensors Actuators B Chem.*, **2020**, *310*, 127828.

[33]. Zeng, X.; Chen, B.; Zhou, S.-F.; Zhan, G. Synthesis of ZIF-8 Nanocrystals Mediated by CO_2 Gas Bubbling: Dissolution and Recrystallization. *Langmuir*, **2020**, *36* (47), 14306–14317.

[34]. Gao, Q.; Xu, S.; Guo, C.; Chen, Y.; Wang, L. Embedding Nanocluster in MOF via Crystalline Ion-Triggered Growth Strategy for Improved Emission and Selective Sensing. *ACS Appl. Mater. Interfaces*, **2018**, *10* (18), 16059–16065.

[35]. Hu, G.; Li, Y.; Li, L.; Xu, S.; Wang, L. Fluorinated ZnFeIII Hollow Metal-Organic Framework as a 19F NMR Probe for Highly Sensitive and Selective Detection of Hydrogen Sulfide. *ACS Omega*, **2020**, *5* (14), 8373–8379.

[36]. Zhou, X.; Lin, X.; Yang, S.; Zhu, S.; Chen, X.; Dong, B.; Bai, X.; Wen, X.; Geyu, L.; Song, H. Highly Dispersed Metal–Organic-Framework-Derived Pt Nanoparticles on Three-Dimensional Macroporous ZnO for Trace-Level H_2S Sensing. *Sensors Actuators, B Chem.*, **2020**, *309*, 127802.

[37]. Wu, X.; Xiong, S.; Gong, Y.; Gong, Y.; Wu, W.; Mao, Z.; Liu, Q.; Hu, S.; Long, X. MOF-SMO Hybrids as a H_2S Sensor with Superior Sensitivity and Selectivity. *Sensors Actuators, B Chem.*, **2019**, *292*, 32–39.

[38]. Winarta, J.; Shan, B.; McIntyre, S. M.; Ye, L.; Wang, C.; Liu, J.; Mu, B. A Decade of UiO-66 Research: A Historic Review of Dynamic Structure, Synthesis Mechanisms, and Characterization Techniques of an Archetypal Metal-Organic Framework. *Cryst. Growth Des.*, **2020**, *20* (2), 1347–1362.

[39]. Li, Y.; Zhang, X.; Zhang, L.; Jiang, K.; Cui, Y.; Yang, Y.; Qian, G. A Nanoscale Zr-Based Fluorescent Metal-Organic Framework for Selective and Sensitive Detection of Hydrogen Sulfide. *J. Solid State Chem.*, **2017**, *255*, 97–101.

[40]. Guo, L.; Wang, M.; Cao, D. A Novel Zr-MOF as Fluorescence Turn-On Probe for Real-Time Detecting H_2S Gas and Fingerprint Identification. *Small*, **2018**, *14* (17), 1703822

[41]. Nagarkar, S. S.; Desai, A. V.; Ghosh, S. K. A Nitro-Functionalized Metal-Organic Framework as a Reaction-Based Fluorescence Turn-On Probe for Rapid and Selective H_2S Detection. *Chem. - A Eur. J.*, **2015**, *21* (28), 9994–9997.

[42]. Surya, S. G.; Bhanoth, S.; Majhi, S. M.; More, Y. D.; Teja, V. M.; Chappanda, K. N. A Silver Nanoparticle-Anchored UiO-66(Zr) Metal-Organic Framework (MOF)-Based Capacitive H_2S Gas Sensor. *CrystEngComm*, **2019**, *21* (47), 7303–7312.

[43]. Nagarkar, S. S.; Saha, T.; Desai, A. V.; Talukdar, P.; Ghosh, S. K. Metal-Organic Framework Based Highly Selective Fluorescence Turn-on Probe for Hydrogen Sulphide. *Sci. Rep.*, **2015**, *4* (1), 7053.

[44]. Yang, X. L.; Ding, C.; Guan, R. F.; Zhang, W. H.; Feng, Y.; Xie, M. H. Selective Dual Detection of H_2S and Cu^{2+} by a Post-Modified MOF Sensor Following a Tandem Process. *J. Hazard. Mater.*, **2021**, *403*, 123698.

[45]. Nandi, S.; Reinsch, H.; Biswas, S. A Vinyl Functionalized Mixed Linker CAU-10 Metal-Organic Framework Acting as a Fluorescent Sensor for the Selective Detection of H_2S and Palladium(II). *Microporous Mesoporous Mater.*, **2020**, *293*, 109790.

[46]. Das, A.; Banesh, S.; Trivedi, V.; Biswas, S. Extraordinary Sensitivity for H_2S and Fe(III) Sensing in Aqueous Medium by Al-MIL-53-N_3 Metal-Organic Framework: In Vitro and in Vivo Applications of H_2S Sensing. *Dalt. Trans.*, **2018**, *47* (8), 2690–2700.

[47]. Zhang, X.; Zhang, Q.; Yue, D.; Zhang, J.; Wang, J.; Li, B.; Yang, Y.; Cui, Y.; Qian, G. Flexible Metal–Organic Framework-Based Mixed-Matrix Membranes: A New Platform for H_2S Sensors. *Small*, **2018**, *14* (37), 1801563.

[48]. Zhu, Z.; He, X.; Wang, W. N. Unraveling the Origin of the "Turn-On" Effect of Al-MIL-53-NO_2 during H_2S Detection. *CrystEngComm*, **2019**, *22* (2), 195–204.

[49]. Tan, J.; Hussain, S.; Ge, C.; Wang, M.; Shah, S.; Liu, G.; Qiao, G. ZIF-67 MOF-Derived Unique Double-Shelled Co_3O_4/$NiCo_2O_4$ Nanocages for Superior Gas-Sensing Performances. *Sensors Actuators, B Chem.*, **2020**, *303*, 127251.

[50]. Wang, Z.; Wang, X.; Li, J.; Li, W.; Li, G. Eu^{3+}/TFA Functionalized MOF as Luminescent Enhancement Platform: A Ratiometric Luminescent Sensor for Hydrogen Sulfide in Aqueous Solution. *J. Inorg. Organomet. Polym. Mater.*, **2019**, *29* (6), 2124–2132.

[51]. Huang, Y.; Chen, W.; Zhang, S.; Kuang, Z.; Ao, D.; Alkurd, N. R.; Zhou, W.; Liu, W.; Shen, W.; Li, Z. A High Performance Hydrogen Sulfide Gas Sensor Based on Porous α-Fe_2O_3 Operates at Room-Temperature. *Appl. Surf. Sci.*, **2015**, *351*, 1025–1033.

[52]. Deng, J.; Ma, J.; Mei, L.; Tang, Y.; Chen, Y.; Lv, T.; Xu, Z.; Wang, T. Porous α-Fe_2O_3 Nanosphere-Based H_2S Sensor with Fast Response, High Selectivity and Enhanced Sensitivity. *J. Mater. Chem. A*, **2013**, *1* (40), 12400–12403.
[53]. Li, Z.; Huang, Y.; Zhang, S.; Chen, W.; Kuang, Z.; Ao, D.; Liu, W.; Fu, Y. A Fast Response & Recovery H_2S Gas Sensor Based on α-Fe_2O_3 Nanoparticles with Ppb Level Detection Limit. *J. Hazard. Mater.*, **2015**, *300*, 167–174.
[54]. Li, Z.; Lin, Z.; Wang, N.; Huang, Y.; Wang, J.; Liu, W.; Fu, Y.; Wang, Z. Facile Synthesis of α-Fe_2O_3 Micro-Ellipsoids by Surfactant-Free Hydrothermal Method for Sub-Ppm Level H_2S Detection. *Mater. Des.*, **2016**, *110*, 532–539.
[55]. Wu, K.; Zhang, C. Facile Synthesis and Ppb-Level H_2S Sensing Performance of Hierarchical CuO Microflowers Assembled with Nano-Spindles. *J. Mater. Sci. Mater. Electron.* **2020**, *31* (10), 7937–7945..
[56]. Alev, O.; Sarıca, N.; Özdemir, O.; Arslan, L. Ç.; Büyükköse, S.; Öztürk, Z. Z. Cu-Doped ZnO Nanorods Based QCM Sensor for Hazardous Gases. *J. Alloys Compd.*, **2020**, *826*, 154177.
[57]. Ali, F. I. M.; Awwad, F.; Greish, Y. E.; Abu-Hani, A. F. S.; Mahmoud, S. T. Fabrication of Low Temperature and Fast Response H_2S Gas Sensor Based on Organic-Metal Oxide Hybrid Nanocomposite Membrane. *Org. Electron.*, **2020**, *76*, 105486.

18 MOF-based Sensors for Volatile Organic Compounds

Shuvendu Tripathy[1] *and Santimoy Khilari*[2]
[1]Department of Physics, Vikash Group of Institutions, Bargarh Odisha, India
[2]Department of Chemistry, Guru GhasidasVishwavidyalaya, Bilaspur, Chhattishgarh, India

18.1 INTRODUCTION

The existence and survival of human civilization strongly depends on clean air with good oxygen levels and the lowest life-threatening gas levels. Levels of air contamination are a significant concern and its monitoring has become a priority area for sustainable public health. Volatile organic compounds (VOCs) are considered to be a major source of air pollution as these compounds are emitted into the atmosphere very easily. Most of the VOCs are organic molecules/compounds with a very low boiling point, such as aldehyde, alcohol, benzene, acetone, and toluene. The release of such VOCs into the air increases the contamination level. These compounds not only contaminate the environment but also directly affect public health. For example, benzene is a widely used solvent that causes serious health problems such as headaches, drowsiness, confusion, unconsciousness, rapid heart rate, and tremors, even at low concentrations [1]. Moreover, benzene is considered to be a potential candidate for causing severe health problems such as cancer. In contrast to benzene, acetone is accepted as slightly less harmful but its danger lies in its intense flammability. Similarly, other VOCs have their adverse effect on the environment (Table 18.1). Thus, the monitoring of such VOC's levels in the atmosphere becomes very important to make the environment sustainable to the existence of life. To tackle this issue, researchers engage themselves in developing innovative and convenient techniques for the detection and monitoring of hazardous air contaminants. The development of such a monitoring system primarily relies upon the sensing capacity of the sensor. Further, the sensing/detection of such VOCs is achieved solely by the active sensing element of the designed sensor. The active sensing elements of the VOC sensors are chemical compounds that can easily interact with the analyte, namely, the VOC and deliver a characteristic signal in terms of its physiochemical change. Thus, the scientific community made a significant effort to design and develop efficient sensing materials for VOC detection in the last few decades. In past years, several materials have been explored for VOC sensing and monitoring which include metal oxides, metals, polymers, etc. In the course of sensor development, metal oxide-based gas sensors are considered to be one of the key technologies for the rapid detection of different toxic gases, volatile organic compounds, and chemical vapors. However, lack of selectivity is the major limitation for metal oxide-based gas sensors' practical applications. Furthermore, most of the oxide-based gas sensors are operated at very high temperatures which limits their practical implication.

Functionalized organic conducting polymers are introduced as one of the alternative sensing materials to metal oxides which are used to detect toxic gases. However, the strong affinity of conducting polymers for VOC analytes causes permanent changes in the polymer structure, resulting in poor sensitivity. In the past few years, a considerable amount of effort has been made to design a new class of composite sensing materials with improved sensitivity and selectivity towards analytes. It is well known that the sensitivity of sensors strongly depends on their structural properties and porosity of the active material. Therefore, the high surface area and tunable porosity remains an important parameter for the design of new sensor materials. In particular, the metal-organic frameworks (MOFs) are considered to be emerging materials for sensing applications because of their high specific surface area, unique morphology, and a high degree of structural and chemical tunability. Additionally, the presence of ample metal sites and functional linkers makes MOFs selective adsorbers of analyte molecules which helps to achieve a high sensing capacity towards desired analytes.

An effective sensor must satisfy several essential parameters for practical applicability. First, it should have a very rapid response/recovery time. Secondly, the sensor must be capable of delivering a broad sensing range, and high repeatability. In addition, the sensor materials must be low-cost, abundant, and easily be fabricated on large scale. Due to unique properties and chemical stability, MOFs satisfy essential requirements for an effective active sensor material. Taking into account different working standards, various MOFs have been developed for possible use as chemiresistive sensors, capacitive sensors, colorimetric sensors, and luminescent sensors. Among the sensors, luminescent and chemiresistive MOFs are extensively used for detection for a wide range of VOC analytes in the vapor phase. Moreover, the MOF-based illumination sensors are most effective in solution. The main purpose of this chapter is to discuss the detection of VOCs in vapor as well as solution phases using different types of MOF-based sensors.

DOI: 10.1201/9781003188148-18

TABLE 18.1
Exposure to various concentrations of VOCs and their harmful effects

VOCs	Threshold limit value	Health issues	References
Benzene	0.1 to 1 ppm	Leukemia	1
Formaldehyde	20 ppm	Eye irritation to acute myeloid leukemia	2
Methanol	4000–13,000 mg/l	Migraines, vomiting, and blindness.	3
Toluene	20 ppm	Visual impairment; female reproductive; pregnancy loss	2
Chloroform	10 ppm	Liver and embryo/fetal damage; CNS impairment	2
Ethyl benzene	20 ppm	kidney damage cochlear damage, Cancer	2,4

18.2 TYPES OF SENSORS USED FOR THE DETECTION OF VOCs

Chemical sensors are widely used to monitor air quality in real-time. Several traditional techniques such as gas chromatography and solid-phase ion extraction have been used to detect toxic gases. Unfortunately, these traditional techniques are complex, time-consuming, and expensive. On the other hand, the chemical sensors are low-cost, flexible, and portable and provide real-time data. A chemical sensor is a measurement device that accounts for the chemical or physical characteristics of a particular analyte-active element interaction. Changes in such physical or chemical characteristics are interpreted with an optical or electrical sign by utilization of a suitable transduction unit. The magnitude of the signal depends upon the concentration of the analytes and the efficacy of the active sensing element, namely, the sensing material. MOF-based VOC sensors may be divided into three types: electrical, mass, and optical signal-based sensors. In the next section, we will discuss some of the MOF-based sensors frequently used for detection of VOCs which work on different sensing platforms/mechanisms.

18.2.1 Chemiresitance Sensor

Electrical signals are widely used to sense different kinds of analytes in chemical sensors. In this context, the electrical conductivity or resistance is considered a potential parameter for analyte monitoring. The basic operation of chemiresistive MOF sensors makes use of a change within the resistance of the active sensing material as a result of adsorption and interaction with the tested gaseous analyte [5]. A simple chemiresistive sensing arrangement is depicted in Figure 18.1. The active sensing material is placed on a suitable substrate (thin film, self-standing materials, and the like.). Further, two electrodes are connected to the sensing material for resistance measurement. The chemiresistive sensing capacity of a sensing material changes with temperature. Thus, programmed heating facilities have to be attached to a substrate for suitable heating purposes. Finally, the data of resistance change is recorded on a digital device. Previously, porous metal oxides were explored as chemiresistive sensors for the detection of VOCs. However, low surface area and high operating temperature remain as a major bottleneck for real-time application. Compared to metal oxide-based porous materials, MOFs have massive surface areas for interacting with gaseous analytes. However, the majority of the MOFs have poor electrical conductivity due to their inherent porosity. Therefore, the poor electrical conductivity of MOFs becomes a major restriction for their electronic device application. Previous studies indicate that insufficient mobility in charge carriers leads to insulating properties in MOFs. The scientific community has devoted its efforts to developing conduction/semiconducting MOFs for suitable electronic applications. The electrical conductivity in MOFs has been improved by various methods including the inclusion of guest molecules/conducting media and post-synthetic modification. Electrically conducting MOFs have many valuable uses in different applications like light-harvesting, fuel cells, thermoelectric, sensor, and energy storage, and so on. In the next section, we will discuss the MOF-based chemiresistive sensor for VOC detection and monitoring.

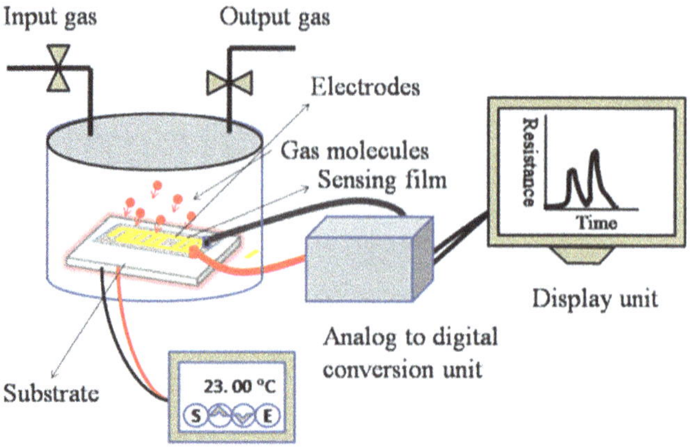

FIGURE 18.1 Schematic diagram of chemiresitance based gas/vapor sensor set up, it determines the change in electronic/ionic resistance of sensing material upon interaction with guest molecules.

The efficacy of a sensing material/element is estimated from its sensor response characteristics to a definite analyte. The sensor response of a chemiresistive MOF-based sensor is attributed to the absorption or interaction of analyte molecules with active sensor materials. This interaction causes an electron or hole transfer between host (sensing material/element) and analyte molecules, leading to a change in electrical or ionic conductivity in the active sensing materials. The sensor response (R) is defined by the following equation:

$$R = \frac{R_g}{R_a} \tag{1}$$

Where

R_a = sensor resistance in the absence of analyte

R_g = sensor resistance in the presence of an analyte.

A high 'R' value signifies a better sensing ability of a sensor and vice versa. Owing to the low cost, and simple fabrication method, chemiresistive MOF-based sensors have been widely studied for the detection of different gas/vapor molecules.

18.2.2 Electrochemical Sensor

Electrochemical techniques are widely used to design and develop various kinds of sensors such as glucose sensors, H_2O_2 sensors, heavy metal detection sensors, and the like. Electrode material, in other words, sensing material, plays a vital role in the efficacy of the electrochemical sensor. The electrochemical sensor works on the principle of electrochemical interaction between electrode material (sensing element) and analyte molecules [6]. Most of the electrochemical interactions are either Faradic or non-Faradic in nature. Faradic interactions involve a redox change of analyte (oxidation/reduction) at a certain potential at the sensing material electrode. Non-Faradic change may appear due to adsorption or other weak interaction between sensing element and analyte. An electrochemical MOF sensor has been established on the principle of the redox response of the analytes in an electrochemical system. The presence of metal ions in MOFs makes them very efficient for redox interactions with analytes. Further, the high surface area, and the tunable porosity of MOFs remain as additional advantages for high-performance electrochemical sensor development. The electrochemical estimation is typically completed utilizing a three-electrode system comprising a working electrode, a counter electrode, and a reference electrode (Figure 18.2). Here, the concentration of the analyte can be estimated by measuring physical quantities such as current, potential, and impedance. These methods are known as amperometry (current response measurement), potentiometry (potential/voltage response measurement), and impedimetric (impedance spectroscopy measurement).

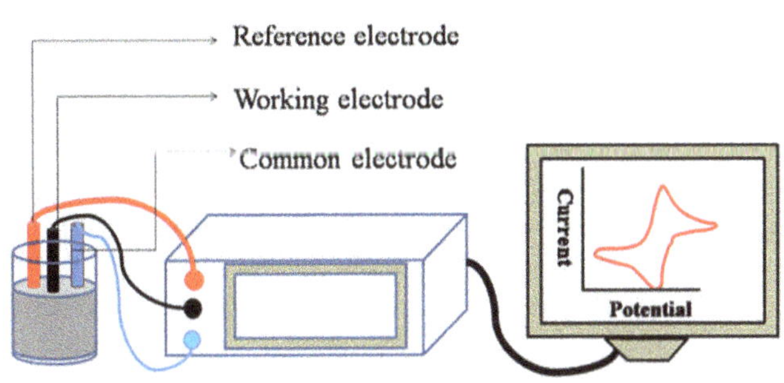

FIGURE 18.2 Schematic of the electrochemical sensor. Electrochemical sensor applies a potential between the working and reference terminals that leads to oxidation or reduction of the analyte, creates an electrical current.

18.2.3 Impedimetric Sensors

Impedance spectroscopy analysis is a useful technique to study the electrical properties of insulator material like MOFs. This electrical property can be used to understand and analyze MOF-analyte interaction which further offers a way to estimate the analyte's concentration. Thus, the impedance analysis tool is widely used to detect toxic gases. The functioning standard of impedimetric sensing depends on the estimation of the electrical impedance of the sensing element as a function of variation in the frequency of the applied current [7]. MOF-based sensors exhibited remarkable changes in the impedance response upon contact with VOCs such as methanol, and ethanol (Figure 18.3). The data obtained from the impedance response is graphically represented by a complex plot (real and imaginary), known as the Nyquist plot. The impedance parameter Z is a complex quantity containing two parts, namely, real part Z' and imaginary part Z''. Furthermore, the argument of Z [arg(z)] gives the phase difference between the current and voltage. MOFs may prove to be good candidates for impedance-based detecting because of their guest-dependent conductivity. There are clear benefits of the impedance technique as far as simplicity of fabrication in contrast with other detecting modules. MOF-based impedance sensors may prove to be good candidates for detecting VOCs due to their guest-dependent conductivity.

18.2.4 Capacitive Sensors

Most of the VOC sensors make use of the electrical conductivity of active material for the detection of gases. Unfortunately, because the majority of MOFs are insulators, their utility as active materials in sensing applications is limited. Low electrical conductivity of MOFs makes them

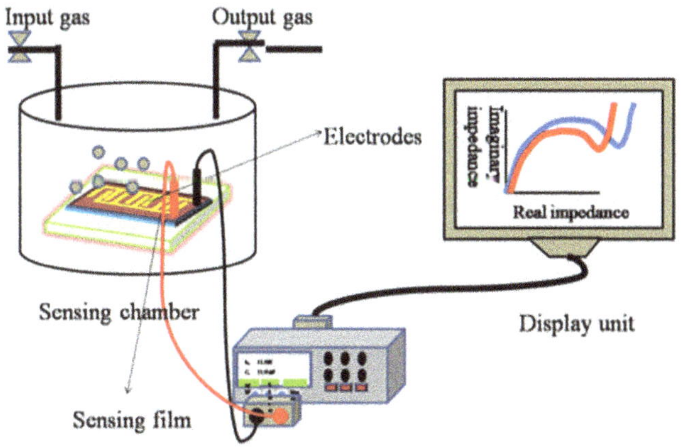

FIGURE 18.3 Schematic illustration of the impedimetric sensor which measures the change in impedance of sensing material upon interaction with guest molecules.

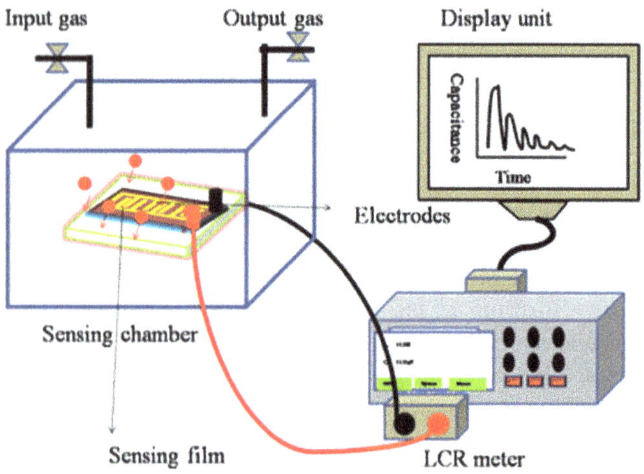

FIGURE 18.4 Schematic illustration of the capacitive sensor which measures the change in capacitance of sensing material upon interaction with guest molecules.

insufficient for electrochemical or electrical sensors for VOC detection where electrical conductivity plays a key role. However, the insulating property can be important in creating dielectric layers for capacitive sensors. The operational principle of such a capacitive device relies on the extent of the analyte molecules' absorption on the MOF [8]. As a result of the interaction between MOF and analyte, a change in capacitances is observed that is a function of the concentration of analyte molecules (Figure 18.4). Typically, the capacitive sensors are of two types: parallel plate capacitor type and interdigitated electrode configuration (IDE). These configurations contrast with each another in the positioning of the MOF-sensor sensing element and in the geometry of the sensor. The MOF-based capacitive sensor technology is similar to the metal oxide-semiconductor-based sensor. The important advantages of capacitive sensors are low fabrication cost, active sensor material can be deposited on a variety of substrates and it only requires a small amount of active material to construct a sensor. The sensitivity of the capacitor sensor can be defined by equation 2

$$S(\%) = \frac{C - C_0}{C_0 \times C_g} \tag{2}$$

Where C_0 in the sensor capacitance in the absence of analyte , C is the sensor capacitance in the presence of analyte, C_g is the analyte concentration and S is the sensitivity.

18.2.5 Luminosity-Based Optical Sensors

Luminescent metal-organic frameworks (LMOFs) are a unique class of MOFs with several potential applications in the field of catalysis, sensing, and optical communications. The luminescence originates in LMOFs primarily from two sources, namely, from organic ligands and inorganic metal-centered. Primarily, three types of luminescence occur in MOFs, namely, chemiluminescence, photoluminescence, and cathodoluminescence. Nevertheless, MOFs with photoluminescent properties have been successfully employed as VOC sensing materials in the last few years. Photoluminescence happens when electrons in energized singlet states revert back to the ground state through photon emission in IR or the visible region. The interaction between analyte with the framework results in changes in the luminescence of the MOF and can be used for detecting analyte species. MOF-based sensors work by using 'turn-on' and 'turn-off' responses [9]. Once the analyte molecules interact with the MOFs, the fluorescence intensity signal is either enhanced or quenched. 'Turn on' signifies the fluorescence intensity signal is enhanced. On the other hand, 'Turn off' means the fluorescence intensity signal is quenched. The better way to evaluate the sensor signal is by using the ''Turn on' method. The overall approach for introducing or enhancing luminescence in intrinsically non-luminescent MOFs is by the inclusion of various types of luminous molecules or ions, organic dyes, and quantum dots. Ln^{3+} ions are a suitable candidate for this purpose because Ln^{3+} ions emit strongly in the visible/ infra-red region. Especially, Eu^{3+} and Tb^{3+} ions are broadly employed for the development of LMOFs because of their strong luminescence in the red and green portions of the visible spectrum. Another straightforward route is to identify VOCs using MOF-based colorimetric sensors. Here, upon the interaction of the analyte with MOFs, a color change will be detected which may be seen with the naked eye.

18.2.6 Quartz Crystal Microbalance-Based Sensors

The quartz crystal microbalance (QCM)-based detector relies upon inheriting piezoelectric effect of quartz crystal. The active MOF film has been deposited on the QCM surface for the chemical sensing of toxic gases. A variation of mass has been observed once the analyte gas interacts with the MOF film. The change in mass (Δm)is a function of the change in oscillatory frequency (Δf)of a quartz crystal which is known as Sauerbrey equation 3 [10, 11].

$$\Delta f = -\left(\frac{2f_0^2}{A\sqrt{\mu\rho}}\right)\Delta m \tag{3}$$

Where f_0= resonant frequency of quartz crystal, Δf = frequency change, Δm= mass change, ρ = density of quartz crystal, μ = shear modulus of quartz, and A= active crystal area.

Thus, QCR sensors determine the change in the mass per unit area by measuring the variation in their oscillatory frequency. Here, the physicochemical property of the MOF (for example, specific surface area) plays an important part in the detection of the analyte. The major advantages of the QCM-based sensing technique are simple, low cost, operate at room temperature, have a rapid response time, and have excellent sensitivity.

18.2.6.1 The Sensing of VOCs

Aldehyde Sensing

Among different organic VOCs, aldehyde group (-CHO) bearing formaldehyde (HCHO) is one of the most harmful indoor pollutants. It can lead to many health complications, possibly leading to conditions ranging from eye irritation to acute myeloid leukemia. The World Health Organization recommends a formaldehyde level of 20 ppm as a dangerous life-threatening issue [2]. As a result, detecting and monitoring of HCHO at low concentrations in the air has become highly significant. ZnO, a typical n-type semiconductor, has typically been used as a sensor to detect VOCs such as HCHO [12]. However, ZnO sensors are operating at high temperatures, followed by rapid hole-electron recombination in the ZnO matrix which made it insufficient for practical implementation. As a result, a novel gas sensor material with high formaldehyde selectivity at room temperature is desperately needed. MOFs have surfaced as the next-generation materials for gas-sensing applications due to their large surface area (up to 8000 $m^2 g^{-1}$) and abundant porosity [13]. In latest years, primary research interests have been involved in the development of durable MOF-based sensors for low-level HCHO detection. Several kinds of HCHO sensors were developed on different feedback/signal mechanisms such as chemiresistive, chemical sensors, and the like. Chen et al. proposed the chemiresistive sensing properties of pure MOFs for the first time [14]. The authors developed a porous metal-organic framework zeolitic imidazolate framework (ZIF) [ZIF-67; (Co(mim)2; mim = 2-methylimidazolate)] with a surface area of 1832.2 $m^2 g^{-1}$ and a bandgap of 1.98 eV. The developed ZIF-67 is highly stable in air and has delivered a sensitivity of 5-500 ppm for formaldehyde at 150 °C. Further, the zeolitic ZIF-8 is considered an important class of MOF. The ZIF-8 is an excellent choice for a wide range of sensor applications due to its excellent chemical stability, porous structure, high crystallinity, and large surface area [15]. The ZnO@ZiF-8 core-shell structure has been synthesized using the self-template method and its response to VOCs has been investigated. The selectivity of the core-shell structure for detecting formaldehyde has improved. It has been observed that the change in sensor resistance is caused by a reaction between reducing gas and oxygen species present on the core-shell structure. Formaldehyde molecules can easily pass through the ZiF-8 pores and are efficiently oxidized by the oxygen species present on ZnO at 300 °C. The formaldehyde oxidation reaction which is responsible for the sensor response is as follows;

$$HCOH + 2O^- \rightarrow CO_2 + H_2O + 2e^-$$

It is proposed that the excellent selectivity toward formaldehyde in the presence of other VOCs is due to formaldehyde's size. Because of its small size in comparison to other VOCs, formaldehyde molecules can easily pass through ZiF-8 pores and react with oxygen species present on the surface of ZnO. The conductivity of the sensor material is an important parameter in the design of efficient resistive sensor material. Typically, ZIF-8 as a sensor material has low electrical conductivity. As a result, there are few reports of ZIF-8 being used as a sensor material in resistive gas sensors. Combining ZIF-8 with functionalized multiwalled CNTs improves its conductivity significantly which can improve the sensing activity [16]. A resistive sensor based on ZIF-8/MWCNT has been created by depositing a thin layer of the nanocomposite onto conducting electrodes. The resistive sensors are notable for their low detection limits for formaldehyde at room temperature. A sensitivity of 4.83 ppm detection limit was reported with ZIF-8/MWCNT-based resistive VOC sensor. In addition to having a low detection limit, they have a high selectivity for formaldehyde in the presence of other VOC gases such as ethanol, methanol, acetone, and acetonitrile. This note-worthy selectivity appears due to the suitable porous structure of the composite.

Consequently, porous bimetallic metal-organic frameworks have been synthesized, and their gas sensing properties have also been investigated [17]. The experimental results show that the composite (Co/Zn =4:1 At.) is showing 5 times more sensitivity to 10 ppm HCHO at 120 °C than pristine Co_3O_4. The sensor's high sensitivity is due to the large surface area and porous architecture of the developed MOFs pyrolyzed framework, which is useful for gas adsorption and surface-assisted reactions.

MOF-based luminescent sensing platforms have been widely used in recent years to detect cations, anions, temperatures, pH, and harmful gases, and so on. Changes in the MOF's luminescence property can be caused by metal ions, organic ligands, or guest species. These changes appear because of their unique photophysical properties. In this respect, lanthanide-based MOFs (Ln-MOFs) have emerged as potential candidates for luminescent VOC sensing. The forbidden f-f transition is responsible for the fluorescent sensing properties of Ln-MOFs. Furthermore, the f-f transition signals are extremely weak. However, by incorporating different types of metal ions into the frameworks of MOFs, the intensity signals can be improved. Li et al. synthesized dual-emitting mixed MOFs by incorporating Eu^{3+} into NH_2-UiO-66 frameworks [18]. The Eu^{3+} and Zr^{4+} ions are responsible for the dual-emitting luminescent property of the MOF. Due to the presence of intrinsic porous architecture and open sites of amine groups that can easily bind the organic molecules, Eu/Zr-MOF can be a promising material for formaldehyde sensors. It has been discovered that the incorporation of formaldehyde results in a significant enhancement of fluorescence at 465 nm. The intensity variation is caused by the transfer of lone pair electrons from the amino group to the positively charged formaldehyde. As a result, a ratiometric fluorescent sensing method for formaldehyde has been developed, based on the intensity ratio of two emission bands at 465 and 615 nm for Eu/Zr-MOF. The formaldehyde detection limit of the Eu/Zr-MOF sensor has been determined to be 0.2 mg/L in this study. Similarly, an Ag^+/Eu^{3+} co-doped MOF has been synthesized and used as a luminescent self-calibrated sensor for indoor formaldehyde detection at room temperature [19]. They discovered that the incorporation of Ag+ ions can result in instantaneous luminescence of both Eu^{3+} and organic ligands. When formaldehyde interacts with Ag+ ions in the composite, it causes an alternation of luminescent intensities in both emitters. As a result, a ratiometric fluorescent luminescent self-calibrated sensor based on the intensity alternation of both emitters has been developed. Furthermore, the Ag^{+}/Eu^{3+} co-doped MOF-based sensor has high sensitivity and a low detection limit (51 ppb). Min et al. created a Tb-functionalized La-MOF. Tb@La-MOF (Tb@[La(hisp)(Hhisp)$(H_2O)_2$]$\cdot 2H_2O_n$) has been developed using a post-synthetic modification technique [20]. The hybrid exhibits high acetaldehyde sensitivity and low detection limits (0.00058% for acetaldehyde in water, 0.66% for acetaldehyde in ethanol).

Colorimetric naked-eye and luminescent activity can be a useful tool for practical application. This type of sensor can be developed by the incorporation of suitable guest species onto the MOF matrix. The color of the doped MOFs can change during the encapsulation/adsorption of the analytes on their surfaces which offers an inference about the analyte. As a result, these types of materials are emerging as potential candidates for VOC chemical sensing. The advantages of this technique are (i) it does not necessitate any expensive equipment, (ii) it is quick, and (iii) it does not necessitate high temperatures. Cu (I)-MOF was synthesized by Yu et al. [21] from CuI and benzimidazolyl-3, 5-bis (4-pyridyl) benzene which has been used to detect formaldehyde. When exposed to different concentrations of formaldehyde vapors (1.6, 0.16, and 0.016 ppm, respectively), a significant color change has been observed, namely orange, deep-orange, and light brown.

18.2.6.2 Alcohols

Alcohols are fascinating because of their various uses, including medicine, the food industry, and as a precursor material in the production of ethylene and propylene. Furthermore, alcohol has significant harmful effects such as blindness, depression, anxiety, and long-term health problems. As a result, precise real-time quantitative detection of alcohols is extremely important and required. The use of MOFs has also been extended to the detection of alcoholic beverages. Its ability to change physical properties in response to analyte interaction is primarily responsible for this.

ZIF-8 is a thermally stable porous MOF structure formed by the direct coordination interaction of Zn^{2+} ions with organic ligands (2-methyimidazolate). Ren et al. investigated the chemiresistive strategy for detecting ethanol using ZIF-8 as an active sensing material [22]. ZnO@ZIF-8 core-shell framework has been developed by encapsulating ZIF-8 shells on ZnO hollow spheres. It has been discovered that ZnO@ZIF-8 core-shells have superior ethanol sensing properties when compared to ZnO hollow spheres. At an operating temperature of 160 °C, the ZnO@ZIF-8 core-shell exhibited selective sensing properties for ethanol ((R_a/R_g = 35.90 to 100 ppm). On the other hand, ZnO hollow spheres show a sensing response of 14.54 at the same ethanol concentration and operating temperature. Because of the high porosity, a large amount of vapor can be absorbed on the surface of ZnO@ZIF-8 which exhibits high sensitivity. Despite significant advancements in chemiresistive VOC sensors, their high operating temperature remains an issue for practical implementation. Although, significant advancements in chemiresistive VOC sensors have been reported but their high operating temperature remains a big challenge for researchers. Carbon nanotubes (single and multi-walled carbon nanotubes) have emerged as a promising material for developing VOC sensors due to their large surface area, good thermal stability, high electron mobility, and additional active absorption sites. The incorporation of carbon nanotubes in MOFs improves sensitivity and making it useful as a room temperature chemiresistive VOC sensor. Ghanbarian et al. used a chemiresistive sensor based on MIL-53(Cr-Fe)/Ag/CNT to detect VOCs such as methanol, ethanol, and isopropanol at room temperature [23]. A sonochemical method has been used to generate a bimetallic framework MIL-53(Cr-Fe). Further, Ag nanoparticles and carbon nanotubes (CNTs) have been incorporated into the bimetallic framework to improve its selectivity and conductivity. The composite structure then is exposed to various VOC concentrations at room temperature. The change in resistance of the sensor before and after exposure to VOCs has been measured using a homemade sensing chamber setup. It is estimated that it has a higher sensitivity in methanol that can measure 30.5 ppm methanol. The indium-imidazoledicarboxylate (SNNU-50) is a significant MOF material with a bandgap of 3.41 eV, comparable to the bandgap of In_2O_3 [24]. The SNNU-50 polyhedral crystals have been developed by a 5-day chemical reaction between $In(NO_3)_3 \cdot xH_2O$ and 4,5-imidazoledicarboxylic acid (IMDC) in the presence of NMF/ 1,4-dioxane/DMI at 100 °C. The sensor array has been fabricated by coating SNNU-50 on Ag-Pd electrodes. SNNU-50 exhibits sensing of butanol isomers at room temperature.

Alternatively, the incorporation of the carbon nanostructure material graphene in MOFs not only improves the sensitivity of chemiresistive VOC sensors but also allows for the fabrication of ambient temperature-based sensors. Tung et al. created MOF and graphene-based composite materials to study their sensing response. These materials are used to detect VOCs like acetone, methanol, ethanol, tetrahydrofuran (THF), acetonitrile, and chloroform [25]. Further, three graphene hybrid composite materials are being used in this study: copper–benzene-1, 3, 5-tricarboxylate (pG-Cu BTC), zirconium 1, 4-dicarboxybenzene (pG-UiO66), and 2-methylimidazole zinc salt (pG-ZIF 8). It has been found that the absorption capacity and selectivity of different MOFs differ. Cu-BTC, for example, is hydrophobic and has a high absorption limit for moisture and alcohol. It is reported that the aromatic ring in UiO 66 makes it more sensitive to aromatic analytes. Furthermore, the hydrophobic nature of the ZIF 8 allows for excellent absorption of a wide range of gases. The pG-Cu BTC sensor has high sensitivity and selectivity for chloroform and methanol. It can detect chloroform and methanol at very low levels ranging from 2.82 to 22.6 ppm.

MOFs' inability to conduct electricity limits their use in electronic devices. Conductive two-dimensional (2D) MOFs, on the other hand, are an important class of materials for chemiresistive VOC detection. Campbell et al. developed many structurally identical 2D MOF materials to investigate their sensing response for VOC detection and monitoring [26]. In this study, three 2D MO materials, $Cu_3(HHTP)_2$, $Cu_3(HITP)_2$, and $Ni_3(HITP)_2$, have been used to design a chemiresistive sensor array. MOFs exhibit a different response for each analyte. When compared to Cu-based MOFs, Ni containing MOF $Ni_3(HITP)_2$ exhibits the opposite relative conductance response. Although $Cu_3(HHTP)_2$and $Ni_3(HITP)_2$ have the same relative response, their magnitudes are different. Authors hypothesize that charge transfer and hydrogen bonding are two important contributors to the observed sensor response in the 2D MOFs chemiresistive sensor.

Most of the MOFs are non-conducting or semiconducting. However, the dielectric property of MOFs can be a key parameter to be used for sensing purposes. The frequency-dependent dielectric response of MOF materials has been extended to construct a VOC sensor at room temperature. Hosseini et al. described a Cu-BTC-based alcohol sensor device that uses capacitance as its operating principle [27]. The sensing device has been developed using an electrochemical method. The Cu-BTC was made on a Cu substrate by combining 1-methyl-3-octylimidazolium chloride and ionic liquid. To make the upper electrode of the capacitor, Ag paste has been applied to the MOF film, while the copper substrate served as the back terminal. The LCR meter has been used to estimate the capacitive detecting data. Furthermore, the sensing test is carried out at a frequency of 1 MHz. The fabricated device has a good sensing activity for detecting ethanol and methanol. The sensor has a low detection limit of 130.0 ppm for ethanol and 39.1 ppm for methanol. Further, De Smet et al. developed a sensor device by layering composite MOFs on top of patterned aluminum electrodes [28]. The composite MOF is made by incorporating various weight ratios (1-100 wt percent) of MIL-53-NH_2(Al) MOF nanoparticles in a polymer (Matrimid) matrix. The impedimetric variation of the deposited composite films after exposure to gases such as water vapor, and methanol has been studied. Impedance spectroscopy has been used to monitor the sensing activity at room temperature in a frequency range of 20 Hz-1 MHz. It has been found that incorporating MOFs into the polymeric matrix increased capacitive response to methanol and water. The film with a polymer composition of 40% wt NH_2-MIL-53(Al) is the best responsive to methanol and water. They have been able to detect methanol in the presence of 6000 ppm of water. This suggests that the MOF-based sensor is highly selective for methanol. MOFs' high surface areas and porosity allow for the absorption of guest molecules, thereby supporting MOFs as active sensing materials.

Furthermore, it has been discovered that the MOF's effective refractive index is sensitive to the interaction of guest molecules and pores. Hupp et al. created a refractometric optical-based sensor that uses HKUST-1 as the active sensing material [29]. Importantly, the refractive index of MOF has been changed by exposing it to volatile chemical compounds such as ethanol, ethane, and ethylene. This optically based MOF sensor is capable of measuring ethanol concentrations of up to 0.3 ppm. The refractive index of spin-coated MIL-101 thin films varied significantly when exposed to different analyte molecules such as ethanol and isopropanol [30]. They discovered an increase in the refractive index of up to 30% after exposing MOFs to various VOC molecules using spectroscopic ellipsometry.

LMOFs have been synthesized rapidly in recent years, transforming into a promising material with potential applications in fluorescence-based detection of different analytes including VOCs. Lee et al. synthesized $[In(OH)(bdc)]_n$ and $Zn_4O(bdc)_3$(MOF-5) MOFs and investigated their luminescence response after ethanol exposure [31]. The main component of the emission spectra of [In(OH)(bdc)]n and MOF-5 upon excitation at 270 nm is 326 nm and 366 nm, respectively. Furthermore, when exposed to ethanol vapor, the spectra of $[In(OH)(bdc)]_n$ changes significantly. The single peak of the $[In(OH)(bdc)]_n$ emission characteristic is changed to a triplet peak emission characteristic with maximum intensity at 389 nm. In the case of MOF-5, however, a blue shift in the emission spectra is observed, and the maximum intensity of the spectra is found to be identical to water. As a result, MOF-5 has been unable to distinguish between water and ethanol vapors.

Methanol is a toxic alcohol to humans. Methanol consumption will result in migraines, vomiting, and blindness. As a consequence, detecting methanol with considerable sensitivity and selectivity is extremely important. Lanthanide-organic frameworks (Ln-MOFs) have long been used to detect methanol in ethanol solutions. The Ln-MOFs exhibit sharp and powerful luminescent properties, which are important for sensor development. The fluorescence properties of Ln ions, as well as the luminescence processes of organic ligands, have been investigated in optical sensors based on Ln-MOFs. In this direction, Chen et al. utilized MOF-76 as a sensor to detect methanol, ethanol, 1-propanol, 2-propanol, acetonitrile, chloroform, and N,N-dimethylformamide [32]. In this study, they have used the solvothermal method to create Eu(BTC) MOF, which uses the luminescent properties of Eu^{3+} to detect VOCs. They discovered luminescent open metal sites within MOFs that act as sensing sites for analyte binding. Similarly, Fonseca and colleagues created a modified MOF-76 and investigated its photoluminescence response for detecting methanol in ethanol medium [33]. It has been discovered that the intensity of Tb^{3+} luminescence gradually increases as the methanol content of the ethanol fuel increases. The deactivation of the Tb^{3+} emitting state as a result of interaction with the C-H and OH oscillators upon methanol addition is one possible explanation for the increase in intensity. Wang et al. reported the synthesis of a novel luminescent Eu-MOF via a solvothermal process [34]. In contrast to n-propanol and ethanol, the Eu-MOF demonstrated a distinct luminescent response to methanol. In this demonstration, Eu-MOF shows size-selective sensing of methanol over aliphatic alcohols in a turn-on luminescence mode.

18.3 CONCLUSION

VOC sensors have recently received a lot of attention as a result of the expanding and strict environmental security guidelines on indoor and open-air quality monitoring. MOFs are among the most promising sensor materials due to their large surface areas, flexible pore sizes, tunable functional sites, and unique physical and chemical properties. With the advancement of electrically conducting MOFs, chemiresistive detection using MOFs is gaining popularity due to its low manufacturing cost and flexibility. This research area, nevertheless, is still very much in the initial phases of development. There is more than enough potential for improving the sensing performance of these MOF materials by experimenting with new ligands and guest molecules. Colorimetric sensors are generally more delicate, have lower identification limits, and have more evident selectivity due to the explicit change in color in the presence of analytes induced by host-guest interactions. Colorimetric sensors are typically more sensitive and have lower detection limits. They have even more visible selectivity because of the explicit change in color caused by host-guest interactions in the presence of analytes. Owing to these advantages, a colorimetric sensor for quick identification of complex gases or VOC mixtures would be a promising future course of MOF-based colorimetric sensors. Furthermore, MOF-based luminescent sensors exhibit excellent selectivity due to the specific organic reactions with VOCs. The piratical MOF-based luminescent sensors' application is limited because a majority of the reactions between MOF and VOC molecules are irreversible. The practical application of MOF-based luminescent sensors is limited because a majority of the reactions between MOF and VOC molecules are irreversible. As a result, MOF-based luminescent sensors with novel sensing mechanisms must be investigated to improve sensing performance and reduce costs. A detailed theoretical and experimental investigation is required to understand the complexity of the possible physical and chemical processes in real-world systems involving MOF, analytes, and water molecules.

REFERENCES

[1]. Infante, Peter F. "Benzene and leukemia: The 0.1 ppm ACGIH proposed threshold limit value for benzene." *Applied Occupational and Environmental Hygiene* 7, no. 4 (1992): 253–262.

[2]. Tsai, Wen-Tien. "An overview of health hazards of volatile organic compounds regulated as indoor air pollutants." *Reviews on environmental health* 34, no. 1 (2019): 81–89.

[3]. Moon, Chan-Seok. "Estimations of the lethal and exposure doses for representative methanol symptoms in humans." *Annals of occupational and environmental medicine* 29, no. 1 (2017): 1–6.

[4]. Harati, Bahram, SeyedJamaleddinShahtaheri, Ali Karimi, Kamal Azam, Alireza Ahmadi, Maryam Afzali Rad, and Ali Harati. "Cancer risk analysis of benzene and ethyl benzene in painters." *Basic & clinical cancer research* 8, no. 4 (2016): 22–28.

[5]. Koo, Won-Tae, Ji-Soo Jang, and Il-Doo Kim. "Metal-organic frameworks for chemiresistive sensors." *Chem* 5, no. 8 (2019): 1938–1963.

[6]. Kumar, Pawan, Ki-Hyun Kim, Parveen Kumar Mehta, Liya Ge, and GrzegorzLisak. "Progress and challenges in electrochemical sensing of volatile organic compounds using metal-organic frameworks." *Critical Reviews in Environmental Science and Technology 49*, no. 21 (2019): 2016–2048.

[7]. Chidambaram, Arunraj, and Kyriakos C. Stylianou. "Electronic metal–organic framework sensors." *Inorganic Chemistry Frontiers* 5, no. 5 (2018): 979–998.

[8]. Kukkar, Deepak, Kowsalya Vellingiri, Rajnish Kaur, Sanjeev Kumar Bhardwaj, Akash Deep, and Ki-Hyun Kim. "Nanomaterials for sensing of formaldehyde in air: Principles, applications, and performance evaluation." *Nano Research* 12, no. 2 (2019): 225–246.

[9]. Kukkar, Deepak, Kowsalya Vellingiri, Ki-Hyun Kim, and Akash Deep. "Recent progress in biological and chemical sensing by luminescent metal-organic frameworks." *Sensors and Actuators B: Chemical 273* (2018): 1346–1370.

[10]. Sauerbrey, G. J. Z. P. "The use of quarts oscillators for weighing thin layers and for microweighing." *Z. Phys. 155* (1959): 206–222.

[11]. Wang, Luyu. "Metal-organic frameworks for QCM-based gas sensors: A review." *Sensors and Actuators A: Physical* 307 (2020): 111984.

[12]. Chu, Xiangfeng, Tongyun Chen, Wangbing Zhang, Banqiao Zheng, and HengfuShui. "Investigation on formaldehyde gas sensor with ZnO thick film prepared through microwave heating method." *Sensors and Actuators B: Chemical* 142, no. 1 (2009): 49–54.

[13]. Czaja, Alexander U., Natalia Trukhan, and Ulrich Müller. "Industrial applications of metal–organic frameworks." *Chemical Society Reviews* 38, no. 5 (2009): 1284–1293.

[14]. Chen, Er-Xia, Hui Yang, and Jian Zhang. "Zeoliticimidazolate framework as formaldehyde gas sensor." *Inorganic chemistry* 53, no. 11 (2014): 5411–5413.

[15]. Tian, Hailin, Huiqing Fan, Mengmeng Li, and Longtao Ma. "Zeoliticimidazolate framework coated ZnOnanorods as molecular sieving to improve selectivity of formaldehyde gas sensor." *ACS sensors* 1, no. 3 (2016): 243–250.

[16]. Jafari, Nasim, and Sedigheh Zeinali. "Highly Rapid and Sensitive Formaldehyde Detection at Room Temperature Using a ZIF-8/MWCNT Nanocomposite." *ACS omega* 5, no. 9 (2020): 4395–4402.

[17]. Zhou, Wei, Ya-Pan Wu, Jun Zhao, Wen-Wen Dong, Xiu-Qing Qiao, Dong-Fang Hou, Xianhui Bu, and Dong-Sheng Li. "Efficient gas-sensing for formaldehyde with 3D hierarchical Co_3O_4 derived from Co_5-based MOF microcrystals." *Inorganic chemistry* 56, no. 22 (2017): 14111–14117.

[18]. Li, Chunmei, Jinping Huang, Huilin Zhu, Lili Liu, Yaomiao Feng, Gang Hu, and Xibin Yu. "Dual-emitting fluorescence of Eu/Zr-MOF for ratiometric sensing formaldehyde." *Sensors and Actuators B: Chemical* 253 (2017): 275–282.

[19]. Hao, Ji-Na, and Bing Yan. "A dual-emitting 4d–4f nanocrystalline metal–organic framework as a self-calibrating luminescent sensor for indoor formaldehyde pollution." *Nanoscale* 8, no. 23 (2016): 12047–12053.

[20]. Min, Jie, Xiang-Long Qu, and Bing Yan. "Tb post-functionalized La (III) metal organic framework hybrid probe for simple and highly sensitive detection of acetaldehyde." *Sensors and Actuators B: Chemical* 300 (2019): 126985.

[21]. Yu, Yang, Xiao-Meng Zhang, Jian-Ping Ma, Qi-Kui Liu, Peng Wang, and Yu-Bin Dong. "Cu (I)-MOF: naked-eye colorimetric sensor for humidity and formaldehyde

in single-crystal-to-single-crystal fashion." *Chemical Communications* 50, no. 12 (2014): 1444–1446.

[22]. Ren, Guojian, Zhimeng Li, Weiting Yang, Muhammad Faheem, Jianbo Xing, Xiaoqin Zou, Qinhe Pan, Guangshan Zhu, and Yu Du. "ZnO@ ZIF-8 core-shell microspheres for improved ethanol gas sensing." *Sensors and Actuators B: chemical* 284 (2019): 421–427.

[23]. Ghanbarian, Maryam, SedighehZeinali, and Ali Mostafavi. "A novel MIL-53 (Cr-Fe)/Ag/CNT nanocomposite based resistive sensor for sensing of volatile organic compounds." *Sensors and Actuators B: Chemical* 267 (2018): 381–391.

[24]. Bai, Xiao-Ying, Wen-Juan Ji, Shu-Ni Li, Yu-Cheng Jiang, Man-Cheng Hu, and Quan-GuoZhai. "Nonlinear optical rod indium-imidazoledicarboxylate framework as room-temperature gas sensor for butanol isomers." *Crystal Growth & Design* 17, no. 2 (2017): 423–427.

[25]. Tung, Tran Thanh, ManhTrung Tran, Jean-François Feller, Mickael Castro, Truc Van Ngo, Kamrul Hassan, Md J. Nine, and DusanLosic. "Graphene and metal organic frameworks (MOFs) hybridization for tunable chemoresistive sensors for detection of volatile organic compounds (VOCs) biomarkers." *Carbon* 159 (2020): 333–344.

[26]. Campbell, Michael G., Sophie F. Liu, Timothy M. Swager, and MirceaDincă. "Chemiresistive sensor arrays from conductive 2D metal–organic frameworks." *Journal of the American Chemical Society* 137, no. 43 (2015): 13780–13783.

[27]. Hosseini, M. S., S. Zeinali, and M. H. Sheikhi. "Fabrication of capacitive sensor based on Cu-BTC (MOF-199) nanoporous film for detection of ethanol and methanol vapors." *Sensors and Actuators B: Chemical* 230 (2016): 9–16.

[28]. Sachdeva, Sumit, Sander JH Koper, AnahidSabetghadam, Dimitri Soccol, Dirk J. Gravesteijn, FreekKapteijn, Ernst JR Sudhölter, Jorge Gascon, and Louis CPM De Smet. "Gas phase sensing of alcohols by Metal Organic Framework–polymer composite materials." *ACS applied materials & interfaces* 9, no. 29 (2017): 24926–24935.

[29]. Lu, Guang, Omar K. Farha, Lauren E. Kreno, Paul M. Schoenecker, Krista S. Walton, Richard P. Van Duyne, and Joseph T. Hupp. "Fabrication of Metal-Organic Framework-Containing Silica-Colloidal Crystals for Vapor Sensing." *Advanced Materials* 23, no. 38 (2011): 4449–4452.

[30]. Demessence, Aude, Patricia Horcajada, Christian Serre, Cédric Boissière, David Grosso, Clément Sanchez, and Gérard Férey. "Elaboration and properties of hierarchically structured optical thin films of MIL-101 (Cr)." *Chemical communications* 46 (2009): 7149–7151.

[31]. Lee, Tu, Zheng Xin Liu, and Hung Lin Lee. "A biomimetic nose by microcrystals and oriented films of luminescent porous metal–organic frameworks." *Crystal growth & design 11*, no. 9 (2011): 4146–4154.

[32]. Chen, Banglin, Yu Yang, Fatima Zapata, Guannan Lin, Guodong Qian, and Emil B. Lobkovsky. "Luminescent open metal sites within a metal–organic framework for sensing small molecules." *Advanced Materials* 19, no. 13 (2007): 1693–1696.

[33]. Fonseca, Ramon RF, Rafael DL Gaspar, Ivo M. Raimundo Jr, and Priscilla P. Luz. "Photoluminescent Tb^{3+}-based metal-organic framework as a sensor for detection of methanol in ethanol fuel." *Journal of Rare Earths* 37, no. 3 (2019): 225–231.

[34]. Wang, Jian, Min Jiang, Ling Yan, Ren Peng, MengjieHuangfu, XinxinGuo, Yang Li, and Pengyan Wu. "Multifunctional luminescent Eu (III)-based metal–organic framework for sensing methanol and detection and adsorption of Fe (III) ions in aqueous solution." *Inorganic chemistry* 55, no. 24 (2016): 12660–12668.

19 Metal-Organic Frameworks for Organic Dye Adsorptions

Strategic Design and Interaction Aspects

Nabakrushna Behera[1], *Sumit Mohapatra*[1], *Tankadhar Behera*[1] *and Sipun Sethi*[1]
[1]School of Chemistry, Sambalpur University, Jyoti Vihar, Sambalpur, Odisha, India

19.1 INTRODUCTION

With the rapid progression of technology, organic dyes are widely used for the manufacture of colored goods in various industries, such as textile, leather, pharmaceutical, food, paper, and cosmetics (Figure 19.1). This has created the problem of generating contaminated waste water which is very difficult to purify. About 100 tons of dyes are discharged every year as waste materials from industry, making the aquatic environment very toxic and hence, lowering drinking water quality [1]. Generally, organic dyes possess considerable stability in light, heat, and oxidizing agents along with having a low biodegradability. As a consequence, these can have the potential danger of bioaccumulation [2]. Thus, it is essential to remove organic dyes from the waste water, and for this, extensive attention has been devoted by the scientific communities to find out the best possible methods. Moreover, toxic waste is a global challenge that seems to be escalating with time. Recently several methods of removing organic dyes from industrial waste water have been suggested that include photocatalysis, membrane filtration, adsorption, chemical coagulation, and flocculation, and so forth. Among these methods, adsorption is considered to be the most viable technology for dye removal. The process of adsorption involves easy operation and economic feasibility, having the capability to perform well with high efficiency. The synthesis of novel materials (adsorbents) with high dye adsorption and removal efficiencies is essential. For this, metal-organic frameworks (MOFs) are receiving increased attention for their efficient performance in the field of adsorption chemistry [3]. MOFs are porous coordination polymers (PCPs). They have always been of interest to researchers in all fields of chemistry since their inception. Due to their useful intrinsic features, they have emerged as one of the most fascinating advanced materials for both industry and academia. Classically, MOFs are formed from metal nodes or metal precursors (namely, metal ions or clusters) that are connected by organic multifunctional ligands (organic linkers) to create periodic frameworks with well-defined porosity and a high degree of functionality (Figure 19.2) [4]. A large variety of synthetic approaches which include solvothermal, mechano-chemical, sonochemical, electro-chemical, microwave processes, and so forth, have been developed [5]. Fine-tuning of the parameters, namely, reagent concentrations, reagent ratios, solvent compositions, temperature, and reaction times are vital for optimizing the synthesis of MOF materials. To date, a large number of crystalline MOFs have been produced with diverse structural designs and promising efficacy. A few examples of porous MOFs synthesized by several research groups have been represented in Figure 19.3 [6]. MOFs find applications in different fields as illustrated in Figure 19.4 [7] and are considered to be leading-edge materials. The versatile applications of MOFs are undoubtedly attributed to their distinct structural features that include robustness, exceptional internal surface areas (up to ~ 10,000 m^2/g), low densities (0.21–1.00 g/cm^3), ample void spaces (55–90%), and high thermal and chemical stability [7].

19.2 STRATEGIC DESIGN OF MOFs FOR ORGANIC DYE ADSORPTIONS

Microscopic observation of MOFs reveals that their compositions are very straightforward. However, it is quite hard to precisely predict the structure, topology, and porosity of the ensuing framework, even after having detailed knowledge of a specified organic linker and metal source, for which, several possible reasons have been suggested. Firstly, the attainment of several conformations of the organic ligands facilitates the formation of polymorphic MOFs having entirely different structures and properties. Secondly, metal ions participate in the construction of varied metal clusters that result in different MOF structures. Thirdly, while using an extended form of ligand, often the resulting MOF suffers from undesired and inconsistent framework interpenetration. Because of these limitations, it is challenging to design MOFs with desirable shapes, sizes, and functions. However, recently structure topologies have been developed to create extremely porous MOFs [8]. Based on a given topology, one can easily find out many key properties of a framework, namely, shape and size of the cavities, the connectivity

DOI: 10.1201/9781003188148-19

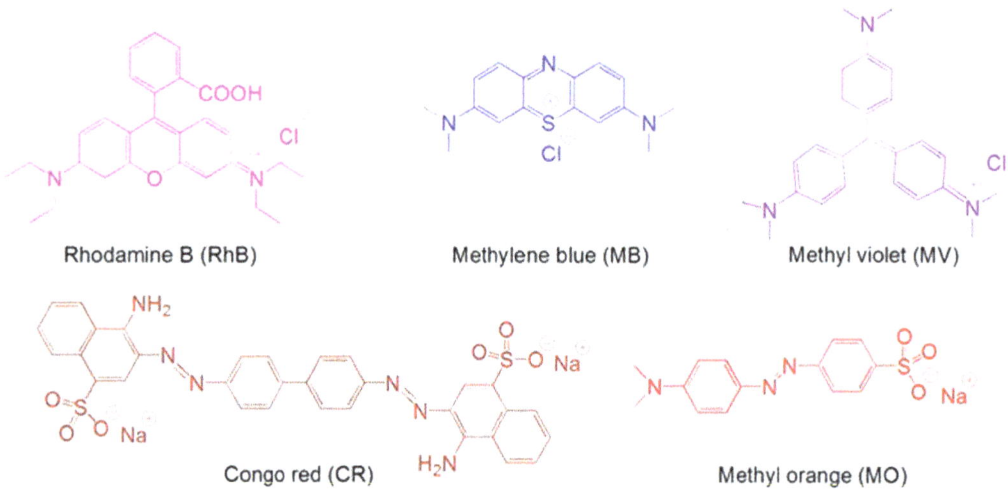

FIGURE 19.1 Schematic structural illustration of some dyes.

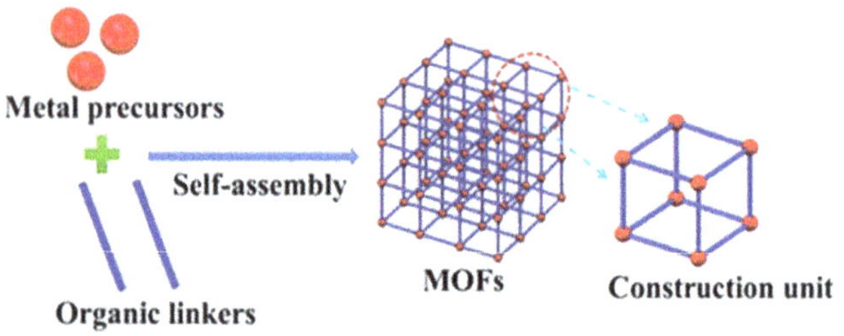

FIGURE 19.2 Schematic diagram of MOFs generated from interconnected organic ligands and metal nodes. Reprinted with permission from ref. [4]. Copyright 2018 The Royal Society of Chemistry.

of its nodes, and the possibility of this framework being able to tolerate self-interpenetration [9]. In the past, MOF topochemistry normally relied on the study of framework topology after its structure was determined. On the other hand, the symmetry-guided design helps to realize a bottom-up design with the preferred network and topology before the building of MOFs [8].

Apart from the use of different metal nodes along with organic linkers for the synthesis of a large variety of MOF structures, the post-synthetic modification of MOFs by different functionalities rather offers a better platform to create modified MOFs with versatile properties and applications. Because of the functionalization, the degree of application of MOF-based materials is enhanced appreciably. Here, our discussion is based on the adsorption of toxic dyes by the recently reported structural MOFs such as optimized, functionalized, composite, derived, and defectives (Figure 19.5). These are identified as very superior dye adsorbents. Optimized MOFs with distinctive dimensions, namely, one dimensional (1D) chains, two dimensional (2D) layers, and three dimensional (3D) supraframeworks have gained significant research attention. Metal node and organic linker functionalizations are another way of getting better functionalized MOFs. The combination of MOFs with other functional materials generates MOF-based composites with better adsorption capacity. The direct carbonization of MOFs and the introduction of defects into MOFs, results in the formation of derived as well as defective MOFs, respectively.

19.3 USE OF MOFs FOR ORGANIC DYE ADSORPTIONS

19.3.1 Dye Adsorption Using Optimized MOFs

Dye waste water is considered one of the major obstacles to sustainable development. In recent times, MOFs with their exceptional adsorptive properties have been used as adsorbents in adsorption techniques for dye removal. For instance, Fe-based MOF (MOF-235) has been found to adsorb both cationic (MB) and anionic (MO) dyes effectively from the contaminated water with adsorption capacity (Q_0) 252.0 mg/g and 477.0 mg/g, respectively [10]. In contrast, MIL-100(Fe) adsorbs MB at 736.2 mg/g and MO at 1045.2 mg/g [11]. However, the adsorption capacity of its chromium analogue, MIL-100(Cr) of MB and MO dyes is about 645.3 mg/g and 211.8 mg/g, respectively. This large difference in adsorption capacities is due to the nature of the framework metal ions, suggesting that the replacement of metal ions may be an efficient way to tailor MOFs for effective adsorption applications [11]. Mesoporous MOFs derived from a copper–gallic acid combination has been found to possess higher adsorption capacity for the removal of MB (124.64 mg/g) and CR (344.54 mg/g) [12]. Excellent adsorption of CR (12,000 mg/g in 20 min) by a cadmium-based MOF has recently been reported [13]. To date, this MOF has proved itself as the best adsorbent with the highest proficiency and fastest adsorption of CR. Al-based MOFs (Al-FA and Al-SA) which use fumaric acid and succinic acid for their framework construction adsorb mono-azo dye (Acid orange 7) and di-azo dye (Acid black 1) with maximum adsorption capacities of 559.28 and 332.48 mg/g, respectively [14].

High specific surface area and well pollutant matched porosity are the characteristic phenomena of the

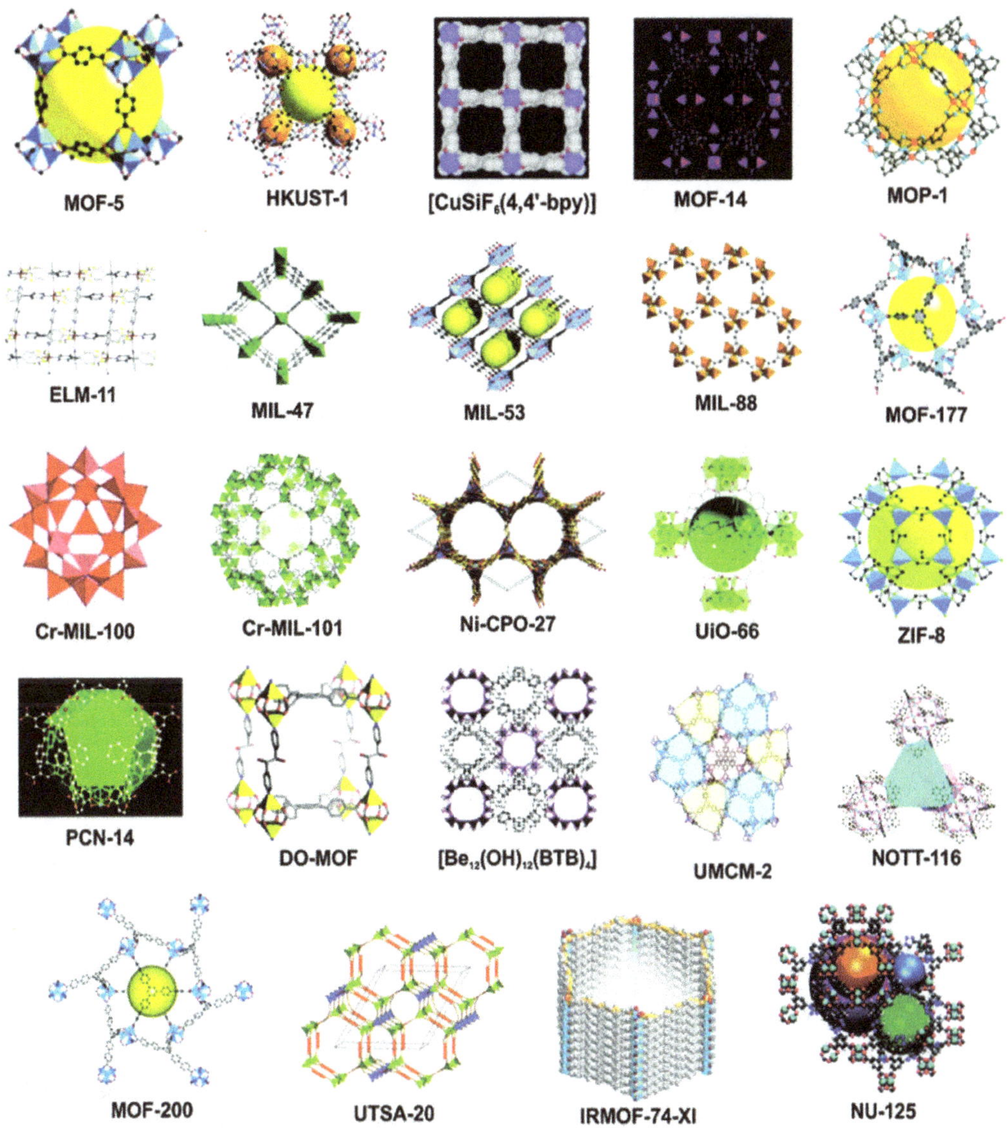

FIGURE 19.3 Different porous MOFs synthesized by several research groups with an aim to accommodate chemical species in their pores/channels. Reprinted with permission from ref. [6]. Copyright 2015 The Royal Society of Chemistry.

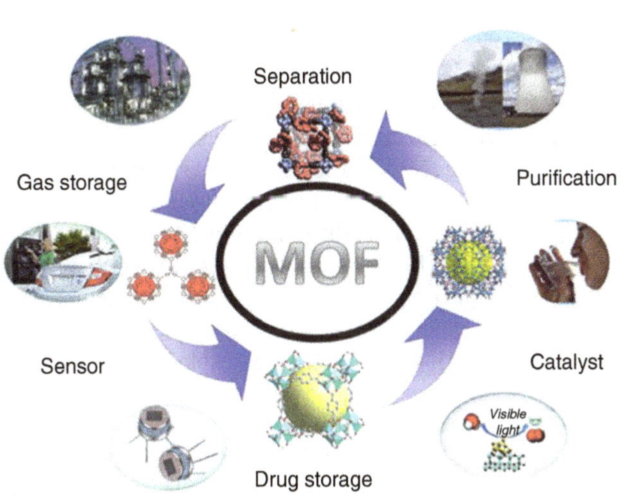

FIGURE 19.4 Wide-ranging applications of MOFs. Reprinted with permission from ref. [7]. Copyright 2013 The Royal Society of Chemistry.

multidimensional MOFs. This includes zero-dimensional (0D) clusters, one-dimensional (1D) chains, two-dimensional (2D) layers, and three-dimensional (3D) pillar structures that exhibit huge adsorption performance. Fundamentally, the use of highly directional ligands plays a crucial role in the construction of coordination-driven self-assembling for these pertinent dimensions and provides a well systematic geometrical topology. Very often, multitopic (linear or bent ditopic, tritopic, tetratopic, and the like.) carboxylic ligands are preferred for MOFs, whereas angularly bent ligands are a common choice for discrete metal-organic polyhedral cage-like assemblies. The bridging linkers based on isophthalic acid and its 5-substituted derivatives are used to design discrete 0D metal-organic self-assemblies which can then transform into 2D MOF structures [15].

An example of a 0D discrete structure coordination compound is $[Co(tib)(Hadc)_2]$ (BUC-60) [16]. A cobalt center in BUC-60 is tetrahedrally connected through two nitrogen atoms and two oxygen atoms from two tib ligands and two

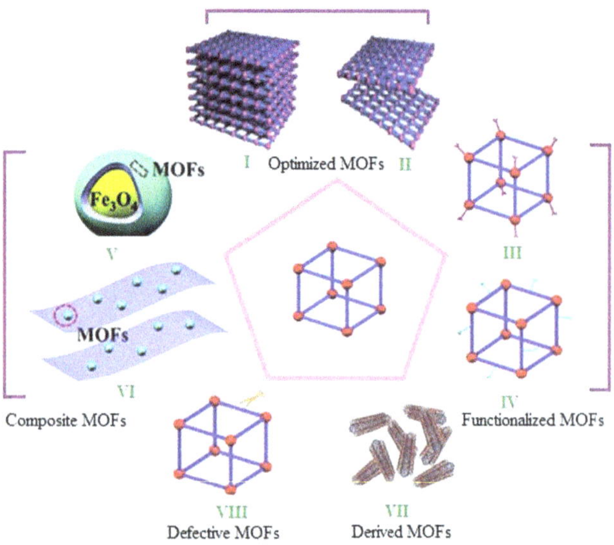

FIGURE 19.5 Strategies adopted to modify MOF-based adsorbents for better performances. Reprinted with permission from refs. [3,4]. Copyright (2019 and 2018) Elsevier & The Royal Society of Chemistry.

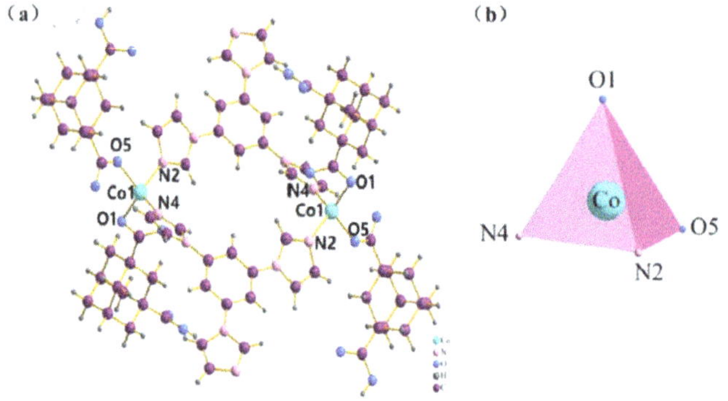

FIGURE 19.6 (a) Discrete framework of BUC-60; (b) Coordination polyhedron of cobalt center. Reprinted with permission from ref. [16]. Copyright 2018 Elsevier.

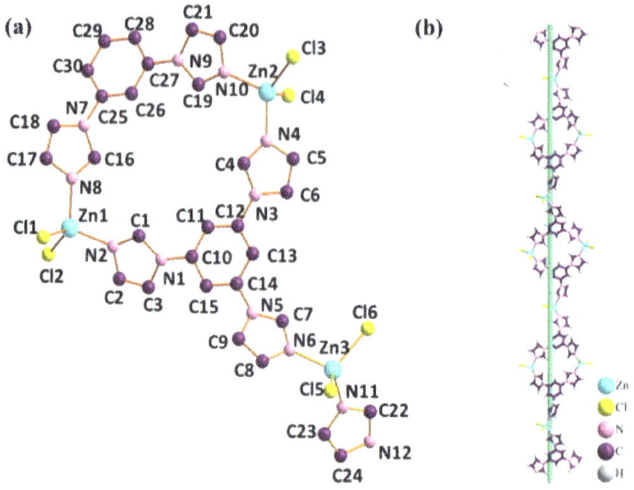

FIGURE 19.7 (a) The asymmetric unit of BUC-61 and the coordination environments around the Zn(II) atoms. H atoms are omitted for clarity; (b) The 1D helix chain in BUC-61. Reprinted with permission from ref. [16]. Copyright 2018 Elsevier.

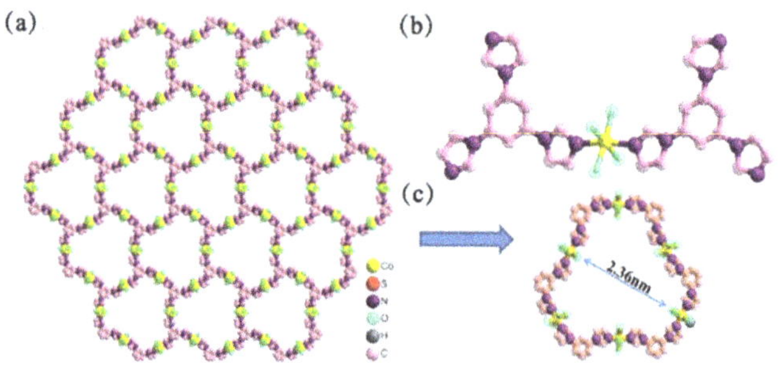

FIGURE 19.8 (a) The 2D graphene-like structure of BUC-17; (b) A symmetric unit of BUC-17 and octahedral environments of Co(II), Yellow = Co(II) node; (c) The hexagonal hole in BUC-17 MOF. Reprinted with permission from ref. [18]. Copyright 2017 The Royal Society of Chemistry.

$Hadc^-$ ligands, respectively (Figure 19.6). The bidentate nature of tib ligand bridges two Co(II) centers providing the third imidazole group to be in the terminal, whereas $Hadc^-$ acts as a monodentate ligand. The mode of coordination of both the tib and $Hadc^-$ ligands results in a 0D discrete unit of $[Co(tib)(Hadc)_2]$ (Figure 19.6a). This BUC-60 exhibits very good adsorption performance for CR being 1949 mg/g and for mordant blue 13 (MB13) being 564 mg/g [16].

A 1D coordination framework, $[Zn_3(tib)_2Cl_6]$ (BUC-61) has been reported to have a tetrahedral zinc center, in which two Cl^- ligands and two nitrogen atoms from two tib ligands occupy the tetrahedral sites (Figure 19.7a) [16]. Each tib ligand tridentately joins with Zn(II) centers into a 1D helix chain running parallel to the a-axis (Figure 19.7b), which is very different from the coordination mode observed for the tib ligand in BUC-60. Further, it has been shown that all the Zn1, Zn2 and Zn3 centers in BUC-61 are nearly identical. The highest adsorption capacities of BUC-61 for CR and MB13 are 1992 and 209 mg/g, respectively.

A novel 1D discrete single-walled metal-organic nanotube (JLU-MONT1) with a rare armchair (3,3) carbon nanotube topology is reported as a very good adsorbent by Zhou et al. for the carcinogenic basic red 9 (BR9) and basic violet 14 (BV14). Two promising reasons may be given for this remarkable adsorption: (a) the framework has enough space to hold the dyes and (b) the channels of JLU-MONT1 are well-matched with the size of these triangular dye molecules. This also has an excellent adsorption capacity of 1615 mg/g for the similar shaped MV dye [17]. A water-stable 2D graphene-like MOF (BUC-17) based on a cobalt system ($[Co_3(tib)_2(H_2O)_{12}](SO_4)_3$) incorporating 1,3,5-tris(1-imidazolyl)benzene (tib) ligands has been reported recently [18]. Structurally, Co(II) in BUC-17 enjoys octahedral geometry (Figure 19.8b). Two nitrogen atoms from two different tib ligands and four oxygen atoms from four coordinated water molecules occupy the axial and equatorial planes, respectively. Further, six Co(II) atoms are tethered by six tib ligands to generate a graphene-like hexagonal pore subunit (Figure 19.8a) with an aperture size of ca. 2.36 nm (Figure 19.8c). This BUC-17 exhibits ultra-high adsorption capacity, particularly for CR up to 4923.7 mg/g at room temperature [18]. Examples of other 2D MOFs which effectively act as adsorbents for MB and MO include Ni(II)-based and Zn(II)-based MOFs [3].

MOFs having pillared-layer structures (PL-MOFs) are three-dimensionally extended and are built from several two-dimensional layers and parallel pillars. The use of both dual-ligand and mixed-ligand is a common strategy for the construction of PL-MOFs in which one ligand contributes to building the layer and the other acts as the pillar [19]. Out of two distinctive classes of PL-MOFs, one involves the polycarboxylate ligands link $[M_2(COO)_4]$ paddlewheel secondary building units for the formation of the layer structure and N-contained ligands (bipyridine or pyrazine) as the pillar bridge adjacent layers to form the ultimate framework [19]. Another one is the SIFSIX series, in which the coordination of N-contained ligands to the single metal centers forms layers and inorganic anions, for example, SiF_6^{2-} acts as the pillars [20]. Besides, other strategies are also used for designing large varieties of structurally good 3D PL-MOFs. For instance, a Co-based MOF, $[Co(biimb)(BPDC)(H_2O)_2]\cdot DMA\cdot 10H_2O$ based on tripodal imidazole-containing ligand (biimb) and linear dicarboxylic acid (H_2BPDC) selectively adsorbs MO (20 mg/g) over other dyes due to the size matching effect. Structurally, two imidazole groups of biimb ligand attach to one of two crystallographically independent Co(II) centers forming a 2D layer and the 3D pillar is accomplished by the coordination of the third imidazole group with another Co(II) center [21].

The Zn-based 3D self-penetrated framework, $[Zn_2(mtrz)_2(azdc)]\cdot DMA\cdot CH_3OH\cdot H_2O$ (Hmtrz = 3-methyl-1,2,4-triazole & H_2azdc = azobenzene-4,4′-dicarboxylic acid), featuring Zn-triazolate layers pillared by dicarboxylate ligands has been reported (Figure 19.9) [22]. In this framework, a distorted tetrahedral environment around each Zn(II) center is observed with one carboxylate O-atom from the dicarboxylate ligand and three N-atoms from three individual triazolate ligands (Figure 19.9a). Each triazolate ligand links to three Zn(II) centers via its three ring N-atoms. Triazolate spacers bridge the adjacent Zn(II) centers to form a dimeric unit and its further extension results in a Zn–triazolate 2D layer (Figure 19.9b). Finally, the dicarboxylate pillars interlink the 2D layers to generate the self-penetrated networks (Figure 19.9c, d). This pillared framework efficiently adsorbs 99% of cationic MB in 360 min. Although RhB is a cationic dye, its proficient adsorption is not observed which may be attributed to the large size of RhB, indicating the involvement of the shape and size exclusion effect. The Cu-based 3D pillared-layer MOF, $[Cu_4(1,4\text{-}ndc)_4(L)_2]\ H_2O\cdot 2DMF$(1,4-ndc = 1,4-naphthalenedicarboxylate; L = 1-(1-(1H-imidazol-1-yl)naphthalen-4-yl)-1H-imidazole) has been reported to efficiently adsorb RhB. It has a 6-connected two-fold interpenetrating pcu (primitive cubic) framework generated from binuclear Cu(II) clusters along with the coexistence of hydrogen-bonding acceptors and donors [23].

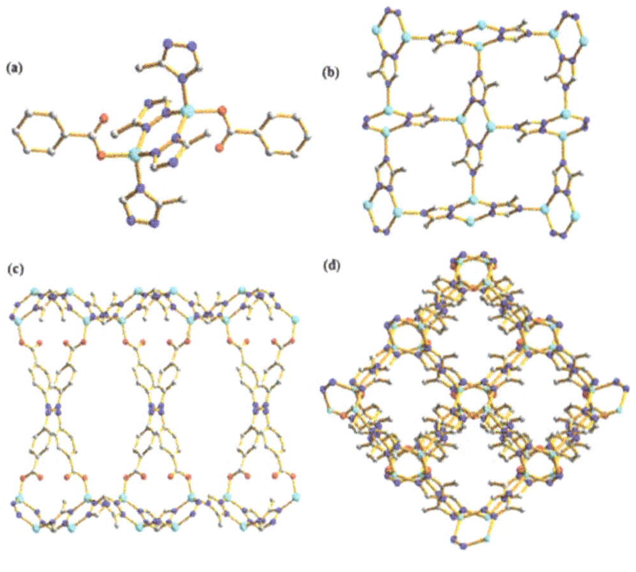

FIGURE 19.9 (a) The coordination environment of the Zn(II) center; (b) 2D Zn-triazolate layers; (c,d) 3D frameworks showing the channels along *a* and *c* direction, respectively. Reprinted with permission from ref. [22]. Copyright 2019 Elsevier.

19.3.2 Dye Adsorption Using Functionalized MOFs

The important components that fundamentally constitute MOF in a periodic mode are metal nodes and polydentate organic ligands (linkers). The functionalization of these components is an effective strategy to tune the physical and chemical properties of MOFs at the molecular level. The majority of synthetic modification of MOFs entails the complete replacement of either the metal or the ligand. On the other hand, MOFs formed via a mixed-component approach involve different metals or linkers with the same structural character, and these are considered to be 'molecular substitutional alloys' [24]. These types of MOFs can be synthesized directly with the use of more than one metal or ligand or prepared by post-synthetic modification. For instance, the partial doping of Cu-BTC MOF with a metal Ni(II) ion results in Ni/Cu-BTC which in turn enhances the adsorption property for CR from waste water because of the synergetic effect. Here the Ni(II) ion plays an important role even with its possible incorporation into unexposed metal sites that are less reachable to dye molecules [25]. Further, the addition of titanium has a considerable effect on the morphology and crystal size of UiO-66, a zirconium-based MOF. The resulting hybrid material, Ti-UiO-66 has better adsorption capacity for an organic dye CR in comparison to the parent UiO-66. The Ti-UiO-66 having 2.7% doped titanium exhibits the highest adsorption capacity of 979 mg/g, which is three times higher than that of the parent UiO-66 [26]. The scope of this study provides an opportunity to dope varied MOFs with different metal ions, improving their adsorptive properties for environmental applications.

Sequential structural modification of MOFs can be achieved by the complete substitution of one linker by another, known as isoreticular synthesis. For instance, 1,4-benzenedicarboxylate linkers can be replaced with more extended dicarboxylates, leading to a significant increase in internal surface area. Linkers can also be modified through different approaches to providing frameworks with a chosen chemical functionality. For instance, the introduction of amine functionalities into a range of MOFs can be performed directly in several ways that include: (a) post synthetically 'grafting' amines onto vacant coordination sites of metal, (b) use of ligands having pendant amines bonded covalently, and (c) using ligands in which the amines are embedded within the core backbone of the bridging ligands [3]. Many of the amine-functionalized MOFs such as NH_2-MIL-125, NH_2-MIL-101 and NH_2-UiO-66, and the like, have been reported. In particular, the maximum adsorption capacity by NH_2-MIL-125 (405.61 mg/g) for MB is more than the parent MIL-125 (321.39 mg/g). The specific surface area of NH_2-MIL-125 is 1028 m^2/g which is much higher than MIL-125 (470 m^2/g) [3]. Furthermore, NH_2-UiO-66 has a very good adsorption capacity for cationic dyes. Favorable adsorption can also be possible by MOFs with other functional groups: ($-SO_3H$) modified ligands [3]. Most importantly, the linker functionalization affects the specific surface area along with the surface charge that plays an important role in the selective adsorption of dyes.

19.3.3 Dye Adsorption Using Composite MOFs

Although several selected porous MOFs exhibit good dye adsorbent properties, nevertheless, the hybrid composites of MOFs have received extensive attention due to their versatile applications including the adsorption process. The combination of MOFs with appropriate materials can greatly improve their morphologies, stabilities, chemical, and physical properties. Their derivative composite materials show a uniform distribution of MOF entities. As a typical example, materials like graphene oxide (GO) and carbon nanotubes (CNT) have been found suitable while both of them were used to formulate hybrid composites with a ZIF-8 MOF, forming ZIF-8@GO and ZIF-8@CNT, respectively. These composites adsorb malachite green (MG), a cationic dye, from the colored waste water, very successfully. As compared to ZIF-8 on its own, which acquires maximum adsorption capacity 1667 mg/g for MG, composite ZIF-8@GO and ZIF-8@CNT exhibit adsorption capacities of 3300 and 2034 mg/g, respectively [27]. Carbon nitride (g-C_3N_4), which has a graphene-like 2D functional material can be coupled with suitable MOFs to provide better dye adsorptive properties [3].

Composites formed by mixing MOFs with magnetic metal nanoparticles serve as excellent dye adsorbents for the treatment of waste waters. Generally, these composites

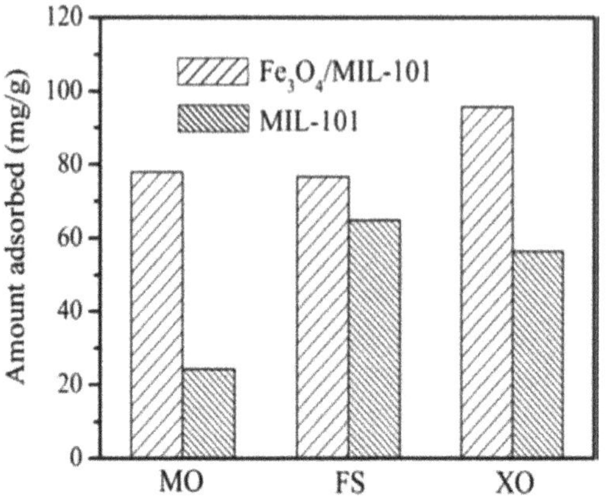

FIGURE 19.10 Comparison of the adsorption capacity of Fe_3O_4/MIL-101 composite and MIL-101 MOF for different anionic dyes. Reprinted with permission from ref. [28]. Copyright 2016 Elsevier.

feature good magnetic separation characteristics, chemical inertness, mechanical stabilities, and a high porosity of MOF. For instance, $Fe_3O_4/Cu_3(BTC)_2$ composite has an adsorption capacity of 244 mg/g for MB with respect to the initial concentration of 300 mg L^{-1} for MB [3]. Another magnetic hybrid material Fe_3O_4@MIL-101(Fe) formed through heterogeneous coprecipitation assembly has a faster absorption rate and higher adsorption capacity than sole MIL-101, for anionic dyes such as MO, xylenol orange (XO), and fluorescein sodium (FS) [28]. Figure 19.10 provides a comparison between MIL-101 and Fe_3O_4/MIL-101 for their adsorbing capacities.

The treatment of dye waste water with composite bioadsorbents involves low ecological toxicity. The bioadsorbent (CS/Fe_3O_4/MIL-101) based on chitosan (CS) wrapping nanoscale Fe_3O_4 and MIL-101 displays remarkable adsorption performance for MO [29]. The presence of chitosan has led to an increase in MO adsorption by about 40 % in comparison with pure MIL-101.

The introduction of new functionality for preparing MOF-based composites is always a challenge to enhance their adsorptive properties. Recently, MOF composites containing polyoxometalate (POM) have received more attention due to their ultrahigh porosity, incredibly high surface area, rich topological structure, and highly electronegative properties. For instance, the removal rate of $H_6P_2W_{18}O_{62}$/MOF-5 composite for MB is better (85%) than that of MOF-5 (nearly zero), indicating that the adsorption performance of pure MOF-5 can be enhanced by modifying with $H_6P_2W_{18}O_{62}$. The composite exhibits selective adsorption ability for the cationic dyes in an aqueous solution with a much faster rate of adsorption. In contrast to anionic dye MO (removal rate 10%), the removal rate has been up to 97% for MB and 68% for RhB within 10 min [30]. It is noteworthy to mention here that POM with many negative charges plays an important role in the adsorption process of cationic dyes. An analogous effect is observed due to the $H_6P_2Mo_{15}W_3O_{62}$ component being present in the $H_6P_2Mo_{15}W_3O_{62}@Cu_3(BTC)_2$ composite while it is used for the selective adsorption of MB and MO [3].

19.3.4 Dye Adsorption Using Derived MOFs

Apart from their direct use, the MOFs can be employed as ideal templates/precursors for the construction of a variety of functional nanostructured materials such as metal-free porous carbon, metal dispersed porous carbon, and so forth. These MOF-derived materials which remarkably serve as adsorbents are capable of inheriting the advantages of MOF precursors, particularly their high specific surface area and tailorable porosity. The porous materials formed under the heat treatment of MOF templates involve the modification of the pore size distribution along with the high surface area, resulting in the enhancement of the adsorption capacity of the carbon network for dye molecules. Nano porous carbons derived from MOF-5, MIL-100, ZIF-67, and so forth, have been found to have improved adsorption capacity for the removal of MB from wastewater. The porous carbon made by carbonization of ZIF-8 at 1000 °C exhibits adsorption capacities of 84.3, 153.5, 200, and 186.3 mg/g for RhB, MO, brilliant green (BG), and MB dyes, respectively [31]. It has also been observed, albeit not always, that the MOF carbonized at higher temperature acquires a huge surface area which indicates that the carbonization temperature is very important for the structural configuration of the resulting porous carbon materials.

Interestingly, the adsorption capacity of nanostructured porous material can be further improved by introducing hetero atoms, namely, N, O, and S. As an example, the metal-free N-doped porous carbons prepared through the carbonization of ZIF-8/dicyandiamide composite have high adsorption capacity for MB (1148.2 mg/g) [32]. What is more, when the wastewater containing MB is treated with excessive adsorbents, the removal percent is almost 100%. This shows that N-doped porous carbons can serve as better adsorbents for removing toxic dyes from wastewater. The strategic synthesis of N-doped porous carbon and its adsorption for MB has been schematically described in Figure 19.11.

Since the MOF involves well-coordinated metal-linker associations, its metal entity can be inherently directed to be uniformly dispersed over the porous carbon matrix through

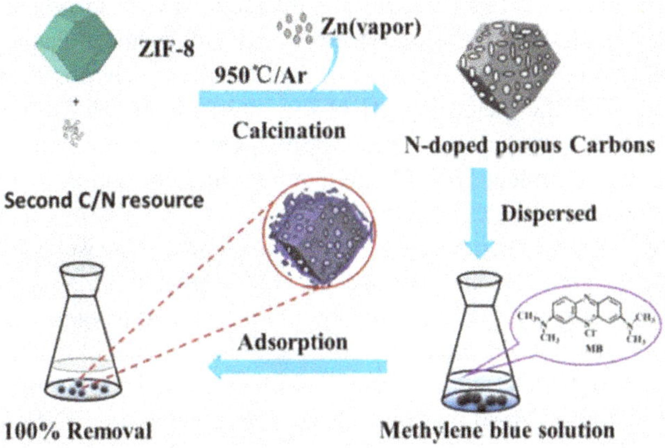

FIGURE 19.11 Synthesis of N-doped porous carbon and its adsorption for MB removal. Reprinted with permission from ref. [32]. Copyright 2017 Elsevier.

a controlled carbonization process. For instance, the ZIF-67 MOF, isostructural with ZIF-8, has been carbonized to result in Co-nano porous carbon (Co/NPC) of open nanoporosity with the uniform distribution of Co nanoparticles of high content [33]. This Co/NPC is ideal for the fast adsorption and removal of MB dye with large capacities. Interestingly, this Co/NPC material also exhibits a strong magnetic response due to dispersed Co nanoparticles, for which it involves easy magnetic separation.

19.3.5 Dye Adsorption Using Defective MOFs

MOFs have been engineered in several ways to fine-tune their properties. One of the emerging ways which have been very instrumental for inculcating improved adsorption properties in MOFs is defect engineering. The introduction of defects brings structural disorder and heterogeneity within MOFs, and hence, modifies the atomic arrangements and surface properties of the resulting materials. Defects within MOFs could be generated by the use of acid modulators, namely, acetic acid and benzoic acid or post-synthetic treatment with inorganic acids, or the mixed-linker approach [34]. Wang et al. have reported the synthesis of defective UiO-66 in which benzoic acid is used as modulator and HCl for post-synthetic treatment [35]. UiO-66 exhibits over nine times higher adsorption capacity for safranine T (ST) (366 mg/g) dye than its defect-free analog UiO-66 (39 mg/g). This defective UiO-66 is also capable of adsorbing ST selectively over crystal violet dye due to a size-exclusion effect. Furthermore, the missing-linker defects generated by the acid modulator make the framework more hydrophilic, providing countless possibilities for dye adsorption [3]. Thus, the construction of defective MOFs by using benzoic acid as a modulator along with post-synthetic acid treatment seems a promising strategy for large molecule separation. Defective MOFs having missing-linker defects are generally produced from the mixed-linker MOFs which primarily contain thermo-stable as well as thermo-labile linkers. Upon treatment under pyrolysis, the thermo-labile linkers in MOF undergo decomposition to produce a defective MOF. In 2017, Bueken et al. reported how the combination of a thermo-stable linker (1,4-benzenedicar-boxylate) with a thermo-labile linker (*trans*-1,4-cyclohexanedicarboxylate) is a feasible strategy to exclusively generate missing-linker defects in mixed-linker UiO-66 MOF post synthetically [36]. The occurrence of missing-linker defects that are experimentally found to be homogeneously distributed throughout the lattice in defective UiO-66 MOF is shown in Figure 19.12 [36].

19.4 INTERACTION ASPECTS OF MOFs WITH DYES

It has been well established that MOFs as adsorbents play a very crucial role due to their enormous surface areas and tunable porosities. What is more, the MOFs can be decorated with various active functionalities which in turn have made them relatively more competitive as adsorbents. Further, it is

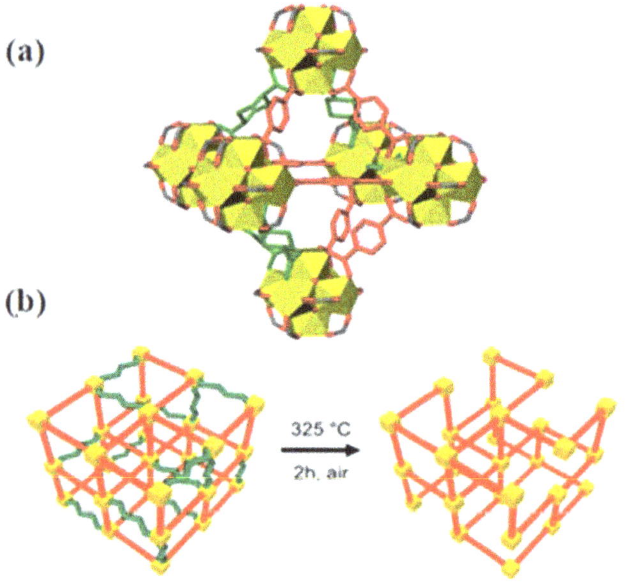

FIGURE 19.12 (a) Structure of the octahedral cage of mixed-linker UiO-66 containing thermo-stable (red) and thermo-labile (green) linkers; (b) Depiction of missing-linker defect creation by the decomposition of thermo-labile linkers in the thermo-stable matrix. Yellow = Zr_6 nodes. Reprinted with permission from ref. [36]. Copyright 2017 American Chemical Society.

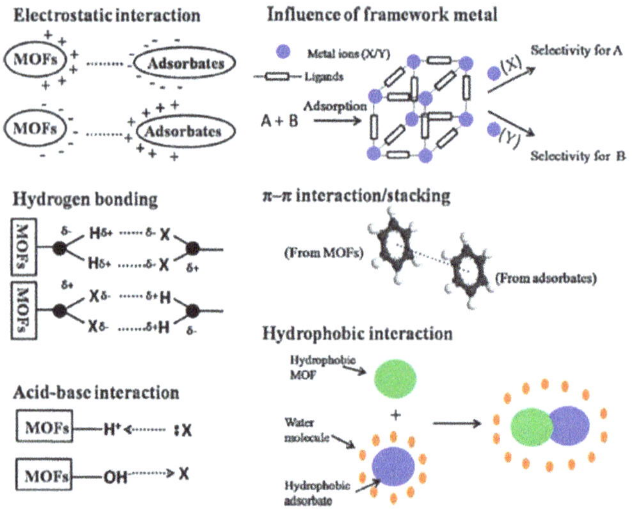

FIGURE 19.13 Schematic diagram of possible mechanisms for adsorptive removal of harmful materials over MOFs. Reprinted with permission from ref. [37]. Copyright 2015 Elsevier.

essential to know about the mechanistic interactions between the MOFs and dye molecules, as this will pave the way to rationally design MOF materials for future applications. Several key supramolecular interactions such as electrostatic, acid-base, hydrogen bonding, hydrophobic, π–π stacking, and so forth, exist between the MOF adsorbents and dye adsorbates during the selective uptake of dyes have been described in Figure 19.13 [37]. Strictly speaking, a particular adsorption process is not always restricted to a single type of interaction, and hence, the possibility of multiple interactions might take place.

19.4.1 Electrostatic Interactions

Very often, electrostatic interactions play an important role during the adsorptive removal of toxic dyes by MOF materials. Depending on the pH of the medium, MOFs fine-tune their surface charges (electric charges) which are present at the interface and dispersed in the polar media. Using the protonation and deprotonation process, the net charge on a MOF can be altered. As a result, the charge (cationic or anionic) on the MOF will be easily capable of interacting with the diverse charges of the dye (adsorbate), and this is known as electrostatic interaction. The study of the adsorptive removal of toxic dye using MOFs was first made by Haque et al. in 2010 [38]. Two porous MOFs, namely, MIL-101(Cr) and MIL-53(Cr) were used by them for the adsorptive removal of MO dye from aqueous solutions and were found to be more adsorbent than the activated carbon. Interestingly, after the functionalization of MIL-101(Cr) by grafting with ethylenediamine (ED) and protonated ethylenediamine (PED), the functionalized MOFs show better adsorption for MO, despite the decrease in porosity and pore size after grafting. Since MO normally exists in the sulfate form, it interacts strongly with positively charged adsorbents in the electrostatic mode. The distribution of positive charges among MOFs follows the order MIL-101(Cr) < ED-MIL-101(Cr) < PED-MIL-101(Cr), as shown in Figure 19.14a. However, upon increasing the solution pH, the positive charge on PED-MIL-101(Cr) decreases due to its deprotonation. Hence, the more positive charge on the adsorbent may be achieved by increasing acidity, especially for PED-MIL-101 at low pH [38]. Haque et al. have also reported how the MOF-235 efficiently adsorbs both MO and MB dyes from contaminated water [10]. The high adsorption of MB or MO on MOF-235 has been described in terms of electrostatic interactions between the adsorbents and dyes (Figure 19.14b). The negative and positive forms of MO and MB dyes make both suitable to establish electrostatic interactions with adsorbents having charges of positive (framework) and negative (charge-balancing anion), respectively. It has also been observed that with the increasing pH of the MO solution, adsorption decreases owing to the decrease of positive charge density on the adsorbent [10]. On the contrary, the concentration of adsorbed MB goes on intensifying with an increase of pH, as the negative charge density on the adsorbent increases with increasing pH. Very recently, Yu et al. prepared a Zn-based MOF containing a large number of nitrogen and oxygen sites projected from the diazene, ether, and ester groups [39]. This Zn-based MOF, upon treatment with aqueous HCl, forms a protonated Zn-MOF with rich positive charges (–N=NH$^+$–, –OH$^+$–, –COOH$^+$–). The resultant protonated Zn-MOF exhibits ultrahigh uptake capacities for different anionic dyes such as Amido Black 10 B (402.82 mg/g), MO (744.02 mg/g), Orange

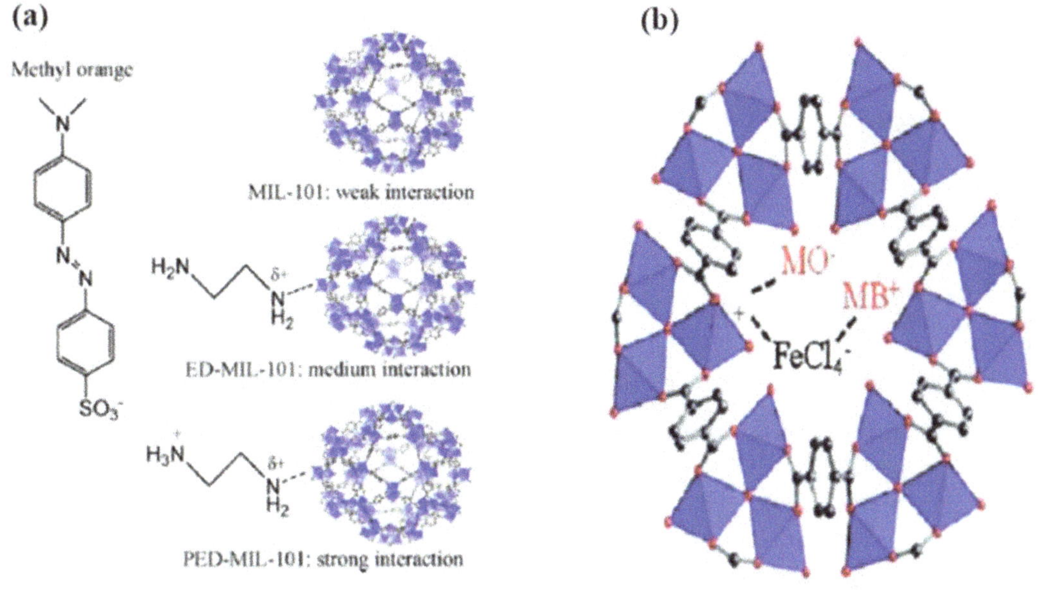

FIGURE 19.14 Possible adsorption mechanisms of dyes: (a) Electrostatic interaction between anionic MO and cationic PED-MIL-101(Cr). Reprinted with permission from ref. [38]. Copyright 2010 Elsevier; (b) Electrostatic interactions of MO and MB dyes with an adsorbent MOF-235 having counter positive charge (frameworks) and negative charge (charge-balancing anion), respectively [10]. Reprinted with permission from ref. [10]. Copyright 2011 Elsevier.

II (522.83 mg/g), and direct red 80 (1496.34 mg/g) [39]. The most plausible adsorption mechanism explains that there exists strong affinity due to the greater electrostatic attraction between the positively charged surfaces of the protonated Zn(II)-MOF and the $-SO_3^-$ group of the anionic dyes.

19.4.2 Acid-Base Interactions

Generally, MOF-dye adsorption only based on acid-base interactions is hardly seen. However, a cobalt-based 3D-supramolecular MOF constructed from mixed ligand systems has been reported to have selective adsorption for orange IV (OIV) and MO dyes through Lewis acid-base interactions [40]. In the interaction process, the cobalt ions (Lewis acid) as open active sites can coordinate with the sulfonate salt (Lewis base) contained in OIV/MO. Proper functionalization of the MOF can lead to establishing such interactions with the dye molecules. As an example, the functionalization of the MIL-100(Fe) MOF has been achieved with thioglycolic acid, which has resulted in a fresh thiol functionalized analog material (TH-MIL-100) for efficient removal of eosin B dyes [41]. The effective adsorption of eosin B dyes by TH-MIL-100 has been attributed to the acid-base interaction between the acidic thiol group (–SH) and the alkaline medium of eosin B dye solution. Recently, with the help of Fe^{3+} (hard acid), Ahamad et al. have modified the surface properties of the Cu- and Ni-based MOFs containing free carboxylic acid groups [42]. The oxygen (hard base) of the carboxylate group binds Fe^{3+} strongly and generates many unsaturated coordination sites in the MOF. These post synthetically modified MOFs, due to Fe^{3+} ions, effectively adsorb MB and MO molecules more effective in comparison to the parent MOFs. The unsaturated metal sites easily interact with the hard donors of the dyes (N and/or O) in accordance with the hard-soft acid-base (HSAB) principle (Figure 19.15) [42].

19.4.3 Hydrogen Bonding Interactions

By and large, the adsorption phenomenon anchored in hydrogen bonding between MOF and dye molecules seldom happens. Such interactions frequently happen during the adsorption of harmful organic molecules over MOFs [37]. Nevertheless, strategic functionalities on both the MOF and the dye molecule could offer appropriate acceptor and donor pairs for hydrogen bonding interactions, which in turn promotes adsorption. This interaction is one of the well-studied interactions in the context of dye adsorption. Li et al. have used MIL-53(Al) and its modified analog MIL-53(Al)-NH_2 to remove MB and MG dyes from an aqueous solution [43]. While MIL-53(Al) exhibits extremely low adsorption capacity, MIL-53(Al)-NH_2 selectively adsorbs the same with a high rate. The presence of amine functions in MIL-53(Al)-NH_2 facilitates strong hydrogen bonding interaction with N-atoms of dye molecules (Figure 19.16), although there exist weak π-π interactions (to be discussed later) between the benzene rings of MIL-53(Al)-NH_2 and the dye molecules. This adsorption process is controlled by the pH of the medium. Similar modifications of titanium-based MOFs by the amino groups have enhanced the adsorption capacity of modified MOFs for different organic dyes [44].

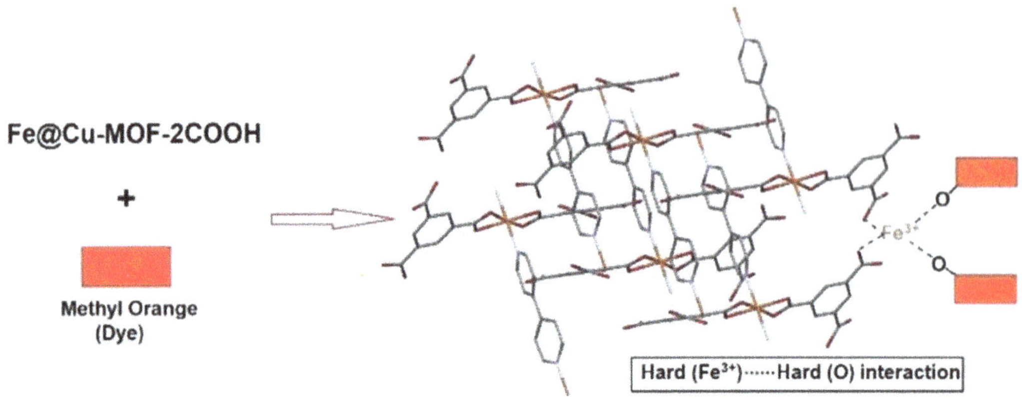

FIGURE 19.15 Plausible adsorption mechanisms of dyes through Fe---O interactions. Reprinted with permission from ref. [42]. Copyright 2020 The Royal Society of Chemistry.

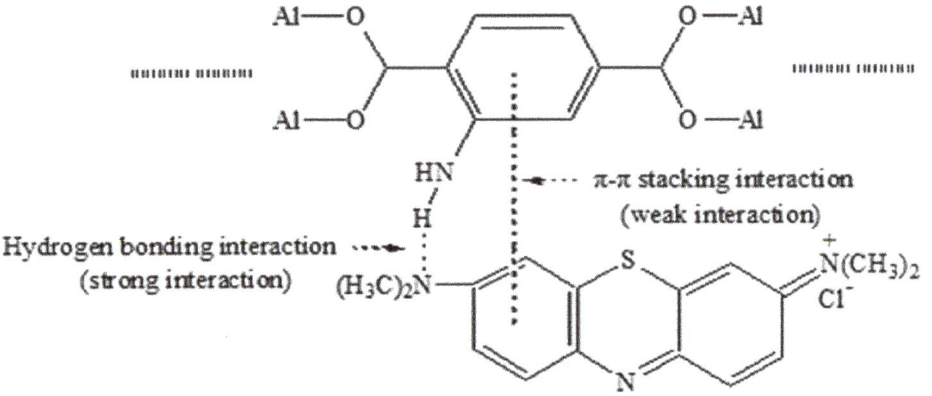

FIGURE 19.16 Possible adsorption mechanism of MB dyes with MIL-53(Al)-NH_2. Reprinted with permission from ref. [43]. Copyright 2015 American Chemical Society.

19.4.4 π–π Interactions

Apart from the interactions discussed in the preceding sections, another important feature which strongly contributes to the high adsorption process is π–π interaction/stacking. The reason behind this interaction is the presence of rich aromatic factions between the MOF and dye molecules. For example, the functionalization of UiO-66 MOF by Ce(III) doping increases the adsorption sites and promotes adsorbent–adsorbate π–π interactions, overriding the possible effects of electrostatic interactions [45]. The strong π–π interactions between the benzene rings of Ce(III)-doped UiO-66 and the aromatic spine of cationic MB relatively enhance the adsorption uptake than that of pure UiO-66. Anionic dyes such as MO and CR, which contain aromatic backbones also interact with the benzene rings of functionalized UiO-66 through π–π interactions to make the adsorption process effective [45]. Recently, Mao et al. reported a Zn-based MOF that possesses an ultrahigh adsorption capacity (2348 mg/g) for effective removal of CR, involving mostly π–π stacking interactions [46]. Even though both MOF and dye molecules contain aromatic moieties as observed in most of the cases, the adsorption process is controlled by multiple interactions [47-49] and the collaborative effects of all interactions contribute to the enhancement of the adsorption capacity of the MOF. An example of a Cu-based MOF engaged in manifold interactions with various dye molecules is depicted in Figure 19.17 [49].

19.5 CONCLUSIONS

Our chapter has highlighted some of the most skilful strategies to obtain MOF-based materials for the organic dye adsorptions from wastewater. The existence of unique textural properties of MOFs provides ample design criteria for the development of good adsorbents. Fabrication of optimized MOFs, functionalization of organic linkers of MOFs, the introduction of defects into MOFs, and the construction of MOF-based composite materials are some key strategies through which the surface properties of MOFs could readily be customized for the adsorptive removal of organic dyes. In addition, rational design of functional MOFs with higher adsorption values genuinely demands a clear understanding of the MOF-dye interactions and/or mechanisms. In many cases, as observed, the major driving forces that make MOFs able to adsorb dyes effectively from wastewater have been attributed to multiple interactions, namely, electrostatic interactions, π–π interaction/stacking, hydrogen bonding interactions, and so forth. Apart from this, the adsorption

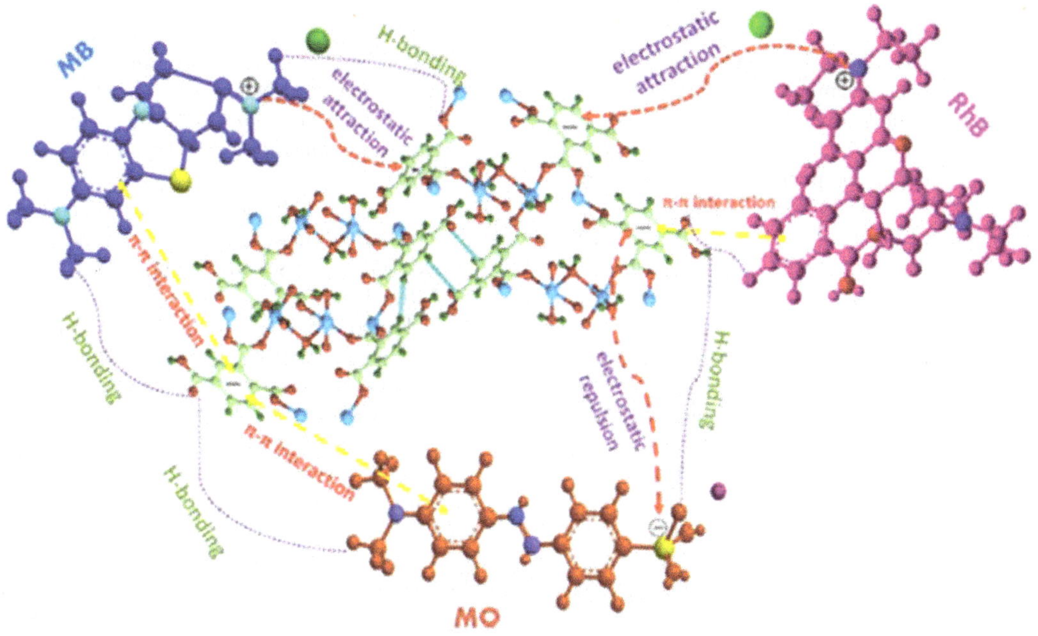

FIGURE 19.17 Cu-based MOF is involved in multiple interactions with RhB, MO and MB dyes. Reprinted with permission from ref. [49]. Copyright 2020 Elsevier.

process also depends upon precise conditions such as pH of the solution, rich adsorption interaction sites of MOFs, and the size and shape of the organic dyes. Especially, MOFs that are resistant to structural damage induced by moisture, redox agents, acids, bases, and radiations are preferred for future applications.

REFERENCES

[1]. J. Fan, D. Chen, N. Li et al., Adsorption and biodegradation of dye in waste water with Fe_3O_4@MIL-100 (Fe) core-shell bio-nanocomposites, *Chemosphere* 191 (2018) 315–323.

[2]. Q. Yang, R. Lu, S.S. Ren, C. Chen, Z. Chen, X. Yang, Three dimensional reduced graphene oxide/ZIF-67 aerogel: effective removal cationic and anionic dyes from water, *Chem. Eng. J.* 348 (2018) 202–211.

[3]. D. Jiang, M. Chen, H. Wang et al., The application of different typological and structural MOFs-based materials for the dyes' adsorption, *Coord. Chem. Rev.* 380 (2019) 471–483.

[4]. J. Li, X. Wang, G. Zhao et al., Metal–organic framework-based materials: superior adsorbents for the capture of toxic and radioactive metal ions, *Chem. Soc. Rev.* 47 (2018) 2322–2356.

[5]. D.Y. Heo, H.H. Do, S.H. Ahn, S.Y. Kim, Metal-organic framework materials for perovskite solar cells, *Polymers* 12 (2020) 2061.

[6]. P. Silva, S.M.F. Vilela, J.P.C. Tomé, F.A.A. Paz, Multifunctional metal–organic frameworks: from academia to industrial applications, *Chem. Soc. Rev.* 44 (2015) 6774–6803.

[7]. S. Chaemchuen, N.A. Kabir, K. Zhou, F. Verpoort, Metal–organic frameworks for upgrading biogas *via* CO_2 adsorption to biogas green energy, *Chem. Soc. Rev.* 42 (2013) 9304–9332.

[8]. M. Zhang, Y.-P. Chen, M. Bosch et al., Symmetry-guided synthesis of highly porous metal-organic frameworks with fluorite topology, *Angew. Chem. Int. Ed.* 53 (2014) 815–818.

[9]. M. Zhang, M. Bosch, T. Gentle III, H.-C. Zhou, Rational design of metal-organic frameworks with anticipated porosities and functionalities, *CrystEngComm* 16 (2014) 4069–4083.

[10]. E. Haque, J.W. Jun, S.H. Jhung, Adsorptive removal of methyl orange and methylene blue from aqueous solution with a metal-organic framework material, iron terephthalate (MOF-235), *J. Hazard. Mater.* 185 (2011) 507–511.

[11]. M. Tong, D. Liu, Q. Yang, S. Devautour-Vinot, G. Maurin, C. Zhong, Influence of framework metal ions on the dye capture behavior of MIL-100 (Fe, Cr) MOF type solids, *J. Mater. Chem. A* 1 (2013) 8534–8537.

[12]. B. Azhar, A.E. Angkawijaya, S.P. Santoso et al., Aqueous synthesis of highly adsorptive copper–gallic acid metal–organic framework, *Sci. Rep.* 10 (2020) 19212.

[13]. D.-D. Guo, B. Li, Z.-P. Deng, L.-H. Huo, S. Gao, A rational design of layered metal–organic framework for high-performance adsorption of hazardous organic dye, *Dalton Trans.* 50 (2021) 7818–7825.

[14]. K.-W. Jung, B.H. Choi, C.M. Dao et al., Aluminum carboxylate-based metal organic frameworks for effective adsorption of anionic azo dyes from aqueous media, *J. Ind. Eng. Chem.* 59 (2018) 149–159.

[15]. N. Ahmad, H.A. Younus, A.H. Chughtai et al., Synthesis of 2D MOF having potential for efficient dye adsorption and catalytic applications, *Catal. Sci. Technol.* 8 (2018) 4010–4017.

[16]. J.-J. Li, C.-C. Wang, J. Guo, J.-R. Cui, P. Wang, C. Zhao, Three coordination compounds based on tris(1-imidazolyl)benzene: hydrothermal synthesis, crystal structure and adsorption performances for organic dyes, *Polyhedron* 139 (2018) 89–97.

[17]. Y. Zhou, S. Yao, Y. Ma, G. Li, Q. Huo, Y. Liu, An anionic single-walled metal–organic nanotube with an armchair (3,3) topology as an extremely smart adsorbent for the effective and selective adsorption of cationic carcinogenic dyes, *Chem. Commun.* 54 (2018) 3006–3009.

[18]. J.-J. Li, C.-C. Wang, H.-F. Fu et al., High-performance adsorption and separation of anionic dyes in water using a

chemically stable graphene-like metal–organic framework, *Dalton Trans.* 46 (2017) 10197–10201.
[19]. Y.-H. Tang, F. Wang, J. Zhang, Construction of unprecedented pillar-layered metal organic frameworks via a dual-ligand strategy for dye degradation, *Dalton Trans.* 47 (2018) 4032–4035.
[20]. K.A. Forrest, T. Pham, B. Space, Comparing the mechanism and energetics of CO_2 sorption in the SIFSIX series, *CrystEngComm* 19 (2017) 3338–3347.
[21]. Y. Deng, Y. Zhao, P. Wang, Z.-Y. Yao, X.-D. Zhang, W.-Y. Sun, Coordination polymers with 1,3-bis(1-imidazolyl)-5-(imidazol-1-ylmethyl)benzene and biphenyl-4,4′-dicarboxylate ligands: Selective adsorption of gas and dye molecules, *Microporous Mesoporous Mater.* 241 (2017) 192–201.
[22]. J. Liu, Y. Wei, F. Bao, G. Li, H. Liu, H. Wang, Pore-size tuning in pillared-layer metal–organic framework with self-penetrated rob net for selective gas adsorption and efficient dyes adsorption in aqueous solution, *Polyhedron* 169 (2019) 58–65.
[23]. L. Yang, X. Li, C.-Y. Sun, H. Wu, C.-G. Wang, Z.-M. Su, A stable pillared-layer Cu(II) metal–organic framework with magnetic properties for dye adsorption and separation, *New J. Chem.* 41 (2017) 3661–3666.
[24]. M.I. Breeze, G. Clet, B.C. Campo et al., Isomorphous substitution in a flexible metal–organic framework: mixed-metal, mixed-valent MIL-53 type materials, *Inorg. Chem.* 52 (2013) 8171–8182.
[25]. J. Hu, H. Yu, W. Dai, X. Yan, X. Hu, H. Huang, Enhanced adsorptive removal of hazardous anionic dye "congo red" by a Ni/Cu mixed-component metal–organic porous material, *RSC Adv.* 4 (2014) 35124–35130.
[26]. Y. Han, M. Liu, K. Li et al., In situ synthesis of titanium doped hybrid metal–organic framework UiO-66 with enhanced adsorption capacity for organic dyes, *Inorg. Chem. Front.* 4 (2017) 1870–1880.
[27]. J. Abdi, M. Vossoughi, N.M. Mahmoodi, I. Alemzadeh, Synthesis of metal-organic framework hybrid nanocomposites based on GO and CNT with high adsorption capacity for dye removal, *Chem. Eng. J.* 326 (2017) 1145–1158.
[28]. Z. Jiang, Y. Li, Facile synthesis of magnetic hybrid Fe_3O_4/MIL-101 via heterogeneous coprecipitation assembly for efficient adsorption of anionic dyes, *J. Taiwan Inst. Chem. Eng.* 59 (2016) 373–379.
[29]. L. Liu, J. Ge, L.T. Yang, X. Jiang, L.G. Qiu, Facile preparation of chitosan enwrapping Fe_3O_4 nanoparticles and MIL-101(Cr) magnetic composites for enhanced methyl orange adsorption, *J. Porous Mater.* 23 (2016) 1363–1372.
[30]. X. Liu, W. Gong, J. Luo, C. Zou, Y. Yang, S. Yang, Selective adsorption of cationic dyes from aqueous solution by polyoxometalate-based metal–organic framework composite, *Appl. Surf. Sci.* 362 (2016) 517–524.
[31]. Z. Abbasi, E. Shamsaei, S.K. Leong, B. Ladewig, X. Zhang, H. Wang, Effect of carbonization temperature on adsorption property of ZIF-8 derived nanoporous carbon for water treatment, *Microporous Mesoporous Mater.* 236 (2016) 28–37.
[32]. S. Xu, Y. Lv, X. Zeng, D. Cao, ZIF-derived nitrogen-doped porous carbons as highly efficient adsorbents for removal of organic compounds from waste water, *Chem. Eng. J.* 323 (2017) 502–511.
[33]. N.L. Torad, M. Hu, S. Ishihara et al., Direct synthesis of MOF-derived nanoporous carbon with magnetic Co nanoparticles for efficient water treatment, *Small* 10 (2014) 2096–2107.
[34]. V.K.-M. Au, Recent advances in the use of metal-organic frameworks for dye adsorption, *Front Chem.* 8 (2020) 708.
[35]. K. Wang, C. Li, Y. Liang et al, Rational construction of defects in a metal-organic framework for highly efficient adsorption and separation of dyes, *Chem. Eng. J.* 289 (2016) 486–493.
[36]. B. Bueken, N.V. Velthoven, A. Krajnc et al., Tackling the defect conundrum in UiO-66: a mixed-linker approach to engineering missing linker defects, *Chem. Mater.* 29 (2017) 10478–10486.
[37]. Z. Hasan, S.H. Jhung, Removal of hazardous organics from water using metal-organic frameworks (MOFs): plausible mechanisms for selective adsorptions, *J. Hazard. Mater.* 283 (2015) 329–339.
[38]. E. Haque, J.E. Lee, I.T. Jang et al., Adsorptive removal of methyl orange from aqueous solution with metal-organic frameworks, porous chromium-benzenedicarboxylates, *J. Hazard. Mater.* 181 (2010) 535–542.
[39]. C.-X. Yu, J. Chen, Y. Zhang et al., Highly efficient, and selective removal of anionic dyes from aqueous solution by using a protonated metal-organic framework, *J. Alloys Compd.* 853 (2021) 157383.
[40]. Z. Zong, C. Fan, X. Zhang, X. Meng, F. Jin, Y. Fan, Four Co(II) coordination polymers based on 4,4′-(1H-1,2,4-triazol-1-yl) methylenebis (benzoic acid): syntheses, structural diversity, magnetic properties, dye adsorption and photocatalytic properties, *CrystEngComm* 21 (2019) 673–686.
[41]. A.C. Tella, J.T. Bamgbose, V.O. Adimula et al., Synthesis of metal–organic frameworks (MOFs) MIL-100(Fe) functionalized with thioglycolic acid and ethylenediamine for removal of eosin B dye from aqueous solution, *SN Appl. Sci.* 3 (2021) 136.
[42]. M.N. Ahamad, M.S. Khan, M. Shahid, M. Ahmad, Metal organic frameworks decorated with free carboxylic acid groups: topology, metal capture and dye adsorption properties, *Dalton Trans.* 49 (2020) 14690–14705.
[43]. C. Li, Z. Xiong, J. Zhang, C. Wu, The strengthening role of the amino group in metal–organic framework MIL-53 (Al) for methylene blue and malachite green dye adsorption, *J. Chem. Eng. Data* 60 (2015) 3414–3422.
[44]. Y.-H. Fan, S.-W. Zhang, S.-B. Qin, X.-S. Li, S.-H. Qi, An enhanced adsorption of organic dyes onto NH_2 functionalization titanium-based metal-organic frameworks and the mechanism investigation, *Microporous and Mesoporous Mater.* 263 (2018) 120–127.
[45]. J.-M. Yang, R.-J. Ying, C.-X. Han et al., Adsorptive removal of organic dyes from aqueous solution by a Zr-based metal–organic framework: effects of Ce(III) doping, *Dalton Trans.* 47 (2018) 3913–3920.
[46]. Y. Mao, Q. Wang, L. Yu et al., A 2-fold interpenetrated nitrogen-rich metal–organic framework for rapid and selective adsorption of congo red, *Inorg. Chem.* 59 (2020) 8213–8219.
[47]. Y. Sun, M. Chen, H. Liu, Y. Zhu, D. Wang, M. Yan, Adsorptive removal of dye and antibiotic from water with functionalized zirconium-based metal organic framework and graphene oxide composite nanomaterial UiO-66-$(OH)_2$/GO, *Appl. Surf. Sci.* 525 (2020) 146614.
[48]. X. Sang, Q. Zha, X. Nie et al., Interfacial growth of metal-organic framework on carboxyl-functionalized carbon nanotube for efficient dye adsorption and separation, *Anal. Methods* 12 (2020) 4534–4540.
[49]. I. Mantasha, H.A.M. Saleh, K.M.A. Qasem, M. Shahid, M. Mehtab, M. Ahmad, Efficient and selective adsorption, and separation of methylene blue (MB) from mixture of dyes in aqueous environment employing a Cu(II) based metal organic framework, *Inorg. Chim. Acta* 511 (2020) 119787.

20 MOF-based Electrochemical Sensors for Pesticides

Yong Wang[1,2], *Qin Xiao*[1] *and Qianfen Zhuang*[1]
[1] College of Chemistry, Nanchang University, Nanchang, China
[2] Jiangxi Province Key Laboratory of Modern Analytical Science, Nanchang University, Nanchang, China

20.1 INTRODUCTION

In the agri-food industry, pesticides are widely used for the control, prevention, and elimination of pests to increase crop yields [1-3]. However, pesticides usually possess high toxicity and long half-lives and are related to many health and environmental problems [1-3]. Therefore, it is imperative to establish a fast, efficient, and sensitive analytical method for the detection of pesticide residue. At present, chromatography-based methods are regarded as the standard for the detection of pesticides [4-7]. Despite their high sensitivity, these chromatography-based methods require skilled personnel, expensive equipment, long pretreatment times, and a lot of toxic solvents, which limit their applications. In the last few years, electrochemical sensors, coupled with advanced materials, have become promising alternatives due to their outstanding advantages such as simple operation, rapid analysis, high sensitivity, good miniaturization, and low cost [8-14]. Among advanced materials, metal-organic frameworks (MOFs), that are a kind of highly ordered crystalline material composed of metal nodes and organic linkers, have recently attracted wide attention from researchers [15, 16]. The MOFs and their derived materials possess excellent physicochemical properties such as high porosity, large specific surface area, high thermal stability, strong chemical stability, excellent catalytic activity, and are thus well-situated as the electrode modifier for the construction of electrochemical sensors [15, 16].

In this chapter, depending on the electrochemical signal transduction mechanism, different types of electrochemical (voltammetric, impedimetric, conductometric, electro-chemiluminescent, and photoelectrochemical) sensors are respectively outlined, and some typical work on the use of MOFs in the aforementioned electrochemical sensors is presented.

20.2 APPLICATION OF MOF-BASED ELECTROCHEMICAL SENSORS FOR PESTICIDES

20.2.1 MOF-based Voltammetric Sensor

Among the electrochemical sensors, the voltammetric sensor is one of the most widely used sensors because of its high sensitivity and high selectivity. A series of stable MOFs such as UiO series (UiO stands for the University of Oslo), MIL series (MIL refers to Materials of Institute Lavoisier), ZIF series (ZIF denotes Zeolitic Imidazolate Frameworks), are used to construct various voltammetric sensors for pesticide detection. Based on different MOF families, the following section will introduce their application in the voltammetric sensor for pesticides.

20.2.1.1 UiO-based Voltammetric Sensor

In 2018, Meng et al. [17] pyrolyzed Pt/UiO-66 MOFs in one step to synthesize Pt/ZrO_2/porous carbon tri-component nanohybrids for the first time. It was found that the composite material possessed high binding toward the phosphate group and strong electrocatalytic ability for methyl parathion. The composite material was then used for the construction of a highly sensitive voltammetric sensor for methyl parathion. The good performance of the sensor was confirmed to originate from the outstanding electrocatalytic ability of Pt species, the good conductivity of porous carbon, and the strong adsorption properties of ZrO_2 for methyl parathion. In 2019, Mahmoudi et al. [18] prepared UiO-66 MOFs by a hydrothermal method and introduced Ce metal into the UiO-66 MOFs to produce Ce-doped UiO-66 (Ce/UiO-66). Subsequently, they added multi-walled carbon nanotubes (MWCNTs) into Ce/UiO-66 to form the composite material, and coated acetylcholinesterase (AChE) to construct a biosensor for the detection of organophosphate pesticide (OPs) (Figure 20.1). The biosensor presented good performance due to the large surface area and good conductivity of the MWCNTs together with appropriate oxophilicity of Ce to acetylthiocholine. Bagheri's group [19] reported the use of TiO_2 functionalized graphene oxide@UiO-66 to construct an enzymeless electrochemical sensor for simultaneous detection of paraoxon and chlorpyrifos. The developed sensor had high sensitivity and a low detection limit because the modified electrode had good electrical conductivity and a unique structure with the increased electron transfer rate. Yang et al. [20] firstly prepared two UiO-66-based electrochemical signal tags by the adsorption of Cd(II) or Pb(II) ions onto the surface of UiO-66-NH_2 and the immobilization of the antigen of triazophos or thiacloprid, and then fabricated two capture probes by coupling the antibody of

DOI: 10.1201/9781003188148-20

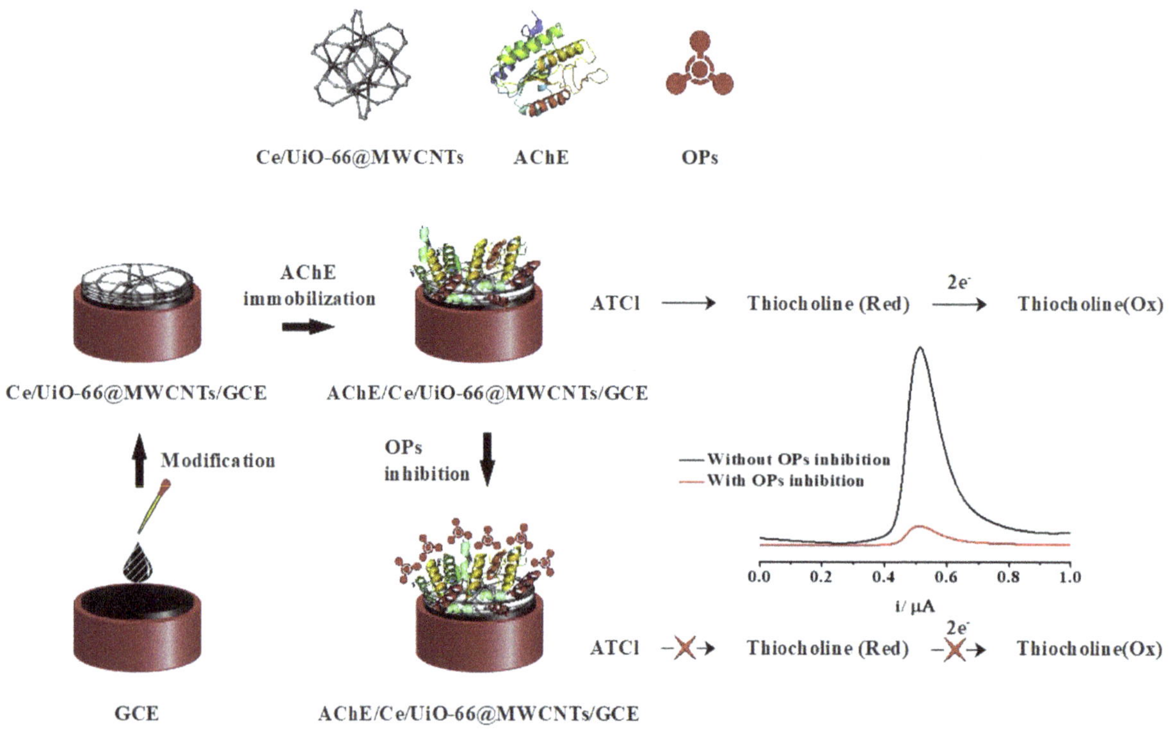

FIGURE 20.1 Schematic diagram of the AChE/Ce/UiO-66@MWCNTs/GCE-basedvoltammetric sensing of organophosphate pesticide. Adapted with permission from [18]. Copyright (2019) Elsevier.

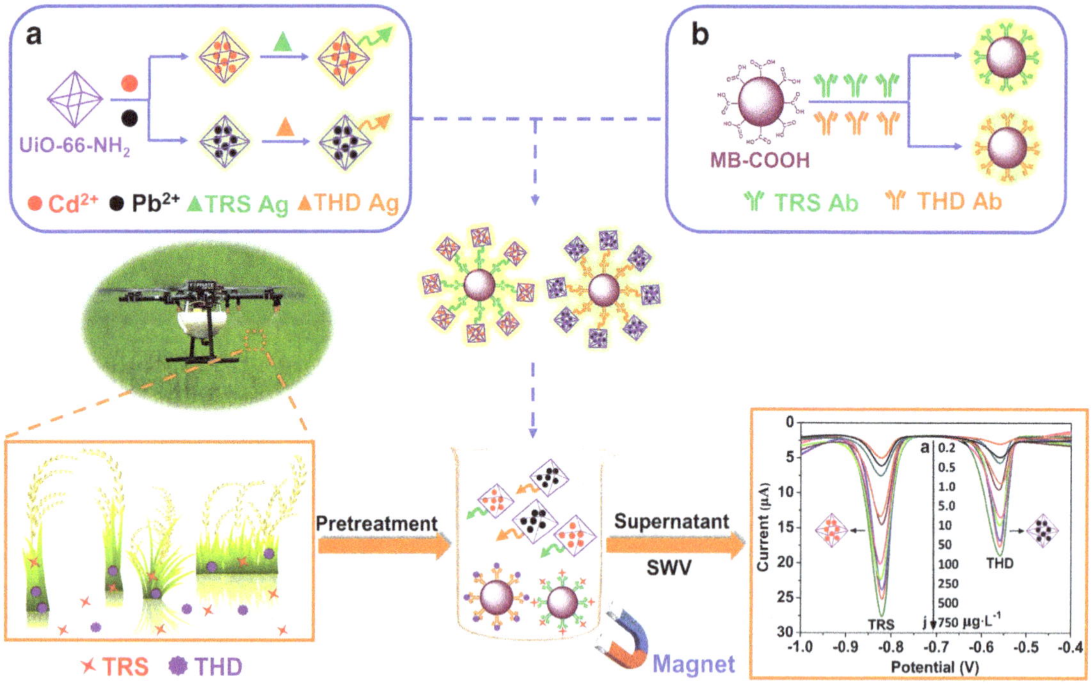

FIGURE 20.2 Schematic diagram of the magnetic bead and UiO-66-NH_2 adsorbed with Cd(II) and Pb(II) ions for voltammetric imunosensing of simultaneous detection of triazophos (TRS) and thiacloprid (THD). Adapted with permission from [20]. Copyright (2019) Springer Nature.

triazophos or thiacloprid with a carboxyl modified magnetic bead (Figure 20.2). Based on the competitive heterogeneous voltammetric immunoassay and magnetic separation, the triazophos and thiacloprid can be simultaneously determined with high sensitivity and high selectivity. Ma et al. [21] hydrothermally prepared UiO-66 MOFs and introduced Pt nanoparticles onto the UiO-66 MOF to give Pt nanoparticle-anchored UiO-66 (Pt@UiO-66). Subsequently, they modified the Pt nanoparticle-anchored UiO-66 together with chitosan and acetylcholinesterase on the glassy carbon electrode to construct a biosensor for malathion detection. The biosensor for malathion gave high sensitivity due to the combination of excellent conductivity with the increasing adsorption sites and high surface area of Pt@UiO-66. In 2020, Xu et al. [22] reported the use of UiO-66 MOFs with isolated dispersed Pt nanoparticle active sites to modify the carbon paste microelectrode and exploited organophosphorus insecticide phosalone as a template molecule to synthesize the mesoporous structured conductive molecularly imprinted polymer on the aforementioned modified electrode by electropolymerization and a subsequent sol-gel process. Then, they used the modified electrode to construct a voltammetric sensor for phosalone detection. The sensor displayed excellent performance because the modified electrode had high porosity, a large surface area, increased electron-transfer rate, fast diffusion ability, and high molecular recognition specificity. Yao et al. [23] used amine-functionalized UiO-66 MOFs, multiwalled carbon nanotube@reduced graphene oxide nanoribbon, and a kind of covalent organic framework material (the material is produced using melamine and cyanuric acid as monomers through polycondensation) to prepare the nanocomposite, and exploited nanocomposite together with aptamer for the construction of a voltammetric aptasensor for the detection of kanamycin. Interestingly, due to its rich chemical functionality and amino-rich structure, such nanocomposite provided high binding ability toward the aptamer, causing the improved sensor's sensitivity. In 2021, Xu et al. [24] hydrothermally prepared UiO-66 MOFs, and then coupled a MOF with a complementary probe and carboxy-ferrocene to give a nanocomposite probe. After that, the nanocomposite probe can be hybridized with malathion aptamer. On the basis of the strong binding between the aptamer with malathion, the nanocomposite probe can be liberated, giving a decreased voltammetric signal. The constructed sensor showed high sensitivity because of the high specific surface area of UiO-66.

20.2.1.2 MIL-based Voltammetric Sensor

In 2017, Shi et al. [25] synthesized Fe_3O_4/reduced graphene oxide (Fe_3O_4/rGO) and core-shell molecularly imprinted polymer-coated metal-organic framework MIL-101 to construct a sensor for the detection of methamidophos or omethoate (Figure 20.3). The sensor was then used for the detection of methamidophos or omethoate in a rape sample with satisfactory results. In 2018, Hadi et al. [26] prepared the

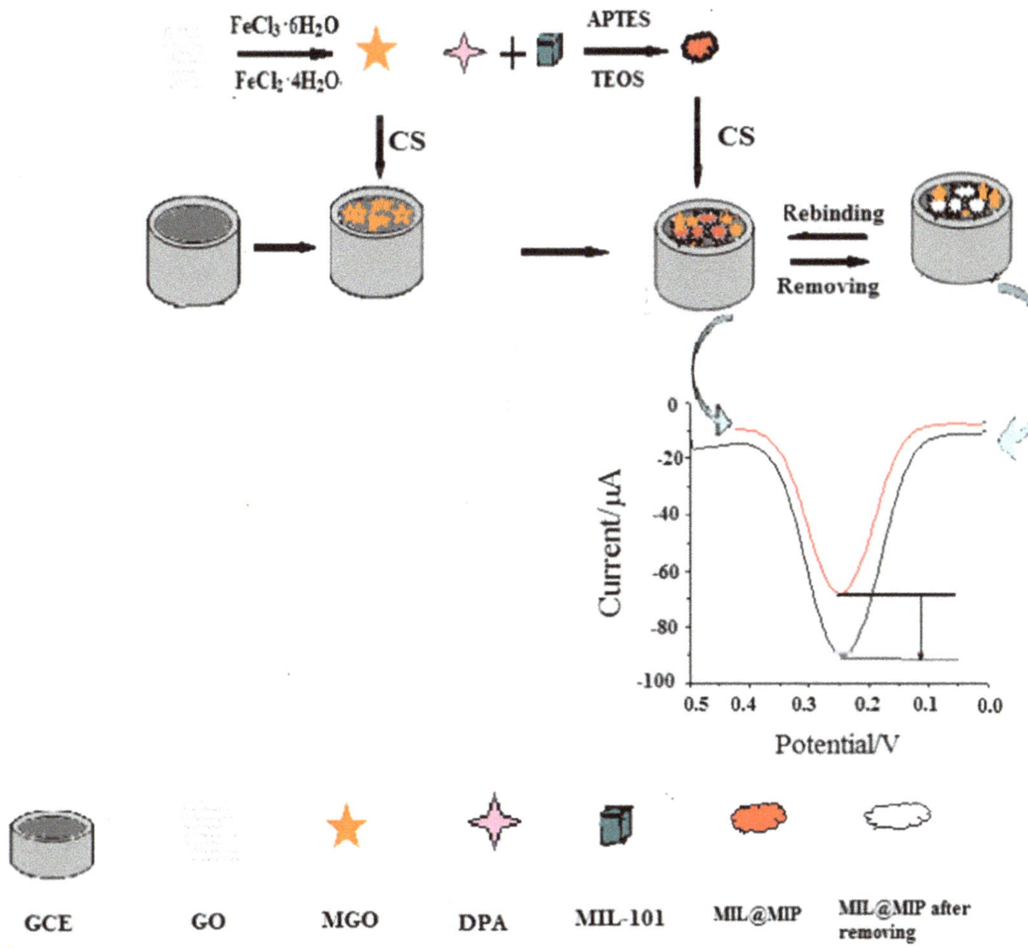

FIGURE 20.3 Schematic illustration of the application of the Fe_3O_4/reduced graphene oxide and core-shell molecularly imprinted polymer-coated metal-organic framework MIL-101 for sensing of detection of methamidophos or omethoate. Adapted with permission from [25]. Copyright (2016) Springer Nature.

nanocomposite consisting of multi-walled carbon nanotube and MIL-101(Cr) for the modification of a glassy carbon electrode, and the modified electrode was further utilized for voltammetric detection of picloram in tap and river water samples. In 2019, Soltani-Shahrivar et al. [27] solvothermally synthesized MIL-101(Fe)-reduced graphene oxide nanocomposite for modification of a glassy carbon electrode, and the modified electrode can be used for simultaneous voltammetric detection of carbofuran and carbaryl.

20.2.1.3 ZIF-based Voltammetric Sensor

In 2018, Ma et al. [28] reported the use of lipase@amine-modified zeolitic imidazolate framework-8 (An-ZIF-8)/chitosan/glassy carbon electrode (GCE) and lipase@ZIF-8/chitosan/GCE to construct sensors for the detection of p-nitrophenyl palmitate or methyl parathion (Figure 20.4). The sensor displayed good performance such as strong stability, wide linear range, and low limit of detection. In 2019, Gan et al. [29] reported the encapsulation of highly dispersed Au nanorods inside the ZIF-8 by epitaxial growth or nucleus coalescence and further used the aforementioned composite to encapsulate graphene oxide nanosheets. The obtained materials showed enhanced chemical stability, avoided the dissolution, aggregation, and migration of Au nanorods, accelerated the mass transfer of reactants to Au nanorods, and displayed high electrocatalytic ability toward niclosamide, dichlorophen, carbendazim, and diuron. Therefore, the composite material was used for the construction of a sensor for the detection of the four pesticides. In 2020, Cheng et al. [30] reported the encapsulation of nano-Burkholderia cepacia lipase onto the hierarchical ordered macro-microporous ZIF-8 structures and found that the composite had good crystallinity, high thermal stability, large surface area, and strong enzyme activity. Therefore, they utilized the composite to fabricate voltammetric biosensors for the detection of nitrogenous diphenyl ether pesticides. Li et al. [31] reported the use of acetylcholinesterase as a recognition element and ZIF-8 as a degradable carrier for encapsulating electroactive methylene blue to construct an immobilization-free homogeneous voltammetric sensor for the detection of paraoxon. The principle of the sensor was as follows: acetylcholinesterase can catalyze hydrolytic reactions to produce the proton. The resultant acidic solution can cause the release of methylene blue in the ZIF-8, giving a strong current. However, in the presence of acetylcholinesterase inhibitors, paraoxon, the release of methylene blue was reduced, resulting in the current decline.

20.2.1.4 Voltammetric Sensor Based on Other MOFs

In 2012, Wang et al. [32] integrated benzoic acid-functionalized single-walled carbon nanotubes with Zn^{2+}

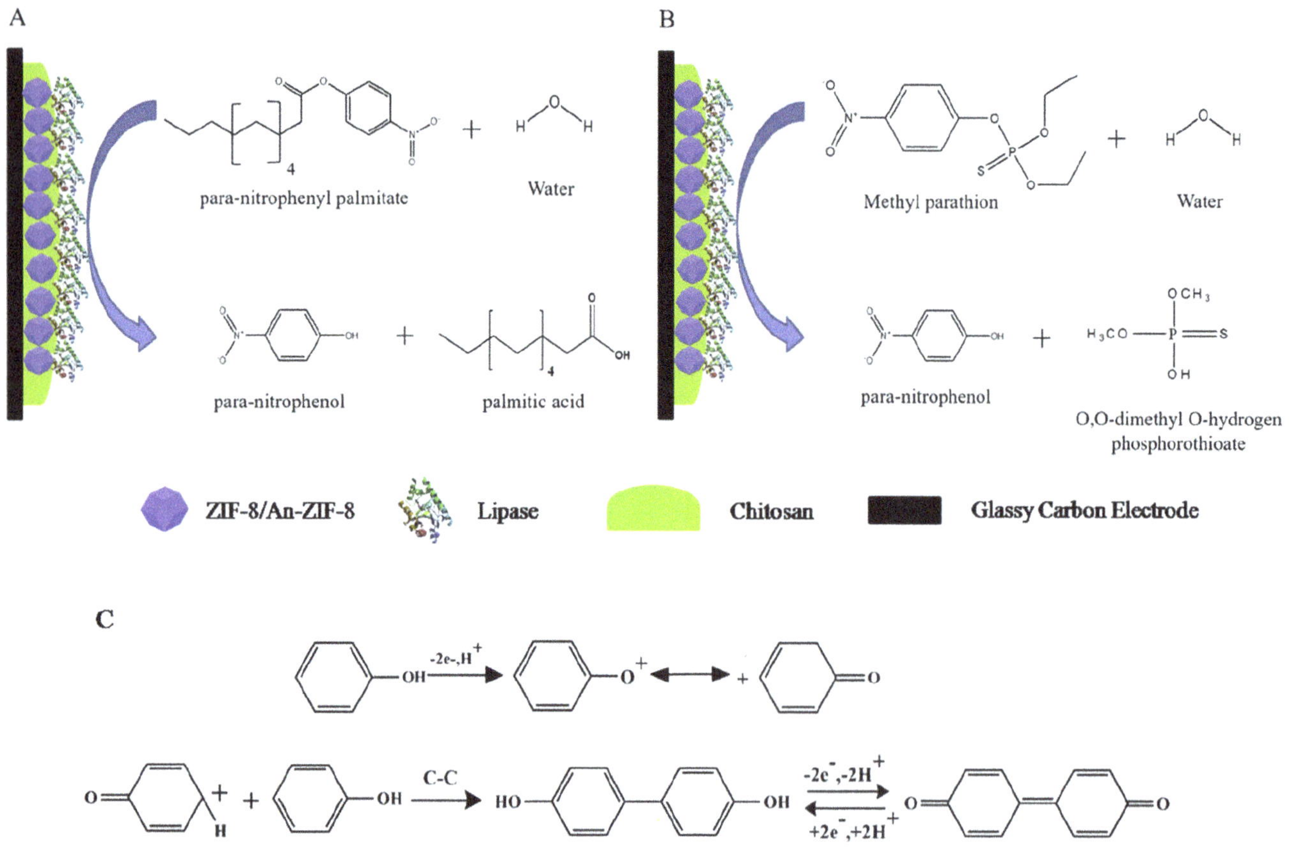

FIGURE 20.4 Schematic diagram of lipase@An-ZIF-8/chitosan/GCE-based sensor for the detection of p-nitrophenyl palmitate (A) and methyl parathion (B). (C) Reaction mechanism of phenol oxidation and its related dimer formation. Adapted with permission from [28]. Copyright (2018) Elsevier.

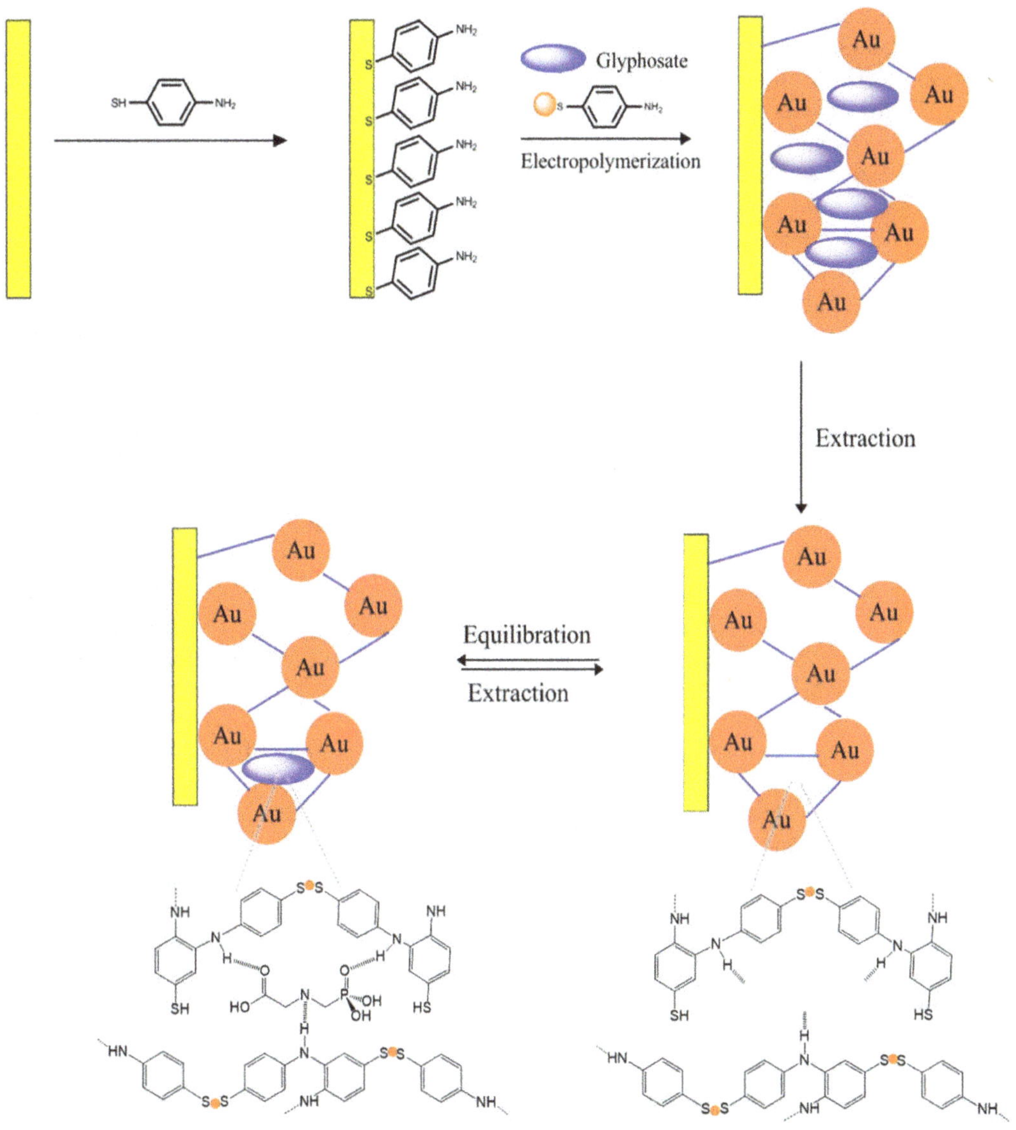

FIGURE 20.5 A fabrication process of the glyphosate sensor based on molecularly imprinted MOF films via the electropolymerization of PATP-functionalized gold nanoparticles. Adapted with permission from [34]. Copyright (2015) Taylor & Francis.

to give three-dimensional porous inorganic-organic hybrid frameworks with large surface areas and high conductivity as well as good dispersibility. And then, they exploited the materials to construct a voltammetric sensor for the sensitive detection of methyl parathion. In 2014, Hu et al. [33] incorporated Fe_3O_4 nanoparticles with Co MOF to give a magnetic nanocomposite with large surface area and good compatibility. The methyl parathion was adsorbed and enriched onto the magnetic nanocomposite and separated from the solution via a magnet. The as-prepared magnetic nanocomposite containing methyl parathion can be transferred onto a magnetic glassy carbon electrode and be detected using the voltammetric technique. In 2015, Do et al. [34] prepared the molecularly imprinted MOF films on a gold electrode by electropolymerization of p-aminothiophenol (PATP)-functionalized gold nanoparticles (AuNPs) in the presence of glyphosate as template molecules. The molecularly imprinted MOF films were found to have high selectivity due to their specific recognition via hydrogen bonds between glyphosate and aniline moieties. And then, they exploited the MOF films to construct a voltammetric sensor for highly sensitive and highly selective detection of glyphosate (Figure 20.5).

In 2018, Song et al. [35] employed a simple room temperature mixing method to prepare the ball-flower-like Cu-hemin MOFs on a flexible three-dimensional nitrogen-containing melamine carbon foam composites and then immobilized acetylcholinesterase onto the as-prepared composites for the detection of trichlorfon. They found that the composite possessed a suitable pore size that avoided the stacking of the protease and contained nitrogen elements that greatly improved the biocompatibility. Dong et al. [36] prepared three kinds of MOFs, based on the reaction of 2-aminoterephthalate and three different metal ions like Fe^{3+}, Zr^{4+}, and La^{3+}, and annealed them under N_2 atmosphere to produce three kinds of carbon-supported metal oxide hybrids. The MOF-derived materials were then used for immobilizing acetylcholinesterase to construct methyl parathion biosensors. It was found that the biosensor based on La-MOF-derived materials had the best electrochemical performance due to rich active sites, fast electron transfer rate, and short diffusion length. In 2019, Xie et al. [37] synthesized the copper-cerium oxide composite by calcination of Cu(II)/Ce(III) MOFs and used the as-prepared materials to construct a non-enzyme voltammetric sensor for the detection of malathion. It was found that the addition of malathion inhibited the current of CuO because of the binding between CuO and sulfur-containing functional groups. Cao et al. [38] used the hierarchical Cu-BTC MOFs to construct a voltammetric sensor for the detection of glyphosate. The sensor had high reproducibility, good stability, and high selectivity toward the major metabolite of glyphosate aminomethylphosphonic acid and other interference. Al'Abri et al. [39] synthesized the copper-based porous coordination polymer to modify the carbon paste electrode and then exploited the modified electrode to construct a voltammetric sensor for the detection of malathion. Wang et al. [40] synthesized MOF nanofibers with good performance to fabricate Burkholderia cepacia lipase@MOF nanofiber biosensors for detection of methyl parathion with high sensitivity. Nagabooshanam et al. [41] modified gold microelectrode with zinc-based metal-organic frameworks (MOF-Basolite Z1200) and acetylcholinesterase enzyme to construct an electrochemical microanalytical device for voltammetric detection of organophosphate pesticide chlorpyrifos in the food chain. The obtained sensor possessed many advantages such as portability, good stability, fast response time, and low cost. Tu et al. [42] prepared a reduced graphene oxide-encapsulated Ce-MOF to construct a voltammetric sensor for the detection of dichlorophen. The structure of the reduced graphene oxide-encapsulated Ce-MOF can prevent the decomposition of the MOF in solution and improve the conductivity via the combination of reduced graphene oxide and MOF. Song et al. [43] respectively synthesized MnO_2/Mn_3O_4 hierarchical micro cuboids derived from a Mn-MOF and Ti_3C_2 MXene/Au NPs composite to modify the glassy carbon electrode for the construction of a voltammetric sensor for the detection of organophosphorus pesticides. The sensor provided excellent electrochemical performances due to the synergistic signal amplification effect of MnO_2/Mn_3O_4 and Ti_3C_2 MXene/Au NPs composites, the large specific surface area, the extraordinary biocompatibility, and good conductivity. Qiao et al. [44] developed a 'signal-on' voltammetric aptasensor for the detection of acetamiprid with high sensitivity and high selectivity. They employed gold nanoparticles-CuMOF to label probe DNA as a signaling element and immobilized the complementary strand with the gold nanoparticles-reduced graphene oxide. The addition of acetamiprid can cause the hybridization of an Au-CuMOF labeled probe DNA with the exposed complementary DNA, and as a result, the CuMOF can approach the electrode and produce a strong current. In 2020, Gu et al. [45] hydrothermally prepared the metal-organic framework HKUST-1, and directly pyrolyzed the MOF to its derivative, CuOx@mC composite. Then, they respectively constructed electrochemical sensors for the detection of glyphosate-based on HKUST-1 or CuOx@mC composite. The results showed that the CuOx@mC-based sensor had better electrochemical performance towards glyphosate than that HKUST-1-based sensor, due to the mesoporous structure and high conductivity of CuOx@mC. Basu's group [46] designed an interface consisting of intertwined interfaces of a metal-organic framework-5 (MOF-5), gold nanorods, cysteamine functionalization, and the acetylcholinesterase (AChE) to construct a voltammetric biosensor for the detection of the organophosphate group of pesticides. They then exploited molecular docking to study the structural aspects of the molecular recognition of organophosphate pesticides. And it was found that the biosensor offered high stability, good specificity, and anti-interference properties. Rashi Bhardwaj et al. [47] concatenated polyclonal antibody (rIgG) based nano immuno-hybrid (Chi-AuNP-rIgG-BSA) on MOF modified ITO substrate to develop a voltammetric sensor for the detection of chlorpyrifos, hexaconazole, bifenthrin, imidacloprid, and dimethoate with high sensitivity and wide linear range. Molecular docking analysis of the binding of the pesticide and rIgG gave an insight into their interaction. The constructed sensor had many merits like short incubation, rapid analysis time, high stability, and good selectivity. Tu et al. [48] used MXene/carbon nano horns/ β-cyclodextrinmetal-organic frameworks to develop an electrochemical sensor for carbendazim pesticide detection. In the construction process of the sensor, β-CD MOFs possessed a host-guest recognition ability and porous structure with high porosity, leading to the enhanced adsorption of carbendazim. In addition, MXene/carbon nano horns had a large specific surface area, good conductivity, and a lot of active sites, which facilitated the mass transfer and electrocatalysis of carbendazim. As a result, the carbendazim sensor was demonstrated to have high selectivity, high sensitivity, and good stability. In 2021, Karimi-Maleh et al. [49] employed a hydrothermal method coupled with thermal annealing to prepare Co_3O_4 nanowire. Then, they prepared core-shell Co_3O_4@MOF-74 nanocomposite by the solvothermal method and modified the nanocomposite onto the electrode. The modified electrode was electropolymerized using pyrrole as a monomer to obtain molecularly imprinted electrodes. It was found that the molecularly imprinted sensor could be used for the detection of organophosphorus insecticide fenamiphos with high sensitivity, high selectivity, and strong stability (Figure 20.6). Duan et al. [50] solvothermal synthesized CoS-nanoparticles attached ZnS rods using cobalt(II)-ion exchanged zinc-based biological metal-organic framework-1 as precursors and then dropped the composite onto the electrode to obtain the modified electrode. After electropolymerization of pyrrole as a monomer, the molecularly imprinting modified electrode was used for the construction of a chloroneb sensor. It was found that CoS and ZnS had electrocatalytic activity, and the porous structure of the material renders electrocatalysts with high surface area and rich porosity, resulting in the improvement of the sensors' sensitivity. Wang et al. [51] employed the solvothermal method coupled, with pyrolysis,

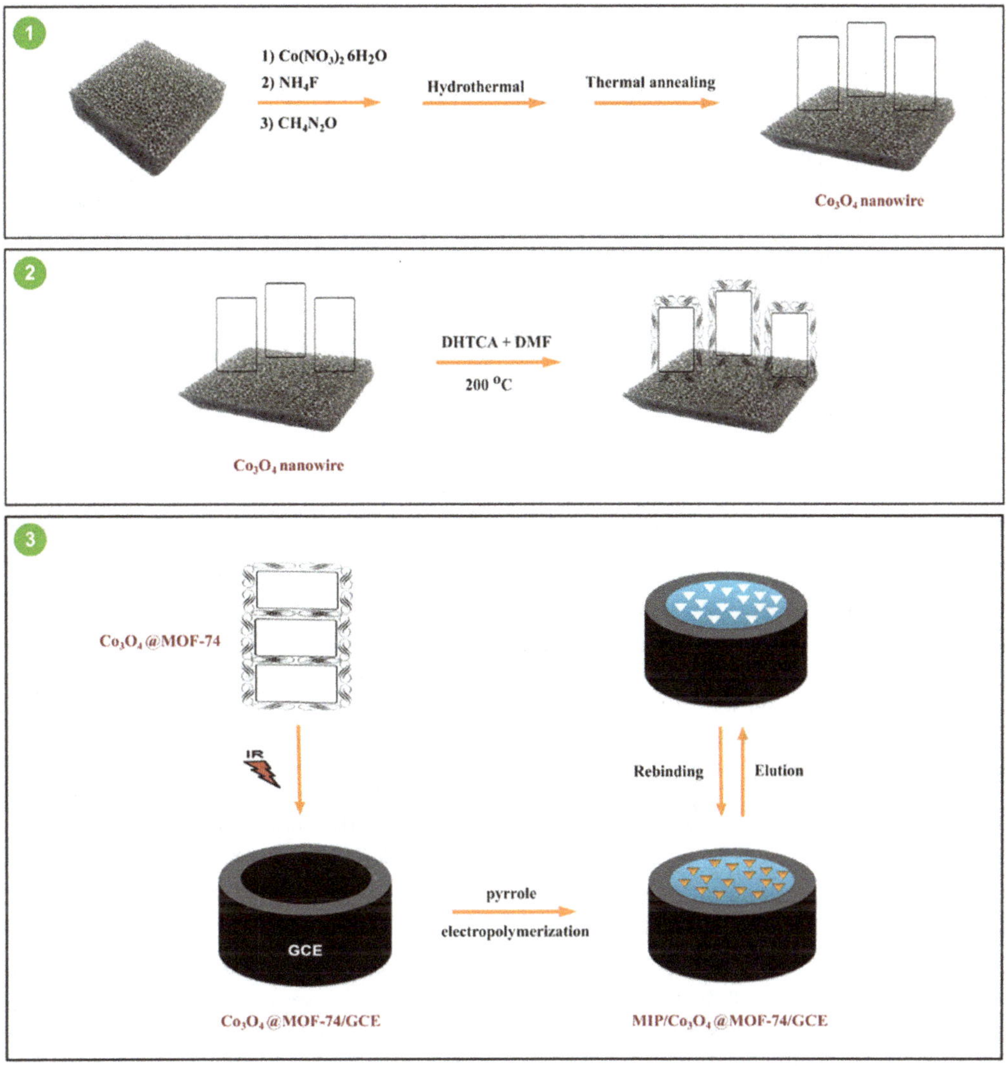

FIGURE 20.6 A fabrication process of the fenamiphos sensor based on the core-shell Co_3O_4@MOF-74 nanocomposite and molecularly imprinted technique. Adapted with permission from [49]. Copyright (2021) Elsevier.

to prepare a derivative of poly-vinylpyrrolidone doped Cu MOFs, three-dimensional nitrogen-doped macro-meso-microporous carbon composites (N/Cu–HPC). The N/Cu–HPC was used for fabricating a voltammetric sensor for the detection of neonicotinoid pesticides. It was reported that the addition of poly-vinylpyrrolidone improved the hydrophilicity of N/Cu–HPC and enhanced the electrocatalytic ability toward neonicotinoids due to a rich electronegative nitrogen species on the N/Cu–HPC. In addition, the macro-meso-microporous structure facilitated the mass and charge transfers between neonicotinoids and the N/Cu–HPC.

20.2.2 MOF-based Impedimetric Sensors

Electrochemical impedance spectroscopy (EIS) is an important label-free biosensing technique for the detection of a broad range of target analytes due to its ability to obtain specific frequency responses based on the combination of the target analyte and molecular recognition element [52-54]. In 2015, Deep et al. [55] obtained nano-metal organic framework films by successive dipping of a 2-aminobenzylamine modified indium tin oxide electrode in a solution containing 2-aminoterephthalic acid and Cd^{2+}. Subsequently, they conjugated the pendent –COOH functional groups on the MOF film with an anti-parathion antibody to develop an impedimetric immunosensor for parathion. It was found that the sensor had superior specificity towards other organophosphate pesticides like malathion, fenitrothion, monochrotophos, paraoxon, and dichlorovos. In 2020, Nagabooshanam et al. [56] used MOFs consisting of zinc metal and an imidazole ligand as a transducing element and acetylcholinesterase as a recognition element to construct an electrochemical micro paper analytical biosensing device for ultrasensitive impedimetric detection of chlorpyrifos. In addition, the integration of portable electronics based on an Arduino microcontroller and artificial intelligence into the biosensing platform gave many advantages like portability, low cost, and user-friendliness.

20.2.3 MOF-based Conductometric Sensors

Conductometry is one of the most widely used techniques of electroanalytical chemistry and is usually used for developing an electrochemical sensor for the detection of an analyte of interest due to its simplicity, label-freeness, and low cost [57]. In 2015, Bhardwaj et al. [58] assembled thin films of a silica-modified copper-metal organic framework, $Cu_3(BTC)_2$, onto a conducting substrate of 2-aminobenzene-1,4-dicarboxylic acid doped polyaniline, and conjugated the film with antiatrazine antibodies to produce an immunosensing conductometric sensing platform for antiatrazine detection. The conductometric sensor provided high sensitivity and high specificity in the presence of paraoxon, malathion, monochrotophos, endosulfan, and parathion.

20.2.4 MOF-based Electrochemiluminescence Sensors

Electrochemiluminescence is the process where species produced at the surface of electrodes transit to excited states via electron-transfer reactions and then emit light [59-62]. The electrochemiluminescence sensor combines the advantages of electrochemistry and luminescence, and thus attracts wide interest from researchers. In 2020, Chen et al. [63] used the NH_2-MIL-88(Fe) MOFs to immobilize CdTe quantum dots on the interior and outer surface. The MOFs containing CdTe quantum dots can be employed to construct an electrochemiluminescence aptasensor for the detection of malathion (Figure 7). It was found that the MOFs played an important role as nanocarriers and signal enhancers to improve the sensitivity of the electrochemiluminescence aptasensor. In addition, the authors established the detailed mechanism of the significantly enhanced electrochemiluminescence.

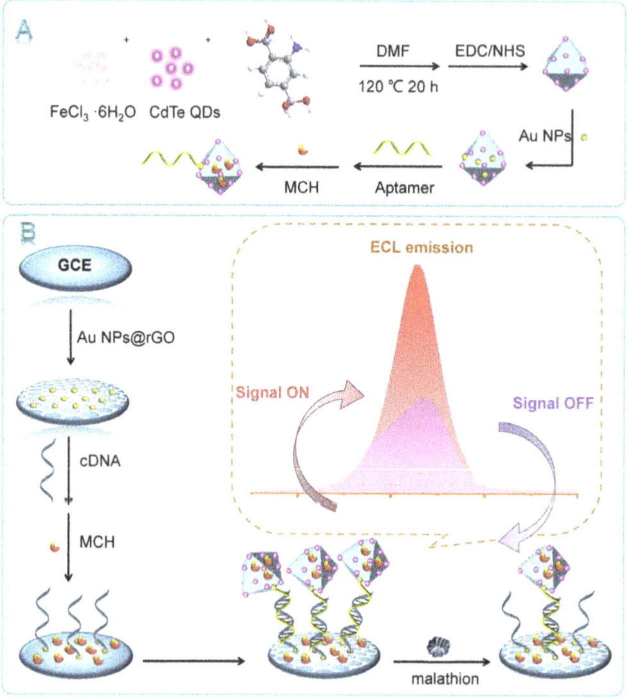

FIGURE 20.7 (A) Preparation of NH_2-MIL-88(Fe) MOFs containing CdTe quantum dots and their corresponding aptamer signal probes. (B) Construction process of the aptasensor sensor for malathion detection and its related sensing mechanism. Adapted with permission from [63]. Copyright (2021) Elsevier.

20.2.5 MOF-based Photoelectrochemical Sensors

A photoelectrochemical sensor involves developing a device with light as excitation source and photocurrent or photovoltage as detection signal [64-66]. The sensor possesses many good characteristics such as high sensitivity, simple equipment, and fast response. In 2015, Jin et al. [67] solvothermal synthesized an amino-functionalized metal-organic framework modified with TiO_2 (amino-MIL-125/TiO_2) and placed the material onto a glassy carbon electrode to construct a photoelectrochemical sensor for the detection of the herbicide, clethodim. The photoelectrochemical mechanism for clethodim oxidation at the modified glassy carbon electrode is displayed in Figure 20.8. Under visible light irradiation, the amino-MIL-125/TiO_2 delivered electrons to the glassy carbon electrode, and the leaving positively charged holes were located on the surface. Subsequently, the hydroxy radicals could be generated via the reaction of the holes with H_2O and were rapidly attacked by clethodim to enhance the efficiency of charge separation. As a result, the photocurrent was noticeably increased. In 2017, Jin et al. [68] used the solvothermal synthesized NH_2-MIL-125(Ti)/TiO_2 composite together with chitosan (CS) for immobilization of glucose oxidase (GOx) to develop a sensitive visible-light activated photoelectrochemical sensor for the detection of acetochlor. It was found that the visible-light activated GOx/CS/NH_2-MIL-125(Ti)/TiO_2 composite could produce a photocurrent in the presence of glucose, but the addition of acetochlor inhibited the GOx activity, leading to a decline in photocurrent. In 2019, Cao et al. [69] synthesized the nanocomposite containing the hierarchically porous Cu-BTC (BTC refers to benzene-1,3,5-tricarboxylic acid) MOF and g-C_3N_4 nanosheet and used the nanocomposites to construct a photoelectrochemical sensor for the detection of glyphosate. Among the nanocomposites, the Cu metal center of the hierarchically porous Cu-BTC can be strongly associated with the glyphosate to form related complexes, causing poor electron transfer, and consequently leading to a remarkable decline in photocurrent. The nanocomposite-based sensor had the advantage of rapid and easy operation.

20.3 CONCLUSION

In this chapter, the use of MOFs and their derived materials to construct an electrochemical sensing platform for pesticide detection are outlined. Depending on the electrochemical signal transduction mechanism, different types of electrochemical

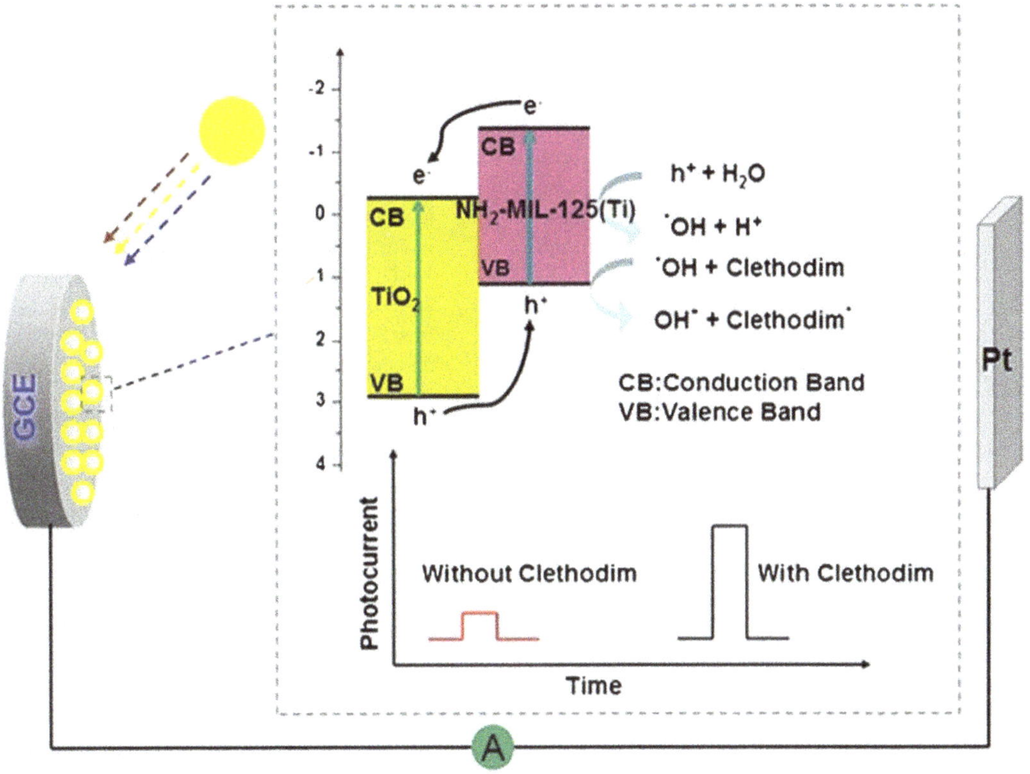

FIGURE 20.8 Schematic diagram of the photoelectrochemical sensing mechanism for the clethodim detection at the amino-MIL-125/TiO_2 modified GCE. Adapted with permission from [67]. Copyright (2015) Springer Nature.

(voltammetric, impedimetric, conductometric, electrochemiluminescence, and photoelectrochemical) sensors are introduced in detail. Although there is a rapid development trend for MOF-based electrochemical sensors for pesticides, some challenges are still to be addressed. Firstly, the present MOF-based electrochemical sensors pay much more attention to the extension of the analytical application and neglect the detailed investigations on the relationship between the structure/composition of MOFs and their electrochemical performances and stability on a microscopic scale. In the future, advanced separation and characterization techniques coupled with theoretical calculations need to be developed to reveal their relationship. Secondly, the performance (sensitivity, selectivity, response time, stability, and reproducibility) of the electrochemical sensor is directly related to the structure/composition of the materials. Detailed insights of the above-mentioned structure/composition-property relationship can give a guide to designing an electrochemical sensor with excellent performance. However, this is not sufficient. Many new electrochemical principles and new advanced materials, as well as many amplification techniques, can further aid the improvement of the electrochemical sensor's performances.

Finally, the application of most MOF-based electrochemical sensors in real samples is limited because the real samples are complex, and they contain many serious matrix effects and background interferences. To this end, it is essential to combine sample pretreatment techniques with multivariate statistic tools like multivariate curve resolution by alternating least-squares and parallel factor analysis.

ACKNOWLEDGEMENTS

This book chapter was supported by the National Natural Science Foundation of China (NSFC–32160600, NSFC–21864017 and NSFC–31960495), and the Science and Technology Innovation Platform Project of Jiangxi Province (20192BCD40001).

REFERENCES

[1]. Sherma J (1995) Pesticides. *Anal. Chem.* 67:1–20
[2]. Samsidar A, Siddiquee S, Shaarani SM (2018) A review of extraction, analytical and advanced methods for determination of pesticides in environment and foodstuffs. *Trends Food Sci. Technol.* 71:188–201
[3]. Sulaiman NS, Rovina K, Joseph VM (2019) Classification, extraction, and current analytical approaches for the detection of pesticides in various food products. *J. Consum. Prot. Food Saf.* 14:209–221
[4]. Van Buuren C, Lawrence JF, Brinkman UA, Honigberg IL, Frei RW (1980) Reversed-phase liquid chromatography of basic drugs and pesticides with a fluorigenicion-pair extraction detector. *Anal. Chem.* 52:700–704
[5]. Tekel J, Hatrik S (1996) Pesticide residue analyses in plant material by chromatographic methods: Clean-up procedures and selective detectors. *J. Chromatogr. A* 754:397–410
[6]. Xu ML, Liu JB, Lu J (2014) Determination and control of pesticide residues in beverages: A review of extraction techniques, chromatography, and rapid detection methods. *Appl. Spectrosc. Rev.* 49:97–120

[7]. Rahman MM, Abd El-Aty AM, Kim SW, Shin SC, Shin HC, Shim JH (2017) Quick, easy, cheap, effective, rugged, and safe sample preparation approach for pesticide residue analysis using traditional detectors in chromatography: A review. *J. Sep. Sci.* 40:203–212

[8]. Skladal P (1996) Biosensors based on cholinesterase for the detection of pesticides. *Food Technol. Biotechnol.* 34:43–49

[9]. Pundir CS, Chauhan N (2012) Acetylcholinesterase inhibition-based biosensors for pesticide determination: A review. *Anal. Biochem.* 429:19–31

[10]. Xia N, Gao YP (2015) Carbon nanostructures for development of acetylcholinesterase electrochemical biosensors for determination of pesticides. *Int. J. Electrochem. Sci.* 10: 713–724

[11]. Arduini F, Cinti S, Scognamiglio V, Moscone D (2016) Nanomaterials in electrochemical biosensors for pesticide detection: advances and challenges in food analysis. *Microchim. Acta* 183:2063–2083

[12]. Zhao FN, Wu J, Ying YB, She YX, Wang J, Ping JF (2018) Carbon nanomaterial-enabled pesticide biosensors: Design strategy, biosensing mechanism, and practical application. *Trends Anal. Chem.* 106:62–83

[13]. Noori JS, Mortensen J, Geto A (2020) Recent development on the electrochemical detection of selected pesticides: A focused review. *Sensors* 20:2221

[14]. Wang WR, Wang XX, Cheng N, Luo YB, Lin YH, Xu WT, Du D (2020) Recent advances in nanomaterials-based electrochemical (bio)sensors for pesticides detection. *Trends Anal. Chem.* 132:116041

[15]. Hou XD, Xu H, Zhen TY, Wu W (2020) Recent developments in three-dimensional graphene-based electrochemical sensors for food analysis. *Trends Food Sci. Technol.* 105:76–92

[16]. Tajik S, Beitollahi H, Nejad FG, Sheikhshoaie I, Nugraha AS, Jang HW, Yamauchi Y, Shokouhimehr M (2021) Performance of metal-organic frameworks in the electrochemical sensing of environmental pollutants. *J. Mater. Chem. A* 9:8195–8220

[17]. Meng TY, Wang L, Jia HX, Gong T, Feng Y, Li RX, Wang H, Zhang YF (2018) Facile synthesis of platinum-embedded zirconia/porous carbons tri-component nanohybrids from metal-organic framework and their application for ultra-sensitively detection of methyl parathion. *J. Colloid Interface Sci.* 536:424–430

[18]. Mahmoudi E, Fakhri H, Hajian A, Afkhami A, Bagheri H (2019) High-performance electrochemical enzyme sensor for organophosphate pesticide detection using modified metal-organic framework sensing platforms. *Bioelectrochemistry* 130:107348

[19]. Karimian N, Fakhri H, Amidi S, Hajian A, Arduini F, Bagheri H (2019) A novel sensing layer based on metal-organic framework UiO-66 modified with TiO_2-graphene oxide: application to rapid, sensitive, and simultaneous determination of paraoxon and chlorpyrifos. *New J. Chem.* 43:2600–2609

[20]. Yang Y, Cheng JL, Wang B, Guo YR, Dong XW, Zhao JH (2019) An amino-modified metal-organic framework (type UiO-66-NH_2) loaded with cadmium(II) and lead(II) ions for simultaneous electrochemical immunosensing of triazophos and thiacloprid. *Microchim. Acta* 186:101

[21]. Ma L, He Y, Wang YR, Wang YH, Li RT, Huang ZH, Jiang YJ, Gao J (2019) Nanocomposites of Pt nanoparticles anchored on UiO66-NH_2 as carriers to construct acetylcholinesterase biosensors for organophosphorus pesticide detection. *Electrochim. Acta* 318:525–533

[22]. Xu L, Li JB, Zhang JJ, Sun JY, Gan T, Liu YM (2020) A disposable molecularly imprinted electrochemical sensor for the ultra-trace detection of the organophosphorus insecticide phosalone employing monodisperse Pt-doped UiO-66 for signal amplification. *Analyst* 145:3245–3256

[23]. Yao X, Shen JH, Liu QY, Fa HB, Yang M, Hou CJ (2020) A novel electrochemical aptasensor for the sensitive detection of kanamycin based on UiO-66-NH_2/MCA/MWCNT@rGONR nanocomposites. *Anal. Methods* 12:4967–4976

[24]. Xu GL, Huo DQ, Hou JZ, Zhang C, Zhao YN, Hou CJ, Bao J, Yao X, Yang M (2021) An electrochemical aptasensor of malathion based on ferrocene/DNA-hybridized MOF, DNA coupling-gold nanoparticles and competitive DNA strand reaction. *Microchem. J.* 162:105829

[25]. Shi XJ, Lu JX, Yin HZ, Qiao XG, Xu ZX (2017) A biomimetic sensor with signal enhancement of ferriferrous oxide-reduced graphene oxide nanocomposites for ultratrace levels quantification of methamidophos or omethoate in vegetables. *Food Anal. Methods* 10:910–920

[26]. Hadi M, Bayat M, Mostaanzadeh H, Ehsani A, Yeganeh-Faal A (2018) Sensitive electrochemical detection of picloram utilising a multi-walled carbon nanotube/Cr-based metal-organic framework composite-modified glassy carbon electrode. *Int. J. Environ. Anal. Chem.* 98:197–214

[27]. Soltani-Shahrivar M, Karimian N, Fakhri H, Hajian A, Afkhami A, Bagheri H (2019) Design and application of a non-enzymatic sensor based on metal-organic frameworks for the simultaneous determination of carbofuran and carbaryl in fruits and vegetables. *Electroanalysis* 31:2455–2465

[28]. Ma BK, Cheong LZ, Wen XC, Tan CP, Shen C (2018) Lipase@ZIF-8 nanoparticles-based biosensor for direct and sensitive detection of methyl parathion. *Electrochim. Acta* 283:509–516

[29]. Gan T, Li JB, Li HX, Liu YX, Xu ZH (2019) Synthesis of Au nanorod-embedded and graphene oxide-wrapped microporous ZIF-8 with high electrocatalytic activity for the sensing of pesticides. *Nanoscale* 11:7839–7849

[30]. Cheng Y, Ma BK, Tan CP, Lai OM, Panpipat W, Cheong LZ, Shen C (2020) Hierarchical macro-microporous ZIF-8 nanostructures as efficient nano-lipase carriers for rapid and direct electrochemical detection of nitrogenous diphenyl ether pesticides. *Sens. Actuators,* B 321:128477.

[31]. Li XY, Gao X, Gai PP, Liu XJ, Li F (2020) Degradable metal-organic framework/methylene blue composites-based homogeneous electrochemical strategy for pesticide assay. *Sens. Actuators,* B 323:128701.

[32]. Wang F, Zhao JB, Gong JM, Wen LL, Zhou L, Li DF (2012) New multifunctional porous materials based on inorganic–organic hybrid single-walled carbon nanotubes: gas storage and high-sensitive detection of pesticides. *Chem. Eur. J.* 18:11804–11810

[33]. Hu LP, Wu N, Zheng J, Xu JL, Zhang M, He PG (2014) Preparation of a magnetic metal organic framework composite and its application for the detection of methyl parathion. *Anal. Sci.* 30: 663–668

[34]. Do MH, Florea A, Farre C, Bonhomme A, Bessueille F, Vocanson F, Tran-Thi NT, Jaffrezic-Renault N (2015) Molecularly imprinted polymer-based electrochemical sensor for the sensitive detection of glyphosate herbicide. *Int. J. Environ. Anal. Chem.* 95: 1489–1501

[35]. Song YG, Shan BX, Feng BW, Xu PF, Zeng Q, Su D (2018) A novel biosensor based on ball-flower-like Cu-hemin MOF

grown on elastic carbon foam for trichlorfon detection. *RSC Adv.* 8 :27008–27015.

[36]. Dong SY, Peng L, Wei WB, Huang TL (2018) Three MOF-templated carbon nanocomposites for potential platforms of enzyme immobilization with improved electrochemical performance. *ACS Appl. Mater. Interfaces* 10:14665–1467

[37]. Xie Y, Tu XL, Ma X, Fang QW, Liu GB, Dai RY, Qu FL, Yu YF, Lu LM, Huang XG (2019) A CuO-CeO_2 composite prepared by calcination of a bimetallic metal-organic framework for use in an enzyme-free electrochemical inhibition assay for malathion. *Microchim. Acta* 186: 567

[38]. Cao Y, Wang LN, Shen C, Wang CY, Hu XY, Wang GX (2019) An electrochemical sensor on the hierarchically porous Cu-BTC MOF platform for glyphosate determination. *Sens. Actuators, B* 283:487–494

[39]. Al'Abri AM, Halim SNA, Bakar NKA, Saharin SM, Sherino B, Nodeh HR, Mohamad S (2019) Highly sensitive and selective determination of malathion in vegetable extracts by an electrochemical sensor based on Cu-metal organic framework. *J. Environ. Sci. Health, Part B* 54:930–941

[40]. Wang ZP, Ma BK, Shen C, Cheong LZ (2019) Direct, selective, and ultrasensitive electrochemical biosensing of methyl parathion in vegetables using Burkholderia cepacia lipase@ MOF nanofibers-based biosensor. *Talanta* 197:356–362

[41]. Nagabooshanam S, Souradeep R, Mathur A, Mukherjee I, Krishnamurthy S, Bharadwaj LM (2019) Electrochemical micro analytical device interfaced with portable potentiostat for rapid detection of chlorpyrifos using acetylcholinesterase conjugated metal organic framework using Internet of things. *Sci. Rep.* 9:19862

[42]. Tu XL, Xie Y, Ma X, Gao F, Gong L, Wang DW, Lu LM, Liu GB, Yu YF, Huang XG (2019) Highly stable reduced graphene oxide-encapsulated Ce-MOF composite as sensing material for electrochemically detecting dichlorophen. *J. Electroanal. Chem.* 848:113268

[43]. Song DD, Jiang XY, Li YS, Lu X, Luan SR, Wang YZ, Li Y, Gao FM (2019) Metal-organic frameworks-derived MnO_2/Mn_3O_4 microcuboids with hierarchically ordered nanosheets and Ti_3C_2 MXene/Au NPs composites for electrochemical pesticide detection. *J. Hazard. Mater.* 373:367–376

[44]. Qiao XY, Xia FQ, Tian D, Chen PP, Liu JH, Gu JS, Zhou CL (2019) Ultrasensitive "signal-on" electrochemical aptasensor for assay of acetamiprid residues based on copper-centered metal-organic frameworks. *Anal. Chim. Acta* 1050:51–59

[45]. Gu CM, Wang Q, Zhang L, Yang PP, Xie YX, Fei JJ (2020) Ultrasensitive non-enzymatic pesticide electrochemical sensor based on HKUST-1-derived copper oxide @ mesoporous carbon composite. *Sens. Actuators, B* 305:127478

[46]. Chansi, Rao RP, Mukherjee I, Basu T, Bharadwaj LM (2020) Metal organic framework steered electrosynthesis of anisotropic gold nanorods for specific sensing of organophosphate pesticides in vegetables collected from the field. *Nanoscale* 12:21719–21733

[47]. Rashi Bhardwaj C, Pragadeeshwara Rao R, Mukherjee I, Agrawal PK, Basu T, Bharadwaj LM (2020) Layered construction of nano immuno-hybrid embedded MOF as an electrochemical sensor for rapid quantification of total pesticides load in vegetable extract. *J. Electroanal. Chem.* 873:114386

[48]. Tu XL, Gao F, Ma X, Zou J, Yu YF, Li MF, Qu FL, Huang XG, Lu LM (2020) Mxene/carbon nanohorn/β-cyclodextrin-Metal-organic frameworks as high-performance electrochemical sensing platform for sensitive detection of carbendazim pesticide. *J. Hazard. Mater.* 396:122776

[49]. Karimi-Maleh H, Yola ML, Atar N, Orooji Y, Karimi F, Kumar PS, Rouhi J, Baghayeri M (2021) A novel detection method for organophosphorus insecticide fenamiphos: Molecularly imprinted electrochemical sensor based on core-shell Co_3O_4@MOF-74 nanocomposite. *J. Colloid Interface Sci.* 592:174–185

[50]. Duan D, Ye JP, Cai X, Li K (2021) Cobalt(II)-ion-exchanged Zn-bio-MOF-1 derived CoS/ZnS composites modified electrochemical sensor for chloroneb detection by differential pulse voltammetry. *Microchim. Acta* 188:111

[51]. Wang QZ, Hui ZS, Zhao YJ, Zhuang YT, Xu ZH, Bu T, Li RX, Wang L (2021) Macro-meso-microporous carbon composite derived from hydrophilic metal-organic framework as high-performance electrochemical sensor for neonicotinoid determination. *J. Hazard. Mater.* 411:125122

[52]. Pejcic B, De Marco R (2006) Impedance spectroscopy: Over 35 years of electrochemical sensor optimization. *Electrochim. Acta* 51:6217–6229

[53]. Lisdat F, Schafer D (2008) The use of electrochemical impedance spectroscopy for biosensing. *Anal. Bioanal. Chem.* 391:1555–1567

[54]. Munoz J, Montes R, Baeza M (2017) Trends in electrochemical impedance spectroscopy involving nanocomposite transducers: Characterization, architecture surface and bio-sensing. *Trends Anal. Chem.* 97:201–215

[55]. Deep A, Bhardwaj SK, Paul AK, Kim KH, Kumar P (2015) Surface assembly of nano-metal organic framework on amine functionalized indium tin oxide substrate for impedimetric sensing of parathion. *Biosens. Bioelectron.* 65:226–231

[56]. Nagabooshanam S, Sharma S, Roy S, Mathur A, Krishnamurthy S, Bharadwaj LM (2020) Development of field deployable sensor for the detection of pesticide from food chain. *IEEE Sens. J.* 21:4129–4134.

[57]. Korotcenkov G, Cho BK (2017) Metal oxide composites in conductometric gas sensors: Achievements and challenges. *Sens. Actuators*, B 244:182–210

[58]. Bhardwaj SK, Bhardwaj N, Mohanta GC, Kumar P, Sharma AL, Kim KH, Deep A (2015) Immunosensing of atrazine with antibody-functionalized Cu-MOF conducting thin films. *ACS Appl. Mater. Interfaces* 7:26124–26130

[59]. Bertoncello P (2011) Nanomaterials for biosensing with electrochemiluminescence (ECL) detection. *Front. Biosci., Landmark Ed.* 16:1084–1108

[60]. Hao N, Wang K (2016) Recent development of electro-chemiluminescence sensors for food analysis. *Anal. Bioanal. Chem.* 408:7035–7048

[62]. Ma C, Cao Y, Gou XD, Zhu JJ (2020) Recent progress in electrochemiluminescence sensing and imaging. *Anal. Chem.* 92:431–454

[63]. Chen PP, Liu Z, Liu JH, Liu HB, Bian WW, Tian D, Xia FQ, Zhou CL (2020) A novel electrochemiluminescence aptasensor based CdTe QDs@NH_2 -MIL-88(Fe) for signal amplification. *Electrochim. Acta* 354:136644

[64]. Zhang XR, Guo YS, Liu MS, Zhang SS (2013) Photo-electrochemically active species and photoelectrochemical biosensors. *RSC Adv.* 3:2846–2857

[65]. Zhao WW, Xu JJ, Chen HY (2014) Photoelectrochemical DNA biosensors. *Chem. Rev.* 114:7421–7441

[66]. Shu J, Tang DP (2020) Recent advances in photoelectrochemical sensing: from engineered photoactive materials to sensing devices and detection modes. *Anal. Chem.* 92:363–377

[67]. Jin DQ, Xu Q, Yu LY, Hu XY (2015) Photoelectrochemical detection of the herbicide clethodim by using the modified metal-organic framework amino-MIL-125(Ti)/TiO_2. *Microchim. Acta* 182:1885–1892

[68]. Jin DQ, Gong QQ, Zhou H (2017) Visible-light-activated photoelectrochemical biosensor for the detection of the pesticide acetochlor in vegetables and fruit based on its inhibition of glucose oxidase. *RSC Adv* 7:17489–17496

[69]. Cao Y, Wang LN, Wang CY, Hu XY, Liu YL, Wang GX (2019) Sensitive detection of glyphosate based on a Cu-BTC MOF/g-C_3N_4 nanosheet photoelectrochemical sensor. *Electrochim. Acta* 317:341–347

21 An Overview of Metal-Organic Frameworks for Detection of Pesticides

Archana Mishra[1,2], Soumya Mukundan[1] and Jitendra Kumar[1,2]

[1]Nuclear Agriculture and Biotechnology Division, Bhabha Atomic Research Centre, Trombay, Mumbai, India

[2]Homi Bhabha National Institute, Anushakti Nagar, Mumbai, India

21.1 INTRODUCTION

Agricultural development has a long history and practice that began about 10,000 years ago in the Fertile Crescent of Mesopotamia, which is today's Jordan, Turkey, Iraq, and Syria. The people who lived in these places generally collected edible seeds by means of fire-stick farming, and forest gardening. After the population became settled and started living on farms, large volumes of cereals, pulses, oilseeds, and millets were cultivated. Rice and sorghum were cultivated in the African area. Rice and millet were also cultivated in China. America independently domesticated corn, squashes, potato, and sunflowers. As per the report by Food and Agriculture Organization (FAO), in 2017, the world population reached 7.55 billion, and at this rate, it may go up to 9.8 billion by 2050. According to the population growth, there is an increase in demand for grain consumption and therefore food production has been increased up to 2.2-fold from 1970 to 2020. In line of required increases in food production to parallel the growth of population, the total cultivated area today is the same as it was in 1965 and also there is no scope of increase in the cultivated area. It seems there will be reduction in the total area of cultivation [1-2].

Pesticide has played a key role in increasing the demand on agricultural productivity per hectare so that food for the growing human population can be supplied. Domesticated crops are often attacked by diseases, weeds, and pests that cause a significant loss in crop yield. If pesticides were not used, there would be a loss of fruit, vegetables, and cereals due to pests and diseases which would be as much as 78%, 54%, and 32%, respectively. This contributed greatly to the extensive use of pesticides. As per the report of a database of FAOSTAT, there has been a 46% increase in global uses of pesticides between 1996–2016 [2-3].

21.2 PESTICIDES

Pesticides have been playing a key role in agricultural growth, reducing the losses of food and improving the agricultural productivity and availability of quality food to the public. During World War II (1939-1945), there was a need to control insect-borne diseases to increase crop production. Thus, synthetic pesticides were introduced and applied to significantly improve agricultural production up to the level of consumption needed to feed the increased population of the world. It was reported that there was an 11% annual increase in worldwide pesticide production from 0.2 million tons in the 1950s to more than 5 million tons by 2000 [1-2].

21.2.1 Classifications of Pesticides

Pesticides are grouped in different classifications depending on the chemical groups, functional groups, modes of action, and toxicity. Depending on the target pest, pesticides are classified into fungicides, insecticides, herbicides, and rodenticides. For example, fungicides are used to control fungi, insecticides are used to kill insects, rodenticides kill rodents, while herbicides are used to kill weeds. In terms of chemical classes, pesticides are classified according to their organic and inorganic ingredients. Among inorganic pesticides, sulfate group pesticides are prominent. Among organic pesticides, the classification is as per their chemical structure, such as organochlorine, organocarbamate, organophosphorus synthetic pyrethroid, synthetic urea, triazine, and benzimidazole, and so forth. [2, 5-7].

21.2.2 The Role of Pesticides in Crop Production

The main benefits of pesticides were seen using different types of pesticides in a range of areas like agriculture and public health. Approximately 30% of agricultural products are saved using pesticides. As per the report, without using pesticides, it is very difficult to control up to 78% loss of fruit production, 54% loss of vegetable production, or 32% loss of cereal production. Therefore, pesticides play a critical role in increasing crop production worldwide and make a significant contribution to removing hunger and provide high-quality food for people. For public health, pesticides are mostly used to kill mice in houses, offices, malls, and streets. They are also used to control pests, such as ticks, rats, and mosquitoes in daily

DOI: 10.1201/9781003188148-21

life. Pesticides help us in providing a better opportunity to reduce the burden of diseases caused by insects and other pest vectors. Uses of pesticides are the most practical approach, to controlling insects and pests, which spread deadly diseases like malaria [2]. Further benefits from pesticide application are the controlling of pests which has brought significant improvement in livestock productivity and, consequent further increases in farming family income and also aiding in the production of nutritious, safe, and affordable food which contributes to increasing life expectancy [2, 8].

21.2.3 Pesticide Behavior in the Environment and Its Toxicity

When pesticides are applied to plants in the agriculture field or disposed of, they have the potential to enter the environment or to enter non-target plants and their surroundings by transfer processes which include volatilization, adsorption, spray drift, leaching, and runoff [2]. In the environment, pesticides undergo degradation by hydrolysis and oxidation thus producing a different, new chemical product which shows different behavior in the environment in comparison to the original compound. For example, methyl paraoxon is the oxidized product of methyl parathion, one of the organophosphate pesticides, and it inhibits acetylcholinesterase enzyme and is more toxic to the neuromuscular system than its parent compound [9-14]. In another case, dichlorodiphenyltrichloroethane (DDT), one of the organochlorine compounds, was widely used all over the world to reduce insect-borne diseases, such as malaria, yellow fever, and typhus. DDT has low acute toxicity, however; it is highly accumulative and persists in fat tissues, causing long-term damage to humans. Therefore, the US in 1972 banned DDT because of its harm to non-target plants and animals and because of this accululation and persistence in fat tissues, causing long-term damage. Many other synthetic pesticides such as captan, chlordane, hexachlorocyclohexane (BHC), aldrin, endrin, parathion, dieldrin, and 2,4-D also show similar behavior in the environment, causing toxicity and therefore, most of these pesticides are now banned. Many new classes of chemical pesticides were introduced after the green revolution, such as chloronicotinyl, spinosyn, triazolopyrimidine, azolone, triketone and isoxazole, strobilurin, and organophosphate insecticides. These have been commercialized and used in lesser concentrations per hectare. Some of these pesticides are now banned in most countries but their residues are still persisting in the environment for a long time and some of these have shown acute toxicity in mammals [2, 7, 11]. Pesticides are causing different types of toxicity as shown in Figure 21.1.

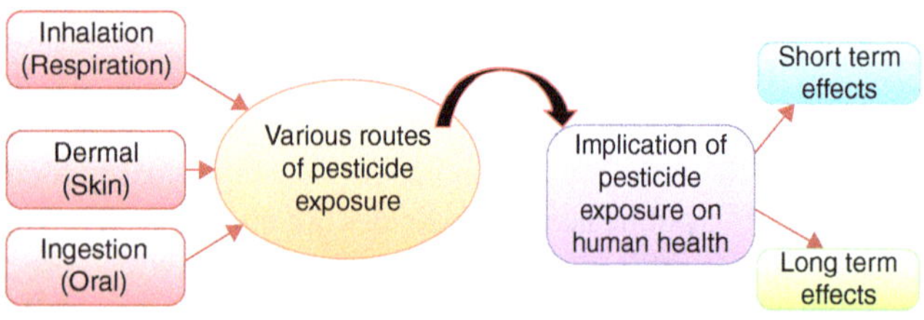

FIGURE 21.1 The schematic diagram for toxicity of pesticides.

21.3 METHODS FOR THE DETECTION OF PESTICIDES

Because of toxicity caused by pesticide residues, there is strict regulation and legislation worldwide to protect consumers. The maximum residue level (MRL) is the maximum permissible amount of an individual pesticide residue present in or on food commodities. MRLs of these pesticides are dependent on their toxicity and the particular food commodities. The MRL also varies from region to region across the world. Food manufacturers are exporting products to various other countries, which must meet all the individual MRLs in their respective target regions. International regulatory agencies (USEPA, EU, and FSSAI) have developed a regulatory mechanism to control pesticides so that the content of residues of individual pesticides should not reach a level exceeding the regulatory maximum residue limits [11]. Pesticides can be detected by using both conventional analytical as well as advance sensor-based methods as shown in Figure 21.2.

21.3.1 Conventional Methods to Detect Pesticides

Among conventional methods, chromatographic methods are commonly used in the determination and separation of target pesticides include gas chromatography (GC) and liquid chromatography (LC) using their specific detector. Mass spectrometry (MS) is also one of the most used additional techniques coupled with GC and LC to enhance detection performance. Initially, pesticide residues were normally analyzed by the following methods: GC with electron capture detection, flame ionization detection, or nitrogen-phosphorus detection, especially for volatile compounds in complex samples and LC ultraviolet, diode array, fluorescence, or electrochemical detection, suitable for non-volatile compounds. Using these conventional methods, analysis of a large number of pesticides in various food commodities is a challenge. The method for analyzing and detecting pesticide residues is typically composed of extracting the residues, clean-up procedures to remove other unwanted components, and an analytical procedure to determine and measure the amount of pesticide residue present in samples. However, these techniques were lacking the required sensitivity as well as selectivity due to the complexity of food matrices. These conventional techniques were replaced by GC and LC coupled to MS techniques, especially with tandem mass spectrometry, LC-MS-MS, and GC-MS-MS [15].

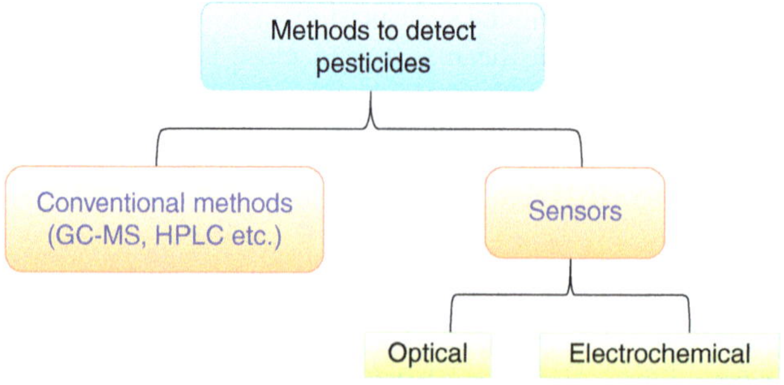

FIGURE 21.2 Methods for detection of pesticides.

TABLE 21.1
Sensor-based Detection of Pesticides

Table 21.1a: Electrochemical sensor for detection of pesticides

S. No.	Detection system	Pesticides	Limit of detection (ppb)	Ref
1.	Electrochemical sensors	Atrazine	2.2	[16]
2.	Electrochemical sensors	Carbendazim	5.736	[17]
3.	Electrochemical sensors	Quinalphos	0.378	[18]
4.	Electrochemical sensors	Fenitrothion	0.036	[19]
5.	Electrochemical sensors	Methyl parathion	50	[20]

Table 21.1b: Optical sensor for detection of pesticides

S. No.	Detection method	Pesticides	Limit of detection (ppb)	Ref
1.	Optical sensor	Thiabendazole	2.8	[21]
2.	Optical sensor	Methyl parathion	300	[14]
3.	Optical sensor	Methyl parathion	300	[10]
4.	Optical sensor	Methyl parathion	300	[22]
5.	Optical sensor	Methyl parathion	100	[23]
6.	Optical sensor	Paraquat	1.6	[24]

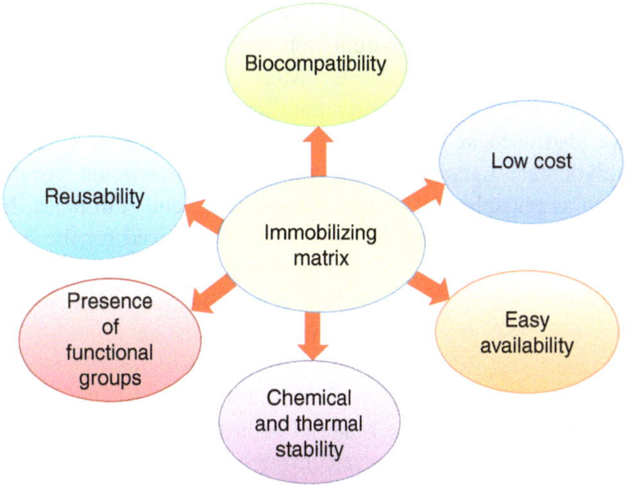

FIGURE 21.3A Characteristics features of an ideal immobilizing matrix.

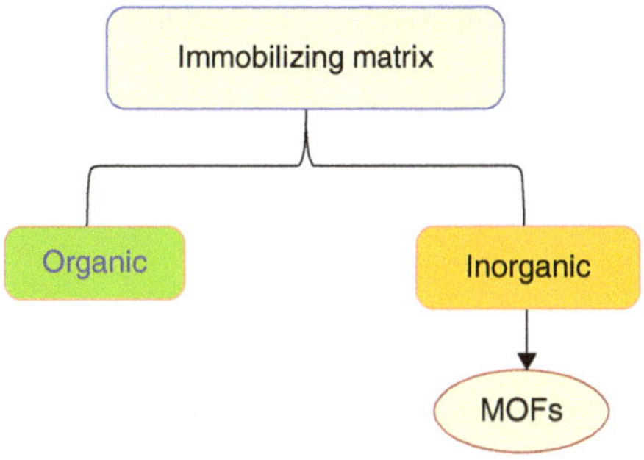

FIGURE 21.3B Types of immobilizing matrices used for sensor applications.

21.3.2 Sensors

Currently, there are numerous types of sensors that have been developed for pesticide detection such as electrochemical (Table 21.1a) and optical (Table 21.1b) as listed in Table 21.1. The electrochemical detection methods include screen-printed electrodes, field effect transistors (FETs), and the capacitive-based method. On the other hand, optical detection methods include, optical-MEMS, colorimetry, surface-enhanced Raman scattering, surface plasmon resonance (SPR), chemiluminescent, and fluorescent methods. Our laboratory has also significantly contributed to this field and developed several electrochemical and optical biosensors for the detection of single to multi-samples of methyl parathion pesticide, using microbial organophosphorus hydrolase enzyme (OPH) [9-14].

21.3.3 The Role of Immobilization in Developing Sensors

Immobilization is a method wherein a molecule of interest like an enzyme is associated with suitable support to improve its applicability [25]. For efficient immobilization, a suitable matrix is required, which plays a very important role in maintaining the catalytic/structural properties of immobilized enzymes. Figure 21.3a shows the various features of an immobilizing matrix. Immobilization plays a key role in developing sensors using various types of matrices for optimum and stable detection.

A wide range of matrices varying from organic/natural to inorganic and hybrid/bio-hybrid matrices haves been applied as immobilizing supports [26-30]. In this regard, MOFs have gained interest due to their beneficial properties like high surface area, tunable pore size, high thermal and chemical stability, and so forth (Figure 21.4).

21.4 METAL-ORGANIC FRAMEWORKS

MOFs are highly ordered crystalline structures with tunable pore sizes which have gained interest as advanced functional porous materials in the field of environmental remediation and monitoring. MOFs can be self-assembled from inorganic moieties using metal ions or metal clusters and bi/multidentate organic ligands through coordination chemistry and its interactions. MOFs are also known as porous coordination polymers. The designable pore structure and easy functionalization of MOFs are generally because of the various connectivities of the inorganic moieties and the features of the organic linkers. During the preparation of MOFs, various inorganic moieties such as Mg ions, Zn ions, Co ions, divalent Cd ions, Ti ions, and Zr ions are used. Based on the coordination number of the metal ions that are present in the center, MOFs can be in the form of various types of geometric coordination structures. MOFs are also used as organic linkers such as neutral or charged polydentate ligands, which span various functional groups such as phosphates, amines, carboxylates, and sulfonates. In 1995, MOFs showing porosity were synthesized for the first time. In one review, it was reported that the building block approach was used to prepare crystalline materials, which possess open-framework structures with cavities or/and channels [31]. Presently, the Cambridge Structural Database has a large number of registered MOFs [32]. The highly used MOFs include the Material of Institute Lavoisier (MIL) series that were synthesized through a lanthanide series or transition metals and trimesic acid or terephthalic acids such as MIL-53, MIL-100, and MIL-101[31]. Zeolitic imidazolate framework (ZIF) series MOFs are engineered from metal ions and imidazole linkers, for example, ZIF-10, ZIF-68, and ZIF-69. The properties of the isoreticular metal-organic framework (IRMOF) series differs due to linker species and the functional groups that are present. Thus, IRMOF has emerged as an outstanding model system for studying the removal of different types of contaminants [31].

21.4.1 Characteristics of MOFs

A material with a high surface area shows high adsorption capacity. Also, adsorption capacity is considered as one of the important parameters for a material that is going to be used for adsorption/binding. In this, MOFs are the materials of choice because of their very high surface area which varies from 200 to 10400 m^2 g^{-1}. MOFs with different spatial structures show different surface areas. Overall, because of their high surface area, MOFs are an important support/sorbent for the adsorption of organic/inorganic pollutants and environmental monitoring also. Another important feature of MOFs is high thermal and chemical stability. MOFs have high thermal stabilities which are above 573 K and can exceed 773 K [31, 32]. The excellent thermal stability of MOF materials allows its use in applications that operate in a high-temperature environment. The chemical stability of ZIF-7 presented interesting features without any structural alteration after it was treated in extreme conditions. The reversible phase transformation of ZIF-7 was investigated through immersion in DMF and ethanol. It was observed that the structural flexibility of the frameworks is maintained under extreme heat and solvent treatment. The stability of the thiol-functionalized MOFs (MIL-101-SH and UiO-66-SH) in water, was studied using inductively coupled plasma–optical emission spectrometry (ICP–OES) and it was observed that Zr- and Cr-based MOFs can retain outstanding chemical stability after thiol-functionalization in comparison to a Cu-based MOF (HKUST-1) [31]. Overall, the tolerability, surface area, high porosity, thermal, chemical, and water stability indicate that MOFs are suitable for various applications like removal of organic pollutants from the environment, drug delivery, storage, and sensors (Figure 21.4).

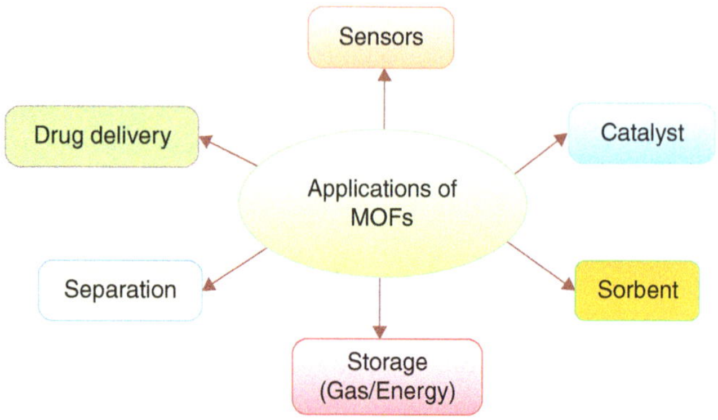

FIGURE 21.4 Various applications of MOFs

21.4.2 Luminescent Metal-organic-frameworks

Luminescent metal-organic frameworks (LMOFs) have high sensitivity, fast response, and easy operation. They have shown great potential as sensors in the field of environmental remediation and monitoring especially for the detection of pesticides. LMOFs have been widely used for the detection of organophosphate groups of pesticides like methyl parathion, parathion, and the like. A LMOF, comprising 1,2,4,5-tetrakis(4- carboxyphenyl) benzene and Zn(II) was synthesized [33]. The developed system was used for the detection of methyl parathion, and it showed low LOD (0.12ppb). LMOFs can also be used for the detection of multiple pesticides in a single sample wherein quenching efficiencies are different. Also, LMOF-based sensors have been used for the detection of other pesticides like paraquat. Paraquat is toxic to humans, however; it is widely used to protect fruit and tea plants against various pests. 3D LMOFs were fabricated by Zhang et al. and used for the detection of paraquat [34]. The study showed a LOD of 9.73×10^{-6} M. The above examples show that LMOFs have wide applicability and can be used for the detection of various classes of pesticides.

21.5 MOF-BASED DETECTION SYSTEMS FOR PESTICIDES

There is an urgent need to synthesize porous materials with novel properties for various ranges of applications. In this respect, MOFs have gained a lot of importance. MOFs are porous coordination polymers that are self-assembled by the combination of metal ions or clusters of metal ions and organic ligands. MOFs have various characteristic features like high surface area, large pore volume, ease of modification, and extraordinary thermal stability. Because of their beneficial properties, MOFs have been widely applied in a wide range of applications like drug delivery, energy, gas storage systems, adsorption, catalysts, separation, and sensing (Figure 21.4). Many sensors using MOFs as the immobilizing materials have been developed for the detection of various pesticides as listed in Table 21.3. Table 21.3a lists the MOF-based biosensors for the detection of organophosphate pesticides and Table 21.3b, lists the detection of other pesticides. In recent years, MOFs have gained a lot of importance as immobilizing support to

TABLE 21.3A
MOF-based Sensors for Detection of Organophosphate Pesticides.

S. No.	MOFs	Sample preparation method/technique	Pesticide	LOD/LOQ	Ref
1	Co/Zn-ZIF	MSPE/GC	Organophosphorus	0.018–0.045 mg L^{-1}	[41]
2	UiO-66	QuEChERS/GC-MS	Organophosphorus	0.16 to 1.56 ng g^{-1}	[42]
3	NH2-MIL-125	SPME/GC–MS	Organophosphorous	0.23 to 7.5 ng g^{-1}	[43]
4	ZIF-8 91.9 to 99.5	DSPE/GC	Profenofos, phosalone, fenitrothion,and fenthion	0.03 to 0.21 ng mL	[44]
5.	Pt/MOF		Methyl parathion	1.45×10^{-9} mM	[45]

TABLE 21.3B
MOF-based Sensors for Detection of Other Pesticides Except for Organophosphate Pesticides.

S. No.	MOFs	Sample preparation method/technique	Pesticide	LOD/LOQ	Ref
1.	ZIF-90	SPME/GC	Pyrethroid	21.2 to 0.5 ng g^{-1}	[46]
2.	Fe-MOF	MSPD GC–MS/MS	Pyrethroids	1.8 to 2.8 ng g^{-1}	[47]
3.	MOF-5	SPME/GC	Triazole	0.05 to 1.58 ng g^{-1}	[48]
4.	MIL-101(Cr)	SPE/LC–MS	Triazine	1.08 to 18.10 pg g^{-1} / 63.60 to 60.20 pg g^{-1}	[49]
5.	SPE/ HPLC	ZIF-8	Metolcarb, isoprocarb, diethofencarb	0.25 to 0.1 ng g^{-1}	[50]

develop sensors/biosensors. Further, MOFs associated with bio components are applied to develop electrochemical and optical biosensors. Developed sensor/biosensors were used for the detection of various pesticides such as methyl parathion, and so on. Organophosphate pesticides are the pesticides that are most commonly used to control various pests to improve yield. Thus, there are various studies wherein the focus is on developing MOF-based sensors for the detection of organophosphate pesticides. In 2019, Wang et al. immobilized *Burkholderia cepacia* lipase (BCL) on a MOF nanofiber and the developed biosensors were used for detection of methyl parathion pesticide. This showed good sensitivity with a wide linear range (0.1-38 μM) and low LOD 0.067 μM [35]. Water–stable zirconium luminescent MOF (Zr-LMOF was used by He et al. to detect parathion methyl and a 78 to 107% recovery range was observed for spiked food samples. This emerged as a non-destructive method for the rapid and in-situ image detection of pesticide residue [36]. To develop a highly sensitive electrochemical sensor, Au nanorods (AuNRs) were embedded with graphene oxide-wrapped microporous ZIF-8 [37]. AuNRs@ZIF-8@GO has shown highly sensitive detection of niclosamide, dichlorophen, carbendazim, and diuron. Zhou et al. used ZIF-8-wrapped urchin-like Au–Ag nanocrystals for surface-enhanced Raman spectroscopy for the detection of traces of pesticide of hexachlorocyclohexane. It was observed that the ZIF-8 shell layer can improve the surface-enhanced Raman spectroscopy performance of urchin-like Au–Ag alloyed nanocrystals to detect hexachlorocyclohexane pesticides wherein LOD was below 1.5 ng mL^{-1} [38]. Cao et al. have developed an electrochemical sensor (Cu-BTC) platform to detect glyphosate and the results showed that the fabricated sensor displays ultralow LOD and a wide detection range. Furthermore, this sensor exhibits acceptable reproducibility and stability as well as sufficient selectivity against other interferences such as trichlorfon, carbendazim, acetochlor, thiram, and various metal ions [39]. There are various reports wherein the focus is on recent advances in the applications of MOFs for sensing pesticides. A pillar-layered entangled luminescent MOF composite was fabricated by Tao et al. in 2017. This showed a blue-green emission with a very high fluorescence quantum yield (99%) and emerged as a simple and reversible method to detect 2,6-dichloro-4-nitroaniline [40]. It was observed that due to the beneficial characteristics of MOFs, MOFs can be applied as an efficient system for the detection of pesticides. An added advantage with MOFs is their luminescent, chromogenic, optical, and conductive properties which further enhances their applicability in the field of sensors including biosensors.

21.6 CONCLUSION AND FUTURE PERSPECTIVES

To feed a continuously increasing population, a lot of pesticides have been used to increase crop production. Because of the extensive use of pesticides, the pesticide residues could be found in water bodies, the environment, food crops, and even animals. These pesticides are toxic to animals as well as human beings. In this respects, efforts have been made to develop efficient, low-cost, environment-friendly systems to detect pesticides with excellent detection ranges and low detection limits. In this chapter, we have summarized the pesticides, their types, applications, and their toxicity effects. Also, we have discussed the progress made in the field of monitoring and detection of pesticides using MOFs. MOFs have emerged as an efficient supporting matrix for developing a sensor for the detection of various pesticides. These MOF matrices can be used to develop a reliable and efficient sensor for the sensitive detection of various toxic pesticides, which are banned or supposed to be banned in the near future. Also, there is a need to explore MOF for fabrications of easy-to-handle and cost-effective technologies for developing economically feasible biosensors for the detection of pesticides.

ACKNOWLEDGMENTS

The authors are grateful for financial support from Bhabha Atomic Research Centre (BARC), Mumbai, India.

REFERENCES

[1]. Unsworth J, *History of Pesticide Use. International Union of Pure and Applied Chemistry (IUPAC)*, 2010.

[2]. Tudi M, Daniel Ruan H, Wang L, Lyu J, Sadler R, Connell D, Chu C, Phung DT Agriculture Development, Pesticide Application and Its Impact on the Environment, *International Journal of Environmental Research and Public Health*, 2021, 18: 1112.

[3]. Food and Agricultural Organization (FAO) of the United Nations. *Advancement of pesticide regulatory management in Asia.* Bangkok, Thailand. Regional office for Asia and the Pacific, 2013.

[4]. Garcia FP, Ascencio SYC, Oyarzun JCG, Hernandez AC, Alavarado PV, Pesticides: Classification, uses and toxicity. Measures of exposure and genotoxic risks, *International Journal of Environmental Science and Toxicology Research,* 2012, 1: 279–293.

[5]. Amaral AFS, Pesticides and Asthma: Challenges for Epidemiology, *Frontier Public Health*, 2014, 2:6.

[6]. Mnif W, Hassine AIH, Bouaziz A, Bartegi A, Thomas O, Roig B, Effect of Endocrine Disruptor Pesticides: A Review, *International Journal of Environmental Science and Toxicology Research*, 2011, 8:2265–2303.

[7]. Kim KH, Kabir E, Jahan SA, Exposure to pesticides and the associated human health effects, *Science of The Total Environment*, 2017, 575:525–535.

[8]. Lamichhane JR, Pesticide use and risk reduction in European farming systems with IPM: An introduction to the special issue, *Crop Protection*, 2017, 97: 1–6.

[9]. Kumar J, Melo JS, Overview on Biosensors for Detection of Organophosphate Pesticides, *Current Trends in Biomedical Engineering and Biosciences*, 2017, 5:555663.

[10]. Kumar J, Jha SK, D'Souza SF, Optical microbial biosensor for detection of methyl parathion pesticide using Flavobacterium sp. whole cells adsorbed on glass fiber filters as disposable biocomponent, *Biosensors and Bioelectronics,* 2006, 21: 2100–2105.

[11]. Mishra A, Kumar J, Melo JS, Sandaka BP, Progressive development in biosensors for detection of dichlorvos pesticide: A review, *Journal of Environmental Chemical Engineering,* 2021, 9:105067.

[12]. Kumar J, Mishra A, Melo JS, Biodegradation of Methyl Parathion and its Application in Biosensors, *Austin Austin Journal of Environmental Toxicology*, 2018, 4: 1024.

[13]. Kumar J, Melo JS, Microbial Biosensors for Methyl Parathion: From Single to Multiple Samples Analysis, T.G. Everett (Ed.), *Advances in Biosensors Research*, New York,: 2015, 89–111.

[14]. Kumar J, D'Souza SF, Immobilization of microbial cells on inner epidermis of onion bulb scale for biosensor application, *Biosensors and Bioelectronics*, 2011, 26:4399–4404.

[15]. Hercegová A, Dömötörová M, Matisová E, Sample preparation methods in the analysis of pesticide residues in baby food with subsequent chromatographic determination, *Journal of Chromatography A*, 2007, 1153: 54–73.

[16]. Švorc Ľ, Rievaj M, Bustin D, Green electrochemical sensor for environmental monitoring of pesticides: determination of atrazine in river waters using a boron-doped diamond electrode, *Sensors and Actuators B: Chemical*, 2013, 181: 294–300.

[17]. Yanke JGM, Dedzo GK, Ngameni E, Solvent effect on the grafting of an organophilic silane onto smectite-type clay: application as electrode modifiers for pesticide detection, *Electroanalysis*, 2017, 29(8):1894–1902.

[18]. Khare NG, Dar RA, Srivastava AK, Adsorptive stripping voltammetry for trace determination of quinalphos employing silicon carbide nanoparticles modified glassy carbon electrode, *Electroanalysis,* 2015, 27(2):503–509.

[19]. Shams N, Lim HN, Hajian R, Yusof N A, Abdullah J, Sulaiman y, Ibrahim I, Huang N M, Electrochemical sensor based on gold nanoparticles/ethylenediamine-reduced graphene oxide for trace determination of fenitrothion in water, *RSC Advances*, 2016, 6(92): 89430–89439.

[20]. Kumar J, D'Souza SF, Microbial biosensor for detection of methyl parathion using screen printed carbon electrode and cyclic voltammetry, *Biosensors and Bioelectronics*, 2011, 26: 4289–4293.

[21]. Piccirilli GN, Escandar GM, A novel flow-through fluorescence optosensor for the determination of thiabendazole, *Analytica Chimica Acta*, 2007, 601(2):196–203.

[22]. Kumar J, D'Souza SF, An optical microbial biosensor for detection of methyl parathion using Sphingomonas sp. immobilized on microplate as a reusable biocomponent, *Biosensors and Bioelectronics*, 2010, 26: 1292–1296.

[23]. Mishra A, Kumar J, Melo JS, An optical microplate biosensor for the detection of methyl parathion pesticide using a biohybrid of Sphingomonas sp. cells-silica nanoparticles, *Biosensor and Bioelectronics*, 2017, 87: 332–338.

[24]. Bian QQ, Liu YF, Yu JS, CdTe/CdS semiconductor quantum dots as a highly sensitive sensor for pesticide paraquat, *Chemical Journal of Chinese Universities*, 2010, 31(6):1118–1125.

[25]. Melo JS, Tripathi A, Kumar J, Mishra A, Bhanu S, Bhainsa KC, *Immobilization: then and now, Tripathi A, Melo JS, Immobilization strategies (Biomedical, bioengineering, and environmental application)*, 2020, Springer Nature, Singapore.

[26]. Mishra A, Melo JS, Sen D, D'Souza SF, Evaporation induced self-assembled microstructures of silica nanoparticles and Streptococcus lactis cells as sorbent for uranium (VI), *Journal of colloid and interface science,* 2014, 414 : 33–40.

[27]. Mishra A, Melo JS, Agrawal A, Kashyap Y, Sen D, Preparation, and application of silica nanoparticles-Ocimum basilicum seeds bio-hybrid for the efficient immobilization of invertase enzyme, *Colloids & Surfaces B: Biointerfaces*, 2020, 188: 110796.

[28]. Mishra A, Kumar J, Melo JS, Silica based bio-hybrids and their relevance to bionanotechnology, *Austin J Plant Biology*, 2020, 6(1): 1024–1028.

[29]. Mishra A, Pandey VK, Shankar BS, Melo JS, Spray drying as an efficient route for synthesis of silica nanoparticles-sodium alginate biohybrid drug carrier of doxorubicin, *Colloids & Surfaces B: Biointerfaces*, 2021,197: 111445.

[30]. Lahiri S, Mishra A, Mandal D, Bhardwaj RL, Gogate PR, Sonochemical reprocessing of uranium from nanosilica based sorbent and its biohybrid, *Ultrasonics Sonochemistry*, 2021, 76: 105667.

[31]. Xu Y, Wang H, Li X, Zeng X, Du Z, Cao J, Jiang W, Metal–organic frameworks for the extraction and detection of pesticides from food commodities, *Comprehensive Reviews In Food Science And Food Safety*, 2021, 20:1009–1035.

[32]. DeCoste JB, Peterson GW, Metal-organic frameworkss for air purification of toxic chemicals, *Chemical Reviews*, 2014, 114(11): 5695–5727.

[33]. Xu XH, Guo YN, Wang XY, Li W, Qi PP, Wang ZW, Wang XQ, Gunasekaran S, Wang Q, Sensitive Fluorescence and Visual Detection of Organophosphorus Pesticides with $Ru(bpy)_3^{2+}$ -ZIF-90-MnO_2 Sensing Platform, *Sensors and Actuators,* 2018, B 260: 339–345.

[34]. Chen HJ, Fan P, Tu XX, Min H, Yu XY, Li XF, Zeng JL, Zhang SW, Cheng P, A bifunctional luminescent metal-organic frameworks for the sensing of paraquat and Fe^{3+} ions in water, *Chemistry: An Asian Journal*, 2019, 14: 3611–3619.

[35]. Wang PL, Xie LH, Joseph EA, Li JR, Su XO, Zhou HC, Metal-organic frameworkss for food safety, *Chemical Reviews*, 2019, 119(18):10638–10690.

[36]. He KY, Li ZS, Wang L, Fu YC, Quan HR, Li YB, Wang W, gunasekharan S, Xu XH, A water-stable luminescent metal-organic frameworks for rapid and visible sensing of organophosphorus pesticides, *ACS Applied Materials and Interfaces*, 2019, 11(29): 26250–26260.

[37]. Mansouriieh N, Sohrabi MR, Khosravi M, Optimization of profenofos organophosphorus pesticide degradation by zerovalent bimetallic nanoparticles using response surface methodology, *Arabian Journal of Chemistry*, 2019, 12(8): 2524–2532.

[38]. Zhou X, Liu GQ, Zhang HW, Li Y, Cai WP, Porous zeolite imidazole frameworks-wrapped urchin-like Au-Ag nanocrystals for SERS detection of trace hexachloro-cyclohexane pesticides via efficient enrichment, *Journal of Hazardous Materials*, 2019, 368: 429–435.

[39]. Cao Y, Wang LN, Shen C, Wang CY, Hu XY, Wang GX, An electrochemical sensor on the hierarchically porous Cu-BTC MOF platform for glyphosate determination, *Sensors and Actuators B-Chemical*, 2019, 283:487–494.

[40]. Tao CL, Chen B, Liu XG, Zhou LJ, Zhu XL, Cao J, Gu ZG, Zhao Z, Shena L, Tang BZ A highly luminescent entangled metal-organic frameworks based on pyridine-substituted tetraphenylethene for efficient pesticide detection, *Chemical Communications*, 2017, 53(72): 9975–9978.

[41]. Li X, Ma W, Li H, Bai Y, Liu H, Metal-organic frameworkss as advanced sorbents in sample preparation for small organic analytes, *Coordination Chemistry Reviews*, 2019, 397: 1–13.

[42]. Mao X, Yan A, Wan Y, Luo D, Yang H, Dispersive solid phase extraction using microporous sorbent UiO-66 coupled to gas chromatography-tandem mass spectrometry: A QuEChERStype method for the determination of organophosphorus pesticide residues in edible vegetable oils without matrix interference, *Journal of Agricultural and Food Chemistry,* 2019, 67(6): 1760–1770.

[43]. Pang Y, Zang X, Li H, Liu J, Chang Q, Zhang S, Wang Z, Solid-phase microextraction of organophosphorous pesticides from food samples with a nitrogen-doped porous carbon derived from g-C_3N_4 templated MOF as the fiber coating, *Journal of Hazardous Materials,* 2020, 384: 121430.

[44]. Amiri A, Tayebee R, Abdar A, Sani FN, Synthesis of a zinc-based metal-organic frameworks work with histamine as an organic linker for the dispersive solid-phase extraction of organophosphorus pesticides in water and fruit juice samples, *Journal of Chromatography A*, 2019, 1597: 39–45.

[45]. Meng TJ, Wang L, Jia HX, Gong T, Feng Y, Li RX, Zhang Y, Facile synthesis of platinum-embedded zirconia/porous carbons tri-component nanohybrids from metal-organic frameworks and their application for ultra-sensitively detection of methyl parathion, *Journal of Colloid, and Interface Science*, 2019, 536:424–430.

[46]. Zhang S, Yang Q, Yang X, Wang W, Li Z, Zhang L, Wang Z, *A zeolitic imidazolate frameworks based nanoporous carbon as a novel fiber coating for solid-phase microextraction of pyrethroid pesticides*, Talanta, 2017, 166:46–53.

[47]. Liang T, Wang SS, Chen LG, Niu N, Metal organic frameworks-molecularly imprinted polymer as adsorbent in matrix solid phase dispersion for pyrethroids residue extraction from wheat, *Food Analytical Methods*, 2019, 12(1), 217–228.

[48]. Liu GY, Li LY, Gao YH, Gao MK, Huang XD, Lv J, Xu DH, A beta-cyclodextrin-functionalized magnetic metal organic frameworks for efficient extraction and determination of prochloraz and triazole fungicides in vegetables samples, *Ecotoxicology and Environmental Safety*, 2019, 183:7.

[49]. Jiang YX, Piao HL, Qin ZC, Li XL, Ma PY, Sun Y, Wang X, Song D, One-step synthesized magnetic MIL-101(Cr) for effective extraction of triazine herbicides from rice prior to determination by liquid chromatography-tandem mass spectrometry, *Journal of Separation Science,* 2019, 42(18): 2900–2908.

[50]. Hao L, Liu XL, Wang JT, Wang C, Wu QH, Wang Z, *Use of ZIF-8-derived nanoporous carbon as the adsorbent for the solid phase extraction of carbamate pesticides prior to high performance liquid chromatographic analysis*, Talanta, 2015, 142:104–109.

22 MOF-based Electrochemical Sensors for Glucose

Ummama Saeed[1], *Rafia Batool*[1], *Dilshad Hussain*[2], *Saadat Majeed*[3], *Muhammad Najam-ul-Haq*[3] *and Batool Fatima*[1]

[1]Department of Biochemistry, Bahauddin Zakariya University, Multan, Pakistan

[2]HEJ Research Institute of Chemistry, International Center for Chemical and Biological Sciences, University of Karachi, Karachi, Pakistan

[3]Institute of Chemical Sciences, Bahauddin Zakariya University, Multan, Pakistan

22.1 INTRODUCTION

Glucose is the simplest form of sugar and a key component to maintain our body in working order and provide energy to continue bodily mechanisms such as regulation of metabolism (gluconeogenesis, fermentation). Normal blood glucose level is 140 mg/dL (7.8 mmol/L). Any kind of disturbance in this level of glucose can lead to worrying complications [1]. The quantification and surveillance of the level of glucose are necessary due to its wide range of applications in biotechnology, in the food industry, and clinical diagnosis [2]. Various methods were used for the detection of glucose such as high-performance liquid chromatography (HPLC), spectrophotometry, optical rotation, and colorimetry, and so on, but all these methods have their limitations. Amongst these, electrochemical sensing of glucose has been applied in many fields because of high sensitivity, selectivity, fast analysis speed, low LOD, facile instrument operation and cost effectiveness [3]. In electrochemical sensing, enzymatic and non-enzymatic sensors are widely utilized for qualitative and quantitative determination of glucose in biological samples [2].

The term 'biosensor' denotes a significant and advanced analytical device containing biological sensing elements with an extensive range of applications, such as food safety, food processing, drug discovery, biomedicine, diagnosis, environmental monitoring, defense, and safety. Sensors are self-contained unified devices that can measure a physical property and alter it into computable and processable signals. Usually, biosensors consist of three working systems: 1) bio receptor, 2) transducer in constricted assembly with bio-receptor and 3) signal-processing component. The binding of the analyte to a specific bio-receptor as the recognition element leads to conversion into a quantifiable output as an electrical signal by the transducer. Sensitivity, selectivity, reproducibility, stability are fundamental characteristics of sensors [4]. Based on the type of biological sensing component, biosensors are classified as enzymatic biosensors, immune biosensors, nuclear biosensors, and receptor biosensors [5]. Based on the transduction method, sensors are classified as piezoelectric, electrochemical, and optical sensors. Electrochemical sensors work by transforming the chemical signal into an electrical signal and provide basic quantitative as well as qualitative information about diagnostic biomarkers of various diseases [6].

Electrochemical sensors classified based on their transducers includes impedimetric parameters in which impedance between the electrodes is measured, amperometry measures current, potentiometry is based on the generated potential while conductometry is related to the conductivity [7]. In electrochemical sensors, some major analytical techniques are used to analyze the samples including cyclic voltammetry (CV), differential pulse voltammetry (DPV), square wave voltammetry (SWV), chronoamperometry, linear sweep voltammetry (LSV), and electrochemical impedance spectroscopy (EIS) [8].

Professor Leland C. Clark Jr. is known as the father of biosensors, as he has constructed the first-ever biosensor in 1956 for sensitive and rapid identification of glucose in blood [9]. Leading-edge advancements in technology have allowed sensors to be developed that are more precise, definitive, and sensitive to biomarkers and analytes study. As these sensors turn out to be more innovative, there is also a drive to miniaturize and transport them by using more advanced materials such as metal-organic frameworks (MOFs) [10].

22.2 METAL-ORGANIC FRAMEWORKS

Metal-organic frameworks or coordination polymers are emerging classes of hybrid materials with high surface area and porosity properties constructed from metallic ions or clusters bridged by organic linkers that have caught the eye of researchers in recent years. They have defined pore aperture, tailorable composition, and structure tenable size, versatile functionality, high loading capacity, and improved biocompatibility which makes them a promising candidate for electrochemical sensing [11]. MOFs can be used as substrates

DOI: 10.1201/9781003188148-22

either in search of various nanostructures or they might act as templates for the formation of nanomaterials. MOFs are central crystalline compounds and have applicability in diverse areas of life because of their realistic design, facile way of synthesis, and elevated porosity [12].

Currently, metal-organic frameworks have gained much attention as sensing materials especially in electrochemical sensing for glucose detection. Among the MOFs hybrids and carbon nanomaterials, carbon nanotubes (CNTs) and graphene oxide (GO) can elevate the electrical conductivity and mechanical strength of the nanomaterials. While the major difficulty is in the low catalytic activity of MOFs which could be dealt with by the use of catalytically active constituents in metal-organic frameworks such as metal nanoparticles (NPs) comprising of copper, ruthenium, palladium, and gold. Because of expeditious advancements in nanotechnology, much progress and development in the preparation and assembly of these nanomaterials has been made especially in electrochemical biosensors. MOFs are solid materials with crystalline nature in which solid-state and coordination chemistry are the two significant areas from which MOFs synthesis have emerged. In 1995 Yaghi and his co-workers found the structure of MOF [17] while in 1997 Kitagawa revealed 3D MOFs [13]. For the last decade, more advancements have been seen in this field.

22.3 STRUCTURE OF METAL-ORGANIC FRAMEWORKS

Metal-organic frameworks contain metal ionic groups and linkers and because of these structures they are also known as organic-inorganic hybrid networks (Figure 22.1). They have different surface areas for micro, mesoporous and microporous materials. Microporous materials have a pore size of 2 nm, for example, zeolites and MOFs. Mesoporous materials have pore sizes fluctuating from 2 nm to 50 nm, for example, oxides of titanium, zirconium, and tin, sometimes silica and alumina are also reported [14]. Macro-porous materials have cavities with pore sizes of more than 50 nm [15]. Organic linkers and inorganic metal ions have the pre-decided agreement of inorganic metal atoms bounded by organic moieties which results in the formation of repeating units and greater surface area of the MOFs [16]. Ligands are used as organic units in which selection of ligand and metal is significant to find out the properties of the metal-organic frameworks. MOFs have well-organized composition, turnability, porousness, flexibility in network structure, and maximum surface area [17]. The nano porosity of the MOFs makes them leading materials over that of metal oxides and other zeolites in which an oxygen atom is replaced by the organic linker which binds with the inorganic ions and is considered to self-assemble into the one-,two-, and three dimensional MOFs [12]. Strong bonding between the metal ions and organic parts is responsible for their 1,2 and 3-dimensional crystalline structure.

22.3.1 Zero Dimension MOFs

Zero-dimension MOFs are mainly nanocrystals/particles which depend on the MOFs' crystal growth and nucleation. Nucleation and crystal growth are controlled by physically confining the MOFs reaction. Prussian blue cubic nanocrystals of diameter: 15 nm, are one of the first 0D nano-MOFs [19].

22.3.2 1D MOFs

1 dimensional metal-organic frameworks are enlarged enormously in one specific direction. 1D -MOFs porosity is due to the presence of an inanity originated upon assembly of (1D) chains. Pores in one-dimension MOFs are predominantly in the form of stretches. Shahhoseini, Leila, et al. [20]

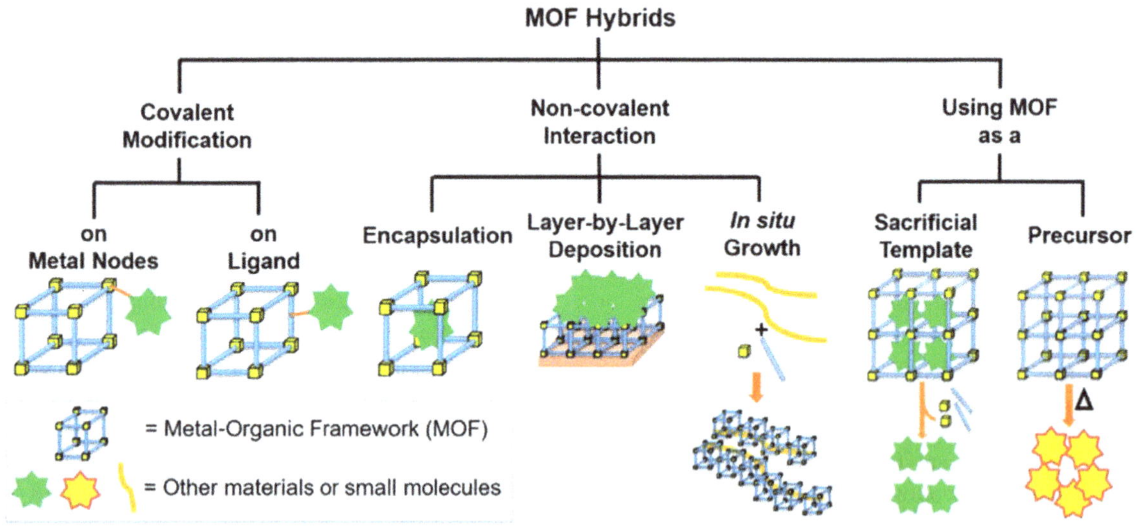

FIGURE 22.1 Schematic representation of the basic structure of metal-organic frameworks [18]. Copyright (2018) The Authors, some rights reserved; exclusive licensee [MDPI]. Distributed under a Creative Commons Attribution License 4.0 (CC BY)

fabricated a sensor for enzyme-free glucose detection by using 1D MOFs. Ni (II)- coordination polymers of Ni are used in this biosensor. Ni(II)-Cp fabrication was done by deliberated diffusion and evaporation of $NiCl_2$ solution consisting of two consistent layers as well diaza-macrocycle enduring of 2 pyridine sidearms in dimethylformamide (DMF) and C_{60} addition to Ni(II)-Cp to improve the chemical as well as electrical stability of the sensor. Ni (II)-CP/C_{60}/GCE-based sensors display a low detection limit (4.3 μM) with sensitivity (614 μA mM^{-1} cm^{-2}).

22.3.3 2D MOFs

2-dimension metal-organic frameworks comprise of indefinite piles of layers which are the main cause of the generation of porosity in the materials. 2D MOFs are one of the excellent and advanced approaches to improving electrochemical biosensor performance due to their porous structure, larger surface area, sensitivity, selectivity, improved electrocatalytic activity, and faster electron transfer. 2D bimetallic metal-organic frameworks are prepared by top-down, exfoliation, bottom-up, and ball milling methods [21]. 2D MOFs can act as one of the best electrochemical sensing materials because we can achieve advanced quantitative as well as qualitative electrochemical-based detection of different disease biomarkers for diagnostics and theragnostic purposes as compared to conventional approaches. 2-dimensional MOFs nanosheets in electrochemical sensing are one of the recent utilizable approaches because of efficient stability, enhanced sensitivity, high porosity for the detection of analytes [22].

Gumilar, et al. [23], synthesized Ni-BDC-MOF in their work for the electrochemical sensing of glucose. Ni-BDC-MOF drives oxidation of glucose in a wide linear range (0.01 mM to 0.8 mM) with 635.9 μA mM^{-1} cm^{-2} of sensitivity, exhibit high selectivity with low LOD of 6.68 μM and fast response time of <5 s. In another study conducted by Shu, Yun, et al. [24], Ni/NiO-NP decorated MOFs used for electroanalytic determination of glucose showed a wide linear range (4 μM to 5664 μM) with a low limit of detection of 0.8 μM and sensitivity of 367.45 mA M^{-1} cm^{-2}.

Chen, Chen, et al. [25] fabricated two-dimensional MOFs and Persian blue analogs as high electronic conductive materials with slow diffusion of MOFs for electrochemical applications. In their work, they synthesized 2D leaflike zeolitic imidazolate frameworks of Cobalt and Iron which were (in situ) converted into PBAs for fabrication of PBA/ZIF nanocomposite on Ni-foam. All the fabricated nanocomposites act as multi-functional sensors which selectively and sensitively detect glucose with a wide range of linearity and good stability. Among all the nanocomposites CoFe-PBA/Co-ZIF/NF show high performance with a linear range from 1.4 μM to 1.5 mM with selectivity and sensitivity towards glucose in human serum samples.

22.3.4 3D porous MOFs

Porosity in 3-dimension metal-organic frameworks is directed by the addition of linker coordination geometry with extensive coordination complexes outstretched all over the three dimensions. Dot dimension MOFs and incarcerated 3D MOF pores (bisecting channels complexes that are identical to the pores present in zeolites), which are produced by building blocks assemblage. MIL-101 and mil-100 are the major examples of the three-dimension metal-organic frameworks which played a significant role in the development of promising MOFs-based sensing materials especially for glucose detection [26].

Liu, Qianshi, et al. [27] originated a new advanced sensor based on 3D graphene foams synthesized by the chemical Vapor deposition method. The 3DG (3DG/PB-AuNPs) based sensor was constructed by the electrode deposition and copolymerization strategies of gold nanoparticles and Prussian blue PB (which was stabilized and protected by the polydopamine film) on 3D graphene (Figure 22.2). CV and DPV were conducted to evaluate the electroactive characters of the newly constructed sensor which shows a linear range of (5 μM–65 μM), a low detection limit of (1.5 μM) with very short reaction time as well as high sensitivity for glucose detection in biological samples. A study conducted by He, Juan, et al, [28] constructed a sensor based on Cu-hemin metal-organic frameworks for glucose biosensing. The resulting sensor showed a linear range (9.10 μM-36.0 mM) and a low limit of detection of 2.73 μM and showed satisfactory detection of glucose from human serum samples.

22.3.5 MOF-derived Nanomaterials

MOFs derived from metallic or carbonaceous nanomaterials have more active metal sites, conductivity, and surface area than simple MOFs. MOF derived nanomaterials are suitable for use as electrode modifying materials in electrochemical sensing [30]. The incorporation of NMs into the MOFs results in enhanced performance, particularly for catalysis and gas storage. Zero dimensional NPs can be combined with MOFs which results in the generation of core-shell nanostructures [31].

Xiao, Xuelian, et al. [32] synthesized metal/metal oxide@ carbon composite by direct carbonization of bimetallic Cu/Ni-based MOFs for detection of glucose. The respective Cu/Ni-MOFs exhibited excellent performance for the sensing of glucose with linearity ranges from 0.1 μM to 2.2 mM and low LOD of 0.06 μM. The respective material shows good reproducibility, sensitivity, and stability for glucose identification in biological samples. In a study conducted by Chen et al. [33]. Ni-MOF microspheres embellished with gold NPs were used for the determination of glucose. The fabricated Au@Ni-BTC sensor showed increased glucose identification performance with linearity (5– 7400 mM), sensitivity (1447.1 μA mM^{-1} cm^{-2}), and low LOD of (1.5 μM). They concluded in their work that this synthesized Metal-based MOF is a promising material for efficient non-enzymatic detection of glucose.

22.3.6 Metal Oxide NPs-based MOFs

Metal oxide nanoparticles are MOFs modifiers and are synthesized due to their high reactivity and productivity.

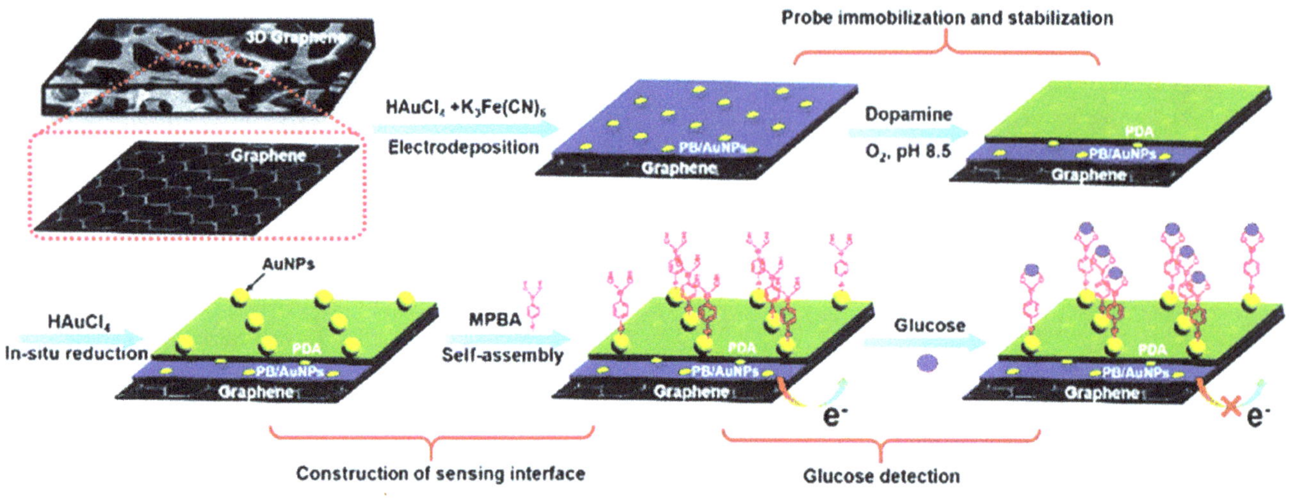

FIGURE 22.2 Illustrative view of fabrication of 3D MOFs based 3DG (3DG/PB-AuNPs) sensor [29]. Copyright (2020) The Authors, some rights reserved; exclusive licensee [Royal Society of Chemistry]. Distributed under a Creative Commons Attribution License 4.0 (CC BY).

Nickel-copper oxide nanowires (Ni-CuO NWs) sensing probe was synthesized by NiO2 as major substrate consisting of hierarchal MOFs structure with CuO nanowires. (Ni-CuO NWs) sensing probe exhibit electroactive properties which were evaluated by analytical techniques such as cyclic voltammetry (CV) and chronoamperometry (CA) and shows that Ni-CuO nanowires displayed admirable electrochemical activity for glucose oxidation with a low detection limit (0.07 μM), linearity (0.2 to 3.0 mM) along with excellent selectivity and sensitivity of 5610.6 μA mM^{-1} cm^{-2}. The coherent strategy based on a metal-organic framework derived from metal oxide composites presents a competent approach for the fabrication of electrochemical non-enzymatic glucose sensing [34].

22.3.7 Carbon Composite-based MOFs

Carbon-derived MOF materials (MOFs-CMs) possess a great chemical stability surface area and electronic conductivity, having a broad range of applications. Zhang, Yaxing, et al. [35] established novel MOF-based porous Ni_2P/graphene composite for enzyme-free glucose detection in an alkaline medium (Figure 22.3). Porous Ni_2P/graphene-based MOF composite displays linear response ranges from 5μM to 1.4 mM with a low LOD of 0.44μM with an excellent linear response of R^2=0.9897 in human serum accompanying glucose concentration from 1 mM to 8 mM.

22.3.8 Enzyme-based Glucose Sensors

Amongst all biosensors, enzymatic biosensors are well-known biosensors due to their high specificity and sensitivity in complex systems [5]. Enzyme-based glucose biosensors generally involve two enzyme families, glucose oxidases GOx and glucose dehydrogenases GDH. Certain enzymes vary in selectivity, turnover rate cofactors, and redox potentials for a glucose level. Glucose oxidase (GOx) is the ideal enzyme because of its sensitivity and selectivity, cost-effectiveness, ionic stability, temperature, and pH withstanding ability. Glucose oxidase (GOx/GOD) catalyzes the conversion of glucose to gluconolactone and hydrogen peroxide with the consumption of oxygen. Two electrons and protons are exchanged during this conversion [5].

$$\text{glucose} + \text{oxygen} \xrightarrow{\text{GOx}} \text{gluconolactone} + \text{hydrogen peroxide}$$

Lee, H., Hong, Y. J et al. [36] fabricated a novel sensor based on SiO_2 (GOx@Au@MagSiO_2) microspheres for enzymatic glucose detection. SiO_2 (GOx@Au@MagSiO_2) microspheres were developed by explicit gold linkage with magnetic SiO2 microspheres and glucose oxidase was directly attached to the gold nanoparticles and magnetic nanozyme. The electrochemical performance of SiO_2 (GOx@Au@MagSiO_2) was determined by the use of evaluation parameters. It has been proven that nanozyme could be proficiently utilized for one-pot, direct glucose determination in real blood samples. At different concentration ranges from 50–700 mg/dL (2.8 mM–38.9 mM) and LOD of 0.02 mg/mL.

22.3.9 Non-enzymatic Glucose Sensing

Non-enzymatic glucose sensing is one of the advanced development techniques that overcome certain drawbacks of conventional enzymatic glucose sensing. Since the involvement of surface electrode modification is significant in the fabrication of glucose sensing materials. Synthesis of more selective and sensitive materials like metal-organic frameworks play an important role in non-enzymatic glucose detection [37]. Chen et al. developed a new electrochemical sensor prepared on metal-organic framework-based microspheres for glucose detection (Ni3N@C) fabricated electrocatalysts performed excellently for non-enzymatic glucose detection by showing a very low LOD of 0.3 μM. Pre-eminent working of Ni3N@C based electrochemical sensor ascribed on the rapid charge transfer capability combinatorial effects of N doped carbon C and equally disseminated Ni3N nanocomposites.

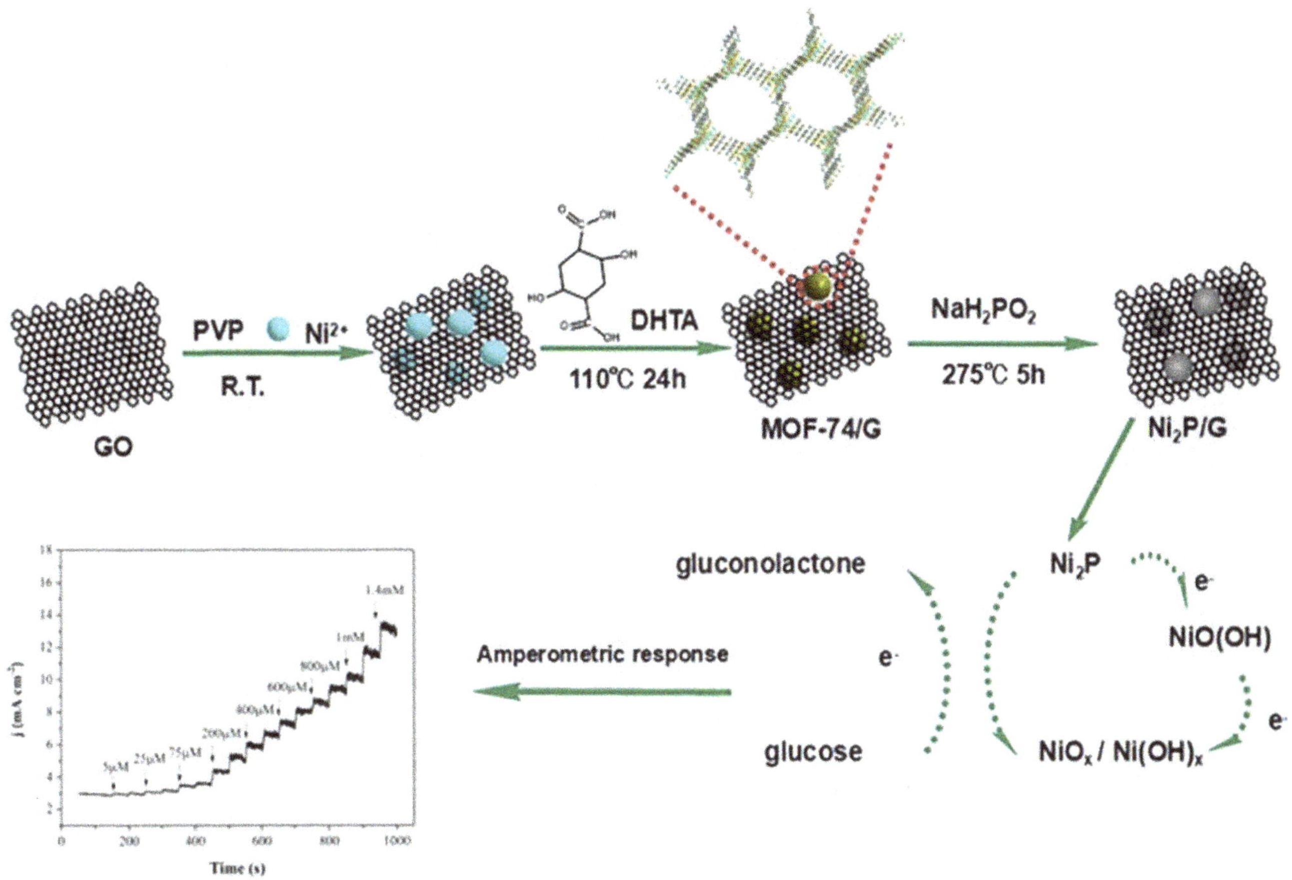

FIGURE 22.3 Amperometric Glucose Sensor MOF-based Porous Ni_2P/Graphene Composites [35]. Copyrights (2018) American Chemical Society.

To evaluate the Ni3N@C sensor characteristics cyclic volumetry CV has been used. Novel Ni3N@C sensor display exceptional viability and reproductivity for the detection of glucose in serum sample [38].

In a study conducted by Sun, Yanmei, et al. [39] they synthesized Cu-based-MOF through isothermal method for non-enzymatic determination of glucose from urine samples in the presence of interferences. The electrochemical behavior of Cu-based-MOF was determined by differential pulse voltammetry. Electrochemical outcomes indicated that Cu-MOF modified electrodes exhibit good electro-catalytic oxidation towards glucose with linearity ranges from 0.06 μM to 5 mM with sensitivity and detection limit of 89 μA/mM cm^2 and 10.5 nM respectively. Lu et al., in their work synthesized core-shell MOF@MOF composite of Ui-67@Ni-MOF under PVP regulation for non-enzymatic detection of glucose in alkaline media. The electrochemical behavior of UiO-67@Ni-MOF was determined by cyclic voltammetry. The MOF nanocomposite shows excellent electrochemical performance towards glucose oxidation. Additionally, the sensor has the ability for selectivity and sensitivity for real-time amperometric detection of glucose in human serum samples with reliability and accuracy. They concluded in their work that this specific sensor can act as an alternative tool for rapid diagnosis of diabetes for monitoring of blood glucose level daily [40].

In another work conducted by Li et al. [41] a Co-MOF nanosheet array on Ni-foam was synthesized for electrocatalytic oxidation of glucose in alkaline media. The sensor exhibited the enzyme-free sensing of glucose with an amperometric response of <5s, good sensitivity of 10,886 μA mM^{-1} cm^{-2} and low LOD of 1.3 nM. They concluded in their work that this sensor has the capability of long-time stability and high reproducibility for glucose detection in human serum and fruit juice. Meng, Wei, et al. [42], sin their work fabricated a Co-based porous MOF-ZIF-67 as the electrochemical sensor of glucose. In their work ZIF-67 modified GCE shows high electrocatalytic activity for glucose oxidation. To enhance the electrochemical performance of ZIF-67, silver-ZIF-67 nanocomposite was constructed through a sequential deposition reduction method. As a result, the reaction time of the modified electrode was shortened by more than two times with two and half times more sensitivity for glucose. The sensor exhibits a wide linear range of 2-1000μM with a low LOD of 0.66 μM.

22.3.10 MOF-based Voltammetric Sensing of Glucose

Voltammetric sensors are based on the system that is used to detect the minor fluctuations in current because of the occurrence of the redox reaction of electroactive substances and the working electrode [43]. Voltammetric biosensors involve various important techniques such as cyclic voltammetry, anodic, cathodic, adsorptive, and catalytic adsorptive stripping voltammetry [44]. Cyclic voltammetry measures the current response of the redox reactions. Cyclic voltammetry uses a three different electrode system. The system consists of working electrodes, reference electrodes, and counter electrodes [45].

Metal-organic framework fabricated voltammetric biosensors have gained much attention from researchers due to their ability to detect even trace amounts' amplified current signal from electroactive species samples [46]. Zhang et al. [47] developed an advanced highly sensitive copper metal-organic framework (Cu MOFs)-based voltammetric biosensor of very low detection limit along with wearisome stability (Figure 22.4). In his study, C/Cu nanocomposites were prepared by in situ growth which results in MOFs pyrolysis on the copper surface. Cu@porous carbon matrix electrode working for glucose detection was analyzed by cyclic voltammetry and chronoamperometry. Novel fabricated electrodes display the low LOD of (0.6 μM), immoderate sensitivity of (10.1 mA cm^{-2} mM^{-1}), and (less than 2 s) of latent period.

In a workshop conducted by Xiao et al. [48] metal-metal oxide nanocomposite was prepared by the direct carbonization of bimetallic Cu/Ni-MOF for cyclic voltammetric and amperometric identification of glucose. The metal-metal oxide nanocomposite exhibits a synergistic effect towards each other for excellent sensing performance of glucose having linearity of 0.1 μM to 2.2 mM with low LOD of 0.06 μM. Besides this, the sensor possesses impressive reproducibility, stability, and selectivity for the detection of glucose in real samples. In their work, they concluded that metal-metal oxide nanocomposite can act as a high-performance glucose sensor in daily life.

Bai et al. [49], in their work, experimented with the formation of Ni-CuO nanowires through oxidation/reduction and replacement reaction procedures. Electrochemical sensing measurements were conducted by cyclic voltammetry and amperometric methods. The sensor exhibits good electrochemical activity for glucose oxidation and has the ability of more sensitivity of 610.6 μA mM^{-1} cm^{-2} with a low limit of detection of 0.07 μM and wide linear range from 0.2 to 3.0 mM with good selectivity and can be used in clinical diagnostics, biotechnology, and food industry.

22.3.11 Amperometric Glucose Sensing

Amperometric sensors work by the generation of a current signal by connecting the electrodes and applying the potential. The analyte is subjected to or takes part in a redox reaction and the current produced in an electrochemical cell can be observed [50]. The amperometric sensor quantifies the current flow of the redox reactions between the connected electrodes. The most familiar and one of the earlier biosensors was constructed by Clark and Lyon in 1962 [51] which was also fabricated for glucose sensing with this leading-edge advancement in technology different amperometric-based sensors on different nanomaterials have been evolved. Metal-organic frameworks-based amperometric sensors are also displaying their worthwhile role in glucose detection. Paul et al. [52] constructed a novel biosensor based on MOF encapsulating TiO_2 for glucose detection in aqueous media through amperometry (Figure 22.5). ZIF-8-MOFs encapsulating TiO_2 nanoparticles and GOx exhibit very small level of glucose with a detection limit of (80 nM) can be measured in aqueous media as well as glucose oxidases concentration GO_x (62 μg in 1 mL). The developed strategy proved to be reliable and promising for both glucose oxidases as well as glucose peroxidase.

In a study conducted by Zhang et al. [53], a non-enzymatic glucose sensor based on carbon/Cu composite was synthesized by in-situ growth and pyrolysis of metal-organic frameworks on Cu-foam. The electrocatalytic activity of the respective sensor was determined by amperometry and cyclic voltammetry. The nanocomposite exhibits ultra-high sensitivity (10.1 mA cm^{-2} mM^{-1}), low limit of detection (0.6μM), a short response time of less than 2 s, and good stability of the modified electrode for glucose sensing in specimens. In another study done by Arul et al. [54], CuO was fabricated on GCE by the electrodeposition method using Cu-MOF. The modified electrode showed the linear amperometric response of glucose by increasing concentration from 500 nM to 5 mM and LOD was found to be 70 nM. The same method was utilized for the determination of glucose in human serum samples, and it was validated by the commercially available glucometer.

22.3.12 MOF-based Ratiometric Sensors

Ratiometric sensors or devices are exercised to express the output signal that delivers measurements/ratios as the result

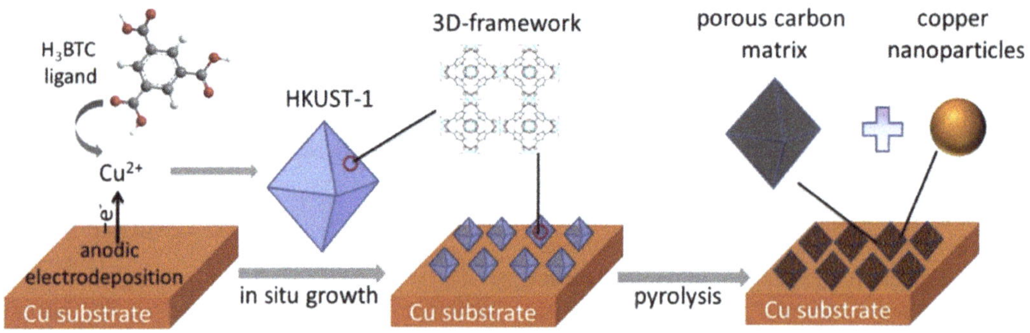

FIGURE 22.4 Representation of fabrication and working of (Cu MOFs) based voltammetric biosensor [47]. Copyrights (2018) Elsevier.

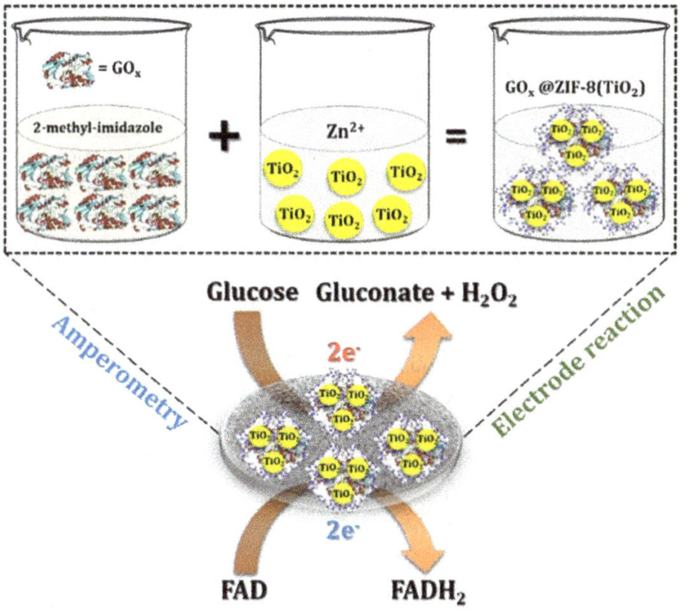

FIGURE 22.5 ZIF-8-MOFs encapsulating TiO_2 nanoparticles for glucose sensing [52]. Copyrights (2018) The Authors, some rights reserved; exclusive licensee [American Chemical Society]. Distributed under a Creative Commons Attribution License 4.0 (CC BY).

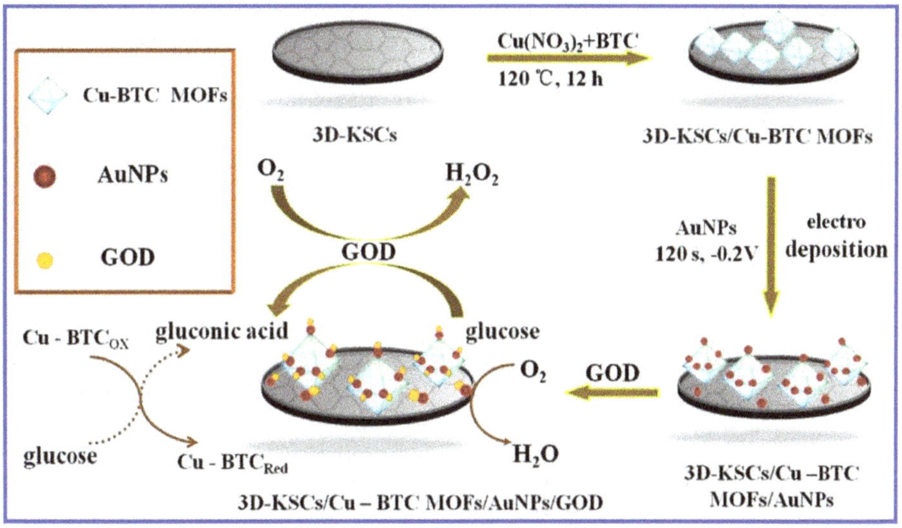

FIGURE 22.6 Electrochemical glucose sensor based on AuNPs/Cu-BTC-MOFs/3D-KSCs [56]. Copyright (2018) Elsevier.

of any variation in the input signal. These sensors hold two-fold electrochemical signals and the quantifiable evaluations of the target are constructed on the ratio of these dual signals [55]. MOF-derived ratiometric sensors are one of the recent approaches for glucose detection. A number of studies have been completed by researchers to fabricate more advance and efficient sensors. Song et al. [56], developed a novel ratiometric sensor for glucose detection. A AuNPs/Cu-BTC-MOFs/3D-KSCs electrode was fabricated to develop a ratiometric sensor to increase the electrical conductivity of copper metal-organic frameworks glucose oxidase GOD was immobilized on AuNPs (Figure 22.6). Cyclic voltammetry, differential pulse voltammetry, amperometry, and EIS were employed to measure response signals. The $j_{O2}/j_{Cu\text{-}BTC}$ is known as the response signal and displayed a linear range from 44.9 μM to 4.0 mM and from 4.0 to 19 mM. The limit of detection was 14.77 μM. The novel ratio-metric electrochemical sensor provided great accuracy, and good reproducibility, immense selectivity and proved to be a consistent and precise approach for glucose detection.

A cobalt carbon nanotube hybrid nanocomplex was constructed for the detection of glucose and hydrogen peroxide-based on the peroxidase-like properties of this nanocomplex. This sensing system changes the color from yellow to blue for hydrogen peroxide detection with a detection limit of 100 nM while glucose shows a detection limit of 150 nM using hydrogen peroxide-related reactions with glucose and glucose oxidase system as a model. This sensor is used for satisfactory recovery detection for glucose in serum samples. The developed ratiometric assay is a cost-effective and reliable detection of hydrogen peroxide and hydrogen peroxide-related reactions in clinical diagnosis and biomedical applications [57].

22.4 CONCLUSION

Metal-organic frameworks (MOFs) are emerging classes of hybrid material having high surface area and porousness constructed from metal ions or clusters bridged by organic linkers that have caught the eye of researchers in recent years. They have defined pore aperture, tailorable composition, and structure tenable size, versatile functionality, high loading capacity, and improved biocompatibility, making them a promising candidate for electrochemical sensing. Amid the MOF hybrids and carbon nanomaterials, carbon nanotubes (CNTs) and graphene oxide (GO), can enhance the electrical conductivity and mechanical strength of the nanomaterials, while the major difficulty placed in the low catalytic activity of MOFs could be dealt with by the use of catalytically active visitant constituents into metal-organic frameworks. The spatial dimension of the porous material makes them capable of an efficient material in electrochemical sensing with the leading-edge advancement in nanotechnology in the last few years proves to be a promising material for glucose sensing.

REFERENCES

[1]. Bruen D, Delaney C, Florea L, Diamond DJS (2017) Glucose sensing for diabetes monitoring: recent developments. *Sensors* 17 (8):1866

[2]. Soomro RA, Ibupoto ZH, Sirajuddin, Abro MI, Willander M (2015) Electrochemical sensing of glucose based on novel hedgehog-like NiO nanostructures. *Sensors and Actuators B: Chemical* 209:966–974.

[3]. Li J, Liu Y, Tang X, Xu L, Min L, Xue Y, Hu X, Yang Z (2020) Multiwalled carbon nanotubes coated with cobalt(II) sulfide nanoparticles for electrochemical sensing of glucose via direct electron transfer to glucose oxidase. *Microchimica Acta* 187 (1):80.

[4]. Metkar SK, Girigoswami KJB, biotechnology a (2019) *Diagnostic biosensors in medicine–A review*. 17:271–283

[5]. Lai J, Yi Y, Zhu P, Shen J, Wu K, Zhang L, Liu J (2016) Polyaniline-based glucose biosensor: A review. *Journal of Electroanalytical Chemistry* 782:138–153.

[6]. Cho I-H, Kim DH, Park SJBr (2020) *Electrochemical biosensors: Perspective on functional nanomaterials for on-site analysis*. 24 (1):1–12

[7]. Ye Y, Ji J, Sun Z, Shen P, Sun XJTTiAC (2020) *Recent advances in electrochemical biosensors for antioxidant analysis in foodstuff*. 122:115718

[8]. Švorc Ľ, Borovska K, Cinková K, Stanković DM, Planková AJEA (2017) *Advanced electrochemical platform for determination of cytostatic drug flutamide in various matrices using a boron-doped diamond electrode*. 251:621–630

[9]. Rakesh P, Pramod P, Sujit PJAJoPR (2019) *Biosensors: Current tool for Medication and Diagnosis*. 9 (1):27–34

[10]. Anik Ü, Timur S, Dursun ZJMA (2019) *Metal organic frameworks in electrochemical and optical sensing platforms: a review*. 186 (3):1–15

[11]. Wang Q, Astruc DJCr (2019) *State of the art and prospects in metal–organic framework (MOF)-based and MOF-derived nanocatalysis*. 120 (2):1438–1511

[12]. Yuan S, Feng L, Wang K, Pang J, Bosch M, Lollar C, Sun Y, Qin J, Yang X, Zhang PJAM (2018) *Stable metal–organic frameworks: design, synthesis, and applications*. 30 (37):1704303

[13]. Qiu W, Wang Q, Yano N, Kataoka Y, Handa M, Gao F, Tanaka HJEA (2020) *Flexible flower-like MOF of Cu2 (trans-1, 4-cyclohexanedicarboxylic acid) 2 as the electroactive matrix material for label-free and highly sensitive sensing of thrombin*. 353:136611

[14]. Zdravkov B, Čermák J, Šefara M, Janků JJOC (2007) *Pore classification in the characterization of porous materials: A perspective*. 5 (2):385–395

[15]. Jing P, Zhang SY, Chen W, Wang L, Shi W, Cheng PJCAEJ (2018) *A macroporous metal–organic framework with enhanced hydrophobicity for efficient oil adsorption*. 24 (15):3754–3759

[16]. Farha OK, Yazaydın AÖ, Eryazici I, Malliakas CD, Hauser BG, Kanatzidis MG, Nguyen ST, Snurr RQ, Hupp JTJNc (2010) *De novo synthesis of a metal–organic framework material featuring ultrahigh surface area and gas storage capacities*. 2 (11):944–948

[17]. Chae HK, Siberio-Perez DY, Kim J, Go Y, Eddaoudi M, Matzger AJ, O'Keeffe M, Yaghi OMJN (2004) *A route to high surface area, porosity, and inclusion of large molecules in crystals*. 427 (6974):523–527

[18]. Sosa JD, Bennett TF, Nelms KJ, Liu BM, Tovar RC, Liu Y (2018) Metal–Organic Framework Hybrid Materials and Their Applications. *Crystals* 8 (8):325

[19]. Stylianou KC, Imaz I, Maspoch D (2011) Metal-Organic Frameworks: Nanoscale Frameworks. *Encyclopedia of Inorganic and Bioinorganic Chemistry*:1–19

[20]. Shahhoseini L, Mohammadi R, Ghanbari B, Shahrokhian SJASS (2019) *Ni (II) 1D-coordination polymer/C60-modified glassy carbon electrode as a highly sensitive non-enzymatic glucose electrochemical sensor*. 478:361–372

[21]. Dowaidar M (2021) *2D MOFs have unique features for biological applications. They can be utilized for gene therapy, bioimaging, biosensing, photodynamic therapy, and tissue engineering*.

[22]. He J, Yang H, Zhang Y, Yu J, Miao L, Song Y, Wang L (2016) Smart Nanocomposites of Cu-Hemin Metal-Organic Frameworks for Electrochemical Glucose Biosensing. *Scientific Reports* 6 (1):36637.

[23]. Gumilar G, Kaneti YV, Henzie J, Chatterjee S, Na J, Yuliarto B, Nugraha N, Patah A, Bhaumik A, Yamauchi Y (2020) General synthesis of hierarchical sheet/plate-like M-BDC (M= Cu, Mn, Ni, and Zr) metal–organic frameworks for electrochemical non-enzymatic glucose sensing. *Chemical science* 11 (14):3644–3655

[24]. Shu Y, Yan Y, Chen J, Xu Q, Pang H, Hu X (2017) Ni and NiO Nanoparticles Decorated Metal–Organic Framework Nanosheets: Facile Synthesis and High-Performance Nonenzymatic Glucose Detection in Human Serum. *ACS Applied Materials & Interfaces* 9 (27):22342–22349.

[25]. Chen C, Xiong D, Gu M, Lu C, Yi F-Y, Ma X (2020) MOF-Derived Bimetallic CoFe-PBA Composites as Highly Selective and Sensitive Electrochemical Sensors for Hydrogen Peroxide and Nonenzymatic Glucose in Human Serum. *ACS Applied Materials & Interfaces* 12 (31):35365–35374.

[26]. Duan D, Yang H, Ding Y, Li L, Ma GJEA (2019) A three-dimensional conductive molecularly imprinted electrochemical sensor based on MOF derived porous carbon/carbon nanotubes composites and prussian blue nanocubes mediated amplification for chiral analysis of cysteine enantiomers. 302:137–144

[27]. Liu Q, Zhong H, Chen M, Zhao C, Liu Y, Xi F, Luo TJRA (2020) *Functional nanostructure-loaded three-dimensional graphene foam as a non-enzymatic electrochemical sensor for reagentless glucose detection.* 10 (56):33739–33746

[28]. He J, Yang H, Zhang Y, Yu J, Miao L, Song Y, Wang L (2016) Smart nanocomposites of Cu-hemin metal-organic frameworks for electrochemical glucose biosensing. *Scientific reports* 6 (1):1–9

[29]. Liu Q, Zhong H, Chen M, Zhao C, Liu Y, Xi F, Luo T (2020) Functional nanostructure-loaded three-dimensional graphene foam as a non-enzymatic electrochemical sensor for reagentless glucose detection. *RSC Advances* 10 (56):33739–33746.

[30]. Liu C-S, Li J, Pang H (2020) Metal-organic framework-based materials as an emerging platform for advanced electrochemical sensing. *Coordination Chemistry Reviews* 410:213222.

[31]. Li B, Ma JG, Cheng PJS (2019) *Integration of metal nanoparticles into metal–organic frameworks for composite catalysts: design and synthetic strategy.* 15 (32):1804849

[32]. Xiao X, Peng S, Wang C, Cheng D, Li N, Dong Y, Li Q, Wei D, Liu P, Xie Z (2019) Metal/metal oxide@ carbon composites derived from bimetallic Cu/Ni-based MOF and their electrocatalytic performance for glucose sensing. *Journal of Electroanalytical Chemistry* 841:94–100

[33]. Chen J, Yin H, Zhou J, Wang L, Gong J, Ji Z, Nie Q (2020) Efficient nonenzymatic sensors based on Ni-MOF microspheres decorated with Au nanoparticles for glucose detection. *Journal of Electronic Materials* 49:4754–4763

[34]. Bai X, Chen W, Song Y, Zhang J, Ge R, Wei W, Jiao Z, Sun YJASS (2017) *Nickel-copper oxide nanowires for highly sensitive sensing of glucose*. 420:927–934

[35]. Zhang Y, Xu J, Xia J, Zhang F, Wang ZJAam, interfaces (2018) *MOF-derived porous Ni2P/graphene composites with enhanced electrochemical properties for sensitive nonenzymatic glucose sensing.* 10 (45):39151–39160

[36]. Hwang D-W, Lee S, Seo M, Chung TD (2018) Recent advances in electrochemical non-enzymatic glucose sensors–a review. *Analytica chimica acta* 1033:1–34

[37]. Hwang D-W, Lee S, Seo M, Chung TDJAca (2018) *Recent advances in electrochemical non-enzymatic glucose sensors–a review.* 1033:1–34

[38]. Chen J, Yin H, Zhou J, Wang L, Ji Z, Zheng Y, Nie QJMT (2021) *Hybrid Ni3N-nitrogen-doped carbon microspheres (Ni3N@ C) in situ derived from Ni-MOFs as sensitive non-enzymatic glucose sensors.* 36 (5):286–295

[39]. Sun Y, Li Y, Wang N, Xu QQ, Xu L, Lin M (2018) Copper-based Metal-organic Framework for Non-enzymatic Electrochemical Detection of Glucose. *Electroanalysis* 30 (3):474–478

[40]. Lu M, Deng Y, Li Y, Li T, Xu J, Chen S-W, Wang J (2020) Core-shell MOF@MOF composites for sensitive nonenzymatic glucose sensing in human serum. *Analytica Chimica Acta* 1110:35–43.

[41]. de Carvalho PAV, Lopes IC, Silva EHC, Bruzaca EES, Alves HJ, Lima MIS, Tanaka AA (2019) Electrochemical behaviour of anticancer drug lomustine and in situ evaluation of its interaction with DNA. *Journal of pharmaceutical and biomedical analysis* 176:112786

[42]. Meng W, Wen Y, Dai L, He Z, Wang L (2018) A novel electrochemical sensor for glucose detection based on Ag@ ZIF-67 nanocomposite. *Sensors and Actuators B: Chemical* 260:852–860.

[43]. Alqarni SA, Hussein MA, Ganash AA, Khan AJB (2020) *Composite Material–Based Conducting Polymers for Electrochemical Sensor Applications: A Mini Review.* 10 (1):351–364

[44]. Sinha A, Lu X, Wu L, Tan D, Li Y, Chen J, Jain RJTTiAC (2018) *Voltammetric sensing of biomolecules at carbon based electrode interfaces: A review.* 98:174–189

[45]. Espinoza EM, Clark JA, Soliman J, Derr JB, Morales M, Vullev VIJJoTES (2019) *Practical aspects of cyclic voltammetry: How to estimate reduction potentials when irreversibility prevails.* 166 (5):H3175

[46]. Elgrishi N, Rountree KJ, McCarthy BD, Rountree ES, Eisenhart TT, Dempsey JLJJoce (2018) *A practical beginner's guide to cyclic voltammetry.* 95 (2):197–206

[47]. Zhang X, Luo J, Tang P, Morante JR, Arbiol J, Xu C, Li Q, Fransaer JJS, Chemical AB (2018) *Ultrasensitive binder-free glucose sensors based on the pyrolysis of in situ grown Cu MOF.* 254:272–281

[48]. Xiao X, Peng S, Wang C, Cheng D, Li N, Dong Y, Li Q, Wei D, Liu P, Xie Z, Qu D, Li X (2019) Metal/metal oxide@carbon composites derived from bimetallic Cu/Ni-based MOF and their electrocatalytic performance for glucose sensing. *Journal of Electroanalytical Chemistry* 841:94–100.

[49]. Bai X, Chen W, Song Y, Zhang J, Ge R, Wei W, Jiao Z, Sun Y (2017) Nickel-copper oxide nanowires for highly sensitive sensing of glucose. *Applied Surface Science* 420:927–934.

[50]. Sahin B, Kaya TJMRE (2019) *Electrochemical amperometric biosensor applications of nanostructured metal oxides: a review.* 6 (4):042003

[51]. Clarke GA, Hartse BX, Niaraki Asli AE, Taghavimehr M, Hashemi N, Abbasi Shirsavar M, Montazami R, Alimoradi N, Nasirian V, Ouedraogo LJJS (2021) *Advancement of Sensor Integrated Organ-on-Chip Devices.* 21 (4):1367

[52]. Paul A, Srivastava DNJAo (2018) *Amperometric glucose sensing at nanomolar level using MOF-encapsulated TiO2 platform.* 3 (11):14634–14640

[53]. Zhang X, Luo J, Tang P, Morante JR, Arbiol J, Xu C, Li Q, Fransaer J (2018) Ultrasensitive binder-free glucose sensors based on the pyrolysis of in situ grown Cu MOF. *Sensors and Actuators B: Chemical* 254:272–281.

[54]. Arul P, Abraham John S (2017) Electrodeposition of CuO from Cu-MOF on glassy carbon electrode: A non-enzymatic sensor for glucose. *Journal of Electroanalytical Chemistry* 799:61–69.

[55]. Yang T, Yu R, Yan Y, Zeng H, Luo S, Liu N, Morrin A, Luo X, Li WJS, Chemical AB (2018) *A review of ratiometric*

electrochemical sensors: From design schemes to future prospects. 274:501–516

[56]. Song Y, Xu M, Gong C, Shen Y, Wang L, Xie Y, Wang LJS, Chemical AB (2018) *Ratiometric electrochemical glucose biosensor based on GOD/AuNPs/Cu-BTC MOFs/ macroporous carbon integrated electrode*. 257:792–799

[57]. Qian P, Qin Y, Lyu Y, Li Y, Wang L, Wang S, Liu Y (2019) A hierarchical cobalt/carbon nanotube hybrid nanocomplex-based ratiometric fluorescent nanosensor for ultrasensitive detection of hydrogen peroxide and glucose in human serum. *Analytical and Bioanalytical Chemistry* 411 (8):1517–1524.

23 MOF-based Electrochemical Sensors for Protein Detection

Yang Liu[1]*, Juanhua Zhou*[1]*, Shiyu Zhang*[1] *and Hongye Wang*[1]
[1]Department of Chemistry, Tsinghua University, Beijing, China

23.1 INTRODUCTION

The content of protein in life is extremely rich, and the normal progress of life activities is inseparable from the activity of protein. Proteins can catalyze reactions, transport substances, transmit signals, and then affect cell proliferation, differentiation, and apoptosis. Therefore, proteins can be used as biomarkers, and their expression and activity can help early diagnosis and process monitoring of disease [1, 2].

At present, many protein detection technologies have been developed, including enzyme-linked immunosorbent assays (ELISA) [3], fluorescence [4], mass spectroscopy (MS) [5], electrochemical methods [2, 6], and so on. The sensitivity of immunofluorescence is very high, but its application in multiple detections is limited due to its inherent shortcomings such as optical background and wide spectral bandwidth [7]. Compared with fluorescence, MS can achieve multiple detections [8]. However, it is difficult in the qualitative and quantitative analysis of low abundance proteins [9]. Moreover, the above two methods are complex, time-consuming, expensive, and require well-trained personnel, which limits their feasibility and portability in practical application [10]. However, the electrochemical method can overcome the above weaknesses. It has high sensitivity, low cost, portability, and simple operation. Therefore, it is widely used in protein analysis of actual samples, especially in the field of point of care and wearable sensing devices [11]. Additionally, combined with appropriate recognition strategies and signal transduction methods, electrochemical sensors based on electrochemical methods have unparalleled advantages in the field of protein sensing.

In electrochemical biosensors, biorecognition elements and transducers are two important components, which determine the selectivity and sensitivity of the sensing. Metal-organic frameworks (MOFs) and their derivatives have attracted extensive attention in the field of electrochemical sensing because of their porous, orderly structure, and adjustable properties in recent years. Thanks to the particular size and shape of their channels and cavities, MOFs have specific selectivity for analytes, which makes them recognition elements [12]. Besides, the introduction of different functional groups in MOFs can also improve selectivity. Also, their large surface area and high porosity enable them to catalyze and amplify signals [13]. On the other hand, by integrating functional molecules or composites with MOFs and MOF derivatives, sensor probes can be constructed for biosensing applications. In general, the MOF is a promising material for electrochemical sensors.

This chapter mainly introduces the role of MOFs and their derivatives in the electrochemical sensor for protein assay. First of all, we give an overview of the structural properties and regulation of MOFs. Then we present different signal transducer and amplification mechanisms adjusted by MOFs. Later, we summarize the application of MOF-based electrochemical sensors for protein detection recently. Finally, we discuss the challenges and future direction of MOF-based electrochemical sensors.

23.2 STRUCTURAL PROPERTIES OF MOFs FOR PROTEIN RECOGNITION

MOFs, an emerging class of materials utilized for electrochemical sensors, have attracted increasingly more attention for their enormous specific surface areas as well as extra-high porosity in recent years [14]. Alterable building blocks and pores guarantee that the functional orientation design of MOFs can meet various sensing demands. In this section, the structural features of MOFs (and their derivates) will be discussed. Meanwhile, several aspects of MOFs' modification and regulation strategies will also be described, to show how, or potentially how they can assist protein recognition.

23.2.1 Porosity and Pore Modulation

Since MOF-5 ($Zn_4O(BDC)_3{\cdot}(DMF)_8(C_6H_5Cl)$) was reported by the Yaghi group in 1999 [15], one of the most essential characteristics of MOFs is their fine porosity. A rational methodology to customize the pores of MOFs is ligand expansion, which means designing ligands of different lengths (prolong carbon chain) or different substituent groups (including changing the number of groups). One landmark work can see the series of IR-MOF-74 reported by Deng et al. [16]. MOF-74, a kind of well-known and classic structure, was firstly reported by the Yaghi group in 2005 with the formula as $Zn_2(DHBDC)(DMF)_2{\cdot}(H_2O)_2$ [17]. The original structure of MOF-74 was assembled with DHBDC (2,5-Dihydroxy-1,4-benzene dicarboxylic acid) as an organic ligand. By expanding the numbers of the phenylene ring from one to

DOI: 10.1201/9781003188148-23

maximally eleven, the length of this series of organic links was altered from 7.0 to 50 Å, resulting in IRMOF-74 with different pore apertures, ranging from 14 to 98 Å, meanwhile, robust frameworks without interpenetrating structures were yielded as well. Before this work, the largest internal pore reported was 47 Å and the largest pore reported was 32 Å [18, 19], meanwhile, longer organic links were usually found and believed to induce interpenetrating structures or fragile frameworks [18, 20, 21]. Moreover, the authors also demonstrated that, when the pore apertures within the MOF structure were sufficient, proteins could be encapsulated into the pores.

As the protein can be encapsulated into the pores of MOF, proteins' performance in MOFs with pore size modulation has attracted great attention from researchers. In 2018, Li et al reported a series of Zr MOF named NU-100x (x ranging from 3 to 7, meaning the MOFs constructed by ligands with a series of expanded aromatic rings) [22]. By regulating the pore of NU-100x, the authors realized the immobilization and reaction of an enzyme (and co-enzyme) into MOFs in the absence of cells. Represented by such work, MOF-based cell-free enzyme immobilization widened the promising future of MOF-protein compatibility, guiding the development of novel MOF-based protein sensing platforms. Notably, the aspect of pore modulation is not only restricted to pore aperture modulation, but also the properties of the pore surface which are also very important. Chen et al. put forward a method to regulate the features of the pore surface resulting in the improvement of the capacity of fluorocarbons and CO_2 adsorption and separation [23].

23.2.2 The Role of Metal Ions and Multivariate MOFs

It is worth mentioning that Zr-based materials were believed to be promising candidates for protein detection, because of the excellent affinity of the Zr cluster with phosphate groups. For example, He et al designed and synthesized a 2D MOF nanosheets called 521-MOF({ $[Zr_6(\mu_3\text{-}O)_8(NBB)_{0.67}(CF_3COO)_6](DEF)_{11}\}n$, NBB= 4′,4‴,4‴″-Nitrilotris [1,1′-biphenyl]-4-carboxylic acid) [24], realizing high selective detection of mucin 1 protein (MUC1), which benefited from the fine affinity of Zr-MOF and the aptamer of MUC1. Additionally, except for MOFs, other Zr-based kinds of materials also showed enough affinity to the phosphate groups [25, 26]. Therefore, the importance of metals in MOFs cannot be overlooked. Different metal species correspond to various functions. For instance, Wang et al prepared Ln-MOF as a self-luminescent electrochemiluminescent probe to detect cytokeratins21-1, adding the property of lanthanide to the design of the luminescent MOF [27].

Since the importance of metal and ligand replacement strategies are emphasized above, we also want to introduce some aspects of mixed metal or mixed ligand MOFs, meanwhile, some generalized ideas of modulation on SBUs (secondary building units) will also be discussed in this paragraph. SBU, a critical factor for the structure of MOFs, indicating the linking between metal ions and the coordinating sites of ligands in certain geometries, can imply the topology of the final structure of the MOF [17]. Since it is the block that will be connected by ligands, SBU is also the key component to providing a potentially porous periodic arrangement [28]. Therefore, the regulation of the SBU may cause a change in the topological modes of MOFs. For example, Rosi et al. constructed several rod-packing MOFs by rod-shaped SBUs [17], demonstrating the design and building process of topology based on the SBU geometry. Noteworthily, the classic structure, MOF-74, was firstly manufactured in this way. Mixing metals (substantially, metal ions) within one SBU is also a kind of modulation. Taking MOF-74 as proto as well, Wang et al synthesized a series of MM(mixed-metal)-MOF-74 [29], containing 2,4,6,8 and at most 10 different metals with a huge range of ionic radii in a single pure phase, while, certainly the SBU remained unchanged, exhibiting possibilities for the design of mixed-metal MOFs to expand their properties. Another classic structure, the MOF-5 type structure, witnessed the feasibility of mixed-ligand MOF design. Deng et al synthesized a series of MTV(multivariate)-MOF-5 [30], containing up to 8 different ligands in one phase while the main framework of MOF-5 (the network connected by Zr-O cluster and phenylene units) was kept unchanged, though the ligands distributions were nonuniform. This pioneering work achieved up to 8 functionalities stacked on a single MOF, while the nonlinearity of the accumulation of group-induced (or to say group-improved) functions was found. Such strategies like mixed metals or ligands and SBU regulation do attribute to the diversity of functions and structures of MOFs. There should be more attempts to investigate the usage of such methods in the electrochemical sensing of protein since they have provided the theoretical basis and potentials for protein recognition.

23.2.3 Hierarchical Structure and MOF-based Recognition Platform

Another crucial structural property of MOF is a hierarchical structure. Besides ligand extension, some new methods which focused on the fine design and regulation of the synthesizing process can help prepare hierarchical MOFs. For instance, Yue et al. have summarized a general template-free strategy to construct hierarchical MOF hybrids [31]. In short, the proposed perturbation-assisted nano fusion synthesis method was to control the disturbance extent in the mother liquor to realize the regulation of the condensation and redissolution process. By such a method, the kinetic-controlled products during the perturbation synthesis can be prepared free from some of the disadvantages of template-assisted strategies like structure collapse. In a different way, Gu et al fabricated NH_2-UiO-66 on NH_2-MIL-125 by the modulation of modular growth [32], forming a novel MOF-on-MOF hierarchical architecture, NH_2-UiO-66@NH_2-MIL-125. Similarly, the authors also fabricated different hierarchical MOFs to achieve target functionalities [33].

Apart from the structural properties and their applications in protein recognition, MOFs themselves can play a role as a substrate, or a protein recognition platform. Considering that a MOF is a kind of crystalline material with ultrahigh specific surface areas, recognition units and electrochemical probes can be fabricated on them simultaneously. Jiang et al utilized Fe-MIL-101 as a substrate to immobilize the antibody and ABEI [34], a kind of electrochemiluminescence probe, realizing the recognition of mucin1 on the MOF platform.

23.3 SIGNAL TRANSDUCER AND AMPLIFICATION MECHANISMS OF MOFs

An electrochemical sensor is a device that provides identical information about its environmental variations. A transducer is an indispensable component of it, which can convert environmental variations into measurable signals on modern instruments by electrochemical technologies. The intensity of the electrochemical output is directly related to the strength of the variation, which makes it possible to qualitatively and quantitatively analyze the variations. Benefiting from the unique structural features and remarkable optoelectronic properties, MOFs and MOF derivatives hold enormous potential for electrochemical signal transduction. In this section, the electrochemical signal transducer and amplification mechanism adjusted by MOFs and MOF derivatives are systematically summarized. In particular, three different signal transduction techniques, including electrogenerated chemiluminescence (ECL) sensors, photoelectrochemical sensors, and impedimetric sensors are discussed, and corresponding amplification mechanisms based on MOFs, MOF derivatives, and MOF-based composites are presented with specific cases.

23.3.1 Electrogenerated Chemiluminescence Sensors

Electrogenerated chemiluminescence, also known as electrochemiluminescence, is a widely used electrochemical technique in different fields. Luminophores like ruthenium complexes, polyaromatic hydrocarbons, luminol, and its derivations, quantum dots and so on are commonly used in the ECL system and developing effective immobilization strategies for these luminophores is a persistent research hotspot in this field [35]. Benefiting from organic-inorganic hybrid extended networks, the combination of ECL luminophores and MOFs/MOF derivatives is believed to be an effective and inspiring immobilization strategy, which can realize the immobilization of numerous luminophores, enhanced ECL signals and efficiency. Furthermore, MOFs and MOF derivatives can be used as electrocatalysts to boost the ECL signal. Thus, numerous MOFs and their derivative-based on ECL sensors have been constructed.

In the first category, the ECL luminophores are incorporated with MOFs and MOF derivatives by in-situ synthesis, encapsulation, doping, or post-synthetic modification processes. These functional ECL emitters are further applied to sensing applications. As MOFs are metal-ligand coordination networks composed of metal-containing nodes and organic linkers [36], ECL luminophores can serve as building blocks to synthesize functional MOFs with ECL activities. In 2015, Xu et al. firstly reported the ECL of Ru-MOF, which was synthesized by the coordination interaction between $[Ru(dcbpy)_2bpy](PF_6)_2$ and a zinc ion. The efficient electron transfer and porous structure of Ru-MOF resulted in intense ECL emission, and the detection of cocaine in a serum sample was further confirmed [37]. Besides ruthenium complexes, carboxyl functionalized polyaromatic hydrocarbons have also been investigated to fabricate MOFs with the ECL property [38, 39]. Huang et al. used 1,1,2,2-tetra(4-carboxylbiphenyl)ethylene(H_4TCBPE) and Hf^{4+} to synthesize a Hf-based metal–organic framework(Hf-TCBPE). It was discovered that the Hf-TCBPE exhibited a stronger ECL emission in comparison with H_4TCBPE aggregates and H_4TCBPE monomers. This phenomenon was attributed to the restricted intramolecular free motions of TCBPE fixed in the framework and the highly porous nature of Hf-TCBPE, thus matrix coordination-induced ECL enhancement was proposed [38]. Also, Jin et al. reported an electroactive MOF(E-MOF) composed of hydroquinone and phenanthroline as oxidative and reductive couples. The surface state-induced ECL of E-MOF was investigated in both co-reactant and annihilation ECL processes [40].

In another category, MOFs and MOF derivatives act as the ECL modifiers to influence (either increase or decrease) ECL intensity, instead of ECL emitters. On one hand, benefiting from the well-ordered porous structure and large surface area, MOFs and MOF derivatives can load with aptamers or antibodies to participate in the ECL immunosensing process with increased selectivity and sensitivity. On the other hand, the catalytic sites of the MOFs and MOF derivatives can enhance the ECL signal. Wu et al. fabricated AuNPs@CuMOFs simply by stirring the mixed solution of AuNPs and CuMOFs. The AuNPs@CuMOFs not only showed brilliant sensitizing effect on the ECL of luminol than that of CuMOFs, but also served as a carrier to load the aptamer (aptamer/AuNPs@CuMOFs) through the Au-S bond to prepare the α-syn oligomer sensor [41].

23.3.2 Photoelectrochemical Sensors

Conversely, to the signal transduction form of ECL, PEC sensing uses the electrical signal as the output with light to excite the photoactive species. For PEC sensing applications, photoactive material plays an indispensable role. MOFs and their derivatives have been applied to improve the photoelectric conversion efficiency, owing to their unique properties.

Utilizing photoactive molecules as a building block, MOFs and MOF derivatives with PEC activity have been synthesized and used to construct PEC sensors. Porphyrinic or metalloporphyrins ligand-based MOFs have received much attention because of the remarkable features of photoactive porphyrin molecules. By incorporation of porphyrin into the framework of MOFs, the smaller energy gap (HOMO-LUMO) of porphyrin MOFs could be achieved and the electron-hole recombination process within the framework is prolonged, which is essential to improve the photoelectric conversion efficiency [42]. Combination with other materials which help to separate the charge carriers would significantly increase the electrical signal. For example, by incorporating the highly conductive RGO with porphyrinic MOF (PCN-224), Peng et al. discovered that the photocurrent of PCN-224/RGO gained ~12 times higher amplification, which should be attributed to the significantly improved separation efficiency of charge carriers [43].

In addition, the incorporation of MOFs and MOF derivatives with photoactive materials is an effective way to synthesize functional, heterogeneous composites with superior PEC activity. Wang et al. reported a highly sensitive PEC biosensor with multiple signal amplification of MOF for protein kinase activity analysis and inhibitor evaluation [44]. Using g-C_3N_4 nanosheets (CN-NS) as a light-harvesting unit, a hierarchically porous Cu-BTC/CN-NS composite has been reported [45]. The hierarchically porous Cu-BTC effectively improved the photocurrent response of CN-NS. In the presence of glyphosate, the Cu ion would coordinate with glyphosate, forming Cu-glyphosate complexes, which hinder the electron transfer process and lead to a decreased photocurrent.

Fabricating MOFs and semiconductor composites with well-designed structures is an attractive way to construct PEC sensors with a specific function. Zhan et al. proposed a self-template strategy to synthesize vertically standing ZnO@ ZIF-8 with core-shell heterostructure [46]. The ZIF-8 shell endowed this composite with molecular-size-selective ability. A 3D nanocomposite, Cu_2O@Cu-MOF/CM, was successfully constructed by an in-situ oxidation process and was believed to improve the conductivity, thus leading to better PEC performance [47].

23.3.3 Impedimetric Sensors

Electrochemical impedance spectroscopy (EIS) is an effective analytical technique to identify and analyze the charge transfer resistance caused by the interaction among the biomolecules and the features of the electrode surface throughout the modification steps [48]. Because of the low dielectric constant and large molecular weight of protein, EIS sensors have been widely used in the sensitive detection of various proteins. Li et al. fabricated a Zr-TAPP complex-based ultrasensitive biosensing layer for the detection of neuron-specific enolase (NSE) [49]. The biosensing layer constructed by Zr-TAPP displayed a large specific area, strong affinity towards $Anti_{NSE}$ by multiple interactions, and excellent electrochemical activity. Wang et al. proposed a MOF-on-MOF strategy to fabricate functional bimetallic MOFs with integrated merits [50]. Specifically, the Tb-MOF-on-Fe-MOF exhibited excellent biocompatibility, strong fluorescence, strong binding affinity with CA125 aptamers, and was applied for simultaneously detecting CA125 and MCF-7 cells with LODs of 58 $\mu U \cdot mL^{-1}$ and 19 $cell \cdot mL^{-1}$. Novel nanohybrids of Cr-based MOF and cobalt phthalocyanine (CoPc) nanoparticles (denoted as Cr-MOF@CoPc) were engineered for colorectal cancer (CT26) cells detection. The presence of CoPc NPs endowed the nanohybrids with enhanced electrochemical activity, excellent biocompatibility and thermostability, fluorescence activity, also improved immobilization ability [51].

In addition to the recognition-detection model, enzyme-catalyzed precipitation techniques have also been proposed, which are considered as an effective approach to amplify the EIS response. Glucose oxidase-initiated cascade catalysis has been coupled with MOF-based EIS sensing. Zhou et al. synthesized a Cu-based MOF, functionalized with Pt nanoparticles, aptamer, hemin and GOx (Pt@CuMOFs-hGq-GOx) to construct a cascade reaction, leading to the formation of nonconductive insoluble precipitates (IPs) on the electrode surface to amplify the EIS signal. H_2O_2 was generated by the oxidation of glucose with the catalysis of GOx, which subsequently reacted with diaminobenzidine (DAB) and resulted in the in-situ formation of nonconductive IPs on the electrode interface with the synergistic peroxide catalysis of Pt@CuMOFs and hGq. Thus, the electron transfer in the resultant sensing interface was effectively hindered and the EIS signal was efficiently amplified, achieving the ultrasensitive detection of Carcinoembryonic antigen (CEA) with a LOD of 0.023 $pg.mL^{-1}$ [52].

23.4 Detection of Different Proteins Based on MOF-Based Electrochemical Sensors

Proteins are involved in almost every physiological process of cells. One of the most important physiological functions of proteins is to catalyze various chemical reactions in cells as enzymes. Also, proteins can participate in intracellular substance and signal transduction, and then affect the life process of cells or the occurrence and evolution of diseases. Here, we summarized MOF-based electrochemical sensors for different kinds of protein detection in recent years.

23.4.1 Antigens

Cancer is one of the major diseases threatening human health in the world [53]. Tumor biomarkers are important for the early diagnosis, treatment, and prognosis of cancer. Therefore, we present some recent work on MOF-based electrochemical sensors for tumor biomarkers detection here. Overexpression of MUC1, a kind of transmembrane mucin, is associated with a variety of cancers, such as lung, pancreatic, and breast cancer, and is often considered as an important tumor biomarker. Jiang et al. reported an ultrasensitive biosensor for detecting MUC1 on MCF-7 cancer cells based on Fe-MIL-101 loaded with the abundant ABEI as a luminophore [34]. Based on electrochemiluminescence immunoassay, the novel metal-organic framework, ABEI@Fe-MIL-101, co-reacting with dissolved oxygen could be utilized as a high-efficiency ECL indicator leading to a detection limit as low as 12 cells for MUC1 on MCF-7 cancer cells. Besides MUC1, alpha-fetoprotein (AFP) is also an important tumor biomarker, which is widely used to indicate early hepatocellular carcinoma [54]. To construct an electrochemiluminescent immunosensor for the detection of AFP, Wang et al. synthesized ultrathin Co/Ni-based MOF nanosheets with micro flower-like structures which were considered as a platform for luminol-functionalized Ag NPs [55]. Benefiting from its superior structure, the composite materials showed excellent catalytic and amplification performance such as resistance to particle agglomeration. Moreover, prostate-specific antigen (PSA) is currently a recognized tumor biomarker for screening prostate cancer [56]. Dai et al. combined palladium nanoparticles (Pd NPs) with electroactive amino-zeolitic imidazolate framework-67 (Pd/NH_2-ZIF-67) to construct a novel redox mediator which could be used in the sandwich-type electrochemical immunosensor [57]. Not only did Pd NPs have outstanding biocompatibility and good catalytic performance, but also the lamellar structure of ZIF-67 increased loading of Pd NPs and catalyzed H_2O_2 decomposition synergistically. The biosensor showed satisfactory selectivity, repeatability, and stability owing to the synergistic amplification of Pd NPs and ZIF-67. Apart from being used to build PSA biosensors, nanomaterials are also widely used in CEA biosensors. Thanks to the ability of MoS_2 quantum dots (MQDs) to catalyze oxygen reactions and adsorption, Xin et al. compounded MoS_2 MQDs with MOF-545-Zn to construct novel composite materials which can co-react with O_2 to achieve high electrochemiluminescent responses [58]. In addition to electrochemiluminescent biosensors, Zhang et al. reported an ultrasensitive photoelectrochemical biosensor based on porous hollow carbon nano bubbles@ZnCdS multi-shelled dodecahedral cages (C@ZnCdSMSDCs) which had stunning photoelectrochemical performance owing to the synergistic effect between the porous shells and the carbon-layer coating [59]. Moreover, a Fe-MOF-based electrochemical aptasensor for CEA detection had been reported as well [60]. In order to improve biocompatibility and amplify the signal, Li et al. coordinated self-polymerized dopamine (DA)-decorated Au NPs and Fe-MOF resulting in great electrochemical performance.

Alzheimer's disease (AD) is a degenerative disease of the central nervous system that usually occurs in the elderly. It will not only seriously affect the physical and mental health of patients but also greatly reduce the quality of life [61]. Amyloid-β (Aβ) is generally considered a diagnostic biomarker and therapeutic target of AD. Han et al. synthesized electrochemical immunosensors by covalently modifying ferrocene onto a porous Zn MOF as an electrochemical signal label (Fc-Zn-MOF) [62]. The Fc-Zn-MOF with flower structure provided enough binding sites and showed enhanced electrochemical performance resulting in a highly sensitive analysis of Amyloid-β (Aβ). In addition to the strategy of introducing electroactive substances into MOF, Zhao et al. constructed an electrochemiluminescent immunosensor by introducing tris (2,2'- bipyridyl) ruthenium (II) [$Ru(bpy)_3^{2+}$] into three-dimensional (3D) zinc oxide metal-organic frameworks [$Ru(bpy)_3^{2+}$/zinc oxalate MOFs] [63]. Based on the sandwich antibody method and quenching electrochemiluminescence strategy, Au@NiFe MOFs coupled with Ab_2 were used as quenchers, and $Ru(bpy)_3^{2+}$/zinc oxalate MOFs coupled with Ab_1 were used as luminophores fixed on the electrode surface. Once Aβ was recognized by the antibody, the resonance energy transfer (RET) would occur between Au@NiFe MOFs and $Ru(bpy)_3^{2+}$/zinc oxalate MOFs, resulting in the decrease of electrochemiluminescence signal.

Cardiovascular diseases are common diseases that seriously threaten human health and rank first among various causes of death. C-reactive protein (CRP) can be an indicator of cardiovascular diseases in clinical diagnosis. Wang et al. synthesized size/morphology adjustable NH_2-Ni-MOFs as electrocatalysts and realized highly sensitive detection of CRP based on an aptamer binding induced DNA walker–antibody sandwich assay [64]. Furthermore, an innovative label-free CRP immunosensor was constructed on biocompatible ZnO/MPC composite which was prepared by pyrolysis of a mixed ligand MOF [65]. This kind of novel carbon material possessed charming electrochemical properties and could be a potential carrier of load identification units. Moreover, galectin-3 (Gal-3) was associated with heart failure closely. There was a sandwich-type electrochemical immunosensor for detecting Gal-3 based on N-doped graphene nanoribbon immobilized Fe-based-Metal-organic frameworks deposited with Au nanoparticles (N-GNRs-Fe-MOFs@Au NPs) [66].

23.4.2 Enzymes

Protein kinase is a kind of enzyme that catalyzes the process of protein phosphorylation. Its activity is closely related to various diseases. Therefore, the detection of protein kinase activity in clinical diagnosis can help diagnosis. Song et al. proposed a low fouling and highly sensitive electrochemical sensor for T4 polynucleotide kinase (PNK) detection based on zwitterionic peptide and self-sacrificial Fe-MOF [67]. The zwitterionic peptide could be assembled into antifouling layers and the Fe-MOF formed Prussian blue (PB) after reacting with $K_4Fe(CN)_6$. Thus, the biosensor prevented the adsorption of nonspecific proteins and had a high sensitivity. Another kinase, protein tyrosine kinase-7 (PTK7) could also be detected by an electrochemical biosensor based on Zn-MOF-on-Zr-MOF architecture [33]. In addition to electrochemical biosensors, photoelectrochemical biosensors attracted lots of interest as well. Our team developed a highly sensitive photoelectrochemical biosensor for the detection of protein kinase A (PKA) by $Ru(bpy)_3^{2+}$ loaded UiO-66 ($Ru(bpy)_3^{2+}$@UiO-66) [44]. UiO-66 not only increased the load of $Ru(bpy)_3^{2+}$ but also provided a large number of phosphorylated kemptide binding sites. Therefore, the biosensor achieved high sensitivity and fast response.

Telomerase is an enzyme responsible for maintaining telomere length in cells. Telomerase activity is tightly regulated in normal human cells, but the telomerase activity is always overexpressed in cancer cells. Thus, the measurement of telomerase's activity is important to tumor screening. Due to the excellent physical and chemical properties of nanomaterials, biosensors based on MOFs combined with nanomaterials for detecting kinase activity has been reported a lot. The nanoscale metalloporphyrin MOF [P-MOF(Fe)] with Pd NPs grown [Pt@P-MOF(Fe)] was used as biomimetic catalyst signal media which was an ideal platform for sensing kinase activity [68]. Moreover, Xiong et al. constructed an electrochemiluminescence biosensor by ZIF-8 doped with ruthenium polyethyleneimine (Ru–PEI) complex which possessed high ECL efficiency [69]. Owing to an enzyme-assisted DNA recycle–amplification strategy, telomerase activity was successfully detected.

23.5 CONCLUSION

Generally, although MOF-based protein electrochemical sensors do exhibit excellent performance and promising futures, some challenges remain. Additionally, we also propose to give some opinions on the key problems which need to be handled in this area. Here, we believe the following directions should be paid more attention:

a) Firstly, more rational design of MOF-based probes, including consideration of protein recognition, signal transduction, and amplification, should be achieved. Make the concept of 'custom-made' referential, in other words, customize the pore, ligands, metals and SBU, and maybe even the sensing platform involving MOFs, according to the analysis demand.
b) Better conductivity should be developed as well, given that electrochemical performance largely relies on the conductivity of electrode materials. Meanwhile, MOFs usually exhibit mediocre conductivity, so more general strategies to improve this are of great relevance.
c) Considering the important target of biomonitoring, the BC (biocompatibility) performance and toxicity of MOF-based sensors or probes should have a general and uniform evaluation criterion. Additionally, stability of sensing performance in different media, during different periods should also be considered.
d) Finally, a critical challenge regarding MOFs, is how to manufacture them on large scale for adaptation to various uses. Although some MOFs have already been commercially produced, this field is still very small. In the near future, outstanding behaviors and performance of MOFs, especially MOF-based probes with applications for protein detection, will be realized more and serve better for health purposes. Increasing the scale of production is of urgency.

REFERENCES

[1]. C. A. Crutchfield, S. N. Thomas, L. J. Sokoll and D. W. Chan, Advances in mass spectrometry-based clinical biomarker discovery, *Clinical Proteomics* **2016**, 13, 1.

[2]. V. Vanova, K. Mitrevska, V. Milosavljevic, D. Hynek, L. Richtera and V. Adam, Peptide-based electrochemical biosensors utilized for protein detection, *Biosensors and Bioelectronics* **2021**, 180, 113087.

[3]. S. K. Vashist, T. van Oordt, E. M. Schneider, R. Zengerle, F. von Stetten and J. H. T. Luong, A smartphone-based colorimetric reader for bioanalytical applications using the screen-based bottom illumination provided by gadgets, *Biosensors and Bioelectronics* **2015**, 67, 248–255.

[4]. X.-P. He, Y. Zang, T. D. James, J. Li and G.-R. Chen, Probing disease-related proteins with fluorogenic composite materials, *Chemical Society Reviews* **2015**, 44, 4239–4248.

[5]. J. R. Yates, C. I. Ruse and A. Nakorchevsky, Proteomics by Mass Spectrometry: Approaches, Advances, and Applications, *Annual Review of Biomedical Engineering* **2009**, 11, 49–79.

[6]. Y.-C. Zhu, L. Zhang, N. Zhang, W.-W. Zhao, Y.-Y. Liang, J.-J. Xu, and H.-Y. Chen, Photoelectrochemical bioanalysis of protein biomarkers, *Current Opinion in Electrochemistry* **2018**, 10, 120–125.

[7]. M. Culha, Surface-enhanced raman scattering: an emerging label-free detection and identification technique for proteins, *Applied spectroscopy* **2013**, 67, 355–364.

[8]. S. Xu, W. Ma, Y. Bai and H. Liu, Ultrasensitive Ambient Mass Spectrometry Immunoassays: Multiplexed Detection of Proteins in Serum and on Cell Surfaces, *Journal of the American Chemical Society* **2019**, 141, 72–75.

[9]. L. Cohen and D. R. Walt, Highly Sensitive and Multiplexed Protein Measurements, *Chemical Reviews* **2019**, 119, 293–321.

[10]. M. Shen, N. Li, Y. Lu, J. Cheng and Y. Xu, An enhanced centrifugation-assisted lateral flow immunoassay for the point-of-care detection of protein biomarkers, *Lab on a Chip* **2020**, 20, 2626–2634.

[11]. S. G. Meirinho, L. G. Dias, A. M. Peres, and L. R. Rodrigues, Voltammetric aptasensors for protein disease biomarkers detection: A review, *Biotechnology Advances* **2016**, 34, 941–953.

[12]. C.-S. Liu, J. Li and H. Pang, Metal-organic framework-based materials as an emerging platform for advanced electrochemical sensing, *Coordination Chemistry Reviews* **2020**, 410, 213222.

[13]. Y. Xu, Q. Li, H. Xue and H. Pang, Metal-organic frameworks for direct electrochemical applications, *Coordination Chemistry Reviews* **2018**, 376, 292–318.

[14]. H. C. Zhou, J. R. Long and O. M. Yaghi, Introduction to metal-organic frameworks, *Chemical Reviews* **2012**, 112, 673–674.

[15]. H. Li, M. Eddaoudi, M. O'Keeffe and O. M. Yaghi, Design and synthesis of an exceptionally stable and highly porous metal-organic framework, *Nature* **1999**, 402, 276–279.

[16]. H. Deng, S. Grunder, K. E. Cordova, C. Valente, H. Furukawa, M. Hmadeh, F. Gándara, A. C. Whalley, Z. Liu and S. Asahina, Large-pore apertures in a series of metal-organic frameworks, *Science* **2012**, 336, 1018–1023.

[17]. N. L. Rosi, J. Kim, M. Eddaoudi, B. Chen, M. O'Keeffe and O. M. Yaghi, Rod Packings and Metal–Organic Frameworks Constructed from Rod-Shaped Secondary Building Units, *Journal of the American Chemical Society* **2005**, 127, 1504–1518.

[18]. Y. K. Park, S. B. Choi, H. Kim, K. Kim, B.-H. Won, K. Choi, J.-S. Choi, W.-S. Ahn, N. Won, S. Kim, D. H. Jung, S.-H. Choi, G.-H. Kim, S.-S. Cha, Y. H. Jhon, J. K. Yang and J. Kim, Crystal Structure and Guest Uptake of a Mesoporous Metal–Organic Framework Containing Cages of 3.9 and 4.7 nm in Diameter, *Angewandte Chemie International Edition* **2007**, 46, 8230–8233.

[19]. L. Ma, J. M. Falkowski, C. Abney and W. Lin, A series of isoreticular chiral metal–organic frameworks as a tunable

platform for asymmetric catalysis, *Nature Chemistry* **2010**, 2, 838–846.

[20]. H. K. Chae, D. Y. Siberio-Perez, J. Kim, Y. Go, M. Eddaoudi, A. J. Matzger, M. O'Keeffe and O. M. Yaghi, A route to high surface area, porosity, and inclusion of large molecules in crystals, *Nature* **2004**, 427, 523–527.

[21]. T. M. Reineke, M. Eddaoudi, D. Moler, M. O'Keeffe and O. M. Yaghi, Large Free Volume in Maximally Interpenetrating Networks: The Role of Secondary Building Units Exemplified by $Tb_2(ADB)_3$ $[(CH_3)2SO]_4{\cdot}16$ $[(CH_3)_2SO]^1$, *Journal of the American Chemical Society* **2000**, 122, 4843–4844.

[22]. P. Li, Q. Chen, T. C. Wang, N. A. Vermeulen, B. L. Mehdi, A. Dohnalkova, N. D. Browning, D. Shen, R. Anderson, and D. A. Gómez-Gualdrón, Hierarchically engineered mesoporous metal-organic frameworks toward cell-free immobilized enzyme systems, *Chem* **2018**, 4, 1022–1034.

[23]. C. X. Chen, Z. Wei, J. J. Jiang, Y. Z. Fan, S. P. Zheng, C. C. Cao, Y. H. Li, D. Fenske and C. Y. Su, Precise modulation of the breathing behavior and pore surface in Zr-MOFs by reversible post-synthetic variable-spacer installation to fine-tune the expansion magnitude and sorption properties, *Angewandte Chemie International Edition 2016,* 128, 10086–10090.

[24]. L. He, F. Duan, Y. Song, C. Guo, H. Zhao, J.-Y. Tian, Z. Zhang, C.-S. Liu, X. Zhang, P. Wang, M. Du and S.-M. Fang, 2D zirconium-based metal-organic framework nanosheets for highly sensitive detection of mucin 1: consistency between electrochemical and surface plasmon resonance methods, *2D Materials* **2017**, 4, 025098.

[25]. G.-Y. Zhang, S.-Y. Deng, W.-R. Cai, S. Cosnier, X.-J. Zhang and D. Shan, Magnetic zirconium hexacyanoferrate (II) nanoparticle as tracing tag for electrochemical DNA assay, *Analytical chemistry* **2015**, 87, 9093–9100.

[26]. W.-F. Ma, C. Zhang, Y.-T. Zhang, M. Yu, J. Guo, Y. Zhang, H.-J. Lu and C.-C. Wang, Magnetic MSP@ ZrO_2 microspheres with yolk–shell structure: designed synthesis and application in highly selective enrichment of phosphopeptides, *Langmuir* **2014**, 30, 6602–6611.

[27]. Y. Wang, G. Zhao, H. Chi, S. Yang, Q. Niu, D. Wu, W. Cao, T. Li, H. Ma and Q. Wei, Self-Luminescent Lanthanide Metal–Organic Frameworks as Signal Probes in Electrochemiluminescence Immunoassay, *Journal of the American Chemical Society* **2020**, 143, 504–512.

[28]. J. Ha, J. H. Lee and H. R. Moon, Alterations to secondary building units of metal–organic frameworks for the development of new functions, *Inorganic Chemistry Frontiers* **2020**, 7, 12–27.

[29]. L. J. Wang, H. Deng, H. Furukawa, F. Gándara, K. E. Cordova, D. Peri and O. M. Yaghi, Synthesis, and characterization of metal–organic framework-74 containing 2, 4, 6, 8, and 10 different metals, *Inorganic chemistry* **2014**, 53, 5881–5883.

[30]. H. Deng, C. J. Doonan, H. Furukawa, R. B. Ferreira, J. Towne, C. B. Knobler, B. Wang and O. M. Yaghi, Multiple functional groups of varying ratios in metal-organic frameworks, *Science* **2010**, 327, 846–850.

[31]. Y. Yue, P. F. Fulvio and S. Dai, Hierarchical metal–organic framework hybrids: perturbation-assisted nanofusion synthesis, *Accounts of chemical research* **2015**, 48, 3044–3052.

[32]. Y. Gu, Y. n. Wu, L. Li, W. Chen, F. Li and S. Kitagawa, Controllable Modular Growth of Hierarchical MOF-on-MOF Architectures, *Angewandte Chemie International Edition* **2017**, 129, 15864–15868.

[33]. N. Zhou, F. Su, C. Guo, L. He, Z. Jia, M. Wang, Q. Jia, Z. Zhang and S. Lu, Two-dimensional oriented growth of Zn-MOF-on-Zr-MOF architecture: A highly sensitive and selective platform for detecting cancer markers, *Biosensors and Bioelectronics* **2019**, 123, 51–58.

[34]. X. Jiang, Z. Wang, H. Wang, Y. Zhuo, R. Yuan and Y. Chai, A novel metal–organic framework loaded with abundant N-(aminobutyl)-N-(ethylisoluminol) as a high-efficiency electrochemiluminescence indicator for sensitive detection of mucin1 on cancer cells, *Chemical Communications* **2017**, 53, 9705–9708.

[35]. W. Miao, Electrogenerated Chemiluminescence and Its Biorelated Applications, *Chemical Reviews* **2008**, 108, 2506–2553.

[36]. S. Liu, C. Lai, X. Liu, B. Li, C. Zhang, L. Qin, D. Huang, H. Yi, M. Zhang, L. Li, W. Wang, X. Zhou, and L. Chen, Metal-organic frameworks, and their derivatives as signal amplification elements for electrochemical sensing, *Coordination Chemistry Reviews* **2020**, 424, 213520.

[37]. Y. Xu, X.-B. Yin, X.-W. He and Y.-K. Zhang, Electrochemistry and electrochemiluminescence from a redox-active metal-organic framework, *Biosensors and Bioelectronics* **2015**, 68, 197–203.

[38]. W. Huang, G.-B. Hu, L.-Y. Yao, Y. Yang, W.-B. Liang, R. Yuan and D.-R. Xiao, Matrix Coordination-Induced Electrochemiluminescence Enhancement of Tetraphenylethylene-Based Hafnium Metal–Organic Framework: An Electrochemiluminescence Chromophore for Ultrasensitive Electrochemiluminescence Sensor Construction, *Analytical Chemistry* **2020**, 92, 3380–3387.

[39]. D. Zhu, Y. Zhang, S. Bao, N. Wang, S. Yu, R. Luo, J. Ma, H. Ju and J. Lei, Dual Intrareticular Oxidation of Mixed-Ligand Metal–Organic Frameworks for Stepwise Electrochemiluminescence, *Journal of the American Chemical Society* **2021**, 143, 3049–3053.

[40]. Z. Jin, X. Zhu, N. Wang, Y. Li, H. Ju and J. Lei, Electroactive Metal–Organic Frameworks as Emitters for Self-Enhanced Electrochemiluminescence in Aqueous Medium, *Angewandte Chemie International Edition* **2020**, 59, 10446–10450.

[41]. Q. Wu, R. Tan, X. Mi, and Y. Tu, Electrochemiluminescent aptamer-sensor for alpha synuclein oligomer based on a metal–organic framework, *Analyst* **2020**, 145, 2159–2167.

[42]. G.-Y. Zhang, Y.-H. Zhuang, D. Shan, G.-F. Su, S. Cosnier and X.-J. Zhang, Zirconium-Based Porphyrinic Metal–Organic Framework (PCN-222): Enhanced Photoelectrochemical Response and Its Application for Label-Free Phosphoprotein Detection, *Analytical Chemistry* **2016**, 88, 11207–11212.

[43]. M. Peng, G. Guan, H. Deng, B. Han, C. Tian, J. Zhuang, Y. Xu, W. Liu, and Z. Lin, PCN-224/rGO nanocomposite based photoelectrochemical sensor with intrinsic recognition ability for efficient p-arsanilic acid detection, *Environmental Science: Nano* **2019**, 6, 207–215.

[44]. Z. Wang, Z. Yan, F. Wang, J. Cai, L. Guo, J. Su, and Y. Liu, Highly sensitive photoelectrochemical biosensor for kinase activity detection and inhibition based on the surface defect recognition and multiple signal amplification of metal-organic frameworks, *Biosensors and Bioelectronics* **2017**, 97, 107–114.

[45]. Y. Cao, L. Wang, C. Wang, X. Hu, Y. Liu and G. Wang, Sensitive detection of glyphosate based on a Cu-BTC MOF/g-C_3N_4 nanosheet photoelectrochemical sensor, *Electrochimica Acta* **2019**, 317, 341–347.

[46]. W.-w. Zhan, Q. Kuang, J.-z. Zhou, X.-j. Kong, Z.-x. Xie and L.-s. Zheng, Semiconductor@Metal–Organic Framework Core–Shell Heterostructures: A Case of ZnO@ZIF-8 Nanorods

with Selective Photoelectrochemical Response, *Journal of the American Chemical Society* **2013**, 135, 1926–1933.
[47]. W. Kong, M.-H. Xiang, L. Xia, M. Zhang, R.-M. Kong and F. Qu, In-situ synthesis of 3D Cu_2O@Cu-based MOF nanobelt arrays with improved conductivity for sensitive photoelectrochemical detection of vascular endothelial growth factor 165, *Biosensors and Bioelectronics* **2020**, 167, 112481.
[48]. L. Liu, Y. Zhou, S. Liu and M. Xu, The Applications of Metal–Organic Frameworks in Electrochemical Sensors, *ChemElectroChem* **2018**, 5, 6–19.
[49]. Y. Li, C. Wang, Z. Li, M. Wang, L. He and Z. Zhang, Zirconium-porphyrin complex as novel nanocarrier for label-free impedimetric biosensing neuron-specific enolase, *Sensors and Actuators B: Chemical* **2020**, 314, 128090.
[50]. M. Wang, M. Hu, Z. Li, L. He, Y. Song, Q. Jia, Z. Zhang and M. Du, Construction of Tb-MOF-on-Fe-MOF conjugate as a novel platform for ultrasensitive detection of carbohydrate antigen 125 and living cancer cells, *Biosensors and Bioelectronics* **2019**, 142, 111536.
[51]. F. Duan, M. Hu, C. Guo, Y. Song, M. Wang, L. He, Z. Zhang, R. Pettinari and L. Zhou, Chromium-based metal-organic framework embedded with cobalt phthalocyanine for the sensitively impedimetric cytosensing of colorectal cancer (CT26) cells and cell imaging, *Chemical Engineering Journal* **2020**, 398, 125452.
[52]. X. Zhou, S. Guo, J. Gao, J. Zhao, S. Xue and W. Xu, Glucose oxidase-initiated cascade catalysis for sensitive impedimetric aptasensor based on metal-organic frameworks functionalized with Pt nanoparticles and hemin/G-quadruplex as mimicking peroxidases, *Biosensors and Bioelectronics* **2017**, 98, 83–90.
[53]. J. Ferlay, M. Colombet, I. Soerjomataram, C. Mathers, D. M. Parkin, M. Piñeros, A. Znaor and F. Bray, Estimating the global cancer incidence and mortality in 2018: GLOBOCAN sources and methods, *International Journal of Cancer* **2019**, 144, 1941–1953.
[54]. Y. Sato, K. Nakata, Y. Kato, M. Shima, N. Ishii, T. Koji, K. Taketa, Y. Endo, and S. Nagataki, Early Recognition of Hepatocellular Carcinoma Based on Altered Profiles of Alpha-Fetoprotein, *New England Journal of Medicine* **1993**, 328, 1802–1806.
[55]. S. Wang, M. Wang, C. Li, H. Li, C. Ge, X. Zhang and Y. Jin, A highly sensitive and stable electrochemiluminescence immunosensor for alpha-fetoprotein detection based on luminol-AgNPs@Co/Ni-MOF nanosheet microflowers, *Sensors and Actuators B: Chemical* **2020**, 311, 127919.
[56]. D. Ilic, M. Djulbegovic, J. H. Jung, E. C. Hwang, Q. Zhou, A. Cleves, T. Agoritsas and P. Dahm, Prostate cancer screening with prostate-specific antigen (PSA) test: a systematic review and meta-analysis, *BMJ* **2018**, 362, k3519.
[57]. L. Dai, Y. Li, Y. Wang, X. Luo, D. Wei, R. Feng, T. Yan, X. Ren, B. Du and Q. Wei, A prostate-specific antigen electrochemical immunosensor based on Pd NPs functionalized electroactive Co-MOF signal amplification strategy, *Biosensors and Bioelectronics* **2019**, 132, 97–104.
[58]. W.-L. Xin, L.-F. Jiang, L.-P. Zong, H.-B. Zeng, G.-F. Shu, R. Marks, X.-j. Zhang, and D. Shan, MoS_2 quantum dots-combined zirconium-metalloporphyrin frameworks: Synergistic effect on electron transfer and application for bioassay, *Sensors and Actuators B: Chemical* **2018**, 273, 566–573.
[59]. X. Zhang, J. Peng, Y. Song, Y. Chen, F. Lu and W. Gao, Porous hollow carbon nanobubbles@ZnCdS multi-shelled dodecahedral cages with enhanced visible-light harvesting for ultrasensitive photoelectrochemical biosensors, *Biosensors and Bioelectronics* **2019**, 133, 125–132.
[60]. J. Li, L. Liu, Y. Ai, Y. Liu, H. Sun and Q. Liang, Self-Polymerized Dopamine-Decorated Au NPs and Coordinated with Fe-MOF as a Dual Binding Sites and Dual Signal-Amplifying Electrochemical Aptasensor for the Detection of CEA, *ACS Applied Materials & Interfaces* **2020**, 12, 5500–5510.
[61]. A. Burns and S. Iliffe, Alzheimer's disease, *BMJ* **2009**, 338, b158.
[62]. J. Han, M. Zhang, G. Chen, Y. Zhang, Q. Wei, Y. Zhuo, G. Xie, R. Yuan and S. Chen, Ferrocene covalently confined in porous MOF as signal tag for highly sensitive electrochemical immunoassay of amyloid-β, *Journal of Materials Chemistry B* **2017**, 5, 8330–8336.
[63]. G. Zhao, Y. Wang, X. Li, Q. Yue, X. Dong, B. Du, W. Cao and Q. Wei, Dual-Quenching Electrochemiluminescence Strategy Based on Three-Dimensional Metal–Organic Frameworks for Ultrasensitive Detection of Amyloid-β, *Analytical Chemistry* **2019**, 91, 1989–1996.
[64]. Z. Wang, P. Dong, Z. Sun, C. Sun, H. Bu, J. Han, S. Chen and G. Xie, NH2-Ni-MOF electrocatalysts with tunable size/morphology for ultrasensitive C-reactive protein detection via an aptamer binding induced DNA walker–antibody sandwich assay, *Journal of Materials Chemistry B* **2018**, 6, 2426–2431.
[65]. S. Dong, D. Zhang, H. Cui and T. Huang, ZnO/porous carbon composite from a mixed-ligand MOF for ultrasensitive electrochemical immunosensing of C-reactive protein, *Sensors and Actuators B: Chemical* **2019**, 284, 354–361.
[66]. Z. Tang, J. He, J. Chen, Y. Niu, Y. Zhao, Y. Zhang and C. Yu, A sensitive sandwich-type immunosensor for the detection of galectin-3 based on N-GNRs-Fe-MOFs@AuNPs nanocomposites and a novel AuPt-methylene blue nanorod, *Biosensors and Bioelectronics* **2018**, 101, 253–259.
[67]. Z. Song, Y. Li, H. Teng, C. Ding, G. Xu, and X. Luo, Designed zwitterionic peptide combined with sacrificial Fe-MOF for low fouling and highly sensitive electrochemical detection of T4 polynucleotide kinase, *Sensors and Actuators B: Chemical* **2020**, 305, 127329.
[68]. P. Ling, C. Qian, J. Yu, and F. Gao, Artificial nanozyme based on platinum nanoparticles anchored metal-organic frameworks with enhanced electrocatalytic activity for detection of telomeres activity, *Biosensors and Bioelectronics* **2020**, 149, 111838.
[69]. C. Xiong, W. Liang, Y. Zheng, Y. Zhuo, Y. Chai and R. Yuan, Ultrasensitive Assay for Telomerase Activity via Self-Enhanced Electrochemiluminescent Ruthenium Complex Doped Metal–Organic Frameworks with High Emission Efficiency, *Analytical Chemistry* **2017**, 89, 3222–3227.

24 MOF-based Electrochemical Sensors for Biological Macromolecule Sensing

Peihong Tong[1], *Yongjuan Meng*[1] *and Jianping Li*[1,2]
[1]College of Chemistry and Bioengineering, Guilin University of Technology, Guilin, China
[2]College of Environmental Science and Engineering, Guilin University of Technology, Guilin, China

24.1 INTRODUCTION

In the fields of biological research, medical examination, biomedicine, environmental monitoring, and so on, it is necessary to develop biosensors with high sensitivity, good selectivity, and rapid detection for the qualitative/quantitative analysis of specific biological macromolecules. Due to the growing demand for accurate detection of sensors, it has greatly promoted the design and development of new functional materials. Various functional materials have been developed for the construction of new biosensors, such as metal nanoparticles, graphene oxide, silica nanoparticles, quantum dots, molecularly imprinted polymers, and metal-organic frameworks (MOFs). Among them, biosensors constructed with multi-functional MOF materials have attracted much attention.

MOFs are materials with periodic network structure formed by the self-assembly of organic ligands (such as carboxylate ligands and other negatively charged ligands) with inorganic metals (metal ions and clusters), which have the characteristics of large specific surface area, high porosity, adjustable pore structure and other features [1]. Recently, some researchers have been exploring the use of MOF materials to construct chemical sensors. MOFs have unique luminescent and catalytic properties. In addition, MOFs can be combined with other functional materials to obtain multi-functional MOF composites with new or combined characteristics, which is better than that of single functional materials and better overcomes the limitations of traditional biosensors in detection.

To improve the targeting of MOF materials to biological macromolecules, scientists often functionally modify MOF materials with a variety of methods. For example, polymers were coated on MOFs surfaces, small molecules and antibodies with recognition function were modified, or bimetallic nodes and functional ligands were introduced, to improve the selectivity of MOFs sensors for different biological macromolecules. The function of MOFs is further improved by modification, which gives the sensors based on MOF materials a broad prospect in the detection of biological macromolecules. This chapter summarizes the types of MOFs, their properties, and preparation methods, and reviews the latest progress of the functions and applications of MOF composites in sensors [2].

24.2 THE ROLES OF MOFs IN THE CONSTRUCTION OF BIOSENSORS

There are three roles that MOFs play in biosensors: (a) acting as the supporting platform for biomolecules loaded with targeting recognition units; (b) acting as the signal probe used as a marker in the sandwich detection of biosensors; and (c) used as the sensing elements of biosensors, with the interaction between MOFs material and biomolecules making the fluorescence and other signals of MOFs material change, and the biomolecules being determined directly according to the change of signal.

24.2.1 MOFs as the Supporting Platform for Loading Biomolecules

In the fabrication of biosensors, many materials are used as electrode modifiers and carriers to immobilize biomolecules (such as aptamers, antibodies, and peptides) to achieve the specific analytical performance of the sensor. The stability, the amount, and the method of the immobilizing the biomolecules directly affects the accuracy and sensitivity of the determination of the biosensors. Among the materials used for biomolecule immobilization, MOFs have a large specific surface area, special cavity structure, and good biocompatibility, which makes them an ideal platform for biomolecule immobilization. Tang et al. synthesized a graphene nanoribbon/iron-based MOF@goldnanoparticle composite (N-GNRs-Fe-MOFs@AuNPs) [3], and modified it on a glassy carbon electrode surface, then loaded it with a large amount of galactose lectin-3-Ab_1 to achieve a biosensor. Since the N-GNRs-Fe-MOFs@AuNPs composite has good electrical conductivity and a large specific surface area, the stability and sensitivity of the sensor were improved.

24.2.2 MOFs as the Signal Probe for Markers

In sandwich-like biosensors, materials such as metal nanoparticles, metal oxides, enzymes, and graphene are usually

DOI: 10.1201/9781003188148-24

used as probes to enhance the detection signal or improve the selectivity of the biosensor. Due to their large specific surface areas, high catalytic performances, and excellent optical/electrical performance, MOFs are often developed as the signal probe for sandwich-like biosensors to achieve highly sensitive detection of biological macromolecules. In addition, most MOFs usually need to modify or encapsulate functional nanomaterials to generate or amplify signals such as current, fluorescence, and electrochemiluminescence, and the like. For example, Liu et al. [4] developed an electrochemical immunosensor for the determination of C-reactive protein (CRP) using gold nanoparticles/metal-organic framework (Au-HKUST-1) as signal probes. HKUST-1 materials contain a large number of copper ions, and copper ions have strong electrochemical activity, which can provide strong detection signals. This strategy provides an excellent signal transduction platform for the detection of CRP and can be integrated with other recognition elements to expand its application in bioassay.

24.2.3 MOFs used as the Sensing Elements of Biosensors

The interaction between MOF materials and biomolecules makes the electrochemical, fluorescence or electrochemiluminescence signals of MOF materials change, and biomolecules are detected directly according to the changes of signals. With the regulation and design of MOF materials, many biosensors are constructed to detect biomolecules directly. For example, when a fluorescence-responsive biosensor is used to detect biomolecules based on changes in fluorescence, the interaction between MOFs (host) and biomolecules (guest) enables the guest molecules to induce fluorescence enhancement or quenching of MOF materials, the biomolecule can be quantitatively/qualitatively analyzed according to the change of fluorescence signal. For example, Qu et al. developed a fluorescent Tb-MOF probe with hydrophilicity and good stability. When the target substance (protamine) is combined with Tb-MOF [5], it can effectively inhibit the fluorescence emission of Tb-MOF. The quenching mechanism is attributed to the photoinduced electron transfer, and a simple and rapid method for the detection of protamine has been developed.

24.3 THE FUNCTION OF MOFs IN BIOSENSORS

In biological detection, MOFs usually have the functions of catalysis [6], luminescence [7], specific recognition of target [8], and so on. Based on the above functions of MOFs, a variety of highly sensitive and selective biosensors were constructed to meet the requirements of the detection of trace biological molecules in organisms.

24.3.1 Catalytic Activity of MOFs

Natural enzymes are crucial biological macromolecules with highly efficient and selective catalytic activity. However, the poor stability, high production cost, complex preparation process, and strict catalytic conditions of natural enzymes limit their wide application. Nano-enzyme is a substitute for a natural enzyme, which has attracted wide attention because of its easy production, low cost, and good stability. Among them, MOFs have shown broad development prospects in the field of new nano-enzymes. MOF-type nano-enzymes have the same or stronger catalytic activity as natural biological enzymes, which are good choices for MOFs to replace natural enzymes.

In the field of biosensors, MOFs, as nano-enzymes, can catalyze and amplify signals [9]. The intrinsic catalytic activity of MOFs comes from their abundant active sites and transition metals (such as Fe, Cu, Co, and Zn). The periodic combination of metal nodes and organic ligands provides sufficient catalytic sites. In addition, the catalytic activity of the MOFs modified with functional materials can also be further improved. The way of modification of functional materials can be divided into two: (a) by using the metal nodes of MOF materials and the functional groups of organic ligands (functional molecules were modified on the surface of MOFs to enhance the catalytic activity of MOFs). (b) by combining MOFs with catalytic active materials (such as metal nanomaterials, metal oxides, carbon nanomaterials, enzymes, etc.). The catalytic efficiency of MOF composites was improved. Table 24.1 summarizes the applications of some MOF nano-enzymes in biosensors.

The hollow structure of MOFs with large pore size has a high surface area, good chemical stability, and excellent electrocatalytic activity. Au/HPCN-222 MOF with hollow structure was synthesized by Biswas et al. [20]. An electrochemical sensor constructed by functionalized MOF material was applied to the determination of estradiol. Similarly, Cheng et al. used Hemin@MOF composite material as a signal probe to construct a novel electrochemical sensor for the highly sensitive detection of H_2O_2 and target DNA content [21].

TABLE 24.1
Application of MOFs as Catalysts in Biosensors

Nanozymes name	Sensing target	Linear range	LOD	Refs.
Fe-MIL-88A	Thrombin	10~80 nM	0.8 nM	[10]
Au-SH-SiO_2 @Cu-MOF	L-cysteine	0.02~300 μM	0.008 μM	[11]
MOF-808	Ascorbic acid	30~1030 μM	15 μM	[12]
Co-MOF	Glucose	0.001~3 mM	1.3 nM	[13]
Th-MOF	Uric acid	4.0~70 μM	1.15 μM	[14]
Co-MOF/EG	Glucose	1.0~3330 μM	0.58 μM	[15]
2D Co-MOFs	Alkaline phosphatase	0.5-10 U/L	0.33 U/L	[16]
ZnNi MOF	Adenosine	0.0001~ 100 ng/mL	20.32 fg/ mL	[17]
MIL-68-NH_2/ MoS_2	H_2O_2	10 nM~ 18.3 mM	6.26 nM	[18]
521-MOF	Mucin 1	0.001~0.5 ng/mL	0.12 pg/mL	[19]

24.3.2 The Optical Property of MOFs

Fluorescence, as part of photoluminescence, is an important optical feature of MOFs. The efficiency of fluorescence emission and the wavelength of fluorescence are usually determined by the structure of the material and the interatomic/intermolecular attraction within the material. Recently, researchers have increasingly focused their attention on exploring the field of novel and unique light-emitting MOF materials. The fluorescence of MOFs can originate from metal centers or bridged ligands and can be regulated by the interaction between MOF components [22], in other words: (a) organic ligands with aromatic groups or extended π systems in MOFs produce photoluminescence under exciting light; and (b) metal ions, including lanthanide elements or various inorganic clusters in the MOFs, can also produce photoluminescence. According to the mechanism, the fluorescence generation ability of MOFs can be divided into the metal center/organic ligand-emission, charge transfer from organic ligand to organic ligand, charge transfer from organic ligand to metal, charge transfer from metal center to organic ligand, and from metal center to metal center. In addition, MOF-encapsulated guest molecules (such as organic dyes, transition metal complexes, and quantum dots) can also emit or induce luminescence.

Over the past decade, a variety of luminescent MOF composites have been developed for quantitative and qualitative biosensing. The detection principle is based on the enhancement or quenching of emission intensity or the shift of emission wavelength due to the interaction between MOFs and analytes (for example, non-covalent interaction, coordination bond, π-π interaction, and recognition of the target head). The advantage of MOFs as luminescent probes is that their framework structure can provide many interaction sites for analytes, thus improving detection sensitivity. According to the mode of electron excitation, the sensor can be divided into luminescence biosensors, chemiluminescence biosensors, and electrochemiluminescence biosensors. Jiang et al. developed a novel fluorescence biosensor and developed a dual-signal ratio fluorescence method for the detection of baicalin (BAI) based on an enzyme-catalyzed reaction [23]. The detection principle is shown in Figure 24.1. Black phosphorous quantum dots (BPQD) were prepared by solvothermal reaction and doped into ZIF-8 materials, and then assembled with silver nanoclusters (AgNCs) to obtain (AgNCs)/BPQD/ZIF-8 nanocomposites. In the detection system of AgNCs/BPQD/MOF nanocomposites containing catalase and H_2O_2, BAI can promote the catalytic decomposition of H_2O_2 by catalase, thereby increasing the red fluorescence of AgNCs (630 nm), while the blue fluorescence of BPQD (535 nm) decreases slightly. With the aid of the linear curve between I_{BPQDs}/I_{AgNCs} (I_{535}/I_{630}) and BAI concentration (C_{BAI}), the ratio fluorimetric determination of BAI was achieved by the dual-signal measurement mode.

24.3.3 The Specific Recognition Performance of MOFs

Porous MOF materials have a large surface area and good adsorption capacity, which makes them easy to combine with

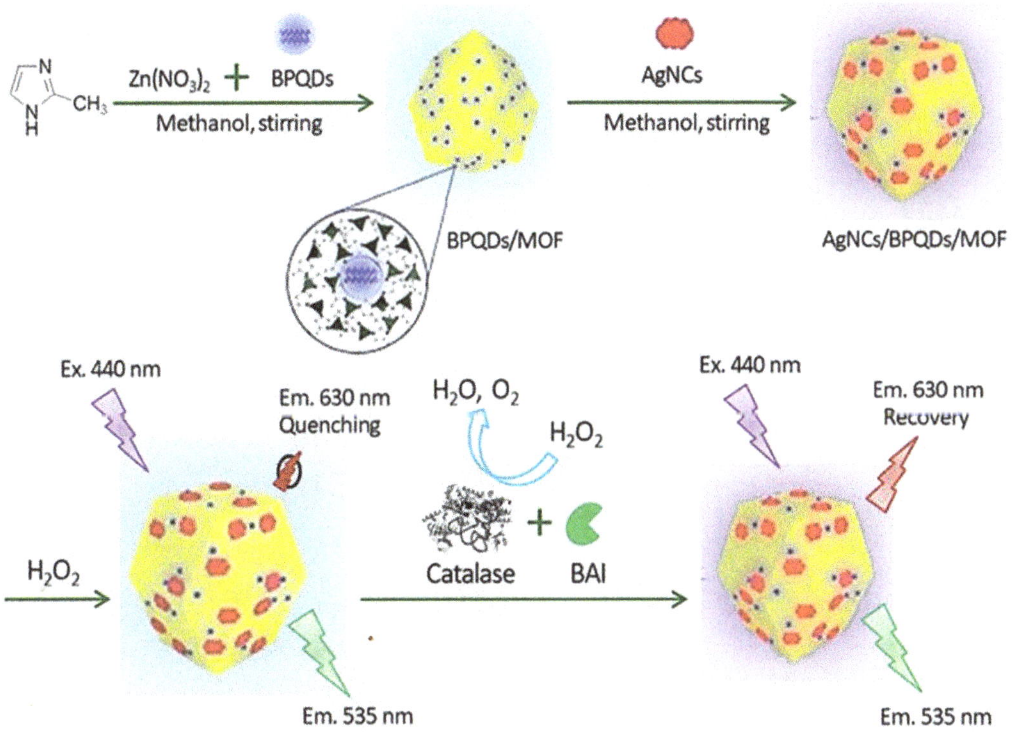

FIGURE 24.1 A schematic diagram of ratio-type fluorescence sensor for visualized detection of baicalin using AgNCs/BPQDs/MOF composite material. Reproduced with permission from Reference [23]. Copyright (2020) Elsevier.

various recognition elements, such as antibodies, aptamers, sugar molecules, peptides, small organic molecules, metal ions, and the like, to achieve functional materials. Therefore, MOF composites with specific targets to various biological macromolecules were prepared by adsorption, intercalation, coupling, and coordination, which can be used to construct various highly selective biosensors. For example, Zhang et al. synthesized an aptamer/Zr-MOF composite using a coordination strategy [24], as shown in Figure 24.2. With the coordination strategy, the aptamer was immobilized on the MOFs surface by utilizing the phosphate group of the aptamer coordinate with the node ion Zr^{4+} of the MOF. Accordingly, a label-free electrochemical aptamer sensor based on the 509-MOF@Apt composite material was developed, which utilizes the change of electrochemical impedance resulting from the combination of the aptamer chain and the target molecule, ultra-sensitive detection of target molecules such as thrombin, kanamycins and carcinoembryonic antigen can be achieved.

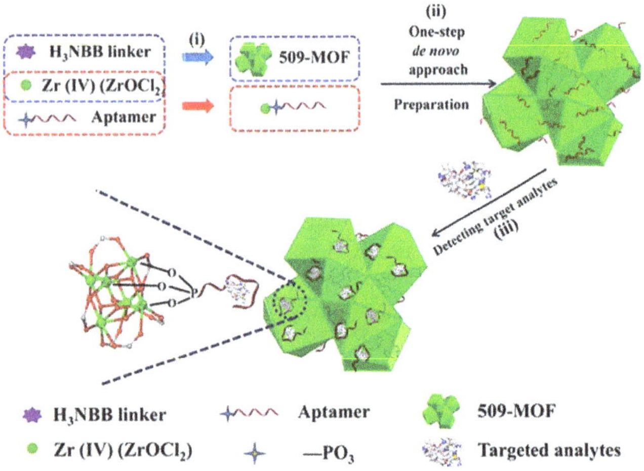

FIGURE 24.2 The preparation process of 509-MOF@Apt composites and sensor for the electrochemical detection. Reproduced with permission from Reference [24]. Copyright (2017) American Chemical Society.

For the modification of antibodies, the free functional groups on the MOF surface are usually used to bind the antibody by a chemical bond with the aid of a coupling agent or cross-linking agent. Many sensors based on this method have been reported. The most commonly used strategy for modifying antibodies on MOF materials is to use an EDC-NHS crosslinker to couple the carboxyl and primary amine groups of MOFs or biomolecules. For example, Liu et al. first prepared $Cu_3(BTC)_2$ materials by a solvothermal method, then linked with the second antibody of prostate-specific antigen (Ab_2) via the amino group of Ab_2, with the carboxyl group of $Cu_3(BTC)_2$ to form an amide bond. An electrochemical immunosensor was constructed, and high sensitivity detection of cancer biomarkers was achieved by using $Cu_3(BTC)_2$-Ab_2 as a signal.

In addition, MOF-based sensors with excellent target recognition performances can also be prepared by combining MOF materials with other recognition units [25, 26]. For example, a molecularly imprinted electrochemical luminescence sensor was constructed with high sensitivity and good stability by using a ZIF-8 composite as a molecularly imprinted film (Figure 24.3) [27]. A ZIF-8-NH_2 imprinted composite was prepared by a layer-by-layer self-assembly method using carboxyl functionalized magnetic $CoFe_2O_4$ nanoparticles as a carrier, ZIF-8-NH_2 precursor solution as functional monomer, and cinchonine as template molecule. Because of the advantages of the good rigid structure, high

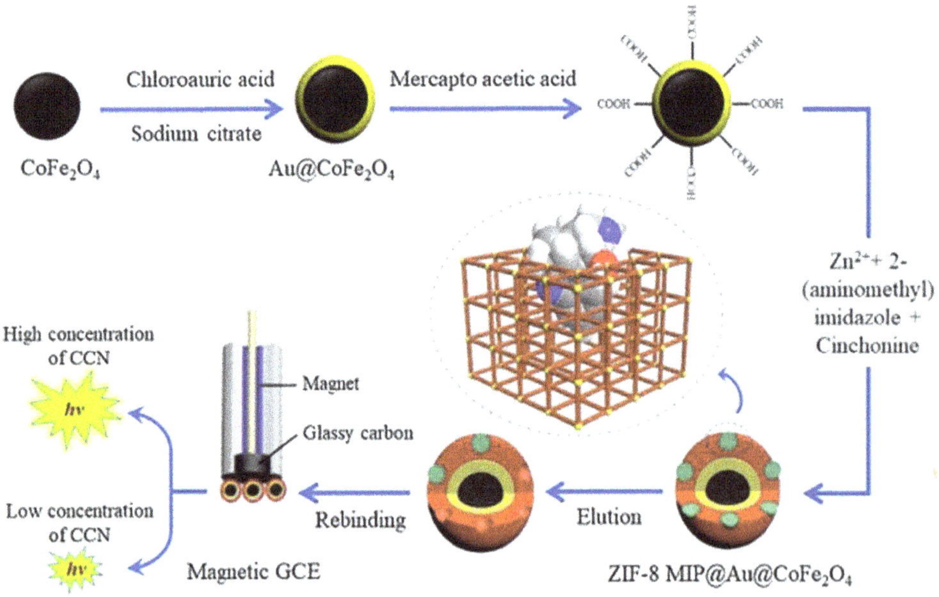

FIGURE 24.3 Construction schematic diagram of molecularly imprinted electrochemiluminescence sensor based on core-shell ZIF-8-NH_2 composite material. Reproduced with permission from Reference [27]. Copyright (2021) Elsevier.

absorption capacity, and the good hydrothermal and chemical stability of the ZIF-8, the conductivity, adsorption capacity, and stability of the imprinting film were improved effectively. The sensor showed good practicability and was used to detect cinchonine in human serum. The strategy provides a reference for the use of MOF materials as imprinted films to construct molecularly imprinted sensors for the detection of biomacromolecules.

24.4 APPLICATION OF MOF-BASED SENSORS IN THE DETECTION OF BIOLOGICAL MACROMOLECULES

Biological macromolecules are indicators that reveal the biochemical processes of organisms and play an important role in life science research, clinical diagnosis, and treatment, and in medical waste pollution monitoring. In recent years, a variety of functional materials have been prepared to be used for constructing biosensors [28]. Because of their unique chemical and physical properties, MOFs have been developed as signal response probes, detection platform substrate material, separation and enrichment carriers, sensor recognition units, and so on. Their applications in the field of biosensors have attracted much attention.

24.4.1 Detection of Antigens

In the early stages of cancer or during the period of suffering from the disease, due to the low content of antigens in the body, it is necessary to develop highly sensitive and selective biosensors for diagnosis. However, when biosensors are used in clinical trials to analyze real samples of patients, there are interferences from other biomolecules in cell/blood/serum, and the concentration of the analytes in samples is very low. It is necessary to overcome the 'false positive' signal caused by non-specific binding. MOFs have recently been considered as an excellent material for enhancing the sensitivity and selectivity of biosensors, and the application of MOF-based biosensors has attracted more and more attention in the detection of antigens. Table 24.2 summarizes some examples of the application of MOF materials in antigen detection.

Carcinoembryonic antigen (CEA), as a tumor marker, plays an important role in the diagnosis and screening of many cancers. The content of CEA in healthy biological samples is less than 5 ng/mL. Therefore, a sensitive method of CEA is crucial for early cancer screening and clinical tumor diagnosis. Guo et al. developed a novel biosensor for the detection of CEA [40]. As shown in Figure 24.4, UiO-66(Zr) was first prepared by a solvothermal method, followed by the addition of $AgNO_3$ solution and an aptamer targeting CEA at room temperature. Then, under vigorous shaking, the freshly prepared ice-cold $NaBH_4$ solution was added to the above mixture. Finally, AgNCs@Apt@UiO-66 nanocomposites were obtained. The material has good biocompatibility, electrochemical activity, and strong affinity. The electrochemical sensor constructed, based on this material, has high sensitivity. CEA was detected by electrochemical impedance spectroscopy and differential pulse voltammetry with detection limits of 8.88 pg/mL and 4.93 pg/mL, respectively. It can be seen that AgNCs@Apt@UiO-66 nanocomposite materials can be used for multifunctional biosensing, and the results show that Zr-MOF-based biosensors have potential applications in clinical diagnosis.

Squamous cell carcinoma antigen (SCCA) is an effective tumor marker for the diagnosis of cervical cancer, and SCCA is usually closely related to the origin of epithelial or endodermal cancers. The concentration of SCCA in the serum of healthy adults is less than 1.5 ng/mL. It makes the task of accurately detecting SCCA difficult. Wei et al. synthesized a Zn-MOF composite to construct an unlabeled photoelectrochemical immunosensor for the detection of SCCA [41]. Au nanoparticles were formed in-situ in Zn-MOF pores by a solution impregnation method to obtain Au-NPs@Zn-MOF, and then the surface of Au-NPs@Zn-MOF was functionalized with 1H-imidazolium salt-1m 3-bis (2-aminoethyl) bromide (IBABr). The IBABr-Au-NPs@Zn-MOF composite was deposited on the surface of a GCE. SCCA antibody was covalently immobilized

TABLE 24.2
Examples of MOF-based Sensors for Antigen Detection

Types of MOFs	Sensing target	Type of sensor	LOD	Refs.
Ce-MoF@HA	Carcinoembryonic antigen	Electrochemical immunosensor	0.2 pg/ml	[29]
MOF/Au/G-quadruplex	Prostatic specific antigen	Electrochemiluminescence	5.8 pg/mL	[30]
ZnS/Zn-MOF/MoS_2	Carcinoembryonic antigen	Photoelectrochemical sensor	1.30 pg/mL	[31]
TCNQ-$Cu_3(BTC)_2$	Prostatic specific antigen	Electrochemical immunosensor	0.06 ng/mL	[32]
Pd/NH_2-ZIF-67	Prostatic specific antigen	Electrochemical immunosensor	0.03 pg/mL	[33]
TbFe-MOFs	Carbohydrate antigen 125	Electrochemical sensor	58 μU/mL	[34]
NH_2-MIL-125(Ti)	Carcinoembryonic antigen	Paper-based analytical device	0.041 ng/mL	[35]
NH_2-MIL-88B (Fe_2Co)-MOF/NF	Cardiac troponin (I)	Electrochemical immunosensor	13 fg/mL	[36]
Ru–PCN-777	Mucin 1	Electrochemiluminescence sensor	33.3 fg/mL	[37]
Zn-MOF-on-Zr-MOF	Protein tyrosine kinase-7	Electrochemical sensor	0.66 pg/mL	[38]
Ab_2@Cu-MOF	α-fetoprotein	Enzyme-linked immunosorbent assay	35 pg/mL	[39]

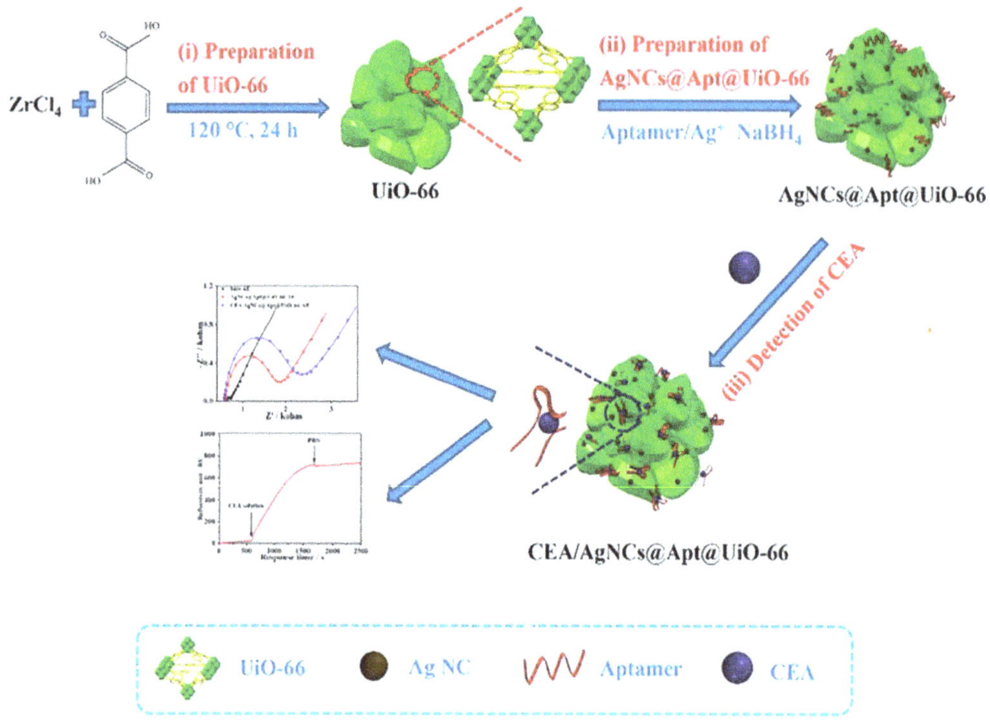

FIGURE 24.4 Schematic diagram of electrochemical sensor for CEA detection based on AgNCs@Apt@UiO-66 composite. Reproduced with permission from Reference [40]. Copyright (2017) American Chemical Society.

on the composite by a glutaraldehyde cross-linking agent to obtain a label-free photoelectrochemical immunosensor. After the combining of SCCA with antibodies, photocurrent caused by the hindrance of electron transfer decreased. The analytical performance is comparable to that of enzyme-linked immunosorbent assay in clinical sample detection. Tuberculosis is becoming more and more popular all over the world. The diagnosis of pulmonary tuberculosis depends on the identification of cultured mycobacterium tuberculosis under a microscope, which is tedious and troublesome. It is encouraging that immunological biomarkers of tuberculosis may be helpful, for example, MPT64 antigen has recently become a rising star in the serological diagnosis of tuberculosis. For example, Chen et al. successfully developed a novel sandwich electrochemical aptamer sensor for rapid and sensitive detection of the MPT64 antigen of mycobacterium tuberculosis by using a dual-signal amplification strategy [42]. Firstly, they coated conductive polyethyleneimine-iron-based metal-organic framework (P-MOF) material on the electrode surface (to promote electron transfer and provide a large specific surface area) to load Au@Pt nanoparticles, and then immobilized MPT64 antigen aptamer I on the modified electrode surface to capture MPT64 antigen. A carbon nanocomposite (C_{60}NPs-N-CNTs/GO) composed of fullerene nanoparticles, nitrogen-doped carbon nanotubes, and graphene oxide was selected as the signal probe, and Au nanoparticles were modified on the composite surface to immobilize MPT64 aptamer Ⅱ. When MPT64 antigen is present, the electrochemical activity of the signal probe of the sandwich sensor is excited by tetraoctylammonium bromide, which produces an obvious current response. The detection limit is as low as 0.33 fg/mL.

The main tumor marker of ovarian cancer is epithelial ovarian cancer antigen (CA 125), so the detection of CA 125 is potentially important in the early detection and prognosis of ovarian cancer. Biswas et al. constructed an ultra-sensitive immunosensor for the detection of CA 125 by taking advantage of the high surface area, electrocatalytic ability, and good protein compatibility of MOF-808 [43]. As shown in Figure 24.5, MOF-808/CNT composites were prepared by in-situ synthesis of MOF-808 on activated carbon nanotubes (CNT), which have excellent electrochemical properties. After the MOF-808/CNT composite was modified on the surface of the glassy carbon electrode, the MOF-808/CNT composite was further functionalized by streptavidin, and then the binding sites of the MOF-808/CNT composite to the antibody against CA 125 were increased. When the antibody-antigen combination is specifically recognized, the diffusion of the redox probe $[Fe(CN)_6]^{4-}/[Fe(CN)_6]^{3-}$ is obviously hindered, and then current change is achieved.

Galactose lectin-3 (GL-3) antigen is a β-galactoside protein. It is an important biomarker for monitoring the risk of cardiac fibrosis, heart failure, and death. Continuous monitoring of the concentration of GL-3 is important for the assessment of cardiovascular risk. Yola et al. prepared a novel sandwich electrochemical immunosensor for the detection of GL-3 based on the graphite carbon nitride nanosheet functionalized with gold nanoparticles (g-C_3N_4@Au NPs/GCE) composite as a sensor platform [44], and Ti-MOF@COFs composite as signal probe.

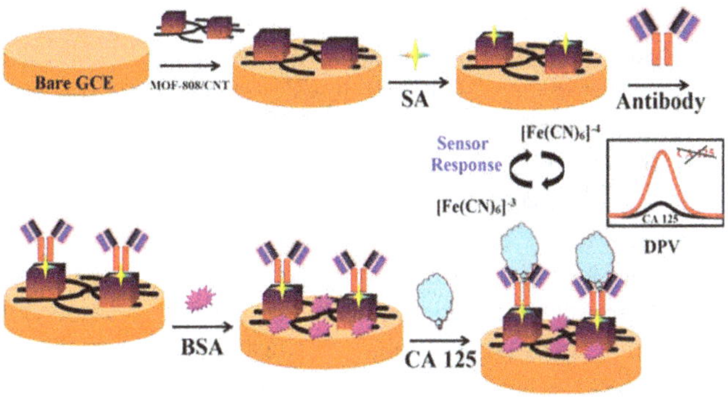

FIGURE 24.5 Construction of a non-labeled electrochemical immunosensor with MOF-808/CNT composite material for the detection of CA 125. Reproduced with permission from Reference [45]. Copyright (2021) American Chemical Society.

g-C_3N_4@AuNPs/GCE is used as an effective carrier for GL-3-Ab_1 antibodies, which provides a binding site for the antibody via amino-gold affinity and also provides effective surface conductivity. The Ti-MOF@COFs composite has good catalytic performance, which can catalyze the decomposition of H_2O_2 to produce a current response. Furthermore, the Ti-MOF@COFs composite can not only bind to GL-3-Ab_2 antibody but also infiltrate GL-3-Ab_2 antibody into the porous Ti-MOF@COFs by π-π stacking and electrostatic action. The detection limit of the sensor is 0.025 pg/mL. Prostate-specific antigen (PSA) is secreted by prostate epithelial cells. The increase of serum PSA concentrations indicates pathological changes or trauma to the prostate. Therefore, Feng et al. constructed a sandwich electrochemical sensor for the detection of PSA [45]. To improve the sensitivity of the electrochemical sensor, Fe-MIL-88B-NH_2 (Fe-MOF) with peroxidase-like activity was designed as the signal probe. Fe-MOF was prepared by hydrothermal method with ferric chloride (III) and 2-aminoterephthalic acid, and then Au nanoparticles were modified on the surface of Fe-MOF for immobilization and labeling of antibodies. The Fe-MOF signal probes catalyze H_2O_2 to form ·OH through its own peroxidase-like activity and then degrades methylene blue (MB) on the surface of graphene oxide-Au nanocomposites (Au-rGO) by inducing the Fenton reaction to generate an electrical signal.

24.4.2 Detection of Antibodies

The detection of antibodies is helpful for clinical diagnosis, and the antibody is also an important indicator of cure in some diseases. There are few reports on the application of MOF-based biosensors to detect antibodies, and it is still in the early stage of development. The p53 antibody was found in human serum after p53 gene mutation. p53 antibodies were detected in the sera from patients with different cancers, such as breast, liver, ovarian, and lung cancer. Therefore, p53 has been considered as an important cancer marker, and highly sensitive detection of p53 antibodies is important in the early diagnosis of cancer.

Wei et al. constructed an ultra-sensitive electrochemiluminescence biosensor using Zn-MOF/GO nanocomposites to detect p53 antibodies [46]. In this work, Zn-MOF was synthesized by a hydrothermal reaction of 5, 10, 15, 20-tetra (4-carboxylphenyl) porphyrin (TCPP) with Zn^{2+}. The Zn-MOF compound has excellent electrochemiluminescence performance. After a Zn-MOF was combined with graphene oxide (GO), the electrochemiluminescence performance was further improved. The Zn-MOF/GO composite was used as the substrate, and then a large number of p53 antigens were modified on its surface to capture the target p53 antibody. The p53 antibody was added to incubate for one hour, and then streptavidin-goat anti-rabbit IgG was added to continue incubation. The modified electrode obtained by the above process was combined with rich-C DNA, and the signal was amplified by capturing the co-reactant Ag (I). The ECL response signal increased with the increase of p53 concentration, which was due to the effect of co-reactant Ag (I). When the concentration of p53 antibody is in the range of 0.1 fg/mL ~ 0.01 ng/mL, there is a linear relationship between the ECL response signal of the biosensor and p53 concentration, and the detection limit is 0.03 fg/mL.

Due to the high infection rate and lethality of H5N1 virus, it is considered to be one of the most prevalent influenza diseases in the world. Wei et al. used a DNA fluorescent probe and Cu-MOF material to construct a fluorescent sensor for the detection of antibody H5N1 [47]. In this detection system, the 5 'end of the short-stranded oligonucleotide was modified with 6-carboxyfluorescein (6-FAM), then it was added to the MOF solution; the DNA probe was adsorbed on the surface of the MOF, and the fluorescence of 6-FAM was effectively quenched by the MOF. Subsequently, the 3 'end of the DNA probe was specifically cleaved with exonuclease I (Exo I), and the fluorescent FAM was released from the MOF, thus fluorescence was recovered. If the H5N1 antigen at the 3 'end of the DNA probe captures the H5N1 antibody, the fluorescence of the system cannot be recovered because the existence of macromolecules inhibits the hydrolysis of Exo I. The method has been applied to the detection of H5N1 in blood samples with a detection limit 1.6×10^{-9} mol/L.

24.4.3 Detection of Polypeptides

As biomarkers, polypeptides have shown important research and application value in immune regulation, information transmission, cell secretion, precursor signaling, disease occurrence, and therapy [48]. Amyloid β oligomers (AβO) have been widely recognized as biomarkers for the early diagnosis of Alzheimer's disease (AD). Qin et al. used ZIF-8 material to encapsulate ferrocene (Fc) in-situ to obtain ZIF-8/Fc composite material for quantitative detection of AβO [49]. Ferrocene is a kind of optical/electrochemical signal probe. In the presence of AβO, the competitive coordination between Zn^{2+} and AβO destroys the original ZIF-8 structure, thus releasing Fc probe molecules from the lattice pores of the MOF. The content of ferrocene released was monitored by ultraviolet/visible spectrophotometry and cyclic voltammetry, and the indirect detection of AβO was achieved. The detection range of AβO by this method is 10^{-5} ~100 μM. The method is selective and feasible in the detection of human cerebrospinal fluid samples.

In addition, the increase of the concentration of N-terminal pro–B-type natriuretic peptide (NT-proBNP) is considered to be an objective indicator for the diagnosis of heart failure (HF). Li et al. constructed a label-free electrochemiluminescence immunosensor using a cerium metal-organic framework/graphite phase carbon nitride nanosheet/Au nanoparticle composite material (Ce-MOF@g-C_3N_4/Au) for sensitive detection of NT-proBNP [50]. Ce-MOF has a large specific surface area and excellent adsorption performance, which not only improves the ECL efficiency of g-C_3N_4, but also significantly improves the interface stability of g-C_3N_4. In addition, a large number of Au NPs are loaded on the surface of Ce-MOF@g-C_3N_4, which not only improves the conductivity of Ce-MOF@g-C_3N_4, but also facilitates the immobilization of NT-proBNP antibody and obtains an excellent immunosensor platform. The ECL immunosensor showed excellent stability and specificity. Polypeptide glycosylation is closely related to many biological and cellular functions, including signal transduction, cell proliferation, and differentiation. At present, mass spectrometry (MS) is an important and effective method for the identification of glycopeptides. Ma et al. loaded Au nanoparticles in-situ on MIL-101(NH_2), and then immobilized L-cysteine (Cys) on Au nanoparticles by Au-S bond to obtain Cys-functionalized MIL-101(NH_2) complex (MIL-101(NH_2)@Au-Cys) [51]. Due to the large specific surface area and ultra-high hydrophilicity of the nanocomposite, it exhibits excellent performance for the adsorption and enrichment of glycopeptides. N-glycopeptide in HeLa cell lysate was enriched by MIL-101(NH_2)@Au-Cys, and then determined by MS. The results are shown in Figure 24.6. MIL-101(NH_2)@Au-Cys has the characteristics of short incubation time, strong binding ability, good selectivity, high recovery (more than 80%), and low detection limit (1 fM) for glycopeptide enrichment. This work not only opens up a new way for the rapid functionalization of MOFs but also provides a new idea for the design of new sample-enrichment materials, especially in the correlation analysis of glycopeptideomics, which has a good application prospect.

Polypeptide phosphorylation is the most important and ubiquitous manifestation of post-translational modification of peptides, which plays an important role in controlling many complex biological processes, such as cell growth and division, metabolic pathways, and signal transduction. MOFs and their coordination compounds have been used as novel affinity probes for the selective enrichment of phosphopeptides before MS analysis due to their controllable framework structure and abundant metal sites. Yang et al. prepared a kind of Ti-MOF nano-tablet for efficient extraction of phosphopeptides from samples [52]. Ti-MOF nanosheets have clear two-dimensional morphology, high active sites, large surface area, and ultra-thin structure. Even in the mixture of phosphorus protein and non-phosphorus protein with a very low molar ratio (1:10000), phosphopeptide can be effectively extracted, with the detection limit of 0.1 fmol/mL by MS. The authors also determined the polypeptides by combining the Ti-MOF nanosheet with LC-MS/MS. Phosphopeptides were selectively preconcentrated from trypsin digestion samples including mouse brain tissue lysate, mouse spinal cord lysate, and mouse testicular lysate by Ti-MOF nanosheets, and then LC-MS/MS analysis was carried out. A total of 2601, 3208, and 2866 phosphopeptides were successfully identified from the three samples.

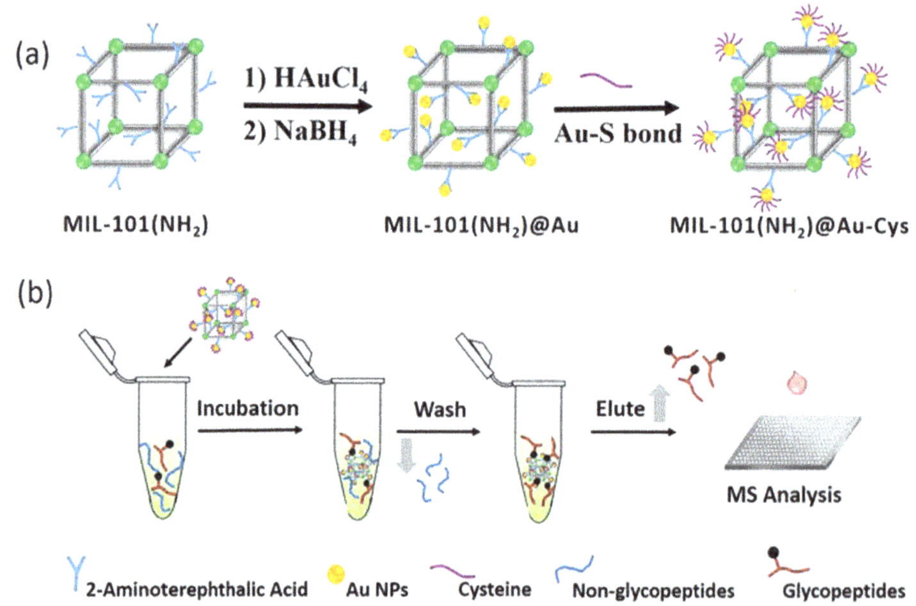

FIGURE 24.6 (a) Synthesis of MIL-101(NH_2)@Au-Cys and (b) enrichment and detection procedures of glycopeptide. Reproduced with permission from Reference [51]. Copyright (2017) American Chemical Society.

24.4.4 Detection of other Biological Macromolecules

In addition to the detection of antigens, peptides, and antibodies described above, there are also studies on the use of MOFs as sensing materials to detect enzymes, polysaccharides, cytokines, hormones, and other biological macromolecules. Table 24.3 summarizes some of the analytical applications of MOF materials for the detection of biological macromolecules.

Thrombin is a serine protease, which is closely related to the blood coagulation process. In addition, thrombin can be used as a biomarker for the diagnosis of lung metastasis, arterial thrombosis, and many other diseases. Qiu et al. developed a label-free electrochemical sensor based on Cu_2(trans-1,4-cyclohexanedicarboxylic acid)$_2$ ($Cu_2(CHDC)_2$), a kind of electroactive MOF, for the detection of thrombin [61]. First, a simple hydrothermal method was used to synthesize $Cu_2(CHDC)_2$ material, and then the AuNPs/$Cu_2(CHDC)_2$ modified electrode was prepared by an electrodeposition method, which was used as the electroactive matrix of thrombin aptamer immobilization. Then the thrombin aptamer chain was immobilized on the surface of AuNPs by an Au-S bond to form a biorecognition layer. When different quantities of thrombin were captured by the aptamer, the current would change accordingly to achieve the purpose of detection.

T4 polynucleotide kinase (PNK) can catalyze the phosphorylation of the 5'-hydroxyl ends of polynucleotides, which is associated with diseases such as Werner syndrome and Bloom syndrome. Song et al. [62] constructed a highly sensitive electrochemical sensor for the detection of T4 PNK based on zwitterionic peptides and Fe-MOF materials. Firstly, the peptides and capture DNA1 were immobilized on the gold electrode by the self-assembly method. In the presence of a target T4 PNK, the 5'- hydroxyl terminal of the hairpin DNA was phosphorylated, then it was cleaved by λ-Exo to release single-stranded DNA2, which triggered the sandwich hybrid chain reaction, and the AuNPs/Fe-MOF nanocomposites connected to DNA3 were selectively attached to the surface of the electrode. Finally, Fe^{3+} in Fe-MOF reacted with $K_4Fe(CN)_6$ to form electroactive $Fe_4[Fe(CN)_6]_3$, which then generated a strong electrical signal for indirect detection of T4 PNK. The detection limit was as low as 3.5×10^{-4} U/mL.

There is evidence that lipopolysaccharide is directly involved in inflammatory and pathological processes associated with AD. Therefore, the development of specific and sensitive methods to detect lipopolysaccharide (LPS) is very important for the diagnosis of AD. Li et al. developed an electrochemical sensor based on bifunctional Cu^{2+} doped MOF nanoparticles (Cu^{2+}-NMOFs) for sensitive detection of LPS in bacteria [63]. After reaction with carbohydrate portions of LPS molecules, Cu^{2+}-NMOFs was labeled as a probe, and then catalyze dopamine oxidation to generate aminochrome, resulting in a strong electrochemical oxidation signal.

24.5 CONCLUSION AND PROSPECTS

The MOF-based sensor detection platform has the advantages of good selectivity, high sensitivity, low detection limit, and fast response for biological macromolecule analysis. To further enhance the analytical performance of MOF-based sensors and expand its application fields, the research and development of MOF-based sensors should focus on several aspects.

Firstly, for the electrochemical and optical biosensors, the low conductivity and optical efficiency of MOF materials limit the improvement of their performance since MOF materials mainly act as signal probes. It is necessary to design some redox-active ligands, dope with functional materials or develop novel optical elements to obtain functional MOF materials with higher electrical activity/optical efficiency. Secondly, it is necessary to intelligently design the structure of MOFs, for example, the two-dimensional structure of MOF material is more favorable as the substrate of the sensor. Thirdly, MOF probes with precise targets for biological systems could be developed to directly diagnose diseases or cancers in vivo. Fourth, to promote the industrialization of MOF-based sensors in the field of detecting biological macromolecules, such as the development of portable MOF sensors make it more convenient and rapid for use in biological analysis.

TABLE 24.3
Application of MOF Materials in the Analysis of other Biological Macromolecules.

Types of MOFs	Function of MOFs	Sensing target	LOD	Refs.
2D MOF nanosheets	Catalytic	Heparin	15 ng/mL	[53]
Tb-MOF	Fluorescent	Protamine	1.21 ng/mL	[5]
Tb-OBBA-Hemin	Catalytic	Estrogen	50 pM	[54]
NU-1000-CMCD	Fluorescent	Cholesterol	0.4 μM	[55]
Pr-MOF-NFs	Luminescent	Prolactin	0.838 ng/mL	[56]
Cu-MOF-NPs	Photoluminescence	Triiodothyronine hormone	0.198 ng/dL	[57]
Cu-MOF@TpBD	Catalytic	Platelet-derived growth factor-BB	0.034 pg/mL	[58]
Ni/Cu-MOFs	Catalytic	Tumor necrosis factor-alpha	2.00 fg/mL	[59]
Cu_2O@Cu-MOF/CM	Photoelectrochemical	Vascular endothelial growth factor 165	2.3 fM	[60]

REFERENCES

[1]. Zhou HC, Long JR, Yaghi OM (2012) Introduction to metal-organic frameworks. *Chem Rev* 112: 673–674

[2]. Tong P, Liang J, Jiang X, Li J (2020) Research Progress on Metal-Organic Framework Composites in Chemical Sensors. *Crit Rev Anal Chem* 50: 376–392

[3]. Tang Z, He J, Chen J, Niu, Y, Zhao Y, Zhang Y, Yu C (2018) A sensitive sandwich-type immunosensor for the detection of galectin-3 based on N-GNRs-Fe-MOFs@AuNPs nanocomposites and a novel AuPt-methylene blue nanorod. *Biosens Bioelectron* 101: 253–259

[4]. Liu TZ, Hu R, Zhang X, Zhang KL, Liu Y, Zhang XB, Bai RY, Li D, Yang YH (2016) Metal-organic framework nanomaterials as novel signal probes for electron transfer mediated ultrasensitive electrochemical immunoassay. *Anal Chem* 88: 12516–12523

[5]. Qu F, Li X, Lv X, You J, Han W (2019) Highly selective metal-organic framework-based sensor for protamine through photoinduced electron transfer. *J Mater Sci* 54: 3144–3155

[6]. Jiao L, Wang Y, Jiang HL, Xu Q (2018) Metal-organic frameworks as platforms for catalytic applications. *Adv Mater* 30: 1703663

[7]. Allendorf MD, Bauer CA, Bhakta RK, Houk RJT (2009) Luminescent metal-organic frameworks. *Chem Soc Rev* 38: 1330–1352

[8]. Kumar P, Deep A, Kim KH (2015) Metal organic frameworks for sensing applications. *TrAC-Trend Anal Chem* 73: 39–53

[9]. Li S, Liu X, Chai H, Huang Y (2018) Recent advances in the construction and analytical applications of metal-organic frameworks-based nanozymes. *TrAC-Trend Anal Chem* 105: 391–403

[10]. Wang Y, Zhu Y, Binyam A, Liu M., Wu Y, Li F (2016) Discovering the enzyme mimetic activity of metal-organic framework (MOF) for label-free and colorimetric sensing of biomolecules. *Biosens Bioelectron* 86: 432–438

[11]. Hosseini H, Ahmar H, Dehghani A, Bagheri A, Tadjarodi A, Fakhari AR (2013) A novel electrochemical sensor based on metal-organic framework for electro-catalytic oxidation of L-cysteine. *Biosens Bioelectron* 42: 426–429

[12]. Zheng HQ, Liu CY, Zeng XY, Chen J, Lü J, Lin R.G, Cao R, Lin ZJ, Su JW (2018) MOF-808: a metal–organic framework with intrinsic peroxidase-like catalytic activity at neutral pH for colorimetric biosensing. *Inorg Chem* 57: 9096–9104

[13]. Li Y, Xie M, Zhang X, Liu Q, Lin D, Xu C, Xie F, Sun X (2019) Co-MOF nanosheet array: a high-performance electrochemical sensor for non-enzymatic glucose detection. *Sensor Actuat B-Chem* 278: 126–132

[14]. Badoei-Dalfard A, Sohrabi N, Karami Z, Sargazi, G (2019) Fabrication of an efficient and sensitive colorimetric biosensor based on Uricase/Th-MOF for uric acid sensing in biological samples. *Biosens Bioelectron* 141: 111420

[15]. Liu B, Wang X, Liu H, Zhai Y, Li L, Wen H. (2020) 2D MOF with electrochemical exfoliated graphene for nonenzymatic glucose sensing: Central metal sites and oxidation potentials. *Anal Chim Acta* 1122: 9–19

[16]. Wan H, Wang Y, Chen J, Meng HM, Li Z (2021) 2D Co-MOF nanosheet-based nanozyme with ultrahigh peroxidase catalytic activity for detection of biomolecules in human serum samples. *Microchim Acta* 188: 1–8

[17]. Tian K, Ma Y, Liu Y, Wang M, Guo C, He L, Song Y, Zhang Z, Du M (2020) Hierarchically structured hollow bimetallic ZnNi MOF microspheres as a sensing platform for adenosine detection. *Sensor Actuat B-Chem* 303: 127199

[18]. Wei P, Sun D, Niu Y, Lu X, Zhai H (2020) Enzyme-free electrochemical sensor for the determination of hydrogen peroxide secreted from MCF-7 breast cancer cells using calcined indium metal-organic frameworks as efficient catalysts. *Electrochim Acta* 359: 136962

[19]. He L, Duan F, Song Y, Guo C, Zhao H, Tian JY, Zhang Z, Liu CS, Zhang X, Wang P (2017) 2D zirconium-based metal-organic framework nanosheets for highly sensitive detection of mucin 1: consistency between electrochemical and surface plasmon resonance methods. *2D Mater* 4: 025098

[20]. Biswas S, Chen Y, Xie Y, Sun X, Wang Y (2020) Ultrasmall Au (0) inserted hollow PCN-222 MOF for the high-sensitive detection of estradiol. *Anal Chem* 92: 4566–4572

[21]. Cheng D, Xiao X, Li X, Wang C, Liang Y, Yu Z, Jin C, Zhou N, Chen M, Dong Y (2018) A non-enzymatic electrochemical sensing platform basedonhemin@MOFcompositesfor detecting hydrogen peroxide and DNA. *J Electrochem Soc* 165: B885

[22]. Dong J, Zhao D, Lu Y, Sun WY (2019) Photoluminescent metal-organic frameworks and their application for sensing biomolecules. *J Mater Chem A* 7: 22744–22767

[23]. Jiang X, Jin H, Sun Y, Sun Z, Gui R (2020) Assembly of black phosphorus quantum dots-doped MOF and silver nanoclusters as a versatile enzyme-catalyzed biosensor for solution, flexible substrate, and latent fingerprint visual detection of baicalin. *Biosens Bioelectron* 152: 112012

[24]. Zhang Z H, Duan F H, Tian J Y, He JY, Yang LY, Zhao H, Zhang S, Liu CS, He LH, Chen M, Chen DM, Du M (2017) Aptamer-embedded zirconium-based metal-organic framework composites prepared by de novo bio-inspired approach with enhanced biosensing for detecting trace analytes. *ACS sensors* 2: 982–989

[25]. Wu T, Wei X, Ma X, Li J (2017) Amperometric sensing of L-phenylalanine using a gold electrode modified with a metal organic framework, a molecularly imprinted polymer, and β-cyclodextrin-functionalized gold nanoparticles. *Microchim Acta* 184: 2901–2907

[26]. Wei X, Wu T, Yuan Y, Ma X, Li J (2017) Highly sensitive analysis of organometallic compounds based on molecularly imprinted electrochemical sensors. *Anal Methods* 9: 1771–1778

[27]. Tong P, Meng Y, Liang J, Li J (2021) Molecularly imprinted electrochemical luminescence sensor based on core–shell magnetic particles with ZIF-8 imprinted material. *Sensor Actuat B-Chem* 330: 129405

[28]. Ma X, Li J, Wang C, Xu G (2016) A review on bio-macromolecular imprinted sensors and their applications. *Chinese J Anal Chem* 44: 152–159

[29]. Li W, Ma C, Song Y, Hong C, Qiao X, Yin B (2020) Sensitive detection of carcinoembryonic antigen (CEA) by a sandwich-type electrochemical immunosensor using MOF-Ce@HA/Ag-HRP-Ab_2 as a nanoprobe. *Nanotechnology* 31: 185605

[30]. Shao K, Wang B, Nie A, Ye S, Ma J, Li Z, Lv Z, Han, H. (2018) Target-triggered signal-on ratiometric electrochemiluminescence sensing of PSA based on MOF/Au/G-quadruplex. *Biosens Bioelectron* 118: 160–166

[31]. Wei Q, Wang C, Li P, Wu T, Yang N. Wang X, Wang Y, Li C (2019). ZnS/C/MoS_2 Nanocomposite Derived from Metal-Organic Framework for High-Performance Photo-Electrochemical Immunosensing of Carcinoembryonic Antigen. *Small* 15: 1902086

[32]. Bhardwaj SK, Sharma AL, Bhardwaj N, Kukkar M, Gill AA, Kim KH, Deep A (2017) TCNQ-doped Cu-metal organic framework as a novel conductometric immunosensing platform for the quantification of prostate cancer antigen. *Sensor Actuat B-Chem* 240: 10–17

[33]. Dai L, Li Y, Wang Y, Luo X, Wei D, Feng R, Yan T, Ren X, Du B, Wei Q (2019) A prostate-specific antigen electrochemical immunosensor based on Pd NPs functionalized electroactive Co-MOF signal amplification strategy. *Biosens Bioelectron* 132: 97–104

[34]. Wang M, Hu M, Li Z, He L, Song Y, Jia Q, Zhang Z, Du M (2019) Construction of Tb-MOF-on-Fe-MOF conjugate as a novel platform for ultrasensitive detection of carbohydrate antigen 125 and living cancer cells. *Biosens Bioelectron* 142: 111536

[35]. Lv S, Tang Y, Zhang K, Tang D (2018) Wet NH_3-triggered NH_2-MIL-125 (Ti) structural switch for visible fluorescence immunoassay impregnated on paper. *Anal Chem* 90: 14121–14125

[36]. Palanisamy S, Senthil Raja D, Subramani B, Wu TH, Wang YM (2020) Label-Free Bimetallic In-situ-Grown 3D Nickel-Foam-Supported NH_2-MIL-88B (Fe_2Co)-MOF-based Impedimetric Immunosensor for the Detection of Cardiac Troponin I. *ACS Appl Mater Inter* 12: 32468–32476

[37]. Hu GB, Xiong CY, Liang WB, Zeng XS, Xu HL, Yang Y, Yao LY, Yuan R, Xiao DR (2018) Highly stable mesoporous luminescence-functionalized MOF with excellent electrochemiluminescence property for ultrasensitive immunosensor construction. *ACS Appl Mater Inter* 10: 15913–15919

[38]. Zhou N, Su F, Guo C, He L, Jia Z, Wang M, Jia Q, Zhang Z, Lu S (2019) Two-dimensional oriented growth of Zn-MOF-on-Zr-MOF architecture: a highly sensitive and selective platform for detecting cancer markers. *Biosens Bioelectron* 123: 51–58

[39]. Hu X, Wei Z, Sun C, Long Y, Zheng H (2020) Bifunctional antibody and copper-based metal-organic framework nanocomposites for colorimetric α-fetoprotein sensing. *Microchim Acta* 187: 1–9

[40]. Guo C, Su F, Song Y, Hu B, Wang M, He L, Peng D, Zhang Z (2017) Aptamer-templated silver nanoclusters embedded in zirconium metal-organic framework for bifunctional electrochemical and SPR aptasensors toward carcinoembryonic antigen. *ACS Appl Mater Inter* 9: 41188–41199

[41]. Wei Q, Wang C, Zhou X, Wu T, Wang Y, Li C, Yang N (2019) Ionic liquid and spatially confined gold nanoparticles enhanced photoelectrochemical response of zinc-metal organic frameworks and immunosensing squamous cell carcinoma antigen. *Biosens Bioelectron* 142: 111540

[42]. Chen Y, Liu X, Guo S, Cao J, Zhou J, Zuo J, Bai L (2019) A sandwich-type electrochemical aptasensor for Mycobacterium tuberculosis MPT64 antigen detection using C_{60} NPs decorated N-CNTs/GO nanocomposite coupled with conductive PEI-functionalized metal-organic framework. *Biomaterials* 216: 119253

[43]. Biswas S, Lan Q, Xie Y, Sun X, Wang Y (2021) Label-Free Electrochemical Immunosensor for Ultrasensitive Detection of Carbohydrate Antigen 125 Based on Antibody-Immobilized Biocompatible MOF-808/CNT. *ACS Appl Mater Inter 13*: 3295–3302

[44]. Yola M L, Atar N (2020) Amperometric galectin-3 immunosensor-based gold nanoparticle-functionalized graphitic carbon nitride nanosheets and core–shell Ti-MOF@ COFs composites. *Nanoscale* 12: 19824–19832

[45]. Feng J, Wang H, Ma Z (2020) Ultrasensitive amperometric immunosensor for the prostate specific antigen by exploiting a Fenton reaction induced by a metal-organic framework nanocomposite of type Au/Fe-MOF with peroxidase mimicking activity. *Microchim Acta* 187: 1–8

[46]. Wei Y, Zhang Y, Chen JS, Mao CJ, Jin BK (2020) An electrochemiluminescence biosensor for p53 antibody based on Zn-MOF/GO nanocomposite and Ag^+-DNA amplification. *Microchim Acta* 187: 1–9

[47]. Wei X, Zheng L, Luo F, Lin Z, Guo L, Qiu B, Chen G (2013) Fluorescence biosensor for the H5N1 antibody based on a metal-organic framework platform. *J Mater Chem B 1*: 1812–1817

[48]. Laverman P, Sosabowski JK, Boerman OC, Oyen WJ (2012) Radiolabelled peptides for oncological diagnosis. *Eur J Nucl Med Mol I* 39: 78–92

[49]. Qin J, Cho M, Lee Y (2019) Ferrocene-encapsulated zn zeolitic imidazole framework (ZIF-8) for optical and electrochemical sensing of amyloid-β oligomers and for the early diagnosis of alzheimer's disease. *ACS Appl Mater Inter* 11: 11743–11748

[50]. Li L, Zhao Y, Li X, Ma H, Wei Q (2019) Label-free electrochemiluminescence immunosensor based on Ce-MOF@g-C_3N_4/Au nanocomposite for detection of N-terminal pro-B-type natriuretic peptide. *J Electroanal Chem* 847: 113222

[51]. Ma W, Xu L, Li X, Shen S, Wu M, Bai Y, Liu H (2017) Cysteine-functionalized metal–organic framework: facile synthesis and highly efficient enrichment of N-linked glycopeptides in cell lysate. *ACS Appl Mater Inter* 9: 19562–19568

[52]. Yang SS, Chang YJ, Zhang H, Yu X, Shang W, Chen GQ, Chen DD, Gu ZY (2018) Enrichment of phosphorylated peptides with metal-organic framework nanosheets for serum profiling of diabetes and phosphoproteomics analysis. *Anal Chem* 90: 13796–13805

[53]. Cheng H, Liu Y, Hu Y, Ding Y, Lin S, Cao W, Wang Q, Wu J, Muhammad F, Zhao X, Zhao D, Li Z, Xing H, Wei H (2017) Monitoring of heparin activity in live rats using metal-organic framework nanosheets as peroxidase mimics. *Anal Chem* 89: 11552–11559

[54]. Wang L, Chen Y (2020) Luminescence-sensing Tb-MOF nanozyme for the detection and degradation of estrogen endocrine disruptors. *ACS Appl Mater Inter* 12: 8351–8358

[55]. Gong M, Yang J, Li Y, Zhuang Q, Gu J (2019) Substitution-type luminescent MOF sensor with built-in capturer for selective cholesterol detection in blood serum. *J Mater Chem C* 7: 12674–12681

[56]. Sheta S M, El-Sheikh S M, Abd-Elzaher M M (2019) A novel optical approach for determination of prolactin based on Pr-MOF nanofibers. *Anal Bioanal Chem* 411: 1339–1349

[57]. Sheta S M, El-Sheikh S M, Abd-Elzaher M M (2019) Promising photoluminescence optical approach for triiodothyronine hormone determination based on smart copper metal-organic framework nanoparticles. *Appl Organomet Chem* 33: e5069

[58]. Li Y, Liu Z, Lu W, Zhao M, Xiao H, Hu T, Ma J, Zheng Z, Jia J, Wu H (2021) A label-free electrochemical aptasensor based on the core-shell Cu-MOF@TpBD hybrid nanoarchitecture for the sensitive detection of PDGF-BB. *Analyst* 146: 979–988

[59]. Yola M L, Atar N (2021) Novel voltammetric tumor necrosis factor-alpha (TNF-α) immunosensor based on gold nanoparticles involved in thiol-functionalized multi-walled carbon nanotubes and bimetallic Ni/Cu-MOFs. *Anal Bioanal Chem* 413: 2481–2492

[60]. Kong W, Xiang M H, Xia L, Zhang M, Kong RM, Qu F (2020) In-situ synthesis of 3D Cu_2O@Cu-based MOF nanobelt arrays with improved conductivity for sensitive photoelectrochemical detection of vascular endothelial growth factor 165. *Biosens Bioelectron* 167: 112481

[61]. Qiu W, Wang Q, Yano N, Kataoka Y, Handa M, Gao F, Tanaka H (2020) Flexible flower-like MOF of Cu_2(trans-1, 4-cyclohexanedicarboxylic acid)$_2$ as the electroactive matrix material for label-free and highly sensitive sensing of thrombin. *Electrochim Acta* 353: 136611

[62]. Song Z, Li Y, Teng H, Ding C, Xu G, Luo X (2020) Designed zwitterionic peptide combined with sacrificial Fe-MOF for low fouling and highly sensitive electrochemical detection of T4 polynucleotide kinase. *Sensor Actuat B-Chem* 305: 127329

[63]. Li Z, Dai G, Luo F, Lu Y, Zhang J, Chu Z, He P, Zhang F, Wang Q (2020) An electrochemical sensor for bacterial lipopolysaccharide detection based on dual functional Cu^{2+}-modified metal-organic framework nanoparticles. *Microchim Acta* 187: 1–10

25 MOF-based Electrochemical Sensors for DNA/RNA/ATP

Saadat Majeed[1], Muhammad Umer Farooq[1], Sayed Tayyab Raza Naqvi[1], Batool Fatima[2], Muhammad Najam-ul-Haq[1], Sabahat Majeed[3], Fahad Ali[1] and Naeem Akhtar Khan[4]

[1]Division of Analytical Chemistry, Bahauddin Zakariya University, Multan, Pakistan

[2]Department of Biochemistry, Bahauddin Zakariya University, Multan, Pakistan

[3]Department of Biosciences, COMSAT University, Islamabad, Pakistan

[4]IRCBM, COMSAT University, Islamabad, Lahore Campus, Pakistan

25.1 INTRODUCTION

No one can deny the importance of nucleic acids and energy in living organisms as these play a vital role in the metabolism of living organisms. Nucleic acids include deoxyribonucleic acid (DNA) and ribosomal nucleic acid (RNA). DNA is a hereditary material that regulates the growth, development, and reproduction of organisms and many viruses. Improper functioning and/or mutation in the structure of DNA leads to various infectious diseases such as heart disease, Parkinson's disease, Alzheimer's disease, diabetes mellitus, myotonic dystrophy, Huntington's disease, and cancer [1]. RNA damage due to change in its structure, oxidative stress, ultraviolet light, reactive oxygen, and nitrogen species causes neurodegenerative and cardiovascular diseases such as Alzheimer's, Parkinson's, oral and systemic diseases [2]. Mitochondria are mainly responsible for energy production in the form of adenosine triphosphate (ATP). Mutation in mitochondrial genes is the main reason for mitochondrial disorders that result in dysfunctions of more energy-consuming organs such as muscle, kidneys, brain liver, and heart [3]. Qualitative, as well as quantitative analysis of DNA, RNA, and ATP, plays a vital role in early diagnosis and treatment of these diseases and patient management. The source of these nucleic acids can be the whole blood, cells, serum, or plasma [1].

Various spectral, chromatographic, electrophoretic, fluorescent, electrochemical, and colorimetric methods have been used for the detection of nucleic acids and ATP. Among them, electrochemical methods have many advantages such as ease of operation, low cost, high selectivity, more sensitivity, and short time requirement for operation [4]. Currently, electrochemical biosensors are receiving more attention than other traditional methods because of their promising sensitivity and selective capability. Nanocomposites and nanohybrids of MOFs enhance the properties of MOFs [5]. Recently, MOF-based biosensors have gained an additional level of interest against biosensors based on other materials such as carbon compounds and nanoparticles or nanohybrid materials. These MOF-based biosensors possess the following potential advantages:

1. A conjugated π electron system is the inherent property of MOF-ligands. It allows π-π stacking to build up between MOF-ligands and capture probe molecules.
2. A network structure of MOFs provides high loading capacity but low degradation for a probe.
3. They have large surface areas.
4. They exhibit flexible pore sizes.
5. They have various organic functionalities.
6. They have more detection sensitivity due to the quenching properties of the metals present on the MOF skeleton.
7. Tailorable material: MOF properties can be changed by changing the metal to ligand ratio.
8. They have the ability to sustain their properties in harsh internal and external conditions such as acids, high temperature, and extreme pH.
9. There is a low cost for fabrication of MOFs.
10. MOFs' one-step synthesis is easy and simple.
11. There exists a large variety of MOF-material that has been synthesized due to the increased variety of metals and ligands.[6]

An up-to-date variety of MOF-based composites and hybrids have been designed which includes MOF/metal nanoparticles, MOF/metal oxide nanoparticles, MOF/polymers, MOF/quantum dots, MOF/enzymes, MOF/biomolecules, and so on. This chapter discusses the electrochemical sensing of DNA, RNA, and ATP based on MOF-based material. The materials include carbon-based MOF composites such as: MOF/carbon nanotubes, MOF/graphene oxide, MOF/polymer-based composites, nanoparticles-based MOFs (silver nanoparticles, gold nanoparticles, copper nanoparticles, platinum nanoparticles, and palladium nanoparticles), and metal or metal oxide MOFs. Graphene-based MOFs are highly selective, sensitive, and stable on the electrode surfaces for sensing DNA, RNA, and ATP. Carbon nanotubes and quantum dots/MOFs show enhanced signals with rapid responses and low limits of detection. This material enhances the properties of the electrode, for example, it increases

DOI: 10.1201/9781003188148-25

surface area, gives a fast response, has a low overvoltage, has thermal stability, chemical stability, and mechanical strength. Polymers are integrated with MOFs due to their attractive properties such as softness, chemical, and thermal stability. Nanoparticle-based MOFs that provide large surface area and more attachment sites for nucleic acids and ATP have been used as electrochemical sensors [7].

25.2 DNA/RNA/ATP

25.2.1 DNA

DNA is a hereditary material that is helpful for genetic and biological development. It is present in all living organisms, including prokaryotes, eukaryotes, and many viruses. Two spirals are wrapped around each other in a large double-helical-shaped polymer. Each spiral consists of large numbers of monomers known as a nucleotide. Each nucleotide consists of three types of building block: deoxyribose sugar, phosphoric acid, and complementary two base pairs. Base pairs include guanine, cytosine, adenine, and thymine. Nucleotides are covalently attached through the phosphate group of one nucleotide and the sugar of another nucleotide. Two spirals of DNA are connected through hydrogen bonds between nitrogenous bases. There is a specific sequence of bonding of these nitrogenous bases, namely, adenine attaches with thymine through double bonds and guanine with cytosine through triple bonds. High stability in the configuration of DNA molecules is necessary for the replication of new DNA as well as RNA molecules. Each strand of DNA acts as a template for the production of new DNA. DNA is a dense complex of proteins that is present in the nucleus in eukaryotes and cytoplasm in prokaryotes. There is a specific concentration of each base pair in the DNA molecule. Change in concentration of base pairs decreases the immunity that results in many various genetic diseases, tumors, and cancer. Therefore, for clinical purposes detection of DNA is important [8].

25.2.2 RNA

RNA is a single-stranded macromolecule that plays an important role in protein synthesis in many organisms and replaces DNA in some viruses for genetic information. It consists of a ribose sugar, nitrogen bases, and phosphate group in its structure. Nitrogenous bases include guanine, cytosine, uracil, and adenine. Of the many types of RNA, three types of RNAs are well known, namely: messenger RNA (mRNA), transfer RNA (tRNA), and ribosomal RNA (rRNA). mRNA carries the information from DNA in the nucleus to the ribosomes. rRNA and protein make up the ribosomes, where the codes carried by rRNA are translated. tRNA carries the amino acids to the ribosomes according to the information carried by the mRNA from DNA where protein synthesis occurs by specific linkage of amino acids. MicroRNA (miRNA) is a small and single-stranded RNA that plays an important role in apoptosis, cell differentiation, and cell proliferation in plants, animals, and some viruses. It is an important biomarker of cancer. Cancer is the result of abnormal expression and function of miRNA. The development of biosensors for the detection of miRNA can help to diagnose and treat these malignant diseases. There are various traditional methods for the detection of miRNAs such as array technology, polymerase chain reaction, and the miRNA analysis method. However, these methods have drawbacks, such as the need for expensive detection kits, interference from other RNAs, and the need for a high concentration of miRNA for detection. Only electrochemical sensors based on MOFs will be discussed in this context [9], [10]. Viruses cause various diseases in animals, especially humans. Many of these viruses contain RNA as hereditary material. For example, Ebolaviruses such as Sudan and Ebola viruses cause hemorrhagic fever in humans. Therefore, to cure this fever and other diseases that are linked with viruses RNA, earlier diagnosis is necessary to control infections and for the management of the diseases. To diagnose these diseases, cost-effective, rapid, specific, selective, and sensitive method development is the need of this era [11].

25.2.3 ATP

Adenosine 5′-triphosphate (ATP) is a naturally occurring nucleotide. It is made up of adenine, ribose, and 3 phosphate groups. It is the primary source of energy that regulates cellular metabolism and performs various biological functions such as enzyme catalysis, DNA replication, neurotransmission, liver glycogen metabolism, vasodilation, cardiac functions, and muscle contraction. It can be released from the cytoplasm of cells and interacts with receptors present at the surface of cells. Changes in the concentration of ATP can lead to various diseases such as Alzheimer's, malignant tumors, hypoglycemia, and Parkinson's. In mammalian cells, an average level of ATP is 3152 μmol/L. The level of ATP is higher in tissue cells as compared to blood cells. ATP level in human erythrocytes is 1500 to 1900 μmol/L [12], [13]. Many receptors and enzymes can undergo phosphorylation and dephosphorylation in a cellular communication network by protein kinase. Phosphorylation and dephosphorylation are also termed as 'on' or 'off' switched. In the phosphorylation reaction transfer of one phosphate group happens to one of the specific residues of protein that may be tyrosine, serine, or threonine. This type of modification in the structure of protein causes the improper functioning and delocalization of proteins that results in the generation of oncoproteins. Abnormal phosphorylation causes many diseases such as cancer and diabetes. To diagnose these diseases there is a need to diagnose and quantify the activity of protein kinase. For this purpose, many analytical methods have been used that include radiolabeled ATP, fluorescence, and the like. There is, however, a need to modify the peptides with an electroactive site to make the procedure easier and low cost [14].

Earlier diagnosis of diseases can help us to stop the transmission of many diseases from person to person or even in a single person. As a point of clinical diagnosis of ATP, it may lead to determining the early signs of many diseases. Various materials have been designed to determine ATP through colorimetric as well as spectroscopic methods. There are, however, many challenges still to be overcome [15].

25.3 MOF-BASED ELECTROCHEMICAL SENSORS

Electrochemical sensors are simple, fast, specific, selective, and portable for biosensing. Here, a probe is modified onto the surface of the electrode that plays an important role in the detection of nucleic acids and ATP. Due to the catalytic properties of the MOF, it is used as a signal unit. MOF-based composites overcome the conductivity problem that is faced in electrochemical biosensors [16]. There are two essential components in sensors, one is active sensing material, and the other is a transducer. The transducer transmits the signal from the reaction center or sensing material. Sensors originate a thermal or electrical signal that is converted into a digital signal. There are various categories of sensors based on the output signal. Among them, electrochemical sensors are better relative to other sensors because of reusability and high sensitivity as well as selectivity for analytes. These show a low limit of detection and good linear response. Among the materials, MOFs show good potential applications due to their porous nature, tunability, various chemical functionalities, and the like [17]. Electrochemical biosensing of DNA based on MOFs is a newly emerging field. It is a simple, fast, and economical assay for the testing of genes. The high specificity of the DNA biosensor enables it to detect genetic diseases, infections caused by viruses and bacteria, and for forensic applications. DNA-sensing has been carried out via fluorescent, colorimetric, and electrochemical methods. Here, we will discuss only electrochemical detection of DNA based on MOF material [18].

25.3.1 MOF-Based Carbon Nanomaterial Sensors

Carbon-based nanomaterials such as carbon nanotube and graphene have been extensively used as electrochemical biosensors to enhance the electrical as well as mechanical properties of composite material. MOF composites of graphene, reduced graphene, and carbon nanotubes have been employed in biomedical, food, environmental, and industrial fields.

25.3.1.1 MOF/Carbon Nanotube Sensors

Based on structure, carbon nanotubes are divided into 3 types: 1) single-walled carbon nanotubes (SWCNT), 2) double-walled carbon nanotubes (DWCNT), and 3) multi-walled carbon nanotubes (MWCNT). These carbon nanotubes are hollow and cylindrical. These have been used in sensor and biosensor applications due to their unique electrical, mechanical, thermal, and optical properties. Large surface area, good thermal stability, controllable pore size, and crystalline structure have allowed the carbon nanotubes to be combined with MOFs to increase the thermal and electrical properties of their composites [19, 20].

A group of researchers studied the DNA of the human immune deficiency virus by using a paper-based electrode made by the DNA hybridization method. This flexible electrode was modified by using nickel metal-organic framework composite/gold nanoparticles/carbon nanotubes/polyvinyl alcohol (Ni-Au composite/CNT/PVA). Methylene blue (MB) was used as an indicator. Two types of electrodes were prepared. One electrode was modified by CNT/PVA, while the second electrode was modified with Ni-Au composite/CNT/PVA. The second electrode was named as the CCP film electrode. CCP film was achieved by deposition of CNT/PVA and Ni-Au composite onto a large network of cellulose membranes by vacuum filtration and drop coating. The CCP film electrode was more advantageous as compared to the CNT/PVA film electrode due to its higher surface area and conjugated π-electron system that facilitates the higher loading of probe DNA. The schematic illustration for sensing of DNA of a modified MOF electrode is presented in Figure 25.1. The sensitivity of target DNA increases due to the hydrogen bond interaction between the MOF and single-stranded DNA. Variation in current was monitored before and after adsorption of MB molecules with DNA. The stable electrochemical property was observed even after stretching from 0%-20% or bending 200 times. LOD and linear range were observed at 0.13 nM and 10 nM-1 μM, respectively. In vitro study of HIV DNA was performed in a serum sample [21].

25.3.1.2 MOF/Graphene Sensors

Graphene, another carbon-based nanomaterial, has good conductivity, thermal stability, and electrocatalytic properties. It has a sheet-like 2-dimensional structure with a large surface area that makes it easier for electron transfer at its edges. In the MOF/Graphene composite electron transfer becomes fast because graphene acts as a conductive bridge. These properties concern the use of MOF/graphene composites in electrochemical sensing and biosensing applications [22]. A group of researchers synthesized the metal-organic framework (MOF)/graphene oxide (GO) composite by dispersing the monolayers of UiO-66-NH_2 onto the GO surface. The electrochemical signal was strong due to the strong π–π interactions between aptamer and composite. The LOD value for the detection of ATP by using this composite was 3.3fM while the linear range was 19.5-97.63fM via electrochemical impedance spectroscopy [23].

25.3.2 MOF-Based Polymer Sensors

The combination of MOFs with polymers provided improved properties which include, chemical and thermal stability, biocompatibility, dispersibility, with a large network of porous composite material. Fabrication of MOF-based polymers is a new, simple, and emerging approach for sensing and biosensing applications [24]. FeTCPP@HKUST-1 is a porphyrin-encapsulated MOF. It is a highly efficient electrochemical signal indicator. After treatment with streptavidin (SA), this composite is used as a signal

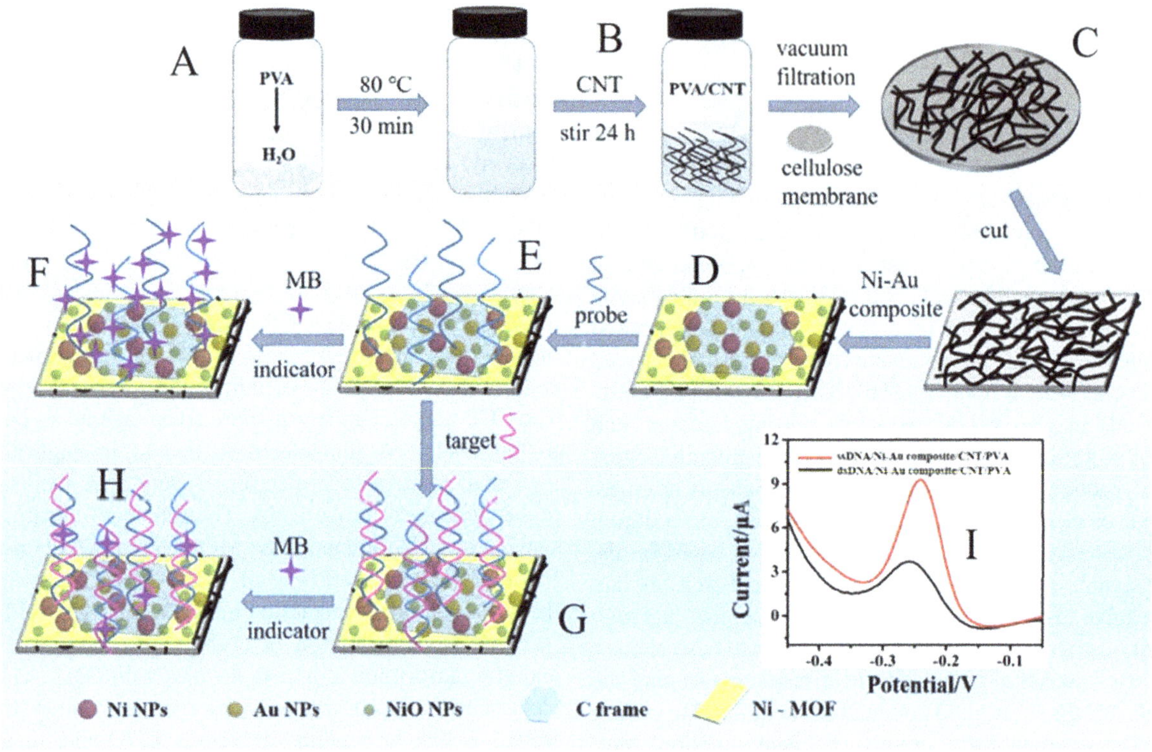

FIGURE 25.1 Fabrication of flexible Ni–Au composite/CNT/PVA film electrode and electrochemical response of target DNA. Adapted with permission from [21]. Copyright (2021) Elsevier.

unit for the electrochemical sensing of DNA. Hairpin DNA is immobilized onto the surface of a glassy carbon electrode acting as a capture unit. Upon introduction of target ssDNA, hairpin DNA is unfolded followed by a free aptamer formation. As the signal-free unit is added, the interaction between streptavidin and aptamer brings FeTCPP@MOF-SA to the electrode surface. Cyclic voltammetry (CV) and differential pulse voltammetry (DPV) measurements show the limit of detection (LOD), 0.48fM. This method shows good feasibility of DNA in a spiked serum sample [25]. Another sensitive and selective electrochemical sensor was designed with zirconium porphyrin. Zirconium-porphyrin MOF (PCN-222) was treated with SA that acts as a signal probe. PCN-222 composite enhances the electrochemical signal (Figure 25.2). Biotin-labeled triple helix immobilized on the GCE acts as a capture unit. Assistant DNA hybridizes with target DNA as the ssDNA is added. End biotin is activated that catches the PCN-222@SA via a streptavidin-biotin link, and a signal probe attaches at the surface of GCE. This method shows LOD of 0.29fM, and specificity to detect DNA in a spiked serum sample [26].

25.3.3 MOF-Based Metal/Metal Oxide Sensors

The introduction of nanomaterials enhances the signal in electrochemical sensing and biosensing systems that depend on transducer material. Composites of metals and metal oxides with MOFs are biocompatible that provide large surface area and good conductivity [27].

25.3.3.1 MOF-based Gold Nanoparticle Sensors

Gold nanoparticles (Au NPs) are one of the most used biocompatible nanomaterials for sensing and biosensing applications due to their unique properties such as large surface area, good electrical properties, and so forth. In a composite of MOF and gold nanoparticles, gold is used as a modifier for other MOF-based metallic structures [28]. A composite of S1-AuNPs@Cu-MOF was prepared and used as a signal unit with catalytic oxidation properties for the detection of miRNA-155. A paper-based electrode was used as a capturing unit that was treated with DNA probe H1 and gold nanoparticles. Non-specific active sites were blocked by MCH. The addition of miRNA-155 is accompanied by opening the hairpin probe H1. The addition of an H2 probe displaces the target from the H1 target hybrid. As a result, a stable H1-H2 hybrid is formed. Another signal probe (S1) can attach with H2 due to its higher length as compared to H1 upon the addition of S1-AuNPs@Cu-MOFs. LOD and linear ranges were 0.35 fM and 1 fM to 10 nM respectively [29].

25.3.3.2 MOF-based Silver Nanoparticle Sensors

Due to enhanced electrical conductivity, silver nanoparticles (Ag NPs) are used for photoelectrochemical as well as antimicrobial activity. The combination of MOF and silver nanoparticles makes it a unique material for the biosensing of DNA, RNA, and ATP [30].

A new approach was proposed by researchers for the detection of HCV-RNA based on an Ag/Zn/MOF. In this

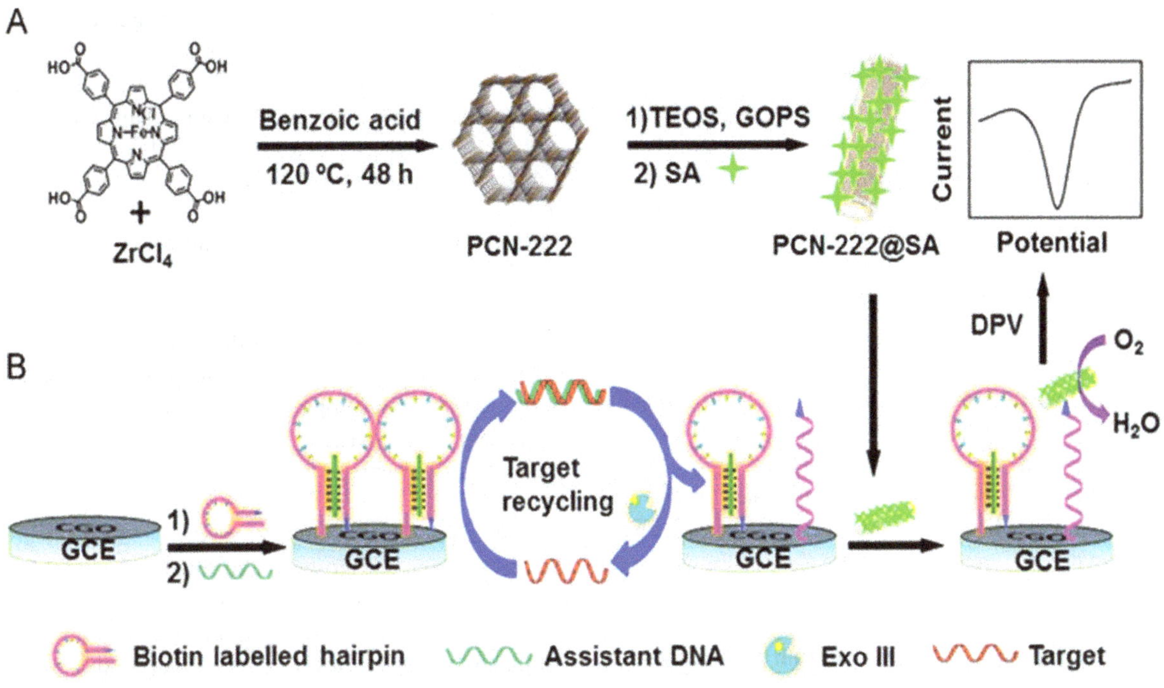

FIGURE 25.2 (A) Preparation of Zirconium-porphyrin MOF (PCN-222@SA) composite (B) Electrochemical DNA sensing-based coupled with target recycling amplification. Adapted with permission from [26]. Copyright (2015) Elsevier.

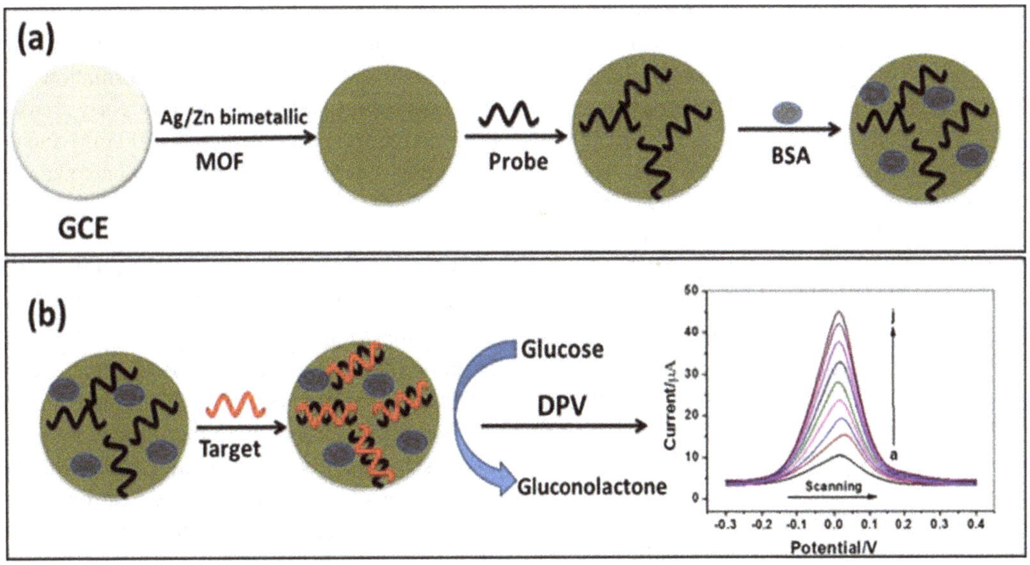

FIGURE 25.3 Fabrication of the HCV-RNA biosensor (A) Synthesis of BSA/probe/bimetallic MOF@GCE (B) Basic principle for HCV target detection and its electrochemical response. Adapted with permission from [31]. Copyright (2021) Elsevier.

method, a bimetallic material (Ag/Zn/MOF) was prepared to confirm the morphology and structure of RNA. MOF composite was coated onto the GCE and then treated with an HCV-capture probe. For the hybridization of HCV-RNA with GCE, the surface of GCE was treated with bovine serum albumin. A step-by-step synthesis of the biosensor was confirmed by using cyclic voltammetry and impedance spectroscopy (Figure 25.3). The activity of the biosensor was checked by adding glucose that was converted into gluconolactone that enhanced the electrochemical signal. The LOD of this biosensor reached down to 0.64fM, which shows its greater sensitivity. The linear detection range was from 1 fM to 100 nM [31].

25.3.3.3 MOF-based Copper/Copper Oxide Nanostructure Sensors

Copper nanoparticles (Cu NPs) are the most extensively used nanomaterials as composites with MOFs for the biosensing of nucleic acids, due to their wide linear range [32]. A novel electrochemical sensor has been developed by researchers that is based on hemin. In this study, Cu-MOF-199 was doped with hemin to synthesize the hemin@MOF. This material was coated onto the surface of GCE that exhibits good electrocatalytic activity for the detection of DNA. Hemin@ MOF was used as an electrochemical signal indicator due to its good electrochemical performance. Probe DNA (pDNA) has a stem-like probe structure that consists of 3 amino acids and 5'-biotin. In this procedure, pDNA was anchored into the network structure of graphene through its carboxylic functional groups via an amidation reaction. Matched target DNA (tDNA) was used to unfold the pDNA by hybridization. Recognition between streptavidin (SA) on the Hemin@MOF-SA and biotin at the pDNA led the Hemin@MOF-SA to access the electrode surface. Hence, Hemin@MOF-SA and matched tDNA also showed the catalytic properties for hydrogen peroxide. This novel electrode shows the linear range and limit of detection 1fM to 1μM and 0.69 fM, respectively. It has one mismatch differential ability up to one base and good stability [33].

25.3.3.4 MOF-based Platinum Nanoparticle Sensors

Platinum nanoparticles (Pt NPs), a noble metal nanoparticle, can be easily aggregated and well dispersed onto a MOF's porous structure that makes composite pores easily accessible for analytes. Pt NPs take part in a multi-electron transfer reaction that enhances the electrochemical signal as compared to one or two-electron transfer metals [34]. FGFR-3 is a mutation gene. It is also a biomarker of achondroplasia. In this study, a signal unit is composed of hemin, PtNPs, bovine serum albumin, signal probe, and a composite of FE-MIL-88. A GCE electrode was modified with gold nanoparticles, reduced graphene oxide- tetraethylenepentamine (rGO-TEPA), a biotin-modified ssDNA probe, and streptavidin. This modified electrode acted as a capture unit. ssDNA was captured by the capture probe through hybridization. An increased electrical signal was observed upon the addition of hydrogen peroxide as hemin-MOFs/PtNPs have a unique ability to catalyze the H_2O_2. In this case, PtNPs were also replaced with gold nanoparticles, but there was a good response with PtNPs as the sensitivity reached down to 0.033fM and linear range from 0.1 fM to 1 nM [35].

The adrenoceptor beta-1 (ADRB1) gene was detected with an electrochemical biosensor that might potentially cause the gene mutation. In other methodologies, only one unit acts as an electrochemiluminescence sensor that works as a signal and capture unit. In this methodology, GCE was modified with PtNPs, Cu (II) ions, and Fe-MIL-88-NH2 nanocomposite. Hydrogen peroxide is catalyzed by Pt/hemin@FeMIL-88NH2/Cu^{2+}. The addition of a probe onto the electrode surface does not affect the reduction signal whereas ssDNA addition decreases the reduction signal. The sensitivity of ADRB1 was recorded down 0.21 fM. ADRB1 was also detected even in a spiked complex serum sample [36].

Detection of exosomal miRNA was performed with MOF@Pt@MOF enzyme through a cascade premier exchange reaction (PER). Target-triggered PER contains three parts: a gated hairpin, DNA polymerase, and a primer. Single strand production is carried out by this target-triggered PER through thermal and autonomous methods. As a result, protector B is used to blockade capture probes that are released from the nascent strand. The LOD of this method reaches down to 0.29 fM which shows good sensitivity. This method is so specific that can even detect the mismatching of a single base unit within homologous miRNA. This method can be used to differentiate between cancer cells and tumor cells by detecting exosomal miRNA-21 [37].

25.3.3.5 MOF-based Palladium Nanoparticle Sensors

Palladium nanoparticles have unique properties and get more attention for the development of electrochemical sensors and biosensors. Composites of palladium and MOFs have been exploited in biosensors for the detection of DNA as well as RNA [38].

miRNA-122 is a biomarker of drug-induced liver injury. An electrochemical biosensor for this marker was developed by a team of researchers. A nanocomposite material of Fe-MIL-88- NH_2, Streptavidin, hairpin signal probe (H2), and palladium nanoparticles were synthesized due to its good catalytic properties which act as a signal unit. While glassy carbon electrodes (GCE) The GCE was modified with nitrogen-doped graphene sheets and a heparin capture probe (H1). Upon addition of miRNA-122, H1 present at the surface of the GCE will open up and become hybridized with miRNA. The addition of MOF-based material leads to the release of miRNA-122 from H1 and the formation of a new stable H1-H2 hybrid takes place. In this assay, sensitivity and the linear range were recorded up to 0.003 fM and 0.01 to 10 pM, respectively. This study was applied on spiked serum samples as well as blood with excellent recovery and selectivity [39].

25.3.3.6 Other MOF or Metal/Metal Oxide/MOF Nanostructures in Sensing Platforms

A new approach for the detection of ATP has attained the attention of researchers that is based on amino-functionalized MOF. After synthesis of Ce-MOF through a one-step process, it was further modified onto the surface of the gold electrode. Then, subsequently, it was conjugated with ATP aptamer. In the presence of target ATP, a strong electrochemical impedance has been observed due to the structural changes in the strands of the aptamer. In this study, sensitivity for the detection of ATP approaches down to 5.6 nM with a linear range of 10 nM to 1 nM. This is the first report in which sensitivity has reached up to nM level. This study was applied for cancer patients in which serum was used as a sample [40].

A CoNi-based metal-organic framework, a novel and efficient material, has been used for the detection of miRNA-126. 2,2′-bipyridine-5,5′-dicarboxylic acid (dcbpy) acts as the building block for the preparation of this MOF. CoNi-MOF displays excellent electrochemical activity as compared to Co-MOF or Ni-MOF. Many forces such as van der Waals, hydrogen bonding, π-π stacking, and metal bindings are involved for immobilization of the DNA probe onto the CoNi-MOF that provides an excellent electrochemical response. It shows an LOD value of 0.14 fM along with good stability, selectivity,

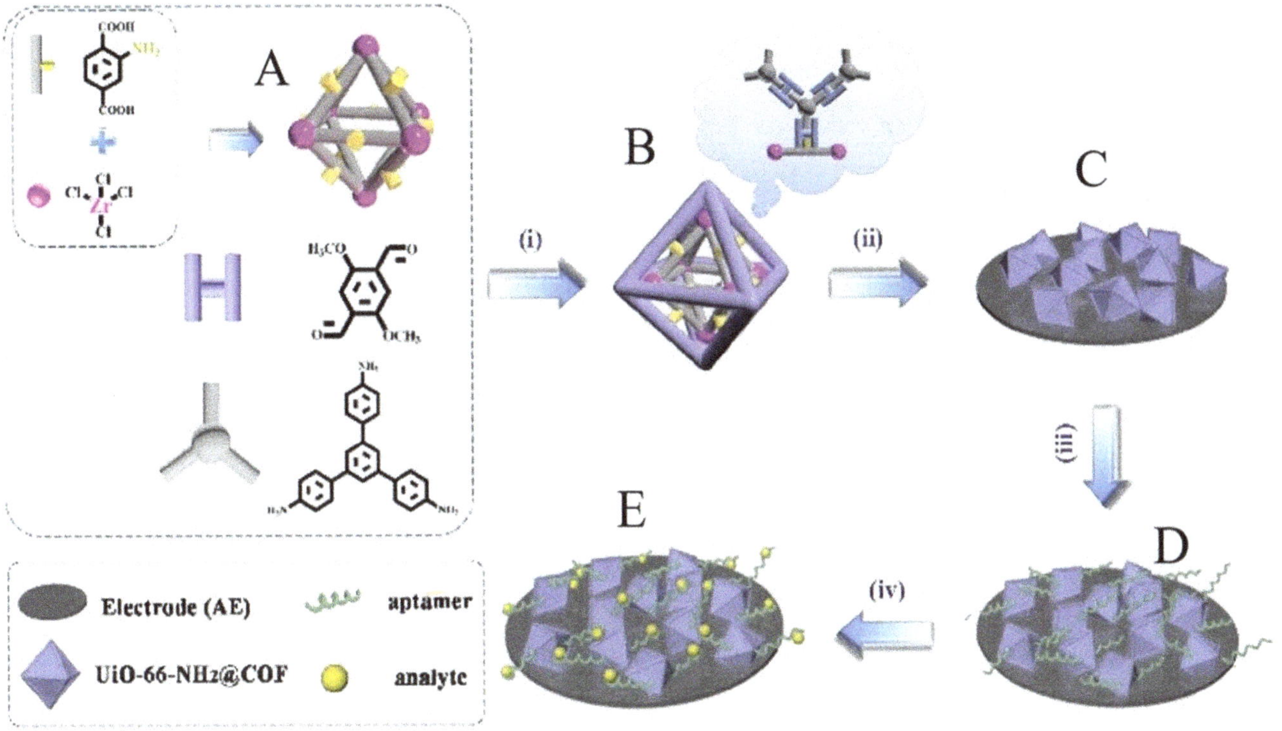

FIGURE 25.4 Step-by-step synthesis of electrochemical aptasensor (MOF@COF) for detection of ATP. Adapted with permission from [42]. Copyright (2021) Elsevier.

and reproducibility. This material shows good applicability in rat glioma cells for the detection of miRNA-126 [41].

Increased porosity in core-shell MOF-based composite enhances the performance and significance of composite in many applications such as biosensing. In this, study UiO-66-NH_2@MOF was synthesized by coating a microporous core of UiO-66-NH_2 onto a mesoporous TAPB-DMTP–COF shell through covalent bonding. Crystal and porous structures of composites increase the electrochemical response for different analytes; ATP as well as antibiotics. The reason behind this strong signal or response is the strong π-π stacking between composite and aptamer. The low limit of detection for this composite was 9.9 fM with a linear range of 20-99 fM [42].

Abbreviations: HIV: Human immune deficiency virus; DNA: Deoxyribonucleic acid ssDNA: single-stranded DNA; FGFR-3: fibroblast growth factor receptor 3; ADRB1 gene: β-1 adrenergic receptor; miRNA: micro ribonucleic acid; HCV: Hepatitis C virus; ATP: Adenosine triphosphate; CNT: Carbon nanotubes; PVA: Polyvinyl alcohol; FETCPP: Iron(III) meso-5,10,15,20-tetrakis(4-carboxyphenyl) porphyrin chloride; MOF: Metal-organic framework; NPs: Platinum nanoparticles; COF: Covalent organic framework; GO: Graphene oxide.

25.3.4 MOF-Based Electrochemiluminescence (ECL) Sensors

The chemiluminescence technique combined with electrochemistry has many advantages in biosensing. As we know, pure MOFs have poor electrical conductivity in comparison to MOF composites. For this reason, composite MOFs are used for the sensitive detection of nucleic acids. Only the ruthenium MOF has good electrical conductivity and intrinsic luminescence properties in its pure MOF form, so it is used in ECL biosensors. ECL biosensors have the same principle of working as electrochemical biosensors.

A group of researchers detected miRNA-155 by ECL biosensors based on a MOF modified by mercury. Here, the signal unit is the combination of Ru-MOF attached with a signal probe (I). The capture unit is made up of gold nanoparticles modified GCE and a hairpin capture probe. Unspecific active sites in the capture unit were blocked by treating with MCH. The formation of the H1-target hybrid takes place through hybridization after the addition of miRNA-155. The addition of probe H2 replaces the target from its hybrid. As a result, new more stable hybrid H1-H2 forms and the free target enters into a new hybridization cycle. Upon the addition of mercury (II), the mildly red colored solution changed into bright red. Mercury (II) triggered the release of $[Ru(bpy)_3]^{2+}$ from Ru-MOFs. The sensitivity of miRNA-155 was observed up to 0.3 fM. miRNA was successfully detected in a spiked serum sample [43]. miRNA was also detected by an ECL biosensor that was based on a Faraday-cage in which the signal probe or signal unit was immobilized onto Ru-MOF. Herein, the glassy carbon electrode was magnetically modified with Fe_3O_4@SiO_2@Au. miRNA-141 addition is accompanied by opening the hairpin cDNA probe. A Faraday structure was formed as an unhybridized fragment was combined with a signal probe. The sensitivity of miRNA increased to 0.3 fM with a linear range from 1 fM to 10 pM. There were 110% recoveries as it is applied for a real serum sample [44].

TABLE 25.1
Analytical Performances of MOF-based Electrochemical and Electrochemiluminescence Biosensors

Analytes	MOF composites	LOD	Linear range	References
HIV DNA	Ni-Au composite/CNT/PVA	0.13 nM	10 nM-1 mM	[21]
ssDNA	FeTCPP@MOF	0.48 fM	--	[25]
DNA	hemin@MOF	0.69 fM	1 fM-1 μM	[33]
FGFR-3 gene	hemin-MOFs/PtNPs	0.033 fM	0.1 fM to 1 nM	[35]
ADRB1 gene	Pt/hemin@FeMIL-88NH_2/Cu^{2+}	0.21 fM	--	[36]
miRNA-155	S1-AuNPs@Cu-MOF	0.35 fM	1 fM to 10 nM	[29]
HCV-RNA	Ag/Zn/MOF	0.64 fM	1 fM to 100 nM	[31]
Exosomal miRNA	MOF@Pt@MOF	0.29 fM	--	[37]
miRNA-122	PdNPs@Fe-MOFs	0.003 fM	0.01 to 10 pM	[39]
miRNA-126	CoNi-MOF	0.14 fM	--	[41]
miRNA-155	Ru-MOF	0.3 fM	--	[43]
miRNA-141	Fe_3O_4@SiO_2@Au	0.3 fM	1 fM to 10 pM	[44]
ATP	Ce-MOF	5.6 nM	10 nM to 1 nM	[40]
ATP	MOF@COF	9.9 fM	20- 99 fM	[42]
ATP	UiO-66-NH_2/GO	3.3 fM	19.5-97.63 fM	[23]

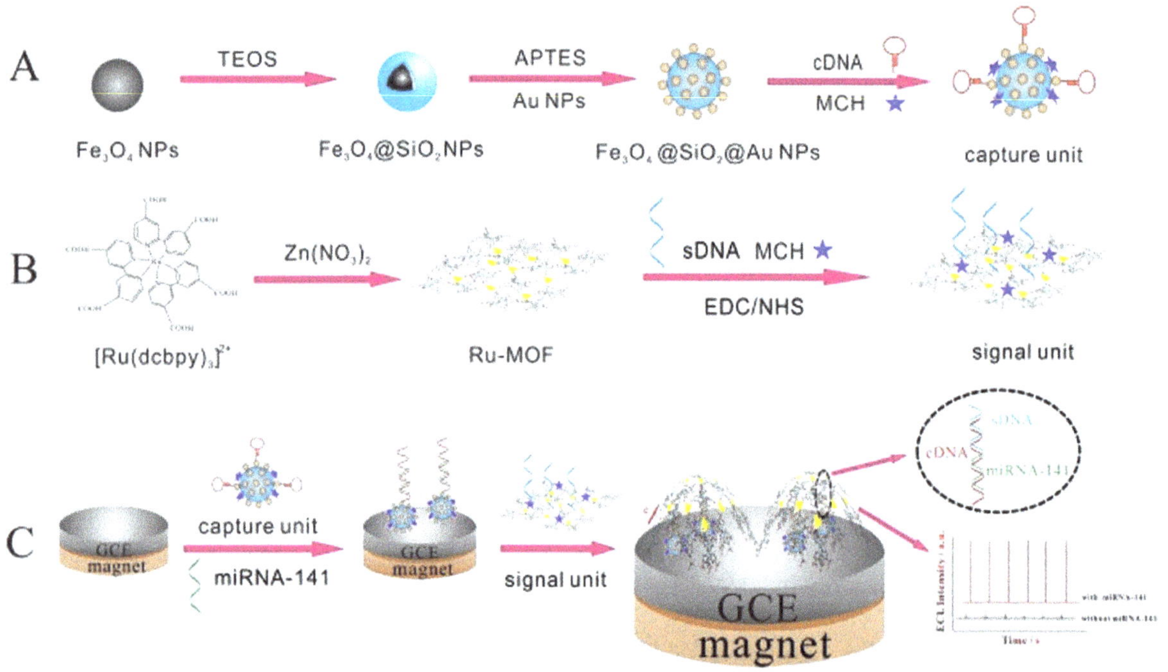

FIGURE 25.5 Electrochemical biosensor for detection of miRNA-141 (A) Synthesis of capture unit (Fe_3O_4@SiO_2@Au-cDNA) (B) Synthesis of Ru-MOF-sDNA (C) Fabrication of Faraday cage. Adapted with permission from [44]. Copyright (2018) Elsevier

25.4 COMPARISON BETWEEN VARIOUS SENSING MATERIALS

It is well established that electrochemical sensors are simple, fast, efficient, and cost-effective because of their high selectivity, sensitivity, and stability toward the detection of nucleic acids and ATP. Carbon-based materials such as graphene and carbon nanotubes have potential over other nanomaterials due to their electrical and thermal conductivity as well as good mechanical strength. Carbon nanotubes, graphene oxide, nanoparticles, and metal/metal oxide have been used for the detection of nucleic acid and/or ATP. Table 25.1 summarizes a few relevant research studies related to these materials. In the case of DNA or gene detection, the hemin-based electrochemical sensor exhibited a LOD of 0.69 fM making it more sensitive than the carbon nanotube-based biosensor. Thus, the hemin-based biosensor works as an advanced MOF-based material for the biosensing of DNA.

Polymers have the inherited advantages of large surface area, high porosity, and more attachment sites that gained significant consideration for the fabrication of electrochemical biosensors. Sensors based on polymer-MOFs exhibited good electrochemical activities. An LOD of 0.48 fM was achieved using a polymer-based biosensor through the electrochemical sensing technique. Nanoparticle-based biosensors exhibited the highest sensitivity, selectivity, and a good linear range. Among them, Pt NPs based hemin-MOF material has the lowest sensitivity (0.033 fM) for the detection of genes. For detection of RNA, MOF-composite based on single and double metals and nanoparticles as well as polymer-based MOFs have been studied. Bimetallic MOF-composites have a lower LOD value that shows their high sensitivity than single metal-based composites. An LOD value achieved through CoNi-MOF was lower (0.14 fM) in comparison to Ag/Zn/MOF (0.64 fM), and Ru-MOF (0.3 fM). Hemin-based electrodes have more sensitivity than single metal-based sensors but less than bimetallic-MOF sensors. A LOD value reported by Pt/hemin@FeMIL-88NH_2/Cu^{2+} material was 0.21 fM. The sensitivity achieved by using MOF-based nanocomposite depends upon the nanoparticles combined with the MOF. Palladium nanoparticle-based MOF material showed the lowest LOD value (0.003 fM) and high sensitivity among all the material reported in this chapter for the biosensing of RNA. But gold nanoparticles showed the least sensitivity.

ATP biosensing was achieved by using three types of materials based on graphene, core-shell composite, and metal MOFs. Among the three materials, graphene-based composite has the lowest LOD value (3.3 fM), and MOF@COF exhibited the highest value. Thus graphene-based MOF composites are the best materials for the detection of ATP. Table 25.1 shows biosensors based on MOFs, which has their structural morphology: Ni-Au composite/CNT/PVA, FeTCPP@MOF, hemin@MOF, hemin-MOFs/PtNPs, Pt/hemin@FeMIL-88NH_2/Cu^{2+}, S1-AuNPs@Cu-MOF, Ag/Zn/MOF, MOF@Pt@MOF, PdNPs@Fe-MOFs, CoNi-MOF, Ru-MOF, Fe_3O_4@SiO_2@Au, Ce-MOF, MOF@COF, and UiO-66-NH_2/GO. All the sensors are reproducible with high sensitivity, and selectivity. Unique features of MOFs and their composites make them biocompatible for a wide range of applications in the field of biosensing, clinical diagnosis, and managing the treatment of various diseases.

25.5 CONCLUSION

There have been great advancements in metal-organic framework sensors associated with a variety of functional MOFs, and their composites, including carbon-based, nanoparticles, polymers, and metal/metal oxide. Their potential has been investigated in the diagnosis of diseases and clinical systems. Their unique features such as large surface area, various functionalities, and conjugates system, flexible pore size, and the like, made them capable of use as electrochemical sensors. These sensors have been used for the analysis of various biological molecules such as nucleic acids and ATP. In this chapter, various MOF-based materials have been summarized with their performance for electrochemical biosensing of DNA, RNA, and ATP. MOF-based composite has a clear and bright future for the prognosis and diagnosis of diseases. MOFs and their composites have gained more interest due to their high sensitivity, selectivity, and thermal stability. Despite many advantages of MOF-based electrochemical sensors, these face many challenges in real samples. These challenges include low LOD with high sensitivity, reproducibility and selectivity, more stability of biosensor, multiple target analysis with high resolution, and adsorption of proteins onto the electrode surface. These challenges can be resolved by using serum samples to avoid protein adsorption, the addition of specific and selective functionalities on material, the addition of catalysts that increase the speed of reaction, and the modification of electrodes to resist bio-fouling. Further advancements might have great potential for electrochemical biosensing of nucleic acids and ATP to diagnose genetic, neurodegenerative, cardiovascular, and cancer diseases.

REFERENCES

[1]. D. C. Wallace, "Diseases of the mitochondrial DNA," *Annual Review of Biochemistry* vol. 61, pp. 1175–1212, 1992.

[2]. Z. Liu, X. Chen, Z. Li, W. Ye, H. Ding, P. Li, and L. H. H. Aung, "Role of RNA oxidation in neurodegenerative diseases," *International Journal of Molecular Sciences,* vol. 21, pp. 5022, 2020.

[3]. T. A. Johnson, H. Jinnah, and N. Kamatani, "Shortage of cellular ATP as a cause of diseases and strategies to enhance ATP," *Journal of Frontiers in pharmacology,* vol. 10, pp. 98, 2019.

[4]. A. Azzouz, K. Y. Goud, N. Raza, E. Ballesteros, S. E. Lee, J. Hong, A. Deep, and K. H. Kim, "Nanomaterial-based electrochemical sensors for the detection of neurochemicals in biological matrices," *Journal of Trends in Analytical Chemistry,* vol. 110, pp. 15–34, 2019.

[5]. S. Liu, L. Wang, J. Tian, Y. Luo, G. Chang, A. M. Asiri, A. O. Al-Youbi, and X. Sun, "Application of zeolitic imidazolate framework-8 nanoparticles for the fluorescence-enhanced detection of nucleic acids," *Journal of ChemPlusChem,* vol. 77, pp. 23, 2012.

[6]. J. Lei, R. Qian, P. Ling, L. Cui, and H. Ju, "Design and sensing applications of metal–organic framework composites," *Journal of Trends in Analytical Chemistry,* vol. 58, pp. 71–78, 2014.

[7]. Ü. Anik, S. Timur, and Z. Dursun, "Metal organic frameworks in electrochemical and optical sensing platforms: a review," *Journal of Microchimica Acta,* vol. 186, pp. 1–15, 2019.

[8]. X. Wang, J. Zhang, Y. Wei, T. Xing, T. Cao, S. Wu, and F. Zhu, "A copper-based metal–organic framework/graphene nanocomposite for the sensitive and stable electrochemical detection of DNA bases," *Journal of Analyst,* vol. 145, pp. 1933–1942, 2020.

[9]. C. Zhang, P. Miao, M. Sun, M. Yan, and H. Liu, "Progress in miRNA Detection Using Graphene Material–Based Biosensors," *Journal of Small,* vol. 15, pp. 1901867, 2019.

[10]. Y. Ren, H. Deng, W. Shen, and Z. Gao, "A highly sensitive and selective electrochemical biosensor for direct detection of microRNAs in serum," *Journal of Analytical chemistry,* vol. 85, pp. 4784–4789, 2013.

[11]. I. Messaoudi, G. K. Amarasinghe, and C. F. Basler, "Filovirus pathogenesis and immune evasion: insights from Ebola virus and Marburg virus," *Journal of Nature Reviews Microbiology,* vol. 13, pp. 663–676, 2015.

[12]. H. J. Agteresch, P. C. Dagnelie, J. W. O. van den Berg, and J. P. Wilson, "Adenosine triphosphate," *Journal of Drugs,* vol. 58, pp. 211–232, 1999.

[13]. P. Xu, and G. Liao, "A novel fluorescent biosensor for adenosine triphosphate detection based on a metal–organic framework coating polydopamine layer," *Journal of Materials,* vol. 11, pp. 1616, 2018.

[14]. H. Song, K. Kerman, and H. B. Kraatz, "Electrochemical detection of kinase-catalyzed phosphorylation using ferrocene-conjugated ATP," *Journal of Chemical communications*, pp. 502–504, 2008.

[15]. M. Wang, L. Yang, B. Hu, J. Liu, L. He, Q. Jia, Y. Song, and Z. Zhang, "Bimetallic NiFe oxide structures derived from hollow NiFe Prussian blue nanobox for label-free electrochemical biosensing adenosine triphosphate," *Journal of Biosensors and Bioelectronics,* vol. 113, pp. 16–24, 2018.

[16]. J. Wang, "Survey and summary: from DNA biosensors to gene chips," *Journal of Nucleic acids research,* vol. 28, pp. 3011–3016, 2000.

[17]. S. Tajik, H. Beitollahi, F. G. Nejad, K. O. Kirlikovali, Q. Van Le, H. W. Jang, R. S. Varma, O. K. Farha, and M. Shokouhimehr, "Recent electrochemical applications of metal–organic framework-based materials," *Journal of Crystal Growth,* vol. 20, pp. 7034–7064, 2020.

[18]. D. W. Kimmel, G. LeBlanc, M. E. Meschievitz, and D. E. Cliffel, "Electrochemical sensors and biosensors," *Journal of Analytical chemistry,* vol. 84, pp. 685–707, 2012.

[19]. Ü. Anik, and S. Çevik, "Double-walled carbon nanotube based carbon paste electrode as xanthine biosensor," *Journal of Microchimica Acta,* vol. 166, pp. 209–213, 2009.

[20]. S. Timur, U. Anik, D. Odaci, and L. Gorton, "Development of a microbial biosensor based on carbon nanotube (CNT) modified electrodes," *Journal of Electrochemistry Communications,* vol. 9, pp. 1810–1815, 2007.

[21]. Q. Lu, T. Su, Z. Shang, D. Jin, Y. Shu, Q. Xu, and X. Hu, "Flexible paper-based Ni-MOF composite/AuNPs/CNTs film electrode for HIV DNA detection," *Journal of Biosensors and Bioelectronics,* vol. 184, pp. 113229, 2021.

[22]. C. W. Kung, Y. S. Li, M. H. Lee, S. Y. Wang, W. H. Chiang, and K. C. Ho, "In situ growth of porphyrinic metal–organic framework nanocrystals on graphene nanoribbons for the electrocatalytic oxidation of nitrite," *Journal of Materials Chemistry A,* vol. 4, pp. 10673–10682, 2016.

[23]. Q. Q. Zhu, H. W. Zhang, R. Yuan, and H. He, "Ingenious fabrication of metal–organic framework/graphene oxide composites as aptasensors with superior electrochemical recognition capability," *Journal of Materials Chemistry C,* vol. 8, pp. 15823–15829, 2020.

[24]. T. Kitao, Y. Zhang, S. Kitagawa, B. Wang, and T. Uemura, "Hybridization of MOFs and polymers," *Journal of Chemical Society Reviews,* vol. 46, pp. 3108–3133, 2017.

[25]. P. Ling, J. Lei, L. Zhang, and H. Ju, "Porphyrin-encapsulated metal–organic frameworks as mimetic catalysts for electrochemical DNA sensing via allosteric switch of hairpin DNA," *Journal of Analytical chemistry,* vol. 87, pp. 3957–3963, 2015.

[26]. P. Ling, J. Lei, and H. Ju, "Porphyrinic metal-organic framework as electrochemical probe for DNA sensing via triple-helix molecular switch," *J ournal of Biosensors and Bioelectronics,* vol. 71, pp. 373–379, 2015.

[27]. Y. Haldorai, S. R. Choe, Y. S. Huh, and Y. K. Han, "A composite consisting of microporous carbon and cobalt (III) oxide and prepared from zeolitic imidazolate framework-67 for voltammetric determination of ascorbic acid," *Journal of Microchimica Acta,* vol. 185, pp. 1–10, 2018.

[28]. M. Çubukçu, F. N. Ertaş, and Ü. Anık, "Centri-voltammetric determination of glutathione," *Journal of Microchimica Acta,* vol. 180, pp. 93–100, 2013.

[29]. H. Wang, Y. Jian, Q. Kong, H. Liu, F. Lan, L. Liang, S. Ge, and J. Yu, "Ultrasensitive electrochemical paper-based biosensor for microRNA via strand displacement reaction and metal-organic frameworks," *Journal of Sensors and Actuators B: Chemical,* vol. 257, pp. 561–569, 2018.

[30]. S. Mohammadi, and G. Khayatian, "Highly selective and sensitive photometric creatinine assay using silver nanoparticles," *Journal of Microchimica acta,* vol. 182, pp. 1379–1386, 2015.

[31]. S. M. El-Sheikh, D. I. Osman, O. I. Ali, W. G. Shousha, M. A. Shoeib, S. M. Shawky, and S. M. Sheta, "A novel Ag/Zn bimetallic MOF as a superior sensitive biosensing platform for HCV-RNA electrochemical detection," *Journal of Applied Surface Science,* vol. 562, pp. 150202, 2021.

[32]. L. Zhang, C. Ye, X. Li, Y. Ding, H. Liang, G. Zhao, and Y. Wang, "A CuNi/C nanosheet array based on a metal–organic framework derivate as a supersensitive non-enzymatic glucose sensor," *Journal of Nano-micro letters,* vol. 10, pp. 28, 2018.

[33]. D. Cheng, X. Xiao, X. Li, C. Wang, Y. Liang, Z. Yu, C. Jin, N. Zhou, M. Chen, and Y. Dong, "A non-enzymatic electrochemical sensing platform based on hemin@ MOF composites for detecting hydrogen peroxide and DNA," *Journal of The Electrochemical Society,* vol. 165, pp. B885, 2018.

[34]. P. Ling, J. Lei, L. Jia, and H. Ju, "Platinum nanoparticles encapsulated metal–organic frameworks for the electrochemical detection of telomerase activity," *Journal of Chemical Communications,* vol. 52, pp. 1226–1229, 2016.

[35]. J. Chen, C. Yu, Y. Zhao, Y. Niu, L. Zhang, Y. Yu, J. Wu, and J. He, "A novel non-invasive detection method for the FGFR3 gene mutation in maternal plasma for a fetal achondroplasia diagnosis based on signal amplification by hemin-MOFs/PtNPs," *Journal of Biosensors and Bioelectronics,* vol. 91, pp. 892–899, 2017.

[36]. G. Yuan, L. Wang, D. Mao, F. Wang, and J. Zhang, "Voltammetric hybridization assay for the β1-adrenergic receptor gene (ADRB1), a marker for hypertension, by using a metal organic framework (Fe-MIL-88NH 2) with immobilized copper (II) ions," *Journal of Microchimica Acta,* vol. 184, pp. 3121–3130, 2017.

[37]. X. Li, X. Li, D. Li, M. Zhao, H. Wu, B. Shen, P. Liu, and S. Ding, "Electrochemical biosensor for ultrasensitive exosomal miRNA analysis by cascade primer exchange reaction and MOF@ Pt@ MOF nanozyme," *Journal of Biosensors and Bioelectronics,* vol. 168, pp. 112554, 2020.

[38]. T. R. Soreta, J. r. Strutwolf, O. Henry, and C. K. O'Sullivan, "Electrochemical surface structuring with palladium nanoparticles for signal enhancement," *Journal of Langmuir,* vol. 26, pp. 12293–12299, 2010.

[39]. Y. Li, C. Yu, B. Yang, Z. Liu, P. Xia, and Q. Wang, "Target-catalyzed hairpin assembly and metal-organic frameworks mediated nonenzymatic co-reaction for multiple signal amplification detection of miR-122 in human serum," *Journal of Biosensors and Bioelectronics,* vol. 102, pp. 307–315, 2018.

[40]. P. Shi, Y. Zhang, Z. Yu, and S. Zhang, "Label-free electrochemical detection of ATP based on amino-functionalized metal-organic framework," *Journal of Scientific reports,* vol. 7, pp. 1–7, 2017.

[41]. M. Hu, L. Zhu, Z. Li, C. Guo, M. Wang, C. Wang, and M. Du, "CoNi bimetallic metal–organic framework as an efficient biosensing platform for miRNA 126 detection," *Journal of Applied Surface Science,* vol. 542, pp. 148586, 2021.

[42]. H.-W. Zhang, Q.-Q. Zhu, R. Yuan, and H. He, "Crystal engineering of MOF@ COF core-shell composites for ultrasensitively electrochemical detection," *Journal of Sensors and Actuators B: Chemical,* vol. 329, pp. 129144, 2021.

[43]. Y. Jian, H. Wang, F. Lan, L. Liang, N. Ren, H. Liu, S. Ge, and J. Yu, "Electrochemiluminescence based detection of microRNA by applying an amplification strategy and Hg (II)-triggered disassembly of a metal organic frameworks functionalized with ruthenium (II) tris (bipyridine)," *Journal of Microchimica Acta,* vol. 185, pp. 1–8, 2018.

[44]. H. Shao, J. Lu, Q. Zhang, Y. Hu, S. Wang, and Z. Guo, "Ruthenium-based metal organic framework (Ru-MOF)-derived novel Faraday-cage electrochemiluminescence biosensor for ultrasensitive detection of miRNA-141," *Journal of Sensors and Actuators B: Chemical,* vol. 268, pp. 39–46, 2018.

26 MOF-based Electrochemical Sensors for Neurochemicals

Suma B Patri[1], *Supritha M Karekuladh*[1], *Pandurangappa Malingappa*[1]
[1]Department of Chemistry, Bengaluru City University, Central College Campus, Bengaluru, India

26.1 INTRODUCTION

Neurochemicals (NCs) are chemical substances present in brain fluid that will regulate thoughts, emotions, transmit signals, and promote growth, reconstruction, and restoration of cells of the central nervous system (CNS). These are vital components and their regulated levels in the blood flow affect the proper functioning of the glands and organs in the body. The role of NCs is mainly interrelated with their structure and concentration in the bloodstream [1].

NCs are categorized into various types depending on their structure, chemical composition, and their role in metabolism. Some neurochemicals are neurotransmitters (NTs) and neuropeptides which control brain functions [2]. NTs are endogenous molecules distributed in the CNS, brain cells, tissues, and body fluids of mammals. They help in transmitting electrical impulses between nerve cells, muscles, organs, or other tissue, thereby acting as messengers of neurological information [3]. NTs are packed and stored inside the synaptic vesicles and synapses are generally released upon electrical stimulation from a neuron into the synaptic cleft through exocytosis, activating postsynaptic receptors and thus kicking off a series of signaling pathways in neurons [4].

Acetylcholine was the first neurotransmitter discovered by Otto Lowei in 1921. After that, around 100 different NTs were discovered [5]. One of the important groups of NTs is catecholamines which include histamine, dopamine, serotonin, epinephrine, and norepinephrine [2].

Dopamine (DA) belongs to catecholamine-based NTs present in the central nervous system of mammals. It regulates numerous brain-related functions such as memory, cognition reward, motivation, proper metabolism, and functioning of the CNS, cardiovascular, hormonal, and renal systems [6, 7]. It is produced in the substantia nigra, the section of the brain which is positioned deep in the brain stem [5] and adrenal medulla. It is important to monitor the levels of dopamine in the body fluids because, increase in its level can cause depression, whereas the deficiency can cause various neurological related disorders [8]. The concentration of dopamine levels could be measured in various biological fluids such as saliva, plasma, urine, or bloodstream. The normal concentration is approximately 10-50 ng/L [9]. Among the various protocols, the first electrochemical studies on the oxidation mechanism of NTs, in other words, catecholamines (dopamine, norepinephrine) were studied by Ralph Adams and his team in 1967, at the University of Kansas [6]. Dopamine can be easily oxidized to form dopamine o-quinone by 2 electron transfer; these electrons will generate a current response depending upon dopamine concentration [10]. Undesirable for electroanalysis, due to the slow interfacial transfer of electrons at the electrode interface as well as the existence of interferents like uric acid (UA) and ascorbic acid (AA) which coexist with the target analyte, dopamine has a poor and sluggish response at the conventional electrode surface. A variety of modified electrodes fabricated using materials such as metal nanoparticles [6], carbon nanomaterials [7], and its composites [11] have been reported in the trace level measurement of DA [12].

Norepinephrine (NE) is another significant neurotransmitter belonging to the catecholamine group secreted in the mammalian CNS. It is also essential for learning, memory, attention, focus and the sleep cycle, hence it is consumed as drugs by athletes to improve their performance in competitive games. It helps to increase the energy production in the body by the conversion of glycogen to glucose and fats into fatty acids in the liver. Abnormalities of NE concentration levels may lead to many severe health conditions [13].

Epinephrine (EP) is another neurotransmitter synthesized in the adrenal medulla and sympathetic nerve terminals and is essential for the functioning of the nervous system, hormonal, renal, and cardiovascular system. It was first discovered in 1901 by Takamine and Aldrichand which was synthesized by Stolz and Dalkin in 1904 for the first time [13, 14]. Measurement of EP is essential in the diagnosis and treatment of Parkinson's disease. Hence, the quantitative measurement of EP in human body fluids has become even more significant.

Serotonin is a necessary NT related to anxiety and behavior. It facilitates several non-neuro-related processes [15]. An abnormal level of serotonin is linked with health ailments. Electrochemical measurement of serotonin includes voltammetric techniques like CV, DPV, and SWV. The redox mechanism of serotonin involves the transfer of

DOI: 10.1201/9781003188148-26

TABLE 26.1
Structure of Selective Neurochemicals

Analyte	IUPAC name	Structure
Dopamine	4[2-Amino ethyl benzene 1, 2-diol]	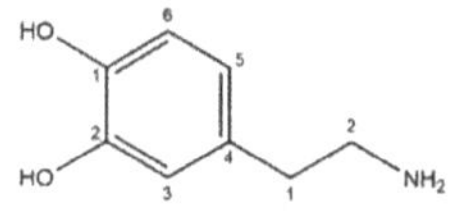 Dopamine
Serotonin	[3 (2-Aminoethyl)-1H-indol-5-ol]	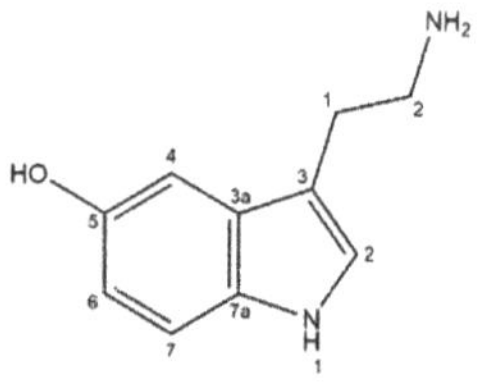 Serotonin
Epinephrine	4[1-R(1-Hydroxy-2-(methylamino) ethyl) benzene-1,2-diol]	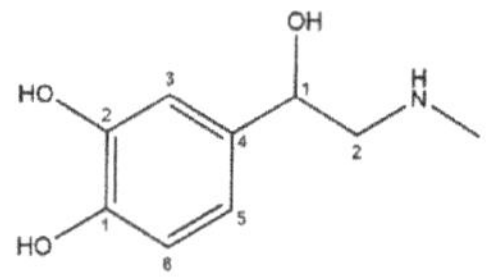 Epinephrine
Norepinephrine	4[1-R(2-amino-1-hydroxyethyl) benzene-1, 2- diol	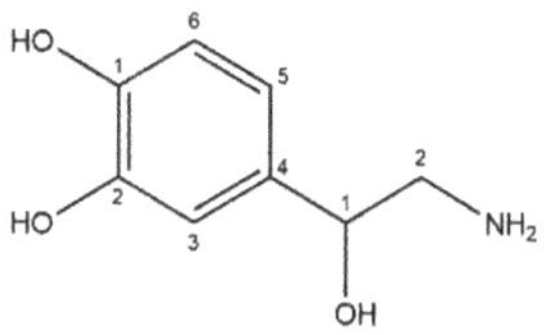 Norepinephrine
Xanthine	[3,7-Dihydro-1H-purine-2,6-dione]	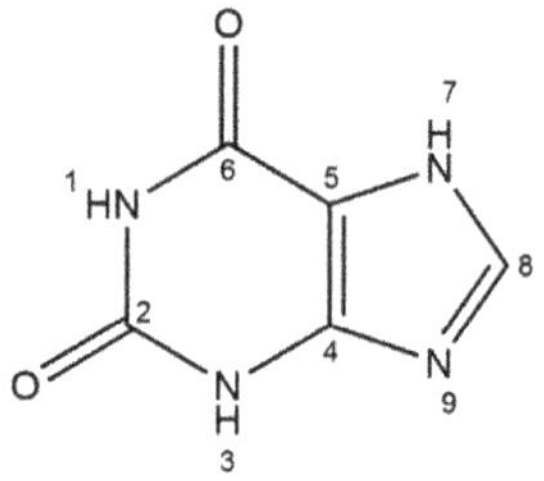 Xanthine
Hypoxanthine	[1,9-Dihydro-6H-purin-6-one]	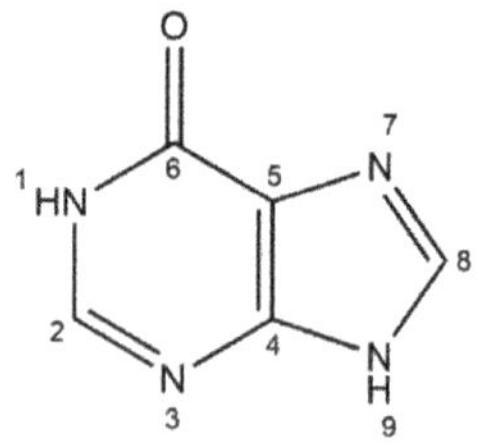 Hypoxanthine

TABLE 26.1 (Continued)

Analyte	IUPAC name	Structure
Choline	[2-Hydroxy-N,N,N-trimethylethan-1-aminium]	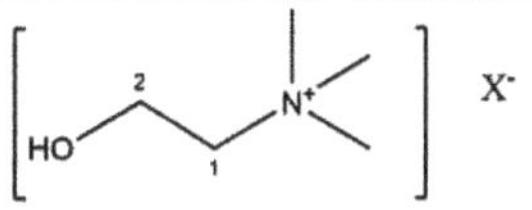 Choline
Acetylcholine	[2-Acetoxy-N,N,N-trimethylethanaminium]	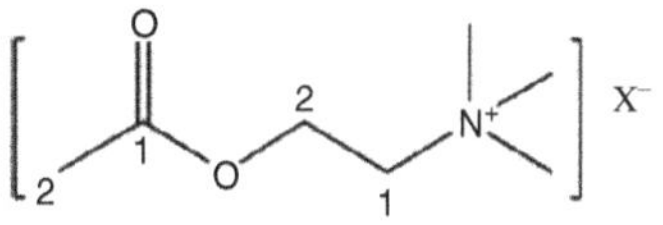 Acetylcholine

2 electrons and 2 protons to form a quinone derivative regardless of the substrate used [16].

Xanthine (XA) and **Hypoxanthine** (HX) are collectively known as oxypurines and obtained using the purine (ATP) degradation pathway and are precursors of uric acid [17]. Hence monitoring XA as well as HX levels in human body fluids is crucial for early clinical diagnosis [18]. It is very challenging to identify NTs in biological fluids because of their presence in trace level in blood [5].

In analytical chemistry, many tools are available for the analysis of neurotransmitters which include chromatography [19], capillary electrophoresis [20], flow injection analysis [21], fluorimetry [22], chemiluminescent [23], and spectroscopic [24] techniques. These conventional methods are laborious, time-consuming, require skilled personnel to operate the instruments which involve prolonged sample preparation before analysis. Because of these limitations, electrochemical sensing platforms offer superior advantages in terms of their simplicity, rapid, highly sensitive, and selective abilities, as alternatives in the trace level quantification of neurotransmitters [25].

An electrochemical sensor is a device that is capable of producing a useful quantitative analytical response in the form of a signal transformed from electrochemical information. A typical working electrochemical sensing device contains two major components: a chemical identification system, and a physicochemical transducer, which transforms the chemical responses into analytical signals that could be recognized easily and transformed by the instrument [1, 26]. In all of these sensing devices, the working electrode is the heart of the sensor. It should generate fast, selective, accurate, and sensitive responses quantitatively of the target analyte. This can be achieved by modification of the surface of the electrode by various functional materials which show better analytical performance. Many approaches are available for the modification of electrodes such as the deposition of nanomaterials which will act as redox mediators, compounds containing specific groups that can enable the charge transfer, and nanostructures with a large surface area to further amplify the sensitivity of the electrode [10].

Metal-organic frameworks (MOFs) which are referred to as coordination polymers or coordination networks are the hybrid molecules containing both inorganic ions and organic linkers. MOFs possess various structural uniqueness such as tunable pore size, large surface area, structural flexibility, controllable synthesis, thermal and mechanical stability [27-29]. Due to these characteristics, they find applications in several domains such as gas storage [30], separation [31], heterogeneous catalysis, ion exchange, supercapacitor [6], sensing [6] and drug delivery [32]. The first MOF was reported by Yaghi et al. in 1999. Modification of these can be carried out by using 3 strategies: dopant alteration, entrapping of nanoparticles in-situ, and post-modification methods which expand the potential for electrochemical application. MOF-based composites have high conductivity and catalyzing ability owing to the combined synergistic effects of both MOF as well as of guest or dopant molecules, thereby overcoming the defects of single MOFs [33].

Chemically modified electrodes (CMEs) with nanoparticles, enzymes, ionic liquids, polymers, or carbon substrates have permitted electrochemical measurement of neurochemicals with very low detection limits, namely, in the nM to µM range. Simultaneous determination of neurochemicals through electroanalytical techniques still marks a challenge, which has successfully been fulfilled through wide-ranging surface functionalization of recognized electrode materials and the development of hybrid functional composite materials.

The chemically modified electrode to sense neurochemicals should possess some characteristics, such as: i) inertness of modified electrode surface in the potential window under study, and ii) the electrode must demonstrate tunable

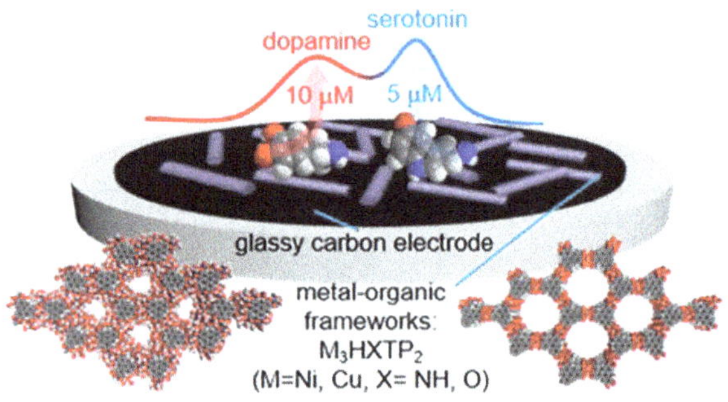

FIGURE 26.1 Voltammetric measurement of electroactive neurochemicals in a multianalyte sample solution using MOFs. Adapted with permission from [25]. Copyright (2020) American Chemical Society.

interfacial chemistry which could be readily optimized to receive fast and reversible electron transfer rates with target analytes which are quite sensitive to interface [34].

MOFs have several features that make them excellent substrate materials in electroanalysis. Despite having all these features, the use of pristine MOFs has limited applications due to their poor conductivity and instability in aqueous solutions. Hence functionalized MOFs can improve the redox kinetics by reducing the reduction/oxidation potentials of biomolecules in support of electrochemical detection strategy. To develop key chemical properties such as conductivity, mechanical stability and catalytic performance, various functional materials such as carbon and its allotropes, polymers, and metal nanoparticles have been utilized in combination with MOFs to prepare sensitive electrode materials and to fabricate electrodes in sensing applications [35, 36].

In this chapter, we have highlighted the utilization of MOFs and their composites in the electrochemical sensing of neurochemicals (NC). The sensing mechanism and quantification parameters of selective neurotransmitters have been discussed in detail. This chapter helps the readers to gain an overall view of various strategies used in the fabrication of electrochemical sensors.

26.2 DIFFERENT SENSING MATERIALS USED IN NEUROCHEMICAL QUANTIFICATION

26.2.1 MOF-based Carbon Paste Electrodes (CPE)

Carbon paste electrodes are usually prepared by mixing appropriate quantities of carbon powder (graphite) and binder using mortar and pestle into a paste. The modified CPE can be prepared either by mixing modifier molecules with graphite powder maintaining the optimum ratio of modifier, binder, and graphite powder, or by surface modification through electrochemical techniques. These CPEs are ideal working electrodes because of their low costs, easy availability, lower background current, and very broad potential window range. A range of indicator molecules can be added into carbon paste in the fabrication of CPEs [37]. The first cyclic voltammetric study of the catecholamine as a function of pH was carried out by R. N Adams et al. [38].

Yuzhi Li et al. have developed dopamine (DA) and uric acid (UA) biosensors comprising a CPE modified with metal-organic framework MIL-101 based on Cr(III) MOF with terephthalate (MIL-CPE). The peak current response was amplified with an increasing concentration of MIL-101 with carbon paste. The optimum ratio for mixing was 1:9 (w/w). The EIS studies showed a lower Rct value at the MIL modified CPE electrode. This increase in the conductance is due to the MOF consisting of l3-oxo bridged chromium(III)-trimers with a terephthalate cross-linking lattice structure. These linking groups provide the channelized way for electron transfer and thereby the conductivity of the MIL modified CPE increases. The electrocatalytic behavior of MIL-101 in the oxidative reaction of DA and UA showed a linear analytical response in the concentration range 5-250 μmol L^{-1} (R = 0.997) and 30-200 μmol L^{-1} (R = 0.995), for DA and UA molecules, respectively. The authors concluded that the reported Cr-terephthalate MOF which has 2 types of mesoporous cages with free internal diameters of 29 and 34 Å, an accessible pore size of 12 Å and 16 Å. This shows that the MOF has a high surface area and numerous coordination unsaturation sites (CUS). However, in aqueous media, the water molecules occupy these CUS, and the large pore volume of these materials may suggest a site in the electro-oxidation of DA and UA molecules. Additionally, the nodes present in the framework could act as catalytically active sites for sensing and biosensing applications [39].

The mesoporous Cr(III) MOF with superior constancy in an aqueous medium and its framework structure can bear either pre or post-synthetic alteration with a different functional motif at the place of the organic terephthalic acid linker. In this regard, Lu-Lu Gao et al. have reported sulfo functionalized MIL-101-graphite paste electrodes (GPE) for electrochemical sensing of dopamine. The electrode fabrication was carried out by doping the graphite paste electrode along with sulfo group-containing MIL-101. The functionalized MOF modified GPE electrode showed enhanced current response and superior sensitivity in the oxidation of DA with a very low limit of detection, namely, 43 nM. These interactions could further support DA assimilation and activation and the MIL-101, which forms hydrogen bonds with DA through the carboxylic groups of the organic linker molecule and aqua/hydroxyl groups present on metal clusters. The sulfo groups present in MOFs serve as mediators for the transfer of protons and thus provide easily accessible sites which favor bonding interactions with the target dopamine molecule and hence promote the reaction process [40].

26.2.2 Metal-metal oxide MOF Composites

Metal and metal oxide (MO) nanoparticles with different sizes and morphologies have been prepared through various methods. These metal oxide nanoparticles demonstrate

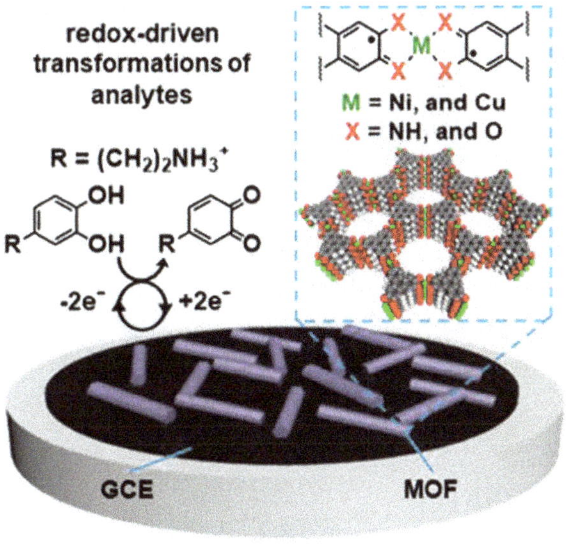

FIGURE 26.2 Schematic representation of MOF modified layered device architecture. Adapted with permission from [25]. Copyright (2020) American Chemical Society.

different electrical and photochemical properties owing to their size, morphology, stability, and high active surface area. This incorporation of metal nanoparticles into substrate provides hardening of the sensing interface which is conductive and enhanced catalytic properties allowing their expansion with other metals. The presence of metal or metal oxide nanoparticles enables the electrical contact of redox-centers in MOFs with the features of the transducer molecules [41].

Researchers are striving hard to develop greener alternatives in material chemistry. In this direction, biomass-derived carbon (BC) material has attracted a lot of interest which is attributed to its virtues of being environmentally friendly as well as its simplistic preparation procedures. They are cost-effective, have natural abundance, a very large surface area, and chemical stability to match their fit in various electrochemical applications. These materials are prepared by straightforward heating of biomass to very high temperatures to obtain a highly porous carbon structure. This bio-derived carbon contains other multifarious elemental particles which provide a highly conducive environment and also facilitate in exposing a large number of catalytically active sites for facile transfer of electrons across the interface.

Zhiwei Lu has reported a multianalyte electrochemical sensor modified with biocarbon/Co_3O_4/$FeCo_2O_4$ based composite for same time measurement of DA, AC, and XA. In this method, the carbonization of pinecones was first carried out to get biocarbon materials. Consequently, the porous biocarbon matrix aids in the in-situ growth of ZIF-67 to obtain the precursor material, and pyrolysis produces BC/Co_3O_4 nanocomposite. Later, the composite preparation was done by hydrothermal reaction followed by the process of calcination at elevated temperature. The authors reported that the composite modified electrode demonstrated exceptional electrical conductivity at reduced overpotential and higher oxidative peaks at BC/Co_3O_4/GCE due to multiple oxidative valence states, namely, Co^{3+}, Co^{2+}, Fe^{3+}, and Fe^{2+}. The large electrode surface area is due to the combined synergistic property of BC, MOF- derived Co_3O_4, and $FeCo_2O_4$. The fabricated sensor displayed good linearity with an increase in its peak currents against an increasing concentration range from 0.1 - 250 µM, 0.1 - 220 µM, and 0.5 - 280 µM of DA, AC, and XA, respectively. The limits of detection are 0.045, 0.028, and 0.12 µM for dopamine, acetylcholine, and xanthine respectively [42].

Feroz Ahmad et al. have synthesized a Cu containing 3D MOF HKUST-1 which exhibits exceptional dispersion ability and sensing action in the electrochemical sensing of two biologically significant analytes such as acetaminophen (AC) and dopamine (DA). The presence of DA showed a pair of redox peaks at 0.250 V and a reduction peak was observed at 0.125 V at a potential scan rate of 100 mVs^{-1}, when scanned in the reverse direction. The limit of detection of 0.113 µM was obtained at neutral pH [43].

Jingjing Guo et al. have reported a simple, environmentally benign, and label-free fluorescent approach for the highly responsive and selective measurement of choline and acetylcholine. The method relies on bifunctional MIL-101(Fe) nanozyme (nanomaterials with enzyme-like activity). The nanozyme had twin functions of mimeting peroxidase and fluorescent emission. The fluorescence sensing approach engages catalyzing the hydrolysis of ACh into choline by the reaction of acetylcholine esterase (AChE) which sequentially gets oxidized by choline oxidase (ChOx) to generate H_2O_2. The unstable peroxide generated due to the catalytic effect of MIL-101(Fe) nanozyme undergoes decomposition into reactive hydroxyl radicals. Hence the organic ligand terephthalic acid which is non-fluorescent in MIL-101(Fe) undergoes oxidization by the generated –OH radicals to form fluorescent 2-hydroxyterephthalic acid. Due to this, a label-free detection mechanism in the determination of choline and ACh via a multiple enzyme cascade reaction was recognized with very low detection limits of 20.0 and 8.9 nM, correspondingly. The proposed biosensor was validated by measuring choline and ACh level present in milk and human plasma sample matrices [44].

Shuisheng Hu et al. have synthesized two-dimensional amino-functionalized Cu-MOF nanosheets (NH_2-CuMOF) via a bottom-up approach. The NH_2-CuMOF nanosheet exhibited a peroxidase mimic property and also displayed an exceptional fluorescence property. Hence, the MOF was used in the detection of HXA to evaluate the freshness of fish. In the quantitative measurement, the fluorescence intensity steadily reduced with the increase of HXA concentration in the range 10 - 2000 µM and the limit of detection (LOD) was 3.93 µM (S/N = 3) [45].

In another method, Cr MOF, MIL-101(Cr) having a permeable cage structure and large specific surface area was chosen as the host framework for loading Pt nanoparticles to construct a non-enzymatic XA sensor. The electrochemical characterization of bimetallic MOF, Pt@MIL-101(Cr)

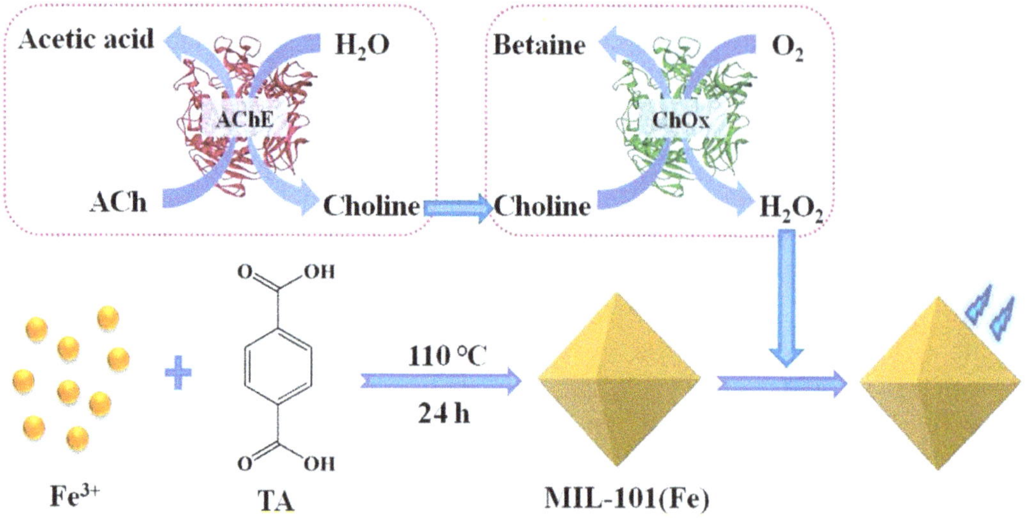

FIGURE 26.3 Schematic representation of fluorescence sensor using bifunctional MIL-101(Fe) nanozyme in the determination of choline and ACh. Adapted with permission from [44]. Copyright (2020) Elsevier.

bespoke electrode revealed the superiority of the modified electrode in terms of conductivity and also its efficiency in the oxidation of XA at physiological pH range. The actual sensing mechanism involved is the Pt@MIL-101(Cr) modified interface has abundant unsaturated Cr(III) sites. These sites may preferably act as catalytically active sites to further progress the electrochemical activity and also the large active electrode surface area of MIL-101(Cr) modified electrode supplies additional active sites for the immobilization of XA molecule. The Pt@MIL-101(Cr) has homogeneously discrete Pt nanoparticles with large surface area and the enhanced electrocatalytic performance is also due to the smaller sized nanoparticles with a rise in the number of edges and corner atoms. The sensitivity of the sensor has been reported as 1.88 μA μM^{-1} cm^{-2} and the detection limit was found to be 0.42 μM [18].

The assimilation of noble metal nanoparticles (NPs) upon MOF lattices for the sensing application utilizes the inherent properties of the MOF, namely, highly porous structures, which restricts the agglomeration of metal NPs, thereby increasing electrocatalytic performance. Shamim Ahmed Hira et al. have fabricated a dopamine sensor by incorporating AgPd bimetallic alloy nanoparticles into a Zr-MOF to prepare an AgPd@Zr-MOF modified electrode. The MOFs together with metal NPs embedded into their structure are admitted to synergistically boost the electrocatalytic action of the individual constituents.

Metal nanoparticles adorned MOFs significantly exhibit superior performance which includes superior electrical conductivity and exceptional electrocatalytic behavior. Shun Lu et al. have successfully fabricated the Au@ZIF-8 based nanocomposite drop coated onto a pre-treated glassy carbon electrode (GCE) in the measurement of dopamine. The good selectivity, repeatability, and stability of the MOF modified electrode in the determination of the DA molecule are attributed to the synergistic effect of ZIF- 8 combined Au NPs with a larger surface area [21].

Kaiming Ge et al. have reported a dopamine sensor that is based on a luminescent Tb-functionalized MOF. The possible sensing mechanism by the MOF was investigated in detail for the detection of DA. The Ln-MOFs luminescence quenching process shows that the competitive mechanism of excitation by light absorption is ascribed to the antagonism between the analyte and the ligand molecule for light energy absorption lacking electron exchange thereby lowering the efficiency of transfer of energy from ligand to lanthanide and quenching the luminescence property of Ln^{3+}. The competition processes rely upon the overlapping of absorption bands in the analyte and ligand. The detection limit was found to be 0.06 μM for DA [46].

Similarly, Zixu Huang et al. reported a strategy for the electrochemical detection of dopamine based on an Ni-based MOF (Ni-MOF). In this method, the Nickel MOF was prepared by using ionic liquid as a template via an ionothermal method. Initially, dopamine gets engrossed onto the Ni-MOF surface. During the process of oxidation, Ni(II)-MOF oxidizes to Ni(III)-MOF. Then, the pre-adsorbed dopamine undergoes oxidation by the reduction of the produced Ni(III)-MOF, and it is regenerated to Ni(II)-MOF continuously. The analytical current responses at the Ni-MOF coated GCE for different concentrations of dopamine showed proportionality in the concentration range 0.2 - 100 μmol L^{-1}, with a limit of detection 60 nmol L^{-1}. The feasibility of the sensor has been evaluated by measuring dopamine levels in dopamine hydrochloride injections with a recovery in the range 95.9 - 104.3 % showing good probability and consistency of the Ni-MOF modified GCE in trace level measurement of dopamine [47].

Ke-Yang Wu et al. have proposed a visible and label-free fluorescent sensor for DA detection. This facile strategy utilizes a water-stable MOF, in other words, ($H_3CmdcpBr$ =

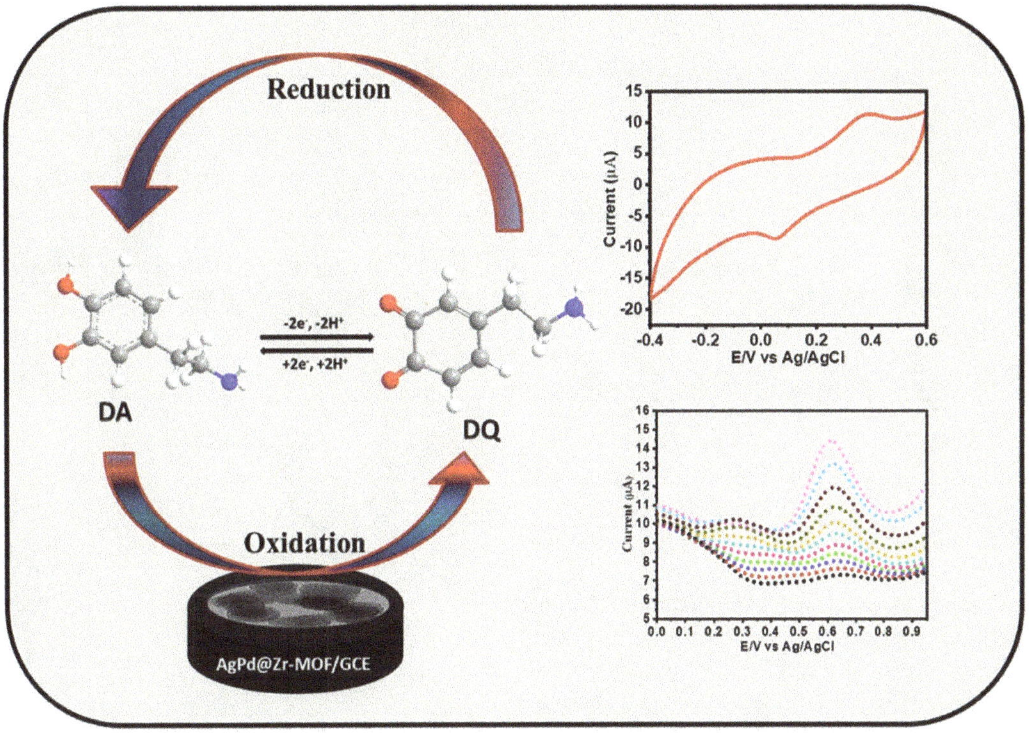

FIGURE 26.4 Voltammetric detection of dopamine at bimetallic alloy MOF composite electrode. Adapted with permission from [6]. Copyright (2021) Elsevier.

N-carboxymethyl-(3, 5-dicarboxyl) pyridinium bromide) for dopamine sensing. The sensing mechanism showed that DA undergoes the polymerization process to give polydopamine (pDA). This pDA binds to the MOF's surface and thus gets quenched its green luminescence. Due to the partial overlapping of the excitation spectrum with the absorption spectrum of pDA, the sensing procedure is visible to our naked eyes under UV light irradiation at 365 nm. The proposed strategy has been applied to biological fluids analysis with a detection limit of 0.41 μM [48].

26.2.3 Carbon Modified MOFs

Several MOF-related carbon composite materials have been developed to increase the electrochemical behavior as well as the electrocatalytic performance of the target analyte through the combination of MOFs along with substrate materials with good conductivity and high mechanical stability. Hence the carbon substrate containing MOF-related electrochemical biosensor may expand the utility in the quantitative measurement of a variety of neurochemicals from real sample matrices.

Yang et al. in a typical approach, have functionalized carbon spheres with MOF (C/Al-MIL-53-$(OH)_2$) and immobilized them onto the surface of GCE using Nafion polymer. Nafion/C/Al-MIL-53-$(OH)_2$ modified electrode has been explored as a new electrocatalyst in dopamine (DA) oxidation using phosphate buffer (PBS). The explanation for the improved electrochemical performance of this sensor may be attributed due to the C/Al-MIL-53-$(OH)_2$/GCE which was drop cast with a negatively charged Nafion film. These modifications might fascinate and accumulate the cationic DA to some extent through an ion-exchange model, caused in the enhancement of peak current response and selectiveness towards the target analyte, namely, dopamine. Under optimized conditions, the peak current increased linearly with increasing DA level, in the range 3.0×10^{-8}–1.0×10^{-5} mol L^{-1} with a limit of detection 0.8×10^{-8} mol L^{-1} [7].

The direct use of sole component MOFs is limited, as they display lower conductivity, less mechanical stability, and lower electrocatalytic activity. To prevail over these limitations, Min-Qiang Wang et al. have manganese-based metal-organic frameworks (Mn-BDC) embedded with MWCNTs through a one-pot solvothermal synthetic route. They revealed that the incorporation of MWCNTs into the composite can instigate the fragmentation of bulky Mn-BDC into thin layers by increasing the surface area. This sensor displayed excellent sensing performance and the practicability of Mn-BDC and Mn-BDC@MWCNT/GCE were evaluated using the differential pulse voltammetry (DPV) technique for multianalyte measurement of ascorbic acid, dopamine, and uric acid. The linear responses were observed in the concentration range 1- 600, 0.1- 420, and 0.1- 600 μM with detection limits of 0.25, 0.024 and 0.023 μM, respectively [49].

Among various carbon substrate materials reported, macroporous carbon (MPC) has better thermal and mechanical stability which is used for several applications. MPC possesses good electrical conductivity with a large specific surface area providing room to load MOFs. In this

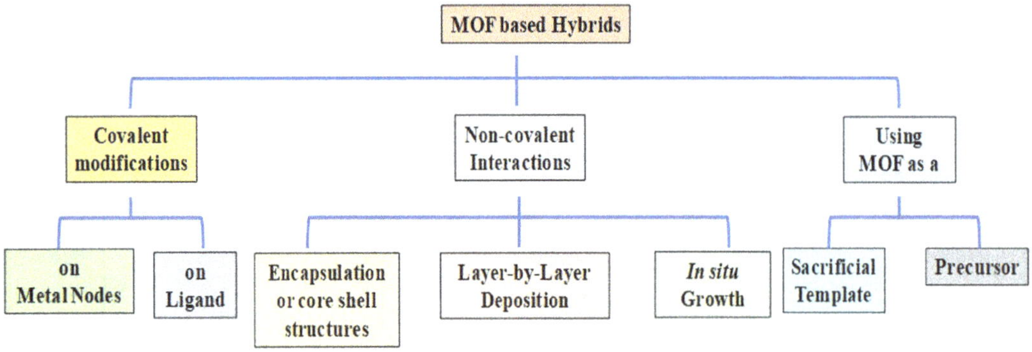

SCHEME 26.1 Different modes of MOF hybrid material preparation.

direction, a new Zr-Porphyrin MOF/MPC composite was synthesized by a single-step solvothermal reaction for the first time by Duanduan Yin et al. The Zr-Por MOF/MPC/GCE composite modified electrode was carried out using uric acid (UA), xanthine (XA) and hypoxanthine (HX) as target molecules to check the electrocatalytic oxidative behavior of this composite at physiological pH conditions. The incorporation of MPC might avert the aggregation of Zr-Por MOF particles thereby increasing the specific surface area of the resulting composite. This combination of the Zr-Por MOF with MPC greatly improves the electrocatalytic competence and overcomes the limitations of MOFs in the detection of neurochemicals. Good linearity for these molecules 5-350 μM (XA) and 5-340 μM (HX) with detection limits of 5.6×10^{-8} and 4.5×10^{-7} M were obtained respectively [50].

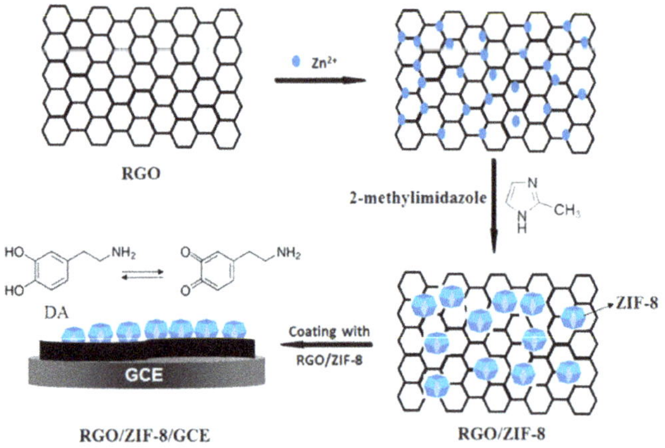

FIGURE 26.5 Synthesis and fabrication of RGO/ZIF-8/GCE for DA detection by voltammetry. Adapted with permission from [54]. Copyright (2017) Elsevier.

26.2.4 Hybrid MOF Composites

Hybridizing nanostructured materials with metal-organic frameworks is a great idea and is promising for the improvement of electrochemical characteristics of the sensor [51]. To date, various hybrid composites have been reported in the literature. The different methods of preparing hybrid materials have been shown in Scheme 1. [52].

Srinivasan et al. have reported an electrochemical sensing platform comprising mesoporous nanohybrids along with core-shell MOFs decorated with mesoporous Ag-doped TiO_2-SnO_2 nanohybrids on g-C_3N_4 nanosheets in the simultaneous measurement of ascorbic acid, dopamine, and uric acid. The hierarchical g-C_3N_4/NC@GC/h-ATS modified ITO electrode has been exploited to fabricate an electrochemical biosensor. The graphitic-C_3N_4 is a layered polymeric material that acts as a metal-free semiconductor. This material can efficiently increase the rapid charge-carrier transportability between inner and outer surfaces along with other large surface areas and higher nitrogen content. The oxidative peak currents of the target analyte such as ascorbic acid, dopamine, and uric acid increased linearly with an increase in the concentration. All these molecules provided good linearity ranges comprising 0.1- 100, 2.5 - 500, and 2.5 - 625 μM for ascorbic acid, dopamine, and uric acid, respectively. The g-C_3N_4/NC@GC/h-ATS functionalized indium tin oxide electrode showed detection limits of 0.01, 0.05, and 0.06 μM for ascorbic acid, dopamine, and uric acid molecules under optimized experimental conditions by a differential pulse voltammetric study [53].

Among different conducting materials reported, graphene is one of the most widely used materials owing to its high conductivity, high electron transfer rate, excellent physicochemical, thermal, and mechanical stability.

Gege Yu et al. have synthesized a hybrid nanocomposite by the RGO-templated growth of ZIF-8 (reduced graphene oxide/zeolitic imidazolate framework-8). The RGO/ZIF-8 modified glassy carbon electrode showed enhanced oxidative and reductive peak currents for dopamine which may be ascribed to the synergistic effects of high electrical conductivity of reduced graphene oxide and large surface area with good porosity of ZIF-8 in the composite. RGO/ZIF-8/GCE showed improvised selectivity and sensitivity in the electrochemical measurement of dopamine in the concentration range 1.0×10^{-7}–1.0×10^{-4} M with a detection limit of 3.0×10^{-8} M and sensitivity of 0.1527 μA μM^{-1} [54].

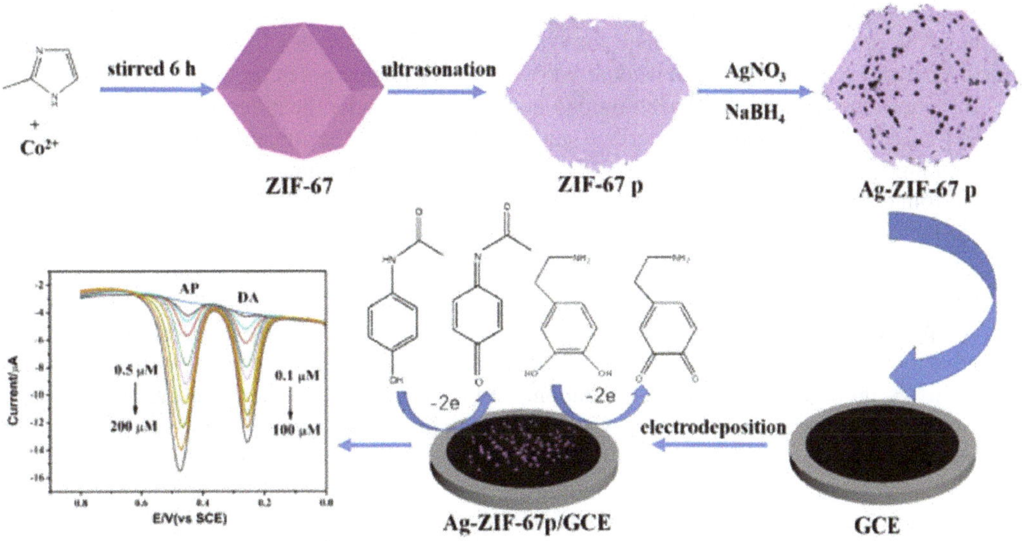

FIGURE 26.6 Schematic representation of fabrication of AP and DA sensing electrode modified with Ag-ZIF-67 nanopinnas. Adapted with permission from [57]. Copyright (2020) Elsevier.

Fariba et al. have fabricated a selective electrochemical platform to measure catechol using Cu-MOF/ZnTeNRs/AuNPs hybrid nanocomposite material. The electrode was fabricated using multilayer modification on the surface of a glassy carbon electrode with a copper metal-organic framework (Cu-MOF), ZnTe nanorods, and gold nanoparticles (AuNPs). ZnTe nanorods and gold metal nanoparticles were integrated into a copper MOF material. This composite comprising Cu-MOF/ZnTe NRs/Au NPs on the surface of the glassy carbon electrode demonstrated better electrocatalytic character and high electron transfer rate in the measurement of catechol. In this composite, Cu-MOF provides the required porosity, and the analyte catechol gets adsorbed onto the EDTA functionalized ZnTe nanorods through the bond formation with –COOH groups of EDTA. Then gold nanoparticles act as accelerators for the transfer of an electron between the analyte molecule and electrode interface. The fabricated sensor showed wide linearity in the concentration range 2.5×10^{-7}–3.0×10^{-4} M with a detection limit of 16 nM [55].

Kai Kang et al. have reported a novel type of artificial nanozyme in the measurement of dopamine molecules using Hemin-doped-HKUST-1 (HKUST-1, also known as MOF-199, a face-centered-cubic MOF containing nanochannels) as a redox mediator. Hemin-doped-HKUST-1 sensor was fabricated by a single-step hydrothermal method, which has been subsequently sandwiched with electrodeposited reduced graphene oxide (rGO) on the surface of GCE. It also serves as a support that prevents the collapse of Hemin-doped MOFs through π–π stack interaction during prolonged electrocatalytic measurement under an aqueous medium [56].

Jing Tang et al. have developed silver metal nanoparticles-ZIF-67 nano pinnas composite (Ag-ZIF-67p) by sonication and reducing $AgNO_3$ within ZIF-67 nano pinnas template. This platform exhibited excellent electrocatalytic performance during dopamine quantification which might be credited to the large surface area, the total number of active sites, and increased electrical conductivity of the Ag-ZIF-67p composite. This structure facilitates more active sites by the dispersion of silver nanoparticles in enhancing an electrocatalytic property as well as increased electron transfer rate at the electrode interface [57].

26.2.5 Polymer Blended MOFs

Most of the MOFs are insulative in nature, which hampers their electrochemical sensing applications. Hence incorporation of conductive polymers into MOFs has been an area of interest to study the enhanced analytical performance and improved material properties in these composites. Kunfeng et al. have prepared a polymer MOF composite using polypyrrole (PPy) which is a conductive polymer with good stability and acts as interconnected wires throughout the matrix thus increasing its conductivity in the composite. Molybdenum oxide-based 3D metal-organic frameworks with twisted helical channels $[Cu_3(trz)_2(MoO_4)_2]\cdot H_2O$ ($CuTRZMoO_4$), was prepared by using ammonium molybdate as a source for molybdenum and 1,2,3 triazole as a linker molecule. Mo exists in multiple oxidative states namely +2, +4, and +6. This enhancement may be attributed to the combined activity of $CuTRZMoO_4$ in facilitating several metallic sites within the structural motifs. The increased active surface area and the microenvironment for the movement of dopamine towards active centers in the composite, namely, $CuTRZMoO_4$@PPy-2 improves the magnitude of analytical signal intensity. [12].

Shamim A H et al. have fabricated a sensor in a facile manner using a copolymer-grafted metal-organic framework like UiO-66-NH_2@PANI-co-ANA synthesized by the polymerization route. This type of combination of MOF along with a polymer boosted electrochemical property was confirmed through an electrochemical characterization study. The square-wave voltammograms for dopamine at varying concentrations at UiO-66-NH_2@PANI-co-ANA/GCE modified interface gave good linearity in the range of 10-100 µM. The detection limit and sensitivity were found to be 0.3 µM and 1102.2 µAµM^{-1} cm^{-2} respectively [58].

Redox-active polyoxometalate like $V_{10}O_{28}$, has been immobilized onto a water-stable Zr metal-organic framework (NU-902). The composite $V_{10}O_{28}$@NU-902 has exhibited redox hopping behavior in an aqueous medium with an enhanced electrocatalysis property in dopamine oxidation. This kind of redox hopping behavior has been observed between the immobilized polyoxometalate units of $V_{10}O_{28}$@ NU-902 thin films under aqueous electrolyte conditions showing that these are electrochemically active in dopamine estimation [59].

26.2.6 Other MOF-based Approaches

Xing Ze Wang et al. have developed a highly sensitive protocol for the monitoring of Alzheimer's disease biomarkers like presenilin-1, amyloid β-protein (Aβ), and acetylcholine (ACh) cerebrospinal fluid. In this study, the authors have applied laser scanning confocal microscopy (LSCM) for the first time to explore the pathway through fluorescence resonance energy transfer (FRET) to propose a neurochemical sensing mechanism. A fluorescent dye like thioflavine T (ThT) has been embedded into an Er-MOF structure to fabricate the dual-emission ThT@Er-MOF ratiometric fluorescent sensor by a post-synthetic modification protocol. This composite enhanced the measurement of acetylcholine (ACh) with a limit of detection of 0.03226 nM [60].

Xie et al. have proposed a protocol to introduce luminescent nanoparticles as guest molecules with high luminous intensity into the non-luminescent MOFs to prepare highly luminescent MOFs (LMOFs) for selective application study. Due to the synergistic effects of carbon dots (CDs) and ZIF-8, biomolecules like dopamine can be accumulated in the sustainable pores prevailing within the ZIF-8 framework structure. Under the influence of basic sites present within the ZIF-8 moiety, the dopamine molecule separately gets oxidized to produce dopamine-quinone moieties. This dopamine-quinone moiety can be engrossed on the ZIF-8 framework through π-π interaction within an aromatic ring of ZIF-8. The electrostatic interaction among the dopamine-quinone molecules and carbon dots causes fluorescence quenching of carbon dots. The prepared LMOFs (CDs@ZIF-8) exhibited a significant fluorescence activity along with the accumulation of the target analyte species providing low detection levels of dopamine even in the presence of potential interferents like serum and urine from complex sample matrices [61].

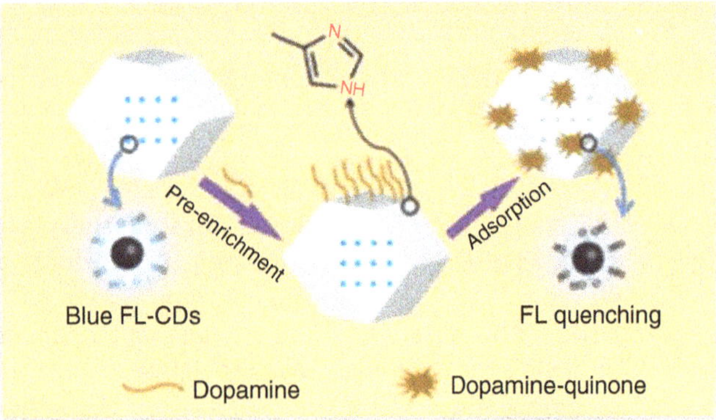

FIGURE 26.7 Schematic representation of carbon dots @ZIF-8 preparation and its application towards dopamine measurement. Adapted with permission from [61]. Copyright (2021) Elsevier.

26.3 CONCLUSIONS

Metal-organic framework based materials provide a sensitive approach in the electrochemical sensing of neurochemicals at very low concentration due to structural features, diversified oxidative states of metal ions as well as the porosity of the frameworks. The incorporation of conductive materials into these frameworks greatly enhances the analytical signal intensity of the electrochemical sensing platform in the measurement of different neurochemicals at trace levels.

ACKNOWLEDGMENT

Authors acknowledge the financial support from Vision Group on Science and Technology (VGST), Government of Karnataka, Bengaluru, India, and also the award of research fellowship to Supritha M. Karekuladh.

ABBREVIATIONS

NR-Not reported, [a]CDs- carbon dots, ZIF-8 –zeoliticimidazolate framework-8, GCE – glassy carbon electrode, BDC-benzene dicarboxylic acid, MWCNT- multiwalled carbon nanotubes, ZrPor /MOF- zirconium porphyrin /metal organic framework, MPC –macroporous carbon, ERG-electrochemically reduced graphene, bpy-4,4' biphenyldicarboxylic acid, CPE-carbon paste electrode, GPE-graphite paste electrode, NPs –nanoparticles, MOG-metal organic gel, IG-isophthalate functionalized graphene, ITO-indium- tin oxide, NC – nonporous nitrogen doped carbons, GC-graphitic carbon, SAMs –self assembled monolayers, BC-biomass carbon, TRZ- 1,2,3 triazine, PPy-polypyrrole, Cmdcp- N-carboxymethyl-(3,5-dicarboxyl) pyridinium bromide, ThT- thioflavine T, ANI-aniline, ANA-anthrnillic acid,

[b]DPV- differential pulse voltammetry, SWV- square wave voltammetry, CV- cyclic voltammetry,

[c]AA-ascorbic acid, DA-dopamine, UA-uric acid, XA-xanthine, HX-hypoxanthine, AP –acetaminophen, NA-noradrenalin, EPI-epinephrine

TABLE 26.2
An Overview of Various MOF-based Electrochemical Approaches Reported in the Literature.

Indicator material[a]	Technique[b]	Analyte[c]	pH range	Oxidation Potential (V)	Linear range (μM)	LOD (μM)	Real samples	[Ref]
CDs@ZIF-8	FL/ quenching	DA	7.0	NR	0.1–200	1.66×10^{-2}	Serum and urine	[61]
Nafion/C/Al-MIL-53-$(OH)_2$/ GCE	DPV	DA	5.0	0.2	0.03–10	0.008	Urine and serum	[7]
Mn-BDC@MWCNT/GCE	DPV	AA, DA, UA	7.5	–0.09 0.12 0.27	1–600, 0.1–420 & 0.1–600	0.25, 0.024 and 0.023	urine	[49]
ERG-UiO-67-bpy/GCE	DPV	NE, UA	6.0	0.256 0.35	4–40 0.5–70	0.030 0.035	urine	[62]
MIL-101/CPE	DPV	DA,UA	7.0	0.272 0.347	5–250 30-200	NR	NR	[39]
ZrO_2/graphene/chitosan/CPE	SWV	DA	7.4	NR	NR	11.03×10^{-3}	NR	[63]
Cu-MOF/ZnTe/Au/GCE	DPV	CC	1.09	0.501	2.5×10^{-7}– 3×10^{-4} m	0.016	Water & Tea	[55]
Ag-ZIF-67	DPV	DA, AC	7.0	0.257 0.47	0.1–100 0.5–200	0.05 0.2	NR	[57]
MOG@IG	SWV	DA	7.0	NR	NR	0.439	NR	[64]
UIO-66-NO_2@XC-72	DPV	AA,DA,UA	6.0	0.015 0.23 0.38	0.2–3.5 0.03–2 0.8–22	0.12 0.005 0.03	injection & urine	[65]
ITO/gC_3N_4/NC@GC/h-ATS	DPV	AA,DA,UA	7.0	NR	0.1–200 2.5–100 2.5–500	0.02 0.01 0.06	Urine	[56]
Au/SAMs/AuNPs	NR	NA, EPI, DA	NR	NR	10–60	NR	NR	[66]
CeO_2/siloxene/GCE	DPV	DA	7.0	NR	0.292–7.8	0.292	NR	[28]
HKUST-1/GCE	DPV	AC, DA	7.0	0.425 0.250	12.5–275 12.5–175	0.092 0.113	NR	[43]
Tb functionalized MOF	LU	DA	NR	NR	NR	0.06	Plasma & urine	[46]
BC/Co_3O_4 /$FeCo_2O_4$/GCE	DPV	DA AC XA	7.0	0.3	0.1–250 0.1–220 0.5–280	0.045 0.028 0.120	Human serum	[42]
CuTRZMoO_4@PPy-2/GCE	DPV	DA	2.5	NR	1–100	0.08	Human serum	[12]
Ni-MOF/GCE	DPV	DA	7.0	0.1	0.2–100	0.06	Dopamine. HCl injection	[47]
[Tb(Cmdcp) $(H_2O)_3$]2$(NO_3)_2 \cdot 5H_2O$}n	LU	DA	7.4	NR	NR	0.41	Urine & serum	[48]
ThT@Er-MOF	FL	Ach	7.0	NR	NR	3.22×10^{-5}	NR	[60]
Pt@MIL-101(Cr)	DPV	XA	7.4	0.68	0.5–162	0.42	Human serum	[18]
UiO-66-NH_2@P(ANI-co-ANA)	CV, SWV, CA	DA H_2O_2	7.02	0.38 -0.35	10–100 25–500	0.3 0.6	Urine& milk	[58]
$V_{10}O_{28}$@NU-902	CV, AM	DA	4.5	0.6	25–400	2.1	NR	[59]

REFERENCES

[1]. A. Azzouz, K.Y. Goud, N. Raza, E. Ballesteros, S.E. Lee, J. Hong, A. Deep, K.H. Kim, Nanomaterial-based electrochemical sensors for the detection of neurochemicals in biological matrices, *TrAC Trends in Analytical Chemistry*, (2019) 110 15–34.

[2]. T. Heinbockel, A.B. Csoka, Introductory Chapter: The Chemical Basis of Neural Function and Dysfunction, in: Neurochemical basis of brain function and dysfunction, *IntechOpen*, 2019.

[3]. N.F. Atta, H. Ekram, Y.M. Ahmed, A. Galal, Determination of some neurotransmitters at cyclodextrin/ionic liquid crystal/graphene composite electrode, *Electrochimica Acta*, (2016) 199, 319–331.

[4]. T. Pradhan, H.S. Jung, J.H. Jang, T.W. Kim, C. Kang, J.S. Kim, Chemical sensing of neurotransmitters, *Chemical Society Reviews*, (2014), 43, 4684–4713.

[5]. B. Si, E. Song, Recent advances in the detection of neurotransmitters, *Chemosensors*, (2018),6, 1.

[6]. S.A. Hira, S. Nagappan, D. Annas, Y.A. Kumar, K.H. Park, NO_2-functionalized metal–organic framework incorporating bimetallic alloy nanoparticles as a sensor for efficient electrochemical detection of dopamine, *Electrochemistry Communications*, (2021) 125, 107012.

[7]. Y. Wang, H. Ge, G. Ye, H. Chen, X. Hu, Carbon functionalized metal organic framework/Nafion composites as novel electrode materials for ultrasensitive determination of dopamine, *Journal of Materials Chemistry B*, (2015), 3, 3747–3753.

[8]. Y. Cheng, J. Wu, C. Guo, X.-G. Li, B. Ding, Y. Li, A facile water-stable MOF-based "off–on" fluorescent switch for label-free detection of dopamine in biological fluid, *Journal of Materials Chemistry B*, (2017), 5, 2524–2535.

[9]. W. Nawrot, K. Drzozga, S. Baluta, J. Cabaj, K. Malecha, *A fluorescent biosensors for detection vital body fluids' agents, Sensors*, (2018), 18, 2357.

[10]. S. Lakard, I.-A. Pavel, B. Lakard, *Electrochemical Biosensing of Dopamine Neurotransmitter: A Review, Biosensors*, (2021), 11, 179.

[11]. A. Amini, S. Kazemi, V. Safarifard, Metal-organic framework-based nanocomposites for sensing applications. *A review, Polyhedron*, (2020), 177, 114260.

[12]. K. Zhou, D. Shen, X. Li, Y. Chen, L. Hou, Y. Zhang, J. Sha, Molybdenum oxide-based metal-organic framework/polypyrrole nanocomposites for enhancing electrochemical detection of dopamine, *Talanta,* (2020), 209, 120507.

[13]. N.G. Mphuthi, A.S. Adekunle, E.E. Ebenso, Electrocatalytic oxidation of Epinephrine and Norepinephrine at metal oxide doped phthalocyanine/MWCNT composite sensor, *Scientific reports*, (2016), 6, 1–20.

[14]. M.D. Tezerjani, A. Benvidi, A.D. Firouzabadi, M. Mazloum-Ardakani, A. Akbari, Epinephrine electrochemical sensor based on a carbon paste electrode modified with hydroquinone derivative and graphene oxide nano-sheets: Simultaneous determination of epinephrine, acetaminophen and dopamine, *Measurement*, (2017), 101, 183–189.

[15]. K. Khoshnevisan, E. Honarvarfard, F. Torabi, H. Maleki, H. Baharifar, F. Faridbod, B. Larijani, M.R. Khorramizadeh, Electrochemical detection of serotonin: a new approach, *Clinica Chimica Acta*, (2020), 501, 112–119.

[16]. D. Dăscălescu, C. Apetrei, Nanomaterials based electrochemical sensors for serotonin detection: A review, *Chemosensors*, (2021), 9, 14.

[17]. S. Sen, P. Sarkar, A simple electrochemical approach to fabricate functionalized MWCNT-nanogold decorated PEDOT nanohybrid for simultaneous quantification of uric acid, xanthine, and hypoxanthine, *Analytica chimica acta*, (2020), 1114, 15–28.

[18]. X. Jin, H. Zhang, Y.T. Li, M.M. Xiao, Z.L. Zhang, D.W. Pang, G. Wong, Z.Y. Zhang, G.-J. Zhang, A field effect transistor modified with reduced graphene oxide for immunodetection of Ebola virus, *Microchimica Acta*, (2019), 186, 1–9.

[19]. L. Yang, M.F. Beal, Determination of neurotransmitter levels in models of Parkinson's disease by HPLC-ECD, in: *Neurodegeneration*, Springer, 2011, 401–415.

[20]. S. Lu, M. Hummel, S. Kang, Z. Gu, Selective voltammetric determination of nitrite using cobalt phthalocyanine modified on multiwalled carbon nanotubes, *Journal of The Electrochemical Society*, (2020) 167, 046515.

[21]. S. Lu, M. Hummel, K. Chen, Y. Zhou, S. Kang, Z. Gu, Synthesis of Au@ZIF-8 nanocomposites for enhanced electrochemical detection of dopamine, *Electrochemistry Communications*, (2020) 114, 106715.

[22]. M. Hsu, Y. Chen, C. Lee, Y.; Chiu H.-T, Gold nanostructures on flexible substrates as electrochemical dopamine sensors. *ACS Appl. Mater. Interfaces*, (2012), 4, 5570–5575.

[23]. L. Ma, T. Zhao, P. Zhang, M. Liu, H. Shi, W. Kang, Determination of monoamine neurotransmitters and metabolites by high-performance liquid chromatography based on Ag (III) complex chemiluminescence detection, *Analytical biochemistry*, (2020), 593, 113594.

[24]. R.T. Kennedy, C.J. Watson, W.E. Haskins, D.H. Powell, R.E. Strecker, In vivo neurochemical monitoring by microdialysis and capillary separations, *Current opinion in chemical biology*, (2002), 6, 659–665.

[25]. M. Ko, L. Mendecki, A.M. Eagleton, C.G. Durbin, R.M. Stolz, Z. Meng, K.A. Mirica, Employing conductive metal-organic frameworks for voltammetric detection of neurochemicals, *Journal of the American Chemical Society*, (2020), 142, 11717–11733.

[26]. D.R. Thevenot, K. Toth, R.A. Durst, G.S. Wilson, Electrochemical biosensors: recommended definitions and classification, *Pure and applied chemistry,* (1999) 71, 2333–2348.

[27]. H.-S. Wang, Metal–organic frameworks for biosensing and bioimaging applications, *Coordination Chemistry Reviews,* (2017) 349, 139–155.

[28]. C. Ge, R. Ramachandran, F. Wang, CeO_2-Based Two-Dimensional Layered Nanocomposites Derived from a Metal-Organic Framework for Selective Electrochemical Dopamine Sensors, *Sensors,* (2020) 20, 4880.

[29]. A. Pandikumar, P. Rameshkumar, *Graphene-based electrochemical sensors for biomolecules*, Elsevier, 2018.

[30]. H. Li, K. Wang, Y. Sun, C.T. Lollar, J. Li, H.-C. Zhou, Recent advances in gas storage and separation using metal–organic frameworks, *Materials Today*, (2018) 21, 108–121.

[31]. X. Zhao, Y. Wang, D.S. Li, X. Bu, P. Feng, Metal-organic frameworks for separation, *Advanced Materials*, (2018) 30, 1705189.

[32]. H.D. Lawson, S.P. Walton, C. Chan, Metal-Organic Frameworks for Drug Delivery: A Design Perspective, *ACS Applied Materials & Interfaces*, (2021) 13, 7004–7020.

[33]. X. Liao, H. Fu, T. Yan, J. Lei, Electroactive metal-organic framework composites: Design and biosensing application, *Biosensors and Bioelectronics*, (2019) 146, 111743.

[34]. S. Rojas, A. Arenas-Vivo, P. Horcajada, Metal-organic frameworks: A novel platform for combined advanced therapies, *Coordination Chemistry Reviews*, (2019) 388, 202–226.

[35]. M. Ding, X. Cai, H.-L. Jiang, Improving MOF stability: approaches and applications, *Chemical Science*, (2019), 10, 10209–10230.

[36]. F.B. Ajdari, E. Kowsari, M.N. Shahrak, A. Ehsani, Z. Kiaei, H. Torkzaban, M. Ershadi, S.K. Eshkalak, V. Haddadi-Asl, A. Chinnappan, A review on the field patents and recent developments over the application of metal organic frameworks (MOFs) in supercapacitors, *Coordination Chemistry Reviews*, (2020) 422, 213441.

[37]. K. Kalcher, Chemically modified carbon paste electrodes in voltammetric analysis, *Electroanalysis*, (1990), 2, 419-433.

[38]. M. Hawley, S. Tatawawadi, S. Piekarski, R.N. Adams, Electrochemical studies of the oxidation pathways of catecholamines, *Journal of the American Chemical Society*, (1967) 89, 447-450.

[39]. Y. Li, C. Huangfu, H. Du, W. Liu, Y. Li, J. Ye, Electrochemical behavior of metal–organic framework MIL-101 modified carbon paste electrode: an excellent candidate for electroanalysis, *Journal of electroanalytical chemistry*, (2013),709, 65–69.

[40]. L.-L. Gao, W.-J. Sun, X.-M. Yin, R. Bu, E.Q. Gao, Graphite paste electrodes modified with a sulfo-functionalized metal-organic framework (type MIL-101) for voltammetric sensing of dopamine, *Microchimica Acta*, (2019),186, 1–9.

[41]. J.M. George, A. Antony, B. Mathew, Metal oxide nanoparticles in electrochemical sensing and biosensing: a review, *Microchimica Acta*, (2018) 185, 1–26.

[42]. Z. Lu, J. Zhong, Y. Zhang, M. Sun, P. Zou, H. Du, X. Wang, H. Rao, Y. Wang, MOF-derived Co_3O_4/$FeCo_2O_4$ incorporated porous biomass carbon: Simultaneous electrochemical determination of dopamine, acetaminophen and xanthine, *Journal of Alloys and Compounds*, (2021) 858, 157701.

[43]. F.A. Sofi, M.A. Bhat, K. Majid, Cu^{2+}-BTC based metal-organic framework: a redox accessible and redox stable MOF for selective and sensitive electrochemical sensing of acetaminophen and dopamine, *New Journal of Chemistry*, (2019), 43, 3119-3127.

[44]. J. Guo, S. Wu, Y. Wang, M. Zhao, A label-free fluorescence biosensor based on a bifunctional MIL-101 (Fe) nanozyme for sensitive detection of choline and acetylcholine at nanomolar level, *Sensors and Actuators B: Chemical*, (2020) 312, 128021.

[45]. S. Hu, J. Yan, X. Huang, L. Guo, Z. Lin, F. Luo, B. Qiu, K.-Y. Wong, G. Chen, A sensing platform for hypoxanthine detection based on amino-functionalized metal organic framework nanosheet with peroxidase mimic and fluorescence properties, *Sensors and Actuators B: Chemical,* (2018) 267, 312–319.

[46]. K. Ge, X. He, Z. Xu, R. Chu, A Luminescent Lanthanide-Functionalized Metal-Organic Framework as a Highly Selective and Sensitive Chemical Sensor for Dopamine, *ChemistrySelect,* (2019) 4, 12573–12579.

[47]. Z. Huang, L. Zhang, P. Cao, N. Wang, M. Lin, Electrochemical sensing of dopamine using a Ni-based metal-organic framework modified electrode, *Ionics*, 27 (2021) 1339–1345.

[48]. K.-Y. Wu, M. Chen, N.H. Huang, R.T. Li, W.L. Pan, W.H. Zhang, W.H. Chen, J.X. Chen, Facile and recyclable dopamine sensing by a label-free terbium (III) metal-organic framework, *Talanta*, (2021) 221, 121399.

[49]. M.-Q. Wang, C. Ye, S.J. Bao, Y. Zhang, Y.N. Yu, M.W. Xu, Carbon nanotubes implanted manganese-based MOFs for simultaneous detection of biomolecules in body fluids, *Analyst*, (2016) 141, 1279–1285.

[50]. D. Yin, J. Liu, X. Bo, M. Li, L. Guo, Porphyrinic metal-organic framework/macroporous carbon composites for electrocatalytic applications, *Electrochimica Acta*, (2017) 247, 41–49.

[51]. C. Liu, J. Wang, J. Wan, C. Yu, MOF on MOF hybrids: Synthesis and applications, *Coordination Chemistry Reviews*, (2021) 432, 213743.

[52]. J.D. Sosa, T.F. Bennett, K.J. Nelms, B.M. Liu, R.C. Tovar, Y. Liu, Metal–organic framework hybrid materials and their applications, *Crystals*, (2018), 8, 325.

[53]. S. Krishnan, L. Tong, S. Liu, R. Xing, A mesoporous silver-doped TiO_2-SnO_2 nanocomposite on gC_3N_4 nanosheets and decorated with a hierarchical core- shell metal-organic framework for simultaneous voltammetric determination of ascorbic acid, dopamine, and uric acid, *Microchimica Acta,* (2020) 187, 1–9.

[54]. G. Yu, J. Xia, F. Zhang, Z. Wang, Hierarchical and hybrid RGO/ZIF-8 nanocomposite as electrochemical sensor for ultrasensitive determination of dopamine, *Journal of Electroanalytical Chemistry*, (2017) 801, 496–502.

[55]. F. Mollarasouli, S. Kurbanoglu, K. Asadpour-Zeynali, S.A. Ozkan, Preparation of porous Cu metal organic framework/ZnTe nanorods/Au nanoparticles hybrid platform for nonenzymatic determination of catechol, *Journal of Electroanalytical Chemistry*, (2020) 856, 113672.

[56]. K. Kang, B. Wang, X. Ji, Y. Liu, W. Zhao, Y. Du, Z. Guo, J. Ren, Hemin-doped metal–organic frameworks based nanozyme electrochemical sensor with high stability and sensitivity for dopamine detection, *RSC Advances*, (2021) 11, 2446–2452.

[57]. J. Tang, Y. Liu, J. Hu, S. Zheng, X. Wang, H. Zhou, B. Jin, Co-based metal-organic framework nanopinnas composite doped with Ag nanoparticles: A sensitive electrochemical sensing platform for simultaneous determination of dopamine and acetaminophen, *Microchemical Journal*, (2020) 155, 104759.

[58]. S.A. Hira, M. Nallal, K. Rajendran, S. Song, S. Park, J.M. Lee, S.H. Joo, K.H. Park, Ultrasensitive detection of hydrogen peroxide and dopamine using copolymer-grafted metal-organic framework based electrochemical sensor, *Analytica chimica acta,* (2020) 1118, 26–35.

[59]. W.H. Ho, T.Y. Chen, K.I. Otake, Y.C. Chen, Y.S. Wang, J.H. Li, H.Y. Chen, C.W. Kung, Polyoxometalate adsorbed in a metal-organic framework for electrocatalytic dopamine oxidation, *Chemical Communications,* (2020) 56, 11763–11766.

[60]. X.Z. Wang, J. Du, N.N. Xiao, Y. Zhang, L. Fei, J.D. LaCoste, Z. Huang, Q. Wang, X.R. Wang, B. Ding, Driving force to detect Alzheimer's disease biomarkers: application of a thioflavine T@ Er-MOF ratiometric fluorescent sensor for smart detection of presenilin 1, amyloid β-protein and acetylcholine, *Analyst*, (2020) 145, 4646–4663.

[61]. S. Xie, X. Li, L. Wang, F. Zhu, X. Zhao, T. Yuan, Q. Liu, X. Chen, High quantum-yield carbon dots embedded metal-organic frameworks for selective and sensitive detection of dopamine, *Microchemical Journal*, (2021) 160, 105718.

[62]. L.L. Gao, J.J. Fang, N.N. Yang, R. Bu, X.M. Yin, L.J. Zhou, W.J. Sun, E.Q. Gao, Synergetic effects between a bipyridyl-functionalized metal-organic framework and graphene for sensitive electrochemical detection of norepinephrine, *Journal of The Electrochemical Society*, (2019) 166, B328.

[63]. H. Devnani, S. Ansari, S.P. Satsangee, R. Jain, ZrO_2-Graphene-Chitosan nanocomposite modified carbon paste sensor for sensitive and selective determination of dopamine, *Materials Today Chemistry*, (2017) 4, 17–25.

[64]. E.C. Vermisoglou, P. Jakubec, O. Malina, V. Kupka, A. Schneemann, R.A. Fischer, R. Zboril, K. Jayaramulu, M. Otyepka, Hierarchical Porous Graphene-Iron Carbide Hybrid Derived From Functionalized Graphene-Based Metal-Organic Gel as Efficient Electrochemical Dopamine Sensor, *Frontiers in chemistry,* (2020) 8, 544.

[65]. W. Zhang, J. Chen, Y. Li, W. Yang, Y. Zhang, Y. Zhang, Novel UIO-66-NO_2@ XC-72 nanohybrid as an electrode material for simultaneous detection of ascorbic acid, dopamine and uric acid, *RSC advances,* (2017) 7, 5628–5635.

[66]. S. Ramirez, N. Silva, M.P. Oyarzun, J. Pavez, J.F. Silva, Gold nanostructures on self-assembled monolayers activity for epinephrine, noradrenaline and dopamine, *Journal of Electroanalytical Chemistry*, (2017) 799, 349–357.

27 Recent Developments in MOF-based Sensors for Pharmaceutical Compounds

Engin Er[1]
[1]Department of Analytical Chemistry, Faculty of Pharmacy, Ankara University, Ankara, Turkey

27.1 WHY METAL-ORGANIC FRAMEWORKS?

Metal-organic frameworks (MOFs) are crystalline porous materials constructed by metal ions/clusters and organic-based ligands with strong coordination bonds. Pristine MOFs and MOF-based composites have been extensively utilized in a wide range of applications including catalysis, energy and gas storage, bio-imaging, drug delivery, and chemical sensing [1–5]. In recent years, the research has been focused on the exploration of the sensing potential of MOFs due to their distinguishing chemical and electrical properties such as high specific surface area (reaches up to 10^4 m^2g^{-1}), adjustable porosity, tailorable structure design and tunable conductivity (Figure 27.1) [6, 7]. These properties make MOFs promising candidates for electroanalytical studies. However, some specific properties of MOFs such as conductivity, size, and morphology, biocompatibility and stability, and the like, are still considered as the main challenges to fabricating the desired electrochemical sensors for potential analytical applications [8]. In this section, we will discuss the crucial parameters affecting the electrochemical performance of MOFs in sensing applications with the following aspects: (i) selection of metal cluster and ligand, (ii) morphology and porosity, and (iii) electronic conduction.

27.1.1 Selection of Metal Cluster and Ligand

Metal nodes and organic ligands have served as the connection and bridging points to constitute the 'primary building units' in the crystalline structure of MOFs (Figure 27.2). The selection of metal types and organic linkers significantly affects the coordination geometry of MOFs. The various metal clusters/ions including alkali metal ions (Na, K, Li, and so forth.), transition-metal ions (Zn^{2+}, Cu^{2+}, Zr^{3+}, Fe^{3+} and Cr^{3+}, and so forth.) and alkaline-earth metals ions (Mg^{2+}, Ca^{2+}, Sr^{2+}, and so forth.) are extensively utilized in the construction of MOFs [9, 10]. However, the crystal geometry of MOFs fundamentally depends on the types of organic ligands. Lewis acidic/basic sites of ligands and the angle of their binding sites with metal nodes are the main components to design the variable metal-ligand coordination geometries (cubic, tetrahedral, octahedral, square-pyramidal, and so forth) [10]. The combination of organic ligands, which generally contain electron-donating and -withdrawing aromatic groups, with metal clusters/ions enables the formation of donor/acceptor structures in the MOFs. These structures lead to improvement in the selectivity and catalytic activity towards the specific analyte [11].

27.1.2 Morphology and Porosity

MOFs are one of the most popular materials, having abundant porosity, cavity, and large surface area in sensing applications. These properties can be controlled by the selection of suitable organic linkers and metal clusters [12]. MOF-based materials with high porosity and large surface area provide an improved electrocatalytic activity towards the various electroactive molecules by accelerating the mass transfer of the analytes into their surfaces. In addition, the porosity and cavity in MOFs significantly improve the detection sensitivity towards the target molecules owing to the high interaction possibilities between analytes and their surfaces [6, 7] (Figure 27.3). These two unique properties enable the preparation of a new generation three-dimensional (3D)-based MOF nano/micro-structures as the electrocatalyst for quantitative purposes [13].

The current studies have demonstrated that 3D MOF-based structures are highly effective in electrocatalysis and are functional materials for the sensitive detection of various small biological molecules and amino acids. For instance, W. Meng et al. fabricated a porous Co-based zeolitic imidazolate framework (ZIF-67), which is synthesized by cobalt salt and 2-methylimidazole, modified with silver nanoparticles (AgNPs) for non-enzymatic detection of glucose. AgNPs/ZIF-67 composite, including porous channels, demonstrated an enhanced electrocatalytic activity towards the oxidation of glucose in alkaline conditions. The obtained results confirmed that AgNPs/ZIF-67 composite has promising analytical performance in terms of sensitivity, selectivity, and stability towards the glucose [14]. J. Tang et al. synthesized the microporous ZIF-67 structure growth on the graphene oxide (GO) layers by an in-situ growth method. The prepared MOF-based composite was electrochemically deposited on the glassy carbon electrode (GCE) and used as an electrochemical sensor (ZIF-67/GO/GCE) for the simultaneous determination of dopamine (DA) as an important neurotransmitter and uric acid (UA) as a main metabolic product in the urinary system.

DOI: 10.1201/9781003188148-27

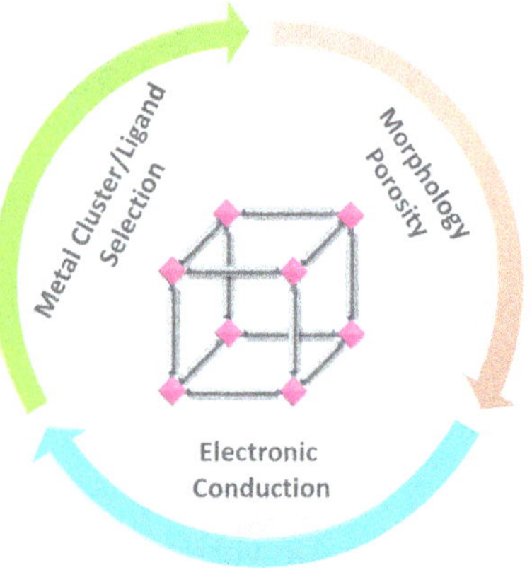

FIGURE 27.1 Tunable MOFs properties for electrochemical applications.

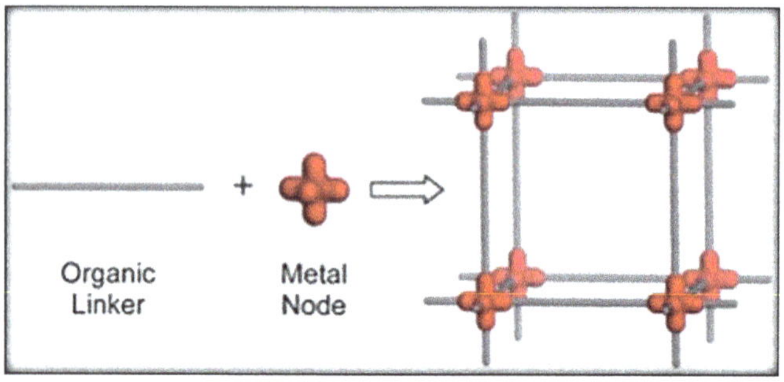

FIGURE 27.2 Schematic design of MOFs in a cubic geometry by the connection of metal nodes and organic linkers. Adapted with permission from reference [9]. Copyright (2012) Royal Society of Chemistry.

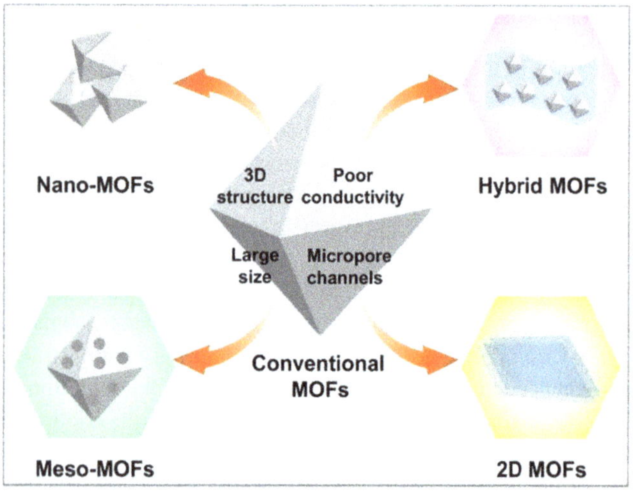

FIGURE 27.3 Different strategies to improve the sensitivity of the conventional MOF-based electrochemical sensors. Adapted with permission from reference [6]. Copyright (2020) Elsevier.

Consequently, ZIF-67/GO/GCE exhibited a challenging electroanalytical performance towards the detection of DA and UA. In addition, ZIF-67/GO/GCE was successfully applied to the simultaneous detection of DA and UA in urine samples with satisfactory recovery results [15]. In another study, Zhao et al. prepared a porous ZIF-67 microcrystal in a rhombic dodecahedral geometry as an electrocatalyst for the sensing of glutathione. The electrochemical results by cyclic voltammetry and amperometry confirmed an enhanced electrocatalytic activity of the ZIF-67 microcrystals towards glutathione detection. The modified electrochemical sensor having wide linearity (0.125-400 μM) and low detection limit (0.125 μM) could be utilized as an alternative platform for glutathione detection in clinical samples [16]. These electrochemical studies have indicated how 3D-MOFs play a critical role in electroanalytical applications.

In addition to the MOFs' porosity, their morphologies, including particular shape and size, are another essential factor affecting their sensitivity and selectivity. Channels and cavities in the particular-shaped MOFs not only possess an enhancement in detection sensitivity but also fabricate a selective sensing surface against the specific analytes depending on the size exclusion effects [6]. Ezzati et al. synthesized bimetallic-based MOFs by a highly controllable electrochemical method for non-enzymatic glucose sensing. Cubic- shaped bimetallic MOFs prepared from nickel-cobalt (Ni-Co) and benzene tricarboxylic acid (BTC) exhibited a good analytical performance (including linear working range, LOD, repeatability, reproducibility, and stability) and an acceptable selectivity against the electro-oxidation of glucose in the presence of AA, DA and UA as the interfering substances, which are naturally found in human blood. The proposed bimetallic composite modified on GCE (E-NiCo-BTC/GCE) is considered an effective electrocatalyst for the detection of glucose in clinical samples [17]. Liu et al. developed an electrochemical sensor based on 3D flower-like copper (II)-porphyrin framework as a MOF structure for the quantitative detection of glutathione. To improve the electrical conductivity and adsorption capability of the glutathione on the 3D MOF-based electrode surface, AgNPs were modified on the prepared Cu-based MOF nanoflowers (Ag/Cu-TCPP). Ag/Cu-TCPP composite exhibited a wide linear dynamic range from 1 to 100 μM with a detection limit of 0.066 μM against the oxidation glutathione. The recoveries of glutathione obtained from human serum samples at Ag/Cu-TCPP modified GCE confirm the applicability of the developed electrochemical sensor for the detection of glutathione in clinical samples [18]. Molybdenum oxide-based 3D MOFs with helical channels ($CuTRZMoO_4$) have been reported by Zhou et al. as the potential electrode materials for electrochemical sensing applications. In this work, to improve its sensitivity towards the DA, 3D MOF nanostructures were modified by polypyrroles (PPy) as electrically conductive polymers. The resulting nanocomposite indicated a remarkable catalytic activity against DA with good electrochemical performance [19]. All of the above reports suggest that the electrocatalytic activity of MOFs strongly depends on their morphologies.

27.1.3 Electronic Conduction

The electrical conductivity of MOFs is formed by charge transport between metal and ligand molecules (metal-to-ligand

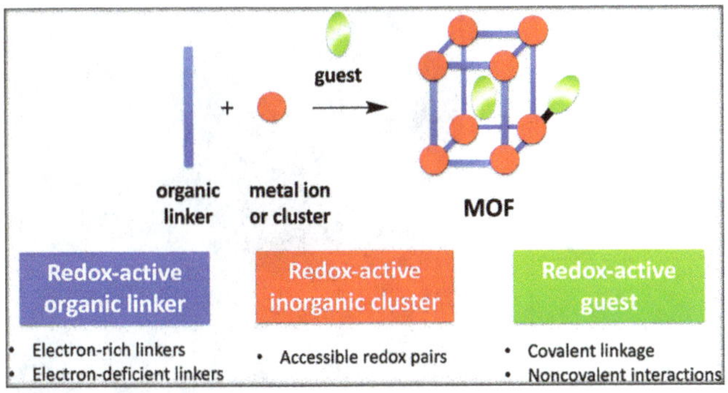

FIGURE 27.4 Modification types to achieve the redox activity in MOFs. Adapted with permission from reference [20]. Copyright (2019) Royal Society of Chemistry.

or ligand-to-metal) or π-π^* transitions in the aromatic rings of the ligands. Most MOFs are considered electrical insulators owing to their poor conductivity features and the large bandgap of the organic ligands. This limited electron transferability in MOFs is developed by the preparation of redox-active MOFs (ra-MOFs) via organic ligand/metal cluster engineering (Figure 27.4) [20]. In this strategy, ra-MOFs can be classified into two main sections: (i) redox-active organic ligand or metal cluster-based conjugated systems, and (ii) redox-active molecules modified redox-inactive frameworks by covalent attachment or immobilization as guests. In the first case, tuning of the redox-activity in metal cluster-based MOFs is related to the amount of stable oxidation states in transition metal nodes. On the other hand, ligand engineering in MOFs has been also reported to promote their redox activities [21]. The conjugation of redox-active organic ligands such as benzene, pyridine, and imidazole-derivatives with the metal clusters is a general strategy to construct the ra-MOFs, which possess an improvement in both conductivity and electrocatalytic activity. The above-mentioned electron-donating and withdrawing groups in the ligand structure facilitate an electron transfer process owing to the presence of abundant π bonds [11, 21]. Numerous ra-MOF studies constructed by electrochemically active organic ligands have been reported as electrocatalysts for analytical applications. For instance, Ko et al. described two-dimensional conductive MOFs modified film electrodes based on M_3HXTP_2 (M=Ni,Cu and X=NH,2,3,6,7,10,11-hexaiminotriphenylene (HITP) or O,2,3,6,7,10,11-hexahydroxytriphenylene (HHTP)) for the voltammetric sensing of neurotransmitters such as dopamine and serotonin in the presence of biologically important compounds. The results demonstrated the modification effects on the electron transfer rate of the conductive 2D-MOFs by tuning the organic ligands and metal nodes [22].

Metal cluster/ligand exchange or the incorporation of redox-active guest molecules as the post-synthetic strategies have been also employed to build ra-MOFs with active sites. The incorporation of redox-active guests into redox-inactive MOFs significantly promotes the electronic conductivity of the related frameworks. It is reported that ferrocene as a redox probe molecule could be an important candidate to fabricate ra-MOFs for potential bio- and pharmaceutical sensing applications [23]. Halls et al. synthesized Zn and Al-based MOFs covalently attached with ferrocene-containing functional groups, and these MOFs have as redox-active materials [24]. Amyloid-β protein is one of the primary biomarkers for the diagnosis of Alzheimer's disease. Han et al. developed a MOF-based sandwich immunosensor with a novel design for sensitive detection of Amyloid-β as a model protein. In this design, ferrocene (Fc) was covalently modified onto a Zn-MOF surface by post-synthetic strategy, and a flower-shaped Fc/Zn-MOF porous structure was obtained. As a second step, Fc/Zn-MOF was decorated by AuNPs labeled with anti-Amyloid-β to prepare a sandwich-immunoassay for Amyloid-β sensing. A well-distributed and high content of Fc on the Zn-MOF surface exhibited an enhanced electrocatalytic activity. Considering the analytical performance (including wide linearity in the range of 0.0001 to 100 ng mL^{-1}, an extremely low detection limit of 0.03 pg mL^{-1}) of the MOF-based immunosensor towards the Amyloid-β detection, the proposed sensing strategy seems to be highly creative in the fabrication of immunosensors for sensitive and accurate detection of specific proteins [25]. In another study, Chang et al. designed an electrochemical sensor based on ferrocene-immobilized MOF structures (Fc-MOF) for the detection of acetaminophen as an analgesic and antipyretic drug. According to the redox reaction of acetaminophen at the modified Fc-MOF electrode, ferrocene as a mediator has increased the electron transfer rate between the target molecule and MOF-based electrode. Electrochemical results reveal that ferrocene immobilization on the MOF structure dramatically enhanced the electrochemical performance towards the acetaminophen as a target analyte [26]. As shown in the reported studies, ra-MOFs not only provide an improvement in electrocatalytic activity but also serve as novel and multi-functional designs for bio- and pharmaceutical analysis.

27.2 FABRICATION OF MOFs AND RELATED COMPOSITES

The catalytic and structural properties of the MOFs are undoubtedly dependent on their production strategies in electrochemical applications. Preparation of MOFs and their functional derivative materials for potential applications may be divided into three categories: (i) morphology-controlled MOFs, (ii) construction of MOF-based composites, and (iii) functional derivative MOF structures [27]. In the first strategy, the combination of the various metal nodes and organic ligands by surface engineering provides an innovative design for the preparation of MOFs in the different morphologies and dimensionalities. Zhou et al., for the first time, synthesized one-dimensional (1D) single-crystal cobalt-based MOF nanotubes with a high aspect ratio (diameter: ~70 nm, length: 25-35 μm) by the hydrothermal approach (Figure 27.5), and it has been used as a precursor for fabricating hierarchical carbon nanostructures, which present extraordinary electrocatalytic activity in important electrochemical applications such as oxygen reduction reaction and rechargeable battery [28]. It is known that surfactant plays a key role in producing the MOF nanocrystals with various sizes and shapes in a controlled way.

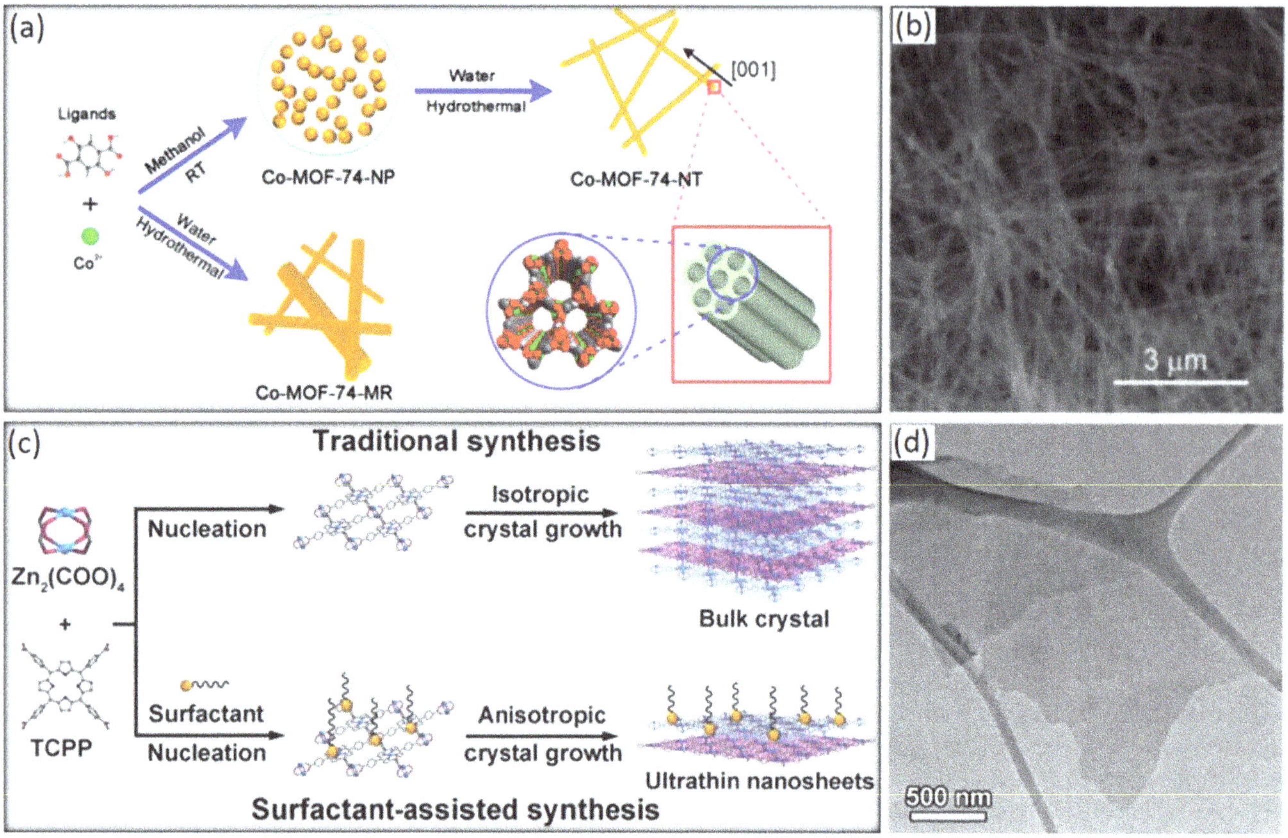

FIGURE 27.5 (a) Schematic procedure of the hydrothermally prepared 1D Co-MOF nanotubes (Co-MOF-74-NT); (b) SEM image of Co-MOF-74-NT. Adapted with permission from reference [28]. Copyright (2018) American Chemical Society. (c) Surfactant-assisted synthesis process of 2D Zn-TCPP nanosheets; (d) TEM image of Zn-TCPP nanosheets. Adapted with permission from reference [29]. Copyright (2018) Wiley-VCH.

Zhao et al. prepared two-dimensional (2D) MOF nanosheets using a surfactant-assisted synthetic approach. Ultrathin 2D MOF nanosheets have been prepared by mixing various metals (M: Zn, Cu, Cd and Co) and porphyrin-based ligand molecules (TCPP=tetrakis(4-carboxyphenyl)porphyrin) in the presence of polyvinylpyrrolidone as a surfactant (Figure 27.5) [29]. Synthesized 2D Cu-based MOF nanosheets could be an effective material for diverse electrochemical applications such as energy storage, batteries, and sensors. On the other hand, three-dimensional (3D) ordered MOF structures are one of the most powerful candidates, especially in sensing applications. For instance, Zhang et al. fabricated an electrochemical sensor based on 3D nanoflower-shaped bimetallic Ce@Zn-MOF microspheres synthesized by an ultrasonic-assisted solvothermal approach for electrochemical detection of uric acid found as a major component in serum and urine. The observed electrocatalytic activity of Ce@Zn-MOF microspheres modified GCE in neutral pH towards the uric acid suggested the potentiality of prepared 3D nanoflower-shaped bimetallic Ce@Zn-MOF microspheres as sensing materials [30]. Despite the fact that MOF structures demonstrate a good catalytic and sensing performance depending on their morphologies, the sensing capability of MOF structures, even at 3D morphology, still needs to be improved.

Construction of MOF-based composites and their derivatives by post-synthesis modification strategies (ii and iii) is a novel design to present an improved sensing performance compared to bare MOF structures in biomedical and pharmaceutical applications. To date, numerous modifications using different nanomaterials containing carbon nanomaterials, metal/metal oxide nanostructures, polymers, and the like, with MOF nanostructures have been performed via different synthesis approaches such as hydro(solvo)thermal, microwave, or ultrasound-assisted synthesis, electrochemical and diffusional methods [31]. The selection of synthesis methods used in the fabrication of MOF-based structures significantly affects the structural and compositional behavior of MOFs. Hydro(solvo)thermal and electrochemical approaches are some of the most effective and common synthesis methods utilized in the fabrication of MOF-based composites for bio- and drug sensing applications. Zhang et al. designed a porous Ni-MOF/carbon nanotube composite by hydrothermal treatment for non-enzymatic glucose sensing. In this process, briefly, nickel salt and 1,4-dicarboxybenzene, as a metal cluster and organic ligand, dissolved in dimethylformamide (DMF) solution were mixed with activated carbon nanotube and kept in a Teflon-lined autoclave at 120 °C for 24 h. After the purification procedure, spheroidal and porous Ni-MOF structures in micron-sizes modified with carbon nanotubes were observed by electron microscopy techniques. The prepared Ni-MOF/carbon nanotube immobilized on GCE exhibited high electrocatalytic activity and analytical performance towards glucose oxidation [32]. In a similar study conducted by Wang et al., a highly stable and self-assembled Cu-MOF/GO nanocomposite has been prepared by a sonication-assisted solvothermal method and used as a sensing material for the detection of acetaminophen and dopamine. A large amount of cubic-shaped Cu-MOF structures with good crystallinities were composed by the modification of wrinkled large GO sheets. The electrochemical activity of Cu-MOF/GO nanocomposite was improved by an electrochemical reduction of GO layers into graphene layers. The synergistic effect of electrochemically reduced graphene and Cu-MOF possesses high electrochemical sensitivity and catalytic activity towards acetaminophen and dopamine [32]. In addition, electrochemical deposition has become an effective method to prepare MOF-based composites for analytical applications. Zhan et al. fabricated a nanocomposite consisting of Zn-MOF modified with AuNPs and electrochemically reduced carbon dots (ErCDs) by a single-step electrodeposition approach. In this self-assembled design, needle-like microstructures of Zn-MOF were successfully anchored with small-sized AuNPs (~ 20 nm) and ErCDs (~ 3-4 nm) on GCE. As a proof of concept application, Zn-MOF/AuNPs/ErCDs/GCE was utilized in the detection of bisphenol A and acted as an endocrine disruptor, and the obtained results showed the efficiency of the fabricated sensor in the sensing applications [33].

In recent years, functional-derived MOF structures have drawn great attention in the fabrication of electrochemical sensors for the detection of pharmaceuticals [6]. Calcination and carbonization processes are typical treatments used in the production of MOF-derived composites, which possess porosity, high surface area, and improved conductivity as well as functionality. In these processes, hydro(solvo)thermally prepared MOF structures have been further carbonized or calcinated at elevated temperatures under nitrogen or argon gas conditions (Figure 27.6). Wang et al. prepared MOF-derived nickel-zinc oxide (NiO@ZnO) hollow microspheres by hydrothermal and calcination treatments. The electrochemical characterization demonstrated that MOF-derivedNiO@ ZnO microspheres have good electrical conductivity and catalytic activity. The fabricated composite has been used as an electrode material for the detection of isoniazid as an anti-tuberculosis drug [34]. Similarly, Yuan et al. successfully synthesized a Fe_3O_4/N/C@MWCNT derived from Fe-MOF by thermal treatment methods including carbonization (Figure 27.6). The electrocatalytic properties and electrochemical performance of the fabricated composite were evaluated by analyzing acetaminophen and metronidazole as the two clinically important pharmaceuticals. This work could be a useful approach for the preparation of other metal oxides @ MWCNT composites as the sensing platforms for potential pharmaceutical analysis in clinical samples [35].

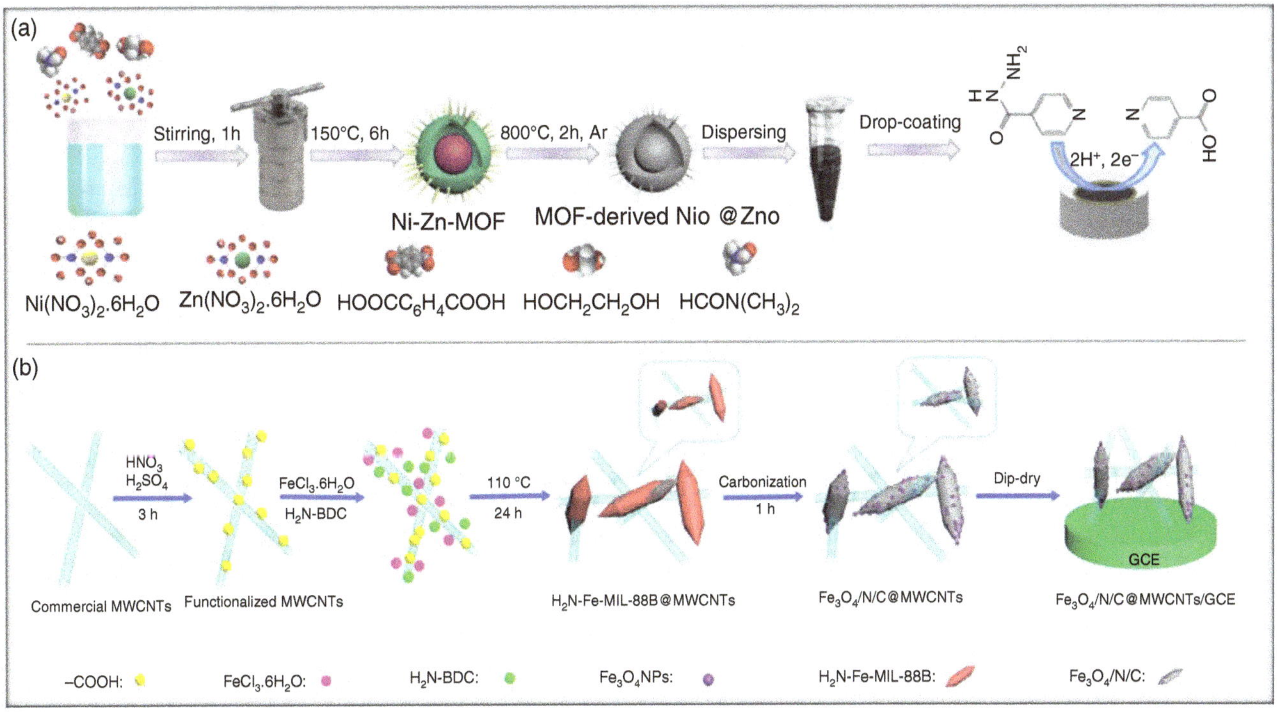

FIGURE 27.6 (a) Schematic illustrations for the production of MOF-derived NiO@ZnO composites prepared by a calcination process. Adapted with permission [34]. Copyright (2020) Springer Nature. (b) Fabrication steps for the modification of Fe_3O_4/N/C@MWCNTs synthesized by carbonization process on GCE. Adapted with permission [35]. Copyright (2019) Elsevier.

27.3 ELECTROANALYSIS OF PHARMACEUTICALS BY MOF-BASED COMPOSITES

MOFs synthesized by traditional routes possess poor conductivities owing to the insulating properties of the organic ligand structures. This phenomenon directly affects the electrochemical efficiency of the crystalline MOFs in sensing applications. Recent developments in the preparation of nanomaterials have offered various opportunities to improve the electrochemical performance of the MOFs towards the detection of a range of molecules. Surface modification strategy and composite design are the main factors to overcoming this electrochemical performance in MOFs. The modification of MOFs with novel carbon-based nanomaterials, conductive polymers, and metal nanoparticles has considerably improved their electrical conductivities, electrocatalytic properties, and durabilities. Recently, MOF-based composites have been extensively used as promising sensing platforms for the determination of clinically important pharmaceutical compounds. Although the MOF-based composites demonstrate high sensitivity towards electroactive compounds including pharmaceuticals, their poor selectivity properties remain a major problem to be overcome. Herein, the molecularly imprinted technique has been come into prominence to fabricate a selective electrochemical sensor based on polymeric materials offering specific molecular recognition ability in a matrix. In this section, we summarize the recent electroanalysis studies performed by MOF-based composites modified electrochemical sensors towards the different-targeted pharmaceutical compounds. For this purpose, MOF-based composites were classified into four different major categories: MOF/carbon-based nanomaterial composites, MOF/metal nanoparticles composites, MOF/conducting polymer composites, and molecularly imprinted MOF-based composites.

27.3.1 MOF/Carbon-based Nanomaterial Composites

Carbon nanomaterials such as graphene, carbon nanotubes, and porous graphitic carbon are promising modifying agents to improve the electrochemical sensing performance of the pristine MOFs by accelerating electron transfer. Graphene, a novel two-dimensional (2D) layered nanomaterial, is one of the most effective candidates to fabricate the various composite materials including MOFs for sensing purposes due to its large surface area, high electrical conductivity, adsorption capability, and mechanical strength. Kemmegne-Mbouguen et al. [36] have used a Zirconium-based MOF composite including a graphene foam (GF/UİO-66) to modify the GCE for simultaneous quantification of acetaminophen and tryptophan. The authors effectively synthesized the GF/UİO-66 composite by an in-situ chemical growth method. Electrochemical characterization of the modified electrode (GF/UİO-66/GCE) was performed by cyclic voltammetry and electrochemical impedance spectroscopy. The DPV results showed good linearity in the concentration ranges of 0.5-200 and 0.5-113 μM towards the AC and TRP. The detection limits of the proposed sensor for AC and TRP were calculated as 0.07 and 0.06 μM, respectively. In another work reported by Ma et al., a nanocomposite based on HKUST-1 (MOF) and electrochemically reduced graphene oxide (ERGO) was developed for simultaneous determination of paracetamol and dopamine. A MOF/ERGO composite prepared on GCE by an electrodeposition technique demonstrated good electrochemical activities towards the paracetamol and dopamine due to their strong synergistic effects. The MOF/ERGO composite modified electrode presented a wide linear range of 0.2-160 μM for paracetamol with a low detection limit of 0.016 μM [37]. Levodopa is a neurotransmitter that is essentially used in the treatment of Parkinson's disease, which is considered to be one of the most widespread neurodegenerative diseases. The detection of levodopa acted as a biomarker at abnormal levels in human serum and urine could give an important signal for the diagnosis of Parkinson's disease at an early stage. In this context, a sensitive electrochemical sensor based on reduced graphene oxide modified with microporous MOFs for the detection of levodopa was proposed by Naghian et al. The obtained sensitivity and electrochemical performance towards the sensing of levodopa have supported a high electrocatalytic activity of the developed MOF-based sensor. [38] Electrochemical deposition or exfoliation is a useful technique to prepare the MOF-based nanostructures for potential sensing applications. This latest study could be a promising design for antidepressant drug analysis. As is known, antidepressant drugs are mainly used in the treatment of common mental disorders such as depression. In this work, sulfur, and nitrogen-doped graphene (SNDGr) modified with Cu-MOFs nanocomposite has been electrochemically prepared on a pencil graphite electrode and utilized in the analysis of an antidepressant agent. The synergic effect of Cu-MOFs and SNDGr enhanced the electron transfer process at the electro-oxidation step of sertraline. The constructed electrochemical sensor presents a wide range of advantages such as sensitivity (0.4557 $\mu A\ \mu M^{-1}\ cm^{2}$), detection limit (0.038 μM), selectivity, and reproducibility towards the detection of sertraline [39].

An overdose of the antineoplastic agents causes high cytotoxicity especially on the neoplastic cells, which may lead to vital chronic effects including lung, liver, kidney, and heart damage with long-term therapy. Hence, the development of analytical-based sensors for the detection of antineoplastic agent amounts in biological samples is a highly desirable issue. Dehdashtian et al. developed an electrochemical sensor based on perlite/cobalt oxide/reduced graphene oxide nanocomposite modified with Cu-MOFs (PC-rGO/MOF-199) for the detection of idarubicin, which belongs to the anthracycline antitumor antibiotics in the treatment of leukemia, in urine and serum samples. The proposed Cu-MOF-based sensor possesses a good sensitivity, reproducibility, and stability towards the electroanalysis of idarubicin. The successful recoveries obtained at PC-rGO/MOF-199 have revealed the potential utilization of the proposed electrochemical sensor for the determination of idarubicin in clinical samples [40].

In a similar study, Jalal et al. [41] reported an electrochemical sensor based on the HKUST-1 framework and graphene oxide nanoribbons (GONRs) modified onto the GCE. The fabricated electrochemical sensor (HKUST-1/GONRs/GCE) demonstrated a good electrochemical performance towards imanitib, which is an anticancer drug used in the treatment of gastrointestinal stromal tumors and chronic myelogenous leukemia (Figure 27.7). The obtained calibration curves at two linear dynamic ranges of 0.04-1 and 1-80 μM with a low detection limit of 6 nM towards the imanitib has proved the superior electrocatalytic performance of the HKUST-1/GONRs/GCE. Considering the recovery results from urine and serum samples for imanitib, the presented MOF-based composite has been considered as an alternative sensing platform for the voltammetric determination of imanitib in real samples.

A study performed by Mahnashi et al. reports a sensitive and selective MIP-based electrochemical sensor for the determination of oxaliplatin using the voltammetric technique. The electrochemical sensor was prepared by the modification of GCE with silver nanoparticle functionalized Cu-MOFs (Ag@Cu-BDC) and nitrogen-doped carbon nanotubes (N-CNTs). The electroanalysis of oxaliplatin at the developed electrochemical sensor was performed by the reduction property of Cu-BDC and its interaction level with target molecules. The MOF-based electrochemical sensor has offered a promising sensitivity towards the oxaliplatin at the concentration range of 0.056-200 ng mL^{-1} with a LOD value of 0.016 ng mL^{-1}. This novel approach provides an opportunity for the fabrication of a sensing platform for the fast analysis of oxaliplatin in clinical samples [42].

The combination of MOFs with porous carbon structures having attractive morphologies enabled the sensitive detection of various pharmaceutical compounds belonging to this group of antibiotics, antitubercular, antihypertensive, anesthetic, and antiarrhythmic drugs. Yan et al. reported a rod-shaped Co-MOF crystal embedded into the mesoporous carbon (MC) structure, which presents an enhanced electrical conductivity and stability. Modification of Co-MOFs on MC probably occurs via defects or carboxyl groups in the composite. The developed Co-MOFs/MC

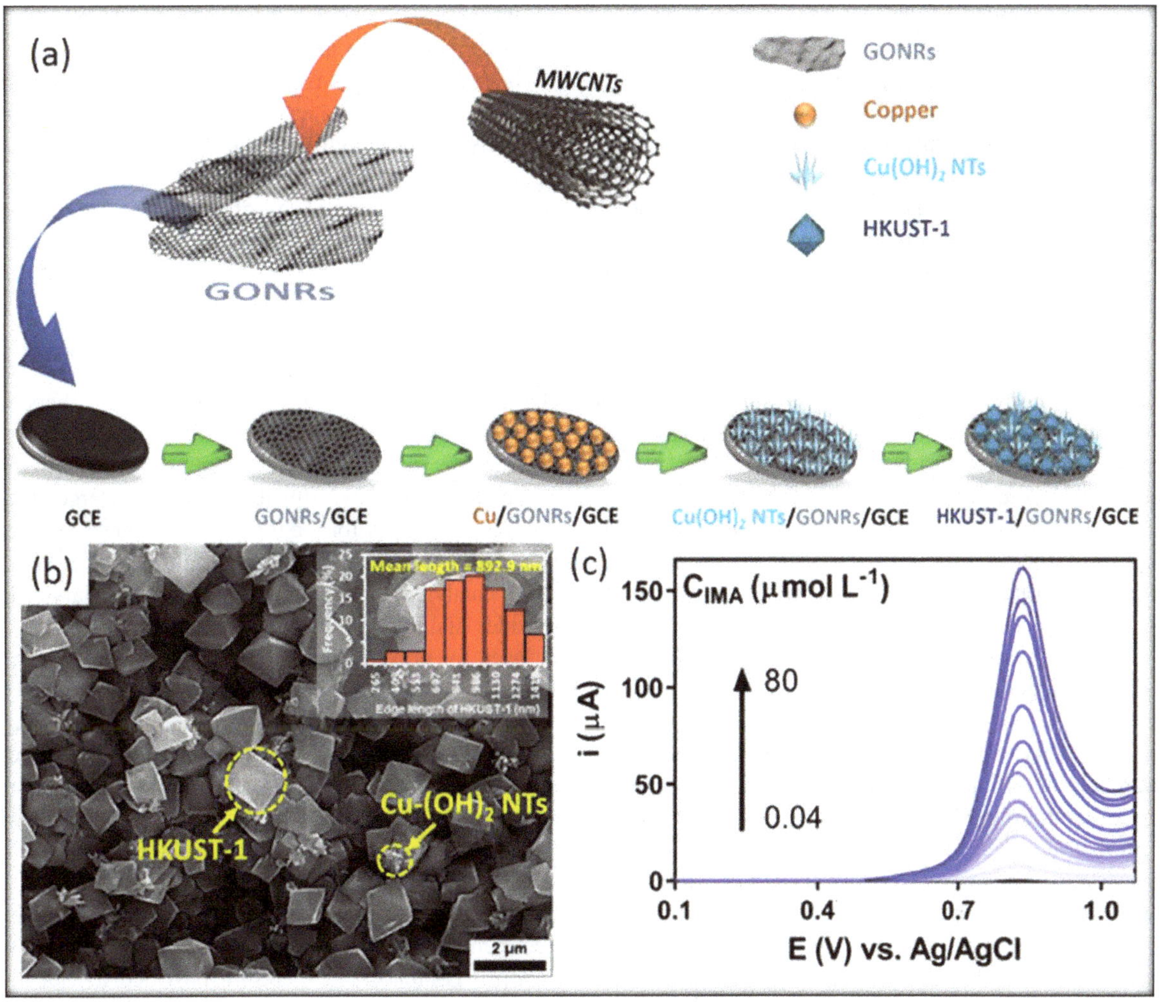

FIGURE 27.7 (a) Fabrication process of HKUST-1/GONRs/GCE; (b) FESEM images of HKUST-1/GONRs/GCE; (c) Differential pulse voltammograms at HKUST-1/GONRs/GCE for different imanitib concentrations (0.04–80 μmol L^{-1}) in pH 7 Britton-Robinson buffer. Adapted with permission from reference [41]. Copyright (2020) American Chemical Society.

TABLE 27.1
Recent electrochemical sensing applications based on MOF/carbon-based nanomaterial composites used in the determination of pharmaceutical compounds

Sensing material	Method	Analyte	Linear working range	LOD	Refs
GF/UiO-66	DPV	Acetaminophen	0.5–200 μM	0.07 μM	[36]
HKUST-1/ERGO	DPV	Paracetamol	0.2–160 μM	0.016 μM	[37]
TMU-22/RGO	SWV	Levodopa	0.1–85 μM	0.025 μM	[38]
Cu-MOF/SNDGr	DPV	Sertraline	0.05–2.67 μM	0.038 μM	[39]
PC-rGO/MOF-199	DPV	Idarubicin	0.005–1 μM	0.0015 μM	[40]
HKUST-1/GONRs	DPV	Imanitib	0.04–1 and 1–80 μM	0.006 μM	[41]
Ag@Cu-BDC/N-CNTs	DPV	Oxaliplatin	0.00014–0.503 μM	0.04 nM	[42]
Co-MOF-74@MC	DPV	Pyrazinamide	0.3–46.5 and 46.5–166.5 μM	0.21 μM	[43]
	Amperometry	Isonicotinyl hydrazide	0.15–1.55 and 1.55–592.55 μM	0.094 μM	
Fe-ZIF-8/NC	LSV	Chloramphenicol	0.1–100 μM	0.031 μM	[44]
	LSV	Metronidazole	0.5–30 μM	0.165 μM	
Fe-MOF/MC	DPV	Amlodipine	0.009–500 μM	1.27 nM	[45]
	DPV	Losartan	0.009–500 μM	2.03 nM	

exhibited a high electrochemical performance for quantitative detection of isonicotinyl hydrazide and pyrazinamide used in the treatment of tuberculosis. The practical applications achieved from human serum and urine samples reveal the efficiency of the electrochemical sensor for isonicotinyl hydrazide and pyrazinamide analysis [43]. A similar electrochemical sensing platform based on a nano porous carbon structure modified with Fe-MOFs has been also proposed for the detection of commonly used antibiotics such as chloramphenicol and metronidazole. Herein, the poor conductivity nature of Fe-MOFs was remarkably improved by the modification of nitrogen-doped nanoporous carbon (NC) particles. Dodecahedron-shaped Fe-MOFs/NC nanostructures possessing a high homogeneity exhibited high sensitivity and selectivity for chloramphenicol and metronidazole owing to the large surface area and enhanced conductivity of Fe-MOFs/NC [44].

Hypertension is considered one of the major diseases over the world triggering death due to cardiovascular complications. The regulation of blood pressure using an antihypertensive drug is a common treatment method, and its overdose in the blood causes cardiotoxicity leading to primary cardiovascular diseases such as heart failure and stroke, and renal damage. Rajpurohit et al. developed an effective and fast electroanalytical approach for simultaneous detection of two important antihypertensive agents, amlodipine, and losartan in pharmaceutical and biological samples. The electrochemical behavior of amlodipine and losartan has been performed by Fe-MOFs/MC nanocomposite using the voltammetric technique. The modification of rod-like shaped Fe-MOFs with small MC nanospheres enhanced the electrocatalytic activity against the amlodipine and losartan due to their active sites, large surface areas, and favorable electrical conductivities. The prepared Fe-MOFs/MC nanocomposite exhibits a challenging electrochemical performance (especially large dynamic range) for the quantitative detection of amlodipine and losartan. The performed analytical validation results confirmed the reliability of the proposed MOF-based nanocomposite for the sensing of amlodipine and losartan in routine pharmaceutical and clinical applications [45].

As seen in previous reports, the synergistic effect of MOF and carbon nanostructures dramatically enhanced the analytical sensitivity for the determination of the above mentioned pharmaceutical compounds. Table 27.1 summarizes the analytical performances of the various pharmaceutical compounds at developed MOF/carbon-based nanomaterial composites.

27.3.2 MOF/Metal Nanoparticle Composites

The decoration of MOFs with differently shaped metal nanoparticles creates novel hybrid materials for sensing applications due to their extraordinary conductivities and surface areas. MOFs with tunable porosity and large surface area act as a host matrix for the integration of metal nanoparticles. This hybrid material does not only show a remarkable electrocatalytic activity towards the specific analytes but also provides stability of the metal nanoparticles on the MOFs surfaces by preventing their possible agglomerations. With these important features, metal nanoparticle-doped MOFs are an attractive nanocomposite used in the detection of pharmaceutical compounds. In a previous report by da Silva et al., gold nanoparticle (AuNPs)/Cu-MOF nanocomposite was immobilized onto GCE to detect the captopril, which is an angiotensin-converting enzyme (ACE) inhibitor, utilized in the regulation of hypertension, in pharmaceutical tablets. The developed nanocomposite significantly increased the electrochemical response towards the oxidation of the Cu-captopril complex and therefore provided a wide dynamic range (0.5 to 2500 μM) and low detection limit of 0.047 μM for captopril [46]. Wang et al. demonstrated a facile approach to synthesize a zeolitic imidazolate framework with a leaf-like morphology (ZIF-L) modified with small-sized

AuNPs by a self-assembly method. AuNPs supported ZIF-L composite structure exhibited a high electrochemical activity against the oxidation of acetaminophen which is a common antipyretic and analgesic agent. The proposed strategy could be a benefit to designing a sensitive electrochemical sensor by the use of various metallic nanoparticles for electroanalytical applications [47]. In a similar work, a hydrothermal route was effectively utilized to synthesize a water-stable MOF-based composite consisting of Cu-MOF and quasi-spherical AuNPs as electrocatalysts. For the electrode modification, diluted Nafion solution as a physical binder was chosen to block the Cu-MOF/AuNP composite on the GCE surface. The synergistic effect of Cu-MOF and AuNPs having a large surface, active sites and high conductivity has dramatically increased the electrochemical activity for the determination of paracetamol known as acetaminophen. The proposed electrochemical performance including wide linear dynamic range (0.01 nM to 100 μM) and extremely low detection limit (0.01 nM) exhibited that Cu-MOF/AuNP composite could be used as a sensitive electrochemical sensing platform for the detection of paracetamol in pharmaceutical samples [48].

Among the metal nanoparticles, AgNPs present an efficient electrochemical performance towards the electroactive analytes owing to their superior electrical conductivity and biocompatibility. Tang et al. prepared a hybridized material based on Co-MOF (ZIF-67p) nanopinnas decorated with AgNPs as an electrochemical sensing platform for the detection of dopamine and acetaminophen (Figure 27.8). AgNPs/ZIF-67p modified GCE remarkably enhanced the electrocatalytic activity towards both target molecules due to their combinations of the large surface area of porous ZIF-67p and excellent conductivity of AgNPs. The obtained analytical performance confirmed the potential applicability of the proposed sensing platform for the detection of acetaminophen in pharmaceutical samples [49].

27.3.3 MOF/Conducting Polymer Composites

Conducting polymers are the favored modification agents used for the fabrication of electrochemical sensors due to their advantages, such as: high electrical conductivity and catalytic activity, chemical stability, and large 3D morphology. Among the conductive polymers, poly(3,4-ethylenedioxythiophene) (PEDOT) has attracted significant attention as an electrocatalyst for potential electrochemical applications such as supercapacitors, solar cells, and sensors. Wang et al. synthesized a Ni-MOF embedded in tubular PEDOT composite by an in-situ growth method for voltammetric sensing of tinidazole utilized as an anti-inflammatory drug. The cross-linked structure of Ni-MOF/PEDOT dramatically improved the electrocatalytic activity towards the reduction of tinidazole. Two linear dynamic ranges of 0.7–35 and 35–255 μM with a LOD value of 0.09 μM for tinidazole were achieved at the surface of Ni-MOF/PEDOT composite. Considering the selectivity, reproducibility, stability, and recovery results obtained for tinidazole, the proposed MOF-based composite confirmed that it has great potential for detection capability towards the tinidazole in real samples [50]. Similarly, Wei et al. developed an electrochemical sensor modified with Zr-MOF (PCN-222) and PEDOT for the sensitive detection of chloramphenicol. The electrocatalytic activity of PCN-222 was enhanced by the modification of

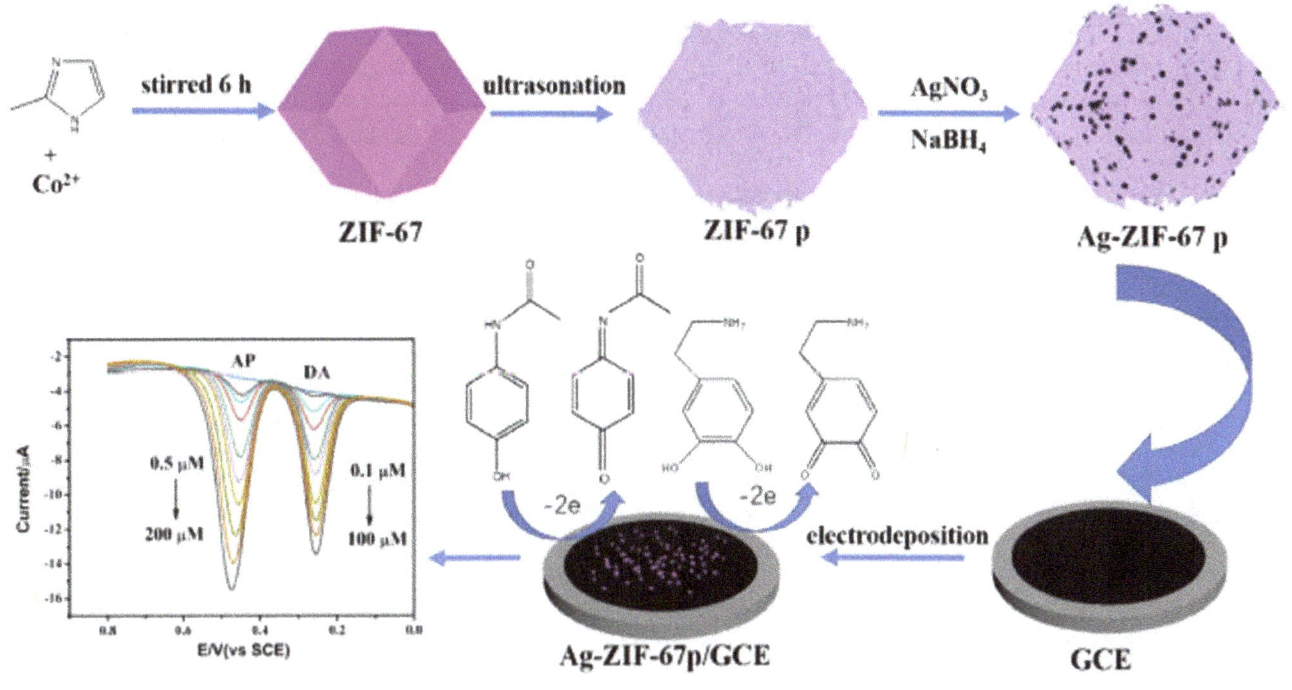

FIGURE 27.8 The fabrication process of Ag-ZIF-67 nanopinnas modified GCE and its application of acetaminophen detection in the presence of dopamine. Adapted with permission from reference [49]. Copyright (2020) Elsevier.

TABLE 27.2
Recent Electrochemical Sensing Applications Based on MOF/Metal Nanoparticle Composites Used in the Determination of Pharmaceutical Compounds

Sensing material	Method	Analyte	Linear working range	LOD	Refs
Cu-MOF/AuNPs	CV	Captopril	0.5–7 and 10–2500 μM	0.047 μM	[46]
ZIF-L/AuNPs	DPV	Acetaminophen	3.5–56 and 56–560 μM	1.02 μM	[47]
Cu-MOF/AuNPs/NFN	DPV	Acetaminophen	0.01 nM–100 μM	0.01 nM	[48]
Ag-ZIF-67p/AuNPs	DPV	Acetaminophen	0.5–200 μM	0.2 μM	[49]

PEDOT on the ITO electrode. In addition, chitosan was used as a binder to make a stable sensor. The synergistic effect of PCN-222 and PEDOT exhibited good electrocatalytic activity against chloramphenicol. A low detection limit for chloramphenicol was obtained as 1.8 nM under optimized voltammetric conditions. The designed MOF-based sensor could be a favorable analytical approach for the detection of chloramphenicol in real samples [51].

27.3.4 Molecularly Imprinted MOF-based Composites

Molecularly imprinted polymer (MIP), which presents a particular size, shape, and functionality, is effectively used as a recognition receptor for specific target molecules. Taking into account the sensitivity, stability, easy fabrication process, and cost-effectiveness in addition to its selectivity, MIP is a powerful candidate for electrochemical sensing applications. The conventional strategy for the preparation of MIP offers some limitations such as long response time, poor site accessibility, and low-rate mass transfer, affecting the electrochemical performance of MIP-based sensors [52]. To overcome these limitations, various supporting nanomaterials such as carbon nanostructures, metal/metal oxide nanoparticles, and porous MOFs have been exploited in the preparation of MIP. For instance, Florea et al. [53] developed a molecularly imprinted microporous MOF prepared by a one-pot electropolymerization method for the sensitive detection of gemcitabine as an antineoplastic drug used in the treatment of non-small cell lung, breast, and ovarian cancers. MIP-based electrochemical sensors were prepared by the electropolymerization of p-amino thiophenol functionalized AuNPs on gold electrodes in the presence of gemcitabine as a template molecule. Considerably good linearity at the range from 3.8 fM to 38 nM with an extremely low detection limit of 3 fM for gemcitabine was obtained in a molecularly imprinted microporous MOF-based electrochemical sensor. In addition, the developed MOF-based MIP sensor was successfully applied to the detection of gemcitabine in drug formulations and spiked serum samples. These attractive analytical performance and recovery results reveal the efficiency of the MIP sensor for biomedical and pharmaceutical analysis involving gemcitabine. In another study conducted by Hatamluyi et al. [54], a well-designed molecularly imprinted 3D MOF-based electrochemical sensor has been reported for the sensitive detection of oxaliplatin, which is a second-generation platinum-based chemotherapeutic agent, at trace level in different biological and pharmaceutical samples. In this context, oxaliplatin imprinted electrode was effectively constructed by the modification of porous 3D Zr-MOF (UiO-66-NH_2) and carbon quantum dots incorporated hexagonal boron nitride nanosheets (CQDs@HBNNS) on GCE. In a MIP-based sensor, the binding capacity is enhanced by a highly porous UiO-66-NH_2, while its sensitivity towards the oxaliplatin has been effectively improved by the use of conductive CQDs@HBNNS. Under optimized electrochemical conditions, CQDs@HBNNS/UiO-66-NH_2/MIP/GCE exhibited two dynamic linear ranges (1 – 20 nM and 20 – 250 nM) with a low detection limit of 0.37 nM for oxaliplatin. The selectivity of the developed MIP sensor was evaluated towards the oxaliplatin in the presence of various biological molecules and ions commonly found in the biological matrix, and the results were found to be highly satisfactory. In addition, acceptable recoveries in different sample matrix including pharmaceutical, serum and urine samples were achieved at CQDs@HBNNS/UiO-66-NH_2/MIP/GCE. The presented analytical results proved that the developed MIP sensor could be a promising method for the monitoring of platinum-based anticancer drugs such as oxaliplatin in clinical and pharmaceutical samples. Norfloxacin, a broad-spectrum antibiotic belonging to the class of fluoroquinolones, has been substantially employed to treat and prevent various inflammations in the human body by inhibiting the activity of DNA gyrase. Norfloxacin concentration directly affects liver enzyme activities. On this point, Ye et al. proposed a polypyrrole-based MIP sensor based on 3D CoFe-MOFs decorated with AuNPs for the detection of trace levels of norfloxacin (Figure 27.9). MIP/CoFe-MOFs/AuNPs modified GCE exhibited an excellent sensitivity, lowest LOD (0.131 pM), selectivity, reproducibility, and stability towards the norfloxacin. To evaluate the applicability of the proposed MIP sensor containing porous MOF structures, the recovery studies for norfloxacin were performed in pharmaceutical and food samples. The obtained results confirmed the feasibility of the MIP-based electrochemical sensor for the detection of norfloxacin in pharmaceutical samples [55].

Embedding the MOF structure with different morphology in a biocompatible polymer is another strategy to fabricate

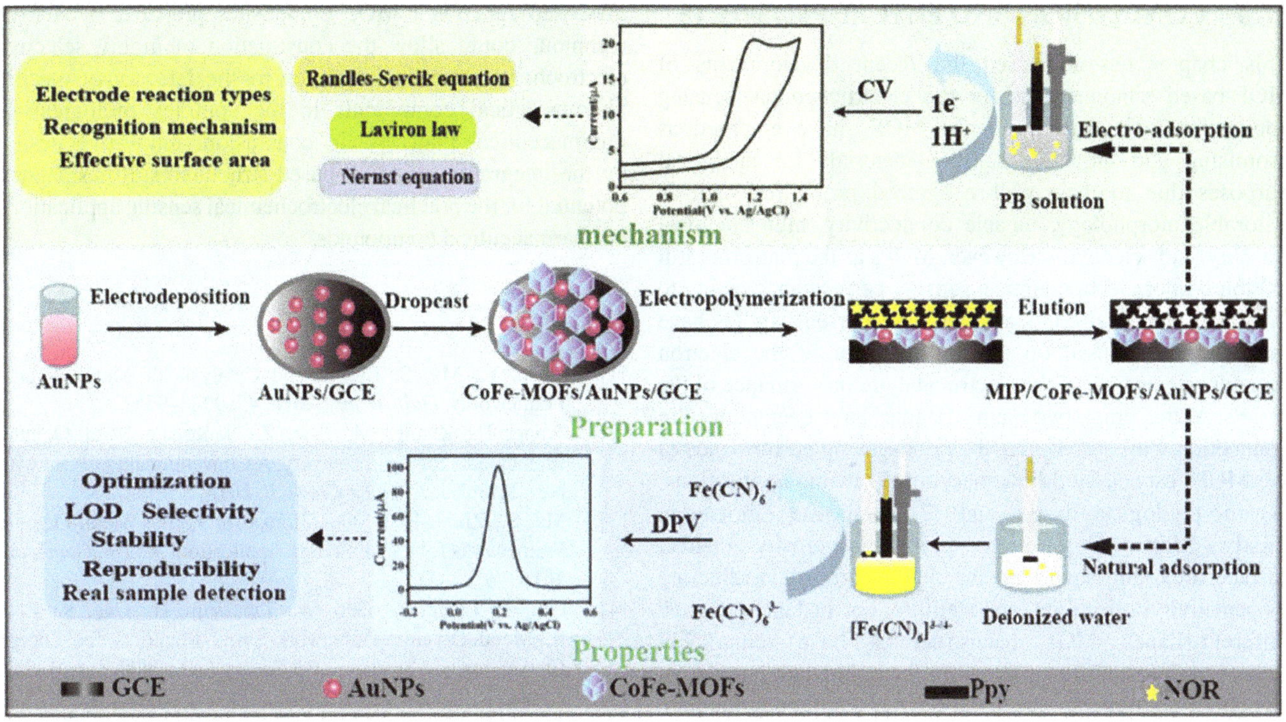

FIGURE 27.9 Schematic representation of the preparation, properties, and determination mechanism process of the MIP/CoFe-MOFs/AuNPs/GCE sensor. Adapted with permission from reference [55]. Copyright (2020) Elsevier.

TABLE 27.3
Recent Electrochemical Sensing Applications Based on Molecularly Imprinted MOF-based Composites Used in the Determination of Pharmaceutical Compounds

Sensing material	Method	Analyte	Linear working range	LOD	Refs
MMOF/MIP	LSV	Gemcitabine	3.8 fM–38 nM	3 fM	[53]
CQDs@HBNNS/UiO-66-NH_2/MIP	DPV	Oxaliplatin	1–20 nM and 20–250 nM	0.37 nM	[54]
MIP/CoFe-MOFs/AuNPs	DPV	Norfloxacin	5 pM–6 nM	0.131 pM	[55]
Co-MOF/MIP	DPV	Ciprofloxacin	0.5–150 μM	0.017 μM	[56]
Cu-MOF/MC/MIP	DPV	Rifampicin	0.08–85 μM	0.28 nM	[57]
		Isoniazid	0.08–85 μM	0.37 nM	

a selective MOF-based composite for sensing applications. Yahyapour et al. [56] synthesized the Co-MOFs nanofibers in the presence of polylactic acid (PLA) as a biodegradable and biocompatible polymer by an electrospinning method. Co-MOFs/PLA nanofibers and ciprofloxacin as a common antibiotic were mixed to construct the MIP-based sensor. For the first time, Co-MOFs/MIP was modified on GCE to detect the ciprofloxacin in pharmaceutical formulations. The linear dynamic range and detection limit for ciprofloxacin at Co-MOFs/MIP sensor were measured as 0.5–150 μM and 0.017 μM, respectively. The presented strategy suggests the Co-MOFs/MIP nanofiber composite as a novel nano sensor for the determination of ciprofloxacin in pharmaceutical applications. Moreover, the MIP-modified electrochemical sensor presents a promising analytical performance for the simultaneous detection of drug molecules. A dual imprinting MIP sensor based on Cu-MOFs/MC was effectively prepared by electro-polymerization for the simultaneous determination of rifampicin and isoniazid as the template molecules. The Cu-MOF/MC/MIP surface has offered a vast quantity of recognition sites for template target molecules, and this feature provided high sensitivity and selectivity towards rifampicin and isoniazid sensing. After optimizing the electro-polymerization cycle and template: monomer ratio, Cu-MOF/MC/MIP on GCE exhibited wide linearity (0.08 - 85 μM), low LOD values (0.28 nM and 0.37 nM for rifampicin and isoniazid), high selectivity, reproducibility, and stability for rifampicin and isoniazid assay. The recovery results for target molecules demonstrated that the developed dual template MIP-based electrochemical sensor could be a favorable analytical tool for the simultaneous determination of rifampicin and isoniazid in pharmaceutical and clinical samples [57].

27.4 CONCLUSION AND FUTURE PROSPECTS

This chapter has addressed the recent developments of MOF-based nanomaterials for the electrochemical sensing applications of pharmaceuticals. MOFs have emerged as promising and attractive sensing materials for analytical purposes due to their well-recognized properties such as tailorable morphology, tunable conductivity, high chemical stability, and selectivity. However, MOFs as the platforms still exhibit a poor electrochemical sensing performance owing to their low electrical conductivities. The current studies have been mainly focused on the improvement of the electron transfer rate as well as the electrocatalytic performance of the MOFs. At this point, two main strategies have been noticed in connection with overcoming the poor sensing performance of the MOFs especially in pharmaceutical sensing applications: (1) morphological design and (2) composite fabrication. Firstly, MOFs morphology including porosity degree, surface area, and shape are the major parameters affecting its sensitivity, selectivity, and stability. For instance, porous/different-shaped MOF nanostructures have remarkably improved their sensitivity owing to their abundant interaction sites between target molecules and MOFs. Although the pore size and shape of the MOFs structures could improve their selectivity towards the specific target molecules (especially for small molecules), a novel strategy has been still required to fabricate the MOF-based sensor for the selective detection of pharmaceuticals. Herein, the molecular imprinting technique offers significant advantages such as a low cost and an easy preparation procedure to fabricate a selective and stable MOF-based electrochemical sensor.

On the other hand, the poor conductive nature of the MOFs has been considered as a second big challenge to improving their electrochemical sensing performance in pharmaceutical analysis. This weak electronic property of the MOFs hindering their potentiality for sensing applications can be easily recovered by the construction of composite materials using noble nanomaterials such as carbon nanostructures, metal/metal oxide nanoparticles, and conductive polymers. Moreover, recent studies have suggested that MOF-based composites facilitate a remarkable sensitivity towards pharmaceutical compounds by accelerating the electron transfer rate between the electrode surface and target analytes. Despite the enormous progress that has been performed in the field of MOF-based sensing, there are still essential points in the development of MOF-based electrochemical sensors for the detection of pharmaceuticals. Therefore, we propose the following potential future research areas for MOF-based electrochemical sensors used in the pharmaceutical assay: (1) biocompatible and biodegradable porous 3D MOFs could be powerful candidates as sensing platforms due to their extremely high surface areas, improved electrical conductivity, and robust skeleton; (2) shape and size-controlled strategy for the preparation of 2D or 3D MOFs structure may be an interesting topic to fabricate uniform sensing platforms for electrochemical applications; (3) 3D-based functional MOF composites prepared by a MIP technique could allow the construction of highly selective electrochemical sensing platforms for the detection of specific pharmaceutical compounds in the complex biological or pharmaceutical matrices. In conclusion, innovative design by the integration of MOF-based structures still has a great potential for the practical electrochemical sensing applications of pharmaceutical compounds.

REFERENCES

[1]. Chen, Y.; Ma, S. Biomimetic Catalysis of Metal-Organic Frameworks. *Dalt. Trans.*, **2016**, *45*, 9744–9753.

[2]. Wang, L.; Zheng, M.; Xie, Z. Nanoscale Metal-Organic Frameworks for Drug Delivery: A Conventional Platform with New Promise. *J. Mater. Chem. B*, **2018**, *6*, 707–717.

[3]. Ma, S.; Zhou, H. C. Gas Storage in Porous Metal-Organic Frameworks for Clean Energy Applications. *Chem. Commun.*, **2010**, *46*, 44–53.

[4]. Qiu, T.; Liang, Z.; Guo, W.; Tabassum, H.; Gao, S.; Zou, R. Metal–Organic Framework-Based Materials for Energy Conversion and Storage. *ACS Energy Lett.*, **2020**, *5*, 520–532.

[5]. Kreno, L. E.; Leong, K.; Farha, O. K.; Allendorf, M.; Van Duyne, R. P.; Hupp, J. T. Metal-Organic Framework Materials as Chemical Sensors. *Chem. Rev.*, **2012**, *112*, 1105–1125.

[6]. Liu, C. Sen; Li, J.; Pang, H. Metal-Organic Framework-Based Materials as an Emerging Platform for Advanced Electrochemical Sensing. *Coord. Chem. Rev.*, **2020**, *410*, 213222.

[7]. Ma, T.; Li, H.; Ma, J. G.; Cheng, P. Application of MOF-Based Materials in Electrochemical Sensing. *Dalt. Trans.*, **2020**, *49*, 17121–17129.

[8]. Chuang, C. H.; Kung, C. W. Metal–Organic Frameworks toward Electrochemical Sensors: Challenges and Opportunities. *Electroanalysis*, **2020**, *32*, 1885–1895.

[9]. Burnett, B. J.; Barron, P. M.; Choe, W. Recent Advances in Porphyrinic Metal-Organic Frameworks: Materials Design, Synthetic Strategies, and Emerging Applications. *Cryst. Eng. Comm.*, **2012**, *14*, 3839–3846.

[10]. Kumar, P.; Deep, A.; Kim, K. H. Metal Organic Frameworks for Sensing Applications. *TrAC - Trends Anal. Chem.*, **2015**, *73*, 39–53.

[11]. Liu, W.; Yin, X. B. Metal-Organic Frameworks for Electrochemical Applications. *TrAC - Trends Anal. Chem.*, **2016**, *75*, 86–96.

[12]. Baumann, A. E.; Burns, D. A.; Liu, B.; Thoi, V. S. Metal-Organic Framework Functionalization and Design Strategies for Advanced Electrochemical Energy Storage Devices. *Commun. Chem.*, **2019**, *2*, 1–14.

[13]. Xue, Y.; Zheng, S.; Xue, H.; Pang, H. Metal-Organic Framework Composites and Their Electrochemical Applications. *J. Mater. Chem. A*, **2019**, *7*, 7301–7327.

[14]. Meng, W.; Wen, Y.; Dai, L.; He, Z.; Wang, L. A Novel Electrochemical Sensor for Glucose Detection Based on Ag@ZIF-67 Nanocomposite. *Sens. Actuators B Chem.*, **2018**, *260*, 852–860.

[15]. Tang, J.; Jiang, S.; Liu, Y.; Zheng, S.; Bai, L.; Guo, J.; Wang, J. Electrochemical Determination of Dopamine and Uric Acid Using a Glassy Carbon Electrode Modified with a Composite

Consisting of a Co(II)-Based Metalorganic Framework (ZIF-67) and Graphene Oxide. *Microchim. Acta*, **2018**, *185*, 486.

[16]. Zhao, J.; Wei, C.; Pang, H. Zeolitic Imidazolate Framework-67 Rhombic Dodecahedral Microcrystals with Porous {110} Facets as a New Electrocatalyst for Sensing Glutathione. *Part. Part. Syst. Charact.*, **2015**, *32*, 429–433.

[17]. Ezzati, M.; Shahrokhian, S.; Hosseini, H. In-situ Two-Step Preparation of 3D NiCo-BTC MOFs on a Glassy Carbon Electrode and a Graphitic Screen Printed Electrode as Nonenzymatic Glucose-Sensing Platforms. *ACS Sustain. Chem. Eng.*, **2020**, *8*, 14340–14352.

[18]. Liu, T.; Zhou, M.; Pu, Y.; Liu, L.; Li, F.; Li, M.; Zhang, M. Silver Nanoparticle-Functionalized 3D Flower-like Copper (II)-Porphyrin Framework Nanocomposites as Signal Enhancers for Fabricating a Sensitive Glutathione Electrochemical Sensor. *Sens. Actuators B Chem.*, **2021**, *342*, 130047.

[19]. Zhou, K.; Shen, D.; Li, X.; Chen, Y.; Hou, L.; Zhang, Y.; Sha, J. Molybdenum Oxide-Based Metal-Organic Framework/Polypyrrole Nanocomposites for Enhancing Electrochemical Detection of Dopamine. *Talanta*, **2020**, *209*, 120507.

[20]. Calbo, J.; Golomb, M. J.; Walsh, A. Redox-Active Metal-Organic Frameworks for Energy Conversion and Storage. *J. Mater. Chem. A*, **2019**, *7*, 16571–16597.

[21]. D'Alessandro, D. M. Exploiting Redox Activity in Metal-Organic Frameworks: Concepts, Trends and Perspectives. *Chem. Commun.*, **2016**, *52*, 8957–8971.

[22]. Ko, M.; Mendecki, L.; Eagleton, A. M.; Durbin, C. G.; Stolz, R. M.; Meng, Z.; Mirica, K. A. Employing Conductive Metal-Organic Frameworks for Voltammetric Detection of Neurochemicals. *J. Am. Chem. Soc.*, **2020**, *142*, 11717–11733.

[23]. Huang, Z.; Yu, H.; Wang, L.; Liu, X.; Lin, T.; Haq, F.; Vatsadze, S. Z.; Lemenovskiy, D. A. Ferrocene-Contained Metal Organic Frameworks: From Synthesis to Applications. *Coord. Chem. Rev.*, **2021**, *430*, 213737.

[24]. Halls, J. E.; Hernán-Gómez, A.; Burrows, A. D.; Marken, F. Metal-Organic Frameworks Post-Synthetically Modified with Ferrocenyl Groups: Framework Effects on Redox Processes and Surface Conduction. *Dalt. Trans.*, **2012**, *41*, 1475–1480.

[25]. Han, J.; Zhang, M.; Chen, G.; Zhang, Y.; Wei, Q.; Zhuo, Y.; Xie, G.; Yuan, R.; Chen, S. Ferrocene Covalently Confined in Porous MOF as Signal Tag for Highly Sensitive Electrochemical Immunoassay of Amyloid-β. *J. Mater. Chem. B*, **2017**, *5*, 8330–8336.

[26]. Chang, Z.; Gao, N.; Li, Y.; He, X. Preparation of Ferrocene Immobilized Metal-Organic-Framework Modified Electrode for the Determination of Acetaminophen. *Anal. Methods*, **2012**, *4*, 4037–4041.

[27]. Liu, C.; Wang, J.; Wan, J.; Yu, C. MOF-on-MOF Hybrids: Synthesis and Applications. *Coord. Chem. Rev.*, **2021**, *432*, 213743.

[28]. Zou, L.; Hou, C. C.; Liu, Z.; Pang, H.; Xu, Q. Superlong Single-Crystal Metal-Organic Framework Nanotubes. *J. Am. Chem. Soc.*, **2018**, *140*, 15393–15401.

[29]. Zhao, M.; Wang, Y.; Ma, Q.; Huang, Y.; Zhang, X.; Ping, J.; Zhang, Z.; Lu, Q.; Yu, Y.; Xu, H.; et al. Ultrathin 2D Metal-Organic Framework Nanosheets. *Adv. Mater.*, **2015**, *27*, 7372–7378.

[30]. Zhang, J.; Gao, L.; Zhang, Y.; Guo, R.; Hu, T. A Heterometallic Sensor Based on Ce@Zn-MOF for Electrochemical Recognition of Uric Acid. *Microporous Mesoporous Mater.*, **2021**, *322*, 111126.

[31]. Safaei, M.; Foroughi, M. M.; Ebrahimpoor, N.; Jahani, S.; Omidi, A.; Khatami, M. A Review on Metal-Organic Frameworks: Synthesis and Applications. *TrAC - Trends Anal. Chem.*, **2019**, *118*, 401–425.

[32]. Zhang, X.; Xu, Y.; Ye, B. An Efficient Electrochemical Glucose Sensor Based on Porous Nickel-Based Metal Organic Framework/Carbon Nanotubes Composite (Ni-MOF/CNTs). *J. Alloys Compd.*, **2018**, *767*, 651–656.

[33]. Zhan, X.; Hu, S.; Wang, J.; Chen, H.; Chen, X.; Yang, J.; Yang, H.; Su, Z. One-Pot Electrodeposition of Metal Organic Frameworks Composite Accelerated by Gold Nanoparticles and Electroreduced Carbon Dots for Electroanalysis of Bisphenol A in Real Plastic Samples. *Sens. Actuators B Chem.*, **2021**, *346*, 130499.

[34]. Wang, J.; Zhao, J.; Yang, J.; Cheng, J.; Tan, Y.; Feng, H.; Li, Y. An Electrochemical Sensor Based on MOF-Derived NiO@ZnO Hollow Microspheres for Isoniazid Determination. *Microchim. Acta*, **2020**, *187*, 380.

[35]. Yuan, S.; Bo, X.; Guo, L. In-Situ Insertion of Multi-Walled Carbon Nanotubes in the Fe3O4/N/C Composite Derived from Iron-Based Metal-Organic Frameworks as a Catalyst for Effective Sensing Acetaminophen and Metronidazole. *Talanta,* **2019**, *193*, 100–109.

[36]. Kemmegne-Mbouguen, J. C.; Tchoumi, F. P.; Mouafo-Tchinda, E.; Langmi, H. W.; Bambalaza, S. E.; Musyoka, N. M.; Kowenje, C.; Mokaya, R. Simultaneous Quantification of Acetaminophen and Tryptophan Using a Composite Graphene Foam/Zr-MOF Film Modified Electrode. *New J. Chem.*, **2020**, *44*, 13108–13117.

[37]. Ma, B.; Guo, H.; Wang, M.; Li, L.; Jia, X.; Chen, H.; Xue, R.; Yang, W. Electrocatalysis of Cu–MOF/Graphene Composite and Its Sensing Application for Electrochemical Simultaneous Determination of Dopamine and Paracetamol. *Electroanalysis*, **2019**, *31*, 1002–1008.

[38]. Naghian, E.; Shahdost-fard, F.; Sohouli, E.; Safarifard, V.; Najafi, M.; Rahimi-Nasrabadi, M.; Sobhani-Nasab, A. Electrochemical Determination of Levodopa on a Reduced Graphene Oxide Paste Electrode Modified with a Metal-Organic Framework. *Microchem. J.*, **2020**, *156*, 104888.

[39]. Habibi, B.; Pashazadeh, S.; Saghatforoush, L. A.; Pashazadeh, A. Direct Electrochemical Synthesis of the Copper Based Metal-Organic Framework on/in the Heteroatoms Doped Graphene/Pencil Graphite Electrode: Highly Sensitive and Selective Electrochemical Sensor for Sertraline Hydrochloride. *J. Electroanal. Chem.*, **2021**, *888*, 115210.

[40]. Dehdashtian, S.; Hashemi, B.; Chegeni, mahdieh; Aeenmehr, A. The Application of Perlite/Cobalt Oxide/Reduced Graphene Oxide (PC-RGO)/Metal Organic Framework (MOF) Composite as Electrode Modifier for Direct Sensing of Anticancer Drug Idarubicin. *IEEE Sens. J.*, **2019**, *19*, 11739–11745.

[41]. Rezvani Jalal, N.; Madrakian, T.; Afkhami, A.; Ghoorchian, A. In-situ Growth of Metal-Organic Framework HKUST-1 on Graphene Oxide Nanoribbons with High Electrochemical Sensing Performance in Imatinib Determination. *ACS Appl. Mater. Interfaces*, **2020**, *12*, 4859–4869.

[42]. Mahnashi, M. H.; Mahmoud, A. M.; Alhazzani, K.; Alanazi, A. Z.; Alaseem, A. M.; Algahtani, M. M.; El-Wekil, M. M. Ultrasensitive and Selective Molecularly Imprinted Electrochemical Oxaliplatin Sensor Based on a Novel Nitrogen-Doped Carbon Nanotubes/Ag@cu MOF as a Signal Enhancer and Reporter Nanohybrid. *Microchim. Acta*, **2021**, *188*, 124.

[43]. Yan, Y.; Ma, J.; Bo, X.; Guo, L. Rod-like Co Based Metal-Organic Framework Embedded into Mesoporous Carbon Composite Modified Glassy Carbon Electrode for Effective Detection of Pyrazinamide and Isonicotinyl Hydrazide in Biological Samples. *Talanta*, **2019**, *205*, 120138.

[44]. Baikeli, Y.; Mamat, X.; He, F.; Xin, X.; Li, Y.; Aisa, H. A.; Hu, G. Electrochemical Determination of Chloramphenicol and Metronidazole by Using a Glassy Carbon Electrode Modified with Iron, Nitrogen Co-Doped Nanoporous Carbon Derived from a Metal-Organic Framework (Type Fe/ZIF-8). *Ecotoxicol. Environ. Saf.*, **2020**, *204*, 111066.

[45]. Rajpurohit, A. S.; Bora, D. K.; Srivastava, A. K. Simultaneous Determination of Amlodipine and Losartan Using an Iron Metal-Organic Framework/Mesoporous Carbon Nanocomposite-Modified Glassy Carbon Electrode by Differential Pulse Voltammetry. *Anal. Methods,* **2018**, *10*, 5423–5438.

[46]. da Silva, D. M.; Carneiro da Cunha Areias, M. Voltammetric Detection of Captopril in a Commercial Drug Using a Gold-Copper Metal-Organic Framework Nanocomposite Modified Electrode. *Electroanalysis*, **2021**, 33, 1255–1263.

[47]. Wang, L.; Meng, T.; Fan, Y.; Chen, C.; Guo, Z.; Wang, H.; Zhang, Y. Electrochemical Study of Acetaminophen Oxidation by Gold Nanoparticles Supported on a Leaf-like Zeolitic Imidazolate Framework. *J. Colloid Interface Sci.*, **2018**, *524*, 1–7.

[48]. Shi, Y.; Zhang, Y.; Wang, Y.; Huang, H.; Ma, J. Amperometric Sensing of Paracetamol Using a Glassy Carbon Electrode Modified with a Composite of Water-Stable Metal-Organic Framework and Gold Nanoparticles. *Int. J. Electrochem. Sci.*, **2018**, *13*, 7643–7654.

[49]. Tang, J.; Liu, Y.; Hu, J.; Zheng, S.; Wang, X.; Zhou, H.; Jin, B. Co-Based Metal-Organic Framework Nanopinnas Composite Doped with Ag Nanoparticles: A Sensitive Electrochemical Sensing Platform for Simultaneous Determination of Dopamine and Acetaminophen. *Microchem. J.*, **2020**, *155*, 104759.

[50]. Wang, H.; Feng, X.; Bo, X.; Zhou, M.; Guo, L. Nickel-Based Metal-Organic Framework/Crosslinked Tubular Poly(3,4-Ethylenedioxythiophene) Composite as an Electrocatalyst for the Detection of Gallic Acid and Tinidazole. *Chem. Electro. Chem.*, **2020**, *7*, 4031–4037.

[51]. Wei, C.; Zhou, H.; Liu, Q. PCN-222 MOF Decorated Conductive PEDOT Films for Sensitive Electrochemical Determination of Chloramphenicol. *Mater. Chem. Phys.*, **2021**, *270*, 124831.

[52]. Liu, Z.; Jin, M.; Lu, H.; Yao, J.; Wang, X.; Zhou, G.; Shui, L. Molecularly Imprinted Polymer Decorated 3D-Framework of Functionalized Multi-Walled Carbon Nanotubes for Ultrasensitive Electrochemical Sensing of Norfloxacin in Pharmaceutical Formulations and Rat Plasma. *Sens. Actuators B Chem.*, **2019**, *288*, 363–372.

[53]. Florea, A.; Guo, Z.; Cristea, C.; Bessueille, F.; Vocanson, F.; Goutaland, F.; Dzyadevych, S.; Səndulescu, R.; Jaffrezic-Renault, N. Anticancer Drug Detection Using a Highly Sensitive Molecularly Imprinted Electrochemical Sensor Based on an Electropolymerized Microporous Metal Organic Framework. *Talanta*, **2015**, *138*, 71 76.

[54]. Hatamluyi, B.; Hashemzadeh, A.; Darroudi, M. A Novel Molecularly Imprinted Polymer Decorated by CQDs@ HBNNS Nanocomposite and UiO-66-NH2 for Ultra-Selective Electrochemical Sensing of Oxaliplatin in Biological Samples. *Sens. Actuators B Chem.*, **2020**, *307*, 127614.

[55]. Ye, C.; Chen, X.; Zhang, D.; Xu, J.; Xi, H.; Wu, T.; Deng, D.; Xiong, C.; Zhang, J.; Huang, G. Study on the Properties and Reaction Mechanism of Polypyrrole@norfloxacin Molecularly Imprinted Electrochemical Sensor Based on Three-Dimensional CoFe-MOFs/AuNPs. *Electrochim. Acta*, **2021**, *379*, 138174.

[56]. Yahyapour, M.; Ranjbar, M.; Mohadesi, A. Determination of Ciprofloxacin Drug with Molecularly Imprinted Polymer/ Co- Metal Organic Framework Nanofiber on Modified Glassy Carbon Electrode (GCE). *J. Mater. Sci. Mater. Electron.*, **2021**, *32*, 3180–3190.

[57]. Rawool, C. R.; Srivastava, A. K. A Dual Template Imprinted Polymer Modified Electrochemical Sensor Based on Cu Metal Organic Framework/Mesoporous Carbon for Highly Sensitive and Selective Recognition of Rifampicin and Isoniazid. *Sens. Actuators B Chem.,* **2019**, *288*, 493–506.

28 MOF-based Electrochemical Sensors for Pharmaceutical Compounds

Atal A.S. Gill[1], *Zondi Nate*[1], *John Alake*[1], *Blessing Wisdom Ike*[1], *Darko Kwabena Adu*[1], *Ruchika Chauhan*[1] *and Rajshekhar Karpoormath*[1]

[1]Department of Pharmaceutical Chemistry, College of Health Sciences, University of KwaZulu-Natal, Westville Campus, Durban, South Africa

28.1 INTRODUCTION

Metal-organic frameworks (MOF) are a crystalline, porous, and polymeric class of materials synthesized by the combination of metal nodes and organic linkers which form 2D or 3D networks of infinite lengths with varying topologies [1, 2]. The majority of the MOFs reported consist of tetrahedral, square, and octahedral geometries. MOFs are synthesized using simple and inexpensive precursors which include inorganic salts like nitrates and chlorides, and organic ligands such as commercially available carboxylic acids. Various methodologies employed for the synthesis of MOFs include electrochemical, microwave-assisted, sonochemical, mechanochemical, solvothermal, and hydrothermal approaches as shown in Figure 28.1, [3, 4]. Due to the porous nature of MOFs, solvents and ligands often get trapped inside the pores which need to be removed by activating the MOF. After activation, the porosity is permanently obtained, and trapped molecules are released. One of the most widely used methods for activation of MOFs is the supercritical drying or vacuum drying method. However, post-activation the electron conductivity is improved in the MOF structure, but it is still poor and can be improved by further modifications.

The low conductivity of electrons in the MOFs can be attributed to their poor overlap between electronic states and frontier orbits of metal ions and ligands [5]. This is the major reason which restricts the use of pristine MOFs in electrochemical devices and electrodes [6, 7]. The most common method employed to overcome the conductivity issue in the MOFs is via incorporating materials with high electro-conductivity in the MOF structure such as metallic nanoparticles, conducting polymers, and conducting nanocrystals. Recently, the use of carbon-based materials and other organic molecules has increased exponentially in the design of electrochemical devices [5–7]. There are various allotropes of carbon-based materials such as graphite, fullerenes, or single-walled carbon nanotubes (SWCNTs) which exhibit amazing properties such as high mechanical strength, high electron conductivity, low cost, thermal stability, and low toxicity. These have been combined with MOFs to form tailored composite materials that have been reported and employed for various applications [8]. Additionally, various organic molecules such as TCNQ and polyacrylic acid have also been used to increase the conductivity of the MOFs and have been employed for electroanalytical applications [9].

Bare working electrodes are usually modified with various nanomaterials such as nanoparticles, nanofibers, and metal-organic frameworks to improve their sensitivity and selectivity to a specific analyte. Among these nanomaterials, MOFs play a major role in providing the modified electrode with tailored properties due to the presence of metal ions and pores. The pore sizes help in the selective uptake of target analytes by increasing the electrochemical response. The variety of topology and pore sizes can be achieved by introducing various combinations of ligands and the geometry can be varied according to the metal ions used. Additionally, the right selection of ligand also aids in achieving better electron transfer capability in the MOF or MOF composite. Various ligands have been used in MOFs. Some of these that are preferred for electrochemical applications include: pyridines, aromatic acids, benzimidazoles, or imidazoles. Further, metal ions are responsible for the selectivity and electrocatalytic activity of the MOF composites. Various transition metals that are used as metal ion cores in MOF composites also function as catalysts for electro-oxidation of various biomolecules such as amino acids, glucose, and secondary metabolites. This is the major reason for the abundant use of MOFs for the modification of electrodes to detect various analytes.

28.2 DIFFERENT TYPES OF STRUCTURES OF MOF FOR ELECTROCHEMICAL DETECTION

28.2.1 2-D MOFs

2-D MOF nanosheets possess a high surface area that results in more accessible active sites and facile diffusion that leads to more efficient response and increased sensitivity of the sensor. Major techniques such as the surfactant-assisted bottom-up synthesis approach, layer-by-layer growth method, and ultrasonication/liquid exfoliation top-down method are used to carry out the synthesis of 2-D MOF nanosheets. Pang et al. managed to synthesize a nanosheet of Ni-MOF with a thickness of 8 nm approximately, using an easy solvothermal

DOI: 10.1201/9781003188148-28

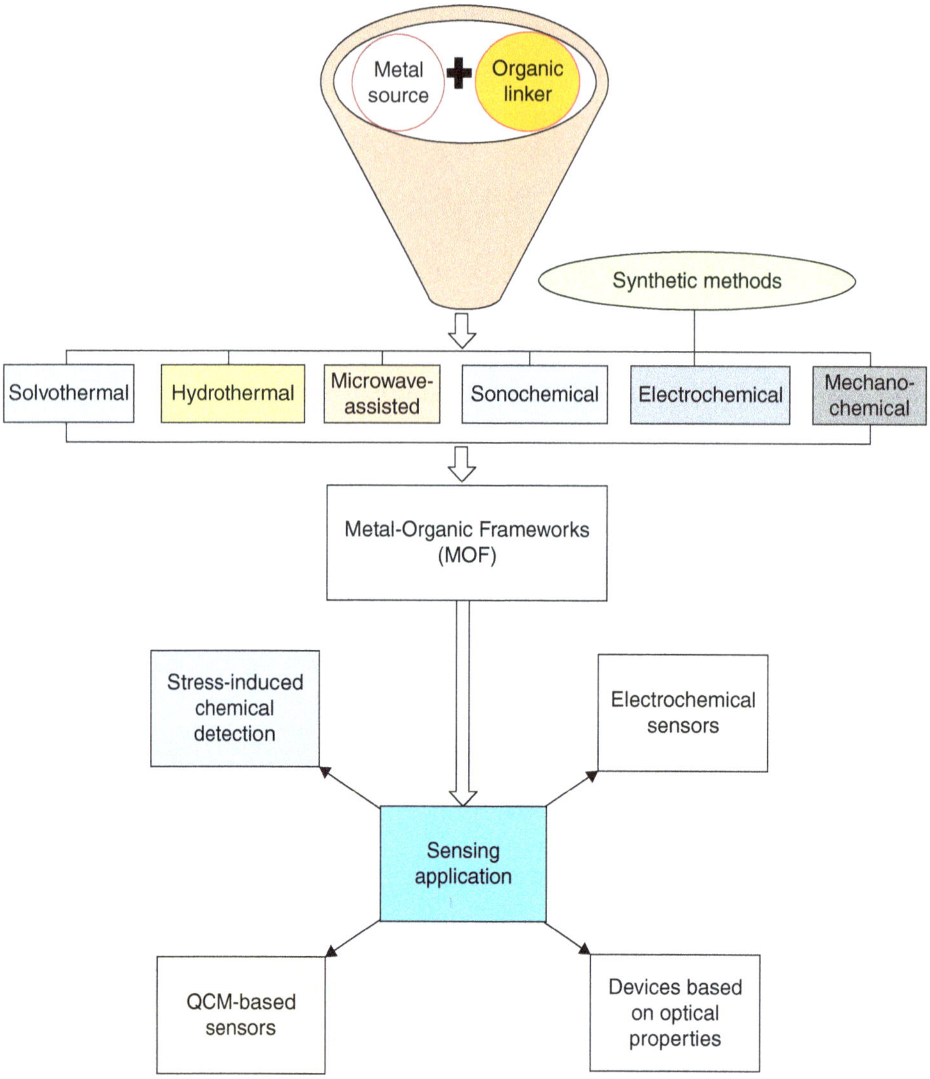

FIGURE 28.1 Schematic diagram of the various synthetic approach and sensing applications for MOFs.

method that was effectively employed for the detection of ascorbic acid [10].

28.2.2 3-D MOFs

The growth of oriented 3-D MOF film on various substrates which are characterized as self-supported and binder-free helps in improving the stability, reproducibility, and sensitivity of the fabricated electrode for sensing applications. The 3-D structure of the MOFs increases the number of electroactive sites which aid in better electron/mass transport in the structure. An example of a self-supported 3-D structure was prepared by Xu et al. [11]. A 3-D Cu-MOF structure was grown over carboxylic acid terminated graphene sheets over a glassy carbon electrode as the substrate. The Cu-MOF crystals were obtained via the solvothermal method and were grown over the graphene sheets using an easy-oriented secondary growth method. This resulted in the growth of 3D porous MOF sheets of 2-5 μm size over the graphene sheets. The growth of the MOF took place according to the orientation of the carboxylic groups on the surface of the graphene sheets.

28.2.3 MOF Thin Films

The growth of electroactive thin films of MOFs over a conductive substrate with strong adhesion, outstanding surface coverage, and controllable thickness are very desirable for the fabrication of an electrode for a highly stable and reproducible electrochemical sensor. The growth of these thin films has been carried out using various techniques which include, layer-by-layer deposition, solvothermal route, and electrochemical deposition techniques. Ho et al. grew thin films of MOF-525 over FTO substrates via a solvothermal method which was used to carry out detection of nitrite using the electrocatalytic activity of the MOF thin films [12]. The only drawback of the method was the intensive preparation steps required to grow the MOF-525 thin films such as pre-treatment of FTO substrate to obtain hydroxyl groups on the surface.

28.2.4 Composite MOF

MOF-based composites are being explored at a high rate due to the diverse tailored properties that can be obtained by the introduction of other nanomaterials to enhance the already extraordinary existing properties of MOFs. Recently, carbon-based materials have been the most used to form composites with MOFs owing to their low cost, high conductivity, stability, and chemical inertness. This is the major reason why MOF/carbon-based composites are preferred over other MOF composites. Various carbon-based materials such as carbon nanofibers, 2-D graphene nanosheets, 3-D graphene aerogels have been employed to prepare MOF composites to fabricate highly efficient materials for electrochemical sensors.

Further, several pristine MOFs have outstanding catalytic activities towards various analytes, however, modification of MOFs with multiple metals could increase the electrocatalytic activity of the overall MOF structure. The porous structure of the MOFs aid in the incorporation of other metal nanoparticles which can act synergistically to increase the electrocatalytic activity of the MOF/nanoparticle composite.

28.3 ELECTROCHEMICAL DETECTION OF RAW MATERIALS USED FOR PHARMACEUTICAL PROCESSES

Pharmaceutical raw materials are usually classified as excipients, active pharmaceutical ingredients, and packaging. These raw materials undergo various safety tests and a series of experiments to prove that they are absorbed in the bloodstream, distributed to the proper site of action in the body, metabolized sufficiently, and demonstrate their non-toxicity and, that they can be considered safe and successful. The quality compliance of pharmaceutical materials cannot be overemphasized. Hence, the compulsion for quality checks of pharmaceutical materials with adherence/and or compliance with several compendia monographs and pharmacopeia methods, categorically established for the testing of raw and finished pharmaceutical products. Amongst some of these many compendia, monographs and pharmacopeia are British Pharmacopeia (BP), Japan Pharmacopeia (JP), United States Pharmacopeia (USP), methods for testing, and validations of pharmaceutical raw materials and final products. Although these methods are used, there is a growing literature that is based on electrochemical detection of pharmaceutical raw materials. The electrochemical detection of raw materials in pharmaceutical processes is essential to the quality control of raw and finished pharmaceutical materials. In achieving quality compliance of pharmaceutical products, it is essential to design and imbibe step-by-step quality checks of items, ranging from the starting material to the finished products.

Analytes undergo oxidative or reducible reactivity in electrochemical studies. The inline testing of each lot of pharmaceutical raw materials is highly important and it is recommended to check preliminary safety and therapeutic records of materials involved in the drug formulations process. Hence the compulsory procedures to be implemented in drug formulation during pharmaceutical processes by companies in quality compliance of raw materials, in productions, handling, transporting, and storage of pharmaceutical products.

Recently, the incorporation of biomolecules in the structure of the MOF such as nucleic acids, antibodies, and enzymes have been explored. These increase the sensitivity, selectivity, and detection range of MOF-based electrochemical sensors. The high specific surface area and porous structure, as well as tailored geometry and functional groups of the MOFs, make them the most promising materials to immobilize biomolecules effectively.

28.4 ANALYTICAL TECHNIQUES INVOLVED IN THE QUALITY CONTROL AND DETECTION OF RAW MATERIALS

There are several analytical techniques involved in the quality control and detection of raw materials. These include: titrimetric, chromatographic, spectroscopic, electrophoretic, and electrochemical methods. This chapter focuses on electrochemical methods. This method is gaining research interest because it is selective, sensitive, and reproducible. It is also easy to set up and cost-effective. Several electrochemical techniques such as cyclic voltammetry, electrochemical impedance spectroscopy, and differential pulse voltammetry have been optimally utilized for real-time study, quantifications, and characterization of a great number of pharmaceutical materials. There are lists of literature available on the use of metal-organic frameworks, for the successful detection of several pharmaceutical raw materials. Some advantages achieved in this field include the use of a copper-based metal-organic framework (Cu BTC) for the detection of ciprofloxacin with a limit detection of 0.47×10^9 mol/L using a differential pulse voltammetry technique [13]. In addition, Sulfadiazine was detected using the electrochemical impedance spectroscopy (EIS) technique with a limit of detection of 0.0013 μM [14]. Electrochemical detection techniques have aided real-time quality control of drugs, with high sensitivity and selectivity and great reproducibility.

28.5 ELECTRO-ACTIVE PHARMACEUTICAL DRUGS

MOFs have been used for many applications, but their use in electrochemical sensor applications is comparatively rare due to their weak electronic conductivity, poor electro-activity, and low stability in aqueous media [15]. However, metal-organic framework composites have been applied in many instances for the electrochemical detection of pharmaceutical drugs. This is because pristine MOFs are poor electron conductors however their composites with other materials such as carbonaceous nanomaterials, noble metal nanoparticles, or metal oxide nanoparticles, show great activity in improving their electrocatalytic activity [15]. Metal and metal oxides are inserted in the porous matrix of MOFs,

TABLE 28.1
MOFs and MOF Composites Used in the Electrochemical Detection of Pharmaceutical Drugs

Nanocomposite	Analyte	Technique	Limit of detection	ref
Cu-MOF/graphene,	Acetaminophen	CV, DPV	0.36 μM	[20]
Cu-MOF/graphene	Dopamine	CV, DPV	0.21 μM	[20]
Cu-MOF/graphene	Caffein			[27]
ZIF-67C@rGO-0.06	Metronidazole	CV	0.05 μM	[24]
Co-MOF derived $Co(OH)_2$ /MWCNT	hydrogen peroxide	amperometry	1.36 μM	[23]
AChE@Ni-MOF	Galantamine hydrobromide	DPV	0.31 pM	[28]
Co-Ni-Cu-MOF	Nilutamide	CV, DPV	0.48 ± 0.02 nM	[29]
Ni-doped nanoporous carbon (Ni/C-400)	Acetaminophen	DPVSS	0.0404 μM	[16]
NiCu-CAT	Paracetamol	CV, DPV	5μM	[30]
Co- MOCP	Glutathione	Amperometry	2.5 μM	[31]
MIL-101	Dopamine	CV	Not stated	[25]
HKUST-CNF	Diclofenac	CA, MPA	3.2 μM	[21]
HKUST-CNF	IBUPROFEN	CA, MPA	6.1 μM	[21]
Fc-MOF	Acetaminophen	CA	0.0064 μM	[32]
poly(acrylic acid)-modified Cu-MOF	Vancomycin	CV	1 nM	[9]
FeMOF/MC	Amlodipine	DPV	1.27 mM	[22]
FeMOF/MC	Losartan	DPV	2.03 mM	[22]
MIL-101(Cr)	Ascorbic acid	CV	0.006 mM	[19]
AgPd@Zr-MOF	Dopamine	CV	0.1 μM	[26]

Abbreviations: Cu-MOF (Copper- Metal-organic frameworks), MIL (Matérial Institut Lavoisier), *HKUST* ([$Cu_3(BTC)_2$] MOF) BTC(1,3,5 benzentricarboxilate), CNF (Carbon Nanofibre), FeMOF (iron- Metal-organic frameworks), MC (Mesoporous carbon), MOCPs (Metal-organic coordination polymers), CV (Cyclic Voltammetry), NiCu-CAT(Nickel-Copper-catecholates), ZIF (zeolite imidazolate framework), rGO (reduced graphene oxide), MWCNT (Multi-wall Carbon Nanotube), AChE@Ni-MOF (*acetylcholinesterase*-Nikel-based Metal-organic frameworks), Fc-MOF (ferrocene metal-organic framework), CA (chronoamperometry), DPV (Differential Pulse Voltammetry), MPA (multiple pulse amperometry) and LOD (Limit of Defection).

which aids in improving their catalytic activity. Therefore, metal and metal oxide-based MOF composites have been used for electrochemical detection of electroactive drugs [16, 17]. MOFs and MOF composites have been shown by several researchers to improve the electrocatalytic activity of several drugs during electrochemical analysis [18]. Modification of GCE with MIL-101(Cr) improved the electro-oxidation of ascorbic acid aiding its detection by cyclic voltammetry [19]. In the research reported by Tashkhourian et al [19], the limit of detection was reported to be 0.006 Mm. Wang et al. [20] also utilized Cu-MOF decorated graphene for the simultaneous detection of acetaminophen in the presence of dopamine using differential pulse voltammetry. Two well-defined peaks of dopamine and acetaminophen were obtained. In another research, ibuprofen and diclofenac were simultaneously detected by chronoamperometry (CA) and multiple pulse amperometry (MPA) using HKUST metal-organic framework-carbon nanofiber composite (HKUST-CNF) with a limit of detection of 6.1 μM and 3.2 μM, respectively [21]. A similar case of simultaneous detection of two species is reported by Rajpurohit et al. [22] who utilized iron-MOF/mesoporous carbon (FeMOF/MC) to detect amlodipine and losartan by differential pulse voltammetry. Furthermore, other metal-organic frameworks have been used in the detection of electroactive pharmaceutical entities such as metronidazole and hydrogen peroxide [23, 24]. This does not only confirm the applicability of MOF composites in drug detection but also suggests that MOF/composite can be applied in the simultaneous detection of two entities.

Physiological compounds such as glutathione and dopamine which are also available as medication have been successfully detected using MOF-based electrochemical sensors. Electrochemical detection of dopamine using cyclic voltammetry has been reported by three different research groups. These groups used Cu-MOF/graphene, MIL-101, and AgPd@Zr-MOF, and a limit of detection ranging between 0.1 μM and 0.21 μM was obtained [20, 25, 26]. According to Yuan et al. (2014), glutathione was successfully detected by Metal-organic coordination polymers (MOCPs) using amperometry. A detection limit of 2.5 μM was obtained. Table 28.1 shows a few examples of some composites of MOFs and pharmaceutical drugs that have been recently used.

28.6 ELECTRO-INACTIVE PHARMACEUTICAL DRUGS

Electro-inactive drugs may not undergo redox reaction or their less stable radical cation may shift outside the working potential of the commonly used electrode systems but the introduction of redox-active groups to convert the redox inactive drugs is promising [17]. Hence with the appropriate redox couple like $[Fe(CN)_6]^{3-/4-}$, Atal et al. [9] utilized a poly(acrylic acid)-modified Cu-MOF to electrochemically determine electro-inactive drug vancomycin. The researchers employed the cyclic voltammetry technique and reported the lowest detectable concentration of 1 nM.

28.7 WASTE FROM PHARMACEUTICAL PROCESSES

Pharmaceutical wastes are waste products from pharmaceutical processes, which include by-products from pharmaceutical processes, expired drugs, and pharmaceutically active drugs in waste water. This work focuses on the detection and determination of by-products from pharmaceutical processes such as hydrogen peroxide, sodium sulfate, heavy metals, nitrite, metal hydride, and cyanide, and pharmaceutical compounds such as diclofenac, dopamine, and ibuprofen. MOF and MOF-based nanocomposites have successfully been applied in photocatalytic detection, adsorption, and removal of pharmaceutical waste [33, 34]. This cannot be said of MOF-based electrochemical detection and determination of pharmaceutical waste. Wang et al. [35] and Chang et al. [36] applied MOF-based nanocomposite, Ni (II)-MOF/CNT and Co@MOF-808/FTO respectively to electrochemically detect hydrogen peroxide. The type of electrochemical mechanism used in the detection of H_2O_2 was a direct electrochemical reduction. In the direct electrochemical reduction method, electron transfer or reduction occurs between the electron and the analyte without any intermediary. A limit of detection of 2.1 μmol/L and 1.3 μM was obtained, respectively. The cyclic voltammetric technique was utilized in the electrocatalytic activity studies of H_2O_2 towards the Ni (II)-MOF/CNT and Co@MOF-808/FTO modified electrodes. In the H_2O_2 detection studies, an amperometric technique was employed in both experiments.

Heavy metals such as Fe, Pb, Zn, Mg, and other metal hydrides used in chemical processes in the pharmaceutical industry end up as waste. Deng et al. [37] designed a graphene-aerogel-metal-organic framework to detect heavy metals, Cd^{2+}, Pb^{2+}, Cu^{2+}, and Hg^{2+} simultaneously and individually in an aqueous solution. The metals were detected at detection limit of 0.02 μM for Cd^{2+}, 1.5 nM for Pb^{2+}, 7 nM for Cu^{2+} and 2 nM for Hg^{2+} for the individual metal detection. In the simultaneous heavy metal detection, the metals were detected at detection limit of 9 nM for Cd^{2+}, 1 nM for Pb^{2+}, 8 nM for Cu^{2+} and 0.9 nM for Hg^{2+}. These results indicate that the synthesized material is good, therefore it can be employed in the industry to detect heavy metals.

Diclofenac (DCF) and ibuprofen (IBP), which are examples of pharmaceutical compounds found in pharmaceutical waste were detected by Motoc et al. [21] in an aqueous solution, using metal-organic framework-carbon nanofiber composite (HKUST-CNT). Cyclic voltammetry (CV), chronoamperometry (CA), and multiple-pulsed amperometry (MPA) were the electrochemical techniques used. The authors reported individual detection limits of 21.70 μg/L for IBP and 100 μg/L for DCF for CV studies. 0.13 μg/L for IBP and 1.5 μg/L for DCF for CA studies. The authors also employed multiple-pulsed amperometry to improve the electroanalytical properties for the detection of the individual drugs and a detection limit of 4.0 μg/L and 2.08 μg/L was obtained for IBP and DCF, respectively. Table 28.2 above shows some examples of MOF and MOF-based nanocomposites of different compositions that have been used to detect H_2O_2, nitrite, heavy metals, and pharmaceutical compounds.

TABLE 28.2
MOF-based Electrochemical Detection of Pharmaceutical Waste

Analyte	Electrode material	Technique	Limit detection (M)	Ref
H_2O_2	Sn-MOF@CNT	CA	4.7×10^{-3}	[38]
H_2O_2	Cu-MOF/MXene	CA	0.35×10^{-6}	[39]
H_2O_2	MIL-53-Cr^{III}	DPV	3.52×10^{-6}	[40]
H_2O_2	AP-Ni-MOF	CV, CA	9×10^{-7}	[41]
Pb^{2+}	MOF-177	CV	4×10^{-9}	[42]
Cd^{2+}	MOF-177	CV	3×10^{-8}	[42]
Nitrite	Cu (II)-MOF	CA	5.0×10^{6}	[43]
Nitrite	MOF-525	Amperometry	2.1×10^{-6}	[44]
Dopamine	MOF nanosheet/Au/ poluxanthurenic acid composite	DPV	1×10^{-6}	[45]

Abbreviations: MOF (Metal-organic frameworks), CNT (Carbon nanotube), H_2O_2 (Hydrogen peroxide), Sn (Tin), Pb^{2+} (Lead ion), Cd^{2+} (Cadmium ion), Au (gold), Cu (II) (Copper (II) ion), Cr(III) (Chronium ion), Ni (Nickel), MOF-177 (Zinc-based metal organic framework), MOF-525 (Zirconium-based metal organic framework), AP-Ni-MOF (Nickel-metal-organic framework based on adipic acid as linker and piperazine as ligand), MIL (Matérial Institut Lavoisier), CA (chronoamperometry), CV (Cyclic Voltammetry), DPV (Differential Pulse Voltammetry).

28.8 CONCLUSION AND FUTURE PERSPECTIVES

The properties of metal-organic frameworks such as high porosity, high crystallinity, large surface-to-volume ratio, and tunable pore size make them suitable to be used in various applications. However, the application of MOFs for electrochemical detection of pharmaceutical drugs appears to be relatively low. This area needs to be further exploited since it has been demonstrated by several research groups that low detection limits and good selectivity are possible. Another area of interest that still needs to be looked into is the detection of pharmaceutical compounds in wastewater using MOF-based electrochemical sensors. The application of less expensive, sensitive, and specific MOF-based electrochemical sensors to the detection of pharmaceutical by-products, will contribute immensely to ridding the environment of these hazardous wastes from pharmaceutical processes.

REFERENCES

[1]. H. He, D. Collins, F. Dai, X. Zhao, G. Zhang, H. Ma, D. Sun, Construction of Metal–Organic Frameworks with 1D Chain, 2D Grid, and 3D Porous Framework Based on a Flexible

Imidazole Ligand and Rigid Benzenedicarboxylates, Cryst. *Growth Des.* 10 (2010) 895–902.

[2]. M.J. Kalmutzki, N. Hanikel, O.M. Yaghi, Secondary building units as the turning point in the development of the reticular chemistry of MOFs, *Sci. Adv.* 4 (2018) eaat9180.

[3]. A.J. Howarth, A.W. Peters, N.A. Vermeulen, T.C. Wang, J.T. Hupp, O.K. Farha, Best Practices for the Synthesis, Activation, and Characterization of Metal–Organic Frameworks, *Chem. Mater.* 29 (2017) 26–39.

[4]. M. Bosch, S. Yuan, W. Rutledge, H.-C. Zhou, Stepwise Synthesis of Metal–Organic Frameworks, Acc. *Chem. Res.* 50 (2017) 857–865.

[5]. W. Liu, X.-B. Yin, Metal–organic frameworks for electrochemical applications, TrAC Trends Anal. *Chem.* 75 (2016) 86–96.

[6]. A. Morozan, F. Jaouen, One-step synthesis of a copper-based metal-organic framework-graphene nanocomposite with enhanced electrocatalytic activity, *Energy Environ. Sci.* 5 (2012) 9269–9290.

[7]. J. Yang, F. Zhao, B. Zeng, One-step synthesis of a copper-based metal–organic framework–graphene nanocomposite with enhanced electrocatalytic activity, *RSC Adv.* 5 (2015) 22060–22065.

[8]. X.-W. Liu, T.-J. Sun, J.-L. Hu, S.-D. Wang, Composites of metal–organic frameworks and carbon-based materials: preparations, functionalities and applications, *J. Mater. Chem. A.* 4 (2016) 3584–3616.

[9]. A.A.S. Gill, S. Singh, N. Agrawal, Z. Nate, T.E. Chiwunze, N.B. Thapliyal, R. Chauhan, R. Karpoormath, A poly(acrylic acid)-modified copper-organic framework for electrochemical determination of vancomycin, *Microchim. Acta.* 187 (2020) 79.

[10]. Q. Li, S. Zheng, X. Hu, Z. Shao, M. Zheng, H. Pang, Ultrathin Nanosheet Ni-Metal Organic Framework Assemblies for High-Efficiency Ascorbic Acid Electrocatalysis, *ChemElectroChem.* 5 (2018) 3859–3865.

[11]. Y. Zhou, C. Li, Y. Hao, B. Ye, M. Xu, Oriented growth of cross-linked metal-organic framework film on graphene surface for non-enzymatic electrochemical sensor of hydrogen peroxide in disinfectant, *Talanta.* 188 (2018) 282–287.

[12]. O. Shekhah, J. Liu, R.A. Fischer, C. Wöll, MOF thin films: existing and future applications, *Chem. Soc. Rev.* 40 (2011) 1081–1106.

[13]. R. Rani, A. Deep, B. Mizaikoff, S. Singh, Copper Based Organic Framework Modified Electrosensor for Selective and Sensitive Detection of Ciprofloxacin, *Electroanalysis.* 32 (2020) 2442–2451.

[14]. S. Chen, C. Wang, M. Zhang, W. Zhang, J. Qi, X. Sun, L. Wang, J. Li, N-doped Cu-MOFs for efficient electrochemical determination of dopamine and sulfanilamide., *J. Hazard. Mater.* 390 (2020) 122157.

[15]. S. Kempahanumakkagari, K. Vellingiri, A. Deep, E.E. Kwon, N. Bolan, K.H. Kim, Metal–organic framework composites as electrocatalysts for electrochemical sensing applications, *Coord. Chem. Rev.* 357 (2018) 105–129.

[16]. L. Guo, L. Hao, Y. Zhang, X. Yang, Q. Wang, Z. Wang, C. Wang, Metal-organic framework precursors derived Ni-doping porous carbon spheres for sensitive electrochemical detection of acetaminophen, *Talanta.* 228 (2021) 122228.

[17]. J. Schram, S.T. Shanmugam, N. Sleegers, A. Florea, N. Samyn, A.L.N. van Nuijs, K. De Wael, Local conversion of redox inactive molecules into redox active ones: A formaldehyde based strategy for the electrochemical detection of illicit drugs containing primary and secondary amines, *Electrochim. Acta.* 367 (2021) 137515.

[18]. A. Mahmood, W. Guo, H. Tabassum, R. Zou, Metal-Organic Framework-Based Nanomaterials for Electrocatalysis, *Adv. Energy Mater.* 6 (2016) 1600423.

[19]. J. Tashkhourian, H. Valizadeh, A. Abbaspour, Ascorbic acid determination based on electrocatalytic behavior of metal-organic framework MIL-101-(Cr) at modified carbon-paste electrode, *J. AOAC Int.* 102 (2019) 625–632.

[20]. X. Wang, Q. Wang, Q. Wang, F. Gao, F. Gao, Y. Yang, H. Guo, Highly dispersible and stable copper terephthalate metal-organic framework-graphene oxide nanocomposite for an electrochemical sensing application, *ACS Appl. Mater. Interfaces.* 6 (2014) 11573–11580.

[21]. S. Motoc, F. Manea, A. Iacob, A. Martinez-Joaristi, J. Gascon, A. Pop, J. Schoonman, Electrochemical Selective and Simultaneous Detection of Diclofenac and Ibuprofen in Aqueous Solution Using HKUST-1 Metal-Organic Framework-Carbon Nanofiber Composite Electrode, *Sensors.* 16 (2016) 1719.

[22]. A.S. Rajpurohit, D.K. Bora, A.K. Srivastava, Simultaneous determination of amlodipine and losartan using an iron metal-organic framework/mesoporous carbon nanocomposite-modified glassy carbon electrode by differential pulse voltammetry, *Anal. Methods.* 10 (2018) 5423–5438.

[23]. N.M. Umesh, K.K. rani, R. Devasenathipathy, B. Sriram, Y.X. Liu, S.F. Wang, Preparation of Co-MOF derived $Co(OH)_d$/ multiwalled carbon nanotubes as an efficient bifunctional electro catalyst for hydrazine and hydrogen peroxide detections, *J. Taiwan Inst. Chem. Eng.* 93 (2018) 79–86.

[24]. H. Chen, X. Wu, R. Zhao, Z. Zheng, Q. Yuan, Z. Dong, W. Gan, Preparation of reduced graphite oxide loaded with cobalt(II) and nitrogen co-doped carbon polyhedrons from a metal-organic framework (type ZIF-67), and its application to electrochemical determination of metronidazole, *Microchim. Acta.* 186 (2019) 623.

[25]. Y. Li, Y. Li, C. Huangfu, H. Du, W. Liu, J. Ye, Electrochemical behavior of metal-organic framework MIL-101 modified carbon paste electrode: An excellent candidate for electroanalysis, *J. Electroanal. Chem.* 709 (2013) 65–69.

[26]. S.A. Hira, S. Nagappan, D. Annas, Y.A. Kumar, K.H. Park, NO2-functionalized metal–organic framework incorporating bimetallic alloy nanoparticles as a sensor for efficient electrochemical detection of dopamine, *Electrochem. Commun.* 125 (2021) 107012.

[27]. A. Venkadesh, J. Mathiyarasu, S. Radhakrishnan, Voltammetric Sensing of Caffeine in Food Sample Using Cu-MOF and Graphene, *Electroanalysis.* 33 (2021) 1007–1013.

[28]. L. Zhang, C. Qiao, X. Cai, Z. Xia, J. Han, Q. Yang, C. Zhou, S. Chen, S. Gao, Microcalorimetry-guided pore-microenvironment optimization to improve sensitivity of Ni-MOF electrochemical biosensor for chiral galantamine, *Chem. Eng. J.* 426 (2021) 130730.

[29]. S. Akhter, N.K. Mohd Zain, M. Shalauddin, V.K. Singh, I.I. Misnon, R.K. Sharma, S. Das, W.J. Basirun, M.R. Johan, R. Jose, Tri-metallic Co-Ni-Cu based metal organic framework nanostructures for the detection of an anticancer drug nilutamide, Sensors Actuators, *A Phys.* 325 (2021) 112711.

[30]. J. Wang, S. Liu, J. Luo, S. Hou, H. Song, Y. Niu, C. Zhang, Conductive Metal-Organic Frameworks for Amperometric Sensing of Paracetamol, *Front. Chem.* 8 (2020) 1–9.

[31]. B. Yuan, R. Zhang, X. Jiao, J. Li, H. Shi, D. Zhang, Amperometric determination of reduced glutathione with a new Co-based metal-organic coordination polymer modified electrode, Electrochem. *Commun.* 40 (2014) 92–95.

[32]. Z. Chang, N. Gao, Y. Li, X. He, Preparation of ferrocene immobilized metal-organic-framework modified electrode for the determination of acetaminophen, *Anal. Methods.* 4 (2012) 4037–4041.

[33]. M. Samy, M.G. Ibrahim, M. Fujii, K.E. Diab, M. ElKady, M. Gar Alalm, CNTs/MOF-808 painted plates for extended treatment of pharmaceutical and agrochemical wastewaters in a novel photocatalytic reactor, *Chem. Eng. J.* 406 (2021) 127152.

[34]. R.M. Rego, G. Sriram, K. V. Ajeya, H.-Y. Jung, M.D. Kurkuri, M. Kigga, Cerium based UiO-66 MOF as a multipollutant adsorbent for universal water purification, *J. Hazard. Mater.* 416 (2021) 125941.

[35]. M.Q. Wang, Y. Zhang, S.J. Bao, Y.N. Yu, C. Ye, Ni(II)-Based Metal-Organic Framework Anchored on Carbon Nanotubes for Highly Sensitive Non-Enzymatic Hydrogen Peroxide Sensing, *Electrochim. Acta.* 190 (2016) 365–370.

[36]. Y.S. Chang, J.H. Li, Y.C. Chen, W.H. Ho, Y. Da Song, C.W. Kung, Electrodeposition of pore-confined cobalt in metal–organic framework thin films toward electrochemical H_2O_2 detection, *Electrochim. Acta.* 347 (2020) 136276.

[37]. M. Lu, Y. Deng, Y. Luo, J. Lv, T. Li, J. Xu, S. Chen, J. Wang, Graphene Aerogel–Metal–Organic Framework-Based Electrochemical Method for Simultaneous Detection of Multiple Heavy-Metal Ions, *Anal. Chem.* 91 (2018) 888–895.

[38]. S. Rani, B. Sharma, R. Malhotra, S. Kumar, R.S. Varma, N. Dilbaghi, Sn-MOF@CNTnanocomposite: An efficient electrochemical sensor for detection of hydrogen peroxide, *Environ. Res.* 191 (2020) 110005.

[39]. D. Cheng, P. Li, X. Zhu, M. Liu, Y. Zhang, Y. Liu, Enzyme-free Electrochemical Detection of Hydrogen Peroxide Based on the Three-Dimensional Flower-like Cu-based Metal Organic Frameworks and MXene Nanosheets, *Chinese J. Chem.* 39 (2021) 2181–2187.

[40]. N. Siraj, M. Rahman, F. Ahmed, S. Chandra, T. Ryu, W. Kim, Electrochimica Acta A base-stable metal-organic framework for sensitive and non- enzymatic electrochemical detection of hydrogen peroxide, *Electrochim. Acta.* 274 (2018) 49–56.

[41]. B. Sherino, S. Mohamad, S.N. Abdul Halim, N.S. Abdul Manan, Electrochemical detection of hydrogen peroxide on a new microporous Ni–metal organic framework material-carbon paste electrode, *Sensors Actuators B Chem.* 254 (2018) 1148–1156.

[42]. S. Sangeetha, G. Krishnamurthy, Fabrication of MOF-177 for electrochemical detection of toxic$Pb^{2+}$and Cd^{2+}ions, *Bull. Mater. Sci.* 43 (2020) 29.

[43]. M.C. Cassani, R. Castagnoli, F. Gambassi, D. Nanni, I. Ragazzini, N. Masciocchi, E. Boanini, B. Ballarin, A Cu(II)-MOF Based on a Propargyl Carbamate-Functionalized Isophthalate Ligand as Nitrite Electrochemical Sensor, *Sensors.* 21 (2021) 4922.

[44]. C.-W. Kung, T.-H. Chang, L.-Y. Chou, J.T. Hupp, O.K. Farha, K.-C. Ho, Porphyrin-based metal–organic framework thin films for electrochemical nitrite detection, *Electrochem. Commun.* 58 (2015) 51–56.

[45]. Z. Qiu, T. Yang, R. Gao, G. Jie, W. Hou, An electrochemical ratiometric sensor based on 2D MOF nanosheet/Au/polyxanthurenic acid composite for detection of dopamine, *J. Electroanal. Chem.* 835 (2019) 123–129.

29 MOF-based Electrochemical Sensors for Endocrine-disrupting Compounds

Yukun Yang[1], Zhuo Shi[1], Wenyan Yan[2], Xiaomin Wang[3], Jinhua Zhang[1], Ligang Yu[1], Caixia Guo[1] and Baoqing Bai[1]
[1]School of Life Science, Shanxi University, Taiyuan, China
[2]College of Biosystems Engineering and Food Science, Key Laboratory of Agro-Products Postharvest Handling, Ministry of Agriculture, Zhejiang University, Hangzhou, China
[3]Institute of Pharmaceutical and Food Engineering, Shanxi University of Chinese Medicine, Yuci, China

29.1 INTRODUCTION

Our endocrine system (ES) is composed of endocrine organs and glands all over the body, including the pancreas, thymus, thyroid gland, and adrenal gland, which can produce, store, and secrete hormones. When functioning normally, the ES works in conjunction with other living systems to regulate the growth, development, metabolism, and reproduction of the whole body. Endocrine-disrupting compounds (EDCs) are generally defined as exogenous chemicals or mixtures of chemicals which can exert a significant effect on ES by mimicking and interfering with the endogenous chemical hormones [1]. Although some EDCs are naturally occurring, the vast majority of EDCs are synthetic chemicals, such as bisphenol A (BPA), organic chlorine pesticides, alkylphenol (AP), alkylphenol polyoxyethylene ether (APE), phthalate (PAE), and polychlorinated biphenyls (PCB). EDCs are extensively present in pesticides, electronic products, personal care products, and cosmetics, as well as food additives or contaminants.

Frequent industrial emissions, agricultural emissions, waste combustion and emissions have led to more and more EDCs entering the environment. Humans may be exposed to EDCs through ingestion of food and water, inhalation of gases and particulates in the air, and skin contact. Moreover, pregnant women and children are the most vulnerable groups to be harmed by EDCs, which can be transferred from pregnant women to developing fetuses or infants through the placenta and breast milk. The effects of exposure to EDCs may have a delayed effect, and more importantly, recent studies have shown that it may increase the susceptibility to non-communicable diseases. Many reports have proved that EDCs exposure has a serious impact on human health, and EDCs have high relevance with diseases including cryptorchidism in young men, breast cancer in women, prostate cancer in men, neurodevelopmental disorders, attention deficit in children, and thyroid tumors.

Nowadays, chemical products are becoming more and more diverse and have become a part of modern life. However, poor chemical management, abuse, or improper handling may pose a threat to the realization of the overall goal of human development and sustainable development. From a global perspective, the pollution of EDCs is a serious environmental problem and global public concern, the degree of EDCs pollution and the scope of EDCs pollution are unprecedented. In order to increase the global awareness of EDCs, reduce the risk of potential diseases, and reduce related medical costs, monitoring of EDCs is very important, especially advanced high-sensitivity and high-selectivity detection methods are highly desired.

29.2 CURRENT RESEARCH OF EDC DETECTION AND MOF-BASED ELECTROCHEMICAL SENSORS

Conventional instrumental techniques are commonly applied for the accurate and sensitive detection of EDCs, including liquid chromatography coupled with mass spectrometry (LC-MS), and gas chromatography coupled with mass spectrometry (GC-MS) [2]. Though these detection methods based on large instruments can offer highly sensitive and accurate detection of EDCs, these conventional methods have some demerits such as tedious tasks of time-consuming sample preparation, requisition of professional technicians, and expensive instruments. In response to the global EDCs pollution status, the development of fast, reliable, and low-cost on-site detection methods for the detection of EDCs has become an important research content in the field of environmental monitoring and food safety. As a promising detection device for the detection of EDCs, electrochemical sensors display the merits of rapid response time, high sensitivity, ease of miniaturization, convenient operation, and low cost [3]. As we all know, the core component of electrochemical sensor construction is the construction of the functional interface of electrochemical sensing. Up to now, various electrode surface modification methods have been explored and developed to improve the performance of electrochemical sensors by

DOI: 10.1201/9781003188148-29

improving electrical conductivity, offering more binding sites, and providing more electrocatalytic sites [4].

Metal-organic frameworks (MOFs) have shown great potential for the construction of electrochemical sensing interfaces to improve the above-mentioned aspects because of their attractive properties, such as high porosity, large specific surface, unsaturated metal sites, tailorable structure, and chemical functionality [5, 6]. Though poor electrical conductivity of most MOFs restricts their application in electrochemical sensors, MOFs with large specific surface could facilitate the loading of various nanomaterials, which is conducive to improving electrical conductivity and amplifying electrical signals, as well as achieving more electrocatalytic sites [7]. Moreover, the functional groups of MOFs are beneficial for the biosensor construction and enrichment of target analytes, which could be achieved strong interaction through hydrogen bonding, π–π stacking, and/or electrostatic force [8]. MOFs with large specific surface area are very suitable as the carriers of molecularly imprinted polymers to increase the imprinting binding sites, thereby improving the detection sensitivity and selectivity of the electrochemical sensor [9, 10]. Additionally, size-controlled manufacturing of nanoscale MOFs with high uniformity has also been implemented to improve the low reproducibility of electrochemical response using micron sizes of MOFs [11].

Recently, studies focusing on the MOF-based electrochemical sensor are increasingly reported due to the advantageous properties of MOF-based electrochemical sensors, expecting to bring revolutionary changes to environmental monitoring, food safety, clinical diagnosis applications. The development of MOF-based electrochemical sensors has also promoted its application in the field of EDCs detection, which provides a unique alternative detection method for EDCs detection. The representative researches on the development of MOF-based electrochemical sensors for EDCs detection are summarized in Tables 29.1-29.4.

29.3 RECENT STUDIES ON MOF-BASED ELECTROCHEMICAL SENSORS FOR EDC DETECTION

29.3.1 Pesticide

Pesticides, as important compounds to ensure the quality and yield of crops, are widely used worldwide. Endocrine-disrupting effects of pesticides have been proved and the exposure to pesticides through food consumption and natural source contamination has raised serious health concerns. Hence, the establishment of ultra-sensitive and accurate determination methods is of great significance to the rapid screening of pesticides.

Due to tailorable chemical characteristics, ordered structure, large surface area, and good absorbability, MOFs are potential surface modifiers for electrochemical sensing. To date, a wide variety of MOF-based electrochemical sensors have been developed for the detection of pesticides (Table 29.1). Unfortunately, the poor electrical conductivity of most MOFs has created a barrier for electrochemical applications due to the large proportion of building organic ligands. Because of this, many functionalized MOFs and MOF-derived materials have been widely used to construct the electrochemical sensing platform for pesticides detection. Song et al. fabricated MOF-derived MnO_2/Mn_3O_4 and Ti_3C_2 MXene/Gold nanoparticles (AuNPs) composite to construct a novel electrochemical sensing platform for ultrasensitive determination of organophsosphorus pesticides (OPs) [12]. Mn-MOFs were used as the precursor of MnO_2/Mn_3O_4 composite. Synergistic signal amplification effects, excellent electrochemical performance, large specific surface area, and long-time stability were achieved using the proposed composite. Under optimal conditions, a wide linear detection range of methamidophos (10^{-12}–10^{-6} M) was achieved using the sensing platform (acetylcholinesterase (AChE)-Chitosan/MXene/AuNPs/MnO_2/Mn_3O_4). Tu et al. exploited Mxene/carbon nanohorns/β-cyclodextrin-MOFs (MXene/CNHs/β-CD-MOFs) with nanoarchitecture to construct an electrochemical sensor for carbendazim (CBZ) detection [13]. The synthetic route of MXene/CNHs/β-CD-MOFs and the sensing strategy are shown in Figure 29.1. Porous MOFs were used as the carrier of β-CD to endow β-CD-MOFs with high adsorption capacity for CBZ. MXene/CNHs had the merits of large specific surface area, abundant available active sites, and high conductivity, enhancing the mass transfer rate and electrochemical catalysis of CBZ on the electrochemical sensing platform. The linear range of the electrochemical sensor using composite (MXene/CNHs/β-CD-MOFs) modified electrode as the working electrode was 3.0 nM-10.0 μM with a low limit of detection (LOD) of 1.0 nM (S/N= 3). Additionally, satisfactory applicability was achieved using the prepared sensor in tomato samples.

Various MOFs and their derived materials with good conductivity and excellent catalytic ability are widely utilized for the construction of electrochemical enzyme-free sensing platforms to detect pesticides. A three-dimensional (3D) nitrogen-doped macro-meso-microporous carbon composite (N/Cu-HPC) derived from polyvinylpyrrolidone (PVP) doped Cu-MOFs was fabricated for the construction of an electrochemical sensing platform for neonicotinoid detection [14]. The preparation process of N/Cu-HPC and the construction process of the electrochemical sensing platform are illustrated in Figure 29.2. The linear detection ranges of imidacloprid (IDP), thiamethoxam (THA), and dinotefuran (DNF) using the N/Cu-HPC modified electrode were 0.5-60 μM, 0.5-60 μM and 1-60 μM, respectively. Low LODs, good anti-interference capacity and reproducibility, and long-term stability were achieved.

As a prospective alternative method for the detection of OPs residue, biosensors have been extensively studied based on the inhibition effect of OPs towards AChE or the hydrolysis effect of organophosphorus hydrolase towards OPs. Li et al. utilized zeolitic imidazolate framework-8 (ZIF-8) to construct a facile immobilization-free electrochemical sensor for sensitive detection of pesticides using AChE as a recognition component and ZIF-8 as a degradable carrier

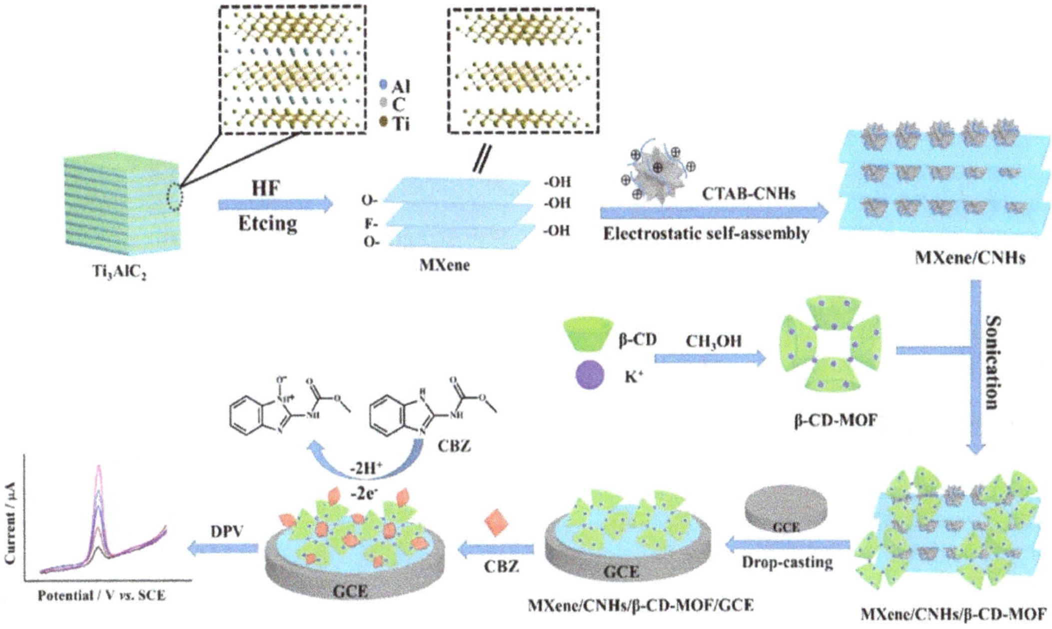

FIGURE 29.1 Synthetic route of MXene/CNHs/β-CD-MOFs and the sensing strategy for CBZ. Adapted with permission from [13]. Copyright (2020) Elsevier.

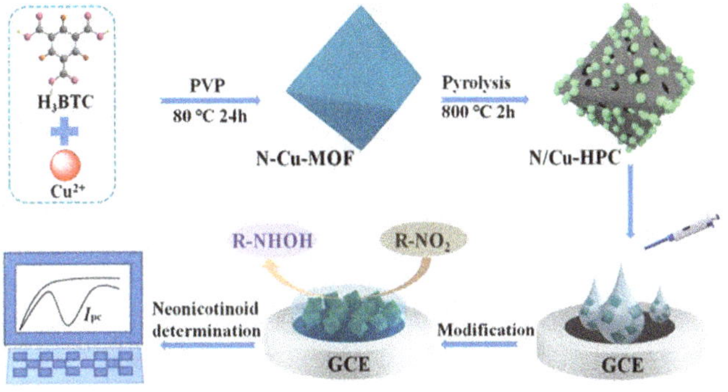

FIGURE 29.2 Preparation of N/Cu-HPC and its application in electrochemical detection of neonicotinoid. Adapted with permission from [14]. Copyright (2021) Elsevier.

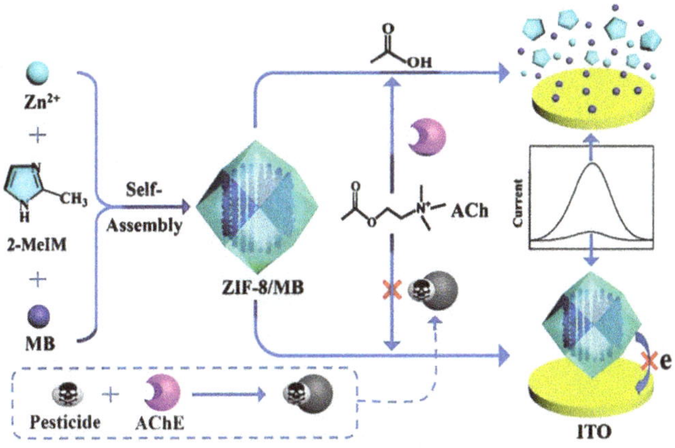

FIGURE 29.3 Schematic illustration of the synthesis of ZIF-8/MB composites and the principle of the ZIF-8/MB composites-based homogeneous electroanalytical strategy for pesticides assay. Adapted with permission from [15]. Copyright (2020) Elsevier.

for encapsulating of AChE [15]. Schematic diagrams of ZIF-8/methylene blue (MB) composites fabrication and the principle of ZIF-8/MB-based electrochemical sensor for the detection of pesticides are shown in Figure 29.3. The AChE-catalyzed hydrolytic reaction of pesticides could induce acidic conditions, causing the dissolution of pH-responsive ZIF-8/MB composite to release the electroactive MB molecules. AChE inhibitors such as organophosphates and carbamates could be detected by the decrease of the signal of MB molecules. Due to the excellent packaging effect of ZIF-8 towards MB molecules, the high analytical performance of the proposed sensing platform for paraoxon detection was achieved with a LOD of 1.7 ng mL^{-1}. In another work, a novel electrochemical biosensing platform of AChE/cysteamine (Cys)/MOF-directed rapid electrochemically grown gold nanorods (aAuNR)/MOFs/ITO was proposed for the detection of OP pesticides by Chan et al. [16]. The programmed self-assembly of electrochemically grown aAuNR was guided by MOFs for enhanced electrochemical sensing efficiency and the aAuNR/MOFs composite acted like a screening mesh for the detection of OP. The proposed AChE/Cys/aAuNR/MOFs/ITO offers good stability, specificity, and anti-interference properties for the detection of OP in real samples.

Although the cholinesterase-based methods show high sensitivity and good selectivity for the detection of pesticides, the enzymes are difficult to obtain and easily inactivated in the actual application process. Thus, other recognition elements that can specifically recognize targets, such as aptamers, molecularly imprinted polymers (MIPs), and antibodies, have been utilized to develop non-enzymatic MOF-based electrochemical sensors for pesticide detection. Xu et al. successfully established an electrochemical aptasensing platform based on nanocomposite probes for malathion detection [17]. The nanocomposite probes were formed by covalently coupling complementary probes (CP), zirconium-based MOFs, and carboxy-ferrocene (Fc-COOH), denoted as CP-MOFs-Fc. The probe CP-MOFs-Fc was assembled on the electrochemical sensing interface by complementary base-pairing reactions through hybridization with malathion aptamer which was modified on the electrode surface in advance. CP-MOFs-Fc was released into the buffer solution after the target recognition of aptamer, resulting in a decrease in the differential pulse voltammetry (DPV) signal of Fc. A linear range of 25-850 ng L^{-1} was achieved by the proposed sensor with a LOD of 17.18 ng L^{-1} (S/N = 3). The introduction of MOFs to load more Fc molecules greatly amplified the electrochemical signal. Additionally, good repeatability and excellent selectivity were realized using this detection strategy. Similarly, Hassan et al. utilized core-shell Co_3O_4@MOFs-74 nanocomposite to prepare a novel molecularly imprinted electrochemical sensor (MIECS) for the detection of the insecticide, fenamiphos (FEN) [18]. The preparation protocol of FEN imprinted electrochemical sensor is described in Figure 29.4. Firstly, Co_3O_4 nanowire was prepared using the hydrothermal method followed by thermal annealing. Then, core-shell Co_3O_4@MOFs-74 nanocomposite was solvothermal prepared in the absence of metal salts. MIP membrane was prepared using cyclic voltammetry on the surface of Co_3O_4@MOFs-74 nanocomposite modified electrode. The developed MIECS had excellent sensing performance, such as high sensitivity, selectivity, stability, and reproducibility, which could be attributed to the fast electron transfer rate, large specific surface area, and the synergistic effect of the Co_3O_4@MOFs-74 nanocomposite.

In another work, Yang et al. described an MOF-based electrochemical immunoassay method for the simultaneous detection of triazophos (TRS) and thiacloprid (THD) [19]. Amino-modified MOFs (UiO-66-NH_2) were utilized for the carrier of cadmium(II) and lead(II) ions loaded with

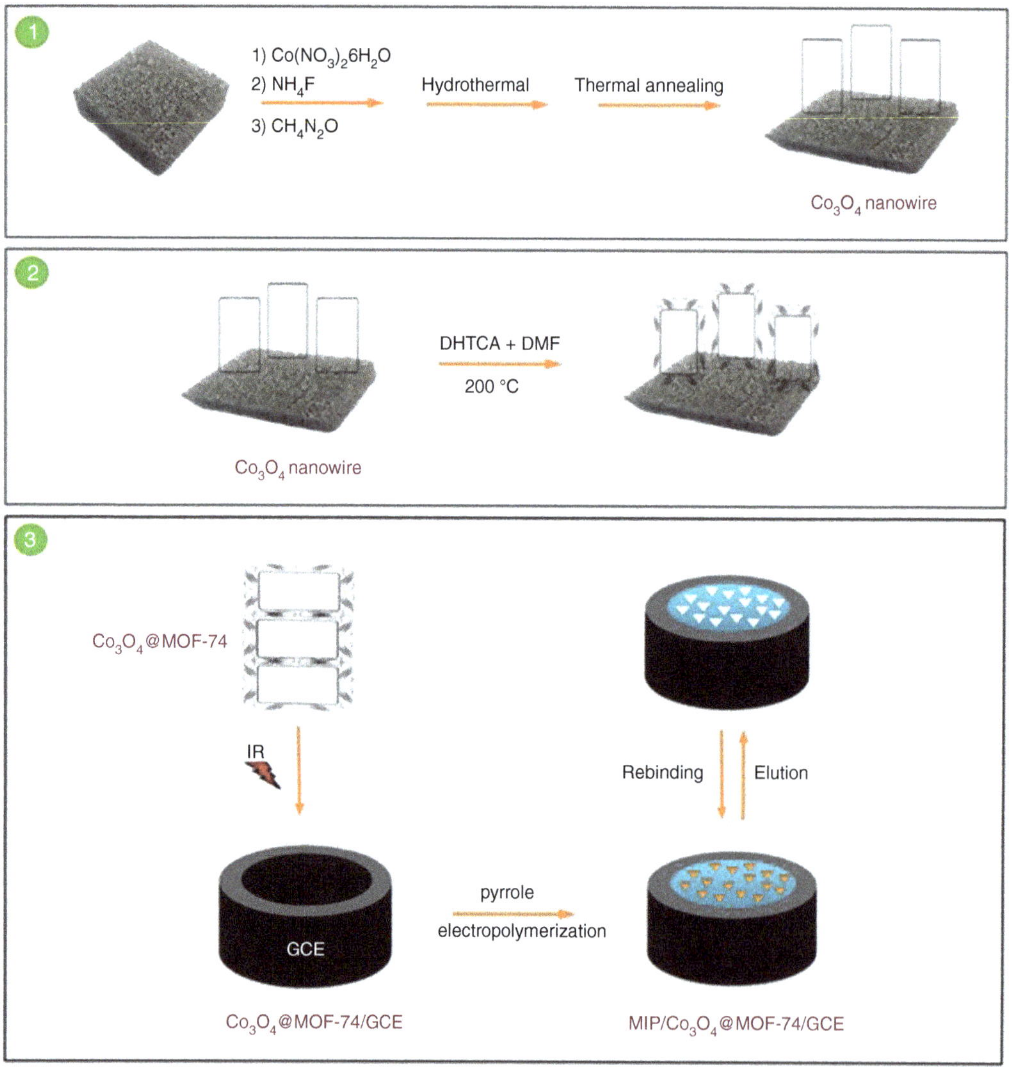

FIGURE 29.4 Preparation protocol of FEN imprinted electrochemical sensor. Adapted with permission from [18]. Copyright (2021) Elsevier.

TABLE 29.1
MOF-based Electrochemical Sensors for Rapid Detection of Pesticides

Target analytes	Modified electrode	Applications	LOD	Linear range	Recovery	Ref.
Methamidophos	AChE-Chit/MXene/Au NPs/MnO_2/Mn_3O_4/ GCE	Fresh fruit	1.34×10^{-13} M	10^{-12}–10^{-6} M	95.2–101.3%	[12]
Carbendazim (CBZ)	MXene/CNHs/β-CD-MOFs/GCE	Tomato	1.0 nM	3.0 nM–10.0 μM	97.77–102.01%	[13]
Imidacloprid (IDP)	N/Cu–HPC/GCE	Oat, corn, and rice	0.026 μM	0.5–60 μM	92.0–100.9%	[14]
Thiamethoxam (THA)			0.062 μM	1–60 μM	92.0–100.9%	
Dinotefuran (DNF)			0.01 μM	0.5–60 μM	92.0–100.9%	
Paraoxon	ZIF-8/MB/ITO	Apple and eggplant	1.7 ng mL^{-1}	5 ng mL^{-1}–10 μg mL^{-1}	93.8–107%	[15]
Neurotoxic organophosphates	AChE/Cys/aAuNR/ MOF/ITO	Vegetable extract	3 ng L^{-1}	30–600 ng L^{-1}	–	[16]
Malathion	CP-MOFs-Fc/GCE	Cucumber and long bean	17.18 ng L^{-1}	25–850 ng L^{-1}	95.00–103.16%	[17]
Fenamiphos (FEN)	MIPs/Co_3O_4@MOF-74/ GCE	Orange juice	3.0×10^{-12} M	1.0×10^{-11}–1.0×10^{-9} M	99.67 ± 0.02–100.45 ± 0.03%	[18]
Triazophos (TRS) and thiacloprid (THD)	Ag-UiO-66-NH_2-based signal tags, Ab-MB-COOH-based capture probes	Rice	0.07 ng mL^{-1} (TRS), 0.1 ng mL^{-1} (THD)	0.2–750 ng mL^{-1}	84.6–92.6%	[19]

Notes: LOD: Limit of detection, AchE: Acetylcholinesterase, Chit: Chitosan, GCE: Glassy carbon electrode, CNHs: Carbon nanohorns, β-CD-MOFs: β-cyclodextrin-metal-organic frameworks, N/Cu–HPC: Nitrogen–doped macro–meso–microporous carbon composites, ZIF-8: Zeolitic imidazolate framework-8, MB: Methylene blue, Cys: Cysteamine, aAuNR: Anisotropic gold nanorod, MOFs: Metal Organic Framework, ITO: Indium tin oxide coated electrode, CP: Complementary probe, Fc: Carboxy-ferrocene, MIPs: Molecularly imprinted polymers, Ag: Antigen, Ab: antibody.

cadmium(II) and lead(II) ions to fabricate electrochemical signal tags, which was bounded with TRS or THD antigens (Ags). Carboxyl modified magnetic bead (MB-COOH) coupled with TRS or THD antibodies (Abs) was utilized as the sensing recognition element. The Ag-UiO-66-NH_2-based signal tags and Ab-MB-COOH-based capture probes were combined as the sensing probe for the target detection. Combined with immune recognition and magnetic separation, fast, simple, sensitive, and selective simultaneous detection of Abs can be achieved with a wide linear range of 0.2-750 ng mL^{-1}. Amplified and distinguishable electrochemical signals corresponding to different targets were achieved using the novel MOF-based signal tags. This method provides universal multiple detection strategies, which can achieve other multi-targets detection by simply changing the loaded signal molecules and different antibodies.

29.3.2 Bisphenol A

Bisphenol A (2,2-bis(4,4′-hydroxyphenyl)propane, BPA) is a typical EDC that is widely utilized as a monomer or additive in plastic industry production. BPA could seriously interfere with hormonal activities at considerably low concentrations and cause all kinds of health concerns, inducing impaired immune functions, cancer development, and reproduction problems. The discharge of wastewater from plastic production plants into surface water, or migration into the food chain by leaching from food plastic packaging and containers can cause BPA pollution. In view of the potential health risks of BPA and the current pollution situation, there is an urgent need to develop a fast and sensitive detection method of BPA. As an important branch of detection methods, MOF-based electrochemical sensors for BPA detection are summarized in Table 29.2.

A copper-centered metal-organic framework (Cu-MOF)-based electrochemical sensor was fabricated by Wang et al. to detect BPA using tyrosinase (Tyr) as a model enzyme [20]. Experimental results showed that the Cu-MOF/GCE had no electrochemical response with the increment of BPA concentration. When Tyr was participated in the sensor construction, well-defined amperometry signals were obtained on the Tyr/GCE and Cu-MOF-Tyr/GCE with fast speed (reaching 95% of steady-state current within 11 s). The 3D structure with a large specific surface area was provided by the modification of Cu-MOF, which is conducive to the immobilization of enzymes and absorption of BPA, thereby improving the sensitivity of the proposed sensor. The LOD and dynamic linear range of the synthesized MOF-based electrochemical sensor were 0.013 and 0.05-3.0 mM, respectively. A MOF-based electrochemical sensor was established for the detection of BPA using amine-functionalized MOFs/reduced graphene oxide (NH_2-MIL-125/rGO) composites [21]. The NH_2-MIL-125/rGO, which was prepared using a one-pot hydrothermal method, was modified on the surface of GCE to fabricate

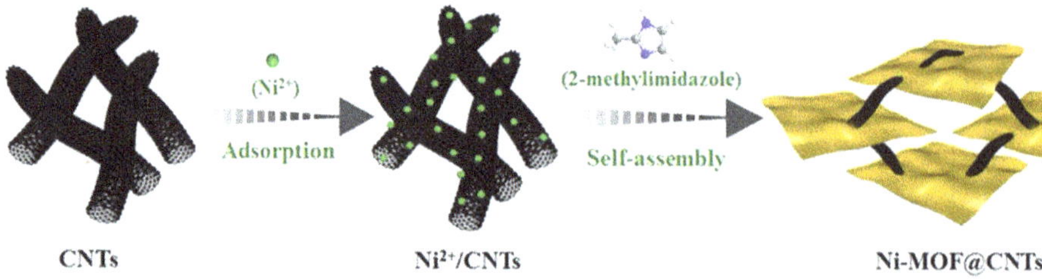

FIGURE 29.5 Illustration for the synthesis of the 3D Ni-MOFs@CNTshybrids. Adapted with permission from [26]. Copyright (2020) Elsevier.

the sensing interface. For the first time, the research group combined the merits of rGO (high electrochemical activity and conductivity, good loading capacity) and NH_2-MIL-125 (high density of hydrophilic active sites, good mechanical properties, and porosity) for the construction of the electrochemical sensor. Due to the synergistic effect of rGO and NH_2-MIL-125, NH_2-MIL-125/rGO/GCE showed high sensitivity and excellent stability for the detection of BPA with an LOD of 0.7966 mM.

Poor conductivity is the main difficulty that limits the application of MOFs in the field of electrochemical sensing. Just like the strategy in the previous articles, coupling conductive materials (metal nanoparticles [22–29], rGO [23, 24], carbon nanotubes (CNTs) [26, 28, 30, 31], cetyltrimethylammonium bromide (CTAB) [22, 27], and multiwall carbon nanotube (MWCNT) [29, 31]) with MOFs has been proved to be an effective method to improve the performance of MOF-based electrochemical sensors. For example, as shown in Figure 29.5, the preparation of 3D Ni-MOFs@CNT hybrids combined the merits of metal nanoparticles, MOFs, and CNTs to achieve the synergistic effect.

Bimetal MOFs are a new kind of MOFs that owns two types of metal irons with unique properties. The ratio of metal ions and the synergy between different components are attributed to their unique properties. Compared with common MOFs, bimetallic MOFs have higher activity, selectivity, and stability in applications [29, 31]. For the first time, bimetallic Ce-Ni-MOFs were utilized for the construction of a MOF-based electrochemical sensor together with MWNTs by Huang et al. [31]. The bimetallic Ce-Ni-MOF owning more catalytic sites and larger specific surface area was added to the traditional conductive material MWCNTs to enhance the performance of the electrochemical sensor. The MOF-based electrochemical sensor had a wider linear range (0.1 μM-100 μM) for BPA detection with a low LOD of 7.8 nM (S/N = 3). The BPA content in different brands of drinking water were detected using the proposed sensor, and satisfactory recoveries (97.4-102.4%) were obtained. Huang et al. also utilized Ce-Zn-MOFs with large specific surface area and multiple catalytic sites to fabricate a bimetal MOF-based electrochemical sensor for BPA detection combined with MWCNTs [29]. The conductivity and catalytic ability of MWCNTs have been improved with the introduction of bimetallic Ce-Zn-MOFs. Under optimal conditions, a wide linear range (0.1-100 μM) and low LOD of 7.2 nM ($S/N = 3$) were achieved by the proposed sensor, which was successfully applied for BPA detection in real samples with desirable recoveries ranging from 96.5 to 103.4%.

Recently, to improve the selectivity of the sensor, MIP was applied widely in the construction of MOF-based electrochemical sensors. A conductive MOF (CMOF)-based MIECS was fabricated for the sensing of BPA [32]. The preparation and detection strategy of this sensor are shown in Figure 29.6. Although CMOFs have better conductivity than traditional MOFs, most CMOFs synthesized with various organic ligands are non-selective for analytes in electrochemical sensors. CMOFs were utilized as a supporter for the fabrication of MIP and ionic liquids (ILs) were used as monomers and cross-linkers in the molecular imprinting process to construct the selective sensing platform. In comparison to traditional MIPs, the MIP prepared using ILs (MIPIL) had better capacity for electrocatalytic activity and target adsorption. The prepared CMOF-based electrochemical sensor had good selectivity for BPA detection, achieving a wide linear range (0.005-5.0 μM) and low LOD (4.0 nM).

29.3.3 Phenolic Compounds

In various samples of coastal and marine environments, phenolic compounds are extensively present, which could accumulate in the human body through the food chain. Even at a very low level of 150 mgL^{-1}, it can cause fainting, itching, anemia and cancer risk, and other harmful effects. As MOFs continue to deepen their application in electrochemical sensors, numerous MOF-based electrochemical sensors are applied for the determination of phenolic compounds [33–39] (Table 29.3). A MOF-based electrochemical sensor was fabricated by Dong et al. for the detection of 2,4-dichlorophenol (2,4-DCP) [34]. As a traditional MOF, $Cu_3(BTC)_2$ had the merits of high adsorption capacity, loose porous structure, enzyme-like characteristics as well as good conductivity. $Cu_3(BTC)_2$ was added in the fabrication process of carbon paste electrode (CPE) to construct $Cu_3(BTC)_2$-CPE, which significantly enhanced the oxidation peak current of 2,4-DCP. Moreover, the proposed MOF-based electrochemical sensor achieved simple, sensitive, and reliable detection of 2,4-DCP. A novel hybrid non-enzymatic electrochemical sensing interface composed of $Cu(OH)_2$ nanosheets and Cu-bipy-BTC MOFs nanoparticles was fabricated by Li et al. for the detection of catechol [39]. This hybrid structure can

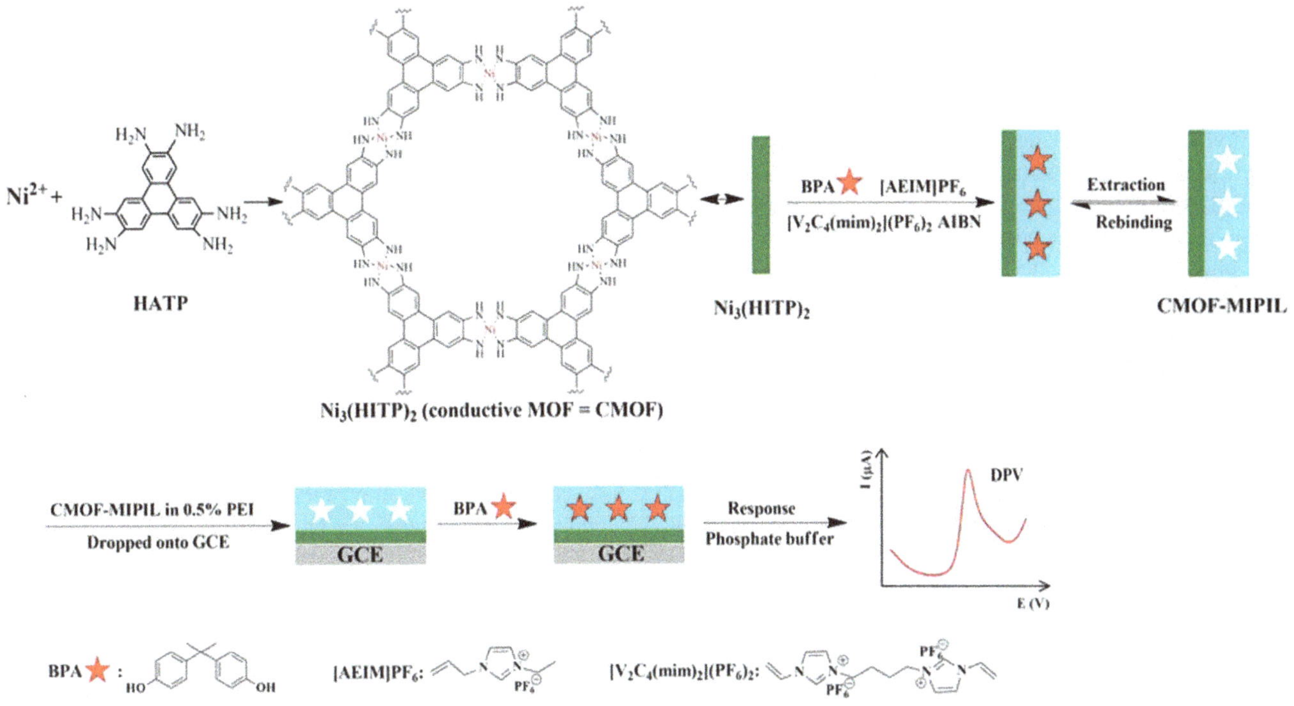

FIGURE 29.6 Preparation of CMOF-MIPIL and determination procedure with CMOF-MIPIL sensor. Adapted with permission from [32]. Copyright (2021) Elsevier.

TABLE 29.2
MOF-based Electrochemical Sensors for Rapid Detection of BPA

Target analytes	Modified electrode	Applications	LOD	Linear range	Recovery	Ref.
BPA	Cu-MOFs/Tyr/GCE	Water bottle, nursing bottle, coffee spoon, mineral water, and bottle	13 nM	5.0×10^{-8}– 3.0×10^{-6} M	94.0–101.6%	[20]
BPA	NH_2-MIL-125/RGO/GCE	Drinking bottles, packaging, and paper cups	0.7966 μM	2.00–200.00 μM	96.85–107.75%	[21]
BPA	Cu-MOFs/ERGO/GCE	Plastic bags, disposable gloves, and water bottle	6.7×10^{-9} M	2.0×10^{-8}– 9.0×10^{-5} M	98.46–105.38%	[23]
BPA	CTAB/Ce-MOFs/GCE	Fresh liquid milk, drinking package and drinking bottle	2.0 nM	0.005–50 M	96.2–104.6%	[22]
BPA	Ce-MOFs-ERGO/GCE	Tap water, mineral water, and pet bottle extract	1.9 nM	3 nM–10 μM	93.2–106%	[24]
BPA	Cu-BTC MOFs/GCE	Urine, Waste water, baby feeding bottle, water bottle, mineral water bottle and fast-food container	0.72 nM	5.0–2000 nM	–	[25]
BPA	Ni-MOFs@CNTs/GCE	Receipts, movie tickets, and polycarbonate water buckets	0.35 nM	0.001–1.0 μM	–	[26]
BPA	CTAB/MIL-101 (Cr)/GCE	E-waste dust samples	9.95 nM	20–350 nM	87.3–111.5%	[27]
BPA	AuNP/mCNT@ZIF-8/GCE	Pond water, tap water and river water	690 nM	1–100 μM	96.5–107.5%	[28]
TBBPA	CNTs@ZIF-67/CPE	Rain and pool water	4.2 nM	0.01–1.5 μM	92.89–106.78%	[30]
BPA	Ce-Ni-MOFs/MWCNTs/GCE	Packaging drinking water	7.8 nM	0.1–100 μM	97.4–102.4%.	[31]
BPA	Ce-Zn-MOFs/MWCNTs/GCE	Packaging drinking water	7.2 nM	0.1–100 μM	96.5–103.4%	[29]
BPA	CMOFs-MIPIL/GCE	Lake and river water samples, plastic bottle and fresh liquid milk	4.0 nM	0.005–5.0 μM	95.3–104.0%	[32]

Notes: LOD: Limit of detection, BPA: Bisphenol A, TBBPA: Tetrabromobisphenol A, Cu-MOFs: Copper-centered metal-organic framework, Tyr: Tyrosinase, GCE: Glassy carbon electrode, NH_2-MIL-125: Amine-functionalized metal–organic framework, RGO: Reduced graphene oxide, ERGO: Electrochemically reduced graphene oxide, Ce-MOFs: Cerium-centered metal-organic framework, Ni-MOFs: Ni-based organic frames, CNTs: Carbon nanotubes, MIL-101 (Cr): Cr-metal organic framework, AuNP: Gold nanoparticle, mCNT: Magnetic carbon nanotubes, ZIF-8: Zeolitic imidazolate framework-8, ZIF-67: Zeolitic imidazole framework-67, Ce-Ni-MOFs: Bimetallic Ce-Ni metal-organic frameworks, MWCNTs: Multiwall carbon nanotubes, Ce-Zn-MOFs: Ce-Zn-metal organic framework, CMOFs: Conductive metal organic framework, MIPIL: Molecularly imprinted poly (ionic liquid).

be in-situ grown directly on the surface of an Au electrode that was electroplated with Cu under specific conditions. Cu-bipy-BTC MOFs nanoparticles, which are embedded in the hydrophobic environment formed by $Cu(OH)_2$ nanosheets, functioned as active sites for catalyzing the oxidation of catechol. This hybrid structure mimicked the active site of laccase, exhibiting high activity and sensitivity. The proposed sensor also showed good selectivity using methylbenzene and phenol as the interferences. Compared with other electrochemical sensors of catechol detection, the proposed MOF-based electrochemical sensor has advantages of better catalytic performance, higher sensitivity, good selectivity, and ease of fabrication.

Most of the MOF-based electrochemical sensors for the detection of phenolic compounds were based on MOF-based hybrid composites composed of graphene aerogels (GAs), reduced graphene oxide (rGO), poly (3,4-ethylenedioxythiophene): poly (styrenesulfonate) (PEDOT:PSS) or MOF-derived materials (N-doped carbon nanotube frameworks (NCNTFs)). An electrochemical sensor was fabricated by modifying the electrode with MOFs and GAs for the detection of 2,2-methylenebis (4-chlorophenol) (DCP) in lake water [35]. The synthesis process of polypyrrole (PPy)@ZIF-8/GAs and the detection strategy of DCP are illustrated in Figure 29.7. ZIF-8 was modified on the GAs decorated with PPy through the coordination interaction between $Zn^{2+}$and amino (-NH-) groups of PPy chains. Improved electron migration rate and electrolyte transportation were achieved by the hybrid PPy@ZIF-8/Gas with 3D architecture, which attributed to the excellent performance of the MOF-based electrochemical sensor for the detection of DCP.

ZIF-67 was employed as a sacrifice template to prepare NCNTFs via simple heat treatment. The NCNTFs were utilized in the fabrication of electrochemical sensor by Zheng et al. for hydroquinone (HQ) and catechol (CC) determination [37]. The NCNTFs with rich nitrogen atoms and pore structures had large specific surface areas. The hydrophilicity and stability of NCNTFs were improved due to the abundant dosage of N atoms, enhancing the ability of electron transfer and catalytic activity. Meanwhile, the mass transfer rate was accelerated effectively by the hollow structure of NCNTFs. The advantages of the MOF-based NCNTFs endowing the proposed electrochemical sensor with excellent analytical performance toward HQ and CC in spiked lake water samples. In another work, Zhang et al. introduced rGO in the construction process of MIL-101(Cr)-based electrochemical sensor [36]. The MIL-101(Cr)@rGO composite with improved dispersibility, stability, and conductivity was obtained by physically mixing GO and MIL-101(Cr) and then followed by a simple chemical reduction. Sensitive electrochemical detection of 4-NP was realized by the MIL-101(Cr)@rGO-based electrochemical sensor.

Due to its good stability, high conductivity, and good solution processability, PEDOT:PSS, has been extensively applied in electrode modification. While in the construction of MOF-based electrochemical sensors, PEDOT:PSS was used for MOF dispersion. ZIF-8@PEDOT:PSS composite was fabricated through a simple and effective ultrasonic method by Gao et al, which had improved stability and conductivity [33]. The ZIF-8@PEDOT:PSS composite was modified on the surface of the electrode for DCP detection with excellent sensing performance. A satisfactory linear ranging from 0.03 μM to 10.0 μM was realized by the ZIF-8@PEDOT: PSS-based electrochemical sensor.

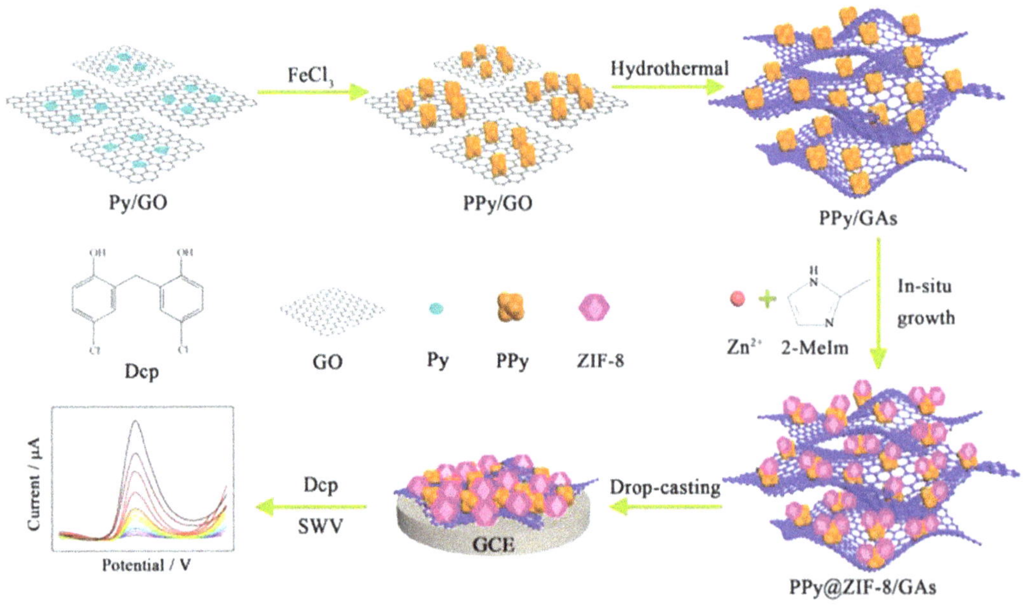

FIGURE 29.7 Synthetic route of GAs/PPy@ZIF-8 and the sensing strategy for DCP. Adapted with permission from [35]. Copyright (2019) Elsevier.

TABLE 29.3
MOF-based Electrochemical Sensors for Rapid Detection of Phenolic Compounds

Target analytes	Modified electrode	Applications	LOD	Linear range	Recovery	Ref.
CT	MOFs-GO/GCE	Tap water	0.1 μM	0.1–566 μM	95.4–103%	[38]
HQ			0.1 μM	0.1–476 μM	100–104%	
2,4-DCP	$[Cu_3(BTC)_2]$-MOFs/CPE	Reservoir raw water	9 nM	0.04–1.0 M	80–91%	[34]
CT	Cu-bipy-BTC MOF/Au	Tap water	4.69 μM	10–1000 μM	99.1–105.6%	[39]
DCP	PPy@ZIF-8/GAs/GCE	Lake water	1.0×10^{-10} M	3.0×10^{-10}–1.0×10^{-5} M	95.96–104.6%	[35]
HQ	MOFs/NCNTFs/GCE	Lake water	0.17 μM	0.8–200 μM	91.6–95.0%	[37]
CC			0.12 μM	0.8–120.0 μM	95.3–105.0%	
4-NP	MIL-101(Cr)@rGO/GCE	Natural water	33 nM	0.1–12.5 μM	95.8–105.7%	[36]
DCP	ZIF-8@PEDOT:PSS/GCE	Lake water	0.01 μM	0.03–10.0 μM	97.78–104.00%	[33]

Notes: LOD: Limit of detection, CT: Catechol, HQ: Hydroquinone, 2,4-DCP: 2,4-dichlorophenol, DCP: 2,2-methylenebis (4-chlorophenol) (dichlorophenol), CC: Catechol, 4-NP: 4-nonylphenol, MOFs: Metal-organic frameworks, GO: Graphene oxide; GCE: Glassy carbon electrode, $Cu_3(BTC)_2$: 1,3,5-benzenetricarboxylic acid copper, CPE: Carbon paste electrode, Cu-bipy-BTC: $[Cu_2(OH)(2,2'\text{-bipy})_2(BTC)_3 2H_2O]_n$, Au: Gold electrode, PPy: Polypyrrole, Gas: Graphene aerogels, NCNTFs: N-doped carbon nanotube frameworks, MIL-101(Cr): Chromium terephthalate MOFs, rGO: Reduced graphene oxide, ZIF-8: Zeolite imidazolate framework-8, PEDOT:PSS: Poly(3,4-ethylenedioxythiophene):poly(styrenesulfonate).

29.3.4 Estrone

Estrogen medical application, industrial and domestic waste discharges will cause estrone accumulation in rivers, lakes, reservoirs, and other water bodies, causing environmental water pollution and then entering the food chain due to bioaccumulation. Estrone can interfere with the human endocrine system at very low concentrations, which may cause declining immunity, decreased sperm count, reproductive and developmental deficiency. Therefore, the monitoring and accurate quantification of estrone are highly desired. Table 29.4 summarizes the MOF-based electrochemical sensors for the detection of estrone. Sun et al. constructed an electrochemical sensor based on Cu-BDC MOFs with regular flaky structures and uniform micropores for the detection of estradiol [40]. The constructed Cu-BDC MOFs modified electrode was more active for the oxidation of estradiol and exhibited remarkable enhancement effects for estradiol detection. The electrochemical signals had a linear relationship with the target concentration in the range of 5.00-650.0 nM with a LOD of 3.80 nM. Good recovery rates ranging from 96.5 to 101% were obtained in water samples.

Due to excellent adsorption performance and good mass transfer rate, hollow MOFs can perform better in the MOF-based electrochemical sensor than the solid structured MOFs, which is an important development direction. In view of this, small-sized AuNPs were anchored into hollow Zr-porphyrin MOFs (HPCN-222) to obtain AuHPCN-222 for the construction of electrochemical sensors by Biswas et al. [41]. The combination of AuNPs and HPCN-222 greatly enhanced the electrochemical properties of conductivity, charge transfer, and catalytic activity. The AuHPCN-222 modified GCE exhibited better electrocatalytic activity toward estradiol compared with AuNPs anchored solid PCN-222/GCE, AuNPs/GCE and HPCN-222/GCE, which was applied in human urine and serum samples for estradiol detection with excellent recoveries. In another work, a hybrid composite of bimetallic MOFs decorated with ultrasmall AuNPs (US AuNPs@AuZn-MOFs) was fabricated for the construction of MOF-based electrochemical sensors using facile sequential ion-exchange and an in-situ chemical reduction method [42]. The bimetallic MOFs were used as the hosts of US AuNPs, greatly enhancing the electron transfer rate and enlarging the electrochemically active surface area. The US AuNPs@AuZn-MOF-based electrochemical sensor exhibited an enhanced electrochemical oxidation current, providing a promising way for the combination of metal nanoparticles and MOFs.

Because of high sensitivity, good selectivity, and rapid binding kinetics, MIP has been widely used in MOF-based electrochemical sensors. Duan et al. fabricated an electrochemical sensor for the detection of 17*β*-estradiol based on MOFs/carbon nanotubes (MIL-53/CNTs) hybrid using Prussian blue (PB) as an electrocatalyst to accelerate electron transfer rate and increase surface area [43]. The preparation process of MIP-PB-MIL-CNTs/GCE is shown in Figure 28.8. The MIL-53/CNTs hybrid can not only increase the specific surface area but also increase the conductivity of the electrochemical sensor. The stability of PB can be further increased by using MIL-53/CNTs hybrid composite as the substrate. The conductive imprinted film was formed by electropolymerization using pyrrole as functional monomer and 17*β*-estradiol as template. Moreover, the overoxidation/dedoping elution method was used to simplify the experimental process. Under optimal conditions, the proposed electrochemical sensor showed good sensitivity and selectivity with a linear range of 10^{-4}-10^{-9} M and an estimated detection limit of 6.19×10^{-15} M.

29.3.5 Phthalate

As plasticizers, phthalates have important industrial applications in food and beverage plastic packages, medical

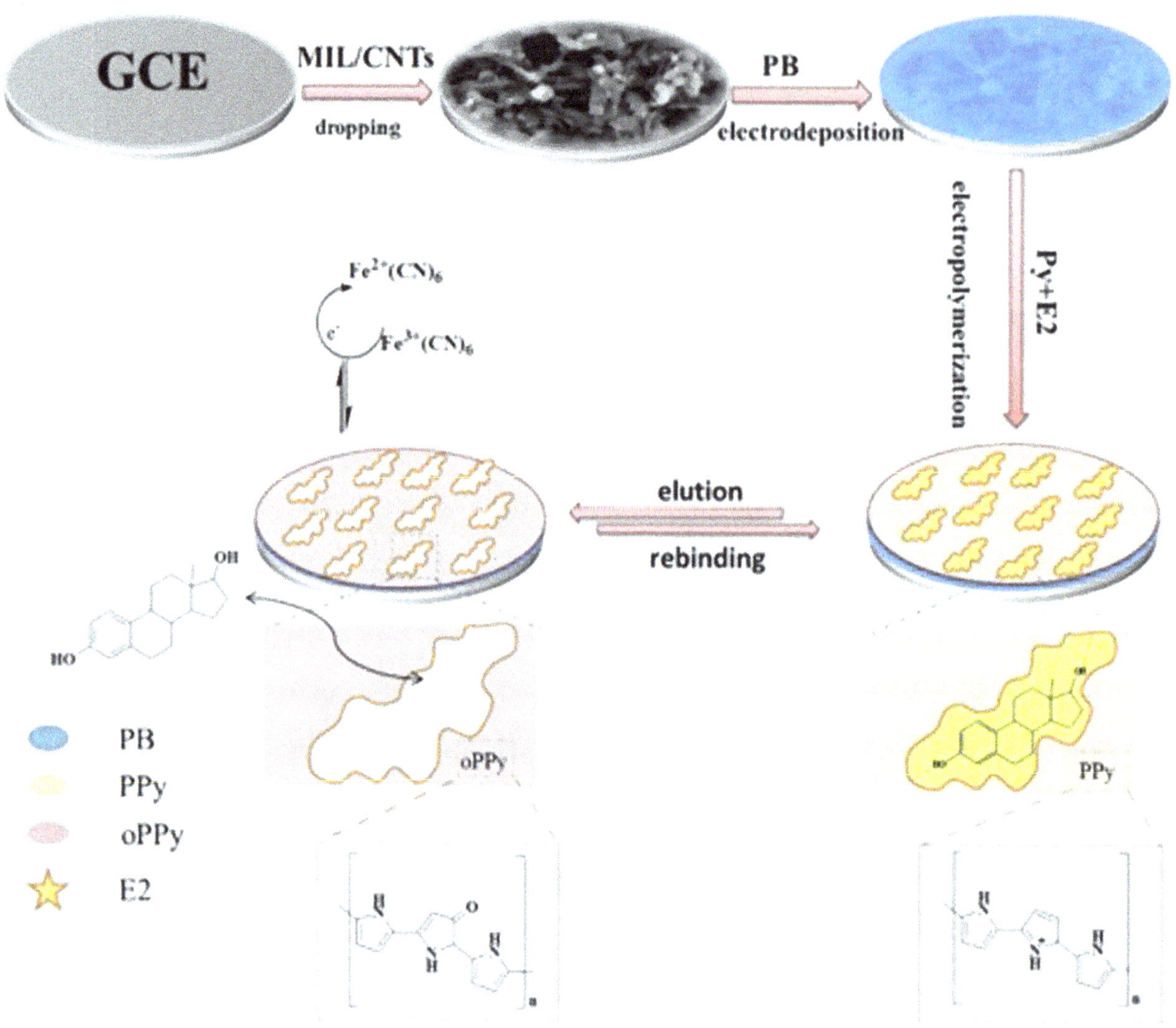

FIGURE 29.8 The preparation process of MIP-PB-MIL-CNTs/GCE. Adapted with permission from [43]. Copyright (2019) Elsevier.

TABLE 29.4
MOF-based Electrochemical Sensors for Rapid Detection of Estrone

Target analytes	Modified electrode	Applications	LOD	Linear range	Recovery	Ref.
Estradiol	Cu-BDC MOFs/CPE	Water	3.8×10^{-9} M	5.0×10^{-9}–6.5×10^{-7} M	96.5–101%	[40]
Estradiol	AuHPCN-222/GCE	Human urine and serum	0.5 nM	0.1–200 μM	98–103%	[41]
Estrone	USAuNPs@AuZn-MOFs)/CPE	Water	12.3 nM	0.05–5 μM	93.0–103.5%	[42]
17*β*-estradiol	MIP-PB-MIL/53-CNTs/GCE	Pond water	6.19×10^{-15} M	10^{-14}–10^{-9} M	96.9–103.9%	[43]

Notes: LOD: Limit of detection, Cu-BDC MOFs: Copper-based metal-organic frameworks, CPE: Carbon paste electrode, AuHPCN-222: Au(0) functionalized hollow PCN-222, GCE: Glassy carbon electrode, US AuNPs: Ultrasmall Au nanoparticles, AuZn-MOFs: Bimetallic metal-organic framework, MIP: Molecularly imprinted polymers, PB: Prussian blue, MIL/53: Metal-organic frameworks, CNTs: Carbon nanotubes.

devices, detergents, electronics and building materials, endowing these industrial products with desirable properties. Because no chemical bonds exist between the products and phthalates, phthalates can be easily released and transferred into the environment and food. In particular, it is common for the phthalates in food packaging and containers to penetrate food through contacting with food, which has been a public concern in recent years. Molecularly surface imprinted MOFs-5 (MIPs@MOF-5) for the recognition of dibutyl phthalate (DBP) was prepared by Kang et al. [44]. By combining the advantages of the ordered pore size and large specific surface area of MOF-5 with high selectivity of MIPs, MIPs@MOF-5 demonstrated higher specific recognition and sensitive detection ability for DBP. In addition, the MIPs@MOF-5 was utilized for the construction of an electrochemical sensor to analyze DBP in tap water samples with satisfactory recovery (89.02-92.27%).

29.3.6 Perfluorooctane sulfonate

Perfluorooctane sulfonate (PFOS) has been widely used in civil, military, commercial, and industrial fields in the past few decades, which is commonly known as perfluoro and polyfluoroalkyl substances (PFAS). Many health concerns, such as obesity, endocrine disruption, immune suppression, and even cancer are related to PFOS exposure. Meanwhile, environmental stability and physiological persistence of PFOS are also serious problems. Cheng et al. constructed a synergistic approach for the targeted affinity-based capture of PFOS using MOFs as porous sorbent probes, which could enhance the detection sensitivity of a microfluidic platform [45]. Schematics of PFOS detection are shown in Figure 28.9. This novel platform functioned as a MOF-based electrochemical sensor to directly measure PFOS concentration through a proportional change in electrical signal (increase in impedance). The nanoporous geometry of MOFs, along with interdigitated microelectrodes, increased the signal-to-noise ratio tremendously. Further, the porose MOF-based capture probes could interact with PFOS at molecular level and effectively enhance the sensitivity. The detection limit of 0.5 ng L^{-1} was achieved unprecedently for in-situ analytical sensors of PFOS.

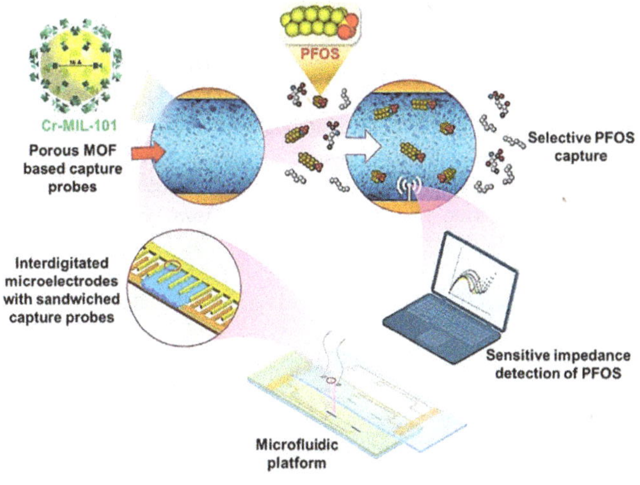

FIGURE 29.9 Schematics of PFOS detection. Adapted with permission from [45]. Copyright (2020) American Chemistry Society.

29.4 CONCLUSION

In recent years, a growing amount of researches has proved that EDCs can exert significant effect on the endocrine system and cause serious health problems. As peoples' awareness of the environment, food, and daily product safety has increased, people have begun to pay more and more attention to EDCs. At the same time, legislative regulations and supervision on EDCs are now more stringent. Highly sensitive, selective, reliable, and on-site detection methods are deeply desired for EDCs monitoring. Due to their special properties, MOFs have been extensively utilized in the construction of sensors, wherein MOF-based electrochemical sensor has aroused the interest of researchers around the world as an alternative detection method of EDCs. The performance of the sensing platform of MOF-based electrochemical sensors could be greatly enhanced by the introduction of MOFs.

In most of the electrochemical sensor construction processes, MOFs are modified on the sensing interface for improving electrical conductivity, providing more electrocatalytic sites, and offering more binding sites. Due to the poor electrical conductivity of MOFs, MOFs are often combined with other nanomaterials for the construction of electrochemical sensors. At the same time, the use of aptamers, antibodies, molecular imprinting materials, and the like, can greatly improve the selectivity of MOF-based electrochemical sensors. However, we should see that there are still many scientific problems with MOF-based electrochemical sensors for EDC detection, such as the stability of electrode modification and low detection efficiency if there are many samples. We believe that with the continuing research efforts on synthesis and optimization of MOFs for electrochemical detection, MOF-based electrochemical sensors are on their way to real applications of EDCs detection.

REFERENCES

[1]. Kaya, S. I., Cetinkaya, A., Bakirhan, N. K. & Ozkan, S. A. (2020) Trends in sensitive electrochemical sensors for endocrine disruptive compounds. *Trends in Environmental Analytical Chemistry* 28:e00106.

[2]. Azzouz, A., Kumar Kailasa, S., Kumar, P., Ballesteros, E. & Kim, K.-H. (2019) Advances in functional nanomaterial-based electrochemical techniques for screening of endocrine disrupting chemicals in various sample matrices. *TrAC Trends in Analytical Chemistry* 113:256–279.

[3]. Hanrahan, G., Patil, D. G. & Wang, J. (2004) Electrochemical sensors for environmental monitoring: design, development, and applications. *Journal of Environmental Monitoring* 6:657–664.

[4]. Liu, J. M., Hu, Y., Yang, Y. K., Liu, H., Fang, G. Z., Lu, X., & Wang, S. (2018) Emerging functional nanomaterials for the detection of food contaminants. *Trends in Food Science and Technology 71*:94–106.

[5]. Liu, L., Zhou, Y., Liu, S. & Xu, M. (2018) The Applications of Metal-Organic Frameworks in Electrochemical Sensors. *ChemElectroChem* 5:6–19.

[6]. Liu, C. S., Li, J., & Pang, H. (2020) Metal-organic framework-based materials as an emerging platform for advanced electrochemical sensing. *Coordination Chemistry Reviews* 410:213222.

[7]. Ma, D. D. & Zhu, Q. L. (2020) MOF-based atomically dispersed metal catalysts: Recent progress towards novel atomic configurations and electrocatalytic applications. *Coordination Chemistry Reviews* 422:213483.

[8]. Jahangiri–Dehaghani, F., Zare, H. R. & Shekari, Z. (2020) Measurement of aflatoxin M1 in powder and pasteurized milk

samples by using a label–free electrochemical aptasensor based on platinum nanoparticles loaded on Fe–based metal–organic frameworks. *Food Chemistry* 310:125820.

[9]. Yang, Y., Yan, W., Wang, X., Yu, L., Zhang, J., Bai, B., Guo, C. & Fan, S. (2021) Development of a molecularly imprinted photoelectrochemical sensing platform based on NH_2-MIL-125(Ti)-TiO_2 composite for the sensitive and selective determination of oxtetracycline. *Biosensors & Bioelectronics* 177:113000.

[10]. Yang, Y., Yan, W., Guo, C., Zhang, J., Yu, L., Zhang, G., Wang, X., Fang, G. & Sun, D. (2020) Magnetic molecularly imprinted electrochemical sensors: A review. *Analytica Chimica Acta* 1106:1–21.

[11]. Xiao, X., Zou, L., Pang, H. & Xu, Q. (2020) Synthesis of micro/nanoscaled metal–organic frameworks and their direct electrochemical applications. *Chemical Society Reviews* 49:301–331.

[12]. Song, D., Jiang, X., Li, Y., Lu, X., Luan, S., Wang, Y., Li, Y. & Gao, F. (2019) Metal–organic frameworks-derived MnO_2/Mn_3O_4 microcuboids with hierarchically ordered nanosheets and Ti_3C_2 MXene/Au NPs composites for electrochemical pesticide detection. *Journal of Hazardous Materials* 373:367–376.

[13]. Tu, X., Gao, F., Ma, X., Zou, J., Yu, Y., Li, M., .Qu, F., Huang, X., & Lu, L. (2020) Mxene/carbon nanohorn/β-cyclodextrin-Metal-organic frameworks as high-performance electrochemical sensing platform for sensitive detection of carbendazim pesticide. *Journal of Hazardous Materials* 396:122776.

[14]. Wang, Q., Zhangsun, H., Zhao, Y., Zhuang, Y., Xu, Z., Bu, T., Li, R., & Wang, L. (2021) Macro-meso-microporous carbon composite derived from hydrophilic metal-organic framework as high-performance electrochemical sensor for neonicotinoid determination. *Journal of Hazardous Materials* 411: 125122.

[15]. Li, X., Gao, X., Gai, P., Liu, X., & Li, F. (2020) Degradable metal-organic framework/methylene blue composites-based homogeneous electrochemical strategy for pesticide assay. *Sensors and Actuators B: Chemical* 323:128701.

[16]. Chansi, R., P. R., Mukherjee, I., Basu, T., & Bharadwaj, L. M. (2020) Metal Organic Framework steered electrosynthesis of anisotropic gold nanorods for specific sensing of organophosphate pesticides in vegetables collected from the field. *Nanoscale* 12:21719–21733.

[17]. Xu, G., Huo, D., Hou, J., Zhang, C., Zhao, Y., Hou, C., Bao, J., Yao, X., & Yang, M. (2021) An electrochemical aptasensor of malathion based on ferrocene/DNA-hybridized MOF, DNA coupling-gold nanoparticles and competitive DNA strand reaction. *Microchemical Journal* 162:105829.

[18]. Karimi-Maleh, H., Yola, M. L., Atar, N., Orooji, Y., Karimi, F., Kumar, P. S., Rouhi, J., & Baghayeri, M. (2021) A novel detection method for organophosphorus insecticide fenamiphos: Molecularly imprinted electrochemical sensor based on core-shell Co_3O_4@MOF-74 nanocomposite. *Journal of Colloid and Interface Science* 592:174–185.

[19]. Yang, Y., Cheng, J., Wang, B., Guo, Y., Dong, X., & Zhao, J. (2019) An amino-modified metal-organic framework (type UiO-66-NH_2) loaded with cadmium(II) and lead(II) ions for simultaneous electrochemical immunosensing of triazophos and thiacloprid. *Microchimica Acta* 186:101.

[20]. Wang, X., Lu, X., Wu, L., & Chen, J. (2015) 3D metal-organic framework as highly efficient biosensing platform for ultrasensitive and rapid detection of bisphenol A. *Biosensors and Bioelectronics* 65:295–301.

[21]. Ling, L. J., Xu, J. P., Deng, Y. H., Peng, Q., Chen, J. H., San He, Y., & Nie, Y. J. (2018) One-pot hydrothermal synthesis of amine-functionalized metal-organic framework/ reduced graphene oxide composites for the electrochemical detection of bisphenol A. *Analytical Methods* 10:2722–2730.

[22]. Zhang, J., Xu, X. & Chen, L. (2018) An ultrasensitive electrochemical bisphenol A sensor based on hierarchical Ce-metal-organic framework modified with cetyltrimethylammonium bromide. *Sensors and Actuators B: Chemical 261*:425–433.

[23]. Li, C., Zhou, Y., Zhu, X., Ye, B. & Xu, M. (2018) Construction of a sensitive bisphenol A electrochemical sensor based on metal-organic framework/graphene composites. *International Journal of Electrochemical Science* 13: 4855–4867.

[24]. Wang, X., Shi, Y., Shan, J., Zhou, H. & Li, M. (2020) Electrochemical sensor for determination of bisphenol A based on MOF-reduced graphene oxide composites coupled with cetyltrimethylammonium bromide signal amplification. *Ionics* 26:3135–3146.

[25]. Hu, P., Zhu, X., Luo, X., Hu, X. & Ji, L. (2020) Cathodic electrodeposited Cu-BTC MOFs assembled from Cu(II) and trimesic acid for electrochemical determination of bisphenol A. *Microchimica Acta* 187:1503–1517.

[26]. Xu, C., Liu, L., Wu, C. & Wu, K. (2020) Unique 3D heterostructures assembled by quasi-2D Ni-MOF and CNTs for ultrasensitive electrochemical sensing of bisphenol A. *Sensors and Actuators, B: Chemical* 310:127885.

[27]. Li, Z., Hu, J., Xiao, Y., Zha, Q., Zeng, L., & Zhu, M. (2021) Surfactant assisted Cr-metal organic framework for the detection of bisphenol A in dust from E-waste recycling area. *Analytica Chimica Acta* 1146:174–183.

[28]. Li, H., Zhu, F., Xiang, J., Wang, F., Liu, Q., & Chen, X. (2021) In-situ growth of ZIF-8 on gold nanoparticles/magnetic carbon nanotubes for the electrochemical detection of bisphenol A. *Analytical Methods* 13:2338–2344.

[29]. Huang, D., Huang, X., Chen, J., Ye, R., Lin, Q., & Chen, S. (2021) An Electrochemical Bisphenol: a Sensor Based on Bimetallic Ce-Zn-MOF. *Electrocatalysis* 12:456–468.

[30]. Zhou, T., Zhao, X., Xu, Y., Tao, Y., Luo, D., Hu, L., Jing, T., Zhou, Y., Wang, P., & Mei, S. (2020) Electrochemical determination of tetrabromobisphenol A in water samples based on a carbon nanotubes@zeolitic imidazole framework-67 modified electrode. *RSC Advances* 10:2123–2132.

[31]. Huang, X., Huang, D., Chen, J., Ye, R., Lin, Q., & Chen, S. (2020) Fabrication of novel electrochemical sensor based on bimetallic Ce-Ni-MOF for sensitive detection of bisphenol A. *Analytical and Bioanalytical Chemistry* 412:849–860.

[32]. Lei, X., Deng, Z., Zeng, Y., Huang, S., Yang, Y., Wang, H., Guo, L., & Li, L. (2021) A novel composite of conductive metal organic framework and molecularly imprinted poly (ionic liquid) for highly sensitive electrochemical detection of bisphenol A. *Sensors and Actuators, B: Chemical* 339:129885.

[33]. Gao, F., Yang, J., Tu, X., Yu, Y., Liu, S., Li, M., Gao, Y., Wang, X., & Lu, L. (2021) Facile synthesis of ZIF-8@ poly(3,4-ethylenedioxythiophene):poly (4-styrenesulfonate) and its application as efficient electrochemical sensor for the determination dichlorophenol. *Synthetic Metals* 277:116769.

[34]. Dong, S., Suo, G., Li, N., Chen, Z., Peng, L., Fu, Y., Yang, Q., & Huang, T. (2016) A simple strategy to fabricate highly

sensitive 2,4-dichlorophenol electrochemical sensor based on metal organic framework $Cu_3(BTC)_2$. Sensors and Actuators, *B: Chemical* 222:972–979.

[35]. Xie, Y., Tu, X., Ma, X., Xiao, M., Liu, G., Qu, F., Dai, R., Lu, L., & Wang, W. (2019) In-situ synthesis of hierarchically porous polypyrrole@ZIF-8/graphene aerogels for enhanced electrochemical sensing of 2, 2-methylenebis (4-chlorophenol). *Electrochimica Acta* 311:114–122.

[36]. Zhang, Y., Yan, P., Wan, Q. & Yang, N. (2018) Integration of chromium terephthalate metal-organic frameworks with reduced graphene oxide for voltammetry of 4-nonylphenol. *Carbon* 134:540–547.

[37]. Zheng, X., Hu, Y., Li, H., Han, B., Lin, R., & Huang, B. (2020) N-doped carbon nanotube frameworks modified electrode for the selective sensing of hydroquinone and catechol. *Journal of Electroanalytical Chemistry* 861:113968.

[38]. Chen, Q., Li, X., Min, X., Cheng, D., Zhou, J., Li, Y., Xie, Z., Liu, P., Cai, W., & Zhang, C. (2017) Determination of catechol and hydroquinone with high sensitivity using MOF-graphene composites modified electrode. *Journal of Electroanalytical Chemistry* 789:114–122.

[39]. Li, Z., Ren, L. & Lu, D. (2020) In-situ fabrication of Cu-bipy-BTC metal-organic framework electrode for catechol detection. *International Journal of Electrochemical Science* 15:7423–7433.

[40]. Sun, D., Deng, Q. & Long, J. (2018) Highly sensitive electrochemical sensor for estradiol based on the signal amplification strategy of Cu-BDC frameworks. *Journal of Solid State Electrochemistry* 22:487–493.

[41]. Biswas, S., Chen, Y., Xie, Y., Sun, X. & Wang, Y. (2020) Ultrasmall Au(0) Inserted Hollow PCN-222 MOF for the High-Sensitive Detection of Estradiol. *Analytical Chemistry* 92:4566–4572.

[42]. Chai, C., Gao, J., Zhao, G., Li, L., Tang, Y., Wu, C., & Wan, C. (2021) In-situ synthesis of ultrasmall Au nanoparticles on bimetallic metal-organic framework with enhanced electrochemical activity for estrone sensing. *Analytica Chimica Acta*, 1152:338242.

[43]. Duan, D., Si, X., Ding, Y., Li, L., Ma, G., Zhang, L., & Jian, B. (2019) A novel molecularly imprinted electrochemical sensor based on double sensitization by MOF/CNTs and Prussian blue for detection of 17β-estradiol. *Bioelectrochemistry* 129:211–217.

[44]. Kang, Y., Zhang, L., Lai, Q., Lin, C., Wu, K., Dang, L., & Li, L. (2020) Molecularly imprinted polymer based on metal-organic frameworks: synthesis and application on determination of dibutyl phthalate. *Polymer-Plastics Technology and Materials* 60:60–69.

[45]. Cheng, Y. H., Barpaga, D., Soltis, J. A., Shutthanandan, V., Kargupta, R., Han, K. S., McGrail, B. P., Motkuri, R. K., Basuray, S., & Chatterjee, S. (2020) Metal-Organic Framework-Based Microfluidic Impedance Sensor Platform for Ultrasensitive Detection of Perfluorooctanesulfonate. *ACS Applied Materials and Interfaces* 12:10503–10514.

30 MOF-based Electrochemical Sensors for Viruses/Bacteria

Hessamaddin Sohrabi[1], *Mir Reza Majidi*[1], *Ahad Mokhtarzadeh*[2] *and Karim Asadpour-Zeynali*[1]

[1]Department of Analytical Chemistry, Faculty of Chemistry, University of Tabriz, Tabriz, Iran

[2]Immunology Research Center, Tabriz University of Medical Sciences, Tabriz, Iran

30.1 INTRODUCTION

Viruses, bacteria, and other microorganisms can be found everywhere in nature and our surrounding environment. Soil, the intestinal tract of animals, waters contaminated with fecal matter, or water (particularly estuarine waters) are the origin of viral and bacterial pathogens. More than 150 kinds of bacteria are living inside and outside of an average person. Many microorganisms are crucial to nature and constructively interact with plants and animals. However, animals and human beings are affected by potentially harmful microorganisms and infected with serious diseases. Generally, infectious diseases are blamed for nearly 40% of the total 50 million annual estimated deaths that occur worldwide. Such microorganisms that generally cause severe and occasionally fatal diseases essentially remain unaffected by environmental conditions in a manner likely to endanger the lives of humans. A large number of such fatal organisms can be grown and preserved for several years. Bacteria outbreaks including the ongoing novel *coronavirus (SARS-CoV-2)* [1], *Dengue virus* [2, 3], *Zika virus* [4, 5], *Ebola virus* [6, 7], *human immunodeficiency virus (HIV)* [8, 9], and viruses like *Clostridium perfringens, Campylobacter jejuni, Listeria monocytogenes, Salmonella, Escherichia coli,* and the like, cause distressing challenges to human health. That is the reason why early detection of them for timely treatment and prevention of the related diseases is so important [10]. Thus far, multiple ways for detecting pathogenic virus and bacteria such as electrochemical sensors [11-14], IgA antibody [15, 16] and antigen testing [17, 18], IgM-IgG combined antibody testing [19], high-throughput sequencing [20], nucleic acid amplification tests [21], real-time RCR (rtPCR) [22], quantitative PCR (qPCR) [23], and so forth, have been planned for their detection. Many assays are still high-priced and require boring sample processing. That is why many scientists are encouraged to embark on devising more cost-effective and quicker diagnosis.

In addition to different classifications, it is possible to categorize biosensors based on sensing mechanisms that are capable of transmuting the interactive bio-analyte responses into identifiable signal forms. Employing nanomaterials in biosensors presents an improved strategy for effectively detecting the target analytes because of their individual properties such as optoelectronic properties, small size, and better surface-to-volume ratio. Moreover, integrating nano-engineered materials and biosensors affects the utility of sensing characteristics with regard to the cost of the device, reagent volumes, response time, specificity, and sensitivity.

The fastest-growing type of chemical sensors realized by the new analytical chemistry has been presented by electrochemical sensing and biosensing analyses [24-26]. Such a chemical sensor is defined as a device, and it prepares fixed details on the sensor's environment. Hypothetically, a particular response can be provided by a chemical sensor, and it is related to the extent of a specific chemical species in a direct manner [27, 28]. Any chemical sensor is composed of a chemically selective layer, which is capable of isolating the response of the analyte from its immediate environment, and a transducer, which is capable of changing the response into an apparent signal on different equipment. Electrical, thermal, mass, or optimal sensors can use these specific properties for their classification and are capable of detecting and responding to the analytes in solid, gaseous, or liquid states. When compared to thermal, optical, and mass sensors, electrochemical sensors are particularly important due to their low cost, exceptional detectability, and simplicity in experiments. Electrochemical sensors have found their special place among the currently-used sensors and have been commercialized thanks to presenting a broad range of new advances in clinical, environmental, agriculture, and engineering gauges. In most of the biosensing and sensing analyses, one single signal tries to determine the quantity of analyte. Nevertheless, other factors like modification by environmental conditions and electrodes also exist, which have effects on the analytical efficiencies. Such parameters can specifically provide poor detection reliability in trace detection. For determining pathogenic viruses and bacteria, electrochemiluminescence, impedance, voltammetry, and amperometry are efficient electrochemical methods and are described below.

Metal-organic frameworks (MOFs) [29, 30] can be considered as porous coordination polymers, which have different metal ions, bonded together via organic bridging ligands, and provide composites with different dimensions. Two main substructures of MOFs can be made up of clusters

DOI: 10.1201/9781003188148-30

of organic molecules and metal ions, they are, therefore, considered as inorganic and hybrid materials. Constructing such frameworks enables us to produce advanced materials, which possess regular porosity in scales from the micro to the nanopore. Functional variations, geometry, and size of metal-organic frameworks were realized as a result of developing more than 20,000 various MOF types. The MOF surface area is broader than typical solids such as carbons and zeolites, nearly 1000 to 10,000 m^2/g. MOFs are desirable for fuel storage and carbon dioxide capturing using this particular feature. Many studies conducted in the area of inorganic-organic hybrid compounds presented novel uses of such particles in modern technology and materials science. In other words, the capacity of attaching and generating super potential compounds with particular features is obtained by inorganic and organic hybrid materials. Other merits of such materials are that they are very stable and highly biocompatible with multifunctional tasks and that they have exceptional mechanical and physicochemical features. In addition, desirable catalytic potential, excellent loading capability, wide surface area, and large volumetric adsorption capacity of such materials increase their bio-connected applications. In applications of pathogenic biosensing, various MOF types binding with metal and organic blocks will be utilized as a matrix of the physical or covalent attachment of particular materials used as a bio-receptor or improving the chemical, optical, electrical signals. It is also possible to use them as a supporter for the deposition of nanoparticles of metals and different species.

In this chapter, electrochemical-based biosensing and sensing assays for highly sensitive viral and bacterial pathogen diagnosis are discussed. Various nanostructured materials and nanoparticles can be used for signal magnification in all electrochemical analysis methods. Multiple benefits including the possibility of electron transfer, immobilizing the antibodies, the capability of protecting biological activity, biocompatibility, and high sensitivity are considered for nanoparticles in electrochemical biosensing/sensing platforms. Originally, different diseases and viral and bacterial pathogenic agents and their harmful impacts on human health were mentioned. Afterward, numerous electrochemical techniques applied to accurately determine the sensitivity of such pathogens are covered. Eventually, a brief and comprehensive overview of the current limitations and forthcoming challenges in this regard is presented.

30.2 ELECTROCHEMICAL SENSING ASSAYS TO DETERMINE DIFFERENT PATHOGENIC BACTERIA SPECIES

There is now a universal concern as to the rapid spreading of pathogenic bacteria and the sudden increase of antibiotic resistance since they are considered as the main foodborne and waterborne infection sources [1-3]. Due to different diseases caused by microbial infection, pathogenic strains of bacteria can be considered as the major concern of the food industry, water supply, hospitals, and environmental biology and some of them can cause death [4]. Contaminated food sources can lead to certain diseases and food distributors have the responsibility to prevent outbreaks [2, 5,6]. Particularly, pathogenic bacterial strains including *Escherichia coli, Staphylococcus, Salmonella,* and *Haemophilus Influenzae* can pollute many water sources and in that case, diseases like typhoid fever, gastroenteritis, cholera, and various diarrheal responses can be observed [2, 7, 8]. According to the report of the World Health Organization (WHO) in 2016, bacterial water contamination annually causes 829,000 deaths from diarrhea [9]. Because of eating contaminated food, approximately 600 million (almost 1 in 10 people) fall sick, and therefore, the annual loss of good health in 33 million people and 420,000 deaths can be observed (disability-adjusted life years; DALYs) [10]. Because of the fatalities due to these microorganisms and their pathogenicity, their quick and sensitive diagnosis is very important using these sensing platforms. The reader can find new analysis on MOF-based identification of pathogenic bacteria.

In a recent study, Sohrabi et al. presented a primary label-free DNA genosensing assay based on DPASV and a direct hybridization in the presence of $[Fe(CN)_6]^{4-/3-}$ to recognize the H. *Influenza* genome in the samples of human plasma (Figure 30.1). Therefore, a Zn-based MOF has been synthesized and conjoined with carboxymethyl cellulose (CMC) and they have been immobilized on the Au electrode surface and AuNPs have been immobilized on the modified electrode surface of the Zn-based MOF/CMC/Au. Desirable performance for determining the L-fuculokinase gene from the *Haemophilus Influenzae* genome, suitable specificity, and high sensitivity can be represented by the fabricated genosensing bio-assay. Correspondingly, LOQ and LOD were equal to 3.23 fM and 1.48 fM. Additionally, t-DNA has obtained a wide linear range of 0.1 pM–10 nM. 98.4-103% and 2.2-3.2 were obtained for the recoveries and RSDs, respectively. The high selective ability of one, two, and three-base mismatched sequences were presented for the fabricated biosensing assay. Afterward, the t-DNAs of *Shigella flexneri and Salmonella typhimurium* bacteria were used for the negative control of the genosensing assay to investigate the selectivity. It was also possible to study the reproducibility and repeatability of the associated bio-assay. Meantime, it should be mentioned that for evaluating the hybridization process, the organized genosensing bio-assay will be reused and regenerated instantly. For diagnosing the infectious agents and on-time provision of effective treatment to patients, the approving aspect of the fabricated genosensing assay can be considered as its singular feature. Such an investigation represents that the hard-growing bacteria can be diagnosed using similar genosensors [14].

According to different research and for the sensitive detection of bacterial lipopolysaccharide (LPS), an electrochemical sensor based on dual-functional Cu^{2+}-modified MOF nanoparticles (Cu^{2+}-MOFs) has been reported. Li et al. [31] have prepared Cu^{2+}-NMOFs. In such an LPS sensor, due to the communication of LPS with the C18 alkyl chains by strong intermolecular interactions (Figure 30.2), LPS originally immobilized in gold nanoparticles/reduced graphene oxide by C18 alkanethiol chains. Thereupon, the

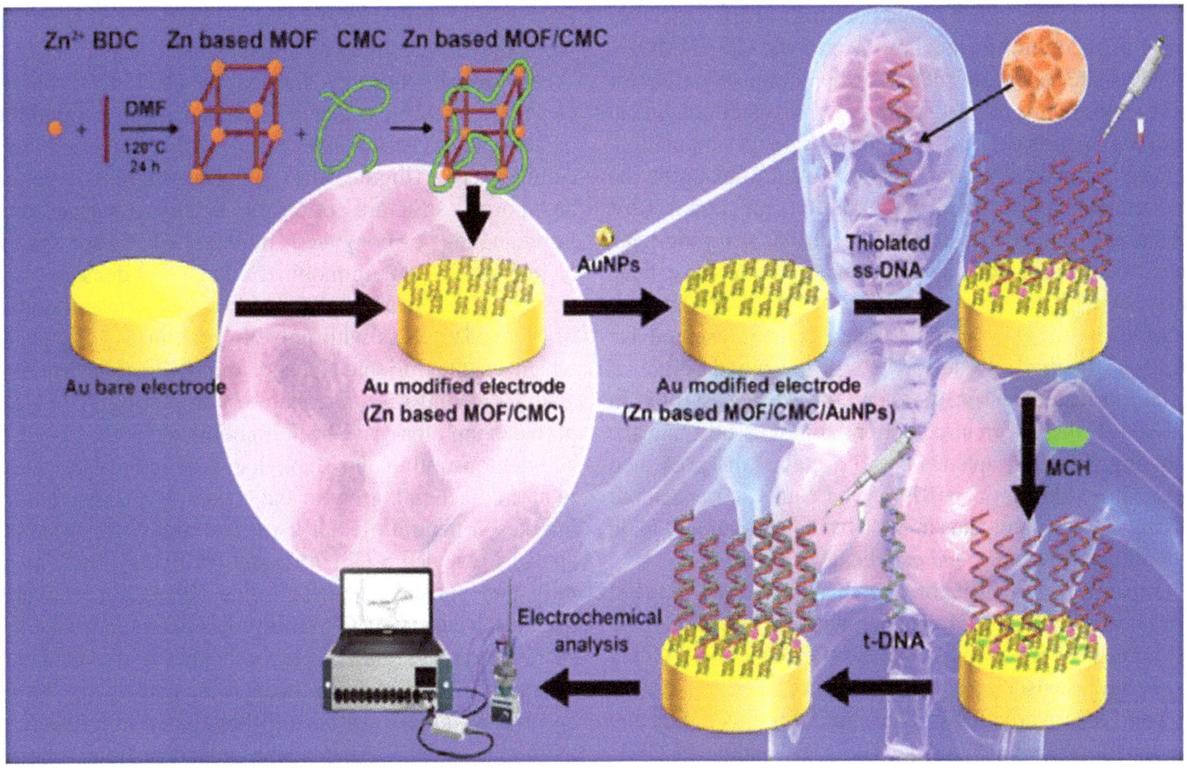

FIGURE 30.1 Graphically representing proposed genosensing bio-assay for sensitive H. *Influenzae*. determination. Adapted with permission from [14]. Copyright (2021) Springer.

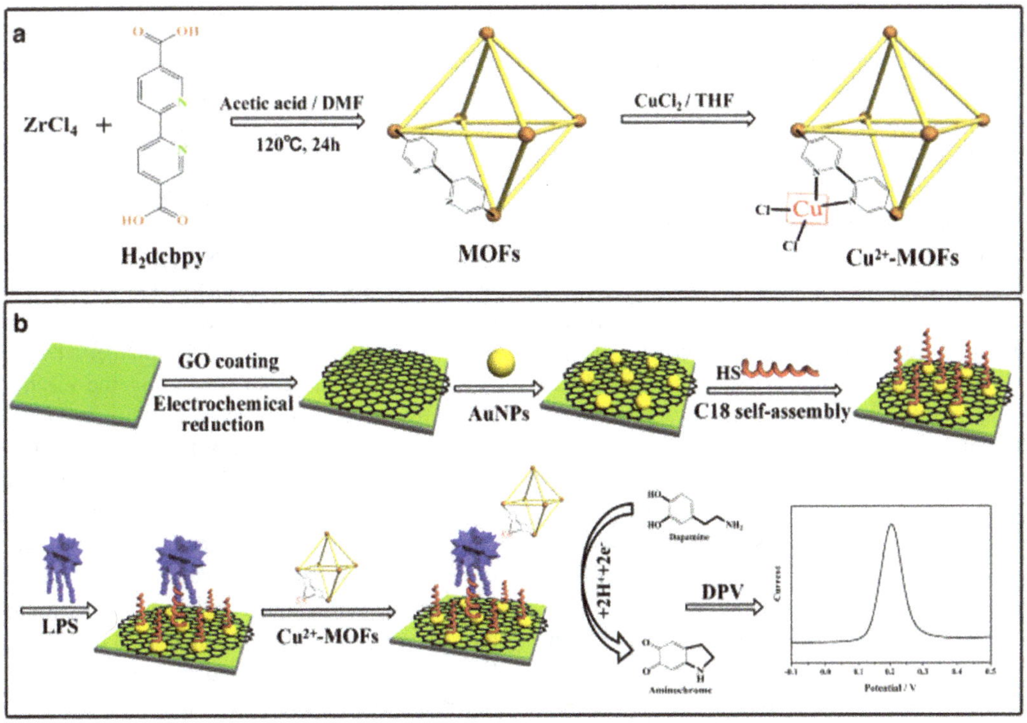

FIGURE 30.2 Schematically representing electrochemical sensing assay. a) Synthesis of Cu^{2+}-NMOFs; b) Fabricating electrochemical LPS sensing assay. Adapted with permission from [31]. Copyright (2021) Springer.

anionic groups of the carbohydrate portions of LPS molecules that serve as recognition units have seized the Cu^{2+}-NMOFs. Cu^{2+} NMOFs are capable of catalyzing dopamine oxidation for producing aminochrome. Generating a strong electrochemical oxidation signal is its result. According to this experimentation, differential pulse voltammetry has been used to examine the electrochemical sensor fabricated on dual-functional Cu^{2+} NMOFs and for screening the LPS level, the stripping peak currents of dopamine oxidized to aminochrome have been employed. A wide linear with a range of 0.0015 - 750 ng/mL and a LOD of 6.1×10^{-4} ng/mL have been represented for the proposed method. To detect LPS in the blood serum of mice, the fabricated sensor has been utilized successfully. In comparison to the different detection structures utilizing the LPS-binding peptides, aptamers, and proteins, the presented LPS determination that depends on the catalytic peroxidase mimicking NMOFs, contains many advantages, namely, good reproducibility, low LOD, and excellent specificity.

By preparing a composite of a MOF, namely, MIL-53 (Fe) (MIL=Materials of Institute Lavoisier) with poly (3,4- ethylene dioxythiophene) polystyrene sulfonate (PEDOT: PSS), Gupta et al. [32] have devised an electrochemical biosensing system for *E. coli*. Afterward, it has been used for modifying the carbon-screen printed electrodes (cSPE). By adding PEDOT: PSS, the beneficial features and the desirable electrochemical conductivity of MOF have been represented by the composite. Anti-*E. coli* antibodies have been used to functionalize the composite modified cSPE to ensure the detection specificity. The presented electrochemical sensor that depends on the measurement of different pulse voltammetry (DPV) signals and electrochemical impedance spectroscopy, made it possible to detect *E. coli* in a wide concentration range of 2.1×10^2– 2.1×10^8 CFU/mL where the superlative LOD is equal to 4 CFU/mL. At the same time, the sensor is capable of tolerating different feasibly interfering species co-present, including *S. arlettae* and *S. aureus*. The outcomes of such a system were validated by standard methods.

According to other research by Zhou et al. [33], for fabricating a MOF-based enzyme (G4-hemin@MOF) (Figure 30.3), a small molecule, hemin/G-quadruplex (G4-hemin), has physically been fixed into NH_2-MIL-53 (Al). Catalyzing a 3,3′,5,5′ -Tetramethylbenzidine (TMB) reaction in the existence of H_2O_2 has been used to examine its catalytic feature and stability. The achieved subsequences represented better stability and activity for such a newly-developed enzyme in comparison to the G4-hemin or old-style hemin, and they are even stronger than hemin@MOF (G4-hemin@MOF > G4- hemin > hemin@MOF > hemin). For at least 24 hours, there was no important loss in the catalytic property. Only a reduction of 13.8 % was obtained after eight days. Using the catalytic feature of G4-hemin@MOF, an electrochemical biosensing platform has effectively been obtained, based on a signal amplification method. The sensing platform had the capability of determining TB (Toluidine Blue) in buffer sensitively and in the existence of many proteins. A better linear response range of 10^{-18} mol/L - 10^{-14} mol/L and a detection limit of 0.33 aM has been obtained for the mentioned method. Furthermore, using this biosensing platform, it is possible to predict great selectivity against mismatch sequences of TB.

An investigation by Farooq et al. [34] discussed an electrochemical biosensor development using an immobilizing bacteriophage in the matrix of surface-modified bacterial cellulose (BC) (Figure 30.4). BC, because of its great fibrous and porous structure, offers an extensive surface area for impregnating carboxylated multiwalled carbon nanotubes (c-MWCNTs) and immobilizes the great-density phage. A positive charge can be generated by the surface modification of BC/c-MWCNTs with polyethyleneimine (PEI) generates, which enables oriented phage immobilization. Microscopic analysis of confocal was capable of presenting 11.7 ± 1.2 phage particles·μm^{-2} that have been immobilized in the matrix of BC and demonstrating anti-staphylococcal activity by generating reduced bacterial growth and clear lytic zone. During 30 minutes at neutral pH, DPV (differential pulse voltammetry) analysis revealed 5 CFU/mL and 3 CFU/mL of *S. aureus* in milk and phosphate buffer saline (PBS), respectively, and during 6 weeks at 4 ^{0}C, stability has been represented. Great specificity for *S. aureus*, in both pure and mixed cultures of non-host bacteria, has been presented by the biosensor, and live *S. aureus* has been differentiated in a live/dead cell mixture effectively. An increase in the deposition time and the phage concentration can additionally upgrade the functioning of the developed biosensor. The method will be used for on-site *S. aureus* detection, particularly in food samples.

One of the greatest inflexible multidrug-resistant bacteria of nosocomial infections is considered to be *Pseudomonas aeruginosa (P. aeruginosa)*. It is difficult to use the presented detection techniques with low detection sensitivity for *P. aeruginosa*. According to an investigation presented by Zhang et al. [35], for rapidly detecting *P. aeruginosa* with higher sensitivity (Figure 30.5), a new enzyme-free electrochemical biosensor has been fabricated. For providing desirable adsorption, the Zr MOF with a very large surface area has first been synthesized. For synthesizing a Cu-Zr MOF possessing great catalytic activity, it has then been joined with a particular Cu^{2+} amount. To function as a signal probe for catalyzing the H_2O_2 decomposition, the nanocomposite of Cu-ZrMOF@Aptamer@DNA has next been composed. Similarly, for increasing the electron transfer for obtaining desirable detection sensitivity, great conductive Super P has been used. To quantify P. aeruginosa where the linearity range is equal to 10-10^6 CFU/mL and LOD is equal to 2 CFU/mL (S/N=3), the fabricated biosensor has been employed. In comparison to the former techniques, the current biosensor is less laborious and more sensitive (only for 120 minutes). Providing advantageous specificity and reproducibility by the biosensor was verified by the analytical functionality evaluation. For quantifying *P. aeruginosa* in spiked urine samples, the biosensor has efficiently been used. These outcomes show that the presented electrochemical biosensor can be a favorable lab device to detect *P. aeruginosa* in clinics. Additional investigations have been done aiming at increasing the stability of storage through simplifying the structure of introduced biosensors and optimizing additional parameters for being easily assembled into a portable device.

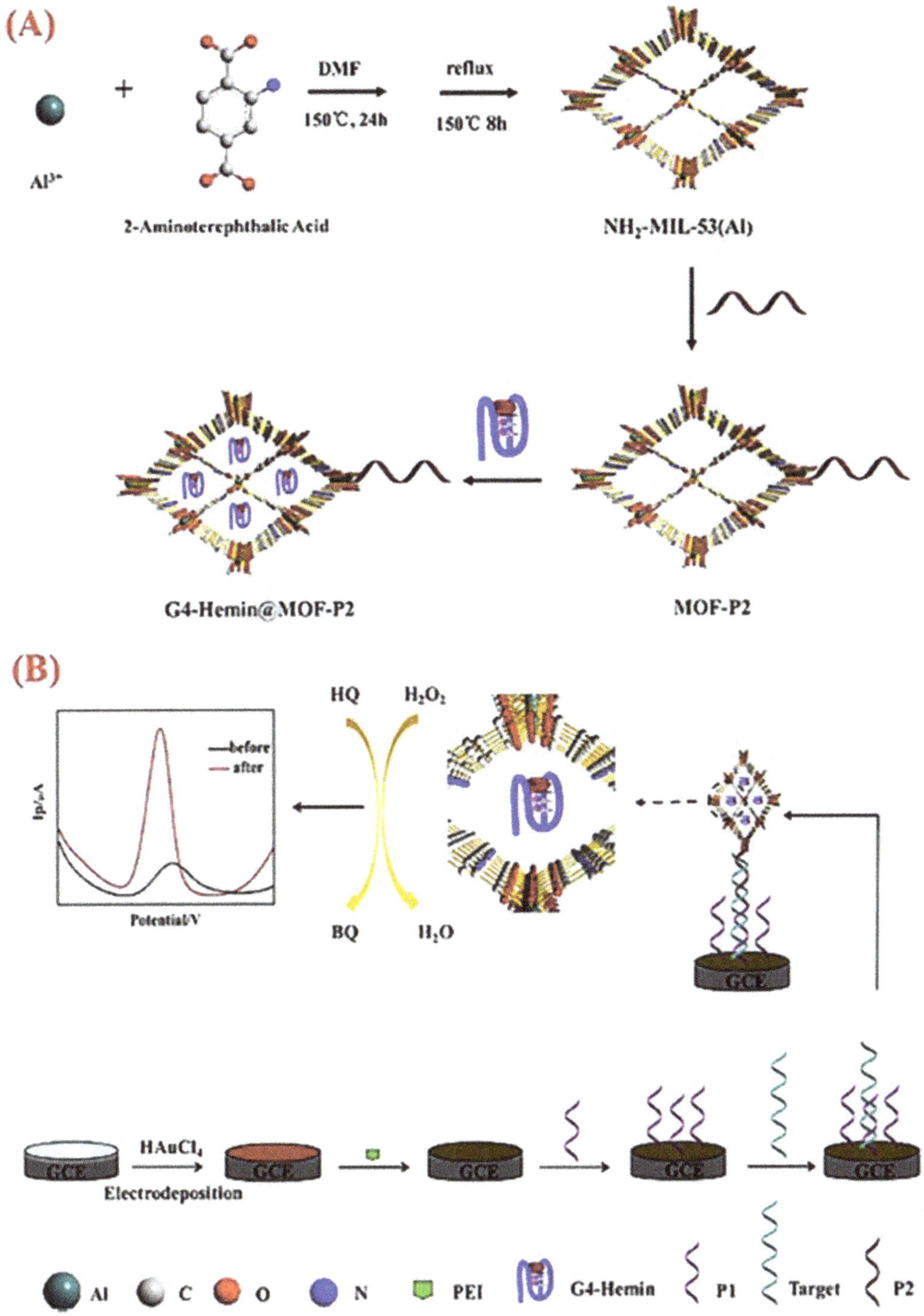

FIGURE 30.3 Graphically outlining that: (A) Synthesis steps of G4-Hemin@MOF-P2 (B) Use of G4-Hemin@MOFin the electrochemical tuberculosis determination. Adapted with permission from [33]. Copyright (2020) Elsevier.

Detecting the pathogenic bacteria sensitively and promptly is challenging in food safety and to prevent epidemics of foodborne disease. For detecting *Vibrio parahaemolyticus* (V.P), the main limitations of standard analytical approaches relate to their tiresome and time-consuming operation and bulky equipment. An electrochemical aptasensor for a quick on-site quantification of V.P in seafood (Figure 30.6) has been represented by Wang et al. [36]. Magnetic nanoscale MOFs (Fe_3O_4@NMOF) with an aptamer label against V.P acted as capture probes, while nanoparticles of gold mixed with ferrocene and phenylboronic acid acted as the nanolabels. When determining V.P, the sandwich-type complex of capture probe-V. P- nano label has been formed and magnetically connected to a screen-printed electrode (SPE) for signal measurement. According to optimized conditions, an increase in ferrocene electrochemical signals were capable of measuring

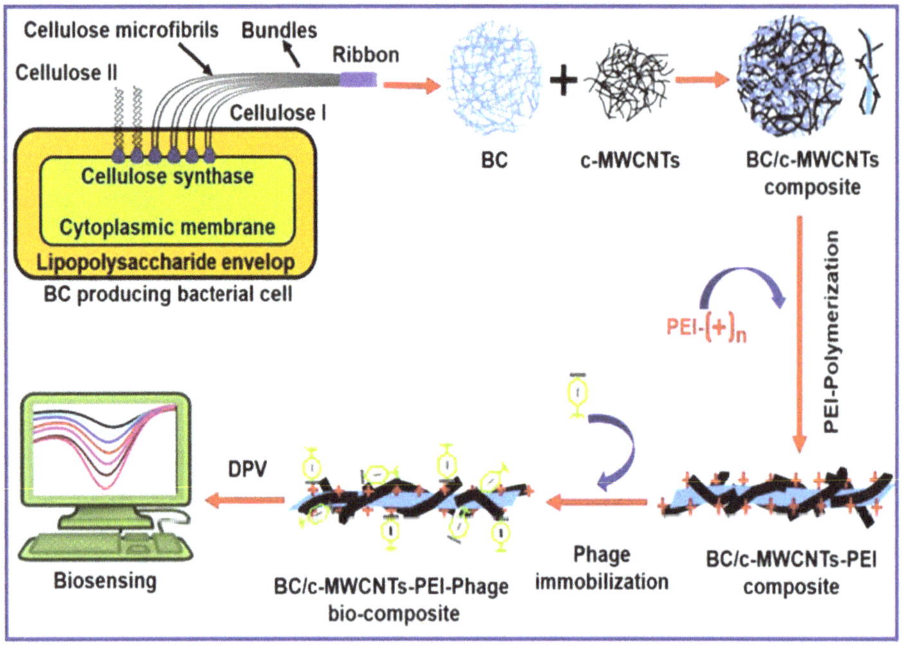

FIGURE 30.4 Schematically representing bacterial cellulose production. Adapted with permission from [34]. Copyright (2019) Elsevier.

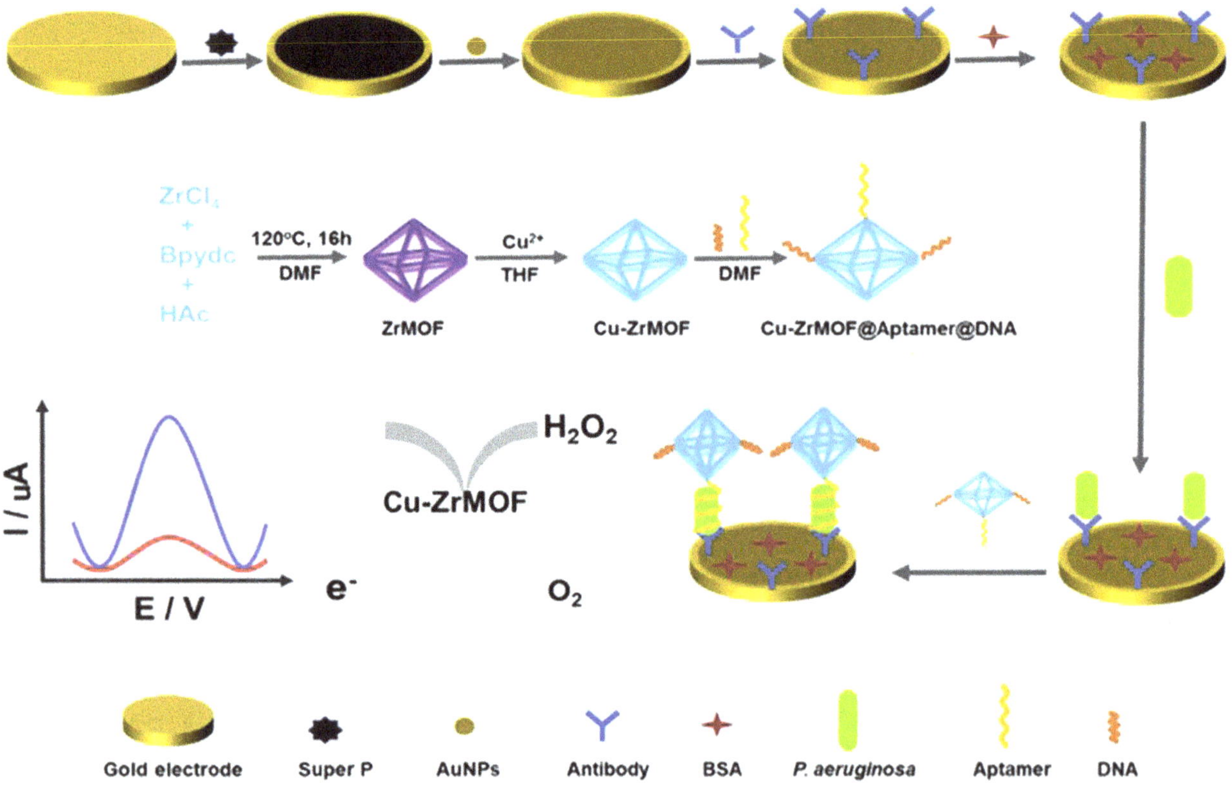

FIGURE 30.5 Presenting the electrochemical sensing assay for determining *P. aeruginosa*. Adapted with permission from [35]. Copyright (2020) Elsevier.

the amount of V.P. The range of quantified concentration has been equal to 10×10^9 CFU/mL. The pretreatment can be prevented by magnetic separation methods on a SPE and the matrix complex can be affected strongly. In addition, 20 minutes were considered for the whole V.P detection time, and for culturing the normal biochemical and agar identification, this time is extremely shorter than the tie essential. For on-site V.P detection, the reusable and portable platform of SPE is favorable. By changing aptamer in aptasensor as a means to monitor the foodborne pathogenic bacteria, an analytical

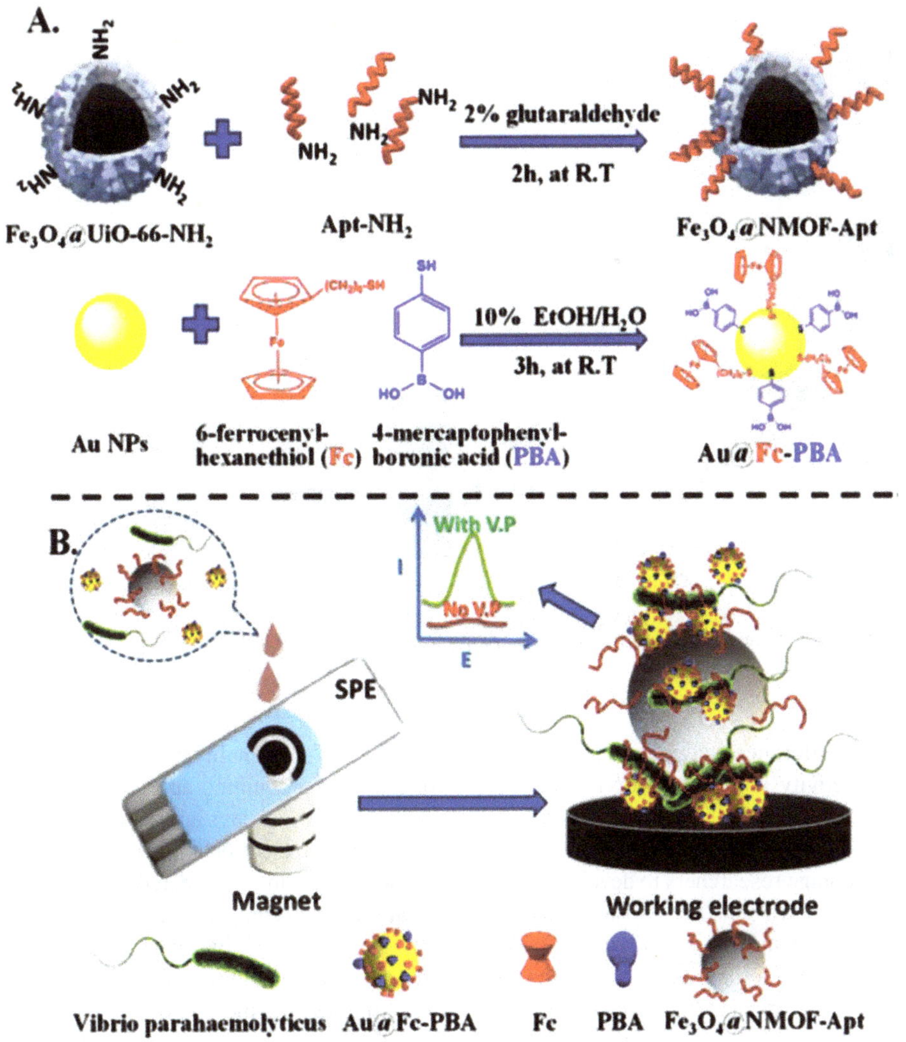

FIGURE 30.6 Schematically representing that (A) the nano label and capture probe preparation and (B) the signal-on type assay for *Vibrio parahaemolyticus* on the platform of SPE with the magnet. Adapted with permission from [36]. Copyright (2019) Elsevier.

strategy will be developed for quantifying different species of Vibrio.

MOFs can be utilized as superb media for developing biosensors for different analytes in environmental media by providing different benefits (for instance, the feasibility of synthesizing them in forms of 3-D, 2-D, and 1-D, flexible framework functionality, and large surface-to-volume ratio). An account of Cu-MOF-based electrochemical biosensor development for *E. coli* bacteria sensitive detection (Figure 30.7) has been presented by Gupta et al. [37]. $Cu_3(BTC)_2$ (BTC = 1,3,5-benzenetricarboxylic acid) has been mainly mixed with polyaniline (PANI) for making a MOF-based electrochemically active platform. Afterward, for serving as a new biosensing electrode, $Cu_3(BTC)_2$-PANI thin film, on a substrate of indium-tin-oxide (ITO), have been bio-interfaced with antibodies of anti-*E. coli*. According to the method of electrochemical impedance spectroscopy (EIS), the sensor showed a beneficial sensitivity to detect lower *E. coli* (2 CFU/mL) concentrations in a shorter response time (~2 minutes). Meanwhile and in the existence of different non-specific bacteria, it was selective. The biosensor fabrication process is uncomplicated (in other words, without using inconvenient steps). Because of a large surface devised transduction platform area (namely, $Cu_3(BTC)_2$-PANI composite), the sensor has the capability of being operated for a wide analyte concentration range at an appropriately low LOD. The represented MOF-based biosensor gives a superior control on surface characteristics and final material using facile fabrication when compared to different electrochemical platforms (for example, CNTs and graphene). From an environmental perspective, there are advantages for the presented MOF-based biosensor according to special conditions such as nontoxicity concerns, which makes it more beneficial for being developed in disposable strip form. Once comparing the performance of Ab/$Cu_3(BTC)_2$-PANI/ITO and previously proposed electrochemical *E. coli* biosensors, it becomes clear that the presented system gives a progressive alternative for biosensing of *E. coli*.

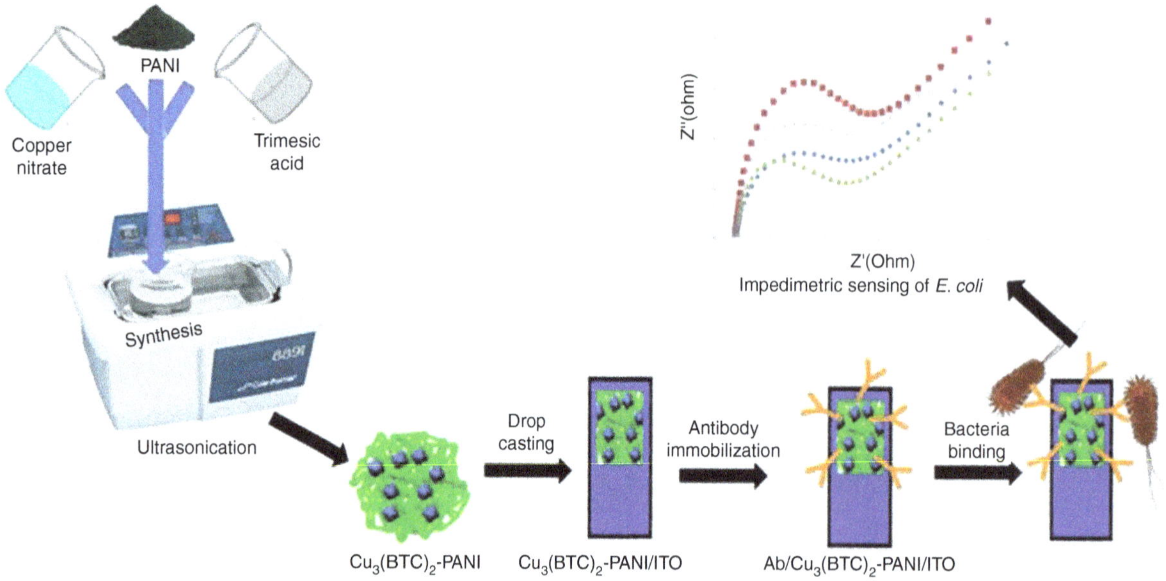

FIGURE 30.7 Illustrating a foreknown sensing bio-assay fabrication. Adapted with permission from [37]. Copyright (2019) American Chemical Society.

For determining the difference between dead/live cells, the limited level of selectivity and sensitivity, inability, and high cost can restrict the usually utilized nucleic acid-based biosensors, antibodies, and enzymes for detecting *S. aureus*. Such disadvantages encourage researchers to develop a stable, electrochemical, selective, ultra-sensitive biosensor that can separate the live *S. aureus* in a live/dead-cell mixture in samples of food. Using antibody (Ab)-hierarchical mesoporous silica (HMS) bio-conjugates, Wang et al. [38] have represented a sensitive antibody-based electrochemical immunosensor for label-free diagnosis of low *S. aureus* concentrations. For the preparation of HMS, a bio-template approach based on butterfly wings has first been used. For conjoining the antibody with glutaraldehyde, the carrier material has then been amino-functionalized. Subsequently, a glassy carbon electrode (GCE) has been used to immobilize the Ab-HMS bio-conjugates and by investigating the alterations in the peak currents after the formation of antigen-antibody complex, it was possible to detect *S. aureus*. DPV (differential pulse voltammetry) has been implemented with bacterial concentration ranges of 10–2 × 10^3 CFU/mL. Using *Salmonella, Listeria monocytogenes (L. monocytogenes),* and *Escherichia coli (E. coli)*, selective tests have been made. Using the sensor, selective assays represented accurate *S. aureus* detection. Additionally, a satisfactory linear relationship between an increase of peak current and logarithmic concentration (R^2= 0.9759) of *Staphylococcus aureus (S. aureus)* with a 20-min detection time and an 11-CFU/mL LOD has been exposed by the immunosensor. Once doing EIS (electrochemical impedance spectroscopy) according to the same terms, a better linear correspondence between the bacterial concentration (R^2= 0.9720) and the impedance change value has been represented by the outcomes and the LOD has been equal to 12 CFU/mL. The sensor functionality has been compared to the functionality of the colony counting approach in the experiment of spiked milk samples. 15 days are required for stabilizing the sensor.

A special electrochemical biosensor based on aptamers immobilized in engineered ZIFs-8 (zeolitic imidazolate Framework-8) employing EDC-NHS chemistry (Figure 30.8) has been presented by Shahrokhian et al. [39]. By observing the special π-π interactions between graphene oxide (GO) and aptamer, the method of DPV has been used with Fc-GO (ferrocene-graphene oxide) as an electroactive marker for detecting *Pseudomonas aeruginosa* (*P. aeruginosa*). In the existence of *P. aeruginosa*, a change has been observed for the aptamer structure and Fc-GO released from the surface of the electrode. According to the signal-off method, a large linear dynamic range (1.2 × 10^1– 1.2 × 10^7 CFU/mL) with a 12-CFU/mL LOQ is displayed for the presented biosensor. When screening disease therapy and monitoring the safety of the clinical sites, the values show that the generated aptasensor enjoys great potential applicability.

A virulence factor concealed by MTB (Mycobacterium tuberculosis) is considered by the 6-kDa early secretory antigenic target named ESAT-6. A voltammetric aptasensor for ultrasensitive ESAT-6 detection (Figure 30.9) has been presented by Linlin et al. [40] . A GCE has been used for placing the reduced graphene oxide doped with the metal-organic framework (MOF-rGO). Immobilizing the electroactive TB (Toluidine Blue) has been augmented by this and the electron has been enabled to transfer from TB to the modified GCE. For bringing the thiolated EBA (ESAT-6 binding aptamer) together on a modified electrode and for strengthening the response to Toluidine Blue, the nanoparticles of Platinum/gold core/shell (Pt@Au) have been used. The conformed GCE acted at −0.36 V (vs. SCE), gives a linear response in ESAT6

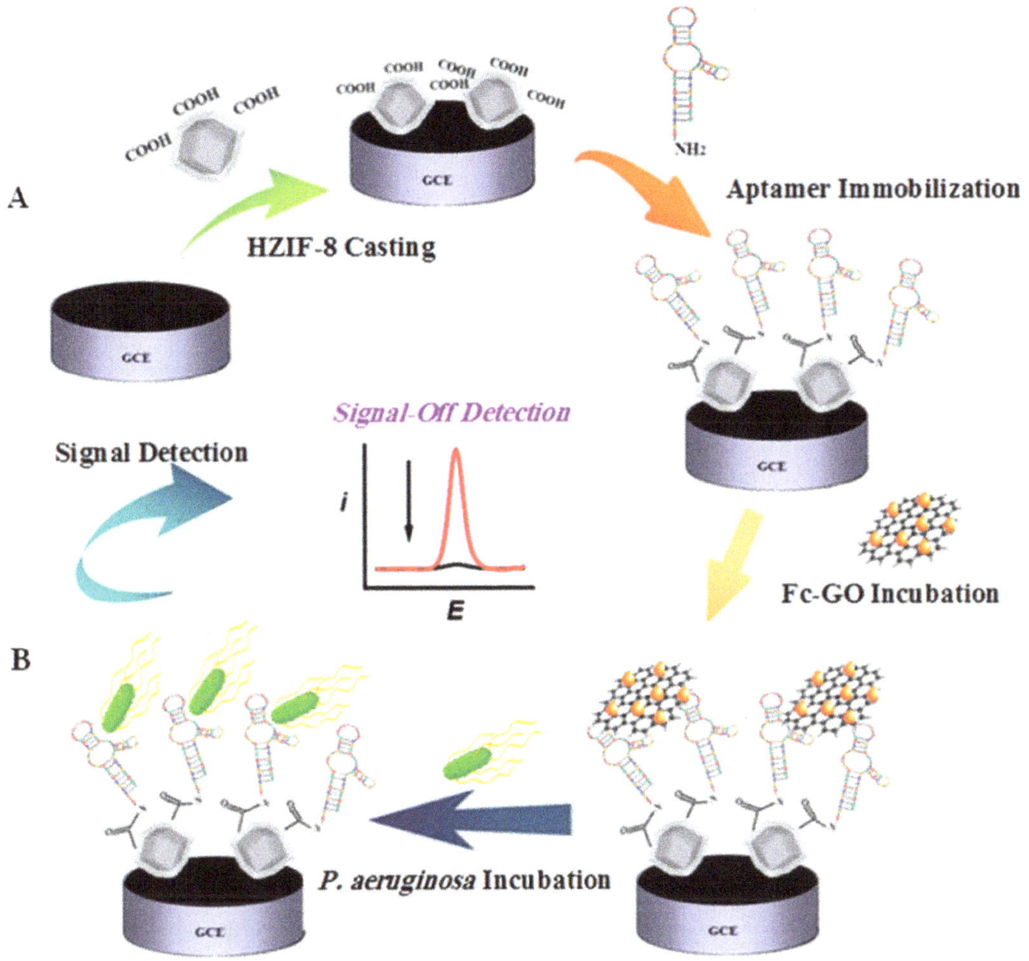

FIGURE 30.8 Schematically illustrating (A) fabrication of aptasensor and (B) signal-off detection of *Pseudomonas aeruginosa*. Adapted with permission from [39]. Copyright (2018) Springer.

concentration range of 1.0×10^{-4} to 2.0×10^{2} ng/mL and the LOD of ESAT-6 is equal to 3.3×10^{-5} ng/mL.

The antibacterial efficacy of three Zn-based nano MOFs (nMOFs), in other words, Zn-BTC, MOF-5, and IRMOF-3 have been investigated by Bhardwa et al. [41], either alone or as mixtures with kanamycin and ampicillin. The mixtures of nMOF/drug presented additive (other combinations of nMOF/drug) or synergistic (kanamycin / IRMOF-3) effects in comparison to nano MOFs or antibiotics alone when testing against *Listeria monocytogenes, Staphylococcus lentus, Staphylococcus aureus* and *Escherichia coli.* By increasing the effectiveness of the available antibiotics, Zn-based nMOFs are capable of alleviating the burden required on novel antimicrobial pharmaceuticals.

According to other research, Li et al. [42] presented an ultrasensitive aptasensor for determining the *Mycobacterium tuberculosis* antigen MPT64 in human serum in a voltammetric manner (Figure 30.10). For serving as the carrier of the aptamers and gold nanoparticles, an amino-modified Zr(IV)-based MOF; type UiO-66-NH_2; composed of 2 amino-terephthalate linkers and Zr_6O_{32} units) with a wide surface has first been synthesized and used. After casting the horseradish peroxidase on the nanomaterials, the signaling nanoprobe has then been engineered. A gold electrode was used to fix two aptamers and their synergistic effect on binding MPT64. If the ratio of two aptamers was equal to 1:1, DPV represented that the peak current will be the greatest. A wide linear response with a range of 0.02 - 1000 pg/mL for MPT64 and a LOD of 10 fg/mL at a working potential of almost −96 mV (in contrast with Ag/AgCl) has been presented by the assay. The values represent the effectiveness of this biosensor for detecting *tuberculosis* at early stages. The difficulty in clinical application is considered for the biosensor. Additionally, different determining factors have effects on the electrode modification process. It is hoped to solve these flaws and facilitate its application in clinical detection by integrating electrochemical detection to different devices like a microfluidic chip.

30.3 ELECTROCHEMICAL SENSING ASSAYS FOR DETERMINING DIFFERENT PATHOGENIC VIRUS SPECIES

Acute diseases which are mainly rooted in viral infections do not require hospitalization in industrialized countries, but

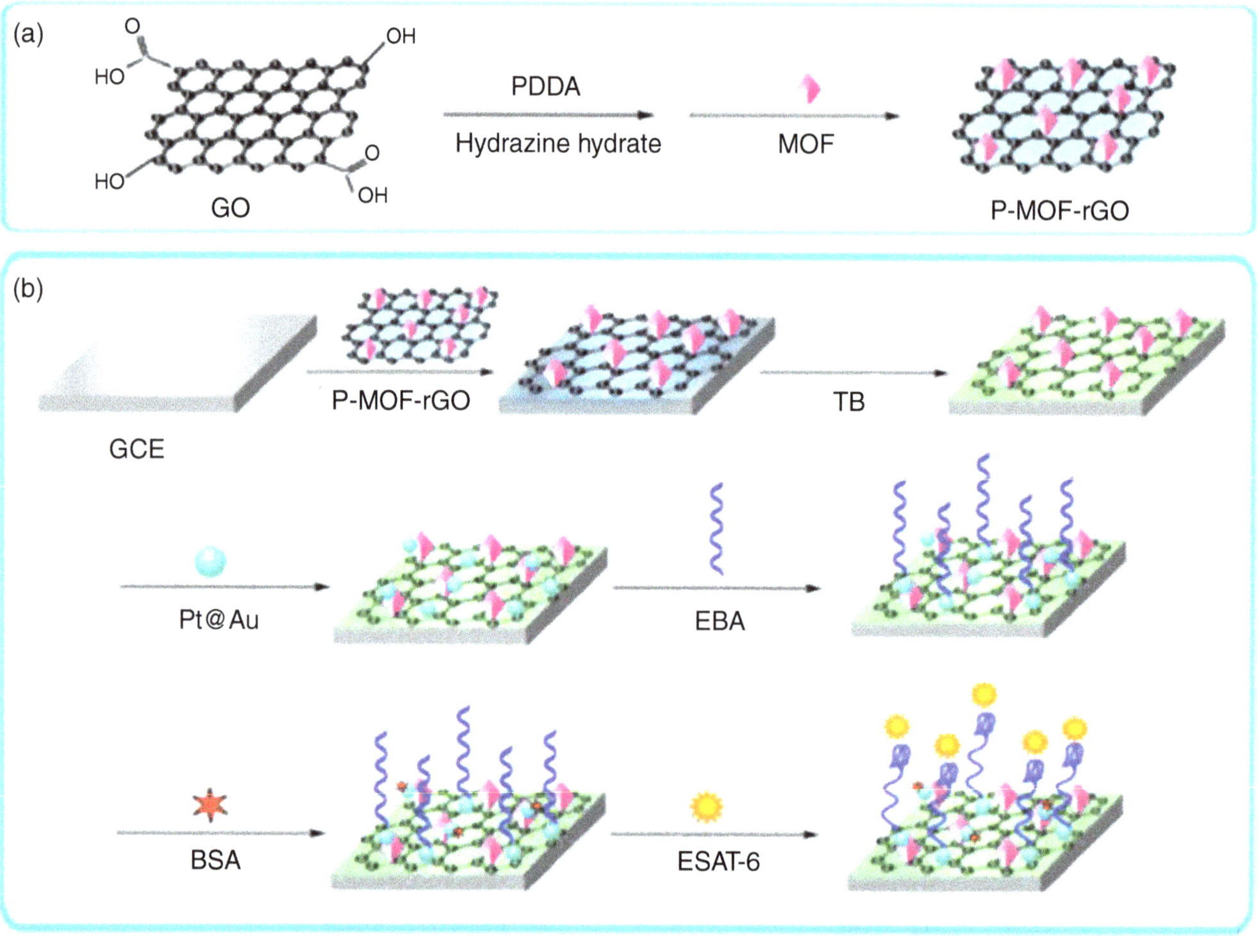

FIGURE 30.9 Schematically representing stepwise electrochemical aptasensor construction procedure. Adapted with permission from [40]. Copyright (2018) Springer.

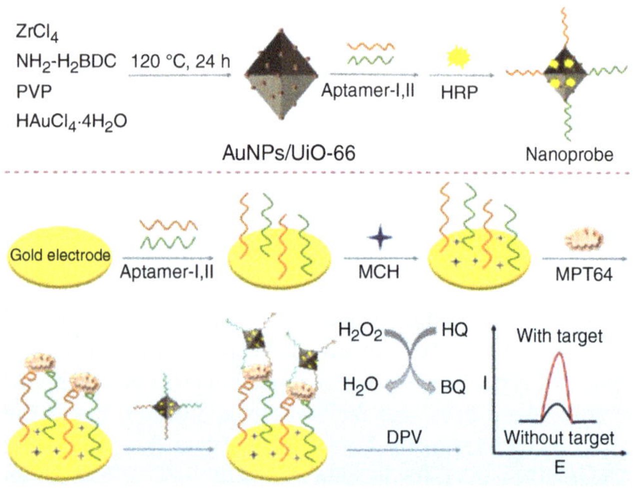

FIGURE 30.10 The fabrication process of UiO-66-NH_2/AuNPs/HRP8/aptamer nanoprobes and electrochemical aptasensing assay. (MCH: 6-Mercapto-1-hexanol; HQ: hydroquinone; HRP: horseradish peroxidase; PVP: poly (vinyl pyrrolidone); BQ: benzoquinone). Adapted with permission from [42] Copyright (2020) Elsevier.

in developing countries, they cause lasting disabilities and sometimes death, particularly in children and infants. People can easily transfer the viral infections to each other, but in many cases, food consumption and/or contaminated water will be considered as other infection sources [7]. A very familiar foodborne infection is considered viral gastroenteritis, which annually causes one million deaths in children aged under five [7]. Other information shows that the current extensive range of previously distinguished viral diseases will be expanded to contain different serious human diseases, including juvenile diabetes, rheumatoid arthritis, tumors, and different immunological and neurological disorders [11].

Sensitive coronavirus detection is very important for avoiding its outbreak. Severe diarrhea and death in neonatal piglets can be by PEDV (porcine epidemic diarrhea virus) which is a highly infectious coronavirus. A PCN-224-based electrochemiluminescence (ECL) system has been presented by Ma et al. [43], which can be used for the detection of PEDV with great sensitivity. PCN-224 is assumed to be used as a reporter of ECL with a strong signal due to its zirconium-based organic porous frame nanomaterial providing a stable structure and a large particular surface area. For the first time, nanoparticles of TiO_2 have been used as an accelerator for supporting the co-reactant potassium peroxydisulfate reduction on the cathode, and therefore, the initial ECL signal of PCN-224 has been improved in a significant manner. Prior to PEDV, the ECL signal has been made weak because of the block effect on electron transfer. Therefore, the new biosensor of 'signal off' obtained a sensitive PEDV detection with a range of 1 pg/mL - 10 ng/mL, where the LOD is equal to 0.4 pg/mL (S/N=3). Particularly, the ECL system has been upgraded in biological analysis by PCN-224 nanomaterial and a novel path for detection of coronavirus has been presented by the projected strategy.

An sMPC (soft metal-phenolic capsule)-based electrochemical immunoassay for extremely sensitive Epstein-Barr virus capsid antigen IgA (EBVCA-IgA) detection, a nasopharyngeal carcinoma biomarker has been presented by Huang et al. [44] (Figure 30.11). Large-sized metal probes contain some metal ions that can provide some benefits to the magnification of signal for anodic stripping voltammetry, but probes comfortably precipitate due to their heavy weight and therefore, can reduce the functionality and recognition efficiency. According to such an investigation, they represented unique surface behavior that could be represented by sMPCs made by metal-coordination interactions in comparison to their solid counterparts, which can make the recognition efficacy better and enlarge sensitivity despite their micrometer size. When sMPCs are used, a greatly improved sensitivity with a 0.46-fM LOD for EBVCA-IgA can be demonstrated by the included electrochemical immunoassay and it is usable in the analysis of the real sample. It can be considered to be the first report of an sMPC based electrochemical method. It can be used for explaining the potential effect of the probe rigidness on the usefulness of an included biosensor and in that case, a chance for designing different effective probes is offered. A satisfactory potential in the point-of-care diagnostics of fatal diseases can be guaranteed by the advantages of such a strategy like good selectivity, ease of fabrication, and exceptional sensitivity.

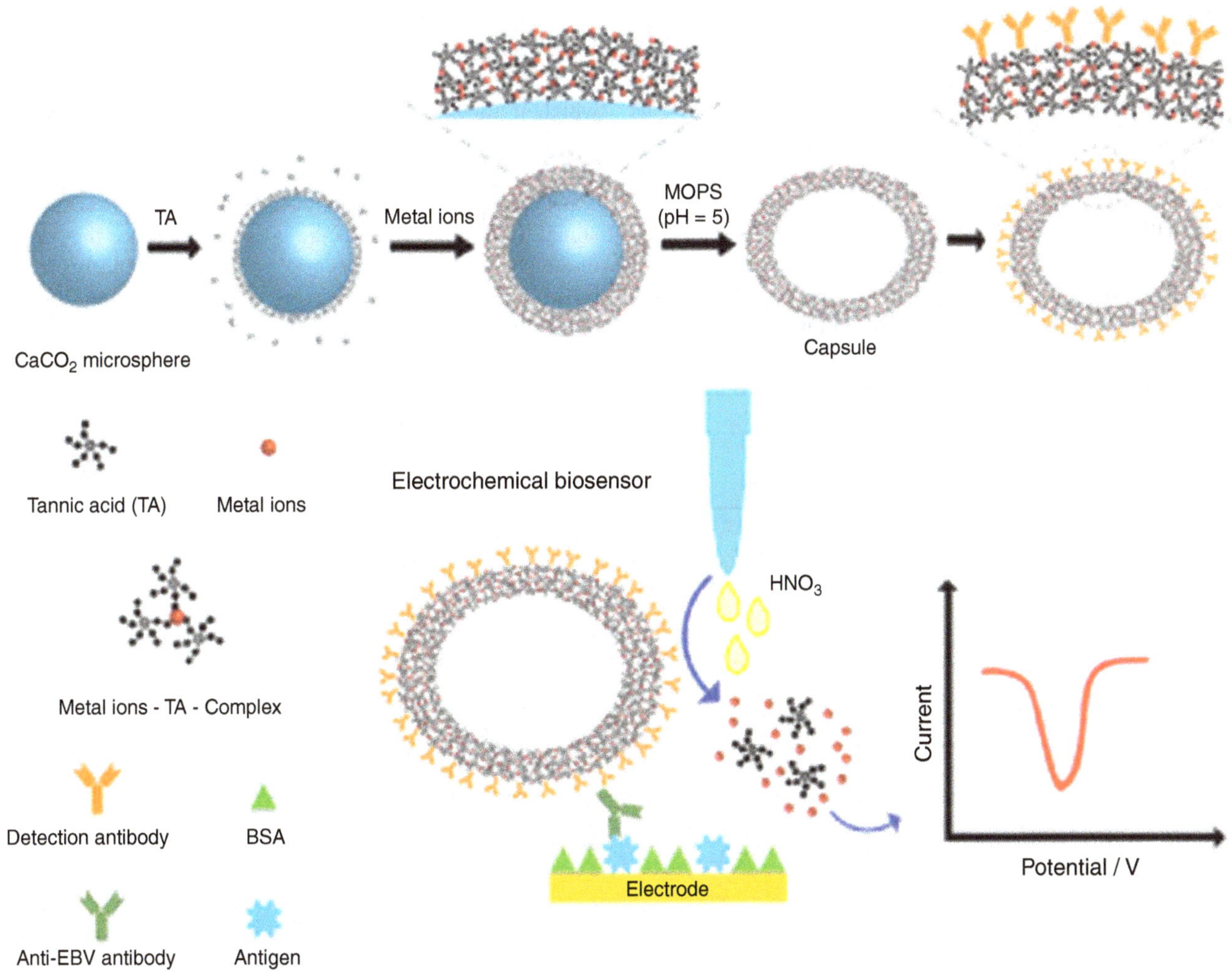

FIGURE 30.11 Schematically representing presented electrochemical immunoassay. Adapted with permission from [44]. Copyright (2020) Elsevier.

For devising novel and more efficient strategies in diagnosing the Hepatitis B virus (HBV), a very challenging task is presented. A new simple electrochemical biosensor with superior stability and sensitivity has been presented by Lin et al. [45], which has been fabricated to detect HBV DNA through integrating the electroactive Cu-MOF as a nanoprobe of signal and multi-site matrix and electro-reduced graphene oxide (ErGO) as magnification material of signal. A better affinity for probing DNA via covalent attachment can be demonstrated by Cu-MOF, which is capable of stabilizing the probe of DNA. 89.9% and 9.48×10^{13} molecules cm^{-2} have been obtained for the hybridization capability of the electrochemical DNA biosensor and the probe DNA density, respectively. In an excellent situation, a functionality toward the determination of HBV DNA with a wide linear range of 50.0 fM - 10.0 nM is considered for the electrochemical DNA biosensor, where LOD is equal to 5.2 fM. Additionally, for quantifying HBV DNA in urine samples and human serum with accurate outcomes, a novel sensor has been utilized.

Real-time polymerase chain reaction (RT-PCR) based Hepatitis-C virus ribonucleic acid (HCV-RNA) identification and measurement is the fundamental element in infection control, administration, and response to remedy because of its sensitivity, specificity, and quantification capabilities. However, some disadvantages are considered to be the requirement of high-tech laboratory equipment, time-consuming and costly which limit the utilization of this approach in rapid diagnosis, point-of-care testing, and blood banks (POCT). According to another research, Sheta et al. [46] fabricated a new label-free electrochemical biosensor that takes advantage of polyaniline@nickel MOF (Ni-MOF) nanocomposite to develop the immediate detection of unamplified HCV nucleic acid (Figure 30.12). Using the smooth layer-by-layer installation of the bovine serum albumin (BSA), deoxyribonucleic acid (DNA), and polyaniline@Ni-MOFnanocomposite onto a GCE, a strong biosensor has been constructed, and then, it has been examined in real-time by EIS and cyclic voltammetry (CV). Using the EIS method, it is possible to observe the specificity and sensitivity of the devised biosensor. The values represented the linear range of 1 fM-100 nM with a 0.75-fM LOD (at an S/N ratio of 3) for the biosensor. By presenting cost-effectiveness and accuracy especially in developing countries, such outcomes have positive effects on the quantitative detection of HCV RNA and other nucleic acids. Furthermore, it is possible to adapt this biosensor for full automation and utilized it in point-of-care testing.

Zhang et al. [47] initiated an ultrasensitive switchable electrochemiluminescence (ECL) RNA sensing platform based on MOF and metal-organic gel (MOG) as nanotag and electrode matrix, respectively. In-situ loading of graphite-like carbon nitride (g-C_3N_4) and assembly of Au nanoparticles (AuNPs) within Zr based MOG (AuNPs&g-C_3N_4@Zr-MOG) has been used for the preparation of the second one, which represents greatly effective solid-state ECL (Figure 30.13). Fe-MIL-88 MOFs have metal active centers to consume co-reactant and can serve as an ECL acceptor in ECL resonance energy transfer system which initiates a double quenching effect on the ECL of AuNPs&g-C_3N_4@Zr-MOG. Subsequently, a DNA probe which is composed of an apyrimidinic/apurinic (AP) site was capable of connecting both, which leads to a turn-off signal. In the existence of the target RNA, since the AP site of the DNA probe has been triggered to be circularly cleaved by endonuclease IV, the ECL has been turned on. The ECL platform represented a wide detection range of 0.3 nM–3 μM using Zika virus (ZIKV) RNA as a model analyte, where the ultralow LOD was equal to 0.1 nM.

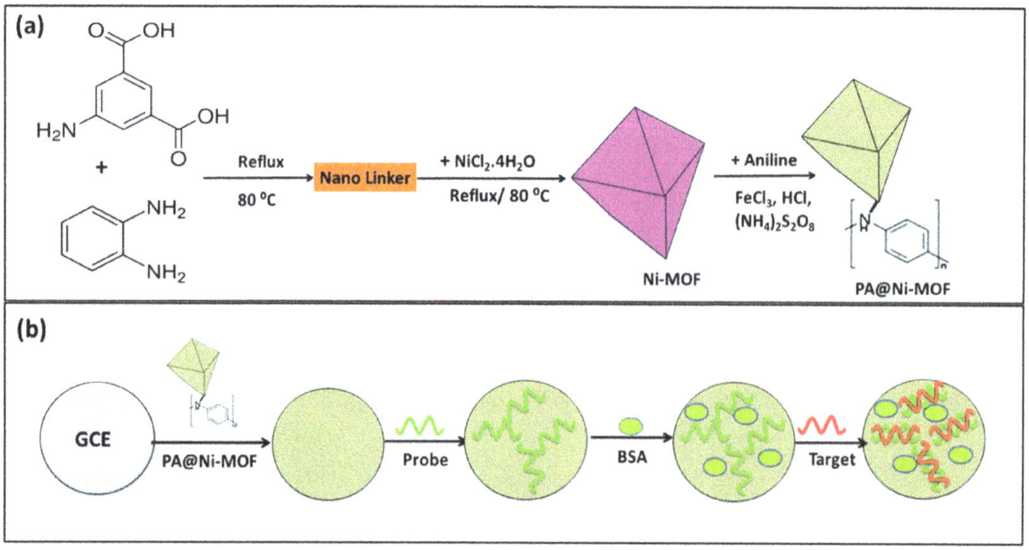

FIGURE 30.12 (a) Graphically representing the synthesis of PA@Ni-MOF composite in the existence of Ni-MOF and (b) Fabricating electrochemical HCV-nucleic acid-sensing assay. Adapted with permission from [46]. Copyright (2020) Elsevier.

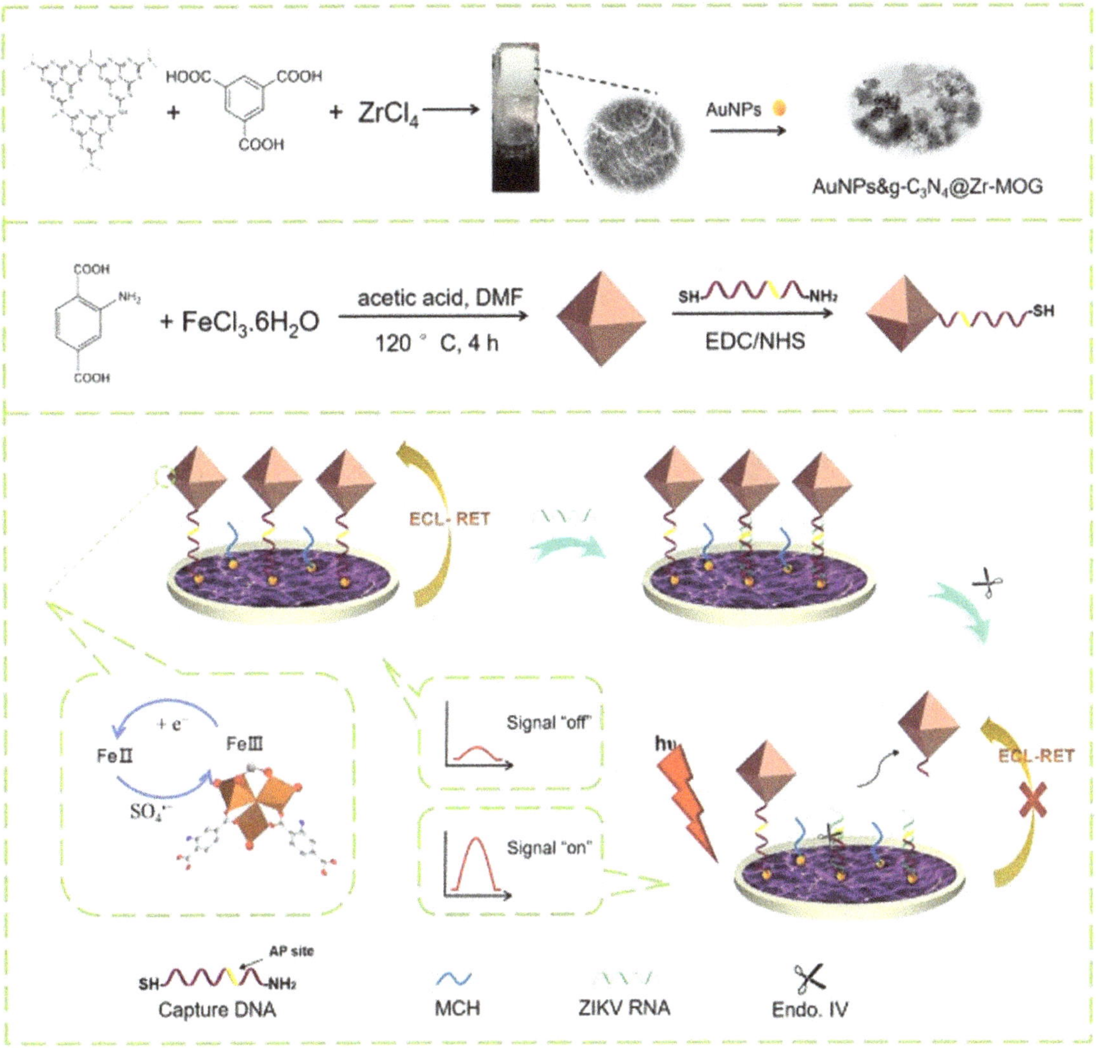

FIGURE 30.13 Graphically representing the fabrication process associated with the presented ECL sensing assay. Adapted with permission from [47]. Copyright (2020) Springer.

Luo et al. [48] were capable of fabricating molecularly imprinted polymer (HM@MIP) nanoprobes from MIL-101 and the pH-responsive polymer (dimethylamino ethyl methacrylate (DMA)). Using the resonance light scattering (RLS) approach, this probe has been used for the selective detection of the hepatitis A virus (HAV). For accelerating the release and capture of the virus by HM@MIPs as expected, DMA regulates the pH of the system, which can enhance or detract from the intensity of RLS. According to the intensity of RLS at 470 nm, it was possible to obtain the LOD of 0.1 pmol/L and a linear concentration of 0.02-2.0 nmol/L during 20 min. A change of 88 - 107% can be observed for the superior recoveries, which indicates the profound HM@MIPs capability for determining HAV in human serum and their potential capability for determining virus in real applications. The outcomes of this investigation are completely satisfactory. Furthermore, the practicability of such a strategy can encourage more investigations on viral molecular imprinting. By pursuing more appropriate functional monomers and imprinting layers, improvement in adsorption kinetics is possible.

In constructing biosensors, a mixture of nanotechnology and supramolecular chemistry has been used, which aims to improve the strength and analytical performance of electron devices. According to this strategy, Qian et al. [48] have fabricated a novel sandwich-type DNA sensor for sensitive detection of HBV (hepatitis B virus) DNA, a marker for chronic hepatitis B (Figure 30.14). The water-soluble pillar arene stabilized Pd NPs, combined with reduced graphene oxide nanosheet (WP5–Pd/RGO), has been synthesized and utilized as supporting material for the modifying electrode surface. Using a new strategy based on the host-guest communication between methylene blue labeled DNA (MB-DNA) and WP5, the probe DNA has been immobilized onto the surface of the electrode. Besides, for anchoring the hydroxylatopillar arena stabilized AuNPs (HP5–Au/CoS), a MOF-derived cobalt sulfide nano box has been developed that contains favorable electrocatalytic performance for H_2O_2 decrease for achieving signal magnification. In a perfect situation, a linear relationship between amperometric currents and the logarithm of tDNA solution from 1 × 10^{-15} mol/L to 1 × 10^{-9}

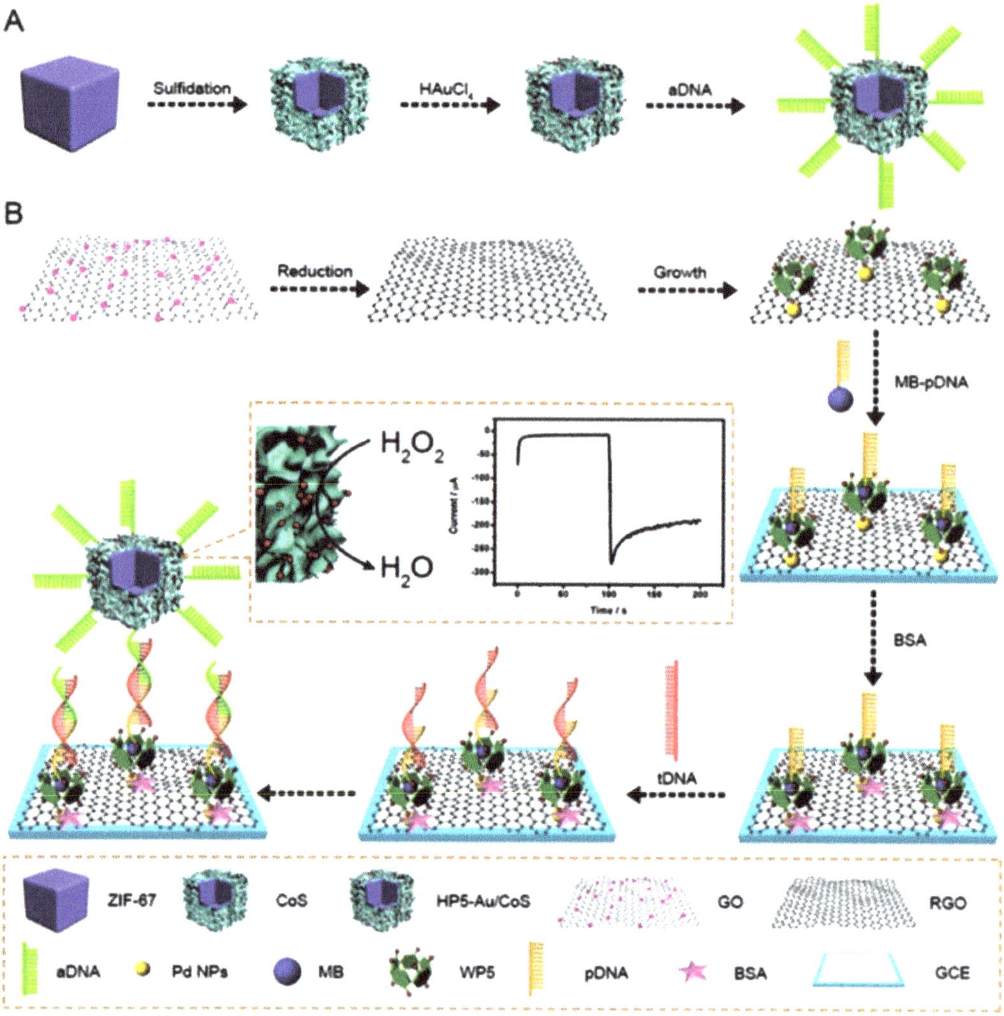

FIGURE 30.14 Demonstrating sandwich-based DNA sensing platform fabrication process. Preparation steps of HP5–Au/CoS-aDNA label (A) and electrochemical assay (B). Adapted with permission from [48]. Copyright (2019) Elsevier.

mol/L, and a lower LOD of 0.32 fmol/L has been presented for the proposed sensor. Additionally, significant behaviors of accuracy, stability, reproducibility, and specificity have been represented by the DNA sensor.

Another research of Jia et al. [49] represented that a bimetallic NiCo-MOF (NiCo-based MOF) has been pyrolyzed into a new composite which contains metallic Co/Ni nanoparticles, CoO, and $NiCo_2O_4$ spinel that has been fixed with nanotubes of carbon under N_2 atmosphere (represented by $NiCo_2O_4$/CoO@CNTs) at high temperature 700 °C, even though organic ligands have been decomposed in the atmosphere of H_2, which leads to lack of carbon nanotubes (indicated by $NiCo_2O_4$/CoO) (Figure 30.15). It has been investigated that the use of composite as an efficient bio platform is possible to immobilize the probe DNA of HIV-1 in the electrochemical HIV-1 DNA diagnosis. Advantageous biocompatibility, strong affinity, and great electrochemical activity for the probe DNA have been presented by $NiCo_2O_4$/CoO@CNTs composite in comparison to the pristine NiCo-MOF and $NiCo_2O_4$/CoO. Looking at the electrochemical measurements, it could be observed that greater sensing performances are considered for the $NiCo_2O_4$/CoO@CNTs-based bioassay, where the exceptionally lower LOD of 16.7 fM has been obtained for HIV-1 DNA and there was a linear range of 0.1 pM - 20 nM, which gives better selectivity even against two-base and noncomplementary mismatch sequences. From that point, the use of MOFs and their derivatives in different fields of biosensing can be extended by the current method. Furthermore, the presented strategy would be believed as a substitute method to diagnose the genes of HIV-1 at trace amounts and expand the MOF sensing applications.

Liu et al. have presented a novel sandwich-type electrochemical immunosensor for ALV J (*Avian leukosis* virus subgroup J). By originally amending a GCE with reduced graphene oxide functionalized with tannic acid and magnetite nanoparticles (rGO-TA-Fe_3O_4), the immunosensor has been fabricated (Figure 16). Subsequently, Ab1s (primary antibodies) have been deposited on the modified GCE. In signal magnification, crystals of hollow eZIF (zeolitic imidazolate framework) functionalized with horseradish

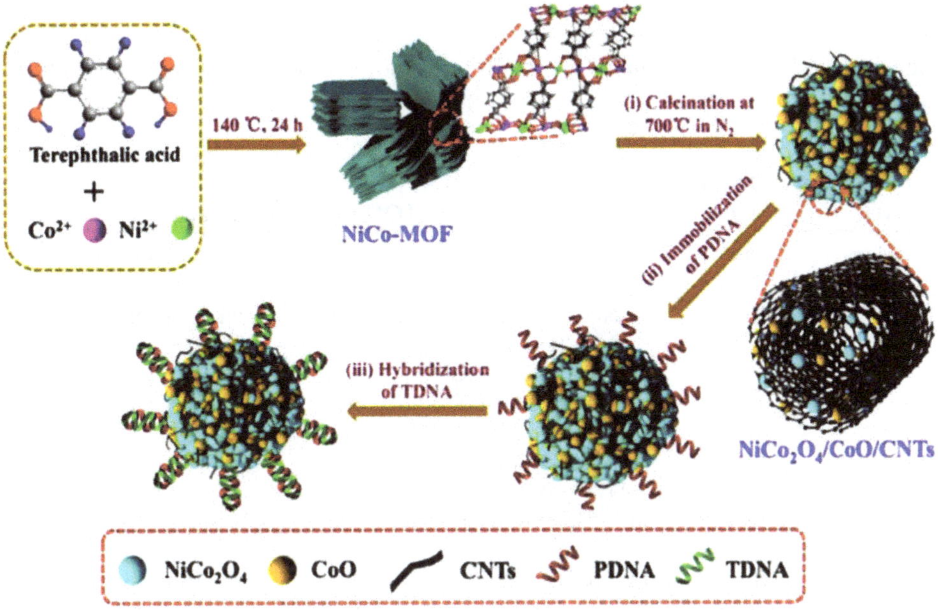

FIGURE 30.15 Schematically representing the construction process of an assay based on CNTs/CoO/$NiCo_2O_4$ for determining HIV-1 DNA, such as (i) constructing platform based on CNTs/CoO/$NiCo_2O_4$, (ii) immobilizing probe DNA, and (iii) determining HIV-1 DNA. Adapted with permission from [49]. Copyright (2018) Springer.

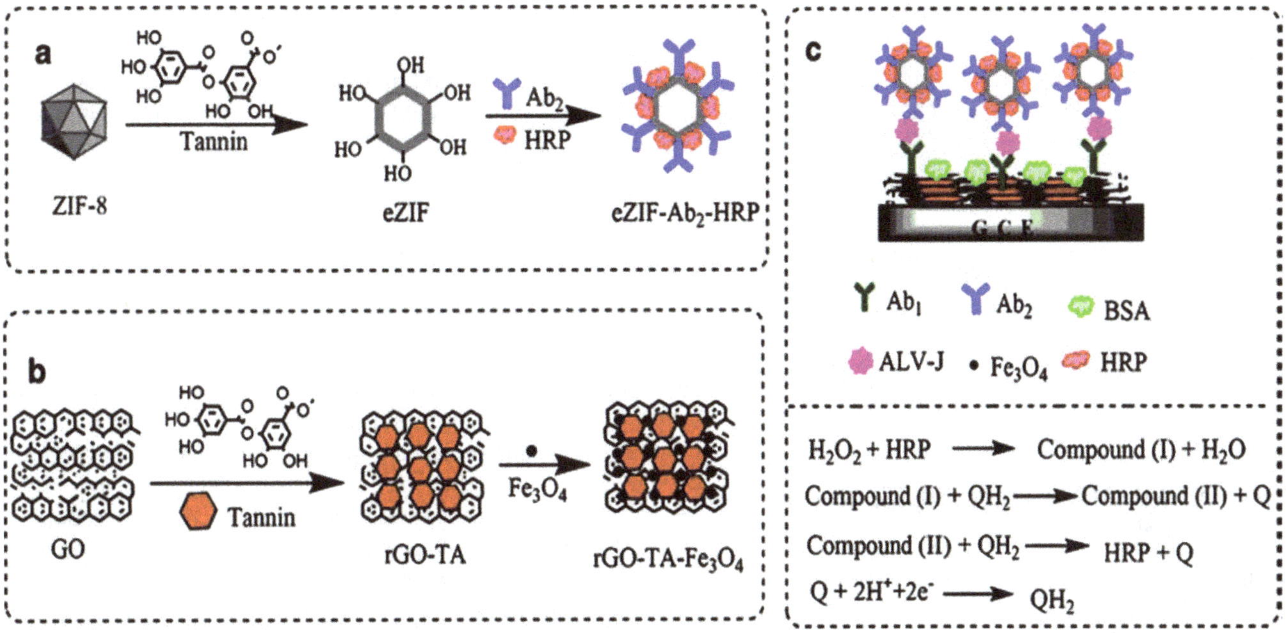

FIGURE 30.16 The strategy used for constructing eZIF-Ab2-HRP (a), rGO-TA-Fe_3O_4 (b), and the sandwich-type immunosensing method (c). It is also possible to demonstrate the performance of the presented immunosensing assay (c). Adapted with permission from [50]. Copyright (2018) Springer.

peroxidase (HRP), carrying secondary antibodies (Ab2) and tannic acid have been used. The crystals of hollow zeolitic imidazolate framework (eZIF) can be used as the exceptional carrier for HRP and Ab2, which encourages the wider MOF employment in electrochemical sensing. According to standard situations, a detection ranges of 152 - 10,000 TCID50 mL^{-1} (where TCID50 is the 50% tissue culture infective dose) has been presented by the immunoassay, where there was a lower LOD of 140 TCID50 mL^{-1} (at S/N = 3). ALV-J has a highly selective immunoassay, and functional stability and excellent reproducibility are considered for it. Experiments with spiked avian serum samples can verify

TABLE 30.1
Electrochemical Sensing Platforms Based on MOFs for Determining Various Viruses and Bacteria Species in Different Real Samples

Pathogen	Bio-recognition layer	Detection strategy	Modifier agent	LOD	Linear range	Recovery (%)	Reference
Pseudomonas aeruginosa	Aptamer	DPV	HZIFs-8/GCE	1 CFU/mL	1.2×10^1–1.2×10^7 CFU/mL	81.2 – 118.3	[39]
Mycobacterium tuberculosis	Aptamer	DPV	C_{60}NPs decorated N-CNTs/GO-PEI - functionalized MOF	0.33 fg/mL	1 fg/mL–1 ng/mL	94.8–102.0	[51]
Mycobacterium tuberculosis	Aptamer	DPV	Zr(IV) based MOF	10 fg/mL	0.02 to 1000 pg/mL	95.9–111.6	[42]
E. coli	Antibody	EIS	Ab/$Cu_3(BTC)_2$-PANI/ ITO	2 CFU/mL	$2 - 2 \times 10^8$ CFU/mL	–	[37]
E. coli	Antibody	EIS, DPV	MIL-53(Fe)/PEDOT composite	4 CFU/mL	$1 \times 10^2 - 2.1 \times 10^8$ CFU/mL	–	[32]
Vibrio parahaemolyticus	Aptamer	SWV	Fe_3O_4@NMOF	3 CFU/mL	10–10^9 CFU/mL	94.0–103.0	[36]
Haemophilus Influenza	DNA	DPAV	Zn based MOF/CMC/ AuNPs	1.48 fM	0.1 pM–10 nM	98.4–103.0%	[14]
Staphylococcus aureus	-	DPV	BC/MWCNTs-PEI	3 CFU/mL	log 10^0–10^7 CFU/mL	–	[35]
Pseudomonas aeruginosa	DNA-Aptamer	DPV	Cu-Zr MOF	2 CFU/mL	10–10^6 CFU/mL	96.9–102.3	[35]
Bacterial lipopolysaccharide	-	DPV	Cu^{2+}-modified MOF	0.61 pg/mL	1.5–7.5 pg/mL	94.4–109.8	[31]
HCV	RNA	CV, EIS	Polyaniline@Ni-MOF	0.75 fM	1 fM–100 nM	–	[46]
HIV-1	DNA	CV, EIS	NiCo-MOF	16.7 fM	0.1 pM to 20 nM	101.8–111.9	[49]
HBV	DNA	CV, EIS	MOF-derived cobalt sulfide nanobox	0.32 fmol/L.	10^{-15}–10^{-9} mol/L,	97.3–105.2	[52]
Zika virus	RNA	CV, EIS	AuNPs and g-C_3N_4 @Zr-MOG	0.1 nM	0.3 nM to 3 μM	–	[47]
HBV	DNA	DPV	ErGO-supported Cu-MOF	5.2 fM	50.0 fM–10.0 nM	95.2–99.8	[45]

the feasibility of the immunoassay for the prompt ALV-J detection.

30.4 SUMMARY, CHALLENGES, AND FUTURE PERSPECTIVES

In theory, a well-designed MOF can perform an essential role in this new scientific area. Meanwhile, developing sophisticated and portable systems based on such particles can offer a route for novel investigations and the way for wider applications is paved. Generally, MOFs can be networked, and porous inorganic-organic hybrid materials accompanied by exceptional catalytic ability, robust adsorption capacity, large specific surface area, and effective loading capability which can attract more consideration by scientists who conduct investigations in bioimaging areas.

Considering biosensing applications based on highly sensitive detection of viruses and bacteria, different MOF types which are bound to metal and organic blocks will be used as matrices for physical or covalent attachment of particular materials to serve as the bioreceptor and/ or improving the chemical, optical and electrical signals. According to biosensing strategies, the accuracy of diagnosis can be upgraded by the use of advanced technologies.

The recent advancements on MOF application in the diagnosis of pathogenic viruses and bacteria have been updated and their present and future problems and perspectives have been mentioned. Different electrochemical methods concerning MOF-based sensing assays have first been summarized for determining different bacteria species. Different investigations concerning sensing assays based on electrochemical MOFs have then been discussed for determining the extensive range of different pathogenic agents. Hitherto, two strategies were presented about the MOF functionalization/modification: 'functionalization during synthesis' and 'post-synthetic modification'. According to such strategies, different approaches are used for modifying/ functionalizing MOFs, including cell membrane coating, polymer MOFs, lipid bi-layer coating, host-guest interaction, covalent binding to pre-functionalized linkers, covalent binding to the linking groups, ligand exchange, coordinative self-assembly, and so on.

To summarize, because of the unique properties of MOFs, they represent appropriate potential in the biosensing area. Nevertheless, some problems and challenges can be observed in improving the functionality and synthesis of MOFs for sensing methods. The considered restrictions are: i) research work on sensing platforms based on MOFs regarding the biological environment, is the challenging task of researchers. The shape and size of MOFs have significant effects on them in-vivo and intracellular applications in the biological system, ii) to establish a favorable biosensing platform, the MOF size has to be smaller to simplify their systemic circulation in the blood and enable their endocytosis by living cells. Nowadays, breaking the large crystals of MOF into nanomaterials can be considered a different important problem for scientists, iii) for producing probes with great sensitivity and selectivity in living systems, it is required to design new MOF structures, iv) for meeting potential clinical demands, regularly and efficiently analyzing them in vivo is needed, such as for the metabolism and accumulation of the residual materials, v) the clinical MOF applications can also provide problems because of their unknown biological activity or toxic side effects, and finally, vi) in clinical investigation fields of recognition, therapy, and bioimaging applications, toxic MOF evaluation is required. The use of low-toxic ligands and metal ions could be an encouraging solution for such a problem.

REFERENCES

[1]. N. Huang, P. Pérez, T. Kato, Y. Mikami, K. Okuda, R.C. Gilmore, C.D. Conde, B. Gasmi, S. Stein, M. Beach, SARS-CoV-2 infection of the oral cavity and saliva, *Nature medicine*, (2021) 1–12.

[2]. A. Rouers, M.H.Y. Chng, B. Lee, M.P. Rajapakse, K. Kaur, Y.X. Toh, D. Sathiakumar, T. Loy, T.-L. Thein, V.W. Lim, Immune cell phenotypes associated with disease severity and long-term neutralizing antibody titers after natural dengue virus infection, *Cell Reports Medicine*, 2 (2021) 100278.

[3]. S.K. Sood, V. Sood, I. Mahajan, An intelligent healthcare system for predicting and preventing dengue virus infection, *Computing*, (2021) 1–39.

[4]. J. Liu, Y. Liu, C. Shan, B.T. Nunes, R. Yun, S.L. Haller, G.H. Rafael, S.R. Azar, C.R. Andersen, K. Plante, Role of mutational reversions and fitness restoration in Zika virus spread to the Americas, *Nature communications,* 12 (2021) 1–12.

[5]. W.A. Russell, Estimating the effect of discontinuing universal screening of donated blood for Zika virus in the 50 US states, *Annals of internal medicine,* 174 (2021) 728–730.

[6]. C. Adaken, J.T. Scott, R. Sharma, R. Gopal, S. Dicks, S. Niazi, S. Ijaz, T. Edwards, C.C. Smith, C.P. Cole, Ebola virus antibody decay–stimulation in a high proportion of survivors, *Nature*, 590 (2021) 468–472.

[7]. J. Lee, A.J. Kreutzberger, L. Odongo, E.A. Nelson, D.A. Nyenhuis, V. Kiessling, B. Liang, D.S. Cafiso, J.M. White, L.K. Tamm, Ebola virus glycoprotein interacts with cholesterol to enhance membrane fusion and cell entry, *Nature Structural & Molecular Biology*, 28 (2021) 181–189.

[8]. M.A. Navia, P.M. Fitzgerald, B.M. McKeever, C.-T. Leu, J.C. Heimbach, W.K. Herber, I.S. Sigal, P.L. Darke, J.P. Springer, Three-dimensional structure of aspartyl protease from human immunodeficiency virus HIV-1, *Nature*, 337 (1989) 615–620.

[9]. M.C. Hosseinipour, C. Innes, S. Naidoo, P. Mann, J. Hutter, G. Ramjee, M. Sebe, L. Maganga, M.E. Herce, A.C. DeCamp, Phase 1 Human Immunodeficiency Virus (HIV) vaccine trial to evaluate the safety and immunogenicity of HIV Subtype C DNA and MF59-adjuvanted subtype C envelope protein, *Clinical Infectious Diseases*, 72 (2021) 50–60.

[10]. L. Luznik, J. Bolaños-Meade, M. Zahurak, A.R. Chen, B.D. Smith, R. Brodsky, C.A. Huff, I. Borrello, W. Matsui, J.D. Powell, High-dose cyclophosphamide as single-agent, short-course prophylaxis of graft-versus-host disease, Blood, *The Journal of the American Society of Hematology*, 115 (2010) 3224–3230.

[11]. R. Gupta, N. Raza, S.K. Bhardwaj, K. Vikrant, K.-H. Kim, N. Bhardwaj, Advances in nanomaterial-based electrochemical biosensors for the detection of microbial toxins, pathogenic bacteria in food matrices, *Journal of Hazardous Materials*, 401 (2021) 123379.

[12]. R. Wang, L. Wang, J. Yan, D. Luan, J. Wu, X. Bian, Rapid, sensitive, and label-free detection of pathogenic bacteria using a bacteria-imprinted conducting polymer film-based electrochemical sensor, *Talanta*, 226 (2021) 122135.

[13]. S.S. Mahshid, S.E. Flynn, S. Mahshid, The potential application of electrochemical biosensors in the COVID-19 pandemic: A perspective on the rapid diagnostics of SARS-CoV-2, *Biosensors and Bioelectronics*, 176 (2021) 112905.

[14]. H. Sohrabi, M.R. Majidi, F. Nami, K. Asadpour-Zeynali, A. Khataee, A. Mokhtarzadeh, A novel engineered label-free Zn-based MOF/CMC/AuNPs electrochemical genosensor for highly sensitive determination of Haemophilus Influenzae in human plasma samples, *Microchimica Acta*, 188 (2021) 1–16.

[15]. J. Mariën, A. Ceulemans, J. Michiels, L. Heyndrickx, K. Kerkhof, N. Foque, M.-A. Widdowson, L. Mortgat, E. Duysburgh, I. Desombere, Evaluating SARS-CoV-2 spike and nucleocapsid proteins as targets for antibody detection in severe and mild COVID-19 cases using a Luminex bead-based assay, *Journal of virological methods*, 288 (2021) 114025.

[16]. I. Iankov, K. Viker, C. Turgeon, D. Matern, E. Galanis, Parameters of immunoglobulin extraction from dried blood spot cards and immunoassays for detection of antibody response to pathogens including the novel SARS-CoV-2, *Journal of immunological methods*, 492 (2021) 112996.

[17]. D. Liu, C. Ju, C. Han, R. Shi, X. Chen, D. Duan, J. Yan, X. Yan, Nanozyme chemiluminescence paper test for rapid and sensitive detection of SARS-CoV-2 antigen, *Biosensors and Bioelectronics*, 173 (2021) 112817.

[18]. B. Diao, K. Wen, J. Zhang, J. Chen, C. Han, Y. Chen, S. Wang, G. Deng, H. Zhou, Y. Wu, Accuracy of a nucleocapsid protein antigen rapid test in the diagnosis of SARS-CoV-2 infection, *Clinical Microbiology, and Infection*, 27 (2021) 289. e281–289. e284.

[19]. H. Wang, X. Li, T. Li, L. Wang, L. Wang, J. Lin, S. Zhang, Y. Xu, W. Wei, Development of a SARS-CoV-2 rapid antibody detection kit and study on dynamic changes in antibodies in infected patients, *The Clinical Respiratory Journal*, (2021).

[20]. M.A. Field, Detecting pathogenic variants in autoimmune diseases using high-throughput sequencing, *Immunology and Cell Biology*, 99 (2021) 146–156.

[21]. H. Liu, J. Wang, H. Zeng, X. Liu, W. Jiang, Y. Wang, W. Ouyang, X. Tang, RPA-Cas12a-FS: A frontline nucleic acid rapid detection system for food safety based on CRISPR-Cas12a combined with recombinase polymerase amplification, *Food Chemistry*, 334 (2021) 127608.
[22]. K.P. Anupama, A. Nayak, I. Karunasagar, I. Karunasagar, B. Maiti, Evaluation of loop-mediated isothermal amplification assay along with conventional and real-time PCR assay for sensitive detection of pathogenic Vibrio parahaemolyticus from seafood sample without enrichment, *Molecular biology reports*, (2021) 1–8.
[23]. V.D. Canh, S. Torii, H. Furumai, H. Katayama, Application of Capsid Integrity (RT-) qPCR to Assessing Occurrence of Intact Viruses in Surface Water and Tap Water in Japan, *Water Research*, 189 (2021) 116674.
[24]. Y. Orooji, H. Sohrabi, N. Hemmat, F. Oroojalian, B. Baradaran, A. Mokhtarzadeh, M. Mohaghegh, H. Karimi-Maleh, An overview on SARS-CoV-2 (COVID-19) and other human coronaviruses and their detection capability via amplification assay, chemical sensing, biosensing, immunosensing, and clinical assays, *Nano-micro letters*, 13 (2021) 1–30.
[25]. H. Sohrabi, A. Hemmati, M.R. Majidi, S. Eyvazi, A. Jahanban-Esfahlan, B. Baradaran, R. Adlpour-Azar, A. Mokhtarzadeh, M. de la Guardia, Recent advances on portable sensing and biosensing assays applied for detection of main chemical and biological pollutant agents in water samples: A critical review, *TrAC Trends in Analytical Chemistry*, (2021) 116344.
[26]. A. Khataee, H. Sohrabi, O. Arbabzadeh, P. Khaaki, M.R. Majidi, Frontiers in conventional and nanomaterials based electrochemical sensing and biosensing approaches for Ochratoxin A analysis in foodstuffs: A review, *Food and Chemical Toxicology*, (2021) 112030.
[27]. H. Sohrabi, O. Arbabzadeh, P. Khaaki, A. Khataee, M.R. Majidi, Y. Orooji, Patulin and Trichothecene: characteristics, occurrence, toxic effects, and detection capabilities via clinical, analytical, and nanostructured electrochemical sensing/biosensing assays in foodstuffs, *Critical Reviews in Food Science and Nutrition*, (2021) 1–29.
[28]. H. Sohrabi, O. Arbabzadeh, P. Khaaki, M.R. Majidi, A. Khataee, S. Woo Joo, Emerging electrochemical sensing and biosensing approaches for detection of Fumonisins in food samples, *Critical Reviews in Food Science and Nutrition*, (2021) 1–16.
[29]. H. Sohrabi, S. Javanbakht, F. Oroojalian, F. Rouhani, Y. Hanifehpour, M. Hashemzaei, A. Shaabani, A. Mokhtarzadeh, A. Morsali, Nanoscale Metal-Organic Frameworks: Recent Developments in Synthesis, Modifications and Bioimaging Applications, *Chemosphere*, (2021) 130717.
[30]. P. Pashazadeh-Panahi, S. Belali, H. Sohrabi, F. Oroojalian, M. Hashemzaei, A. Mokhtarzadeh, M. de la Guardia, Metal-organic frameworks conjugated with biomolecules as efficient platforms for development of biosensors, *TrAC Trends in Analytical Chemistry*, (2021) 116285.
[31]. Z. Li, G. Dai, F. Luo, Y. Lu, J. Zhang, Z. Chu, P. He, F. Zhang, Q. Wang, An electrochemical sensor for bacterial lipopolysaccharide detection based on dual functional Cu^{2+}-modified metal–organic framework nanoparticles, *Microchimica Acta*, 187 (2020) 1–10.
[32]. A. Gupta, A.L. Sharma, A. Deep, Sensitive impedimetric detection of E. coli with metal-organic framework (MIL-53)/polymer (PEDOT) composite modified screen-printed electrodes, *Journal of Environmental Chemical Engineering*, 9 (2021) 104925.
[33]. Y. Zhou, L. Wei, L. Chen, Application of metal-organic framework (MOF)-based enzymatic amplification strategy for the sensitive electrochemical detection of tuberculosis, *Sensors and Actuators B: Chemical*, 324 (2020) 128724.
[34]. U. Farooq, M.W. Ullah, Q. Yang, A. Aziz, J. Xu, L. Zhou, S. Wang, High-density phage particles immobilization in surface-modified bacterial cellulose for ultra-sensitive and selective electrochemical detection of Staphylococcus aureus, *Biosensors and Bioelectronics*, 157 (2020) 112163.
[35]. X. Zhang, G. Xie, D. Gou, P. Luo, Y. Yao, H. Chen, A novel enzyme-free electrochemical biosensor for rapid detection of Pseudomonas aeruginosa based on high catalytic Cu-ZrMOF and conductive Super P, *Biosensors and Bioelectronics*, 142 (2019) 111486.
[36]. W. Wang, L. Tan, J. Wu, T. Li, H. Xie, D. Wu, N. Gan, A universal signal-on electrochemical assay for rapid on-site quantitation of vibrio parahaemolyticus using aptamer modified magnetic metal–organic framework and phenylboronic acid-ferrocene co-immobilized nanolabel, *Analytica Chimica Acta*, 1133 (2020) 128–136.
[37]. A. Gupta, S.K. Bhardwaj, A.L. Sharma, K.-H. Kim, A. Deep, Development of an advanced electrochemical biosensing platform for E. coli using hybrid metal-organic framework/polyaniline composite, *Environmental research*, 171 (2019) 395–402.
[38]. H. Wang, Y. Xiu, Y. Chen, L. Sun, L. Yang, H. Chen, X. Niu, Electrochemical immunosensor based on an antibody-hierarchical mesoporous SiO_2 for the detection of Staphylococcus aureus, *Rsc Advances*, 9 (2019) 16278–16287.
[39]. S. Shahrokhian, S. Ranjbar, Development of a sensitive diagnostic device based on zeolitic imidazolate frameworks-8 using ferrocene–graphene oxide as electroactive indicator for Pseudomonas aeruginosa detection, *ACS Sustainable Chemistry & Engineering*, 7 (2019) 12760–12769.
[40]. L. Li, Y. Yuan, Y. Chen, P. Zhang, Y. Bai, L. Bai, Aptamer based voltammetric biosensor for Mycobacterium tuberculosis antigen ESAT-6 using a nanohybrid material composed of reduced graphene oxide and a metal-organic framework, *Microchimica Acta*, 185 (2018) 1-9.
[41]. N. Bhardwaj, S.K. Pandey, J. Mehta, S.K. Bhardwaj, K.-H. Kim, A. Deep, Bioactive nano-metal–organic frameworks as antimicrobials against Gram-positive and Gram-negative bacteria, *Toxicology research*, 7 (2018) 931–941.
[42]. N. Li, X. Huang, D. Sun, W. Yu, W. Tan, Z. Luo, Z. Chen, Dual-aptamer-based voltammetric biosensor for the Mycobacterium tuberculosis antigen MPT64 by using a gold electrode modified with a peroxidase loaded composite consisting of gold nanoparticles and a Zr (IV)/terephthalate metal-organic framework, *Microchimica Acta*, 185 (2018) 1–7.
[43]. J. Ma, W. Wang, Y. Li, Z. Lu, X. Tan, H. Han, Novel Porphyrin Zr Metal–Organic Framework (PCN-224)-Based Ultrastable Electrochemiluminescence System for PEDV Sensing, *Analytical Chemistry*, 93 (2021) 2090–2096.
[44]. J. Huang, J. Meng, S. Chen, S. Zhang, T. Liu, C. Li, F. Wang, A soft metal-polyphenol capsule-based ultrasensitive immunoassay for electrochemical detection of Epstein-Barr (EB) virus infection, *Biosensors and Bioelectronics*, 164 (2020) 112310.
[45]. X. Lin, X. Lian, B. Luo, X.-C. Huang, A highly sensitive and stable electrochemical HBV DNA biosensor based

on ErGO-supported Cu-MOF, *Inorganic Chemistry Communications*, 119 (2020) 108095.

[46]. S.M. Sheta, S.M. El-Sheikh, D.I. Osman, A.M. Salem, O.I. Ali, F.A. Harraz, W.G. Shousha, M.A. Shoeib, S.M. Shawky, D.D. Dionysiou, A novel HCV electrochemical biosensor based on a polyaniline@ Ni-MOFnano composite, *Dalton Transactions*, 49 (2020) 8918–8926.

[47]. Y.-W. Zhang, W.-S. Liu, J.-S. Chen, H.-L. Niu, C.-J. Mao, B.-K. Jin, Metal-organic gel, and metal-organic framework based switchable electrochemiluminescence RNA sensing platform for Zika virus, *Sensors and Actuators B: Chemical*, 321 (2020) 128456.

[48]. L. Luo, F. Zhang, C. Chen, C. Cai, Molecular imprinting resonance light scattering nanoprobes based on pH-responsive metal-organic framework for determination of hepatitis A virus, *Microchimica Acta*, 187 (2020) 1–8.

[49]. Z. Jia, Y. Ma, L. Yang, C. Guo, N. Zhou, M. Wang, L. He, Z. Zhang, NiCo2O4 spinel embedded with carbon nanotubes derived from bimetallic NiCo metal-organic framework for the ultrasensitive detection of human immune deficiency virus-1 gene, *Biosensors and Bioelectronics,* 133 (2019) 55–63.

[50]. C. Liu, J. Dong, S. Ning, J. Hou, G.I. Waterhouse, Z. Cheng, S. Ai, An electrochemical immunosensor based on an etched zeolitic imidazolate framework for detection of avian leukosis virus subgroup *J, Microchimica Acta*, 185 (2018) 1–9.

[51]. Y. Chen, X. Liu, S. Guo, J. Cao, J. Zhou, J. Zuo, L. Bai, A sandwich-type electrochemical aptasensor for Mycobacterium tuberculosis MPT64 antigen detection using C60NPs decorated N-CNTs/GO nanocomposite coupled with conductive PEI-functionalized metal-organic framework, *Biomaterials*, 216 (2019) 119253.

[52]. X. Qian, S. Tan, Z. Li, Q. Qu, L. Li, L. Yang, A robust host-guest interaction controlled probe immobilization strategy for the ultrasensitive detection of HBV DNA using hollow HP5–Au/CoS nanobox as biosensing platform, *Biosensors and Bioelectronics*, 153 (2020) 112051.

31 MOF-derived Smart Sensors, Challenges and Future Perspectives

Ambika Devi[1], *Prism Bhardwaj*[1] *and Pawan Kumar*[1]
[1]Material Application Research Lab (MARL), Department of Nano Sciences and Materials, Central University of Jammu, Jammu, India

31.1 INTRODUCTION

Metal-organic frameworks (MOFs) are porous coordinated polymers wherein metal ions are connected to multitopic organic linkers or struts through coordinate bonds resulting in 2D or 3D network structures [1]. MOFs have rapidly developed over the past thirty years. Most of the MOF research is associated with their nanoscale framework design and unique functions. MOFs mainly consist of lanthanide elements and transition metals. They also consist of non-identical metal ions having coordination geometry with some beneficial properties [2]. Organic ligands and metal ions have importance in the assembling process of MOFs. The rapid exploration of MOFs so far is due to their structure recombination and tailoring features. The properties and structure of the resultant MOF can be predicted up to a certain extent through the specific tunable mixture of metal ions and linkers. As a result, they have the potential for numerous implications such as luminescence, catalysis, gas storage, drug delivery, gas separation, and so forth. Additionally, MOFs show great responses to external stimuli. Recent studies are mainly based on macroscopical crystalline MOFs which restrict the wide opportunities and various considerable challenges in electronic devices. The circuits and systems fabrication of electronic MOF sensors demands further investigation regarding their sensing performance. Sensitivity, stability, selectivity, the limit of detection (LOD), reproducibility, recovery, and response are commonly the key parameters for their sensing criteria. White sensitivity is considered as the main parameter, and it shows the least change in external stimuli with a detachable signal molecule. The sensitivity of MOFs is commonly affected by the sensor structure, signal unit, and sensing material. In recent years, researchers working in the area of MOF sensing have focused widely on reproductivity, repeatability, and stability of MOFs [3]. They are preferred to other porous substances due to various excellent features such as their open metal sites, high porosity, tunable pore functionality, and pore compositions. Dispersed active sites that have high density can be developed through the incorporation of MOFs by either rational framework designs or post-synthetic methods.

Many attempts have been made to mitigate environmental pollution as well as to detect the materials that pollute the environment with the help of MOF-based smart sensing materials. High-risk materials like volatile organic compounds (VOCs), dyes, pharmaceuticals, nitrogen-based compounds, sulfur-based compounds, and personal care products, NO_x, SO_x, CO_x, H2S, and the like, are found in our environment. Porous MOFs are highly interesting for sensing materials due to their versatile applications [4]. Topologically diverse and attractive structures are one of their distinctive features. Many MOFs can be obtained from minerals through mother nature. Fabrication of target structures for sensing applications with their particular properties and specific functions gives everlasting opportunities for researchers. The initial step is to gain an understanding of the basic geometric principles to replicate their structures from nature. O'Keeffe and Yaghi give a procedure through the demonstration of MOFs into their fundamental topological nets, providing a way for the fabrication and account of other structures in MOFs [5].

In this chapter, we represent current challenges and future directions for MOF-based hybrid sensing materials. This chapter provides detailed insight into MOF-based smart sensors, types of sensors, opportunities, and lastly a conclusion that pays attention towards the future research opportunities. Figure 31.1 shows MOF-based sensors with their characteristics, applications, and future scope.

31.2 MOFs AS SMART SENSORS

Among the microporous materials, MOFs have attained appealing recognition owing to their unique structure and excellent sensing performance. They have emerged as unique, effective sensor materials for monitoring environmental contaminants, pH, temperature, and humidity as well as rapid detection of explosives chemicals and precursors [6]. In the last decade, many MOFs have been shown to have photoluminescence properties that arise either from organic linkers or metal ions. The high selectivity of MOFs towards adsorption with high hydrothermal stability up to 400-450°C and high specific surface area makes their role pivotal in sensing applications. Post-modification of MOFs further increases their sensing ability by improving the stabilization of guest molecules and adsorption strength of adsorbents [7].

DOI: 10.1201/9781003188148-31

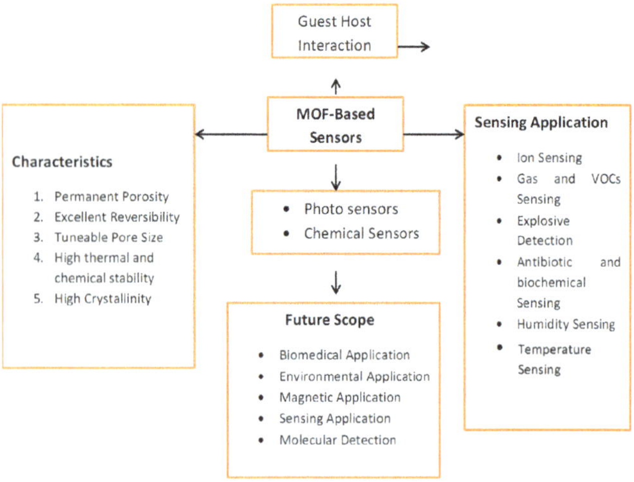

FIGURE 31.1 MOF-based Sensors with their Characteristics, Applications, and Future Scope.

Adapted with permission from [1]. Copyright (2020) Copyright The Authors, some rights reserved; exclusive licensee [EnPress Publisher LLC.]. Distributed under a Creative Commons Attribution License 4.0 (CC BY) https://creativecommons.org/licenses/by/4.0/

31.3 CATEGORIES OF MOF-BASED SENSORS

31.3.1 Luminiscient MOF-based Sensor

In recent years, MOFs have obtained much popularity as a promising luminescent functional material. Luminescent MOFs find application as smart sensors and are extensively explored for detection and separation, exhibiting rapid response time, high sensitivity, and low LOD (limit of detection) value. Post-synthetic modification (PSM) further enhances the sensing ability of luminescent MOFs as PSM will result in either enhancement or weakening of the original fluorescence of MOFs, namely, revealing fluorescence variation and thus allowing for easy detection of the analyte [8]. On absorption of UV-visible radiation, in most of the MOFs, the aromatic unit of the organic linker undergoes excitation from linker LUMO and creates an oxidizing hole in linker HOMO and imparts luminescence. Luminescence can also originate due to metal ion-emission, charge transfer transition from ligand to metal (LMCT), intraligand (LLCT), or metal to ligand (MLCT), guest-emission which involves the inclusion of guest molecules into the MOFs pores, and catalytic-emission [9].

31.3.1.1 Ion Sensing

Anion Sensing

MOF sensors effectively detect anions such as Br^-, I^-, Cl^-, CrO_4^-, PO_4^{3-}, SCN^-, N_3^-,CO_3^{2-}, $Cr_2O_7^{2-}$, MnO_4^-, and SO_4^{3-} through anion exchange, in other words, the exchange of counter anions of MOF with anionic species. The cationic fluorescence-based Cu^{II}-MOF $[CuL^8_2(H_2O)_{0.5}]$ (CH_3OH).$(NO_3)_2$.$0.5H_2O$. $3.25(CH_2Cl_2)$ has been synthesized by Dong et al., for the sensing of particular anionic species such as SCN^-, Br^-, N_3^-, I^-, and Cl^-. This CuII-MOF characterizes colors by exchanging counter anions with these anionic species. Yang et al. constructed Ln-based MOF ($[Tb(H_2O)_2($mucicate$)_{1.5}]5H_2O\}_n]$) to detect CO_3^{2-} in an aqueous medium with a higher luminescence turn-on response than other anions. The hydrogen bond formation between the ligand (mucic acid) OH group and CO_3^{2-} reduces vibrational quenching, resulting in increased fluorescence intensity [10]. Lanthanide-based MOF, [Tb (BTC)]. (CH_3OH) (H_3BTC=1,3,5-benzene tricarboxylic acid) was fabricated by Lobkovsky et al. for selective detection of F^-. There occurs an increase in the intensity of emission when F^- form a hydrogen bond with the methanol molecules at the terminal position. Zhao et al. reported two heterobimetallic cationic luminescent probes $[Ln_2Zn(L^5)_3(H_2O)_4]$ $(NO)_3$.$12H_2O$ where Ln=Tb and Eu, L^5= 4,4′-dicarboxylate-2,2′-dipyridine) that can effectively detect I^- via quick turn off luminescence response (10s) with the counter anions (NO_3^-) present in the MOF channels [11].

Chromate anions and MnO_4^- anions are envisaged with high solubility in water, carcinogenic and mutagenic which results in DNA damage, allergic reactions and at the same time pose a major threat to the global environment. Hence, there is a need for an urgent concerted effort to detect and remove these harmful anions. Zn^{II}-based MOF, $[Zn_2(BDC)_2(TzTz)_2]$ .2DMF, BDC=terephthalic acid, TzTz=2,5-di[4-pyridyl]thiazolo[4,5-d]thiazole, a luminescent sensor with layered framework has been selectively prepared by a solvothermal method which can effectively detect MnO_4^- and $Cr_2O_7^{2-}$. Chromate anion detection can also be achieved by using water-stable Eu-based MOF namely $[Eu_7(mtb)_5(H2O)_{16}]\cdot NO_3\cdot 8DMA\cdot 18H2O$ (**1**),{H_4mtb= 4-[tris(4-carboxyphenyl)methyl] benzoic acid}. This 3D MOF is fabricated using a solvothermal route and has excellent chromate sensing properties in an aqueous solution [12-13].

Cation Sensing

Pollutants produced by heavy metals like Cd^{2+}, Hg^{2+}, As^{3+}, Pb^{2+}, and the like, are capable of imposing a risk to the environment and also adversely affecting human life [14]. To mitigate the adverse impact of pollutants resulting due to heavy metals, it has become a necessity to develop constructive techniques to detect and separate the heavy metals. Fluorescent MIL-101-NH_2(Fe) provides feasibility for effectively detecting and detaching heavy metals (Pb^{2+}, Fe^{3+}, and Cu^{2+}). Fluorescent MIL-101-NH_2(Fe) has been synthesized by the solvothermal method by using NH_2-BDC as an organic linker which contributes to the fluorescence emission. The other four MOFs that are MIL-53- NH_2(Al), UiO-66-NH_2(Zr), MOF-5-NH_2((Zn), and MIL-101-NH_2(Cr) constructed by the similar method, namely, by embellishing organic ligands with -NH_2 groups, show the same performance in terms of heavy metal detection and removal [15]. MOF, $[Me_2\ NH_2]_4[Zn_6(trz)_4(qptc)_3]6H_2O$(**1**) (trz= 1,2,4-triazole, H_4qptc=terphenyl-2,5,2′5′-tetracarboxylate) profitably detects and removes the Cr^{3+} from aqueous solution via the turn-off mechanism, namely, a decrease in the intensity of luminescence [16].

A luminescent MOF $[Co(datz)(oba)(H_2O)]$, {where datz= 3,5-diamino-1,2,4-triazole and oba=4,4'-oxybis(benzoic acid)}, has been synthesized by a hydrothermal method. Aluminum ions (Al^{3+}) in an aqueous solution can be effectively detected by using this fluorescent MOF through luminescence intensity enhancement in combination with the hypsochromic or blue shift of approximately 24 nM [17]. Another example of a water-stable luminescent 3D MOF has been synthesized combining the strong thermal, chemical, and hydrolytic stability of $C\text{-}PO_3$ bonds with the benzene ring π-conjugation and the crystalline character of carboxylate. This turn-off fluorescent MOF sensor $Cd_2(pbdc)(H_2O)_3$ where H_4bdc= 5-phosphonobenzene-1,3-dicarboxylic acid can effectively detect Al^{3+}, Fe^{2+}, and Cu^{2+} in an aqueous medium above other metals while retaining its architecture [18].

31.3.1.2 Gas and Volatile Organic Compound Sensing

Functional metal-organic frameworks exhibit superior performance for the detection of hazardous gases (such as SO_2, H_2S, CS_2, CO, NH_3, NO) and VOCs (such as alcohols, ketones, hydrocarbons, aldehydes) This ultimately overcomes its ill effects on human health as well as the environment [19]. Dinca et al. reported high detection of ammonia via a high-temperature turn-on method with the help of $Mg(H_2DHBDC)$ ($H_2DHBDCA$ =2,5-dihydroxybenzene-1,4-dicarboxylic acid and $Zn_2(TCPE)$ (TCPE=tetrakis(4carboxyphenyl) ethylene) [10]. A 3D TPE-based MOF with 1D rhombus channel, $Zn_2(TCPPE)]$ {H_4TCPPE= tetrakis[4-(4-carboxyphenyl) phenyl]ethene} has been utilized for sensing organic compounds of volatile nature, toxic metals, and explosives [20]. Volatile organic compound formaldehyde imposes a major risk to humans as well as animals. Various products in indoor air can release formaldehyde, including wooden furniture, decoration items, paints, varnishes, plastics, and so forth. Although it has huge applications in various fields, it is considered to be a carcinogenic substance. Its long-term exposure may lead to nausea, inflammation, throat pain, headache, leukemia, and the like. It also causes high pollution in indoor environments and the survey depicted people who spent around 90% time in an indoor atmosphere. Over 2 million workers come into contact with 20 billion kilograms of formaldehyde (CH_2O) in the USA. 0.4 ppm of CH_2O is sufficient to give rise to inflammation in human beings. WHO and EPA, respectively have defined a safe limit of 0.15 mg kg^{-1} of body weight and 0.2 mg kg^{-1} of formaldehyde. High-performance liquid chromatography, spectrophotometry, gas chromatography, and other methods are described for the detection of formaldehyde. However, they are not preferred due to poor suitability for real-time monitoring, their complex operation, and their expense. Metal-organic framework-based sensors are a promising candidate to deal with such hurdles. They have specific selectivity and sensitivity, fast response, easy operation, and cost-effectiveness [21]. Jafari et al. fabricated a resistive gas sensor of ZIF-8 nanoparticles and functionalized SWCNTs to detect formaldehyde at room temperature. The sensor showed 4.83 ppm of LOD in the case of formaldehyde sensing. The high surface area (1412.76 m2 /g) of this composite helped to achieve a higher and fast response [22].

31.3.1.3 Explosive Sensing

Aromatic compounds bearing nitro moieties such as Trinitrophenol (TNP), Trinitrotoluene (TNT), 2, 4-dinitrotoluene (DNT), or even nitroaliphatic compounds are highly explosive and hence detection of such harmful compounds is necessary for anti-terrorism and in the homeland security field. These extremely harmful compounds can be effectively detected using fluorescent MOF Tb (BTC) (BTC= benzene-tricarboxylate) [23]. Another luminescent Zn^{II}-based MOF, HNU-34 formulated as $[Zn_2(tta)_2(TCPE)].2DMF\text{-}4H_2O.2Me_2NH_2$ (1H-tta=1H-tetrazole, H_4TCPE = 1, 1, 2, 2-tetra (4-carboxyphenyl) ethylene,) has been synthesized by Tang et al. which effectively detects nitrophenol, especially mononitrophenol and dinitrophenol [24]. The activated form **1′ of** Zr (IV)-MOF (**1**) having a framework with UiO-68 topology which contains the luminescent BTDB ligand (H_2BTDB = 4,4′ -(benzo[c][1,2,5]thiadiazole-4,7-diyl)dibenzoic acid), selectively detects TNP in methanol [25].

31.3.1.4 Antibiotic Sensing

Antibiotics play a vital role to prevent and cure bacterial infections in human beings. They are also employed in agriculture and livestock farming for the treatment of animal diseases. These antibiotics have been explored for their abuse, misuse, and harsh side effects. They can be harmful to human beings if taken in excessively, as their misuse can cause various health issues, such as allergic reactions, cancer, hereditary genetic infections, and the like. At the same time, their misuse can aggravate environmental pollution. To detect these antibiotics, MOF-based sensors can be deployed or used. Fluorescent Cd-MOF material has been synthesized by hydrothermal methods, using ligands 1,2-phenylenediacetic acid (H_2L), 1,4-bis(2-methyl-imidazol-1-yl)butane (bbi) and metal source $Cd(NO_3)_2.H_2O$, to remove antibiotics present in water. This Cd-MOF probe selectively detects CRO (Ceftriaxone sodium) antibiotic with a detection limit of 55ppb under pH=4-11 via a fluorescence quenching mechanism, namely, a more significant reduction in fluorescence intensity than with other antibiotics. HNU-52, a fluorescent MOF-based sensor with high quenching efficiency was prepared by a solvothermal method and is used in the detection of nitrofuran antibiotics [26-27].

A fluorescent Zn-MOF probe formulated as $[Zn_2(4,4'\text{-}bpy)_2(oba)_4]$ where 4,4′-bpy=4,4′-bipyridine, oba=4,4′-oxybis(benzoic acid), shows high selective sensitivity towards the detection of metronidazole (MET) antibiotic as well as aromatic nitrophenols in aqueous media. Under the stimulation of 325 nM UV light, Zn- MOF emits violet light of 445 nM. In an aqueous solution, aromatic NP and MET antibiotic could selectively quench the violet emission, demonstrating remarkable sensitivity. This fluorescent MOF can be used to monitor the quality of water [28]. BUT-172, a lanthanide-based MOF formulated as $Ln_3(CCTA)_2(OH)_3$, serves as an excellent candidate for detecting and removing

TABLE 31.1
Detection of Various Analytes using MOF-Sensors

S.No.	METAL	LIGANDS	MOF	TARGET/ SENSING	DETECTION LIMIT	REFERENCE
1.	Zn	TzTz, H_2BDC	$[Zn_2(TzTz)_2(BDC)_2]$.2DMF	MnO_4^- and $Cr_2O_7^{2-}$	100 ppb	12
2.	Tb	H_3BTC	[Tb (BTC)]. (CH_3OH)	F^-		11
3.	Zr	TCPP	PCN-221	Hg^{2+}	0.01μM	14
4.	Cr Al Zn	NH_2 –BDC	MIL-101-NH_2 MIL-52-NH_2 MOF-5-NH_2	Fe^{3+}, Cu^{2+}, Pb^{2+}	Fe^{3+}=0.0018 mM Cu^{2+}=0.0016 mM Pb^{2+}=0.0052 mM	15
5.	Ru^{II}-Cu^{II}	tBubpy, Dien	$Ru^{II}(CN)_4$(tBubpy) $[Cu^{II}(dien)]_2(ClO_4)_2$	I^-	1.2 μM	10
6.	Zn	H_4TCPE	HNU-34	2,4-DNP 2,6-DNP TNP	42.55 nM 29.45 nM 36.15 nM	24
7.	Co	2-MIL(2-methylimidazole)	ZIF 67	Formaldehyde Trimethylamine	5 ppm 2 ppm	36
8.	Zr	BDC-NH_2	NH_2-UiO-66	N_2	1 ppm	37
9.	Cd	2-phenylenediacetic acid (H_2L), bbi	Cd-MOF	CRO (Ceftriaxone sodium) antibiotic	55 ppb	25
10.	Zn	4,4′-oba, 4,4′- bpy	$[Zn_2(4,4'\text{- bpy})_2(oba)_4]$	Aromatic nitrofuran, metronidazole antibiotic	NF=0.66 ppm MET=0.81 ppm	28
11.	Cd	H_4pdc	Cd_2 (pbdc)$(H_2O)_3$	Cu^{2+}, Al^{3+}, Fe^{3+}	Cu^{2+}=38.9 nM Al^{3+}=139.0 nM Fe^{3+}=167.6 nM	18
12.	Ln=Eu,	L^5=4,4'- dicarboxylate-2,2'-dipyridine	$[Ln_2Zn(L^5)_3(H_2O)_4]$ $(NO)_3$.12H_2O	CO_3^{2-}	0.001 ppm	11
13.	Co	DATZ, OBA	[Co (DATZ)(OBA) (H_2O)],	Al^{3+}	57.5 ppb	24
14.	Ln	$CCTA^{3-}$	BUT-172, $Ln_3(CCTA)_2(OH)_3$	fluoroquinolone antibiotic	50 ppb	29

antibiotic fluoroquinolone with a low LOD value of around 50 ppb. BUT-172 has high stability in water, including boiling water, as well as in acidic and basic media, that is, stable in the 3-10 pH range whereas another isomeric form BUT-173 decomposes in the same conditions. Both isomers have an identical framework topology, and the only variation is in the $CCTA^{3-}$ arrangement around the one-dimensional secondary building units which in both isomers contain corner-sharing octahedral units $Ln(OH)_2O_4$ [29]. Table 31.1 shows the detection of various analytes using MOF-sensors.

31.3.2 Electrochemical Sensors

Since 2013, the application of MOFs in electrochemical sensors has been a burgeoning field. For electrochemical sensors, modified electrodes comprised of MOFs combined with catalytically active species are considered to be ideal candidates because: i) in comparison to other typical materials, MOFs have a substantially greater SSA(specific surface area), allowing them to sustain high density catalytically active sites, and ii) interlinked porosity of MOFs makes it easy for the targeted analyte to diffuse into these electrochemically active sites. Thus, MOFs are excellent platforms for electrochemical reactions, particularly electrocatalysis [30]. Yuan's group reported electroactive Cu-II MOF as a ratiometric electrochemical sensor. The Cu-II ions in the MOF produce a high electrochemical signal undergoing reduction to Cu-I ions and Cu metals. Lipopolysaccharide(LPS) sensing has been achieved using this MOF [31]. Other aspects that can aid the electrochemical process in MOFs include coupling MOFs with some other functional materials such as polymers, carbon substrates [32], metal nanoparticles, and biomolecules [31]. When compared to pristine MOFs, the MOF composite developed has better physio-chemical features such as catalytical activity, mechanical stability, and electrical conductivity [32]. A variety of composite MOF materials with better sensing properties have been produced recently. A novel composite MOF material cMWCNT/MOF-199 has been constructed by encapsulating carboxylate multiwalled carbon nanotubes (cMWCNT) with MOF-199. In addition, modifying this composite material on the GCE surface (glassy carbon electrode), a molecularity imprinted electrochemical sensor capable of rapid identification of 3-MCPE(3-chloro-1,2-propanediol) was successfully developed [33].

Currently, most of the work on MOF electrochemical activity centers on transition metal materials like Ni-MOF and Co-MOF, however, there are few investigations on lanthanides MOF. MOFs have been used in the fabrication of supercapacitor electrodes because of their superior electroactivity. A novel lanthanide MOF electrode Eu-MOF/CPE (CPE stands for carbon paste electrode) employed for TNP(2,4,6-trinitrophenol) sensing [34]. PBAs (Prussian blue analogues) provide a viable method for conserving the natural or inherent properties of MOFs in composite materials. They also exhibit biocompatibility stability, selectivity as well as redox-active sites for enhanced catalysis. An electrochemical biosensor electrode CoFe-PBA/Co-ZIF/NF (NF stands for nickel foam) having excellent sensitivity, low LOD, quick response time, stability in both basic medium and human blood has been produced, outperforming every single component in terms of sensing H_2O_2 and glucose [35].

31.3.3 Chemiresistive MOF Sensors

The performance of chemiresistive sensors is largely influenced by the interaction between the analyte and the sensing material surface. On adsorption there occurs e^-/hole transfer caused by the surface reaction between analyte and MOF material, resulting in resistance/conductance change. Chen et al. were the first who reported pure chemiresistive Co-MOF sensor ZIF-67 to detect formaldehyde with a detection limit of 5ppm at high temperature. At ambient temperature, this zeolitic imidazole framework lacks conductivity due to poor electron orbital overlap. MOF's electrical conductivity can be enhanced by two methods: i) increasing electron orbital overlap of metal nodes and ligands, namely, through a bond method, and ii) introducing electroactive fragments to the MOFs free space to create electrical pathways via non-covalent interactions, in other words, through a space method. The pore functionalization of UiO-66 with an amino group (UiO-66) has proven to be robust and an effective chemiresistive sensing material for monitoring acidic gases (NO_2, SO_2, CO_2) at low concentration and a temperature of 423K.

Chemiresistive sensor, Cu_3(HITP), where HITP= 2,3,6,7,10,11-hexaminotriphenylene, a conductive 2D MOF sensing material, effectively detects a sub-ppm level of ammonia at ambient temperature [38]. The excellent electrical conductivity of this 2D MOF is due to extended π-conjugation and charge delocalization between metal nodes and linkers [36-37].

Chemiresistive nanofilm Cu-HHTP-10C doped with HITP (a dual ligand strategy), where HHTP=Hexaiminotriphenylene and HHTP=Hexahydroxytriphenylene, demonstrates modulated selective sensitivity with improved response and speedy recovery. The dual ligand method leads to electronic structure modulation due to defect-associated energy levels and the difference in contribution to HOMO levels. This electronically conducting MOF thin film can be carefully synthesized at the nanoscale 20-70 and display 220% better selectivity for benzene (C_6H_6) against ammonia (NH_3). Among the scientific community, carbon nanotubes (CNTs) are regarded as an ideal sensing platform. The composite MOF materials Ni-MOF/-OH-MWNTS and Ni-MOF/-OH-SWNTS have been synthesized by the solvothermal method that includes the inclusion of -OH activated multi-walled (MWNTS) and single-walled carbon nanotubes (SWNTS) into Ni-MOF, and these composite MOFs show a considerable increase in sensing characteristics when compared to pristine MOF. The sensing response of composite materials has been seen by examining the resistance change [40].

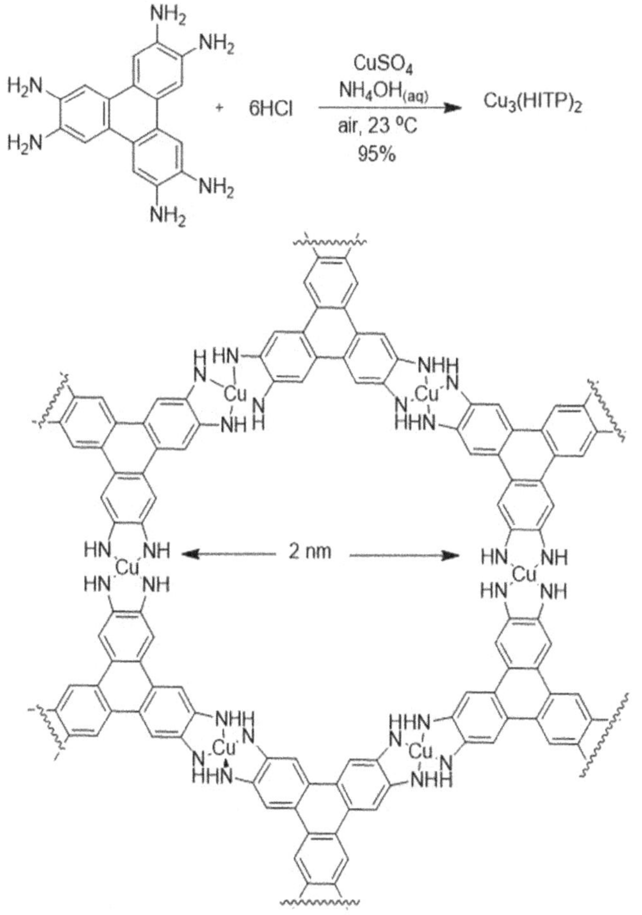

FIGURE 31.2 Synthesis Scheme of Chemiresistive MOF $Cu_3(HITP)_2$ [38].

31.3.4 Electromechanical Sensors

Electromechanical sensors are devices that can convert mechanical stimulus into electric current via the oxidation and reduction process. The primary need for electromechanical sensors is the analyte adsorption on the sensor's surface. MOFs have been used as a sensing material in electromechanical devices due to their remarkable sensing capacity [41]. This is accomplished by coating electromechanical devices such as microcantilevers (MCL), surface acoustic wave sensors (SAWS), quartz crystal microbalance (QCM) with MOF film, which enables the adsorption of targeted analyte species [42]. A novel MOF-14 coated QCM sensor has been developed that can effectively detect benzene vapor. Except for benzene, the other three, namely, xylene, ethylbenzene, and toluene, can be converted to the carboxyl group via oxidation and expelled outside of the body. Furthermore, benzene buildup in the body can cause cancer [43].

31.3.5 Miscellaneous Sensors

PH sensors: Many MOFs are fluorescence sensitive with several pH ranges such as a biological optimum range of 5-7.5. They work beautifully as pH probes in living organisms and aqueous media owing to their potential key features.

Temperature sensors: Temperature sensors that ratiometrically sense the variation in a desirable temperature have been widely studied. Ln-based sensors, mixing with other metals come under this category. Apparently, photoluminescence behavior can be altered by temperature change due to energy change between Ln-based MOF. Chen et al. designed a MOF-thermometer [Eu0.006 9Tb0.9931-DMBDC] (H2DMBDC=2,5-dimethoxy-1,4-benzenedicarboxylic acid) that exhibits a quick response in the 10-300K temperature range [44]. Temperature sensing can also be accomplished by utilizing luminous transition metal MOFs having d^{10} configuration and strongly p-conjugated organic linkers which exhibit intense ligand-based luminescence. Wu et al., for example, constructed a nano thermometer out of luminous 2D Cd -MOF, [Cd(TTAA)] $(H_2NMe_2).2H_2O$ where H_3TTAA=N,N',N"-1,3,-5-triazine-2,4,6-triyltris(4-aminomethylbenzoic acid). The extended conjugation in C_3H_6 ligand units should be primarily responsible for the framework's high fluorescence. As the temperature is reduced from 298Kto 70K, this 2D framework exhibits high sensitivity, short response time, and an 8 nM blue shift [45].

Colorimetric sensors: Colorimetric sensors are most often useful for sensing water contaminants. This sensing is grounded on the chromogenic reaction in colored substances. A change in the color of the solution is helpful to determine the amount and components in the targeted material. Gu et al. reported the RuUiO-67 that is a chromophoric Ru complex-doped MOF. They investigated the performance of the sensing probe and the colorimetric sensing of Hg^{2+} [46].

31.4 CHALLENGES AND OPPORTUNITIES

According to several reports, MOFs exhibit a rapid response to the analyte solvent and have several potential applications focusing on sensing. However, fast detection of the gas phase remains challenging which hampers their excellent opportunities. Mostly reported MOF-based sensors show sensitivity to the guest family where selective detection becomes tough. This is difficult mainly when the analyte possesses identical chemical properties. For instance, the transfer of electrons or holes is due to the reactions between surface or adsorption in a gas molecule with the sensing substances. This is a procedure of the sensing mechanism for chemiresistive sensors. Hence, the conductance change or resistance in the chemiresistive sensors happens when the reaction of sensing layers occurs with the adsorbed target species of gas.

Conversely, the changes in resistance in materials that include carbon, such as graphene, carbon nanotubes, carbonized MOFs, graphene oxides, and the like, are mostly because of the adsorption of targeted analyte species onto their surfaces. The arrangement of adsorption sites gets affected through the adsorbed gas molecules [47]. Most of the reported MOFs are still coping with bad recyclability and short stability. Weak coordination bonds between single metal ions or polynuclear metal clusters and the complete framework of MOFs are the main reasons for this issue. Moreover, transition metals and oxygen that are present in the nodes of MOFs, have a very high affinity towards water. Due to the water adsorption, hydrolysis reactions occur that cause a huge decrease of the binding sites and cause the MOF framework to fail. Consequently, their stability becomes a severe problem for practical applications. The photocatalytic efficiency of MOFs is still very low for removing gaseous pollutants. There has not been any careful study on the photocatalytic reaction mechanisms. The specific and selective adsorption of highly toxic gaseous pollutants is still a constant issue. Lastly, the fabrication cost of MOF-based sensors is still very high [35]. Cost-effective fabrication may pave the way for overcoming various challenges of MOFs. Additionally, one of the main hurdles in exploring the MOFs is their poor stability in water which in turn leads to the breakdown of their framework. This drawback requires them to be explored further [48]. Most of the MOFs possess low conductivity at their intrinsic level which can be overcome by incorporating numerous functional materials or guest molecules like quantum dots, luminol, and so forth [49]. The inefficiency in controlling the final design of frameworks is also a hurdle for the advancement of their applications. It occurs mainly due to mismatching of the prerequisites for a particular function of MOFs [50]. Hence, great efforts are required to design fast-detecting MOF-based smart sensors and in-depth host-guest interaction is needed to be explored computationally as well as thoroughly. Figure 31.3 shows some of the challenges of MOF-based sensors.

31.5 CONCLUSION AND FUTURE PERSPECTIVES

MOFs show eminent sensing performance due to diversity in their structures and the multifunctionality of their pores. MOF-based sensors are, therefore, excellent candidates for the determination of environmental contaminants including anions, heavy metals, organic compounds, toxic gases, and also for the detection of explosives. Luminescent MOF-based sensors operate by the turn-off (quenching) mechanism or turn-on (enhancement) mechanism, namely, magnification in the luminescent intensity of the host framework (MOF) in the existence of a guest molecule. Several studies suggest various ways to modify the specific and sensitive sensing methods of MOFs which include entrapping of functional molecules (in-situ), modification of dopant molecules, and post-modification techniques. Though much reported, MOF-based smart sensors provide great opportunities for selective detection and sensing mechanisms owing to their notable application in high-quality

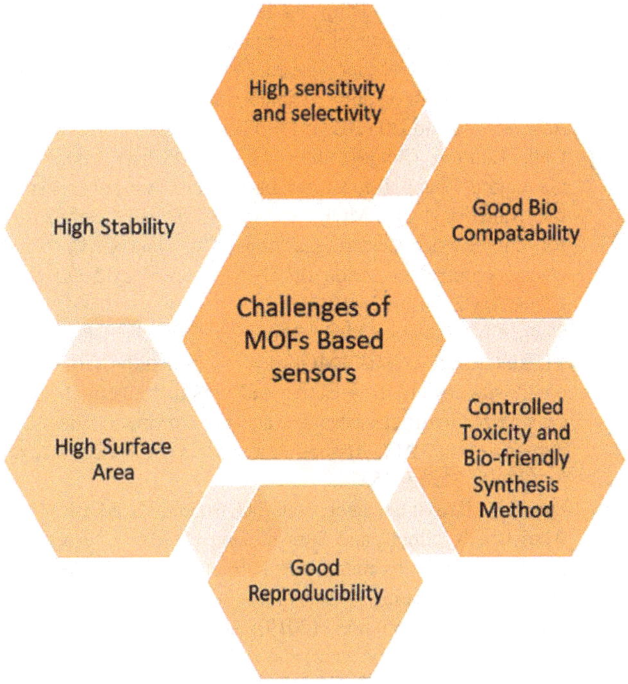

FIGURE 31.3 Challenges of MOF-based Sensors.

smart sensors. However, new hybrid MOF-based sensors are still emerging in research and create future perspectives for further applications, but the fabrication of truly selective and sensitive MOF-based sensors will probably take some more time. Cost-effective, bio-compatible, and high-yield synthesis methods for MOF-based sensors are expected to lead to their preparation on a considerable scale.

ACKNOWLEDGMENTS

We thank Science and Engineering Research Board (SERB), New Delhi, for Research funding under 'Early Carrier Research Grant (Project's reference no. ECR/2018/001716)' to Dr. Pawan Kumar.

REFERENCES

[1]. Soni, Sanju, Parmendra Kumar Bajpai, and Charu Arora. "A review on metal-organic framework: synthesis, properties and application." *Characterization and Application of Nanomaterials* 3, no. 2 (2020): 87–106.

[2]. García, Hermenegildo, and Sergio Navalón, eds. *Metal-Organic Frameworks: Applications in Separations and Catalysis*. John Wiley & Sons, Germany (2018).

[3]. Zhang, Lin-Tao, Ye Zhou, and Su-Ting Han. "The role of metal–organic framework in electronic sensors." *Angewandte Chemie* 133, no. 28(2020):15320-15340.

[4]. Khan, Nazmul Abedin, Zubair Hasan, and Sung Hwa Jhung. "Adsorptive removal of hazardous materials using metal-organic frameworks (MOFs): a review." *Journal of hazardous materials* 244 (2013): 444–456.

[5]. Zhou, Hong-Cai, Jeffrey R. Long, and Omar M. Yaghi. "Introduction to metal–organic frameworks." *Chemical reviews* 112, no. 2 (2012): 673–674.

[6]. Lustig, William P., Soumya Mukherjee, Nathan D. Rudd, Aamod V. Desai, Jing Li, and Sujit K. Ghosh. "Metal–organic frameworks: functional luminescent and photonic materials for sensing applications." *Chemical Society Reviews* 46, no. 11 (2017):

[7]. Kustov, Leonid M., Vera I. Isaeva, Jan Přech, and Kamal Kumar Bisht. "Metal-organic frameworks as materials for applications in sensors." *Mendeleev Communications* 29, no. 4 (2019): 361–368.

[8]. Yu, Qingya, Zheng Li, Qi Cao, Shumei Qu, and Qiong Jia. "Advances in luminescent metal-organic framework sensors based on post-synthetic modification." *TrAC Trends in Analytical Chemistry 129* (2020): 115939.

[9]. Liu, Yang, Xiao-Yu Xie, Chen Cheng, Zhen-Shu Shao, and Huai-Song Wang. "Strategies to fabricate metal–organic framework (MOF)-based luminescent sensing platforms." *Journal of Materials Chemistry C 7*, no. 35 (2019): 10743–10763.

[10]. Yi, Fei-Yan, Dongxiao Chen, Meng-Ke Wu, Lei Han, and Hai-Long Jiang. "Chemical sensors based on metal–organic frameworks." *ChemPlusChem* 81, no. 8 (2016): 675–

[11]. Zhang, Yingmu, Shuai Yuan, Gregory Day, Xuan Wang, Xinyu Yang, and Hong-Cai Zhou. "Luminescent sensors based on metal-organic frameworks." *Coordination Chemistry Reviews* 354 (2018): 28–45.

[12]. Safaei, Shirin, Jun Wang, and Peter C. Junk. "Incorporation of thiazolothiazole fluorophores into a MOF structure: A highly luminescent Zn (II)-based MOF as a selective and reversible sensor for $Cr_2O_7^{2-}$ and MnO_4^- anions." *Journal of Solid State Chemistry* 294 (2021): 121762.

[13]. Liu, Wei, Yanlong Wang, Zhuanling Bai, Yuxiang Li, Yaxing Wang, Lanhua Chen, Lin Xu, Juan Diwu, Zhifang Chai, and Shuao Wang. "Hydrolytically stable luminescent cationic metal organic framework for highly sensitive and selective sensing of chromate anions in natural water systems." *ACS applied materials & interfaces* 9, no. 19 (2017): 16448–16457.

[14]. Moradi, Ehsan, Rahmatollah Rahimi, and Vahid Safarifard. "Porphyrinic zirconium-based MOF with exposed pyrrole Lewis base site as an efficient fluorescence sensing for Hg^{2+} ions, DMF small molecule, and adsorption of Hg^{2+} ions from water solution." *Journal of Solid State Chemistry* 286 (2020): 121277.

[15]. Lv, Shi-Wen, Jing-Min Liu, Chun-Yang Li, Ning Zhao, Zhi-Hao Wang, and Shuo Wang. "A novel and universal metal-organic frameworks sensing platform for selective detection and efficient removal of heavy metal ions." *Chemical Engineering Journal* 375 (2019): 122111.

[16]. Jia, Xiao-Xia, Ru-Xin Yao, Fu-Qiang Zhang, and Xian-Ming Zhang. "A fluorescent anionic MOF with Zn_4 (trz) 2 chain for highly selective visual sensing of contaminants: Cr (III) ion and TNP." *Inorganic chemistry* 56, no. 5 (2017): 2690–2696.

[17]. Singha, Debal Kanti, and Partha Mahata. "Highly selective and sensitive luminescence turn-on-based sensing of Al^{3+} ions in aqueous medium using a MOF with free functional sites." *Inorganic chemistry* 54, no. 13 (2015): 6373–6379.

[18]. Zhang, Chenghui, Huaizhong Shi, Libo Sun, Yan Yan, Bolun Wang, Zhiqiang Liang, Li Wang, and Jiyang Li. "Water stable metal–organic framework based on phosphono-containing ligand as highly sensitive luminescent sensor toward metal ions." *Crystal Growth & Design* 18, no. 12 (2018): 7683–7689.

[19]. Li, Hai-Yang, Shu-Na Zhao, Shuang-Quan Zang, and Jing Li. "Functional metal–organic frameworks as effective sensors of

gases and volatile compounds." *Chemical Society Reviews* 49, no. 17 (2020): 6364–6401.

[20]. Liu, Xun-Gao, Hui Wang, Bin Chen, Yang Zou, Zhi-Guo Gu, Zujin Zhao, and Liang Shen. "A luminescent metal–organic framework constructed using a tetraphenylethene-based ligand for sensing volatile organic compounds." *Chemical Communications* 51, no. 9 (2015): 1677–1680.

[21]. Han, Zejun, Yuan Qi, Zhengyi Yang, Hecheng Han, Yanyan Jiang, Wenjing Du, Xue Zhang et al. "Recent advances and perspectives on constructing metal oxide semiconductor gas sensing materials for efficient formaldehyde detection." *Journal of Materials Chemistry C* 8, no. 38 (2020): 13169–13188.

[22]. Jafari, Nasim, and Sedigheh Zeinali. "Highly Rapid and Sensitive Formaldehyde Detection at Room Temperature Using a ZIF-8/MWCNT Nanocomposite." *ACS omega* 5, no. 9 (2020): 4395–4402.

[23]. Moscoso, Francisco G., José Almeida, Ahmad Sousaraei, Tânia Lopes-Costa, Ana MG Silva, Juan Cabanillas-Gonzalez, Luís Cunha-Silva, and José M. Pedrosa. "A lanthanide MOF immobilized in PMMA transparent films as a selective fluorescence sensor for nitroaromatic explosive vapours." *Journal of Materials Chemistry C* 8, no. 11 (2020): 3626–3630.

[24]. Zhang, Xiaodie, Guojian Ren, Meiling Li, Weiting Yang, and Qinhe Pan. "Selective Detection of Aromatic Nitrophenols by a Metal–Organic Framework-Based Fluorescent Sensor." *Crystal Growth & Design* 19, no. 11 (2019): 6308–6314.

[25]. Mostakim, S. K., and Shyam Biswas. "A thiadiazole-functionalized Zr (IV)-based metal–organic framework as a highly fluorescent probe for the selective detection of picric acid." *CrystEngComm* 18, no. 17 (2016): 3104–3113.

[26]. Xing, Pengcheng, Di Wu, Jingshuai Chen, Jiming Song, Changjie Mao, Yuanhao Gao, and Helin Niu. "A Cd-MOF as a fluorescent probe for highly selective, sensitive and stable detection of antibiotics in water." *Analyst* 144, no. 8 (2019): 2656–2661.

[27]. Yang, Yonghang, Guojian Ren, Weikang Yang, Xudong Qin, Dongxu Gu, Zhiqiang Liang, Dong-Yu Guo, and Pan Qinhe. "A new MOF-based fluorescent sensor for the detection of nitrofuran antibiotics." *Polyhedron* 194 (2021): 114923.

[28]. Wang, Jing, Qingqing Zha, Guoxu Qin, and Yonghong Ni. "A novel Zn (II)-based metal-organic framework as a high selective and sensitive sensor for fluorescent detections of aromatic nitrophenols and antibiotic metronidazole." *Talanta* 211 (2020): 120742.

[29]. Zhong, Wen-Bin, Ru-Xia Li, Jie Lv, Tao He, Ming-Ming Xu, Bin Wang, Lin-Hua Xie, and Jian-Rong Li. "Two isomeric In (III)-MOFs: unexpected stability difference and selective fluorescence detection of fluoroquinolone antibiotics in water." *Inorganic Chemistry Frontiers* 7, no. 5 (2020): 1161–1171.

[30]. Chuang, Cheng-Hsun, and Chung-Wei Kung. "Metal– Organic Frameworks toward Electrochemical Sensors: Challenges and Opportunities." *Electroanalysis* 32, no. 9 (2020): 1885–1895.

[31]. Liu, Lantao, Yanli Zhou, Shuang Liu, and Maotian Xu. "The applications of metal– organic frameworks in electrochemical sensors." *ChemElectroChem* 5, no. 1 (2018): 6–19.

[32]. Kempahanumakkagari, Sureshkumar, Kowsalya Vellingiri, Akash Deep, Eilhann E. Kwon, Nanthi Bolan, and Ki-Hyun Kim. "Metal–organic framework composites as electrocatalysts for electrochemical sensing applications." *Coordination Chemistry Reviews* 357 (2018): 105–129.

[33]. Han, Shuang, Yuxin Ding, Fu Teng, Aixin Yao, and Qiuxue Leng. "Determination of chloropropanol with an imprinted electrochemical sensor based on multi-walled carbon nanotubes/metal–organic framework composites." *RSC Advances* 11, no. 30 (2021): 18468–18475.

[34]. Gao, Loujun, Chunxia Jiao, Hongmei Chai, Yixia Ren, Gangqiang Zhang, Hao Yu, and Long Tang. "A highly sensitive multifunctional Eu-MOF sensor with pentacarboxylate for fluorescence detecting acetone, Cu^{2+} and $Cr_2O_7^{2-}$, and electrochemical detection of TNP." *Journal of Solid State Chemistry* 284 (2020): 121199.

[35]. Chen, Chen, Dengke Xiong, Minli Gu, Chunxiao Lu, Fei-Yan Yi, and Xinghua Ma. "MOF-derived bimetallic CoFe-PBA composites as highly selective and sensitive electrochemical sensors for hydrogen peroxide and nonenzymatic glucose in human serum." *ACS Applied Materials & Interfaces* 12, no. 31 (2020): 35365–35374.

[36]. Dmello, Marilyn Esclance, Nalini G. Sundaram, Akash Singh, Abhishek K. Singh, and Suresh Babu Kalidindi. "An amine functionalized zirconium metal–organic framework as an effective chemiresistive sensor for acidic gases." *Chemical Communications* 55, no. 3 (2019): 349–352.

[37]. Koo, Won-Tae, Ji-Soo Jang, and Il-Doo Kim. "Metal-organic frameworks for chemiresistive sensors." *Chem* 5, no. 8 (2019): 1938–1963.

[38]. Campbell, Michael G., Dennis Sheberla, Sophie F. Liu, Timothy M. Swager, and Mircea Dincă. "Cu_3 (hexaiminotriphenylene) 2: an electrically conductive 2D metal–organic framework for chemiresistive sensing." *Angewandte Chemie International Edition* 54, no. 14 (2015): 4349–4352.

[39]. Wu, Ai-Qian, Wen-Qing Wang, Hong-Bin Zhan, Lin-An Cao, Xiao-Liang Ye, Jia-Jia Zheng, Pendyala Naresh Kumar, Kashi Chiranjeevulu, Wei-Hua Deng, Guan-E Wang, Ming-Shui Yao, and Gang Xu. "Layer-by-layer assembled dual-ligand conductive mof nano-films with modulated chemiresistive sensitivity and selectivity." *Nano Research* 14, no. 2 (2021): 438–443

[40]. Ingle, Nikesh, Pasha Sayyad, Megha Deshmukh, Gajanan Bodkhe, Manasi Mahadik, Theeazen Al-Gahouari, Sumedh Shirsat, and Mahendra D. Shirsat. "A chemiresistive gas sensor for sensitive detection of SO_2 employing Ni-MOF modified–OH-SWNTs and–OH-MWNTs." *Applied Physics A* 127, no. 2 (2021): 1–10.

[41]. Chidambaram, Arunraj, and Kyriakos C. Stylianou. "Electronic metal–organic framework sensors." *Inorganic Chemistry Frontiers* 5, no. 5 (2018): 979–998.

[42]. Kreno, Lauren E., Kirsty Leong, Omar K. Farha, Mark Allendorf, Richard P. Van Duyne, and Joseph T. Hupp. "Metal–organic framework materials as chemical sensors." *Chemical reviews* 112, no. 2 (2012): 1105–1125.

[43]. Ma, Zhiheng, Tongwei Yuan, Yu Fan, Luyu Wang, Zhiming Duan, Wei Du, Dan Zhang, and Jiaqiang Xu. "A benzene vapor sensor based on a metal-organic framework-modified quartz crystal microbalance." *Sensors and Actuators B: Chemical* 311 (2020): 127365.

[44]. Wen, Meicheng, Guiying Li, Hongli Liu, Jiangyao Chen, Taicheng An, and Hiromi Yamashita. "Metal–organic framework-based nanomaterials for adsorption and photocatalytic degradation of gaseous pollutants: recent progress and challenges." *Environmental Science: Nano* 6, no. 4 (2019): 1006–1025.

[45]. Li, Ying. "Temperature and humidity sensors based on luminescent metal-organic frameworks." *Polyhedron* 179 (2020): 114413.

[46]. Fang, Xian, Boyang Zong, and Shun Mao. "Metal–organic framework-based sensors for environmental contaminant sensing." *Nano-micro letters* 10, no. 4 (2018): 1–19.

[47]. Koo, Won-Tae, Ji-Soo Jang, and Il-Doo Kim. "Metal-organic frameworks for chemiresistive sensors." *Chem* 5, no. 8 (2019): 1938–1963.

[48]. Zhou, Jiaojiao, Yun Li, Wenjing Wang, Xuecai Tan, Zhicheng Lu, and Heyou Han. "Metal-organic frameworks-based sensitive electrochemiluminescence biosensing." *Biosensors and Bioelectronics* 164 (2020): 112332.

[49]. Hu, Gui-Bing, Cheng-Yi Xiong, Wen-Bin Liang, Xiao-Shan Zeng, Hui-Ling Xu, Yang Yang, Li-Ying Yao, Ruo Yuan, and Dong-Rong Xiao. "Highly stable mesoporous luminescence-functionalized MOF with excellent electrochemiluminescence property for ultrasensitive immunosensor construction." *ACS applied materials & interfaces* 10, no. 18 (2018): 15913–15919.

[50]. Mandal, Sukhendu, Srinivasan Natarajan, Prabu Mani, and Asha Pankajakshan. "Post-Synthetic Modification of Metal–Organic Frameworks Toward Applications." Advanced Functional Materials 31, no. 4 (2021): 2006291.

Index

R

S

T

U

V

W

Z

For Product Safety Concerns and Information please contact our EU
representative GPSR@taylorandfrancis.com
Taylor & Francis Verlag GmbH, Kaufingerstraße 24, 80331 München, Germany

www.ingramcontent.com/pod-product-compliance
Lightning Source LLC
Chambersburg PA
CBHW080913220326
41598CB00034B/5556
9781032036007